Fluid Mixtures

Research Report

Deutsche Forschungsgemeinschaft

Thermodynamic Properties of Complex Fluid Mixtures

Edited by Gerd Maurer

Research Report

WILEY-VCH Verlag GmbH & Co. KGaA

Deutsche Forschungsgemeinschaft
Kennedyallee 40, D-53175 Bonn, Federal Republic of Germany
Postal address: D-53175 Bonn
Phone: ++49/228/885-1
Telefax: ++49/228/885-2777
E-Mail: postmaster@dfg.de
Internet: http://www.dfg.de

This book was carefully produced. Nevertheless, editor, authors and publisher do not warrant the information contained therein to be free of errors. Readers are advised to keep in mind that statements, data, illustrations, procedural details or other items may inadvertently be inaccurate.

Library of Congress Card No.: applied for

A catalogue record for this book is available from the British Library.

Die Deutsche Bibliothek – CIP Cataloguing-in-Publication Data
A catalogue record for this publication is available from Die Deutsche Bibliothek

ISBN 3-527-27770-6

© 2004 WILEY-VCH Verlag GmbH & Co. KGaA, Weinheim

Printed on acid-free paper.

All rights reserved (including those of translation into other languages). No part of this book may be reproduced in any form – by photoprinting, microfilm, or any other means – nor transmitted or translated into machine language without written permission from the publishers. Registered names, trademarks, etc. used in this book, even when not specifically marked as such, are not to be considered unprotected by law.

Cover Design and Typography: Dieter Hüsken
Composition: ProSatz Unger, Weinheim
Printing: betz-druck gmbh, Darmstadt
Bookbinding: Litges & Dopf Buchbinderei GmbH, Heppenheim

Printed in the Federal Republic of Germany

Contents

I	**Chemical Reactive Systems**	
1	**Experimental and Theoretical Investigations at the System CH_3OH + H_2SO_4 + H_2O**	3
	Matthias Behmann, Sae-Hoon Kim, Hans-Jürgen Koß, and Klaus Lucas	
1.1	Abstract	3
1.2	Introduction	3
1.3	Experimental Methods	4
1.3.1	Apparatus	4
1.3.2	Materials	5
1.3.3	Calibration and Analysis of Raman Signals	5
1.4	Theory	9
1.5	Results and Discussion	12
1.5.1	The System CH_3OH + H_2O	13
1.5.2	The System H_2SO_4 + H_2O	13
1.5.3	The System CH_3OH + H_2SO_4 + H_2O	17
1.5.4	The System CH_3SO_4H + CH_3OH	18
1.5.5	The System CH_3OH + H_2SO_4 + H_2O with CH_3SO_4H	18
1.5.6	The System CH_3OH + H_2SO_4 + H_2O with CH_3SO_4H and $(CH_3)_2O$	21
1.6	Conclusion	22
	References	23
	Nomenclature	24
2	**Measurement of Binary Phase Equilibria of Esterification Systems**	26
	Marko Tischmeyer and Wolfgang Arlt	
2.1	Abstract	26
2.2	Introduction	26
2.3	Fundamentals	27
2.3.1	Calculation of Liquid and Vapor Phase Fugacities	27
2.3.2	Reaction Space and Coordinates Transformation	28
2.3.3	Standard State in VLE Calculation with Chemical Equilibrium	31
2.3.4	Kinetic Modelling	32
2.4	Experimental	34

2.4.1	Chemicals	34
2.4.2	Equipment and Experimental Procedure	35
2.5	Results	39
2.6	Conclusion	45
	References	45
	Nomenclature	46

3	**Solubility of Carbon Dioxide in Aqueous Alkanolamine Solutions in the Temperature Range 313 to 353 K and Pressures up to 2.7 MPa**	48
	Dirk Silkenbäumer, Bernd Rumpf, and Rüdiger N. Lichtenthaler	
3.1	Abstract	48
3.2	Introduction	48
3.3	Experimental	49
3.3.1	Apparatus	49
3.3.2	Procedure	50
3.3.3	Materials	51
3.4	Results	51
3.4.1	Test of Procedure: Results for the System $CO_2 + H_2O$	51
3.4.2	Results for the System CO_2 + MDEA + H_2O	52
3.4.3	Results for the System CO_2 + AMP + H_2O	54
3.4.4	Results for the System CO_2 + MDEA + AMP + H_2O	57
3.5	Modeling	58
3.6	Comparison with Literature Data	64
3.7	Conclusions	66
	References	66
	Nomenclature	67

4	**The Influence of NO_2 on Complex Phase and Reaction Equilibria in Wet Flue Gas Cleaning Processes**	69
	Mohammad A. Siddiqi, Jens Petersen, and Klaus Lucas	
4.1	Abstract	69
4.2	Introduction	69
4.3	Experimental	71
4.3.1	Apparatus	71
4.3.2	Materials	72
4.4	Experimental Results	73
4.4.1	Studies at 298.65 K	73
4.4.2	Studies at 318.45 K and 333.35 K	77
4.5	Model Calculations	78
4.6	Results and Discussion	85
	References	88
	Nomenclature	90

5	**An Infrared Spectroscopic Investigation of the Species Distribution in the System $NH_3 + CO_2 + H_2O$**	92
	Ute Lichtfers and Bernd Rumpf	
5.1	Abstract	92
5.2	Introduction	92
5.3	Phase Equilibria and Quantitative Infrared Spectroscopy in the System $NH_3 + CO_2 + H_2O$	93
5.4	Experimental	98
5.4.1	Apparatus	98
5.4.2	Materials	99
5.4.3	Results	99
5.4.4	Calibration Measurements	101
5.4.5	Evaluation of Spectra	102
5.4.5.1	Spectra Taken in Calibration Measurements	102
5.4.5.2	Spectra in the System $NH_3 + CO_2 + H_2O$	103
5.5	Results	106
5.6	Modeling	107
5.7	Prediction of Caloric Effects	112
5.8	Conclusion	114
	Acknowledgements	115
	References	115
	Nomenclature	117
II	**Associating Mixtures**	121
6	**VLLE for Mixtures of Water and Alcohols: Measurements and Correlations**	123
	Frank Gremer, Gerhard Herres, and Dieter Gorenflo	
6.1	Abstract	123
6.2	Introduction	123
6.3	Experimental	124
6.4	Experimental Results	127
6.5	Conclusions	132
	References	133
	Nomenclature	134
7	**Phase Equilibria in Ternary Systems Containing Phenols, Hydrocarbons, and Water**	135
	Jürgen Schmelzer, Klaus Taubert, Antje Martin, René Meinhardt, and Jochen Kempe	
7.1	Abstract	135
7.2	Introduction	135
7.3	Experimental	136
7.3.1	Liquid-Liquid Equilibrium	136
7.3.2	Vapor-Liquid Equilibrium	137

7.3.3	Infrared Spectroscopic Investigations	139
7.3.4	Materials	139
7.4	Theory	139
7.4.1	UNIFAC Group-Contribution Method	139
7.4.2	NRTL and UNIQUAC Gibbs Excess Energy Models	140
7.4.3	ESD Equation of State (ESD-EOS)	140
7.5	Results and Discussions	142
7.5.1	Prediction Using UNIFAC	142
7.5.2	Correlation of Binary Data with NRTL, UNIQUAC, and ESD-EOS	143
7.5.3	Prediction for Ternary Systems with NRTL, UNIQUAC, and ESD-EOS	143
7.6	Conclusions	146
	Acknowledgements	147
	References	147
	Nomenclature	148

8	**Influence of Additives on Hydrophobic Association in Polynary Aqueous Mixtures. An NMR Relaxation and Self-Diffusion Study**	**150**
	Manfred Holz and Manghaiko Mayele	
8.1	Abstract	150
8.2	Introduction	150
8.3	Methods and Experimental	152
8.3.1	Theoretical Background of the *A*-Parameter Procedure	153
8.3.2	Experimental Determination of the *A*-Parameter	154
8.3.2.1	Measurement of the Intermolecular Relaxation Rate	154
8.3.2.2	PFG NMR for the Determination of Self-Diffusion Coefficients	155
8.3.2.3	Measurement of the Spin Density	155
8.3.3	Experimental	156
8.3.3.1	Apparatus	156
8.3.3.2	Sample Preparation	156
8.3.4	Materials	157
8.4	Results and Discussion	157
8.4.1	The Binary System: Water + *tert*-Butanol	157
8.4.2	The Ternary System: Water + *tert*-Butanol + Urea	159
8.4.2.1	Urea Effect on the Translational Dynamics of *tert*-Butanol and Water	160
8.4.2.2	Influence of Urea on *tert*-Butanol Self-Association	166
8.4.3	Water + Ethylene Glycol Systems	169
8.4.3.1	The Binary System: Water + Ethylene Glycol	170
8.4.3.2	Ternary Systems: Water + Ethylene Glycol + Additive	172
	The Additive Dioxane	172
	The Additive *n*-Butanol	174
	Salts as Additives	175
8.5	Conclusions	177
	References	178
	Nomenclature	181

III	**Aqueous Solutions of Strong Electrolytes**	185

9	**Phase Equilibria of Aqueous Solutions Containing Volatile Electrolytes** .	187
	Jürgen Zipprian, Nils Elm, and Karlheinz Schaber	
9.1	Abstract	187
9.2	Introduction	187
9.3	Experimental	188
9.3.1	Apparatus	188
9.3.2	IR-Spectra and Calibration Curves	190
9.4	Results and Discussion	192
9.4.1	System $NH_3 + H_2O$	192
9.4.2	System $HCl + H_2O$	195
9.4.3	System $HBr + H_2O$	197
9.4.4	System $HCl + CaCl_2 + H_2O$	198
9.4.5	System $HCl + HBr + H_2O$	200
9.5	Theory	202
9.6	Results and Discussion	203
9.6.1	Partial Pressure of HCl	203
9.6.2	Partial Pressure of H_2O	204
9.7	Conclusions	204
	References	205
	Nomenclature	206

10	**Experimental Determination and Prediction of Phase Equilibria in Systems Containing Strong Electrolytes**	208
	Magnus Topphoff, Christian Rose, Jörn Kiepe, and Jürgen Gmehling	
10.1	Abstract	208
10.2	Introduction	209
10.3	Group Contribution Model for the Prediction of Activity Coefficients in Systems Containing Strong Electrolytes	210
10.3.1	Thermodynamic Framework	210
10.3.1.1	Estimation of Parameters	213
10.3.2	Results and Discussion for VLE Data	214
10.3.3	Prediction of Solid-Liquid Equilibria	217
10.3.3.1	Theory	218
10.3.3.2	Results and Discussion	218
10.3.4	Prediction of Gas Solubilities	221
10.3.4.1	Modified PSRK Group Contribution Method	222
10.3.4.2	Results and Discussion	224
10.4	Determination of Vapor-Liquid Equilibria for Systems Containing Strong Electrolytes	225
10.4.1	Materials	226
10.4.2	Headspace Gas Chromatography	226
10.4.3	Ebulliometry	226
10.4.3.1	Experimental Results	227

10.4.4	Measurement of Activity Coefficients at Infinite Dilution in Electrolyte Systems Using the Dilutor Technique	228
10.4.4.1	Experimental and Measurement Procedure	228
10.4.4.2	Principle of the Applied Method	230
10.4.4.3	Experimental Results	231
10.4.5	Measurement of the Mean Activity Coefficients of Ions by Determination of the Electromotive Force	233
10.5	Conclusion	235
	References	236
	Nomenclature	238
11	**Ion Coordination and Thermodynamic Modeling of Molten Salt Hydrate Mixtures**	**241**
	Wolfgang Voigt, Kay Hettrich, and Dewen Zeng	
11.1	Abstract	241
11.2	Introduction	241
11.3	Experimental	243
11.3.1	Solid-Liquid Equilibria	243
11.3.2	UV-Vis Spectroscopy	243
11.3.3	Raman Spectroscopy	243
11.3.4	Chemicals	244
11.4	Results and Discussion	244
11.4.1	System $MgCl_2 + CuCl_2 + H_2O$	244
11.4.1.1	Solid-Liquid Equilibria	244
11.4.1.2	UV-Vis Spectroscopy	247
11.4.1.3	Raman Spectra	250
11.4.2	$MgCl_2 + CsCl + H_2O$	251
11.5	Thermodynamic Modeling	253
11.5.1	Modeling Strategy	253
11.5.2	Computational Problems of Parameter Estimation	254
11.5.3	Model Development for the Systems $MgCl_2 + KCl + H_2O$ and $MgCl_2 + CuCl_2 + H_2O$	255
11.5.3.1	Binary Systems	255
	$MgCl_2 + KCl$	255
	$KCl + H_2O$	257
	$MgCl_2 + H_2O$	258
11.5.3.2	Ternary Systems	260
	$MgCl_2 + KCl + H_2O$	260
	$MgCl_2 + CuCl_2 + H_2O$	261
11.6	Conclusions	263
	Acknowledgements	263
	References	263
	Nomenclature	266

12	**Effects of Salts on Excess Enthalpies of Binary Liquid Mixtures**	268

Peter Ulbig, Thorsten Friese, and Katrin Wagner

12.1	Abstract	268
12.2	Introduction	268
12.3	Experimental/Methods	269
12.3.1	Apparatus	269
12.3.1.1	Calorimetric Measurements	269
12.3.1.2	Solubility Measurements	271
12.3.1.3	Conductivity Measurements	272
12.3.1.4	Measurement of the Excess Volume	273
12.3.2	Materials	273
12.3.3	Analysis	273
12.4	Results and Discussion	275
12.4.1	Excess Enthalpy	275
12.4.1.1	Effect of the Salt Concentration	275
12.4.1.2	Effect of Temperature	275
12.4.1.3	Effect of Different Alcohols	275
12.4.1.4	Effect of Different Salts	278
12.4.2	Solubility	279
12.4.3	Conductivity	279
12.4.4	Excess Volume	280
12.5	Theory	280
12.5.1	The HEACE Model	280
12.5.2	Conductivity Theory	282
12.6	Results and Discussion	283
12.7	Conclusion	286
	Acknowledgements	287
	References	287
	Nomenclature	288

13	**Hydrate Equilibria in Aqueous Solutions Containing Inhibitors**	290

Armin M. Rock and Lothar R. Oellrich

13.1	Abstract	290
13.2	Introduction	290
13.3	Fundamental Hydrate Structural and Phase Behavior	292
13.4	Experimental	296
13.4.1	Apparatus	296
13.4.2	Procedure	298
13.5	Theory	300
13.5.1	Thermodynamic Framework	300
	Solid Hydrate Phase	301
	Fluid Phases	304
13.5.2	Estimation of Hydrate Cavity Interaction Parameters	306
13.6	Results and Discussion	308
13.7	Conclusions	315
	Acknowledgement	316

	References	316
	Nomenclature	319

IV	**Phase Equilibrium of Aqueous Two-Phase Systems**	321

14	**Experimental and Theoretical Investigations on the Precipitation of Polyelectrolytes from Aqueous Solutions by Neutral Polymers**	323
	Thomas Grünfelder and Gerd Maurer	
14.1	Abstract	323
14.2	Introduction	323
14.3	Experimental/Methods	325
14.3.1	Materials	325
14.3.2	Experimental Procedures	326
14.3.2.1	Turbidity Measurements	326
14.3.2.2	Compositions of Coexisting Phases	328
	Freeze Drying	328
	Atomic Absorption Spectroscopy	329
	Polymer Concentrations	329
14.4	Results and Discussion	329
14.4.1	Overview	329
14.4.2	Cloud Points	330
14.4.3	Composition of Coexisting Phases	331
14.4.4	Influence of Molecular Weight on the Phase Behavior	332
14.4.5	Influence of Temperature on the Phase Behavior	332
14.5	Modeling	334
14.5.1	Introduction to VERS-PE	334
14.5.2	Determination of Parameters of the VERS-PE Model	338
14.5.3	Results of Correlations/Predictions with the VERS-PE Model	340
14.6	Conclusions	341
	Acknowledgements	341
	References	342
	Nomenclature	342

15	**Experimental and Theoretical Studies on Partitioning of Native and Unfolded Enzymes in Aqueous Two-Phase Systems**	345
	Maria-Regina Kula and Christian Rämsch	
15.1	Abstract	345
15.2	Introduction	345
15.3	Material and Methods	346
15.3.1	Chemicals and Lysozyme Mutants	346
15.3.2	Determination of Phase Diagrams	347
15.3.3	Protein Conformation	347
15.3.4	Partition Coefficients	348
15.4	Results and Discussion	348
15.4.1	Phase Diagrams of Quaternary Systems	348

15.4.2	Stability of T 4-Lysozyme Variants	350
15.4.3	Partition of T 4-Lysozyme Variants	352
15.4.4	T 4-Lysozyme Unfolding and Refolding in PEG + Na_2SO_4 Systems	354
15.5	Conclusion	356
	Acknowledgements	356
	References	356
	Nomenclature	357
16	**Experimental and Theoretical Investigations on the Partitioning of Proteins in Aqueous Two-Phase Systems**	**359**
	Jochen Brenneisen and Gerd Maurer	
16.1	Abstract	359
16.2	Introduction	359
16.3	Theory	360
16.3.1	Model	360
16.3.2	Gibbs Energy of the Aqueous Solutions	363
16.3.3	Chemical Reactions	363
16.4	Experimental/Methods	364
16.4.1	Liquid-Liquid Equilibrium Measurements	364
16.4.2	Experiments on Subsystems	365
16.4.2.1	Isopiestic Measurements	365
16.4.2.2	Measurement of pH	365
16.4.2.3	Determination of the Protein Net-Charge	366
16.4.3	Materials	366
16.5	Experimental Results/Modelling	367
16.5.1	Parameters for the Gibbs Excess Energy Model	367
16.5.2	Protein Net Charge	371
16.5.3	Protein Partitioning	371
16.6	Conclusions	374
	Acknowledgements	375
	References	375
	Nomenclature	376
V	**Phase Equilibrium of Polymer Systems**	**379**
17	**Phase Behavior of Quaternary Polymer Solutions**	**381**
	Claudia Barth-Wiedmann, Matthias Wünsch, and Bernhard Anton Wolf	
17.1	Abstract	381
17.2	Introduction	381
17.3	Experimental Part	382
17.3.1	Apparatus and Procedures	382
17.3.1.1	Static Light Scattering	382
17.3.1.2	Headspace-Gaschromatography	382
17.3.1.3	Determination of Phase Diagrams	384
17.3.2	Materials	384

17.4	Theoretical Background	385
17.4.1	Flory-Huggins Interaction Parameters	385
17.4.2	Calculation of Phase Diagrams	387
17.4.3	Calculation of Vapor Pressures	389
17.5	Results and Discussion	390
17.5.1	Binary Systems	390
17.5.1.1	Mixed Solvents	390
17.5.1.2	Polymer Solutions	391
17.5.2	Ternary Systems	393
17.5.2.1	Phase Diagrams	393
17.5.2.2	Vapor Pressures	395
17.5.3	Quaternary Systems	398
17.6	Conclusions	399
	Acknowledgement	399
	References	400
	Nomenclature	401
	Abbreviations	402

18	**Calculation of the High Pressure Phase Equilibrium of Mixtures of Ethylene, Poly(ethylene-co-vinylacetate) Copolymers and Vinyl Acetate with a Cubic Equation of State**	403
	C. Browarzik, D. Browarzik, and H. Kehlen	
18.1	Abstract	403
18.2	Introduction	403
18.3	Theory	405
18.3.1	Segment-Molar Quantities and Polydispersity	405
18.3.2	Calculation of the Cloud-Point Curves	407
18.3.3	Calculation of the Coexistence Curves	409
18.3.4	Sako-Wu-Prausnitz Equation of State (SWP-EoS)	411
18.3.5	Fit of Pure-Component Parameters	412
18.4	Results and Discussion	414
18.4.1	The Binary System Ethylene (A) + EVA (B)	414
18.4.2	The Binary System Ethylene (A) + VA (C)	421
18.4.3	The Binary System EVA (B) + VA (C)	422
18.4.4	The Ternary System	423
18.5	Conclusions	425
	References	426
	Nomenclature	428

19	**Modeling of Copolymer Phase Equilibria Using the Perturbed-Chain SAFT Equation of State**	430
	Feelly Tumakaka and Gabriele Sadowski	
19.1	Abstract	430
19.2	Introduction	430
19.3	Theory	431
19.3.1	Copolymer Concept	431

19.3.2	Perturbed-Chain SAFT and its Extension to Copolymer Systems	432
19.4	Pure-Component Polymer Parameters	434
19.5	Results for Copolymer Systems	437
19.5.1	Poly(ethylene-*co*-propylene) (PEP) Systems	437
19.5.2	Poly(ethylene-*co*-1-butene) (EB) Systems	439
19.5.3	Poly(ethylene-*co*-vinyl acetate) (EVA) Systems	441
19.5.4	Poly(ethylene-*co*-methyl acrylate) (EMA) Systems	443
19.6	Conclusions	446
	Acknowledgements	447
	References	447
	Nomenclature	449
20	**Cloud Point Pressures of Ethene + Acrylate + Poly(ethene-*co*-acrylate) Systems**	451
	Michael Buback and Markus Busch	
20.1	Abstract	451
20.2	Introduction	452
20.3	Experimental Methods	455
20.4	Results and Discussion	458
20.5	Conclusions	468
	References	468
	Nomenclature	470
21	**Gas Expanded Polymer Solutions**	472
	Bernd Bungert and Wolfgang Arlt	
21.1	Abstract	472
21.2	Introduction	472
21.3	Fundamentals	474
21.3.1	Separation Processes	474
21.3.1.1	Antisolvent Crystallization	474
21.3.1.2	Liquid-Liquid-Phase Separation of Polymer Solutions	475
21.3.1.3	Liquid-Liquid-Vapor to Vitrified Liquid-Vapor Equilibrium	475
21.3.1.4	Change of Properties by the Dissolution of Compressed Gases	477
21.3.2	Thermodynamic Modeling of Phase Equilibrium	477
21.4	Experimental	478
21.4.1	Apparatus and Chemicals	478
21.5	Results	479
21.6	Conclusion	484
	Acknowledgement	484
	References	484
	Nomenclature	486

22	**Calculation of the Stability and of the Phase Equilibrium on the System Methylcyclohexane + Polystyrene Based on an Equation of State**	488
	Dieter Browarzik and Mario Kowalewski	
22.1	Abstract	488
22.2	Introduction	488
22.3	Theory	490
22.3.1	Calculation of the Cloud-Point and the Shadow Curves	490
22.3.2	Calculation of the Spinodal Curve	492
22.3.3	Calculation of Critical Points	494
22.3.4	Sako-Wu-Prausnitz Equation of State (SWP-EoS)	495
22.4	Results	497
22.4.1	Parameter Fit	497
22.4.2	Formation of the Hour-Glass Curves in the Polydisperse Case	498
22.4.3	Application of the EoS to a Real Polydisperse System MCH + PS	504
22.5	Conclusions	506
	References	506
	Nomenclature	507

VI	**High-Pressure Phase Equilibria**	509
23	**Phase Equilibrium (Solid-Liquid-Gas) in Binary Systems of Poly(ethylene glycols), Poly(ethylene glycol) dimethyl ether with Carbon Dioxide, Propane, and Nitrogen**	511
	Eckhard Weidner and Veronika Wiesmet	
23.1	Abstract	511
23.2	Introduction	512
23.3	Experimental Equipment, Methods, and Substances	512
23.3.1	Polymers	512
23.3.2	Determination of Melting Point	512
23.3.3	Determination of Solubility	513
23.3.4	Correlation of the Experimental Results with SAFT	514
23.4	Results and Discussion	514
23.4.1	PEG + Carbon Dioxide	515
23.4.1.1	Liquid + Gas Systems	515
23.4.1.2	Solid-Liquid Transition	517
23.4.2	PEG + Nitrogen	519
23.4.2.1	Liquid + Gas Systems	519
23.4.2.2	Solid-Liquid Transition	520
23.4.3	PEG + Propane	520
23.4.3.1	Liquid + Gas Systems	520
23.4.3.2	Solid-Liquid Transition	523
23.4.4	Comparison of the Phase Behavior of Liquid and Solid PEGs With Different Gases	524
23.4.5	Influence of Functional Groups	525
23.4.5.1	Short Polymer Chains	526

23.4.5.2	Longer Polymer Chains	528
23.4.6	Classification of Phase Behavior for (Solid and Liquid) PEGs and Pressurized Gases	529
23.5	Application	530
23.5.1	Powder Generation from Poly(ethylene glycols)	531
23.5.2	Polyglycols as Lubricants in Climatisation Systems	531
23.6	Conclusion	532
	Acknowledgements	533
	References	533
	Nomenclature	534
24	**Measurements and Modeling of High-Pressure Fluid Phase Equilibrium of Systems Containing Benzene Derivatives and CO_2**	**535**
	Gerd Brunner, Oliver Pfohl, and Stanimir Petkov	
24.1	Abstract	535
24.2	Introduction	536
24.3	Experimental/Methods	536
24.3.1	Apparatus	536
24.3.2	Procedure	537
24.3.3	Analysis	538
24.3.4	Materials	539
24.4	Results and Discussion	539
24.4.1	Binary Systems: Benzene Derivative + Carbon Dioxide	539
24.4.2	Ternary Systems	543
24.4.3	Discussion of Experimental Results	550
24.4.3.1	Binary Systems Benzene Derivative + Carbon Dioxide	550
24.4.3.2	Ternary System *o*-Cresol + *p*-Cresol + Carbon Dioxide	550
24.4.3.3	Ternary Systems Benzene Derivative + Water + Carbon Dioxide	551
24.4.3.4	Ternary Systems Cresol Isomer + Ethanol + Carbon Dioxide	552
24.5	Theory/Methods	552
24.5.1	Equations of State	552
24.5.1.1	Cubic EoS	552
24.5.1.2	Association EoS	554
24.5.1.3	Mixing Rules	555
24.5.2	Parameter Estimation	556
24.5.2.1	Pure Component Parameters	556
24.5.2.2	Binary Systems: Determining Binary Interaction Parameters	559
24.5.2.3	Ternary Systems	559
24.6	Results and Discussion	563
24.7	Conclusion	565
	References	566
	Nomenclature	567

VII	**Phase Equilibrium in Microemulsion Systems**	569
25	**The Potential of Surfactants Modified Supercritical Fluids for Dissolving Different Classes of Substances**	571
	Uta Lewin-Kretzschmar and Peter Harting	
25.1	Abstract	571
25.2	Introduction	571
25.3	Experimental/Methods	573
25.3.1	Equipment	573
25.3.2	Procedure	573
25.3.3	Analysis	575
25.3.4	Dyeing Experiments	575
25.3.5	Materials	576
25.3.6	Statistical Methods	578
25.4	Correlation and Prediction of the Solubility of Dyes in Supercritical Fluids	578
25.4.1	Method of Mitra/Wilson	579
25.4.2	Method of del Valle/Aguilera	579
25.4.3	Group Contribution Equation of State	580
25.5	Results and Discussion	580
25.5.1	Solubility of Nitroaromatics in Pure and Surfactant-Modified $scCO_2$	580
25.5.2	Solubility of Adamantane in Pure and Surfactant-Modified Supercritical Fluids	584
25.5.3	Solubility of Polar Dyes in Pure and Surfactant-Modified $scCO_2$ and scC_2H_6	585
25.5.4	Dyeing Tests	592
25.6	Conclusion	593
	Acknowledgement	594
	References	594
	Nomenclature	596
26	**Liquid-Liquid Phase Equilibria in Microemulsion Forming Systems Based on Carbohydrate Surfactants**	598
	Sabine Enders, Dirk Häntzschel, Heike Kahl, and Konrad Quitzsch	
26.1	Abstract	598
26.2	Introduction	598
26.3	Experimental	601
26.3.1	Apparatus	601
26.3.1.1	Phase Equilibrium	601
26.3.1.2	Surface Tension	602
26.3.1.3	Interfacial Tension	603
26.3.1.4	Dynamic Light Scattering	603
26.3.2	Materials	603
26.4	Theory	604
26.4.1	Micelle Formation Model	604
26.4.2	G^E-Model	609
26.4.3	Landau-Theory	611

26.5	Results and Discussion	613
26.5.1	Strategy and Concepts of the Investigations	613
26.5.2	Binary Subsystems	614
26.5.3	Ternary Subsystems	615
26.5.4	Quaternary Systems	619
26.6	Conclusion	625
	Acknowledgement	626
	References	626
	Nomenclature	629
27	**Interactions of Polyelectrolytes and Zwitterionic Surfactants in Aqueous Solution**	632
	Heinz Hoffmann, Holger Lauer, and Klaus Redlich	
27.1	Abstract	632
27.2	Introduction	633
27.3	Experimental/Methods	633
27.4	Results and Discussion	634
27.4.1	Aqueous Solutions of Polyacrylic Acids	634
27.4.2	Aqueous Solutions of $C_{14}DMAO$	636
27.4.3	Interactions between PAA and $C_{14}DMAO$ in Aqueous Solution	638
27.4.4	Modified PAAs in Aqueous Solution	642
27.4.5	Interactions Between Modified PAAs and $C_{14}DMAO$ in Aqueous Solution	646
27.5	Conclusion	650
	Acknowledgement	651
	References	651
	Nomenclature	651

Preface

This book presents summaries of the results achieved within a priority program funded by Deutsche Forschungsgemeinschaft (SPP 736) from 1995 to 2002. The main aim of this program was to enable experimental and theoretical work in the area of applied thermodynamics, in particular in phase equilibrium of complex fluid mixtures e.g. in the area of chemical, biotechnological and environmental process sciences. Altogether 27 single research projects were co-ordinated and funded. The research topics range from investigations of fluid systems where the complex behaviour results from chemical reactions (including associating substances) through the phase equilibrium in aqueous solutions of strong electrolytes, aqueous two-phase systems and polymer solutions, to high pressure phase equilibrium phenomena and the properties of microemulsion systems. The priority program was initiated by the editor with support by the late Professor Konrad Bier, Karlsruhe, Professor Klaus Lucas, Aachen, and Professor Karl Stephan, Stuttgart. Professor Dieter Mewes, Hannover, provided strong support within DFG. The project proposals were evaluated and recommended for support by an advisory committee, chaired by Professor Erich Hahne, Stuttgart. Professors Hans-Jürgen Bittrich and Wolfgang Fratzscher, both from Merseburg, Dietrich Wörmann, Köln, Manfred Zeidler, Aachen, and Drs. Bernhard Gutsche, Düsseldorf, and Gerhard Hochgesand, Heusenstamm, served in this committee. Also on behalf of all research grantees, I want to express my gratitude to all those colleagues. Last not least, I appreciate the support by Dr. Walter Lachenmeier of DFG who not only administrated the priority program, but was always open minded to help us to overcome the unavoidable problems during the course of the projects. This book is not only to present the research results, but it is dedicated to all the aforementioned institutions and in particular the persons for their generous and long lasting assistance. Without that dedication, it would have been impossible to achieve the goals. Mrs. Maike Petersen of Wiley-VCH, Weinheim, deserves my thanks both for her advice and her patience with us (the authors and the editor). Last not least I want to thank my secretary Mrs. Monika Reim for assisting me in the time consuming and tedious process of editing/proof reading the various manuscripts and following up the correspondence with the authors.

Kaiserslautern, July 2003　　　　　　　　　　　　　　　　　　　　　　　　　　Gerd Maurer

I Chemical Reactive Systems

1 Experimental and Theoretical Investigations at the System $CH_3OH + H_2SO_4 + H_2O$

Matthias Behmann, Sae-Hoon Kim, Hans-Jürgen Koß, and Klaus Lucas [1] *

1.1 Abstract

The thermodynamic properties of the system $CH_3OH + H_2SO_4 + H_2O$ and their subsystems are studied by Raman spectroscopy, vapor pressure measurement and calorimetry. The degrees of dissociation of the electrolytes in the systems $H_2SO_4 + H_2O$, $CH_3OH + H_2SO_4 + H_2O$, and $CH_3SO_4H + CH_3OH$ are determined at 293.15 K. The amount of CH_3SO_4H produced in the ternary system is measured. With the help of Raman spectroscopy, the partial vapor pressures of CH_3OH, H_2O, and $(CH_3)_2O$ are determined as a function of time in order to observe the formation of $(CH_3)_2O$ in the temperature range of 285 K to 370 K. A solvation model is used to describe the thermodynamic properties of the system $CH_3OH + H_2SO_4 + H_2O$.

1.2 Introduction

The esterification is one of the most important reactions in organic synthesis. This reaction of carbon acids and primary alcohols is often supported by H_2SO_4 as catalyst. The subsystems of $CH_3OH + H_2SO_4 + H_2O$ already show a very complex behavior due to strong electrolytes, chemical reactions and the resulting large number of components in the mixture. In liquid phase, parts of H_2SO_4 dissociate to HSO_4^- and SO_4^{2-}. The acid and CH_3OH react to CH_3SO_4H. Due to the acidic character of CH_3SO_4H, there is an additional dissociation reaction. At higher temperatures, the formation of $(CH_3)_2O$ from CH_3OH is observed.

  Though there are many experimental data available for the thermodynamic properties of the subsystems $H_2SO_4 + H_2O$ and $CH_3OH + H_2O$, only a few measurements are reported for the system $CH_3OH + H_2SO_4 + H_2O$. Most of them are concerned with the formation of

---

[1] RWTH-Aachen, Lehrstuhl für Technische Thermodynamik, Schinkelstr. 8, 52056 Aachen.
* Author to whom correspondence should be addressed.

CH$_3$SO$_4$H. Almost no data are available for the dissociation behavior of H$_2$SO$_4$ and CH$_3$SO$_4$H, which strongly influence the properties of the electrolyte system.

In this work, mixtures of CH$_3$OH + H$_2$SO$_4$ + H$_2$O and their subsystems are studied in liquid and vapor phase. The degrees of dissociation of aqueous sulfuric acid are determined over a wide concentration range by Raman spectroscopy. Using CH$_3$OH and (CH$_3$)$_2$O as standards for calibration, no assumptions on the dissociation behavior of the acid were needed. CH$_3$SO$_4$H and CH$_3$SO$_4^-$ are distinguished by Raman spectroscopy for the first time in this work and their dissociation in CH$_3$OH is studied. In CH$_3$OH + H$_2$SO$_4$ + H$_2$O systems, the amount of CH$_3$SO$_4$H produced and the degrees of dissociation of H$_2$SO$_4$ and CH$_3$SO$_4$H are determined. The partial pressures of CH$_3$OH and H$_2$O are measured for these systems. The formation of (CH$_3$)$_2$O in the liquid phase is observed by its increasing partial pressures at constant temperature. Caloric data on the mixing of CH$_3$OH and H$_2$SO$_4$ + H$_2$O complement the spectroscopic information.

The thermodynamic properties of the systems H$_2$SO$_4$ + H$_2$O and CH$_3$OH + H$_2$O are described by a Wilson (Wilson, 1964) based solvation model (Engels, 1990) over a wide concentration range considering caloric and phase equilibrium data. The behavior of the CH$_3$OH + H$_2$SO$_4$ + H$_2$O system is predicted and compared with experimental data.

1.3 Experimental Methods

1.3.1 Apparatus

For the Raman spectroscopic examinations, the samples are sealed in a quartz cell enabling the optical access to the mixture by laser light and the emitted Raman signal. The temperatures of the samples were hold within 0.2 K. A magnetic stirrer keeps the sample homogeneous.

For Raman measurements in both phases, a 9 W argon ion laser (Spectra Physics) of 514.5 nm wavelength is used. 90% of the laser power is focused into the vapor phase. As shown in Figure 1.1, a part of the scattered Raman signal from the vapor phase is guided to a 500 mm grating spectrometer (B&M).

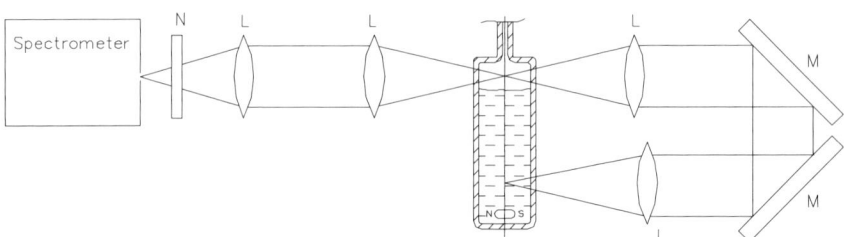

Figure 1.1: Signal path for Raman measurements in liquid and vapor phase (L: lens, M: mirror, N: notch filter).

A notch filter (Kaiser) suppresses the scattering light of the laser wavelength by six magnitudes. The Raman spectrum is detected by a liquid nitrogen cooled backilluminated CCD-camera (S&I). To analyze liquid and vapor phase in rapid succession, the Raman signal of the liquid is guided through the transparent vapor phase and then follows the already described signal path. In order to obtain the vapor pressure of the mixtures, the vapor phase of the sample is connected to a pressure gauge (Burster), which has an accuracy of 0.1 kPa for pressures up to 110 kPa.

The enthalpies of dilution of aqueous sulfuric acid with CH_3OH are determined in an isothermal flow calorimeter (Hart Scientific model 4250). The acid is enclosed in a chemically inert bag of fluorinated ethylene propylene (FEP), which is placed inside a metal vessel connected with the calorimeter. Pumping oil into the vessel, an equivalent amount of the acid flows into the calorimeter.

1.3.2 Materials

The examined mixtures are made of sulfuric acid (Merck, >98%), methanol (Fluka, >99.8%), and twice distilled water. The compositions of the mixtures are determined by weighing the sample after stepwise mixing with a balance, which has an accuracy of 0.1 mg. The concentration of the sulfuric acid is determined by titration with an accuracy of 0.3 mass%. Water-free sulfuric acid is produced by adding SO_3 vapor (Fluka, >98%) to 98% H_2SO_4. Samples for the determination of vapor pressure are mixed from substances degassed in a supersonic bath under vacuum.

To enable spectroscopic studies of CH_3SO_4H, the ester is synthesized from equal amounts of CH_3OH and $ClSO_3H$ (Fluka, >98%) at 232 K as described by Müller (1963) and Klages et al. (1966). The resulting transparent fluid is frozen by liquid nitrogen when not used in the experiments.

1.3.3 Calibration and Analysis of Raman Signals

The concentration ratios of components in a mixture can be determined from the ratios of pure substance areas in the Raman spectrum. The spectra of pure components are reproduced by a series of Gaussian, Lorentzian, and Voigt profiles, each specified by frequency, half width, and height. The ratios of all the profile areas within a pure substance are always kept constant. The experimental Raman spectrum of a mixture is fitted from the reproduced pure component spectra with a concentration dependent factor. The square of the differences between the measured and the reproduced signal is minimized for all frequencies. The Raman signals from various frequency ranges of a pure substance are combined in its reproduced spectrum. This enables the analysis of experimental spectra even with severe overlapping Raman bands, which are common in multi-component mixtures. The uncertainties in concentrations of the mixtures are calculated from the uncertainty of the calibration factors and the possible errors in determining the component areas in the Raman spectrum. The main source of error results from the overlapping of neighbored lines.

1 Experimental and Theoretical Investigations at the System $CH_3OH + H_2SO_4 + H_2O$

CH_3OH, H_2O, and $(CH_3)_2O$ are found in vapor phase Raman spectra of the system $CH_3OH + H_2SO_4 + H_2O$. $(CH_3)_2O$ is formed by liquid phase reaction. The electrolytes H_2SO_4 and CH_3SO_4H in the vapor phase can be neglected due to their small vapor pressures. The concentrations of all three substances in the gas phase are determined by analyzing Raman spectra between 2700 cm^{-1} and 3900 cm^{-1}. Figure 1.2 shows a measured spectrum of the vapor phase.

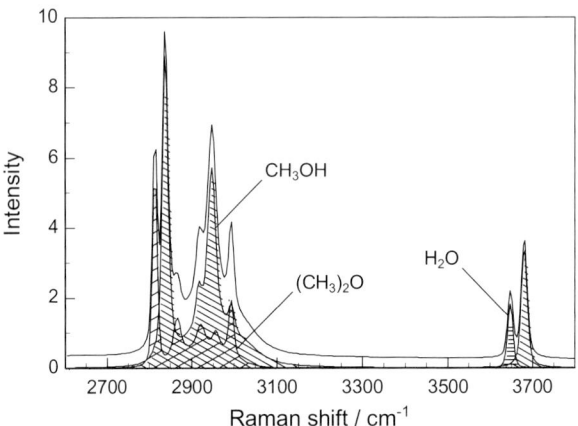

Figure 1.2: Raman spectrum from the vapor phase of the system $CH_3OH + H_2SO_4 + H_2O$ and its partitioning into CH_3OH, H_2O, and CH_3OCH_3 spectrum.

There is only one Raman line of H_2O at 3655 cm^{-1} in the analyzed frequency range (Kohlrausch, 1943; Herzberg, 1964). The Raman signals of CH_3OH are found at 3690 cm^{-1} (OH stretching vibration) and in the range of 2845 cm^{-1} to 3000 cm^{-1} (Furic et al., 1993). Due to the CH_3 groups in the molecule, the Raman spectra of $(CH_3)_2O$ and CH_3OH are seriously overlapping (Taylor and Vidale, 1957). However, the above-described method of analyzing spectra allows the determination of the concentrations within 0.4 mol%. For very small amounts of CH_3OH compared to $(CH_3)_2O$, only the OH stretching line of the alcohol in the upper frequency range is taken into account. No influence of the temperature to the form of Raman spectra was detected in the measured temperature range of 293 K to 363 K.

The system $H_2SO_4 + H_2O$, with CH_3OH and $(CH_3)_2O$ as reference substances, and the system $CH_3SO_4H + CH_3OH$ are studied for the Raman calibration of the electrolytes and their dissociation products. The calibration results of the subsystems are combined for the analysis of the total system. Figure 1.3 shows the Raman spectrum of a mixture of aqueous sulfuric acid and CH_3OH in the analyzed frequency range of 400 cm^{-1} to 1700 cm^{-1}. The broad unspecific Raman signals of H_2O are not given.

The SO_4^{2-} ion is mainly represented by its symmetric stretching vibration found at 982 cm^{-1}. The Raman bands at 890 cm^{-1} and 1040 cm^{-1} are caused by the HSO_4^- ion. The in-phase HO–SO$_2$–OH vibration of H_2SO_4 is found at 910 cm^{-1}. Close to 600 cm^{-1} and 430 cm^{-1} there are two more Raman contours with contributions from H_2SO_4, HSO_4^-, and SO_4^{2-}. All above discussed signals are combined to reproduce the spectra of the acid and its

1.3 Experimental Methods

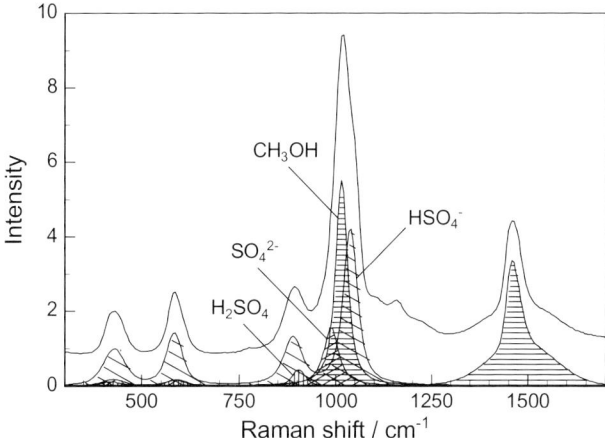

Figure 1.3: Raman spectrum of the liquid phase of the system $CH_3OH + H_2SO_4 + H_2O$ and its partitioning into CH_3OH, H_2SO_4, HSO_4^-, and SO_4^{2-} spectrum.

ions. Due to the strong interactions in the electrolyte mixtures, there are systematic changes in frequency and half width of certain Raman signals (Walrafen et al., 2000). By controlling the parameters of the profiles in the reproduced component spectrum, this phenomen can be taken into account. Nevertheless the ratios of the profile areas are always kept constant for each substance. In the analyzed spectral range, the Raman signals of CH_3OH consist of the lines at 1007 cm^{-1} and at 1460 cm^{-1} assigned to the CO stretching vibration and the deformation vibration of the methyl group. There is another broad band close to the latter frequency. A small Raman line at 1111 cm^{-1} is related to the deformation vibration of the molecule (Furic et al., 1993). Since the band of the CO vibration is found at 1034 cm^{-1} in pure CH_3OH, there is a drastic shift in frequency due to the interaction of CH_3OH with the surrounding liquid.

In the high concentration range of sulfuric acid, $(CH_3)_2O$ is used as reference substance. Due to the basic character of the ether (Arnett et al., 1961), $(CH_3)_2OH^+$ ions have to be taken into account. A Raman spectrum of the $(CH_3)_2O + H_2SO_4 + H_2O$ mixture is shown in Figure 1.4.

The unspecific Raman bands of the OH groups of aqueous sulfuric acid in this frequency range are shown as dashed line in Figure 1.4. The calibration factors of molecular H_2SO_4, HSO_4^-, and $(CH_3)_2O$ with respect to $(CH_3)_2OH^+$ are determined with an accuracy of 2%. No amount of the SO_4^{2-} ion is found at these high concentrations of H_2SO_4.

The calibration factor of CH_3OH with respect to $(CH_3)_2O$ is determined from Raman spectra of the system $CH_3OH + (CH_3)_2O$ in the ether concentration range of 1.4 mol% to 7.5 mol% with an accuracy of 1.2%. The spectra are analyzed in the frequency range of 700 cm^{-1} to 1100 cm^{-1}. Combining the already determined calibration factors, the Raman measuring system is expanded to the second dissociate of sulfuric acid by treating mixtures of aqueous sulfuric acid and methanol (Fig. 1.3). Recording the spectra within a short time after preparation of the mixtures and keeping the sample at 293.15 K, the slow formation of CH_3SO_4H and $(CH_3)_2O$ can be neglected. The calibration factor of SO_4^{2-} is determined from

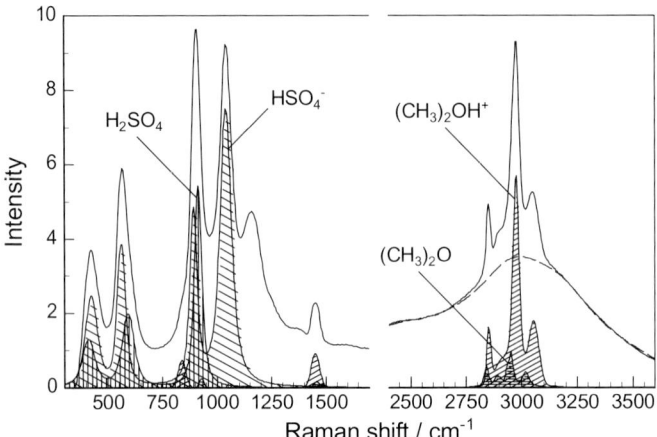

Figure 1.4: Raman spectrum of the system $(CH_3)_2O + H_2SO_4 + H_2O$ and its partitioning into molecular H_2SO_4, HSO_4^-, $(CH_3)_2O$, and $(CH_3)_2OH^+$.

Eq. (1.1) with an accuracy of 3%. The calibration factors of H_2SO_4 and HSO_4^- are already determined from the above described measurements.

$$\left(\frac{n^0_{H_2SO_4}}{n^0_{CH_3OH}}\right)_i = c_{H_2SO_4,CH_3OH}\left(\frac{A_{H_2SO_4}}{A_{CH_3OH}}\right)_i + c_{HSO_4^-,CH_3OH}\left(\frac{A_{HSO_4^-}}{A_{CH_3OH}}\right)_i$$
$$+ c_{SO_4^{2-},CH_3OH}\left(\frac{A_{SO_4^{2-}}}{A_{CH_3OH}}\right)_i \quad (1.1)$$

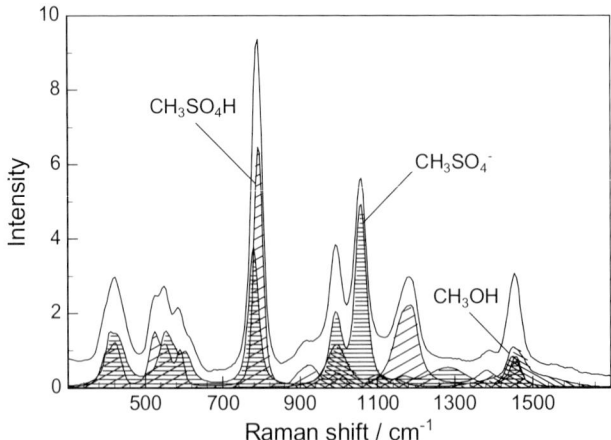

Figure 1.5: Raman spectrum from the system $CH_3SO_4H + CH_3OH$ and its partitioning into CH_3OH, CH_3SO_4H, and $CH_3SO_4^-$ spectrum.

The spectroscopic method of this work allows the quantitative determination of the concentrations of CH_3OH, $(CH_3)_2O$, H_2SO_4, and its two dissociated forms in the liquid phase.

For Raman spectroscopic treatment of CH_3SO_4H, it is synthesized from $ClSO_3H$ and CH_3OH. The intense vibration band of $H_3C-O-SO_3$ at 810 cm^{-1} is most characteristic for the Raman spectrum of molecular CH_3SO_4H, which has not been reported in the literature before (Fig. 1.5).

The characteristic line of $CH_3SO_4^-$ is found at 783 cm^{-1} (Siebert, 1957; Okabayashi et al., 1974; Fischer and Beckenkamp, 1998). There are two more strong Raman lines at 1005 cm^{-1} and 1062 cm^{-1}. The calibration factors of CH_3SO_4H and $CH_3SO_4^-$ with respect to CH_3OH are determined from two series of dilution, where known masses of CH_3OH are consecutively added to a constant amount of CH_3SO_4H.

1.4 Theory

The solvation model of Engels (1990) is used to describe the system $CH_3OH + H_2SO_4 + H_2O$. Rütten et al. (1998) and Kim (1998) already showed that the solvation model is adequate for multi-component electrolyte systems but they considered only first dissociation reaction of H_2SO_4. In this work, the solvation model of Engels (1990) is extended in order to describe the two dissociation reactions of the system $H_2SO_4 + H_2O$. It is well known that the caloric and phase equilibrium data of the system $CH_3OH + H_2O$ cannot be described simultaneously with commonly used local composition models. Katz et al. (1989) concluded that the mixtures of CH_3OH and H_2O act as a ternary system, which consists of CH_3OH, H_2O, and $CH_3OH + H_2O$ (1:1) complex. As the solvation model is a general reaction model, the association of CH_3OH and H_2O can be also taken into account. The reactions between CH_3OH and H_2SO_4 are neglected, as there are no sufficient data to determine the interaction parameters of the model. As most of the existing data lay in low concentration range of H_2SO_4, the thermodynamic properties of the ternary system are predicted only with binary parameters of the systems $CH_3OH + H_2O$ and $H_2SO_4 + H_2O$.

The reactions considered in this model can be formulated as

$$H_2SO_4 + m^{(1)} H_2O \rightleftarrows C^{(1),+} + C^{(1),-} \ (= 2\,C^{(1)}) \qquad m^{(1)} = 1,\ v^{(1)} = 2 \qquad (1.2)$$

$$H_2SO_4 + m^{(2)} H_2O \rightleftarrows 2\,C^{(2),+} + C^{(2),2-} \ (= 3\,C^{(2)}) \qquad m^{(2)} = 1,\ v^{(2)} = 3 \qquad (1.3)$$

$$CH_3OH + m^{(3)} H_2O \rightleftarrows C^{(3)} \qquad m^{(3)} = 1,\ v^{(3)} = 1 \qquad (1.4)$$

with the equilibrium constants defined by

$$K_a^{(1)} = \frac{x_{C^{(1)}}^{v^{(1)}}}{x_W^{m^{(1)}} x_{SA}} \frac{\gamma_{C^{(1)}}^{v^{(1)}}}{\gamma_W^{m^{(1)}} \gamma_{SA}}, \quad K_a^{(2)} = \frac{4 x_{C^{(2)}}^{v^{(2)}}}{x_W^{m^{(2)}} x_{SA}} \frac{\gamma_{C^{(2)}}^{v^{(2)}}}{\gamma_W^{m^{(2)}} \gamma_{SA}}, \quad K_a^{(3)} = \frac{x_{C^{(3)}}^{v^{(3)}}}{x_W^{m^{(3)}} x_M} \frac{\gamma_{C^{(3)}}^{v^{(3)}}}{\gamma_W^{m^{(3)}} \gamma_M} \qquad (1.5)$$

The subscripts W, SA, and M indicate H_2O, H_2SO_4, and CH_3OH, respectively. γ_i is the activity coefficient of component i, $m^{(k)}$ the number of H_2O molecules used in reaction k. $C^{(k)}$ is the complex and $v^{(k)}$ is the number of products of reaction k. The constant 4 in Eq. (1.5) results from the fact that the two cation-complexes of Eq. (1.3) are not distinguishable. The ratio of molar quantities before and after the reactions is given by

$$n^0/n = 1 + \sum_{k=1}^{3}(m^{(k)} + 1 - v^{(k)})x_{C^{(k)}} \tag{1.6}$$

The mole fractions x_i after the reactions can be calculated by

$$x_W = \left(\frac{n^0}{n}\right)x_W^0 - m^{(1)}x_{C^{(1)}} - m^{(2)}x_{C^{(2)}} - m^{(3)}x_{C^{(3)}} \tag{1.7}$$

$$x_{SA} = \left(\frac{n^0}{n}\right)x_{SA}^0 - x_{C^{(1)}} - x_{C^{(2)}} \tag{1.8}$$

$$x_M = \left(\frac{n^0}{n}\right)x_M^0 - x_{C^{(3)}} \tag{1.9}$$

In the above equations the superscript 0 stands for the state before reaction occurs. The activity coefficients of each component are calculated with (Wilson, 1964):

$$\ln \gamma_i = 1 - \ln\left[\sum_j^N x_j \tau_{ij}\right] - \sum_k^N \frac{x_k \tau_{ki}}{\sum_j^N x_j \tau_{kj}} \tag{1.10}$$

The interaction parameters of the complexes $C^{(i),+}$ and $C^{(i),-}$ are set equal in order to reduce the number of model parameters. The temperature dependence of the interaction parameters τ_{ij} of Wilson and the equilibrium constants $K_a^{(k)}$ is formulated in the following forms:

$$\tau_{ij} = \exp[a_{ij} + b_{ij}/T + c_{ij}T] \tag{1.11}$$

$$\ln K_a^{(k)} = A^{(k)} + B^{(k)}/T \tag{1.12}$$

The calculating procedures of thermal properties and the properties of the pure substances used in this model are described elsewhere (Engels, 1990; Kim, 1997; Rütten et al., 1998) in detail and will not be discussed here. The vapor pressures of the complexes are assumed to be zero. In Table 1.1, the experimental data used for the approximation of the model parameters and the calculation results (in form of mean deviation $\delta_m/\%$) are listed. Table 1.2 shows the model parameters obtained from the approximation.

1.4 Theory

Table 1.1: Experimental data used for the approximation.

Reference	No. of data	Type	x_{SA} or x_M	T/K	$\delta_m/\%$
$H_2SO_4 + H_2O$					
Tarasenkov (1955)	76	p	0.0200 – 0.4778	293.15 – 373.15	2.58
Collins (1933)	146	p	0.0727 – 0.2583	298.15 – 408.15	1.97
Burt (1904)	192	p	0.0575 – 0.5833	298.15 – 401.70	2.87
Giauque et al. (1960)	90	H_m^E	0.0177 – 0.8000	298.15	2.38
	90	c_p	0.0177 – 0.8000	298.15	0.75
	90	a_W	0.0177 – 0.8000	298.15	3.24
Kim and Roth (2001)	106	H_m^E	0.0108 – 0.7877	293.15 – 333.15	2.91
this work	22	$\alpha_{SA,1}$	0.0090 – 0.8700	293.15	2.85
	22	$\alpha_{SA,2}$	0.0090 – 0.8700	293.15	1.74
$CH_3OH + H_2O$					
Kooner et al. (1980)[b]	13	p	0.0444 – 0.9361	298.14	0.80
	13	y_M	0.0444 – 0.9361	298.14	0.21
Mc Glashan and	31	p	0.0408 – 0.9514	308.15 – 338.15	0.82
Williamson (1976)[a]	31	y_M	0.0408 – 0.9514	308.15 – 338.15	0.53
Ratcliff and Chao (1969)[a]	10	p	0.2095 – 0.9510	313.05	1.19
	10	y_M	0.2095 – 0.9510	313.05	1.54
Broul et al. (1969)[a]	12	p	0.0343 – 0.7582	333.15	0.95
	12	y_M	0.0343 – 0.7582	333.15	0.65
Abello (1970)[c]	11	H_m^E	0.0780 – 0.7955	298.15	1.54
Benjamin and Benson (1963)[c]	15	H_m^E	0.0292 – 0.9528	298.15	1.98
Bertrand et al. (1966)[c]	9	H_m^E	0.1290 – 0.9120	298.15	1.68
Lama and Lu (1965)[c]	18	H_m^E	0.0287 – 0.9348	298.15	2.19
Tomaszkiewicz et al. (1986)[d]	11	H_m^E	0.0870 – 0.7890	298.15	1.72
Ocon et al. (1959)[c]	21	H_m^E	0.0650 – 0.8650	308.55 – 326.65	2.74
this work	5	H_m^E	0.1003 – 0.4007	298.15	1.24

a) from Gmehling and Onken (1977)
b) from Gmehling et al. (1988)
c) from Christensen et al. (1984)
d) from Gmehling and Holderbaum (1988)

Table 1.2: Adjusted parameters of the model.

i	j	a_{ij}	b_{ij}/K	c_{ij}/K^{-1}
H_2O	H_2SO_4	4.2825	–	–0.0041696
H_2O	CH_3OH	–	12.838	–
H_2O	$C^{(1)}$	3.0902	–	–0.0045159
H_2O	$C^{(2)}$	3.3642	–	–0.0072439
H_2O	$C^{(3)}$	–	150.42	–
H_2SO_4	H_2O	3.1963	–	–0.0050681

11

Table 1.2 (continued)

i	j	a_{ij}	b_{ij}/K	c_{ij}/K^{-1}
H_2SO_4	$C^{(1)}$	0.40552	–	–0.0098070
H_2SO_4	$C^{(2)}$	–	–	–0.036877
CH_3OH	H_2O	–	–589.66	–
CH_3OH	$C^{(3)}$	–	–4.5615	–
$C^{(1)}$	H_2O	4.3129	–	–0.0080645
$C^{(1)}$	H_2SO_4	2.7488	–	–0.0043742
$C^{(1)}$	$C^{(2)}$	–25.839	–	–
$C^{(2)}$	H_2O	–0.10041	–	0.016678
$C^{(2)}$	H_2SO_4	12.354	–	–0.0049475
$C^{(2)}$	$C^{(1)}$	9.458	–	–0.0028993
$C^{(3)}$	H_2O	–	750.98	–
$C^{(3)}$	CH_3OH	–	738.31	–

k	$A^{(k)}$	$B^{(k)}/K$
1	–4.2682	3393.0
2	–33.860	7214.5
3	–4.6008	266.62

1.5 Results and Discussion

The concentrations of the mixtures and the type of the data studied in this work are summarized in Figure 1.6.

The dissociation behavior of aqueous sulfuric acid is determined in the concentration range of 1 mol% to 87 mol% at 293.15 K. At the same temperature, the influence of CH_3OH on the ion equilibrium is studied for acid concentrations below 36 mol%. The enthalpies of dilution of 10.3 mol%, 17.8 mol%, and 25.2 mol% aqueous sulfuric acid with CH_3OH are determined. For mixtures of CH_3OH and concentrated aqueous sulfuric acid of 49 mol%, 70 mol%, 93 mol%, and 100 mol%, the formation of CH_3SO_4H is brought to equilibrium. The formed amounts of CH_3SO_4H and the dissociation behavior of CH_3SO_4H and H_2SO_4 are determined at 293.15 K. For mixtures from 49 mol% H_2SO_4, the vapor pressures of CH_3OH and H_2O and the increasing amounts of $(CH_3)_2O$ are studied between 285 K and 370 K. In addition to Figure 1.6, Raman measurements are carried out on the systems $(CH_3)_2O + H_2SO_4 + H_2O$ and $CH_3SO_4H + CH_3OH$.

1.5 Results and Discussion

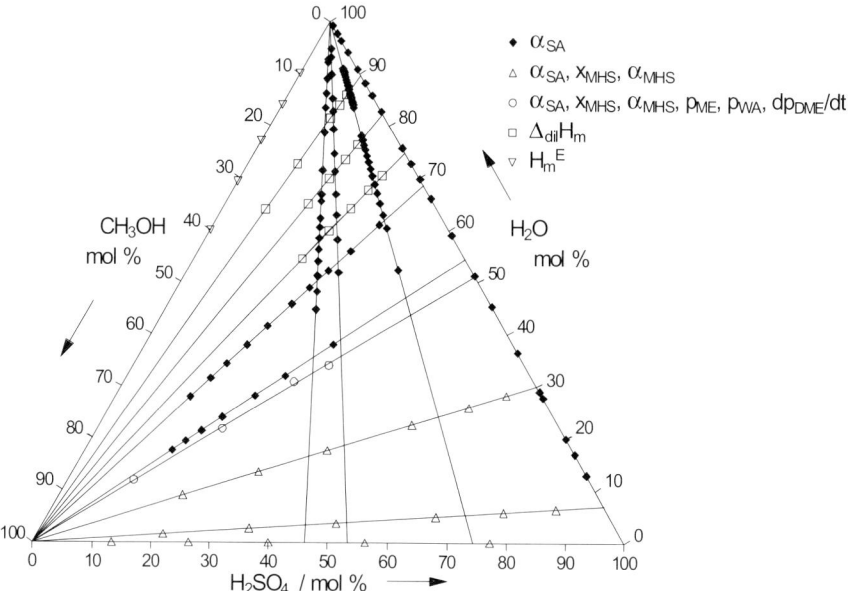

Figure 1.6: Determined properties of mixtures containing CH$_3$OH, H$_2$SO$_4$, and H$_2$O.

1.5.1 The System CH$_3$OH + H$_2$O

The calculation results of the model are shown in Figures 1.7 and 1.8.

By considering the association reaction of CH$_3$OH and H$_2$O, the phase equilibrium and caloric measurements can be described simultaneously in the temperature range of 298.15 K to 338.15 K. The mean deviation, δ_m, of the measured data from calculation for each data set is given in Table 1.1.

The excess molar enthalpy of the system measured in this work agrees very well with the values reported in DECHEMA Chemistry Data Series (Christensen et al., 1984; Gmehling and Holderbaum, 1988).

1.5.2 The System H$_2$SO$_4$ + H$_2$O

The dissociation behavior of aqueous sulfuric acid is studied by Raman spectroscopy. In Figure 1.9, both the measured and calculated results are presented. In concentrated sulfuric acid, the portion of HSO$_4^-$ with respect to the total amount of the acid grows rapidly as H$_2$O is added and reaches its maximum at 41 mol%. Below 50 mol% it is lowered by the portion of SO$_4^{2-}$, which rises up to 33% of the total amount of sulfuric acid and falls again for 10% in the more dilute solutions. The portion of molecular H$_2$SO$_4$ in the concentrated acid solution is continuously diminished with increasing amounts of H$_2$O in the mixtures. The frac-

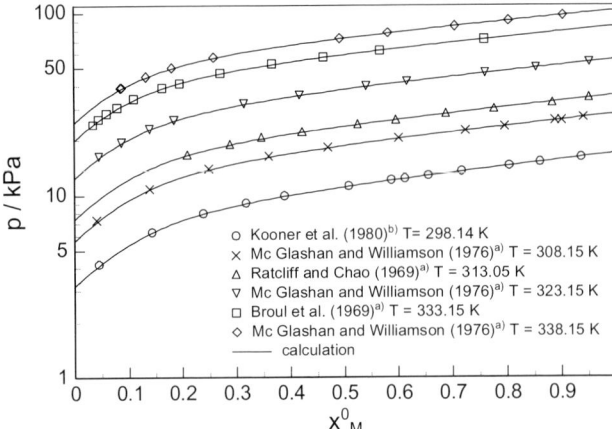

Figure 1.7: Vapor pressure of the system $CH_3OH + H_2O$: a) from Gmehling and Onken (1977); b) from Gmehling et al. (1988).

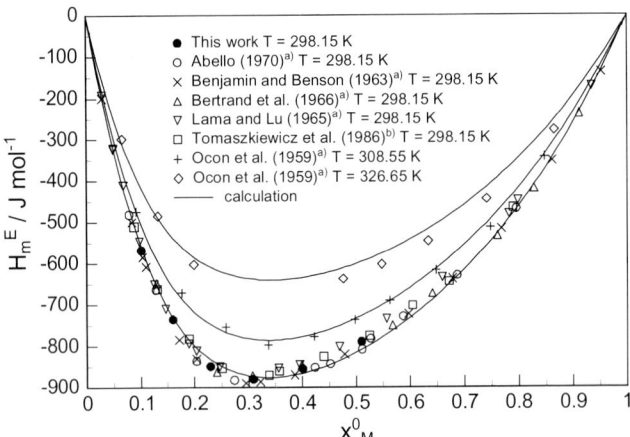

Figure 1.8: Excess molar enthalpy of the system $CH_3OH + H_2O$: a) from Christensen et al. (1984); b) from Gmehling and Holderbaum (1988).

tion of the undissociated acid molecules is almost constant below 40 mol% aqueous sulfuric acid, while decreases again at concentrations smaller than 5 mol%.

In contrast to the common opinion (Young et al., 1959; Walrafen et al., 2000) our Raman measurements yield some small Raman signals of molecular H_2SO_4 left below 40 mol% acid concentration. The large error bars at low concentrations of H_2SO_4 shown in Figure 1.9 result from the serious overlapping of the strong Raman signals of HSO_4^- and SO_4^{2-} and the weak H_2SO_4 signals as shown in Figure 1.3. The degrees of dissociation, $\alpha_{SA,1}$ and $\alpha_{SA,2}$, of H_2SO_4 can be determined within 5% or better at higher concentration range.

1.5 Results and Discussion

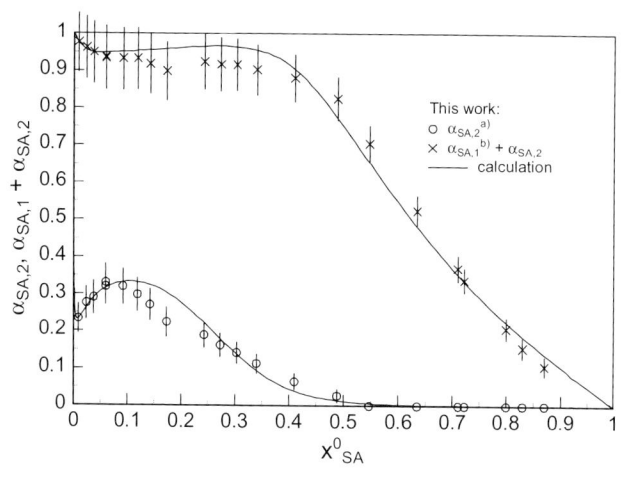

Figure 1.9: Degrees of dissociation of H_2SO_4 in the system $H_2SO_4 + H_2O$ at 293.15 K.
a) $\alpha_{SA,2} = x_{SO_4^{2-}}/(x_{H_2SO_4} + x_{HSO_4^-} + x_{SO_4^{2-}})$; b) $\alpha_{SA,1} = x_{HSO_4^-}/(x_{H_2SO_4} + x_{HSO_4^-} + x_{SO_4^{2-}})$.

At acid concentrations above 45 mol%, $\alpha_{SA,1}$ from this work is smaller than that of Walrafen et al. (2000) and Young at al. (1959). But they are still within the error limits of our results accept for the highest concentration value of Young et al. (1959). In order to compare our measurements with other literature values, the concentration ratio $Q_c = [SO_4^{2-}][H^+]/[HSO_4^-]$ is calculated and given in Figure 1.10.

The uncertainties of our Q_c values represented by the bars are calculated from the uncertainty of the ion concentrations seen in Figure 1.9. Up to an acid concentration of 6 mol l^{-1} the Q_c values of Turner (1974) and Young et al. (1959) are in good agreement with our results. At higher concentrations the values of Young et al. (1959) are smaller. The

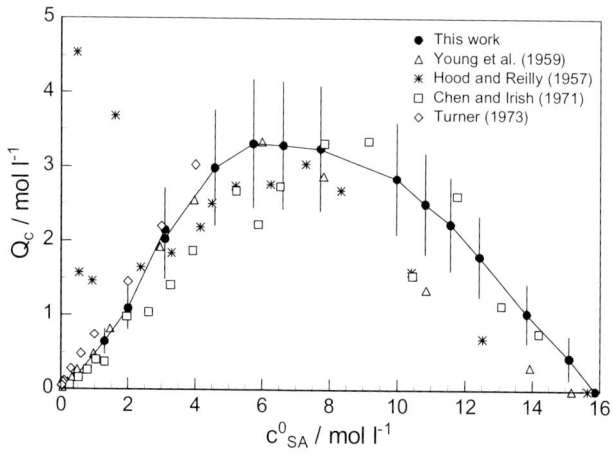

Figure 1.10: Comparison of the ion concentration ratio Q_c of this work and literatures.

1 Experimental and Theoretical Investigations at the System $CH_3OH + H_2SO_4 + H_2O$

NMR values of Hood and Reilly (1957) are similar to that of Young et al. (1959) except for small concentration range below 2 mol l^{-1}. The Q_c values of Chen and Irish (1971) are comparable to our results, but smaller below 7 mol l^{-1}.

The calculated degrees of dissociation of this system are also in good agreement with the measured values within the uncertainty of the measurements (Fig. 1.9). The large deviation at the concentration range of 10 mol% to 25 mol% can be explained by simplification of the dissociation reactions in form of Eqs. (1.2) and (1.3). The other calculation results of the system are given in Figures 1.11 and 1.12. With only one set of parameters, the vapor pressure, water activity, heat capacity, excess enthalpy and degrees of dissociation can be described in large temperature and concentration ranges.

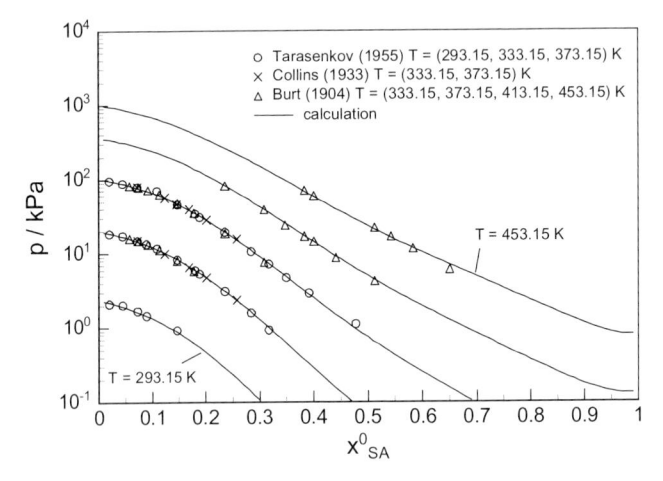

Figure 1.11: Vapor pressure of the system $H_2SO_4 + H_2O$.

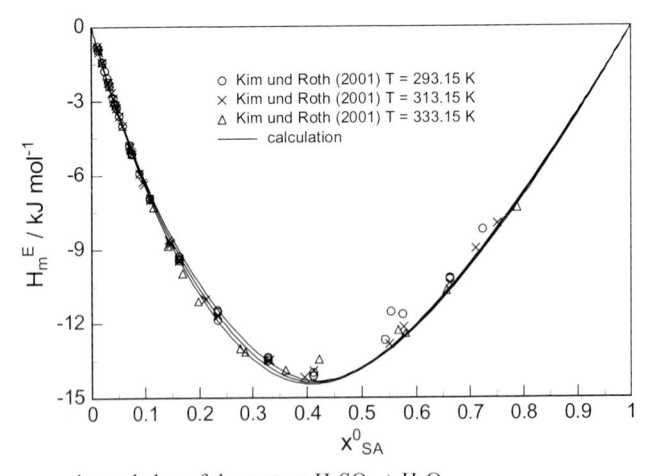

Figure 1.12: Excess molar enthalpy of the system $H_2SO_4 + H_2O$.

1.5 Results and Discussion

1.5.3 The System $CH_3OH + H_2SO_4 + H_2O$

The degrees of dissociation of H_2SO_4 in H_2O and CH_3OH are studied by Raman spectroscopy. The results are shown in Figure 1.13. As mentioned in Section 1.3.3, the formation of CH_3SO_4H and $(CH_3)_2O$ is slow at low acid concentrations.

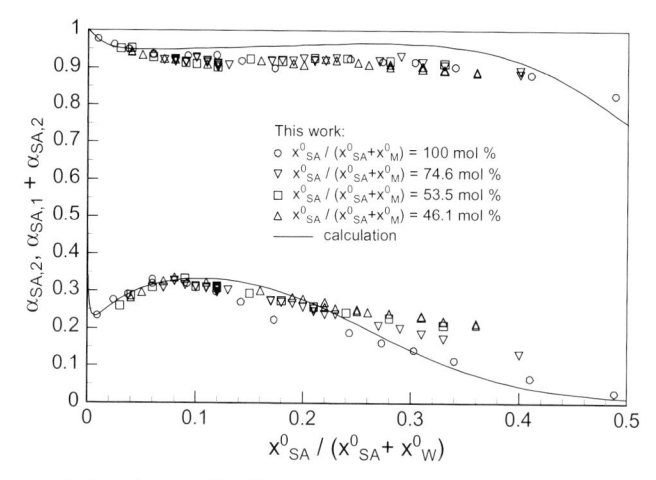

Figure 1.13: Degrees of dissociation of H_2SO_4 in the system $CH_3OH + H_2SO_4 + H_2O$ at 293.15 K.

Adding CH_3OH to 30 mol% aqueous sulfuric acid, the portion of SO_4^{2-} with respect to the amount of acid in the mixture increases. Almost no effect on the fraction of molecular H_2SO_4 is observed within the accuracy of the measurements discussed above. The influence of CH_3OH almost vanishes for increasing water concentrations in the ternary mixtures, turning into the dissociation behavior of the $H_2SO_4 + H_2O$ system.

These spectroscopic studies are complemented by enthalpy of dilution measurement of aqueous sulfuric acid and CH_3OH in an isothermal flow calorimeter at 298.15 K. The mixing process is exothermal. For a given concentration, the determined molar enthalpy of dilution is not influenced by the volume flow rate of the calorimeter (1 µl s^{-1} to 8 µl s^{-1}). Hence, the slow chemical reactions of the mixture can be neglected in the caloric measurements. The accuracy of the measurements is within 2%. In order to obtain the excess molar enthalpy of the ternary system, our results are combined with the excess molar enthalpy data of the system $H_2SO_4 + H_2O$ of Giauque et al. (1960). The determined and the predicted excess molar enthalpies of the ternary system are given in Figure 1.14. The systematic deviations of 4% result from an equivalent deviation in the model of the binary system $H_2SO_4 + H_2O$. The model assumes that there is no dissociation, as well as chemical reaction between CH_3OH and H_2SO_4. The interactions among the CH_3OH molecules and the acid related components are neglected. Even though, the concentration dependence of the excess molar enthalpy of the ternary system is very well represented by the model. This indicates that CH_3OH behaves as an inert component in the H_2O rich mixtures. The negligible influence of CH_3OH on the degrees of dissociation for the diluted systems confirms this assumption.

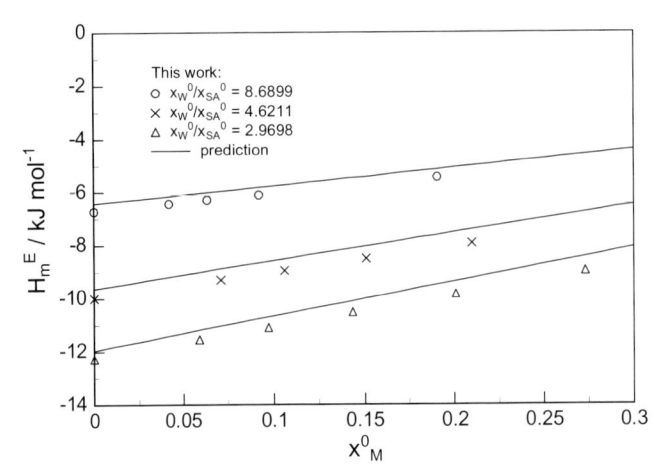

Figure 1.14: Excess molar enthalpy of the system $CH_3OH + H_2SO_4 + H_2O$ at 298.15 K.

At higher amounts of the alcohol, a further dissociation reaction of H_2SO_4 with CH_3OH should be taken into account.

1.5.4 The System $CH_3SO_4H + CH_3OH$

The resulting degree of dissociation of CH_3SO_4H in the system $CH_3SO_4H + CH_3OH$ at 293.15 K is shown in Figure 1.15.

For this system, there are no published data in literature to compare with. As seen by the large fraction of $CH_3SO_4^-$, CH_3SO_4H is a strong acid. At high acid concentrations, its degrees of dissociation are smaller than those of H_2SO_4, but much larger than that of HNO_3, which is also known as strong 1:1 acid (Young et al., 1959). Estimating a pK_a value of –3.4 for the ester, Guthrie (1978) expects CH_3SO_4H to be an even stronger acid than H_2SO_4. Conductometric titration methods confirm this order of acid strength (Chromiak, 1993).

1.5.5 The System $CH_3OH + H_2SO_4 + H_2O$ with CH_3SO_4H

H_2SO_4 and CH_3OH form CH_3SO_4H in an equilibrium reaction according to the equation:

$$H_2SO_4 + CH_3OH \rightleftarrows CH_3SO_4H + H_2O \tag{1.13}$$

Heating the samples for 12 to 24 h at temperatures between 308 K and 318 K depending on the mixture composition, the ester forming reaction is brought to equilibrium, whereas there is almost no formation of $(CH_3)_2O$. The total mole fraction of methyl hydrogen sulfate (x_{MHS}) in Figure 1.16 is determined from the sum of the measured mole frac-

1.5 Results and Discussion

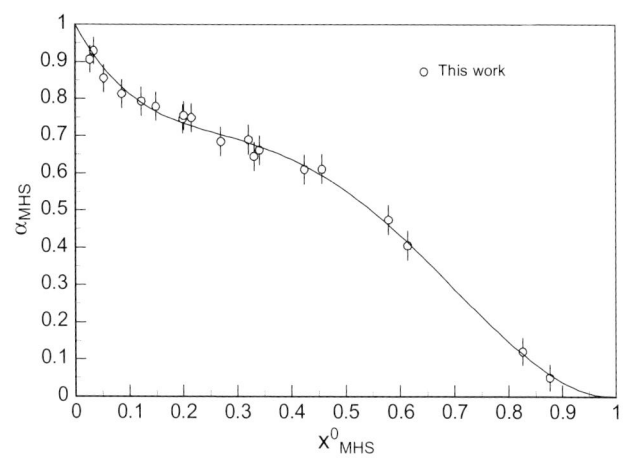

Figure 1.15: Degree of dissociation of CH_3SO_4H in the system $CH_3SO_4H + CH_3OH$ at 293.15 K.

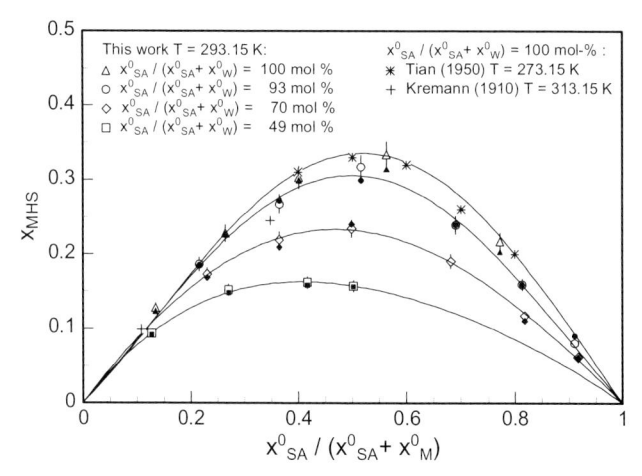

Figure 1.16: Mole fraction of CH_3SO_4H in the system $CH_3OH + H_2SO_4 + H_2O$ determined by Raman spectroscopy (open symbols) and ion chromatography (filled small symbols).

tions of CH_3SO_4H and $CH_3SO_4^-$ by Raman spectroscopy. x_{MHS} is also determined by ion chromatography for the same concentrations. The results are presented in Figure 1.16 as small filled symbols of the same type as used for the spectroscopic results. Both measurements agree very well within the accuracy of the two methods.

There is also a good agreement between the own measurements at 293.15 K and those of Tian (1950) at 273.15 K for the water-free system. The fraction of ester determined by Kremann and Neumann (1910) at 313.15 K in the CH_3OH rich mixture is also comparable to our results. There seams to be no strong dependency of CH_3SO_4H formation upon temperature. Tian (1950) also found that there is no dependency between ester formation and temperature at the system $C_2H_5OH + H_2SO_4$.

1 Experimental and Theoretical Investigations at the System CH₃OH + H₂SO₄ + H₂O

The degree of dissociation of CH_3SO_4H is shown in Figure 1.17.

No difference is found in degree of dissociation among the samples made from 100 mol% and 93 mol% aqueous sulfuric acid. The influence of H_2O to the dissociation behavior of CH_3SO_4H is found to be stronger at small CH_3OH concentrations. With increasing fractions of CH_3OH in the mixture, more and more of H_2SO_4 reacts to the ester and the degree of dissociation converges to the values of the $CH_3OH + CH_3SO_4H$ system presented in Figure 1.15. The degrees of dissociation of H_2SO_4 are shown in Figure 1.18.

The fraction of molecular H_2SO_4 decreases with increasing amounts of H_2O in the samples of small concentrations of CH_3OH. In the limit, where no CH_3OH exists, the acid

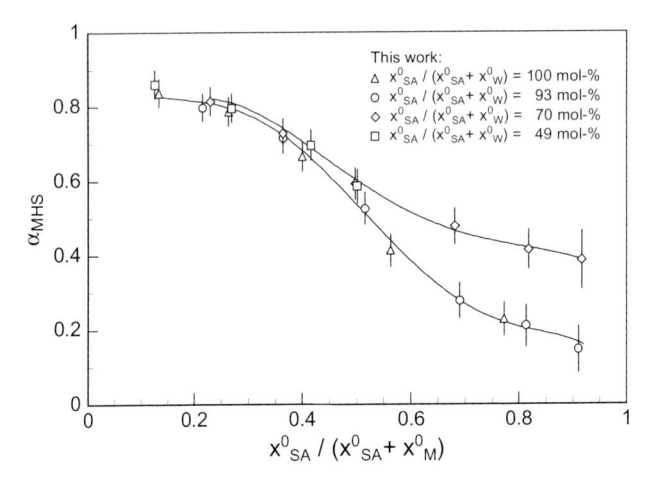

Figure 1.17: Degree of dissociation of CH_3SO_4H in the system $CH_3OH + H_2SO_4 + H_2O + CH_3SO_4H$ at 293 K.

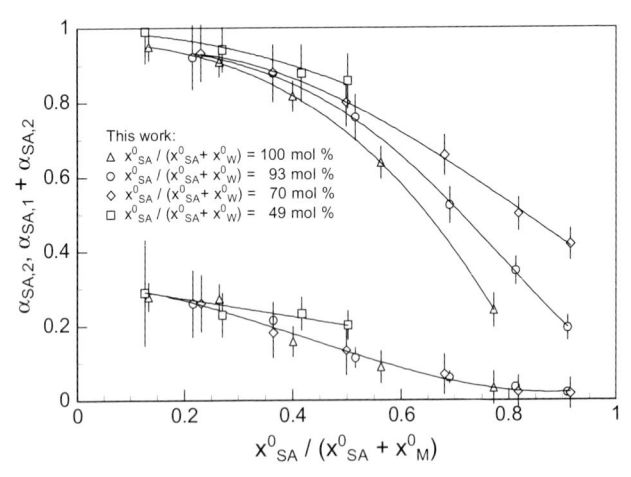

Figure 1.18: Degrees of dissociation of H_2SO_4 in the system $CH_3OH + H_2SO_4 + H_2O + CH_3SO_4H$ at 293 K.

behavior converges to the degrees of dissociation of the $H_2SO_4 + H_2O$ system. By increasing the amounts of CH_3OH in the mixtures, larger fractions of sulfuric acid react into CH_3SO_4H. As the mole fractions of H_2SO_4 and its ions in the CH_3OH rich mixtures are in the range of 1 mol%, the Raman measurement technique reaches its limit in resolving their concentrations.

1.5.6 The System $CH_3OH + H_2SO_4 + H_2O$ with CH_3SO_4H and $(CH_3)_2O$

In the vapor phase of the system $CH_3OH + H_2SO_4 + H_2O$, the concentration of the substances CH_3OH, H_2O, and $(CH_3)_2O$ are determined by Raman spectroscopy. Due to their small vapor pressures, H_2SO_4 and CH_3SO_4H are not found in the vapor phase. CH_3OH undergoes an ion-catalyzed reaction in the liquid phase:

$$2\ CH_3OH \rightarrow (CH_3)_2O + H_2O \tag{1.14}$$

By simultaneous measurement of the vapor pressure, the partial pressures of CH_3OH, H_2O, and $(CH_3)_2O$ are determined. Since the reaction can be seen here as irreversible, the formed amount of $(CH_3)_2O$ is accumulated during the experiment. Four mixtures, which are made by diluting 49 mol% aqueous sulfuric acid with various amounts of CH_3OH, are studied. The ester concentrations in these mixtures are given in Figure 1.16. The partial pressures of methanol and water are plotted against the reciprocal of temperature in Figure 1.19.

It is assumed that the phase equilibrium is established, as the reaction rate of $(CH_3)_2O$ forming reaction is very slow. The partial pressures of CH_3OH and H_2O, determined at different amounts of $(CH_3)_2O$ in the liquid phase, do not vary within the accuracy of the measurement.

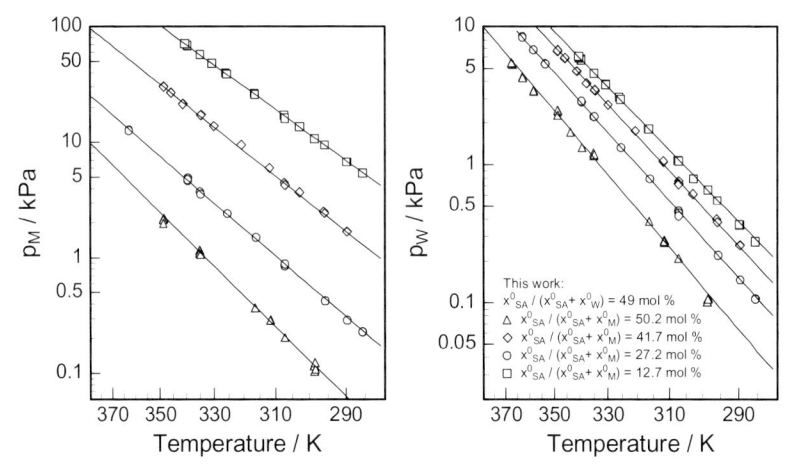

Figure 1.19: Partial pressures of CH_3OH (left) and H_2O (right) over the system $CH_3OH + H_2SO_4 + H_2O + CH_3SO_4H + (CH_3)_2O$.

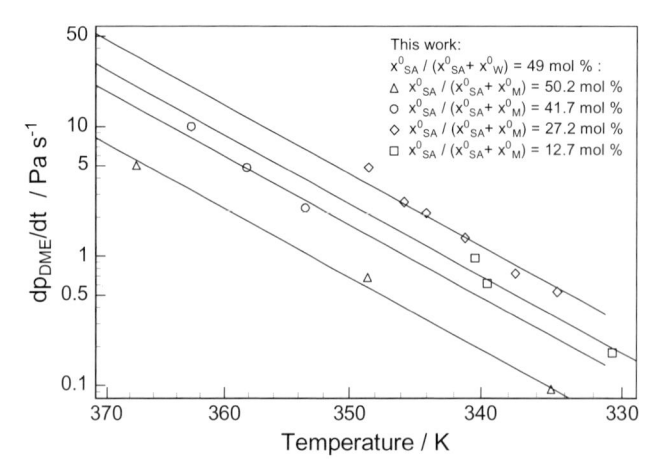

1 Experimental and Theoretical Investigations at the System $CH_3OH + H_2SO_4 + H_2O$

Figure 1.20: Rise of partial pressures of $(CH_3)_2O$ per time over the system $CH_3OH + H_2SO_4 + H_2O + CH_3SO_4H + (CH_3)_2O$.

The rate of partial pressure increase of $(CH_3)_2O$ is shown in Figure 1.20. Since the $(CH_3)_2O$ reaction is far from equilibrium, the observed rates are independent of time.

They grow with increasing CH_3OH concentration, but are reduced again for the mixture with the highest alcohol content. This indicates that the rate of reaction depends not only on CH_3OH but also on H_2SO_4, as one could expect from an ion-catalyzed reaction.

1.6 Conclusion

Mixtures of $CH_3OH + H_2SO_4 + H_2O$ are complex multi component systems due to the formation of CH_3SO_4H and $(CH_3)_2O$ and the dissociation of two electrolytes. The composition of liquid and vapor phase is determined by Raman spectroscopy, especially considering the dissociation behavior of H_2SO_4 and CH_3SO_4H. The formation of $(CH_3)_2O$ in the liquid phase is verified through partial pressure determinations as a function of time.

For the modeling of the system $H_2SO_4 + H_2O$, the data of macroscopic properties are complemented by Raman-determined degrees of dissociation. The properties of the ternary system were predicted from binary subsystem parameters. The importance of data on the compositions of mixtures increases with growing number of components and reactions. Raman spectroscopy is one approach to obtain this information as proven in this work.

References

Arnett, E. M.; Wu, C. Y. (1961): Base Strengths of Some Aliphatic Ethers in Aqueous Sulfuric Acid. J. Am. Chem. Soc. 84, 1680–1684.

Burt, B. C. (1904): The Vapour Pressure of Sulphuric Acid Solutions and the Molecular Condition of Sulphuric Acid in Concentrated Solution. J. Chem. Soc. 85, 1339–1354.

Chen, H.; Irish, D. E. (1971): A Raman Spectral Study of Bisulfate-Sulfate System. II. Constitution, Equilibria, and Ultrafast Proton Transfer in Sulfuric Acid. J. Phys. Chem. 75, 2672–2681.

Christensen, C.; Gmehling, J.; Rasmussen, P.; Weidlich, U. (1984): Heats of Mixing Data Collection. Chemistry Data Series Vol. III, Part 1. DECHEMA Frankfurt/Main (FRG).

Chromiak, E. (1993): Die Untersuchung der Zusammensetzung von Abfallsäuren aus dem Prozeß der Veresterung der Schwefelsäure. Przem. Chem. 72, 60–63.

Collins, E. M. (1933): The Partial Pressures of Water in Equilibrium with Aqueous Solutions of Sulfuric Acid. J. Phys. Chem. 37, 1191–1203.

Engels, H. (1990): Phase Equilibrium and Phase Diagrams of Electrolytes. Chemistry Data Series Vol. XI, Part 1, DECHEMA Frankfurt/Main (FRG).

Fischer, W.; Beckenkamp, K. (1998): Die Zerlegung saurer, methanolischer Karl-Fischer-Lösungen. J. pract. Chem. 340, 52–62.

Furic, K.; Mohacek, V.; Mamic, M. (1993): Methanol in Isolated Matrix, Vapor and Liquid Phase: Raman Spectroscopic Study. Spec. chim. Acta 49A, 2081–2087.

Giauque, W. F.; Hornung, E. W.; Kunzler, J. E.; Rubin, T. R. (1960): The Thermodynamic Properties of Aqueous Sulfuric Acid Solutions and Hydrates from 15 to 300 K. J. Am. Chem. Soc. 82, 62–70.

Gmehling, J.; Hoderbaum, T. (1988): Heats of Mixing Data Collection. Chemistry Data Series Vol. III, Part 3. DECHEMA Frankfurt/Main (FRG).

Gmehling, J.; Onken, U. (1977): Vapor-Liquid Equilibrium Data Collection. Chemistry Data Series Vol. I, Part 1. DECHEMA Frankfurt/Main (FRG).

Gmehling, J.; Onken, U.; Rarey-Nies, J. R. (1988): Vapor-Liquid Equilibrium Data Collection. Chemistry Data Series Vol. I, Part 1b. DECHEMA Frankfurt/Main (FRG).

Guthrie, J. P. (1978): Hydrolysis of Esters of Oxy Acids: pK_a Values for Strong Acids, Brönsted Relationship for Attack of Water at Methyl, Free Energies of Hydrolysis of Esters of Oxy Acids, and a Linear Relationship Between Free Energy of Hydrolysis and pK_a Holding Over a Range of 20 pK Units. Can. J. Chem. 56, 2342–2354.

Herzberg, G. H. (1964): Molecular Spectra and Moleculare Structure: II. Infrared and Raman Spectra of Polyatomic Molecules. D. Van Nostand Company, Inc., New York.

Hood, G. C.; Reilly, C. A. (1957): Ionization of Strong Electrolytes. V. Proton Magnetic Resonance in Sulfuric Acid. J. Chem. Phys. 27, 1126–1128.

Katz, E. D.; Lochmuller, C. H.; Scott, R. P. (1989): Methanol-Water Association and Its Effect on Solute Retention in Liquid Chromatography. Anal. Chem. 61, 349–355.

Kim, S. H. (1998): Experimentelle und theoretische Untersuchungen am Stoffsystem Salpetersäure-Schwefelsäure-Wasser. Shaker Verlag, Aachen, ISBN: 3-8265-3510-3.

Kim, S. H.; Roth, M. (2001): Enthalpies of Dilution and Exess Molar Enthalpies of an Aqueous Solution of Sulfuric Acid. J. Chem. Eng. Data 46, 138–143.

Klages, F.; Jung, H. A.; Hegenberg, P. (1966): Die dynamische Acidität sehr starker Säuren. Chem. Ber. 99, 1704–1711.

Kohlrausch, K. W. F. (1943): Ramanspektren. Heyden & Son Ltd, London, ISBN: 0 85501 071 1.

Kremann, R.; Neumann, H. (1910): Zur Kinetik der Bildung von Methylschwefelsäure und Dimethyläther. Monatsh. 31, 1051–1056.

Müller, E. (1963): Methoden der organischen Chemie Band VI/2. G. Thieme Verlag, Stuttgart.

Okabayashi, H.; Caminiti, M.; Feltrin, A.; Ramondo, F.; Sadun, C. (1974): The Raman Spectra and Molecular Conformations of Surfactants in Aqueous Solution and Christalline States. Bull. Chem. Soc. Jap. 47, 1075–1077.

Rütten, P.; Kim, S. H.; Roth, M. (1998): Measurements of the Heats of Dilution and Discription of the System H₂O/H₂SO₄/HCl wit a Solvation Model. Fluid Phase Equilibria 153, 317–340.

Siebert, H. (1957): Schwingungsspektren einiger Derivate der Schwefelsäure. Z. Anorg. Chem. 289, 15–28.

Tarasenkov, D. N. (1955): Vapor Pressure of Aqueous Sulfuric Acid Solutions. J. Appl. Chem. USSR. 28, 1053–1058.

Taylor, R. C.; Vidale G. L. (1957): Raman Spectrum and Vibrational Assignments of Gaseous Dimethyl Ether. J. Chem. Phys. 26, 122–123.

Tian, A. (1950): Recherches sur l'esterification sulfurique. Compt. rend. 264, 1223–1232.

Turner, D. (1974): Dissociation of the Bisulphate Ion at Moderate Concentrations. J. Chem. Soc. Faraday Trans. I, 1346–1356.

Walrafen, G. E.; Yang, W.-H.; Chu, Y. C.; Hokmabadi, M. S. (2000): Structures of Concentrated Sulfuric Acid Determined from Density, Conductivity, Viscosity, and Raman Spectroscopic Data. J. Sol. Chem. 29, 905–936.

Wilson, G. M. (1964): A New Expression for the Excess Free Energy of Mixing. J. Am. Chem. Soc. 86, 127–130.

Young, T. F.; Maranville, L. F.; Smith, H. M. (1959): The Structure of Electrolytic Solutions. Ed. W.J. Hammer. Wiley, New York, 35–59.

Nomenclature

A	area in the Raman spectrum
A, B	constants
a, b, c	constants
a_w	activity of water
C	complex
c	Raman calibration factor
c	molarity
c_p	heat capacity
$\Delta_{dil}H_m$	molar enthalpy of dilution
H_m^E	excess molar enthalpy
K_a	equilibrium constant of reaction
m	number of solvent molecules in reaction
n	number of moles
p	vapor pressure
Q_c	concentration ratio
T	temperature
t	time
x	mole fraction of liquid phase
y	mole fraction of the vapor phase

Greek letters

α	degree of dissociation
γ	activity coefficient
δ_m	mean deviation
ν	number of complexes formed in reaction
τ	Wilson interaction parameter

Superscripts

0	state before reaction
k	reaction number

Subscripts

1	first dissociation reaction
2	second dissociation reaction
C	complex
DME	dimethyl ether, $(CH_3)_2O$
i, j, k	component index
M	methanol, CH_3OH
MHS	methyl hydrogen sulfate, CH_3SO_4H
SA	sulfuric acid, H_2SO_4
W	water, H_2O

2 Measurement of Binary Phase Equilibria of Esterification Systems

Marko Tischmeyer and Wolfgang Arlt [1]*

2.1 Abstract

The simulation of reactive distillation processes separately requires binary phase equilibrium data and information about the reaction kinetics and equilibrium. A novel apparatus has been constructed to measure these data for reaction rates with half-life of 30 min and below. Experimental data are presented for mixtures of methanol with acetic acid and propionic acid at 332 K to 361 K as a model reaction for reactive distillation. These temperatures represent operation points of technical columns.

2.2 Introduction

Conventional chemical processes which consist of a reaction and a purifying section can in certain cases be replaced by a reactive distillation in order to lower the production costs or to improve the purity of the products. By performing the two steps at once several advantages can be achieved, the most important are:

- reduced downstream processing for separation of the product from residual components,
- the heat of reaction can directly be utilized for evaporation of liquid phase,
- in case of parallel or consecutive reactions the selectivity will be increased by removal of product(s),
- the limitations of azeotropic mixtures can be overcome,
- lower resident times since the concentrations of educts and thus the product yield is higher.

---

[1] Technische Universität Berlin, Institut für Verfahrenstechnik, Fachgebiet Thermodynamik und Thermische Verfahrenstechnik, Straße des 17. Juni 135, 10623 Berlin.
* Author to whom correspondence should be addressed.

The earliest publication concerning the concept of reactive distillation was given by Backhaus (1921). He investigated the methyl acetate process and observed that higher yields of the desired products are achieved by removing them from the reactor during the ongoing reaction. Today there are numerous applications of this technology like the production of esters from acids or acid chlorides with alcohols, their hydrolysis, alkylation processes, the separation of olefines or the formation of ethers from alkenes and alcohols. The latter case is of great interest since the fuel additives methyl *tert*-butyl ether (MTBE) and *tert*-amyl methyl ether (TAME) received a lot of attention in recent times (Rehfinger, 1990) due to clean air legislation (Hall, 1992).

The design of reactive columns needs research in terms of fluid dynamics and phase equilibrium. This chapter is dedicated to the phase equilibria. The slow reaction rate of the systems prohibits the use of a continuous stirred tank reactor, the main tool of chemists. In general, the equilibrium conditions in these systems are not favourable, an equilibrium constant around unity indicates that a multicomponent mixture is formed by the reactants. In order to increase the yield, the removal of one of the products, namely the water in esterification reactions, is the key.

The design of the column needs at least a plate-to-plate calculation. Reaction kinetics or equilibrium constants *and* binary vapor-liquid equilibrium (VLE) data for well-known g^E-models are needed as input. These data are not readily available because the reaction prevents the determination of binary phase equilibrium data.

By a literature search for compoundsmixtures/systems of organic acids with alcohols, 64 isothermal or isobar VLE data sets were found in the DECHEMA data base (Gmehling et al., 1977). Only four of them were found to be thermodynamically consistent with the Redlich-Kister area test. These inconsistencies can have the following reasons:

- when concentrations are determined, the forming of the reaction products was not taken into account,
- if a non-recycle flow still is used, the residence times can be too short for establishing phase equilibrium,
- the calculation of the association effects in the vapor phase has been neglected.

The measuring method applied in this work follows a proposal by Patel and Young (1993) for calorimetry. It is shown that this method is appropriate for those systems used in the reactive distillation.

2.3 Fundamentals

2.3.1 Calculation of Liquid and Vapor Phase Fugacities

When the phase equilibrium is established, the pressure and the temperature as well as the fugacities of the different components show the same values in the liquid and vapor phase. To calculate the vapor-liquid equilibrium, it is necessary to calculate separately the fugacity

2 Measurement of Binary Phase Equilibria of Esterification Systems

f_i of each component i in each phase. For description of mixtures, different approaches are possible, in the present case a γ-φ model was chosen:

$$f_i^L = f_i^V \tag{2.1}$$

$$x_i \gamma_i f_{0i}^L = y_i \varphi_i^V P \tag{2.2}$$

The activity coefficient γ_i is calculated from a model for the Gibbs excess enthalpy g^E. The NRTL (Non-Random-Two-Liquid) model (Renon and Prausnitz, 1968) was chosen according to the strongly non-ideal nature of the system. To calculate the liquid fugacity with this model it is also necessary to include the standard state fugacity f_{0i}^L, which is the fugacity of the pure liquid component i at system pressure and temperature. It can be expressed by

$$f_{0i}^L = \varphi_{0i}^{LV} P_{0i}^{LV} \exp\left(\frac{v_{0i}^L (P - P_{0i}^{LV})}{RT}\right) \tag{2.3}$$

where φ_{0i}^{LV} is the fugacity coefficient of the component i at saturated vapor pressure P_{0i}^{LV}, and $\exp\left(\frac{v_{0i}^L (P - P_{0i}^{LV})}{RT}\right)$ is the Poynting correction, which accounts for the expansion or compression of the liquid from saturated vapor pressure to system pressure.

The fugacity coefficient φ_i^V is calculated with the virial equation of state with the method of Hayden and O'Connell (1975) to calculate the second virial coefficient. This assumes that the vapor is ideal except for association between molecules, usually due to hydrogen bonding. This method incorporates the "chemical theory of dimerization". Therefore, it accounts for strong association and solvation effects, including such as found in systems containing organic acids like in the present system. Association percentage of some occurring components are stated in Table 2.1 (Winkler, 1996):

Table 2.1: Association percentage of acetic acid, methanol, and water.

		Acetic acid T = 391 K	Methanol T = 298 K	Water T = 373 K
Liquid:	% of monomers	37%$_{mol}$	3%$_{mol}$	
Vapor:	% of monomers	61%$_{mol}$	99%$_{mol}$	95%$_{mol}$
	% of dimers	36%$_{mol}$		
	% or trimers	3%$_{mol}$		

2.3.2 Reaction Space and Coordinates Transformation

While a ternary mixture can be visualized easily within a ternary diagram, the possible states of a quaternary mixture can best be depicted in a tetrahedron. All points within the tetrahedron are possible phase equilibrium states for a system without reaction. When one or

2.3 Fundamentals

more reactions occur, the degree of freedom is reduced by the number of independent chemical equilibrium reactions:

Gibbs phase rule: $\quad df = 2 + co - ph - re \quad$ (2.4)

where co is the number of components, ph is the number of phases and re is the number of independent chemical equilibrium reactions.

In a quaternary system with two phases and one reversible reaction (where e.g. the pressure is fixed), the degree of freedom is two. This number is identical to the dimension of the concentration space in which the equilibrium conditions are fulfilled. This part of the concentration space is defined as reaction space (Bessling et al., 1997). The geometrical shape of the reaction space varies with the type of reaction, the position of the chemical equilibrium, and the presence of inert components. The stoichiometric reaction mechanism is

$$\nu_A A + \nu_B B \rightleftarrows \nu_C C + \nu_D D \quad (2.5)$$

where A and B are the educts (alcohol and organic acid), C and D are the products (ester and water) and ν_i are the stoichiometric coefficients. The latter are unity for an esterification reaction. For a value of the equilibrium constant K_a

$$K_a = \frac{a_C \cdot a_D}{a_A \cdot a_B} \quad (2.6)$$

with the activities a_i

$$a_i = x_i \gamma_i \quad (2.7)$$

about unity and ideal mixture behavior (i.e. also $K_x = 1$), the isobaric reaction space is shown in Figure 2.1.

It forms a saddle area that is stretched between the non-reacting binary systems AC, AD, BC, and BD. The dashed lines are the stoichiometric lines which show the change in concentration caused by the reaction. The reaction paths for two initial compositions (50%

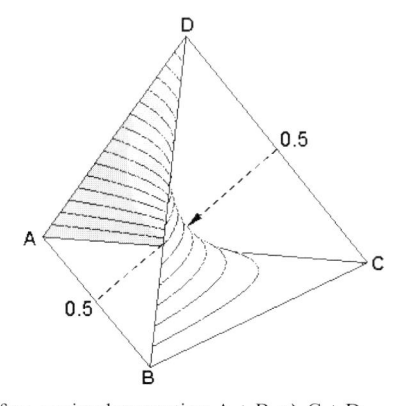

Figure 2.1: Reaction space of an equimolar reaction $A + B \rightleftarrows C + D$.

2 Measurement of Binary Phase Equilibria of Esterification Systems

A/B and 50% C/D) are plotted, which yield the same equilibrium concentration. Since the reaction does not change the number of moles, all other stoichiometric lines (which belong to other initial concentrations) are parallel to the plotted ones.

For process synthesis and design, distillation lines are a valuable means to examine and understand the behavior of a mixture. Separation barriers like azeotropes, border distillation lines or border distillation areas can be identified. The origins or termini of distillation lines are feasible top or bottom fractions of distillation columns, so they can be utilized to verify whether a certain component is producible with this process.

In case of reactive distillation, all points of the (reactive) distillation lines are positioned at the reaction space (see Fig. 2.2/1).

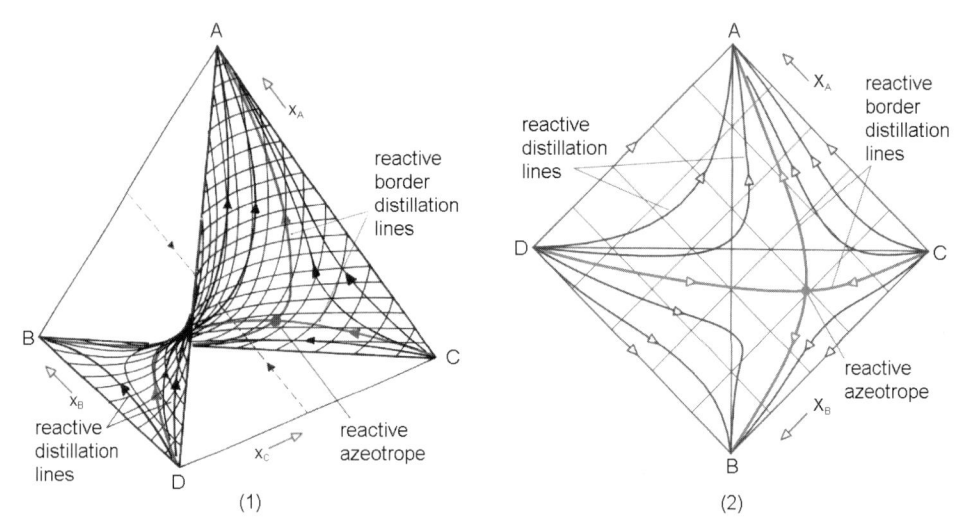

Figure 2.2: (1) Reactive distillation lines of the esterification reaction A + B ⇌ C + D with $K_X = 1$. (2) Reactive distillation lines of the system of (1) in transformed coordinates (Stichlmair, 1998).

A projection of this curved plane along the stoichiometric lines onto a plane orientated vertically to the stoichiometric lines yields a two-dimensional picture of the reactive space. The result of the transfer is a rhombus with component A and B at the top and at the bottom, respectively (see Fig. 2.2/2). The conversion from three to two dimensions can be visualized as a change of the eye position into a point which is located at the main stoichiometric line (initial compositions 50% A/B or 50% C/D, see Fig. 2.3).

This transformation of coordinates proposed by several authors (Barbosa and Doherty, 1988; Espinosa et al., 1995a, 1995b) is a means to depict the conditions of chemical equilibrium.

To carry out the coordinates transformation, the equation set given by Espinosa et al. (1995b) can be applied:

$$X_i = \frac{v_k x_i - v_i x_k}{v_k - v_t x_k} \tag{2.8}$$

2.3 Fundamentals

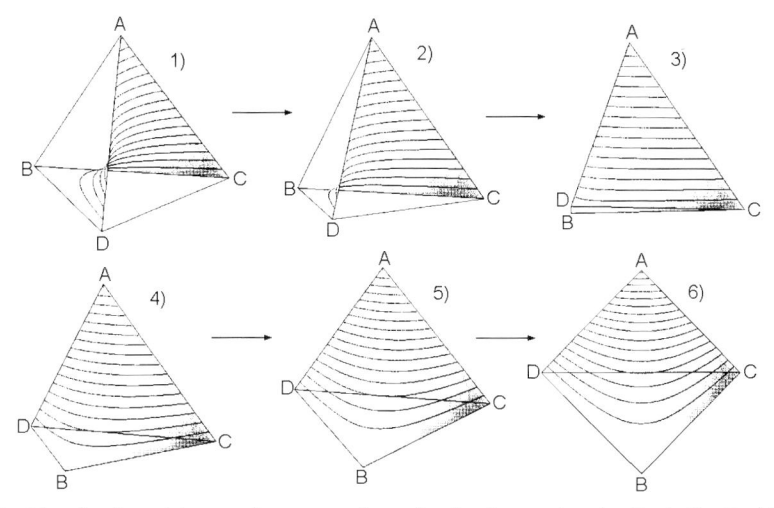

Figure 2.3: Visualization of the coordinates transformation for the reaction A + B ⇌ C + D with $K_X = 1$.

$$Y_i = \frac{v_k y_i - v_i y_k}{v_k - v_t y_k} \quad \text{with} \quad v_t = \sum v_i \tag{2.9}$$

The subscript "k" denotes one arbitrary component which can be chosen and which is not visible in transformed coordinates. The transformed concentrations meet the following conditions:

$$\sum X_i = \sum Y_i = 1 \tag{2.10}$$

This signifies for a quaternary mixture that one component must have transformed concentrations between –1 and 1.

After the transformation of a quaternary mixture, the lever arm rules of the mass balance are valid again since the effect of the reaction is not visible. However, this raises also a major disadvantage: effects caused by pure distillation (i.e. within a non-reactive area of a column) are not visible as the transformed plane shows only the states of chemical equilibrium.

2.3.3 Standard State in VLE Calculation with Chemical Equilibrium

As stated above, for the calculation of the phase equilibrium a standard state fugacity f_i^+ is necessary. Also for the calculation of the chemical equilibrium constant K (Eq. (2.11)), a standard state f_i^+ is required.

$$K = \prod_i \left(\frac{f_i}{f_i^+} \right)^{v_i} \tag{2.11}$$

Two different methods can be used, which yield different results in calculation of the equilibrium conditions. One method refers to the fugacity of the pure liquid component i at system pressure and temperature f_{0i}^L. Then the value of K is calculated as follows:

$$K = \prod_i \left(\frac{f_i}{f_{0i}^L}\right)^{\nu_i} = \prod_i \left(\frac{x_i \gamma_i f_{0i}^L}{f_{0i}^L}\right)^{\nu_i}$$

$$= \prod_i (x_i \cdot \gamma_i)^{\nu_i} = \prod_i x_i^{\nu_i} \cdot \prod_i \gamma_i^{\nu_i} = K_x \cdot K_\gamma \qquad (2.12)$$

$$= K_a = \prod_i a_i^{\nu_i}$$

The subscript "a" denotes that K is calculated from the activities a_i, i.e. from the concentrations and a model for the activity coefficients γ_i (g^E-model). This accounts for the real behavior of the mixture in the liquid state.

Another way is to choose the ideal gas at system temperature and at standard pressure (commonly 1 atm) as standard state. The fugacity of the ideal gas equals the pressure, $f^+ = P^+$. This is appropriate for a chemical reaction in the gas phase and yields the following expression for K:

$$K = \prod_i \left(\frac{f_i}{P^+}\right)^{\nu_i} = \prod_i \left(\frac{y_i \varphi_i P}{1\,\text{atm}}\right)^{\nu_i}$$

$$= \prod_i \varphi_i^{\nu_i} \cdot \prod_i y_i^{\nu_i} \cdot \left(\frac{P}{1\,\text{atm}}\right)^{\sum_i \nu_i} = K_\varphi \cdot K_y \cdot \left(\frac{P}{1\,\text{atm}}\right)^{\sum_i \nu_i} \qquad (2.13)$$

$$= K_f$$

The subscript "f" denotes that the fugacity coefficient (among pressure and concentrations) is used to calculate K. An equation of state can be applied here.

By comparison of Eqs. (2.12) and (2.13) it is obvious, that the results can differ considerably. Since the esterification reaction takes place in the liquid phase and the fugacity here is calculated with f_{0i}^L as standard state, it is advisable to use the first method and thereby the same standard for the equilibrium constant K. This way a consistent model for the chemical and physical equilibrium can be obtained.

2.3.4 Kinetic Modelling

Different approaches are available for description of the reaction kinetics. To obtain a model which is consistent with the equilibrium constant for liquid phase reactions (Eq. (2.12)), activities must be applied for the expression of the reaction rate r. A formulation for a reversible second order reaction like the esterification is given by Eq. (2.14):

2.3 Fundamentals

$$r = \frac{1}{\nu_i} \frac{dx_i}{dt} = -k_{12} \, a_A \, a_B + k_{21} \, a_C \, a_D \tag{2.14}$$

$$da_A \neq da_B \neq -da_C \neq -da_D \tag{2.15}$$

where k_{12} and k_{21} denote the rate constants of the esterification and the hydrolysis reaction, respectively, and a_i is the activitiy of the component i. The temperature dependency of the rate constants is expressed by Arrhenius' law

$$k_{ij} = k_{ij,0} \cdot \exp\left(-\frac{E_{A,ij}}{RT}\right) \tag{2.16}$$

Four adjustable parameters have to be fitted, namely the pre-exponential factors $k_{12,0}$ and $k_{21,0}$ and the energies of activation E_{A12} and E_{A21}.

The computation effort is significant since values for γ_i have to be calculated for each component and each concentration point using a g^E-model. Their values differ for each component and hence the change of activity is not equal for different components even in case of an equimolar reaction. However, the benefit is an improved agreement of the fitted model to experimental data (Pöpken et al., 1999).

At chemical equilibrium ($t \to \infty$), the change of the concentrations and activities is zero. The correlation of the equilibrium constant and the reaction rates then yields

$$0 = -k_{12} \, a_{A\infty} \, a_{B\infty} + k_{21} \, a_{C\infty} \, a_{D\infty} \tag{2.17}$$

$$\frac{k_{12}}{k_{21}} = \frac{a_{C\infty} \, a_{D\infty}}{a_{A\infty} \, a_{B\infty}} = K_a \tag{2.18}$$

which is conform with the expression for K_a stated above.

In the case of the homogeneous esterification the reaction occurs by catalysis of the organic acid. To consider this in the kinetic model an auto-catalysis approach (Pöpken et al., 2000) is utilized:

$$r = \frac{1}{\nu_i} \frac{dx_i}{dt} = a_A^\alpha \left(-k_{12} \, a_A \, a_B + k_{21} \, a_C \, a_D\right) \tag{2.19}$$

The term a_A^α accounts for the activity of the protons in the mixture, which are the catalyst for both the esterification and the hydrolysis reaction. α may show a value of 0.5 or 1 depending on the assumptions about the catalytic mechanism. In this work the best description of the kinetics was achieved with $\alpha = 1$ which assumes catalysis by undissociated acetic acid.

As a special case, if the liquid phase behavior of the mixture is considered to be ideal (i.e. $\gamma_i = 1$ and hence $a_i = x_i$), the reaction rate of the equimolar esterification reaction can be expressed by

$$r = \frac{1}{\nu_i} \frac{dx_A}{dt} = x_A \left(-k_{12} \, x_A \, x_B + k_{21} \, x_C \, x_D\right) \tag{2.20}$$

with

$$dx_A = dx_B = -dx_C = -dx_D \qquad (2.21)$$

When no product C or D is present in the reactor at the start of the reaction and the initial concentrations of A and B are x_{A0} and x_{B0}, then Eq. (2.20) can be simplified to

$$r = \frac{1}{v_i}\frac{dx_A}{dt} = k_{12}\, x_A^2\,(x_{B0}+x_{A0}-x_A) - k_{21}\, x_A\,(x_{A0}-x_A)^2 \qquad (2.22)$$

In this special case the equimolar reaction rate can be determined by measuring the concentration of only one component.

2.4 Experimental

At the moment there are few projects concerning the measurement of reactive phase equilibria. Alsmeyer et al. (2000) have set up a flow-through apparatus for VLE measurements in reactive systems which also allows the determination of reaction kinetics in the liquid phase. The reactive mixture is forced into a steady state which does not represent its chemical equilibrium but reaches phase equilibrium between vapor and liquid. Both phases are then analyzed in situ by mid-infrared (MIR) spectroscopy. While the gas phase is measured in transmission, attenuated total reflection (ATR) is used for the liquid phase.

Pöpken et al. (2000) are using a computer operated static apparatus. They were measuring isothermal *P-x* data of the binary systems required for the methyl acetate production process. Presumption for these measurements is that the VLE is established much faster than the chemical equilibrium which yields operating temperatures up to 323.15 K at max.

Another flow apparatus has been operated until 1998 by Reichl et al. (1998). It was set up as non-recycle flow still and due to the short residence time was used to determine isobaric VLE data of thermally unstable components and of reactive mixtures. Two esterification systems (methyl formate and ethyl acetate) and one etherification system (*tert*-amyl methyl ether) were investigated.

2.4.1 Chemicals

Methanol and ethanol used for experimental procedure were of analytical grade (purity of 99.8% and 99.9%, Merck). Acetic acid was pure (100%, Merck), formic acid was given with 98% purity (Merck). Propionic acid and 1,4-dioxane were >99% pure (Merck).

For calibration of the gas chromatograph, ethyl formate, methyl acetate (both with purity >98%, Merck), methyl propionate (>99%, Merck), and distilled water were used. The chemicals were processed without further purification except for degassing.

2.4.2 Equipment and Experimental Procedure

The experiments were carried out in a 550 cm^3 static equilibrium cell which is located in a thermostatic oil bath at a given temperature with an accuracy of ± 0.1 K. The set-up was modified during the project time, hence two resulting schemes are given in Figures 2.4 and 2.6 and are explained below. Stainless steel 1.4435, Teflon, and glass have been used for construction. The temperature inside the cell was measured by a platinum resistance thermometer PT-100 (Conatex) with an accuracy of ± 0.01 K. The pressure was measured using a calibrated high accuracy sensor head (MKS Baratron Type 121A connected to the signal conditioner type PDR-D-1) attached to the top of the cell. The sensor head and the connecting line from the cell to the sensor were thermostated at 398.15 ± 0.1 K. The sensor was calibrated with the environmental pressure (measured with mercurial manometer), high vacuum (10^{-4} Pa), and the vapor pressure of pure water, methanol, ethanol, and formic acid at a temperature range of 313.15 K to 353.15 K. According to these measurements, the experimental pressure is assessed to be reliable within $\pm 0.5\%$ of the measured value.

Rather slowly reacting esterification systems (i.e. methanol + acetic acid and methanol + propionic acid) were measured with the bypass operated mixing technique as shown in Figure 2.4.

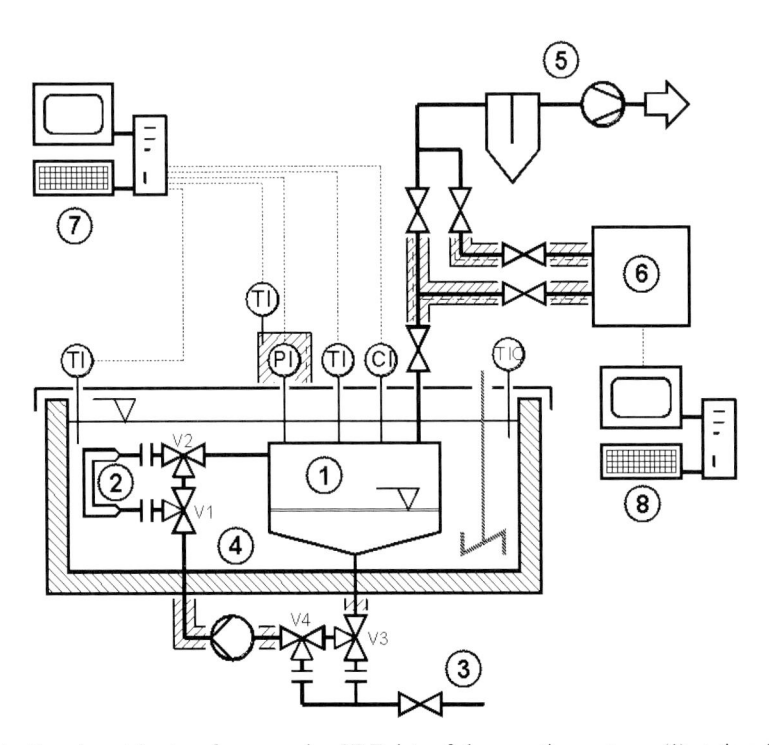

Figure 2.4: Experimental set-up for measuring VLE-data of slow reacting systems, (1) static cell, (2) injection loop (bypass), (3) liquid sample loop, (4) thermostatic oil bath, (5) vacuum pump, (6) gas chromatograph, (7) and (8) personal computers.

2 Measurement of Binary Phase Equilibria of Esterification Systems

When the cell was completely evacuated, the organic acid was fed in as first component via the liquid sample loop (3). The amount of inserted acid was determined gravimetrically (± 0.02 g). The second component was also degassed and loaded into a bypass formed pipe (2). The filled bypass was connected to the circulation flow by quick connections in a way that it was thermostated by the oil bath, too. By switching the valves V1 and V2, the circulation was directed through the bypass and the second component was led into the cell. Mixing occurred only due to the circulation through the pipes and diffusion, no mixer was installed.

The time period needed for complete mixing was above 15 min which was found by measuring the electrical conductivity (Knick laboratory conductometer type 703; see Fig. 2.5, curve (a)).

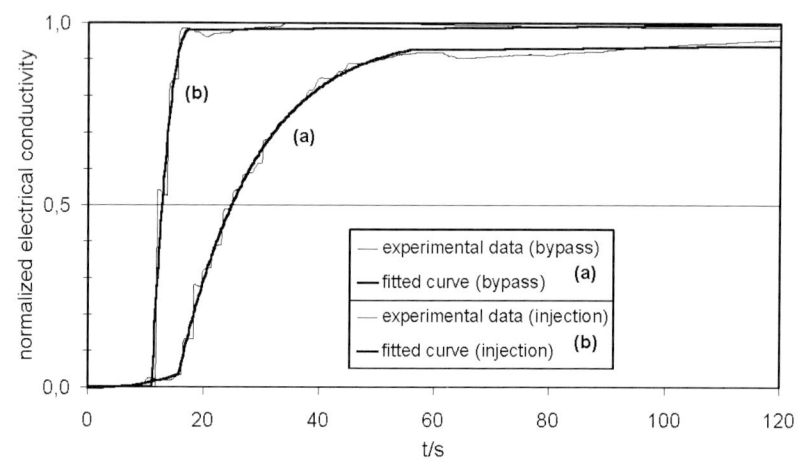

Figure 2.5: Comparison of the residence times of the bypass method, curve (a), and the injection method, curve (b).

This was sufficient for reactions with half life times of several hours but for faster reacting compounds such as ethanol + formic acid and isopropanol + formic acid the apparatus needed major changes. Therefore the bypass system was replaced by an injection pump (see Fig. 2.6).

The recycle flow was reduced to a minimal length which is needed for taking liquid samples outside the oil bath. Phase equilibrium is now reached quickly by mixing the two phases intensively using a stirrer head with magnetic coupling (bmd 075, Büchi GmbH). Herewith the time needed for mixing was reduced to some seconds as can be seen in Figure 2.5 curve (b).

Any necessary amount of alcohol can be injected to obtain a desired mole fraction of the mixture. The alcohol is degassed and filled into the injection cylinder (Fig. 2.6 (2), manufactured by the Laboratory for Thermophysical Properties of the University Oldenburg). When the equilibrium temperature of the oil bath is established throughout, the vapor pressure of pure organic acid is measured. At the time t_0 the predefined amount of alcohol is in-

2.4 Experimental

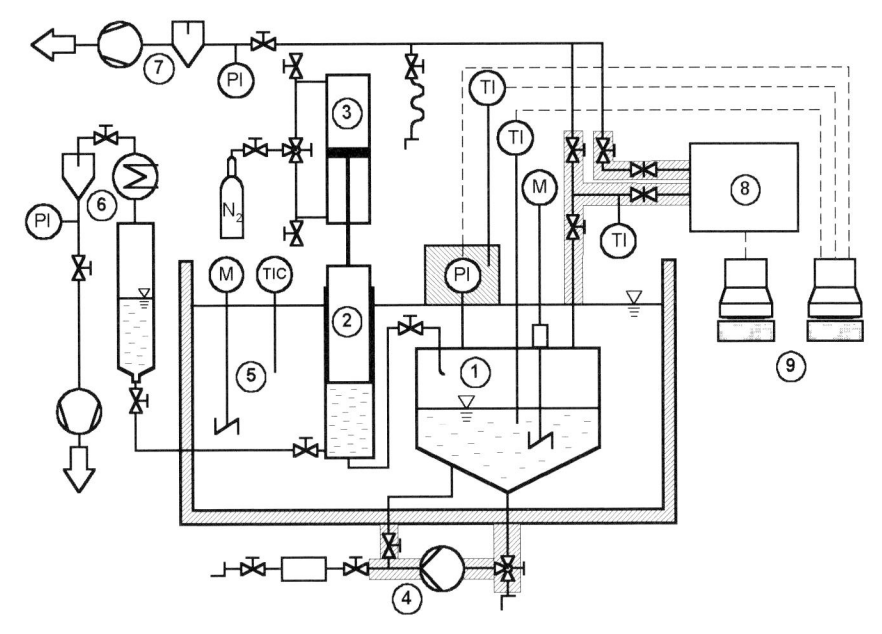

Figure 2.6: Experimental set-up for measuring VLE-data of fast reacting systems, (1) static cell (2) injection cylinder, (3) pneumatic power supply for injection, (4) liquid sample loop, (5) thermostatic oil bath, (6) alcohol reservoir and degassing device, (7) vacuum pump, (8) gas chromatograph, (9) personal computers.

jected into the cell within less than 2 s. Therefore the pressure pump (3) is powered with 1 MPa from a nitrogen vessel. The precision of the injection is controlled by a linear transducer (Haidenhain, type ND221) with an accuracy of ± 10 µm which is calibrated to determine the injected mass of alcohol at a given temperature. During the ongoing reaction, pressure and temperature are monitored continuously, liquid and vapor samples are taken in short intervals and analyzed immediately in case of the vapor samples. The principles of experimental results are given in Figure 2.7.

If the total pressure of the binary mixture is lower than the total pressure at reaction equilibrium P^{VLE+CE}, the pressure over time diagram Figure 2.7 holds. With the addition of a second component, the equilibrium is disturbed due to the heat of mixing and heat of reaction, it still takes some time to re-establish equilibrium conditions. Hence the first minutes after injection cannot be used to determine the total pressure. After the temperature is equilibrated, a polynomial is fitted to pressure data and extrapolated to the time of mixing t_0. We checked with non-reacting mixtures like acetic acid + water and methanol + water that after 60 s of mixing, the cell reaches phase and thermal equilibrium. The gas phase can be analyzed using an on-line gas chromatograph (GC): HP-GC 5890A Series II, split/splitless-injector, TCD and FID detector, six port valve for gas phase sampling. For slow reacting systems a HP-FFAP capillary column, length 25 m, i.d. 0.25 mm, film thickness 0.2 µm was used with helium 5.0 as carrier gas (Messer-Griesheim). To speed up the sample analysis for faster reactions a HP-FFAP capillary column, length 25 m, i.d. 0.53 mm, film thickness 0.1 µm was installed and the time needed for a complete analysis was reduced from 23 min to 12 min.

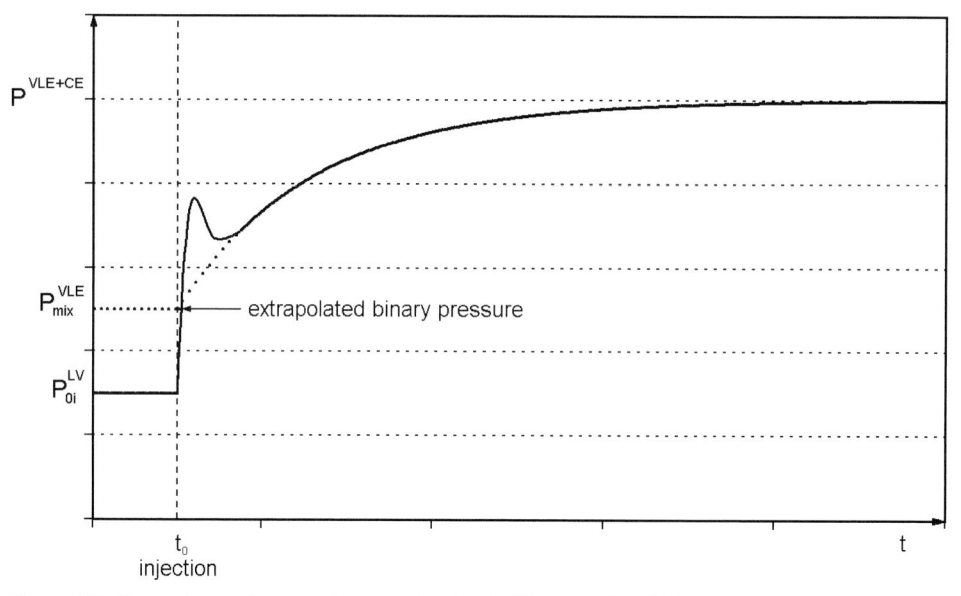

2 Measurement of Binary Phase Equilibria of Esterification Systems

Figure 2.7: Curve shape of an experiment with a low boiling organic acid in the cell and alcohol injected at t_0. P^{VLE+CE} denotes the pressure established in chemical equilibrium, P_{0i}^{LV} the vapor pressure of the pure first component (acid) and P_{mix}^{VLE} the desired binary system pressure obtained by extrapolation.

Gas phase analyses are not necessary to determine the VLE. When x_i and P are known, the values of y_i can also be calculated by use of a g^E-model. Therefore the deviation of measured and calculated pressure data at given x_i is minimized by adjusting the binary interaction parameters of the g^E-model. However, additional y_i data can be used to obtain a more accurate fit by using P and y_i deviations for the fitting procedure. Furthermore, a consistency area test (Gmehling and Onken, 1977) is feasible only by using experimental y_i data.

The GC is also used for the analysis of the liquid samples which can be withdrawn at the pump circuit by connecting an evacuated sample vial. The vapor phase samples are taken every 12 min when a GC run is finished. Liquid samples are withdrawn any few minutes (depending on the reaction rate). The reaction is stopped immediately in the sample vial by addition of 5 g 1,4-dioxane per 1 g sample liquid and cooling down to 283 K. The liquid samples are analyzed after the online vapor measurements by use of an autosampler (type G1895A, Agilent Technologies).

For determination of the binary VLE of the starting system the liquid samples are not necessary since the initial binary composition of alcohol and organic acid are known. Nevertheless, by analyzing the liquid phase during the reaction as well the kinetic information is obtained accessory.

After collecting all data the concentrations of the vapor phase $y_i(t)$ have to be extrapolated to the initial binary values, too. Then the binary VLE diagram is constructed as can be seen in Figure 2.8.

Different polynomial approaches were examined for the extrapolation to t_0. The results for the corresponding binary pressure differ only within 0.06 kPa, which is about the accuracy of the pressure measuring device.

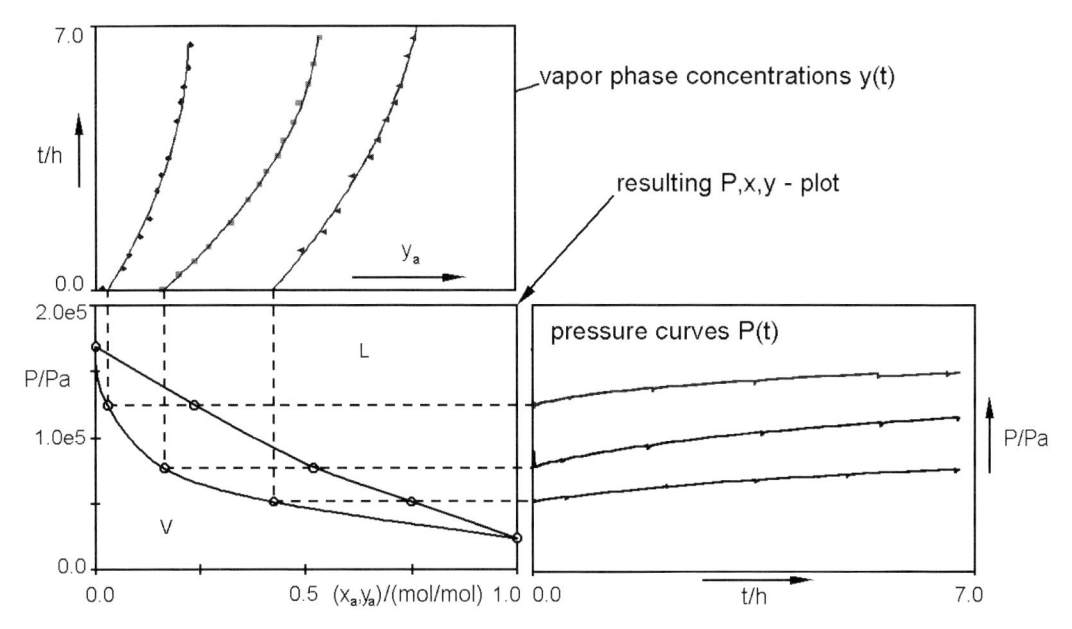

Figure 2.8: Calculation of the VLE-diagram for methanol + acetic acid at $T = 353.15$ K from experimental data; x_a and y_a are the mole fractions of acetic acid in liquid and vapor phase, respectively.

2.5 Results

P, x data were measured for the binary systems methanol + acetic acid and methanol + propionic acid. Measurements for the faster reacting compounds are still under progress (see Table 2.2).

Table 2.2: VLE measurements in progress.

	Ethanol	Isopropanol
Formic acid	303.15 K – 333.15 K	303.15 K – 353.15 K
Propionyl chloride	303.15 K – 323.15 K	303.15 K – 323.15 K

The experimental data obtained so far are listed in Tables 2.7 and 2.8. For all binary systems interaction parameters for the NRTL and UNIQUAC model were fitted to the data at all temperatures. The vapor phase association was considered by the method of Hayden and O'Connell (1975) known as chemical theory. The degree of association is represented by the η_{ij} parameter which is taken from Prausnitz (Prausnitz et al., 1986) according to component i and j (see Table 2.3).

Table 2.3: Hayden and O'Connell association parameters η_{ij}.

	Methanol	Propionic acid	Acetic acid
Methanol	1.63	2.50	2.50
Propionic acid	2.50	4.50	4.50
Acetic acid	2.50	4.50	4.50

The parameters obtained by regression are given in Tables 2.4 and 2.5. The three coefficients A_i, B_i, and C_i of the Antoine equation

$$P_{0i}^{LV} = \exp\left(\frac{A_i - B_i}{C_i + T/K}\right) \tag{2.23}$$

used for calculation of the pure component vapor pressure are listed in Table 2.6. In order to account only for the excess Gibbs energy the parameters A_i were adjusted to the experimental pure component vapor pressures during the g^E-model parameter fitting procedure.

Vapor phase data were measured also but due to problems in the determination of the compositions are not reproduced here.

In Figures 2.9 and 2.10 the experimental data and the results of the NRTL calculation are presented.

During the fitting procedure, the non randomness parameters α_{ij} of the NRTL model were set to 0.47. This value is recommended for mixtures of strongly self-associated substances with non-polar substances (Renon and Prausnitz, 1968). In case of the methanol + propionic acid system at 361.10 K, a good fit could only be achieved by including α_{ij} in the regression. However it was restricted to values between 0.0 and 1.0 to maintain a thermodynamic reasonable magnitude. All systems show a negative deviation from Raoult's Law $P_i = x_i P_{0i}^{LV}$ i.e. $\gamma_1 < 1$ throughout. Kinetic data were also obtained for the first hours of the reaction, but the state still was far from equilibrium after 12 h. Hence complete kinetics were not measured for these systems.

Experimental data were compared with calculated VLE computations from extrapolated NRTL parameters (Apelblat et al., 1976) and estimated parameters by UNIFAC. In case of the methanol + acetic acid system Figure 2.11 shows that UNIFAC overestimates the system pressure and predicts a positive deviation from Raoult's Law. The extrapolated parameters from Apelblat et al. underestimate the system pressure slightly. Both deviations grow with higher temperatures.

In case of the system methanol + propionic acid only a very slight overestimation of the system pressure occurs at both, prediction with UNIFAC and extrapolation from Apelblat et al. The experimental data and calculated curves can be seen in Figure 2.12.

2.5 Results

Table 2.4: Regressed interaction parameters for the system methanol (1) + acetic acid (2), and from Apelblat et al. (1976) at 308.15 K and 318.15 K.

T (K)		a_{12} (J/mol)	a_{21} (J/mol)	α_{12}	ΔP (kPa)
308.15	UNIQUAC	486.1228	−449.5153		0.9
	NRTL	59.7074	−281.3740	0.2472	1.1
318.15	UNIQUAC	−318.9105	334.2756		0.4
	NRTL	−135.0471	−119.2037	0.3307	3.0
332.76	UNIQUAC	255.4752	−483.4497		4.0
	NRTL	243.1955	−285.2002	0.47	0.2
351.94	UNIQUAC	437.5206	−416.9875		3.0
	NRTL	228.4277	−276.7689	0.47	0.9
361.49	UNIQUAC	−503.1419	574.2670		6.0
	NRTL	267.4625	−326.3024	0.47	0.3

Table 2.5: Regressed interaction parameters for the system methanol (1) + propionic acid (2), and from Apelblat et al. (1976) at 308.21 K and 318.16 K.

T (K)		a_{12} (J/mol)	a_{21} (J/mol)	α_{12}	ΔP (kPa)
308.21	UNIQUAC	−502.9222	980.6091		0.9
	NRTL	48.8358	−88.4390	0.6678	1.0
318.16	UNIQUAC	−518.9469	1002.2317		0.9
	NRTL	−374.9271	352.1810	0.1011	1.0
332.51	UNIQUAC	−627.8750	1325.062		1.0
	NRTL	209.9713	−245.0031	0.47	0.4
351.94	UNIQUAC	−539.7068	826.9049		2.0
	NRTL	235.2346	−276.2347	0.47	0.3
361.10	UNIQUAC	−617.0044	1060.9779		2.0
	NRTL	154.1661	−201.3134	1.0	0.5

Table 2.6: Antoine coefficients A_i, B_i, and C_i for pure components.
$$P_{0i}^{LV}/\text{hPa} = \exp\left(A_i - \frac{B_i}{C_i + T/\text{K}}\right) \cdot \frac{1013.25}{760}$$

	A_i	B_i	C_i
Methanol	18.5469	3626.55	34.29
Acetic acid	16.7607	3405.57	56.34
Propionic acid	17.3743	3723.42	−67.48

Table 2.7: Experimental P,x data of the binary system methanol (1) + acetic acid (2).

T/K	x_1/(mol/mol)	P/hPa
332.76	1.0000	798.5
	0.9165	735.1
	0.7547	595.4
	0.4286	369.1
	0.2020	213.3
	0.0000	112.7
351.94	1.0000	1667.5
	0.7422	1248.6
	0.7406	1251.3
	0.4776	785.4
	0.4667	788.1
	0.2247	514.7
	0.2166	530.2
	0.0000	254.8
361.49	1.0000	2327.0
	0.7960	1791.2
	0.7747	1788.8
	0.5271	1225.8
	0.4038	954.2
	0.3952	955.3
	0.1991	628.5
	0.1721	632.3
	0.0000	358.4

Table 2.8: Experimental P,x data of the binary system methanol (1) + propionic acid (2).

T/K	x_1/(mol/mol)	P/hPa
332.51	1.0000	843.2
	0.9738	780.0
	0.7380	563.0
	0.5150	386.0
	0.2410	208.4
	0.1210	134.8
	0.0000	40.4
351.94	1.0000	1653.0
	0.7506	1213.2
	0.7384	1217.9
	0.5003	774.9
	0.4857	776.8
	0.2506	433.3
	0.2423	436.2
	0.0000	94.3
361.10	1.0000	2327.0
	0.5930	1285.6
	0.5647	1283.8
	0.4056	888.3
	0.3975	872.2
	0.1511	433.4
	0.1500	455.4
	0.0000	146.6

2.5 Results

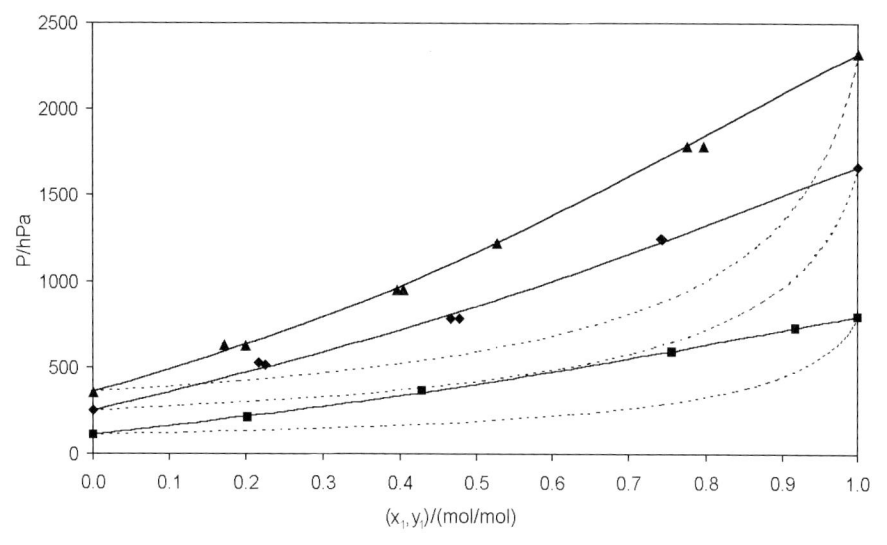

Figure 2.9: Experimental P,x data of the binary system methanol (1) + acetic acid (2) at (v) 332.76 K, (υ) 351.94 K, and (σ) 361.49 K, continuous line = calculated x_i, dashed line = calculated y_i.

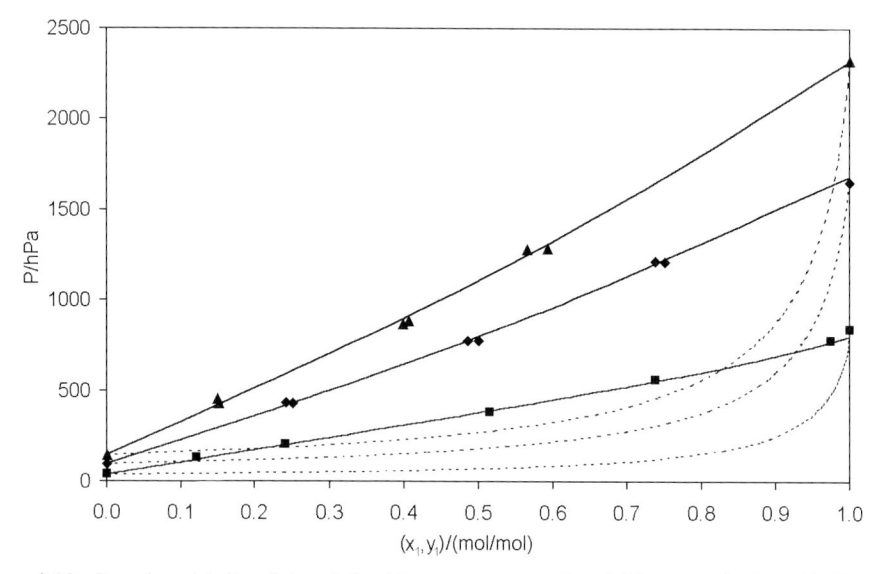

Figure 2.10: Experimental P,x data of the binary system methanol (1) + propionic acid (2) at (v) 332.51 K, (υ) 351.94 K, and (σ) 361.10 K, continuous line = calculated x_i, dashed line = calculated y_i.

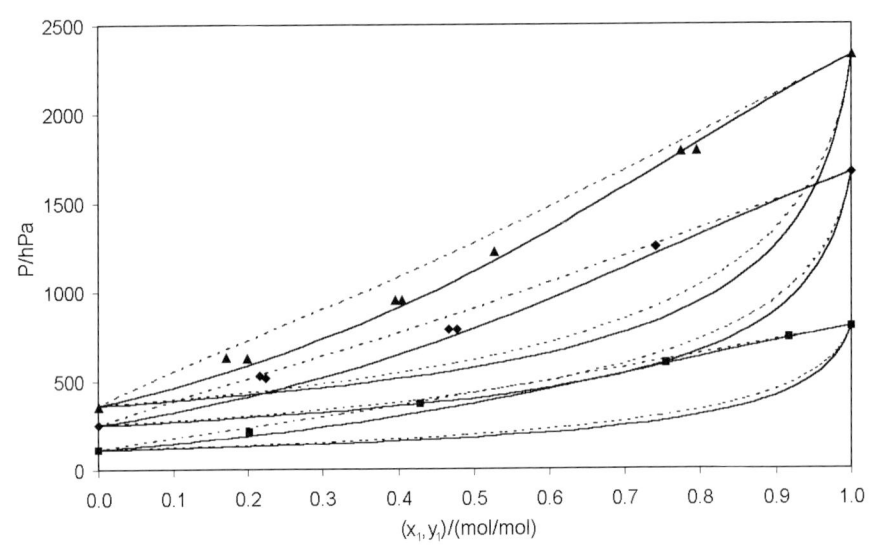

Figure 2.11: Comparison of experimental P,x data of the binary system methanol (1) + acetic acid (2) at (v) 332.76 K, (υ) 351.94 K, and (σ) 361.49 K with estimation by UNIFAC (continuous line) and extrapolation of Apelblat et al. (1976) (dashed line).

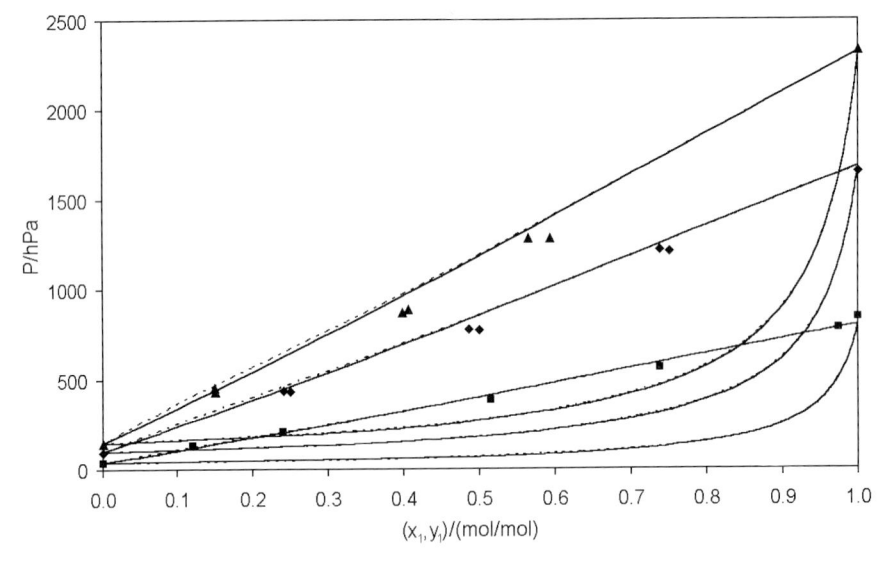

Figure 2.12: Comparison of experimental P,x data of the binary system methanol (1) + propionic acid (2) at (v) 332.51 K, (υ) 351.94 K, and (σ) 361.10 K with estimation by UNIFAC (continuous line) and extrapolation of Apelblat et al. (1976) (dashed line).

2.6 Conclusion

Isothermal binary P,x data have been measured for the reacting systems methanol + acetic acid (332.76 K to 361.49 K) and methanol + propionic acid (332.51 K to 361.10 K). A static cell with the possibility of adding thermostated, degassed components and an extrapolation method to the time of mixing have been used. Parameters for the NRTL and UNIQUAC g^E-models were obtained. Calculated values and experimental data are in good agreement, as can be seen from the diagrams. Also an improvement to the use of estimated or extrapolated NRTL parameters was shown. Hence the new regressed parameters for the chosen g^E-models can be recommended for accurate computation of the equilibrium conditions e.g. in simulation processes.

Further measurements will be carried out on the systems ethanol + formic acid, isopropanol + formic acid and the same alcohols with propionyl chloride. Therefore the vapor phase sample system will be improved in order to obtain complete VLE data sets.

References

Alsmeyer, F.; Marquardt, W.; Koß, H.-J.; Roth, M.; Siesler, H.W.; Müller, D.; Olf, G.; Ronge G. (2000): Measurement of Phase Equilibria in Reacting Systems, Lecture at GVC-Fachausschuß „Thermische Zerlegung von Gas- und Flüssigkeitsgemischen", Wernigerode, 5.–7.4.2000.

Apelblat, A.; Kohler, F. (1976): Excess Gibbs Energy of Methanol + Propionic Acid and of Methanol + Butyric Acid. J. Chem. Thermodyn. 8, 749–756.

Barbosa, D.; Doherty, M.F. (1988): The Simple Distillation of Homogeneous Reactive Mixtures. Chem. Eng. Sci. 43, 541–550.

Backhaus, A.A. (1921): Continuous Pprocess for the Manufacture of Esters. US Patents 1, 400, 849.

Bessling, B.; Schembecker, G.; Simmrock, K.H. (1997): Design of Processes with Reactive Distillation Line Diagrams. Ind. Eng. Chem. Res. 36, 3032–3042.

Espinosa, J.; Aguirre, P.A.; Perez, G.A. (1995a): Some Aspects in the Design of Multicomponent Reactive Distillation Columns Including Nonreactive Species. Chem.-Ing.-Tech. 43, 541–550.

Espinosa, J.; Aguirre, P.A.; Perez, G.A. (1995b): The Product Composition Regions of Single-Feed Reactive Distillation Columns: Mixtures Containing Inerts. Ind. Eng. Chem. Res. 34, 853–861.

Gmehling, J.; Onken, U. (1977): Vapor-Liquid Equilibrium Data Collection, DECHEMA Chemistry Data Series, Vol. I, Part 1, Frankfurt.

Hall, J.R. (1992): Cleaner Products: A Refining Challenge. Hydrocarbon Process. 71 (3), 100C–100F.

Hayden, J.G.; O'Connell J.P. (1975): A Generalised Method for Predicting Second Virial Coefficients, J. Ind. Eng. Chem. Process Des. Dev. 14 (3), 209–216.

Patel, N.C.; Young, M.A. (1993): Measurement and Prediction of Vapor-Liquid Equilibria for a reactive system: Application to Ethylene Oxide + Nonyl Phenol. Fluid Phase Equilib. 83, 79–92.

Pöpken, T.; Geisler, R.; Götze, L.; Brehm, A.; Moritz, P.; Gmehling, J. (1999): Reaktionskinetik in der Reaktivrektifikation – Zur Übertragbarkeit von kinetischen Daten aus einer Rührzelle auf einen Rieselbettreaktor. Chem.- Ing. -Tech. 71 (1+2), 96–100.

Pöpken, T.; Götze, L.; Gmehling, J. (2000): Reaction Kinetics and Chemical Equilibrium of Homogeneously and Heterogeneously Catalyzed Acetic Acid Esterification with Methanol and Methyl Acetate Hydrolysis. Ind. Eng. Chem. Res. 39, 2601–2611.

Prausnitz, J.M.; Lichtenthaler, R.N.; Azevedo, E.G. (1986): Molecular Thermodynamics of Fluid-Phase Equilibria, 2nd ed., Prentice Hall Inc., Englewood Cliffs.

Rehfinger, A.; Hoffmann, U. (1990): Kinetics of Methyl Tertiary Butyl Ether Liquid Phase Synthesis Catalyzed by Ion Exchange Resin; I. Intrinsic Rate Expressions in Liquid Phase Activities. Chem. Eng. Sci. 45 (4), 1605–1617.

Reichl, A.; Daiminger, U.; Schmidt, A.; Davies, M.; Hoffmann, U.; Brinkmeier, C.; Reder, C.; Marquardt, W. (1998): A Non-recycle Flow Still for the Experimental Determination of Vapor-Liquid Equilibria in Reactive Systems. Fluid Phase Equilib. 153 (1), 113–134.

Renon, H.; Prausnitz, J.M. (1968): Local Compositions in Thermodynamic Excess Functions for Liquid Mixtures. AIChE J. 14 (1), 135–144.

Stichlmair, J.G.; Fair, J.R. (1998): Distillation: Principles and Practices, Wiley-VCH, New York.

Winkler, B. (1996): Experimental Setup for the Isothermal Measurement of Binary VLE of Reacting Systems. Presentation at Union Carbide Corporation, South Carolina, USA.

Nomenclature

a	activity
a_{12}, a_{21}	binary interaction parameters for g^E-models
A_i, B_i, C_i	Antoine Coefficients
co	number of components
$E_{A,ij}$	energies of activation for reaction ij
f	fugacity
df	degree of Freedom
g	Gibbs energy per mole
g_{ij}	energies of interaction between an i-j pair of molecules
h_{mix}	enthalpy per mole
k_{12}	reaction rate constants of the esterification
k_{21}	reaction rate constants of the hydrolysis
K_a	chemical equilibrium constant in terms of activities
K_f	chemical equilibrium constant in terms of fugacities
K_x	chemical equilibrium constant in terms of mole concentrations
n	number of components
n_i	number of moles of component i
P	pressure
ph	number of phases
P^{LV}	vapor pressure
r	reaction rate
re	number of independent chemical reactions
R	universal gas constant
t	time
t_0	time of mixing, start of the reaction
T	temperature
v	molar volume

x	liquid mole fraction
y	vapour mole fraction
X	transformed liquid coordinate
Y	transformed vapor coordinate

Greek letters

α	exponent which accounts for the activity of acid protons
α_{ij}	non-randomness constant for binary ij interactions
v_i	stoichiometric coefficient of component i
φ	fugacity coefficient
γ	activity coefficient
η	Hayden O'Connell association parameter

Superscripts

+	standard state conditions
E	excess
L	liquid
V	vapor
LV	in phase equilibrium
VLE	in phase equilibrium
VLE+CE	in phase and chemical equilibrium

Subscripts

0	pure component
A	component A
B	component B
C	component C
D	component D
i0	initial composition of component i
i	component i
j	component j
mix	mixture
∞	after time $\to \infty$

3 Solubility of Carbon Dioxide in Aqueous Alkanolamine Solutions in the Temperature Range 313 to 353 K and Pressures up to 2.7 MPa

Dirk Silkenbäumer[1], Bernd Rumpf[2], and Rüdiger N. Lichtenthaler[1]*

3.1 Abstract

The solubility of carbon dioxide in aqueous solutions containing 2-amino-2-methyl-1-propanol (AMP) was measured in the temperature range from 313 to 353 K at total pressures up to 2.7 MPa using an analytical method. A model taking into account chemical reactions in the liquid phase as well as physical interactions is used to correlate the new data. To test the predictive capability of the model, the solubility of carbon dioxide in an aqueous solution containing AMP and N-methyldiethanolamine (MDEA) was measured at 313 K. Experimental results are reported and compared to literature data and calculations.

3.2 Introduction

Aqueous solutions containing water-soluble amines like monoethanolamine (MEA), N-methyldiethanolamine (MDEA) or sterically hindered amines like 2-amino-2-methyl-1-propanol (AMP) are widely used to separate acid gases like carbon dioxide and hydrogen sulfide from gaseous effluents contaminated with these components (Sartori and Savage, 1983; Xu et al., 1992). Typical examples are the cleaning of raw gases in coal gasification processes or applications in the natural gas industry. Often, aqueous solutions containing "blended" amines like mixtures of MEA and MDEA or MDEA and AMP are used (Jane and Li, 1997). The proper design of absorption columns requires information on vapor-liquid equilibria, caloric effects and also information on the kinetics of mass transfer and chemical reactions. Due to chemical reactions in the liquid phase and a strong deviation from ideality, the thermodynamic descrip-

---

[1] Universität Heidelberg, Physikalisch-Chemisches Institut, Angewandte Thermodynamik, Im Neuenheimer Feld 253, 69120 Heidelberg.
[2] Technische Universität Kaiserslautern, Fachbereich Maschinenbau und Verfahrenstechnik, Lehrstuhl für Technische Thermodynamik, Erwin-Schrödinger-Straße, 67653 Kaiserslautern.
* Author to whom correspondence should be addressed.

tion of aqueous systems containing alkanolamines like MEA, MDEA, or AMP and sour gases like carbon dioxide or hydrogen sulfide is a difficult task. Reliable experimental data at least for the solubility of a single gas in aqueous solutions containing alkanolamines are required to develop and test physico-chemical models to describe vapor-liquid equilibria in these systems. But reliable data on vapor-liquid equilibria in these systems are often scarce or scatter (Xu et al., 1992; Kuranov et al., 1996). Thus, the solubility of carbon dioxide in aqueous solutions containing AMP was measured in the temperature range from 313 to 353 K at pressures up to 2.7 MPa. The physico-chemical model originally developed by Edwards et al. (1978) and recently applied to aqueous solutions containing sour gases and MDEA by Kuranov et al. (1996) was used to correlate the data. To test the predictive capability of the model, the solubility of carbon dioxide in an aqueous solution containing AMP and MDEA was measured at 313 K. Experimental results are reported and compared to model calculations and literature data.

3.3 Experimental

3.3.1 Apparatus

Measurements of the vapor-liquid equilibrium (VLE) were carried out using the apparatus shown schematically in Figure 3.1. The equilibrium cell (EC) is made of stainless steel and has a volume of approximately 950 cm^3. The cell is thermostated in a water bath at the desired temperature with an accuracy of ± 0.1 K.

The mixture is stirred by a propeller stirrer (PS), which is coupled magnetically to the motor drive. When the pressure (LPR, HPR) and temperature readings (TR) remain constant (after approx. 2 h) complete chemical and vapor-liquid equilibrium has been established. The temperature is measured by a resistance thermometer (RT; PT 100, Lauda, Lauda-Königshofen) with an accuracy of ± 0.04 K. The pressure is measured using two different pressure transducers, a low-pressure sensor (LPS; MKS Baratron, München) with an accuracy of $\pm 0.2\%$ of the reading for pressures up to 0.5 MPa and a high-pressure sensor (HPS; Burster, Gernsbach) for pressures up to 10 MPa (accuracy $\pm 0.1\%$ of full scale). The membrane capacitors are located outside the water bath. The capacitors themselves and the tubes connecting the equilibrium cell with LPS and HPS are thermostated at a temperature (Th2) higher than the temperature in the equilibrium cell in order to avoid condensation.

The composition of the liquid and the gaseous phase is analyzed using a gas chromatograph (GC, Siemens) with a computerized integration unit (IU). The liquid phase is circulated through the liquid sample injection valve (LSV) by means of a gear pump (GP). The valve which allows injection of samples of 1 µl is mounted on the top of the gas chromatograph.

In order to analyze the gaseous phase, a Valco six-way valve (6-WV, gaseous sample injection valve) is used, which is located inside the water bath Th1. The sample loop has a volume of 1 ml. Two additional magnetically operated bellows valves (V5, V6) are needed for the injection procedure. The process of analyzing is performed automatically. All the valves are operated on-line by the gas chromatographic control unit. Each registered data

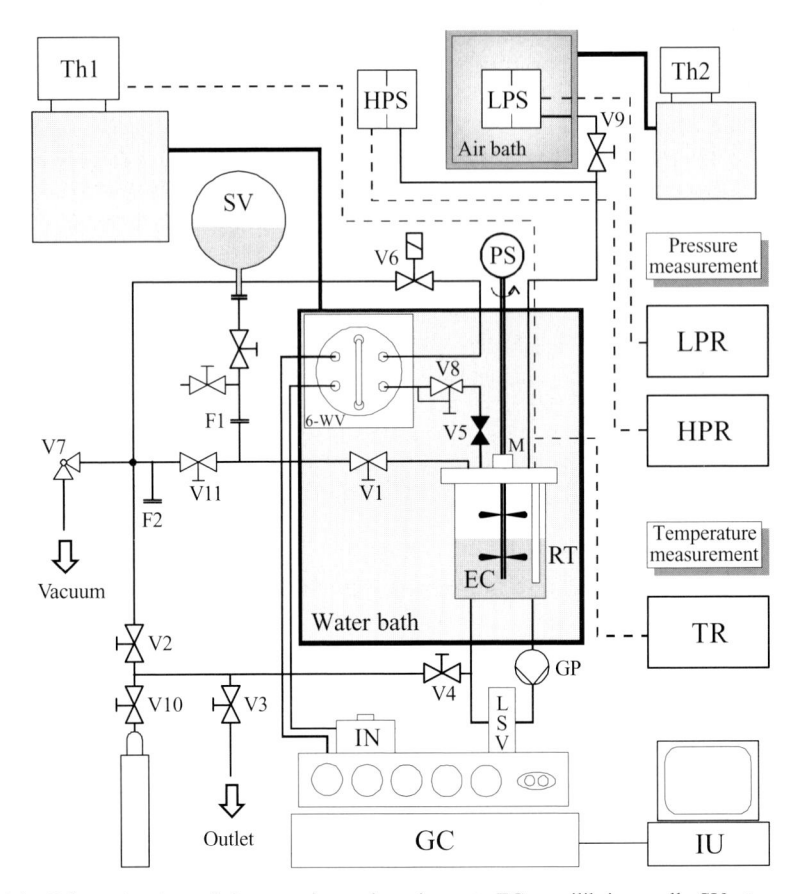

3 Solubility of Carbon Dioxide in Aqueous Alkanolamine Solutions

Figure 3.1: Schematic view of the experimental equipment: EC, equilibrium cell; SV, storage vessel; Th1/Th2, thermostat; GC, gas chromatograph; IN, injector; IU, integrator unit; GP, gear pump; LSV, liquid sampling valve; 6-WV, 6-way valve; RT, resistance thermometer; TR, temperature reading; LPS, low-pressure sensor; HPS, high-pressure sensor; LPR/HPR, pressure readings; PS; propeller stirrer; M, magnetic coupling; V1-V11, bellows valves; F1/F2, adapter.

point is the mean value of at least five samples taken from either of the two phases. The samples are taken alternately.

3.3.2 Procedure

The experimental procedure applied to the system carbon dioxide + water + AMP was as follows. The aqueous degassed AMP solution to be investigated was prepared in the store vessel (SV). The molality of the AMP solution was determined by weighing with an accuracy of ±0.001 mol/kg. The solution was then poured into the equilibrium cell until it was half filled by establishing a pressure gradient from the store vessel (SV) to the equilibrium cell. For this purpose the vessel was thermostated at a temperature higher than the temperature in the equilibrium cell. The gaseous carbon dioxide was added through the valve V4. By filling the cell in the described manner, carbon dioxide is quickly absorbed in the liquid to an extent close to the thermodynamic equilibrium.

The samples of the liquid phase were vaporized during the injection process into the GC. In the vaporized state, the chemical reactions were shifted completely to the educts, thus only carbon dioxide, water, and AMP were detected by the gas chromatograph. The gas chromatograph was equipped with a capillary separation column packed with Poropak Q (10 m × 0.53 mm) and a thermal conductivity detector. The peak areas obtained by GC were calibrated using so-called response factors. The response factor is the ratio of the molar amount to the area of a gas chromatographic peak. The response factor of CO_2 related to that of water was determined by filling the equilibrium cell with the pure substances and taking samples with the LSV (for water) and the 6-WV (for CO_2). The ratio of the volumes of liquid and gaseous sample was known and then the response factor was determined with an accuracy of ±2.7%. The reproducibility of analyzing the liquid phase was better than ±3% (for details cf. Silkenbäumer, 1997).

3.3.3 Materials

Carbon dioxide (≥99.995 mass%) was purchased from Messer-Griesheim, Ludwigshafen, and used without any further purification. Water (≥99.9 mass%) and AMP (≥99 mass%) were purchased from Merck, Darmstadt, and degassed by vacuum distillation. MDEA (Aldrich, ≥99 mass%) was also degassed before usage.

3.4 Results

3.4.1 Test of Procedure: Results for the System $CO_2 + H_2O$

To check the experimental arrangement and procedure, the solubility of carbon dioxide in water was measured at 293.15 K. The experimental results are given in Table 3.1. The molality of CO_2 was about 0.83 mol/kg at maximum, resulting in total pressures up to 2.5 MPa. In Figure 3.2 the results are compared to literature data (Crovetto, 1991; Landolt-Börnstein, 1968). The new data agree well with the literature data. The average relative deviation in the molality of carbon dioxide between the new experimental data and the literature data is 2%.

3 Solubility of Carbon Dioxide in Aqueous Alkanolamine Solutions

Table 3.1: Experimental results for the solubility of carbon dioxide in water at 293 K.

p/MPa	$\bar{m}_{CO_2,\text{literature}}$/mol/kg	$\bar{m}_{CO_2,\text{experimental}}$/mol/kg
0.5215	0.189	0.188
1.012	0.358	0.382
1.507	0.519	0.524
2.020	0.675	0.686
2.533	0.835	0.827

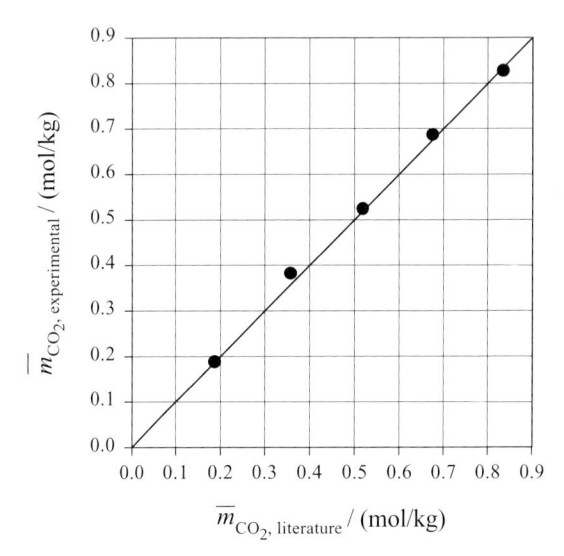

Figure 3.2: Solubility of carbon dioxide in water: Comparison between own measurements at 293 K (●) and literature data (Crovetto, 1991; Landolt-Börnstein, 1968).

3.4.2 Results for the System CO_2 + MDEA + H_2O

The experimental results for the solubility of carbon dioxide in an aqueous solution containing about 2.6 mol/kg of MDEA at 313 K are given in Table 3.2. In Figure 3.3, the experimental results for the total pressure are plotted versus the overall amount of carbon dioxide dissolved in the liquid phase. Furthermore, Figure 3.3 contains results calculated from a correlation by Kuranov et al. (1996) based on their data for the system CO_2 + MDEA + H_2O.

Adding carbon dioxide to a MDEA-containing solution at first results in nearly no change in the total pressure above the aqueous solution as in that range the sour gas is mostly dissolved in nonvolatile, ionic form. When the overall molality of the sour gas surmounts the overall molality of MDEA, the total pressure increases steeply as most of the MDEA has been spent by the chemical reactions and added sour gas can no longer be absorbed chemically, i.e. in nonvolatile, ionic form, but has to be dissolved also physically.

3.4 Results

Table 3.2: Experimental results for the solubility of carbon dioxide in aqueous solutions of MDEA ($\bar{m}_{MDEA}$ = 2.632 mol/kg) at 313 K.

$\bar{m}_{CO_2}$/mol/kg	p/MPa
0.634	0.0120
1.068	0.0182
1.371	0.0252
1.756	0.0445
2.117	0.0868
2.348	0.2230
2.601	0.5440
2.921	1.160
3.114	2.117
3.267	3.029
3.435	4.080

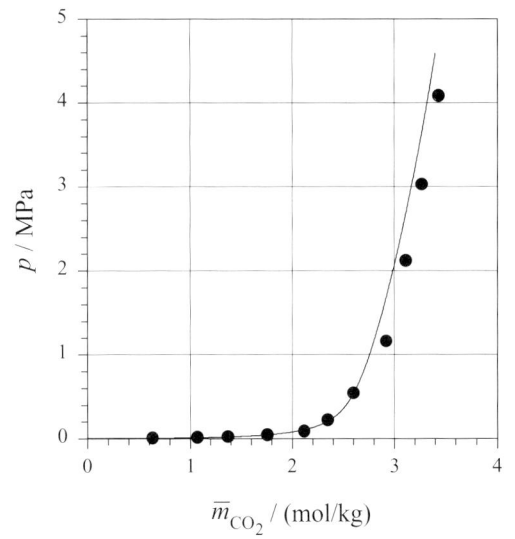

Figure 3.3: Solubility of carbon dioxide in an aqueous solution of MDEA at 313 K. ●: Experimental results, this work ($\bar{m}_{MDEA}$= 2.63 mol/kg); —: results calculated from a correlation by Kuranov et al. (1996).

There is a fair to good agreement with the correlation by Kuranov et al., although at constant pressure the experimental results for the overall molality of carbon dioxide in the liquid phase are slightly, but systematically higher. The average relative deviation in the total pressure of the new data from the correlation by Kuranov et al. (1996) is 19%, the average absolute deviation is 0.23 MPa. However, this comparably large deviations mostly result from a few data points at carbon dioxide molalities where the slope of the total pressure curve is rather steep. For example at $\bar{m}_{CO_2}$ = 3.44 mol/kg, the experimentally determined total pressure is 4.08 MPa whereas the calculated total pressure is 4.85 MPa. Vice versa at

53

$p = 4.08$ MPa, the calculated overall molality of carbon dioxide in the liquid phase is 3.33 mol/kg. Thus, small errors in the overall molality of carbon dioxide lead to comparably large errors in the total pressure.

3.4.3 Results for the System CO_2 + AMP + H_2O

The results for the solubility of carbon dioxide in aqueous solutions of AMP in the temperature range from 313 to 353 K are given in Table 3.3. Two overall molalities of AMP ($\bar{m}_{AMP} \approx 2.4$ and ≈ 6.2 mol/kg) were investigated. The maximum overall molality of carbon dioxide is about 6.3 mol/kg corresponding to a total pressure of about 3 MPa.

Table 3.3: Experimental results for the solubility of carbon dioxide in aqueous solutions of AMP.

$\bar{m}_{AMP}$ = 2.430 mol/kg T = 313.15 K		$\bar{m}_{AMP}$ = 2.451 mol/kg T = 333.15 K		$\bar{m}_{AMP}$ = 2.443 mol/kg T = 353.15 K	
$\bar{m}_{CO_2}$/mol/kg	p/MPa	$\bar{m}_{CO_2}$/mol/kg	p/MPa	$\bar{m}_{CO_2}$/mol/kg	p/MPa
0.557	0.0073	0.672	0.0204	0.375	0.0473
1.447	0.0093	1.002	0.0226	0.748	0.0526
1.906	0.0183	1.268	0.0253	0.892	0.0576
1.999	0.0295	1.874	0.0486	1.231	0.0751
2.082	0.0430	2.022	0.0711	1.640	0.1238
2.229	0.0726	2.183	0.1083	2.038	0.2691
2.416	0.233	2.270	0.1550	2.086	0.2766
2.581	0.615	2.362	0.2244	2.172	0.3836
2.735	1.197	2.420	0.3201	2.265	0.564
2.903	2.052	2.432	0.521	2.504	1.061
		2.681	1.052	2.590	1.633
		2.849	2.023	2.596	1.754
				2.613	1.938

$\bar{m}_{AMP}$ = 6.135 mol/kg T = 313.15 K		$\bar{m}_{AMP}$ = 6.477 mol/kg T = 333.15 K		$\bar{m}_{AMP}$ = 6.242 mol/kg T = 353.15 K	
$\bar{m}_{CO_2}$/mol/kg	p/MPa	$\bar{m}_{CO_2}$/mol/kg	p/MPa	$\bar{m}_{CO_2}$/mol/kg	p/MPa
1.786	0.0073	1.833	0.0221	0.975	0.0498
2.255	0.0076	2.547	0.0278	1.528	0.0585
2.905	0.0083	3.252	0.0380	2.253	0.0785
3.442	0.0102	3.728	0.0532	3.029	0.1055
3.928	0.0138	4.314	0.1045	3.613	0.1650
4.460	0.0258	5.244	0.2500	4.135	0.2738
4.830	0.0447	5.812	0.680	4.756	0.4539
5.386	0.1275	5.932	1.095	5.178	0.745
5.652	0.2867	6.057	1.730	5.447	1.117
5.897	0.6380	6.382	2.743		
6.098	1.2100				
6.210	1.9960				

3.4 Results

In the upper parts of Figures 3.4 and 3.5 the results for the total pressure above aqueous solutions containing about 2.4 and 6.2 mol of AMP per kg of water at 313, 333, and 353 K are plotted versus the overall molality of carbon dioxide dissolved in the liquid phase. Figures 3.4 and 3.5 also show the results of the correlation of the new data (see below). As it was expected, the total pressure shows a similar behavior as in the system CO_2 + MDEA + H_2O.

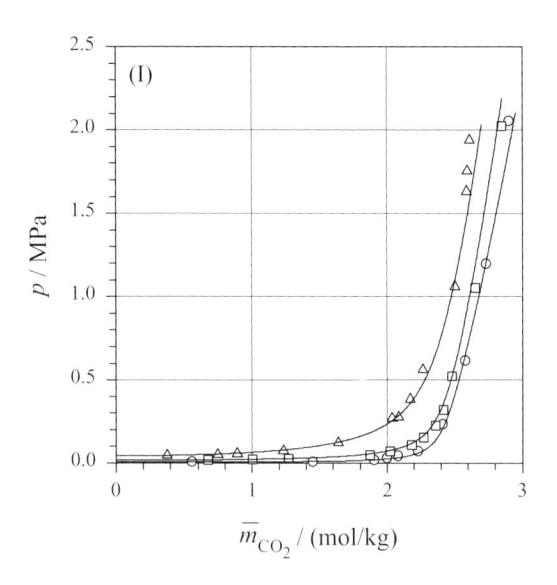

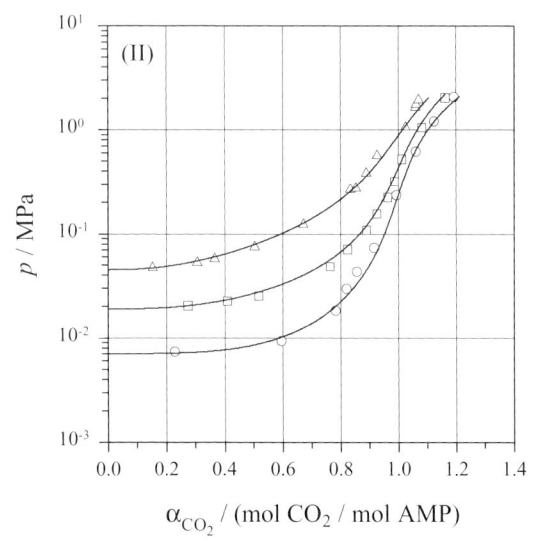

Figure 3.4: Solubility of carbon dioxide in aqueous solutions of AMP. $\bigcirc, \square, \triangle$: Experimental results, this work ($\bigcirc$: $\bar{m}_{AMP}$ = 2.43 mol/kg, T = 313.15 K; $\square$: $\bar{m}_{AMP}$ = 2.45 mol/kg, T = 333.15 K; $\triangle$: $\bar{m}_{AMP}$ = 2.44 mol/kg, T = 353.15 K); —— : correlation, this work.

3 Solubility of Carbon Dioxide in Aqueous Alkanolamine Solutions

In the lower parts of Figures 3.4 and 3.5 the total pressure above the aqueous solution is plotted versus the so-called loading of the AMP solution, i.e. the ratio $\bar{m}_{CO_2}/\bar{m}_{AMP}$. Due to the temperature dependencies of the chemical reactions in the liquid phase and of the physical solubility of carbon dioxide, at constant pressure the overall amount of carbon dioxide in the liquid phase decreases with increasing temperature.

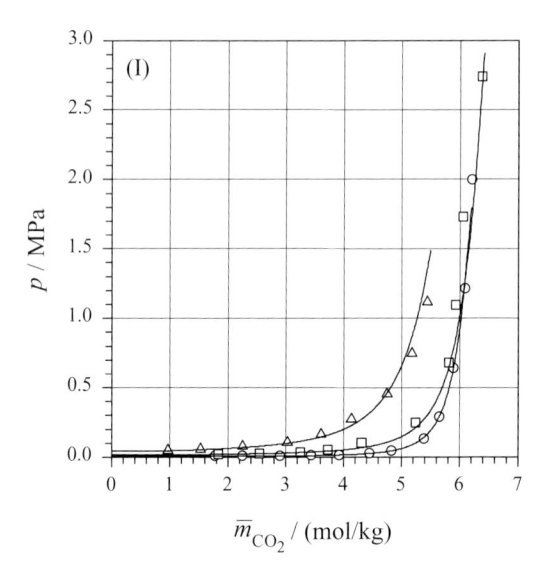

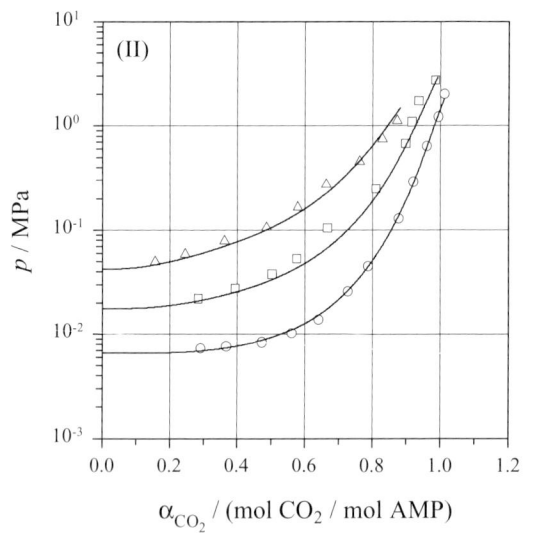

Figure 3.5: Solubility of carbon dioxide in aqueous solutions of AMP. $\bigcirc, \square, \triangle$: Experimental results, this work ($\bigcirc$: $\bar{m}_{AMP}$ = 6.14 mol/kg, T = 313.15 K; $\square$: $\bar{m}_{AMP}$ = 6.48 mol/kg, T = 333.15 K; $\triangle$: $\bar{m}_{AMP}$ = 6.24 mol/kg, T = 353.15 K); —: correlation, this work.

3.4.4 Results for the System CO_2 + MDEA + AMP + H_2O

In order to investigate the simultaneous influence of MDEA and AMP on the solubility of carbon dioxide, the solubility of carbon dioxide in an aqueous solution containing 1.266 mol/kg MDEA and 1.278 mol/kg AMP was measured (cf. Table 3.4). In Figure 3.6, the results for the total pressure are compared to the results for the two ternary systems CO_2 + MDEA + H_2O and CO_2 + AMP + H_2O. Figure 3.6 furthermore contains calculated results for the two ternary systems as well as predicted results for the quaternary system (see below).

Table 3.4: Experimental results for the solubility of carbon dioxide in an aqueous solution containing MDEA and AMP ($\bar{m}_{MDEA}$ = 1.278 mol/kg, $\bar{m}_{AMP}$ = 1.266 mol/kg) at 313 K.

$\bar{m}_{CO_2}$/mol/kg	p/MPa
1.159	0.0125
1.404	0.0197
1.854	0.0443
2.250	0.1100
2.413	0.1890
2.682	0.4550
2.853	1.082
3.087	2.072
3.251	3.012
3.411	4.020

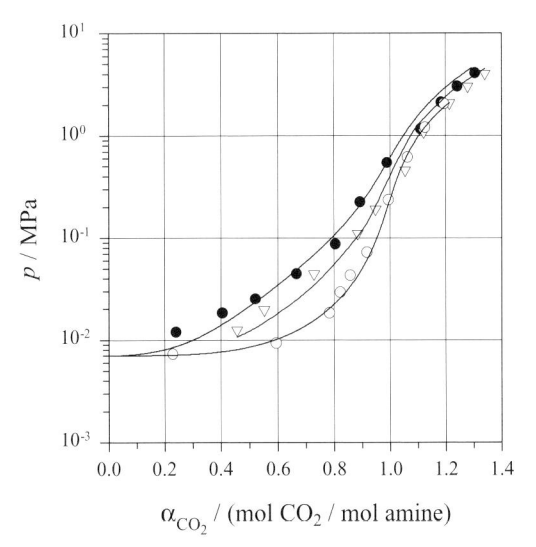

Figure 3.6: Solubility of carbon dioxide in aqueous alkanolamine solutions at 313 K. ●, ○, ▽: Experimental results, this work (●: $\bar{m}_{MDEA}$ = 2.632 mol/kg; ○: $\bar{m}_{AMP}$ = 2.430 mol/kg; ▽: $\bar{m}_{MDEA}$ = 1.266 mol/kg, $\bar{m}_{AMP}$ = 1.278 mol/kg); —: calculated results, this work and Kuranov et al. (1996).

3 Solubility of Carbon Dioxide in Aqueous Alkanolamine Solutions

The overall alkanolamine molality in all solutions is comparable. As AMP is a stronger base than MDEA, at a constant total pressure an aqueous solution containing AMP has a higher carbon dioxide loading than a MDEA-solution. The model based only on data for the ternary systems CO_2 + MDEA + H_2O and CO_2 + AMP + H_2O (see below) well predicts the solubility of carbon dioxide in the aqueous solution containing MDEA and AMP simultaneously.

3.5 Modeling

Figure 3.7 shows a scheme of the model applied to calculate the solubility of carbon dioxide in aqueous solutions containing MDEA and AMP. Due to chemical reactions in the liquid phase, carbon dioxide is dissolved in the liquid phase not only in neutral, but also in nonvolatile, ionic form. The following chemical reactions are considered (R denotes the HO–CH_2–$C(CH_3)_2$-group in AMP):

$$CO_2 + H_2O \rightleftarrows HCO_3^- + H^+ \tag{I}$$

$$HCO_3^- \rightleftarrows CO_3^{2-} + H^+ \tag{II}$$

$$CO_2 + H_2O \rightleftarrows HCO_3^- + H^+ \tag{I}$$
$$HCO_3^- \rightleftarrows CO_3^{2-} + H^+ \tag{II}$$
$$RNH_2 + H_2O \rightleftarrows RNH_3^+ + OH^- \tag{III}$$
$$RNH_2 + HCO_3^- \rightleftarrows RNHCOO^- + H_2O \tag{IV}$$
$$MDEA + H_2O \rightleftarrows MDEAH^+ + OH^- \tag{V}$$
$$H_2O \rightleftarrows H^+ + OH^- \tag{VI}$$

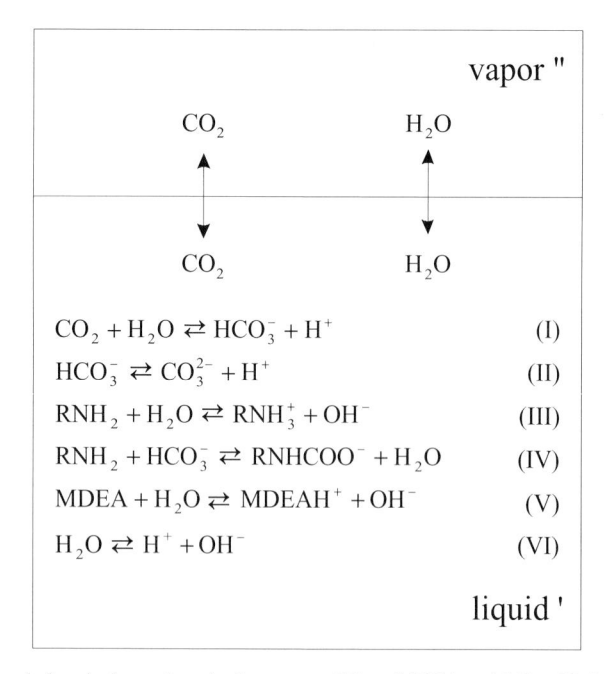

Figure 3.7: VLE and chemical reactions in the system CO_2 + MDEA + AMP + H_2O.

3.5 Modeling

$$RNH_2 + H_2O \rightleftharpoons RNH_3^+ + OH^- \quad \text{(III)}$$

$$RNH_2 + HCO_3^- \rightleftharpoons RNHCOO^- + H_2O \quad \text{(IV)}$$

$$MDEA + H_2O \rightleftharpoons MDEAH^+ + OH^- \quad \text{(V)}$$

$$H_2O \rightleftharpoons H^+ + OH^- \quad \text{(VI)}$$

The condition for chemical equilibrium yields the following equation for a chemical reaction R:

$$K_R(T) = \prod_i a_i^{v_{i,R}} \quad R = \text{I} \ldots \text{VI} \quad (3.1)$$

The balance equation for the number of moles n_i of species i in the liquid phase is

$$n_i = \bar{n}_i + \sum_R v_{i,R} \cdot \xi_R \quad (3.2)$$

where n_i is the overall number of moles of species i, $v_{i,R}$ is the stoichiometric coefficient for species i in reaction R ($v_{i,R} > 0$ for products and $v_{i,R} < 0$ for educts) and ξ_R is the extent of reaction R. Together with an expression for the excess Gibbs energy of the liquid phase, Eqs. (3.1) and (3.2) may be solved in an iterative procedure to yield the true number of moles n_i as well as the true molalities m_i of all species in the liquid phase for given overall molalities of carbon dioxide, MDEA, and AMP and a given temperature.

The condition of vapor-liquid equilibrium is then applied to calculate the total pressure and the composition of the gas phase. For water, the extended Raoult's law is used:

$$p y_w \varphi_w'' = p_w^s \varphi_w^s \exp\left(\frac{v_w (p - p_w^s)}{RT}\right) a_w \quad (3.3)$$

while for carbon dioxide, the extended Henry's law is used:

$$p y_{CO_2} \varphi_{CO_2}'' = H_{CO_2,w}^{(m)}(T, p_w^s) \exp\left(\frac{v_{CO_2,w}^\infty (p - p_w^s)}{RT}\right) m_{CO_2} \gamma_{CO_2}^* \quad (3.4)$$

As the vapor pressures of pure AMP and MDEA are rather small in the temperature range considered here (cf. Xu et al., 1991; Xu et al., 1992), the presence of AMP and MDEA in the vapor phase is neglected.

The calculation requires the knowledge of the six temperature dependent equilibrium constants K_R, the activities a_i of all species present in the liquid phase, Henry's constant $H_{CO_2,w}^{(m)}$ for the solubility of carbon dioxide in pure water, the vapor pressure p_w^s and molar volume v_w of pure water, the partial molar volume $v_{CO_2,w}^\infty$ of carbon dioxide dissolved at infinite dilution in water as well as information on the vapor phase nonideality. Equilibrium constants K_I and K_{II} are taken from Bieling et al. (1989), K_{III} from Littel et al. (1990), K_{IV} from Xu et al. (1992), K_V from Kuranov et al. (1996), and K_{VI} from Edwards et al. (1978) (cf. Table 3.5). Henry's constant for carbon dioxide is taken from Rumpf and Maurer (1993) (cf. Table 3.6).

3 Solubility of Carbon Dioxide in Aqueous Alkanolamine Solutions

The vapor pressure and molar volume of water is taken from Saul and Wagner (1987). A truncated virial equation of state is used to calculate the fugacity coefficients φ_i. Pure component second virial coefficients B_{ij} for carbon dioxide and water are calculated from a correlation based on data recommended by Dymond and Smith (1980) (cf. Table 3.7). The mixed second virial coefficient $B_{CO_2,w}$ is calculated as recommended by Hayden and O'Connell (1975) (cf. Table 3.8). The partial molar volume of carbon dioxide at infinite dilution in water is calculated as recommended by Brelvi and O'Connell (1972) (cf. Table 3.8).

Table 3.5: Equilibrium constants for chemical reactions (I–VI) (Schwabe, 1959; Edwards et al., 1978; Bieling et al., 1989; Littel et al., 1990; Xu et al., 1992). (Note: Numbers for reactions (III) and (V) are given on molarity scale.)

$$\ln K_R = A_R/(T/K) + B_R \ln(T/K) + C_R (T/K) + D_R$$

R	Reaction	A_R	B_R	$C_R \cdot 10^2$	D_R
I	$CO_2 + H_2O \rightleftarrows HCO_3^- + H^+$	−7742.6	−14.506	−2.8104	102.28
II	$HCO_3^- \rightleftarrows CO_3^{2-} + H^+$	−8982.0	−18.112	−2.2249	116.73
III	$RNH_2 + H_2O \rightleftarrows RNH_3^+ + OH^-$	−7261.78	−22.4773	0	142.58612
IV	$RNH_2 + HCO_3^- \rightleftarrows RNHCOO^- + H_2O$	2546.2	0	0	−11.555
V	$MDEA + H_2O \rightleftarrows MDEAH^+ + OH^-$	−13445.9	−22.4773	−4.1447	173.1912
VI	$H_2O \rightleftarrows H^+ + OH^-$	−13445.9	−22.4773	0	140.932

Table 3.6: Henry's constant for the solubility of carbon dioxide in pure water ($273 \le T/K \le 473$) (Rumpf and Maurer, 1993).

$$\ln H_{CO_2,w}^{(m)}(T, p_w^s)/(\text{MPa kg mol}^{-1}) = A_{CO_2,w} + B_{CO_2,w}/(T/K) + C_{CO_2,w}(T/K) + D_{CO_2,w} \ln(T/K)$$

$A_{CO_2,w}$	$B_{CO_2,w}$	$C_{CO_2,w}$	$D_{CO_2,w}$
192.876	−9624.4	0.01441	−28.749

Table 3.7: Pure component second virial coefficients ($273 \le T/K \le 473$).

$$B_{i,i}/(\text{cm}^3/\text{mol}) = a_{i,i} + b_{i,i}(c_{i,i}/(T/K))^{d_{i,i}}$$

i	$a_{i,i}$	$b_{i,i}$	$c_{i,i}$	$d_{i,i}$
CO_2	65.703	−184.854	304.16	1.4
H_2O	−53.53	−39.29	647.3	4.3

Table 3.8: Mixed second virial coefficients and partial molar volume for carbon dioxide at infinite dilution in water.

T/K	$B_{CO_2,w}/(\text{cm}^3/\text{mol})$	$v_{CO_2,w}^\infty/(\text{cm}^3/\text{mol})$
313.15	−163.1	33.4
333.15	−144.6	34.7
353.15	−129	36.3

3.5 Modeling

Activity coefficients of both molecular and ionic species were calculated from the Pitzer (1973) equation for the excess Gibbs energy of an aqueous electrolyte solution:

$$\frac{G^E}{RT n_w M_w} = f_1(I) + \sum_{(i,j)\neq w} m_i m_j (\beta_{i,j}^{(0)} + \beta_{i,j}^{(1)} f_2(I)) + \sum_{(i,j,k)\neq w} m_i m_j m_k \tau_{i,j,k} \quad (3.5)$$

where f_1 is a modified Debye-Hückel term. Both f_1 and f_2 are functions of ionic strength I:

$$I = \frac{1}{2} \sum_i m_i z_i^2 \quad (3.6)$$

In Eq. (3.5), $\beta_{i,j}^{(0)}$, $\beta_{i,j}^{(1)}$, and $\tau_{i,j,k}$ are binary and ternary interaction parameters. The resulting expressions for the activity coefficients of a dissolved species i and for the activity of water are given elsewhere (Bieling et al., 1995). Calculations require the dielectric constant of pure water which was taken from Bradley and Pitzer (1979).

Interaction parameters for the ternary system CO_2 + MDEA + H_2O were taken from Kuranov et al. (cf. Table 3.9) while those for the system CO_2 + AMP + H_2O were determined from the experimental data of the present work as follows (cf. Kuranov et al., 1996): In the system CO_2 + AMP + H_2O eight species (CO_2, RNH_2, RNH_3^+, $RNHCOO^-$, HCO_3^-, CO_3^{2-}, H^+, OH^-) are present in the liquid phase. As the concentrations of hydrogen- and hydroxide ions remain rather small compared to those of the other species, all interaction parameters involving H^+ or OH^- were set to zero. As no reliable experimental information for the binary system AMP + H_2O is available, interaction parameters $\beta_{RNH_2,RNH_2}^{(0)}$ and τ_{RNH_2,RNH_2,RNH_2} had to be set to zero. Furthermore, the ionic strength dependence of the second virial coefficient in Eq. (3.5) is neglected for a neutral species, i.e. $\beta_{RNH_2,j}^{(1)} = \beta_{CO_2,j}^{(1)} = 0$.

Table 3.9: Interaction parameters for the system CO_2 + MDEA + AMP + H_2O.
$f(T) = q_0 + q_1/(T/K)$

Parameter	q_0	q_1	T_{min}/K	T_{max}/K	Subsystem
$\beta_{CO_2,HCO_3^-}^{(0)}$	0.0843	−16.15	313	473	$NH_3 + CO_2 + H_2O$
$\beta_{CO_2,RNH_3^+}^{(0)}$	−0.40384	88.4057	313	353	$CO_2 + AMP + H_2O$
$\beta_{CO_2,RNH_3^+,RNHCOO^-}^{(0)}$	0.40384	−88.4057	313	353	$CO_2 + AMP + H_2O$
$\beta_{RNH_3^+,HCO_3^-}^{(0)}$	0.02118	−30.8522	313	353	$CO_2 + AMP + H_2O$
$\beta_{RNH_3^+,CO_3^{2-}}^{(0)}$	3.58754	−1042.0259	313	353	$CO_2 + AMP + H_2O$
$\beta_{RNH_3^+,HCO_3^-}^{(1)}$	3.79363	−1147.3168	313	353	$CO_2 + AMP + H_2O$
$\tau_{CO_2,RNH_3^+,HCO^-}$	0.00572	0	313	353	$CO_2 + AMP + H_2O$
$\tau_{RNH_3^+,HCO_3^-,HCO_3^-}$	0.00745	0	313	353	$CO_2 + AMP + H_2O$
$\beta_{CO_2,MDEAH^+}^{(0)}$	−0.4147	119.96	313	413	$CO_2 + MDEA + H_2O$
$\beta_{MDEAH^+,HCO_3^-}^{(0)}$	−0.5418	251.43	313	413	$CO_2 + MDEA + H_2O$
$\beta_{MDEAH^+,HCO_3^-}^{(1)}$	1.3284	−787.13	313	413	$CO_2 + MDEA + H_2O$
$\tau_{MDEAH^+,HCO_3^-,HCO_3^-}$	0.0338	−16.164	313	413	$CO_2 + MDEA + H_2O$

The remaining parameters describing interactions between molecular dissolved carbon dioxide and other species were determined following Kuranov et al. (1996): Considering only binary parameters in the following derivation, the activity coefficient of carbon dioxide in an aqueous solution of AMP is

$$\ln \gamma_{CO_2}^* = 2 \left(m_{CO_2} \cdot \beta_{CO_2,CO_2}^{(0)} + m_{RNH_2} \cdot \beta_{CO_2,RNH_2}^{(0)} \right.$$
$$+ m_{RNH_3^+} \cdot \beta_{CO_2,RNH_3^+}^{(0)} + m_{RNHCOO^-} \cdot \beta_{CO_2,RNHCOO^-}^{(0)} \qquad (3.7)$$
$$\left. + m_{HCO_3^-} \cdot \beta_{CO_2,HCO_3^-}^{(0)} + m_{CO_3^{2-}} \cdot \beta_{CO_2,CO_3^{2-}}^{(0)} \right)$$

Neglecting the concentration of hydrogen and hydroxide ions, the condition for electroneutrality of the liquid phase is

$$m_{RNH_3^+} = m_{RNHCOO^-} + m_{HCO_3^-} + 2 \cdot m_{CO_3^{2-}} \qquad (3.8)$$

Introducing that equation into Eq. (3.7) leads to

$$\ln \gamma_{CO_2}^* = 2 \left(m_{CO_2} \cdot \beta_{CO_2,CO_2}^{(0)} + m_{RNH_2} \cdot \beta_{CO_2,RNH_2}^{(0)} \right.$$
$$+ m_{RNHCOO^-} \cdot \left(\beta_{CO_2,RNH_3^+}^{(0)} + \beta_{CO_2,RNHCOO^-}^{(0)} \right)$$
$$+ m_{HCO_3^-} \cdot \left(\beta_{CO_2,RNH_3^+}^{(0)} + \beta_{CO_2,HCO_3^-}^{(0)} \right) \qquad (3.9)$$
$$\left. + m_{CO_3^{2-}} \cdot \left(\beta_{CO_2,RNH_3^+}^{(0)} + \beta_{CO_2,CO_3^{2-}}^{(0)} \right) \right)$$

Following Rumpf and Maurer (1993) and Kuranov et al. (1996), the binary parameters $\beta_{G,C}^{(0)}$ and $\beta_{G,A}^{(0)}$ describing interactions between a dissolved gas G and a strong electrolyte $C_{\nu_C} A_{\nu_A}$ cannot be determined separately. Therefore, introducing

$$B_{G,CA}^{(0)} = \nu_C \beta_{G,C}^{(0)} + \nu_A \beta_{G,A}^{(0)} \qquad (3.10)$$

where C denotes RNH_3^+ and A denotes $RNHCOO^-$ or HCO_3^- or CO_3^{2-} into Eq. (3.9) results in

$$\ln \gamma_{CO_2}^* = 2 \left(m_{CO_2} \cdot \beta_{CO_2,CO_2}^{(0)} + m_{RNH_2} \cdot \beta_{CO_2,RNH_2}^{(0)} \right.$$
$$+ m_{RNHCOO^-} \cdot B_{CO_2,RNH_3^+RNHCOO^-}^{(0)} + m_{HCO_3^-} \cdot B_{CO_2,RNH_3^+HCO_3^-}^{(0)}$$
$$\left. + m_{CO_3^{2-}} \cdot B_{CO_2,(RNH_3^+)_2CO_3^{2-}}^{(0)} \right) \qquad (3.11)$$

where

$$B_{CO_2,RNH_3^+RNHCOO^-}^{(0)} = \beta_{CO_2,RNH_3^+}^{(0)} + \beta_{CO_2,RNHCOO^-}^{(0)} \qquad (3.12)$$

3.5 Modeling

$$B^{(0)}_{CO_2,RNH_3^+HCO_3^-} = \beta^{(0)}_{CO_2,RNH_3^+} + \beta^{(0)}_{CO_2,HCO_3^-} \tag{3.13}$$

$$B^{(0)}_{CO_2,(RNH_3^+)_2CO_3^{2-}} = 2 \cdot \beta^{(0)}_{CO_2,RNH_3^+} + \beta^{(0)}_{CO_2,CO_3^{2-}} \tag{3.14}$$

As can be seen from Eqs. (3.11) to (3.14), in principle only four additional binary parameters ($\beta^{(0)}_{CO_2,RNH_2}$, $\beta^{(0)}_{CO_2,RNHCOO^-}$, $\beta^{(0)}_{CO_2,HCO_3^-}$, $\beta^{(0)}_{CO_2,CO_3^{2-}}$) are required to describe the activity coefficient of carbon dioxide in the system $CO_2 + AMP + H_2O$. A similar argument holds for the ternary parameters describing interactions between molecular dissolved carbon dioxide and other species. The resulting combinations of ternary interaction parameters which directly appear in the expression for the activity coeffcient of carbon dioxide are (Rumpf and Maurer, 1993):

$$\Gamma_{G,G,CA} = \nu_C \cdot \tau_{G,G,C} + \nu_A \cdot \tau_{G,G,A} \tag{3.15}$$

$$\Gamma_{G,CA,CA} = \nu_C^2 \cdot \tau_{G,C,C} + 2 \cdot \nu_C \cdot \nu_A \cdot \tau_{G,C,A} + \nu_A^2 \cdot \tau_{G,A,A} \tag{3.16}$$

As in Eqs. (3.12) to (3.14), only certain "observable" combinations of interaction parameters may be determined.

As molecular dissolved carbon dioxide and AMP are simultaneously present only in very small amounts, the binary parameter $\beta^{(0)}_{CO_2,RNH_2}$ was set to zero. The binary parameters $\beta^{(0)}_{CO_2,HCO_3^-}$ and $\beta^{(0)}_{CO_2,CO_3^{2-}}$ were taken from Kurz et al. (1995). Preliminary calculations showed that the calculated true molality of the carbamate ion remains small compared to the other species. Therefore, parameter $B^{(0)}_{CO_2,RNH_3^+RNHCOO^-}$ was set to zero. As the amount of molecular dissolved carbon dioxide remains small in the pressure range investigated here, all ternary interaction parameters $\tau_{CO_2,CO_2,i}$ were set to zero. A sensitivity study then showed that it is sufficient to include the binary parameter $\beta^{(0)}_{CO_2,RNH_3^+}$ as well as the ternary parameter $\tau_{CO_2,RNH_3^+,HCO_3^-}$.

Besides the parameters describing interactions between neutral carbon dioxide and other species, there are several other parameters in Eq. (3.5) describing interactions between charged species, e.g. $\beta^{(0)}_{RNH_3^+,HCO_3^-}$, $\beta^{(0)}_{RNH_3^+,CO_3^{2-}}$, $\beta^{(0)}_{RNH_3^+,RNHCOO^-}$ as well as ternary parameters like $\tau_{RNH_3^+,RNH_3^+,HCO_3^-}$, $\tau_{RNH_3^+,RNH_3^+,CO_3^{2-}}$ and $\tau_{RNH_3^+,RNH_3^+,RNHCCO^-}$. Although those parameters do not have a direct influence on the activity coefficient of carbon dioxide (cf. Eq. (3.11)), they strongly influence the calculated species distribution in the liquid phase, i.e. the "true" molalities of carbon dioxide and AMP. To further reduce the number of adjustable parameters, it is common practice to neglect all interaction parameters involving only species with the same sign of charge. A sensitivity study then showed that it is sufficient to take into account the following parameters describing interactions between charged species: $\beta^{(0)}_{RNH_3^+,HCO_3^-}$, $\beta^{(0)}_{RNH_3^+,HCO_3^-}$, $\beta^{(0)}_{RNH_3^+,CO_3^{2-}}$ as well as the ternary parameter $\tau_{RNH_3^+,HCO_3^-,HCO_3^-}$.

The influence of temperature on the binary parameters had to be taken into account. It was approximated by

$$f(T) = q_0 + q_1/(T/K) \tag{3.17}$$

whereas the ternary parameters were treated as independent of temperature. Together with the ternary parameters mentioned above, parameters q_0 and q_1 were fitted simultaneously to the new results for the solubility of carbon dioxide in aqueous solutions containing AMP. The resulting set of parameters is given in Table 3.9.

The model correlates the new experimental data in the temperature range from 313 to 353 K with an average relative deviation of 8.5%, the average absolute deviation is 0.04 MPa (cf. Figs. 3.4 and 3.5).

To test the predictive capability of the model, the parameter set as determined from the new experimental data for the system CO_2 + AMP + H_2O was combined with the interaction parameters for CO_2 + MDEA + H_2O as taken from Kuranov et al. (1996) (cf. Table 3.9). As can be seen from Figure 3.6, a good agreement between the experimental data for the total pressure above CO_2 + MDEA + AMP + H_2O and the prediction is observed. However, in the range up to about 0.1 MPa calculated total pressures are slightly, but systematically smaller than the experimental numbers. The average relative deviation in the predicted total pressure is 23%, the average absolute deviation is 0.22 MPa. Vice versa at constant pressure, the average relative deviation between the experimental data for the overall molality of carbon dioxide in the liquid phase and the prediction is only 4.5%, the average absolute deviation in $\bar{m}_{CO_2}$ is 0.09 mol/kg.

3.6 Comparison with Literature Data

Several authors measured the solubility of carbon dioxide in aqueous solutions of AMP (Sartori et al., 1983; Roberts and Mather, 1988; Teng et al., 1989; Teng et al., 1990; Tontiwachwuthikul et al., 1991; Li and Chang, 1994). Some authors used molarity as a concentration scale for AMP. To convert this data to molality scale the reference temperature at which the solutions were prepared must be known as the density required for this conversion depends on temperature. But as that temperature was not given in all of the publications cited above, the molarity given in those publications was converted to molality by using the density of AMP + H_2O at 298.15 K. The density of the system AMP + H_2O was taken from Xu et al. (1991). To determine the overall molality of carbon dioxide in the liquid phase, the authors cited above used wet chemical techniques, i.e. the liquid phase was analyzed.

Some literature data are compared to the results of the present correlation in Figure 3.8. Roberts and Mather (1988) measured the partial pressure of carbon dioxide above aqueous solutions of AMP at 313 K for AMP molarities of 2 and 3 mol/l and partial pressures of carbon dioxide up to about 5.9 MPa. There is a good agreement between the correlation and their data at lower pressures (cf. part I and II of Fig. 3.8). However, at high partial pressures for carbon dioxide the literature data are systematically higher than the correlation based on the data of this work.

Tontiwachwuthikul et al. (1991) determined the solubility of carbon dioxide in 2 and 3 mol/l solutions of AMP in the temperature range from 293 to 353 K and partial pressures of carbon dioxide up to 0.1 MPa. Their results agree well with the present correlation at 333 and 353 K at both concentrations (cf. part I and II of Fig. 3.8), however, at 313 K and 3 mol/l AMP the calculated partial pressures of carbon dioxide are smaller than the experimental data.

3.6 Comparison with Literature Data

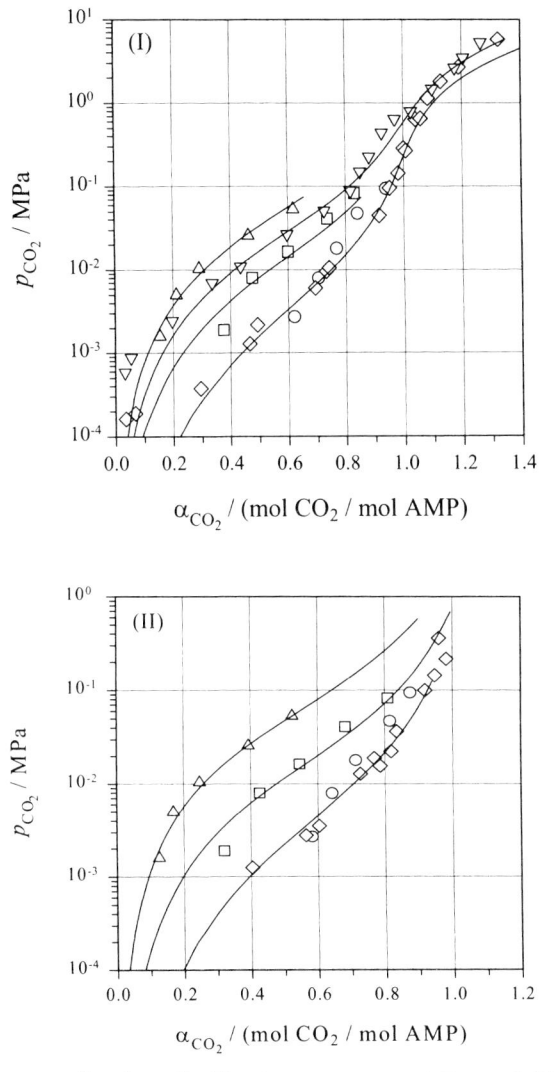

Figure 3.8: Partial pressure of carbon dioxide above aqueous solutions of AMP. Part I: $\bar{m}_{AMP}$ = 2.45 mol/kg: ○, □, △: T = 313, 333, 353 K, experimental results, Tontiwachwuthikul et al. (1991). ◇: T = 313 K, experimental results, Roberts and Mather (1988). ▽: T = 343 K, experimental results, Teng and Mather (1990). —: Calculated results, this work. Part II: $\bar{m}_{AMP}$ = 4.12 mol/kg: ○, □, △: T = 313, 333, 353 K, experimental results, Tontiwachwuthikul et al. (1991). ◇: T = 313 K, experimental results, Roberts and Mather (1988). —: Calculated results, this work.

The data from Teng et al. (1990) for the partial pressure of carbon dioxide above aqueous solutions containing 2 mol/l AMP at 343 K agree well with the results of the present correlation also at high partial pressure of carbon dioxide (cf. part I of Fig. 3.8). The deviation between their data and the prediction is in the range of the combined experimental errors.

3.7 Conclusions

The solubility of carbon dioxide in aqueous solutions containing 2-amino-2-methyl-1-propanol (AMP) was measured in the temperature range from 313 to 353 K at total pressures up to 2.7 MPa. A model taking into account chemical reactions in the liquid phase as well as physical interactions was used to correlate the new data. Using a reasonable number of adjustable interaction parameters, the model is able to correlate the new data mostly within the experimental uncertainty. A comparison with the limited literature data in nearly all cases yields fair to good agreement. To test the predictive capability of the model, the solubility of carbon dioxide in an aqueous solution containing MDEA and AMP was measured at 313 K. The model predictions agree well with the experimental data for this complex, chemically reactive system.

References

Bieling, V.; Rumpf, B.; Strepp, F.; Maurer, G. (1989): An Evolutionary Optimization Method for Modeling the Solubility of Ammonia and Carbon Dioxide in Aqueous Solutions. Fluid Phase Equilibria 53, 251.

Bradley, D. J.; Pitzer, K. S. (1979): Thermodynamics of Electrolytes. 12. Dielectric Properties of Water and Debye-Hückel Parameters to 350 °C and 1 kbar. J. Phys. Chem. 83, 1599.

Brelvi, S. W.; O'Connell, J. P. (1972): Corresponding States Correlations for Liquid Compressibility and Partial Molal Volumes of Gases at Infinite Dilution in Liquids. AIChE J. 18, 1239.

Crovetto, R. (1991): Evaluation of Solubility Data of the System $CO_2 - H_2O$ from 273 K to the Critical Point of Water. J. Phys. Chem. Ref. Data 20, 575.

Dymond, J. H.; Smith, E. B. (1980): The Virial Coefficients of Pure Gases and Mixtures. Oxford University Press: Oxford, 1980.

Edwards, T. J.; Maurer, G.; Newman, J.; Prausnitz, J. M. (1978): Vapor-Liquid Equilibria in Multicomponent Aqueous Solutions of Volatile Weak Electrolytes. AIChE J. 24, 966.

Hayden, J. G.; O'Connell, J. P. (1975): A Generalized Method for Predicting Second Virial Coefficients. Ind. Eng. Chem. Proc. Des. Dev. 14, 209.

Jane, I.-S.; Li, M.-H. (1997): Solubilities of Mixtures of Carbon Dioxide and Hydrogen Sulfide in Water + Diethanolamine + 2-Amino-2-methyl-1-propanol. J. Chem. Eng. Data 42, 98.

Kuranov, G.; Rumpf, B.; Smirnova, N. A.; Maurer, G. (1996): Solubility of Single Gases Carbon Dioxide and Hydrogen Sulfide in Aqueous Solutions of N-Methyldiethanolamine in the Temperature Range 313 – 413 K at Pressures up to 5 MPa. Ind. Eng. Chem. Res. 35, 1959.

Kurz, F.; Rumpf, B.; Maurer, G. (1995): Vapor-Liquid-Solid Phase Equilibria in the System $NH_3 - CO_2 - H_2O$ From Around 310 to 470 K: New Experimental Data and Modeling. Fluid Phase Equilibria 104, 261.

Landolt-Börnstein (1968): Zahlenwerte und Funktionen aus Physik, Chemie, Astronomie, Geophysik und Technik. 6. Auflage, II. Band, 2. Teil, Bandteil b. Springer-Verlag: Heidelberg.

Li, M.-H.; Chang, B.-C. (1994): Solubilities of Carbon Dioxide in Water + Monoethanolamine + 2-Amino-2-methyl-1-propanol. J. Chem. Eng. Data 39, 448.

Li, Y.-G.; Mather, A. E. (1997): Correlation and Prediction of the Solubility of CO_2 and H_2S in Aqueous Solutions of Methyldiethanolamine. Ind. Eng. Chem. Res. 36, 2760.

Littel, R. J.; Bos, M.; Knoop, G. J. (1990): Dissociation Constants of Some Alkanolamines at 293, 303, 318 and 333 K. J. Chem. Eng. Data 35, 276.

Pitzer, K. S. (1973): Thermodynamics of Electrolytes. 1. Theoretical Basis and General Equations. J. Phys. Chem. 77, 268.

Roberts, B. E.; Mather, A. E. (1988): Solubility of CO_2 and H_2S in a Hindered Amine Solution. Chem. Eng. Commun. 64, 105.

Rumpf, B.; Maurer, G. (1993): An Experimental and Theoretical Investigation on the Solubility of Carbon Dioxide in Aqueous Electrolyte Solutions. Ber. Bunsen-Ges. Phys. Chem. 97, 85.

Sartori, G.; Savage, D. W. (1983): Sterically Hindered Amines for CO_2 Removal From Gases. Ind. Eng. Chem. Fundam. 22, 239.

Saul, A.; Wagner, W. (1987): International Equations for the Saturation Properties of Ordinary Water Substance. J. Phys. Chem. Ref. Data 16, 893.

Schwabe, K. (1959): Physikalisch-chemische Untersuchungen an Alkanolaminen. Z. Phys. Chem. Neue Folge 20, 68.

Silkenbäumer, D. (1997): Bestimmung von Dampf-Flüssigkeits-Gleichgewichten multinärer Mischungen: Experimentelle Ermittlung und Modellbeschreibung. Dissertation, Universität Heidelberg.

Teng, T. T.; Mather, A. E. (1989): Solubility of H_2S, CO_2 and Their Mixtures in an AMP Solution. Can. J. Chem. Eng. 67, 846.

Teng, T. T.; Mather, A. E. (1990): Solubility of CO_2 in an AMP Solution. J. Chem. Eng. Data 35, 410.

Tontiwachwuthikul, P.; Meisen, A.; Lim, C. J. (1991): Solubility of CO_2 in 2-Amino-2-methyl-1-propanol Solutions. J. Chem. Eng. Data 36, 130.

Xu, S.; Qing, S.; Zhen, Z.; Zhang, C.; Carroll, J. J. (1991): Vapor Pressure Measurements of Aqueous N-Methyldiethanolamine Solutions. Fluid Phase Equilibria 67, 197.

Xu, S.; Wang, Y.-W.; Otto, F. D.; Mather, A. E. (1992): Representation of the Equilibrium Solubility Properties of CO_2 With Aqueous Solutions of 2-Amino-2-Methyl-1-Propanol. Chem. Eng. Proc. 31, 7.

Nomenclature

$A_{CO_2,w} \ldots D_{CO_2,w}$	coefficients for the temperature dependence of Henry's constant of carbon dioxide in water
$A_R \ldots D_R$	coefficients for the temperature dependence of equilibrium constants
$a_{i,j} \ldots d_{i,j}$	coefficients for the temperature dependence of second virial coefficients
a_i	activity of component i
$B^{(0)}_{G,CA}$	"observable" combination of binary interaction parameters
$B_{i,j}$	second virial coefficient for interactions between species i and j
f	function for the temperature dependence of interaction parameters
f_1, f_2	functions in Pitzer's equation
G^E	excess Gibbs energy
$H^{(m)}_{CO_2,w}$	Henry's constant for the solubility of carbon dioxide in pure water (on molality scale)
I	ionic strength (on molality scale)
K_R	equilibrium constant for chemical reaction R (on molality scale)
M_w	molar mass of water in kg/mol
$\bar{m}_i$	overall molality of component i
m_i	true molality of component i

$\bar{n}_i$	overall number of moles of component i
n_i	true number of moles of component i
p	total pressure
p_i	partial pressure of component i
q_0, q_1	coefficients for the temperature dependence of interaction parameters
R	universal gas constant
v	partial molar volume
y	mole fraction in vapor
z_i	number of charges of component i

Greek letters

α_{CO_2}	loading, i.e. $\bar{m}_{CO_2}/(\bar{m}_{AMP} + \bar{m}_{MDEA})$
$\beta^{(0)}, \beta^{(1)}$	binary interaction parameters in Pitzer's equation
$\Gamma_{G,CA,CA}, \Gamma_{G,G,CA}$	"observable" combination of ternary interaction parameters
γ^*	activity coefficient normalized to infinite dilution (on molality scale)
$\nu_{i,R}$	stoichiometric coefficient of component i in reaction R
τ	ternary interaction parameter in Pitzer's equation
φ	fugacity coefficient
ξ_R	extent of reaction R

Superscripts

m	on molality scale
s	saturation
*	normalized to infinite dilution
∞	infinite dilution
'	liquid phase
''	gas phase

Subscripts

A	anion A
C	cation C
CA	salt $C_{\nu_C} A_{\nu_A}$
G	gas G
i, j, k	component i, j, k
max	maximum
min	minimum
R	reaction R
w	water

4 The Influence of NO_2 on Complex Phase and Reaction Equilibria in Wet Flue Gas Cleaning Processes

Mohammad A. Siddiqi [1], Jens Petersen [1], and Klaus Lucas [2]*

4.1 Abstract

The effect of nitrogen dioxide on the physicochemical processes which take place during the absorption of sulfur dioxide in aqueous solution under conditions similar to those of a wet flue gas cleaning process has been studied. The study is done at 298.65 K, 318.45 K, and 333.35 K and ambient pressure. The results show that after a few hours NO_2 vanishes from the gas phase. This leads to a reproducible frozen (stable) state in which tetravalent sulfur exists in phase and chemical equilibrium. It is found that the simultaneous absorption of SO_2 and NO_2 in water depends on their initial concentrations and the temperature and is governed by the kinetics of a number of concurrent and consecutive reactions in the liquid phase. The gas phase contains SO_2, N_2O, and NO. The liquid phase contains a number of nitrogen-sulfur compounds besides tetravalent and hexavalent sulfur and nitrate. The analysis results together with the known thermodynamic and kinetic data from the literature have been used to develop a model for the description of $SO_2 + NO_2 + N_2 + H_2O$ system. The model has been used to predict not only the stable state composition at various temperatures but also the time dependence of the system composition.

4.2 Introduction

Oxides of sulfur and nitrogen are emitted in large quantities from fossil-fuelled electric power plants and incineration plants. The removal of these oxides mostly takes place via a wet scrubbing process. For the design and the optimization of the flue gas scrubbers a proper description of the physical and chemical processes responsible for the absorption of the

---

1 Gerhard-Mercator-Universität Duisburg, Fakultät für Ingenieurwissenschaften, IVG, Thermodynamik, Lotharstr. 1, 47048 Duisburg.
2 RWTH-Aachen, Lehrstuhl für Technische Thermodynamik, Schinkelstr. 8, 52056 Aachen.
* Author to whom correspondence should be addressed.

components involved is needed. To describe these processes in a thermodynamically consistent way, a knowledge of the phase behavior of the components and the specification of the liquid phase are the prerequisite. We have undertaken a systematic study of sulfur dioxide absorption in aqueous solutions under different conditions (Siddiqi et al., 1996, 1997a; Krissmann et al., 1997, 1998). The dissolution of SO_2 in aqueous solution is a combined phase and chemical equilibrium and may be described in the low pH range (between 1 and 3) by the following reactions:

$$SO_2 (g) \rightleftarrows SO_2 (aq) \tag{4.1}$$

$$SO_2 (aq) + H_2O (l) \rightleftarrows HSO_3^- + H^+ \tag{4.2}$$

The absorption of SO_2 in presence of NO_2 is a much more complex process. After the transfer of NO_2 into the liquid phase a number of concurrent reactions may take place depending on the concentrations of the components, the temperature, and the pH of the solution (Takeuchi et al., 1977; Sato et al., 1979; Lee and Schwartz, 1982; Ellison and Eckert, 1984; Littlejohn et al., 1993; Gutberlet et al., 1996):

$$2\ NO_2 (aq) + H_2O (l) \rightarrow HNO_2 (aq) + NO_3^- + H^+ \tag{4.3}$$

$$2\ NO_2 (aq) + HSO_3^- + H_2O (l) \rightarrow 2\ NO_2^- + SO_4^{2-} + 3\ H^+ \tag{4.4}$$

$$2\ NO_2 (aq) + 2\ HSO_3^- \rightarrow 2\ NO_2^- + S_2O_6^{2-} + 2\ H^+ \tag{4.5}$$

$$2\ NO_2 (aq) + 2\ HSO_3^- \rightarrow NO_2^- + HON(SO_3)_2^{2-} + \tfrac{1}{2}\ O_2 + H^+ \tag{4.6}$$

In acidic solution, which is the case in the present investigation, SO_2 exists mainly as HSO_3^- and interacts with NO_2^- produced in the above mentioned reactions to form hydroxylamine disulfonic acid (HADS), $HON(SO_3)_2^{2-}$, according to the overall reaction (Chang et al., 1982):

$$NO_2^- + H^+ + 2\ HSO_3^- \rightarrow HON(SO_3)_2^{2-} + H_2O(l) \tag{4.7}$$

via the intermediate product nitrososulfonic acid (NSS), $ONSO_3^-$

$$NO_2^- + H^+ + HSO_3^- \rightarrow ONSO_3^- + H_2O(l) \tag{4.8}$$

$$ONSO_3^- + HSO_3^- \rightarrow HON(SO_3)_2^{2-} \tag{4.9}$$

A summary of the reactions that can take place in aqueous solution is given in the literature (Chang et al., 1982; Gutberlet et al., 1996). The experimental work was directed to analyze and quantify all possible compounds which may be formed during the process to obtain a better understanding of the physicochemical processes which accompany the wet flue gas desulfurization. The analysis results have been used to propose a scheme for the reactions that can take place in aqueous solution and used to develop a model for the SO_2 + NO_2 + N_2 + H_2O system. The results of model calculations that give the concentration profile of species produced in this system as a function of time are presented. The effects of temperature and the initial concentration of gas mixture are discussed.

4.3 Experimental

4.3.1 Apparatus

The measurements have been carried out using the apparatus described in previous communications (Siddiqi et al., 1996, 1997b). The whole apparatus was placed in an air thermostat to accomplish the measurements at elevated temperatures. The temperature within the thermostat did not vary more than ±0.5 K. This was checked up to 333.35 K by measuring the temperature at different reference points within the thermostat with the help of pre-calibrated thermocouples. An ancillary part of the apparatus consisted of a Fourier Transform Infra Red (FTIR)-Spectrometer from BioRad Laboratories equipped with a long path gas cell. This assembly was used to measure the concentration of nitrous oxide (N_2O) in the gas phase.

The experimental procedure has already been described in previous communications (Siddiqi et al., 1996, 1997b) and so only a brief description of the modifications will be given here. The basic procedure for filling the gas mixtures is the same except that first SO_2 standard gas mixture was filled to a precalculated pressure in the gas vessel, then the NO_2 standard gas mixture was pressurized up to its precalculated pressure and then the mixture was diluted with pure nitrogen up to the experimental pressure to obtain the (SO_2 + NO_2 + N_2) mixture of desired composition. The initial pressure in the system depended on the desired experimental temperature. It was chosen to furnish a total pressure of nearly 104 kPa after making allowance for the saturated vapor pressure of the liquid phase. It was kept at 101 kPa, 95 kPa, or 84 kPa respectively for studies at 298.65 K, 318.45 K, or 333.35 K. The initial concentrations of SO_2 in the gas mixture are pertinent to its concentration in flue gases. The volume of the gas phase was 22.25 dm^3 and the mass of liquid water 0.4 kg.

After attainment of the experimental pressure and temperature the mixture was brought into contact with water (about 0.4 kg; accurately weighed) in the reaction cell by opening the appropriate valves and starting circulatory pump. The spectra of the gaseous phase and of the liquid were taken by scanning through the wavelength range of 220 nm to 450 nm with the help of the respective spectrophotometer at different time intervals. It was found that after about four hours all the nitrogen dioxide disappeared from the gas phase (no absorption band in 390–410 nm region). This stable state was considered as the state of quasi-equilibrium. After this state the gaseous phase and the liquid phase were analyzed in 30 min intervals. The gas phase and the liquid phase spectra were taken in the UV-Vis region and evaluated to determine the composition of gas phase. Only sulfur dioxide could be detected in the gas phase. The liquid phase spectra were evaluated by taking the molar absorption coefficient for SO_2(aq) from the literature (Siddiqi et al., 1996) at 260 nm and 276 nm. In this way the concentration of molecularly dissolved sulfur dioxide SO_2(aq) could be determined. The samples of the liquid phase were taken from the bottom of the reaction cell and analyzed by ion chromatography for NO_3^-, NO_2^-, SO_3^{2-}, SO_4^{2-} as well as for hydroxylaminemonosufonic acid (HAMS), HADS, hydroxylamine-N,O-disufonic acid (HAODS), hydroxylamine-O-sulfonic acid (HAOMS), hydroxylaminetrisufonic acid (HATS), amine disulfonic acid (IDS), sulfonic acid (AS), and amine trisulfonic acid (NTS) separately.

At the end of each experiment the gas phase was sent through the long path gas cell in the FTIR-spectrometer and the IR-spectra were taken in the region 2180–2260 cm^{-1} for the analysis of nitrous oxide. The spectra were evaluated in the same manner as UV-spectra using the multivariate analysis method described in detail in a previous publication (Krissmann et al., 1997). Before performing the spectroscopic measurements, a calibration was made by taking spectra of (nitrous oxide + nitrogen) gas mixtures of known compositions. The initial concentrations used were 2000 and 100, 200, 500; 4000 and 200, 500, 1000, 2000; 6000 and 500, 1000, 2000, 3000; 8000 and 500, 1000, 2000, 3000; 10,000 and 500, 1000, 2000, 3000 mg/m$_N^3$ respectively for SO_2 and NO_2.

For the ion-chromatographic determination a Dionex DX-100 ion chromatograph equipped with membrane suppressor and a conductivity detector was used. The sample injection loop had a volume of 10 μl. A standard anion-exchanger column Dionex IonPac AS9-SC was used for the determination of sulfate, sulfite, nitrite, and nitrate. The flow rate was kept constant at 2 ml/min. The eluent consisted of an aqueous solution of 1.6 mM sodium carbonate and 2.0 mM sodium bicarbonate. A standard cation-exchanger column Dionex IonPac CS12 was used for the determination of ammonium ion. A 20.0 mM methanesulfonic acid solution was used as eluent. The flow rate was 1.0 ml/min.

For the determination of nitrogen-sulfur compounds (HAMS, HADS, HAODS, HAOMS, HATS, IDS, AS, and NTS) ion-pair chromatography was employed. The standard anion-exchanger column used was a Dionex IonPac NS1–10 μm. Nitrate could also be determined with this column. The flow rate was kept constant at 1 ml/min. The eluent consisted of an aqueous solution of 1.6 mM sodium carbonate and 2.0 mM tetrabutylammoniumhydroxide to which different amounts of acetonitrile were added. The amount of acetonitrile was optimized for the separation and detection of various compounds. The most suitable concentration ratios of acetonitrile for the separation and analysis of various nitrogen-sulfur compounds were found to be: 10 vol% acetonitrile for HAOMS, 15 vol% acetonitrile for HAMS, NO_3^-, HADS, 22 vol% acetonitrile for HAODS, and 30 vol% acetonitrile for HATS, IDS, and NTS.

The calibration curve for the conductivity detector was established by running the standard solutions through the chromatograph.

4.3.2 Materials

The certified standard mixtures ($SO_2 + N_2$ and $NO_2 + N_2$) were supplied by Messer-Griesheim, Germany. The standard mixtures having the volume fraction of 10,000 ppm SO_2, 1000 mg/m^3 NO_2, 10,000 mg/m^3 NO_2 were used to prepare the desired gas mixtures. The sulfur dioxide in the standard mixtures had a mole purity $x(SO_2) \geq 0.9998$ (impurities: $x(CO_2) \leq 30 \times 10^{-6}$; $x(H_2O) \leq 50 \times 10^{-6}$). The nitrogen dioxide standard mixtures had a mole purity $x(NO_2) \geq 0.98$ (impurities: $x(CO_2) \leq 30 \times 10^{-6}$; $x(H_2O) \leq 50 \times 10^{-6}$). Pure nitrogen and the nitrogen in the standard mixtures had a purity $x(N_2) \geq 0.99999$ (impurities: $x(O_2) \leq 0.5 \times 10^{-6}$; $x(H_2O) \leq 0.5 \times 10^{-6}$). Purified and deionized water which had a conductivity <0.06 μS/cm was used. It was freed from dissolved oxygen by bubbling nitrogen sufficiently long through a perforated glass tube with small pores. AS was Merck 99.5% and HAOMS was Fluka purum >97%. The other nitrogen-sulfur compounds, viz. HATS, HADS, HAMS, NTS, IDS, and HAODS, used for calibration purpose in the ion chromatography, were not available commercially. These were synthesized according to the methods given in the literature (Raschig, 1923; Rollefson and Oldershaw, 1932; Sisler and Audrieth, 1938; Brauer, 1975). The other chemicals used for the ion chromatography measurements were "pro-analysis". All the chemicals were dried before use.

4.4 Experimental Results

4.4.1 Studies at 298.65 K

It is found that after four hours practically all NO_2 vanishes from the gas phase. The specification of the system has been undertaken after attaining this reproducible state. This state is marked by the fact that the mass transfer of NO_2 from the gaseous phase to the liquid phase is completed. After mixing the two phases for four hours the spectra in the liquid and in the gas phase were taken at half hour intervals. The liquid phase was analyzed additionally with the help of an ion chromatograph. The gas phase consists of SO_2, N_2O, and N_2 besides small amount of NO. The liquid phase consists of molecularly dissolved SO_2, NO_3^-, HADS, HAMS as well as tetravalent sulfur [S(IV)] $\{HSO_3^-\}$ and hexavalent sulfur [S(VI)] $\{HSO_4^- + SO_4^{2-}\}$. It may be remarked at this point that a small proportion (2–5%) of sulfur measured as S(VI) in solution results from the oxidation of HSO_3^- through atmospheric oxygen during sampling and analysis with the ion chromatograph. A study of the concentration of these components during the time period between four and six hours shows that the concentrations of $SO_2(g)$, NO_3^-, and S(IV) do not change during this period. The concentration of HADS decreases whereas the concentrations of HAMS and S(VI) increase. A typical time dependence is shown in Figure 4.1. The concentrations of SO_4^{2-}, NO_3^- and of HADS and HAMS were determined in two separate runs as different separation columns have to be used for their analysis. This was considered as legitimate as the reproducibility of the measurements is very good.

The variation of concentration with time may be interpreted by considering reaction (4.7) through which HADS is produced. As no NO_2^- was detected in the system it is concluded that this reaction is completed within four hours. The following hydrolysis of HADS to HAMS and HSO_4^- seems to be a slow process and could be measured.

$$HON(SO_3)_2^{2-} + H_2O\,(l) \rightarrow HONHSO_3^- + HSO_4^- \tag{4.10}$$

The observed concentration course of HADS, HAMS, and S(VI) is in accordance with the stoichiometry of the reaction. This quantitative agreement and the fact that an additional detailed chromatographic analysis of the liquid phase did not indicate the presence of $S_2O_6^{2-}$ or any other nitrogen-sulfur compound except HADS and HAMS leads to the conclusion that the sulfonation of HADS and HAMS

$$HON(SO_3)_2^{2-} + HSO_3^- \rightarrow N(SO_3)_3^{3-} + H_2O(l) \tag{4.11}$$

$$HONHSO_3^- + HSO_3^- \rightarrow HN(SO_3)_2^{2-} + H_2O(l) \tag{4.12}$$

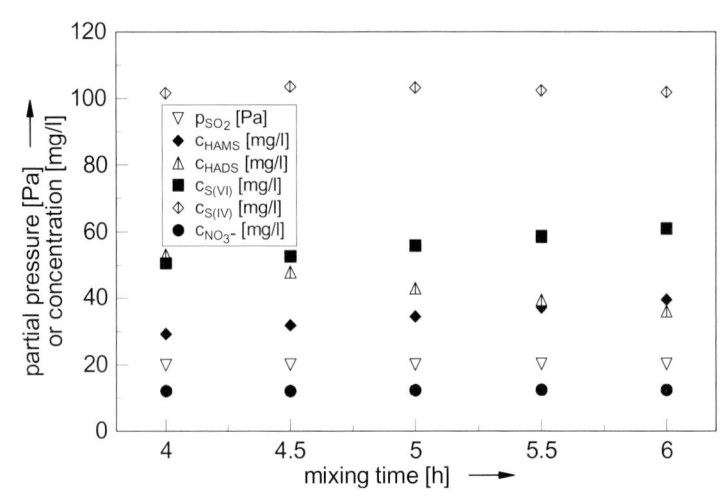

4 *The Influence of NO₂ on Complex Phase and Reaction Equilibria*

Figure 4.1: Time dependence of the concentrations of the components SO$_2$(g), HAMS, HADS, S(IV), S(VI), and NO$_3^-$ at 298.65 K for a gas mixture ($c_{SO_2,ini}$ = 4000 mg/m$_{N^3}$ and $c_{NO_2,ini}$ = 1000 mg/m$_{N^3}$).

and the hydrolysis of HAMS

$$HONHSO_3^- + H_2O(l) \rightarrow NH_2OH + HSO_4^- \qquad (4.13)$$

do not play any significant role for the description of the system at the conditions of the present study.

This scheme is confirmed by performing a comprehensive ion chromatographic analysis. For this the retention times for all possible nitrogen-sulfur compounds and $S_2O_6^{2-}$ were determined. The analysis of the liquid phase showed the presence of HADS, HAMS, NO$_3^-$, SO$_3^{2-}$, and SO$_4^{2-}$ only. No other components could be detected (detection limit 2–3 mg/l). This is also consistent with the overall mass balance for sulfur. 97–98% of the sulfur introduced in the system as SO$_2$ gas could be found out as SO$_2$(g), S(IV), S(VI), HADS, and HAMS. The mass balance is satisfactory as the difference lies within experimental uncertainties or detection limits. In contrast to this the mass balance for nitrogen was not fulfilled. Only 65–75% of nitrogen introduced as NO$_2$ was found in the solution. The other possible nitrogen compounds are NH$_2$OH, NH$_4^+$, NO, N$_2$, and N$_2$O. The presence of NH$_2$OH in any appreciable amount may be excluded from a study of the reactions for its formation and consumption given in the reaction scheme and following considerations. Firstly, that the components HAMS and HAOMS whose hydrolysis may lead to the formation of NH$_2$OH are themselves very stable (Candlin and Wilkins, 1961) and secondly that the NH$_2$OH formed should be further converted to AS, which could not be detected (even 1 mg/l) in a long time experiment of two days. The ion chromatographic analysis of the liquid phase for cations yielded only negligible amounts of NH$_4^+$ (of the order of µg/l). N$_2$O was measured in the gas phase with the help of FTIR-spectroscopy. NO was also found in the gas phase. However, its measurement is associated with greater experimental uncertainties. Due to the overlapping of the sharp peak of NO in the UV region by a broader peak of SO$_2$ the evaluation of small con-

4.4 Experimental Results

centrations of NO (<100 mg/m^3) is not accurate. It is estimated that about 5% of the nitrogen introduced as NO$_2$ to the system exists later as NO in the gas phase. Thus from the analysis of the gas phase and of the liquid phase about 90% of the mass balance for nitrogen is fulfilled. The remaining 10%, depending on the initial concentration of NO$_2$, represents about 4.3×10^{-6} to 1.3×10^{-4} mol and is considered as N$_2$ produced during the reactions. As nitrogen is used for the dilution of initial gas mixtures this additional small amount produced during the reactions is not measurable. The possible reactions for the production of N$_2$ are via NSS (Chang et al., 1982) according to reaction (4.8) which then undergoes hydrolysis to form hyponitrous acid (H$_2$N$_2$O$_2$) and subsequently produces N$_2$:

$$2\ ONSO_3^- + 2\ H_2O(l) \rightarrow H_2N_2O_2(aq) + 2\ HSO_4^- \tag{4.14}$$

$$H_2N_2O_2(aq) + HNO_2(aq) \rightarrow N_2 + H^+ + NO_3^- + H_2O(l) \tag{4.15}$$

The experimental results for the simultaneous absorption of NO$_2$ and SO$_2$ in water at 298.65 K after a mixing period of six hours for various initial concentrations of NO$_2$ in the gas mixtures show that the concentration of SO$_2$ in the gas phase, and the concentration of molecularly dissolved SO$_2$ and of S(IV) in the liquid phase decreases with the increasing initial NO$_2$ concentration. This effect is due to the reaction of tetravalent sulfur with NO$_2$ or one of its resultant products NO$_2^-$/HNO$_2$. The measured values are shown in Figure 4.2 as points.

The concentrations of nitrogen-sulfur compounds ($m_{HADS} + m_{HAMS}$) and NO$_3^-$ in the liquid phase increase with increasing initial NO$_2$ concentration. Typical results for an initial SO$_2$-concentration of 6000 mg/m$_N^3$ and of 10,000 mg/m$_N^3$ are shown as points in Figure 4.3. The NO$_3^-$ may be formed by the disproportion of NO$_2$ according to the reaction (4.3). The

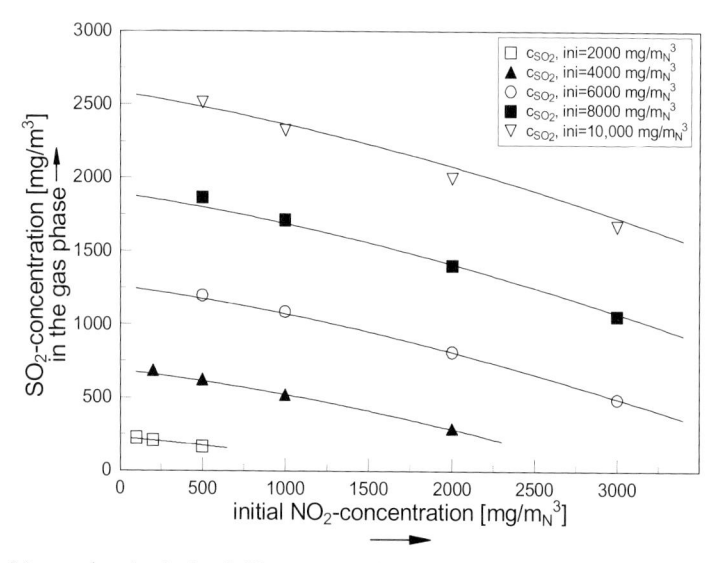

Figure 4.2: Measured and calculated SO$_2$-concentration in the gas phase as a function of initial NO$_2$-concentration at 298.65 K.

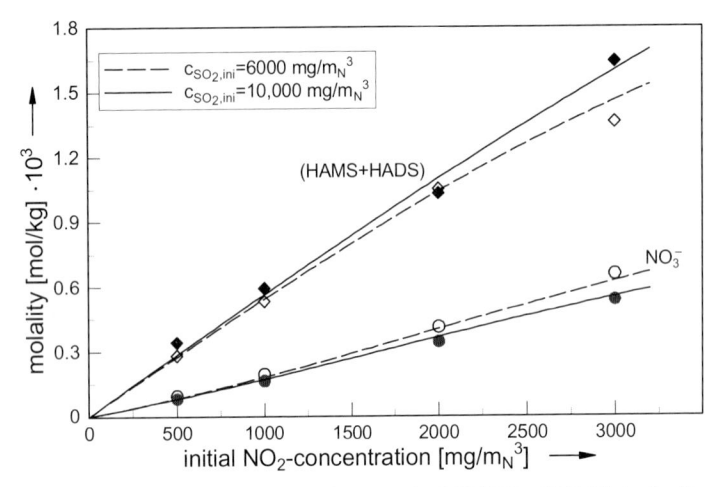

Figure 4.3: Measured and calculated molality of NO_3^- and of (HADS + HAMS) in the liquid phase as a function of initial NO_2-concentration at 298.65 K for some selected initial SO_2-concentrations.

sum of molalities of HADS and HAMS ($m_{HADS}+m_{HAMS}$) is plotted against the initial NO_2 concentration for the simple reason that this sum does not change even when the hydrolysis of HADS takes place beyond six hours.

From a comparison of the ratio of the observed values for molecularly dissolved SO_2 and gas phase SO_2 with the corresponding ratios for the $SO_2 + N_2 +$ water system it may be inferred that a phase and chemical equilibrium for this subsystem according to Eqs. (4.1) and (4.2) exists. However, the equilibrium state is not reached for this complete system. It is in a frozen state which arises due to the kinetically inhibited hydrolysis of HAMS according to Eq. (4.13) which results from a number of concurrent reactions which run completely towards product side.

A study of the individual molalities of HADS and HAMS shows that at the same initial NO_2 concentration the value of HADS decreases and that of HAMS increases with increasing SO_2 concentrations. This effect may be explained by the pH dependence of the HADS hydrolysis, which is favored at low pH.

The kinetic measurements performed between 4–6 h were evaluated (Petersen, 2000) to obtain the reaction rate for the hydrolysis of HADS according to reaction (4.10)

$$r/[\text{mol}/(\text{kg} \cdot \text{s})] = -\frac{dm_{HADS}}{dt} = 1.75 \cdot 10^{-2} \, m_{H^+} \, m_{HADS}$$

which is in reasonable agreement with the value $1.66 \times 10^{-2} \, m_{H^+} \, m_{HADS}$ reported in the literature (Naiditch and Yost, 1941) under similar conditions (pH 3–4, total ionic strength 8×10^{-3} mol/kg).

4.4 Experimental Results

4.4.2 Studies at 318.45 K and 333.35 K

The studies at 318.45 K and at 333.35 K show different results. At these temperatures practically no HADS was detected in the liquid phase over a period of 4–6 h. The values for SO_2(aq), S(IV), S(VI), NO_3^-, and HAMS, however, remained constant between 4–6 h, i.e. the system attains the stable (frozen) state in 4 h. Another nitrogen-sulfur compound HAODS was also detected. The detailed results of the analysis of the gas phase and the liquid phase after six hours of mixing at 318.45 K and at 333.35 K are not tabulated for the sake of conciseness. However, some of the representative results are found in figures which follow. A study of the molalities of (HADS+HAMS) at constant initial SO_2 concentration shows that the slope of ($m_{HADS} + m_{HAMS}$) vs. c_{NO_2} curves at 318.45 K is smaller than that at 298.65 K. At 333.35 K the values of ($m_{HADS} + m_{HAMS}$) even decrease at higher NO_2 concentration (e.g. 2000 mg/m^3) as shown in Figure 4.4.

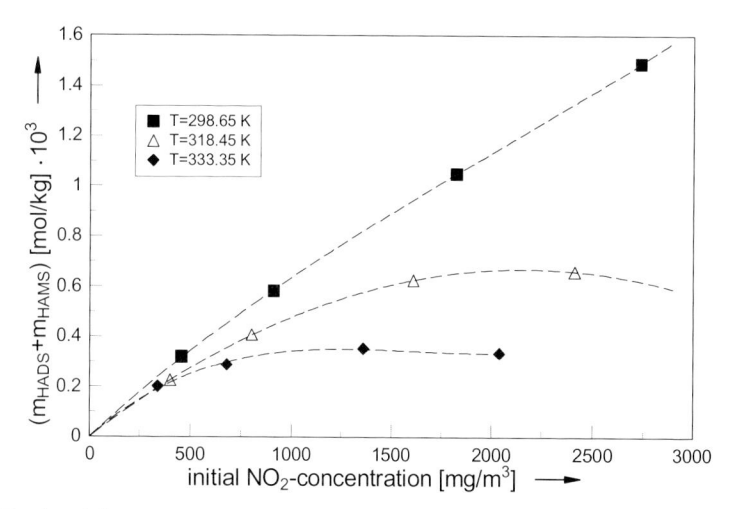

Figure 4.4: Total molality of (HADS + HAMS) as a function of initial NO_2-concentration for a selected initial SO_2-concentration (8000 mg/m$_{N^3}$) at different temperatures.

About 93% of the sulfur introduced in the system as SO_2 gas could be detected as SO_2(g), S(IV), S(VI), HAMS, and HAODS. The production of HAODS may be described by considering the following reactions:

$$HON(SO_3)_2^{2-} + 2\,HNO_2(aq) + HSO_3^- \rightarrow NOSO_3(SO_3)_2^{3-} + 2\,NO(aq) + 2\,H_2O(l) \quad (4.16)$$

$$N(OSO_3)(SO_3)_2^{3-} + H_2O(l) \rightarrow NH(OSO_3)(SO_3)^{2-} + HSO_4^- \quad (4.17)$$

The remaining 7% sulfur is believed to be distributed among the species HADS, HATS, HAOMS or $S_2O_6^{2-}$ which is individually within experimental uncertainties or detection limits (2–3 mg/l). The mass balance for nitrogen is similar to that at 298.65 K. The

4 The Influence of NO₂ on Complex Phase and Reaction Equilibria

amount of N$_2$O found at 318.45 K is about 2–3 times larger than that found at 298.65 K. This may be attributed to arise due to an additional concurrent reaction for HNO$_2$:

$$\text{HONHSO}_3^- + \text{HNO}_2(\text{aq}) \rightarrow \text{N}_2\text{O}(\text{aq}) + \text{HSO}_4^- + \text{H}_2\text{O}(\text{l}) \tag{4.18}$$

which is important at higher temperatures. This reaction explains the observation that at higher temperatures the molality of HAMS decreases and that of hexavalent sulfur increases.

NO formed in reaction (4.16) may undergo the following reaction:

$$\text{NO}(\text{aq}) + \text{NO}_2(\text{aq}) + \text{H}_2\text{O}(\text{l}) \rightarrow 2\,\text{HNO}_2(\text{aq}) \tag{4.19}$$

4.5 Model Calculations

The simultaneous absorption of SO$_2$ and NO$_2$ in water is a very complex process. The physicochemical behavior of aqueous solutions of sulfur dioxide and nitrogen oxides is governed by the kinetics of a number of concurrent and consecutive reactions taking place in the liquid state. In addition to this the phase and chemical equilibrium reactions are also to be considered. The system is modeled on the basis of the following reaction scheme. For the sake of clarity and ready reference some of the reactions already described are reproduced here:

$$\frac{dp_{\text{NO}_2}}{dt} = \frac{dp_{\text{NO}_2(21)}}{dt} + \frac{dp_{\text{NO}_2(22)}}{dt} + \frac{dp_{\text{NO}_2(23)}}{dt} + \frac{dp_{\text{NO}_2(19)}}{dt} \tag{4.20}$$

$$2\,\text{NO}_2(\text{aq}) + \text{H}_2\text{O}(\text{l}) \rightarrow \text{HNO}_2(\text{aq}) + \text{HNO}_3(\text{aq}) \tag{4.21}$$

$$2\,\text{NO}_2(\text{aq}) + 2\,\text{HSO}_3^- \rightarrow \text{HON(SO}_3)_2^{2-} + \tfrac{1}{2}\,\text{N}_2\text{O}(\text{aq}) + \tfrac{1}{2}\,\text{H}_2\text{O}(\text{l}) + \text{O}_2(\text{aq}) \tag{4.22}$$

$$2\,\text{NO}_2(\text{aq}) + \text{HSO}_3^- + \text{H}_2\text{O}(\text{l}) \rightarrow 2\,\text{HNO}_2(\text{aq}) + \text{HSO}_4^- \tag{4.23}$$

$$\text{NO}(\text{aq}) + \text{NO}_2(\text{aq}) + \text{H}_2\text{O}(\text{l}) \rightarrow 2\,\text{HNO}_2(\text{aq}) \tag{4.19}$$

$$3\,\text{HNO}_2(\text{aq}) \rightarrow \text{HNO}_3(\text{aq}) + 2\,\text{NO}(\text{aq}) + \text{H}_2\text{O}(\text{l}) \tag{4.24}$$

$$\text{HNO}_2(\text{aq}) + \text{HSO}_3^- \rightarrow \text{ONSO}_3^- + \text{H}_2\text{O}(\text{l}) \tag{4.8}$$

$$\text{ONSO}_3^- + \tfrac{1}{2}\,\text{H}_2\text{O}(\text{l}) \rightarrow \tfrac{1}{2}\,\text{N}_2\text{O}(\text{aq}) + \text{HSO}_4^- \tag{4.25}$$

$$\text{ONSO}_3^- + \text{HSO}_3^- \rightarrow \text{HON(SO}_3)_2^{2-} \tag{4.9}$$

$$\text{HON(SO}_3)_2^{2-} + \text{H}_2\text{O}(\text{l}) \rightarrow \text{HONHSO}_3^- + \text{HSO}_4^- \tag{4.10}$$

$$\text{HON(SO}_3)_2^{2-} + 2\,\text{HNO}_2(\text{aq}) + \text{HSO}_3^- \rightarrow \text{NOSO}_3(\text{SO}_3)_2^{3-} + 2\,\text{NO}(\text{aq}) + 2\,\text{H}_2\text{O}(\text{l}) \tag{4.16}$$

$$\text{NOSO}_3(\text{SO}_3)_2^{3-} + \text{H}_2\text{O}(\text{l}) \rightarrow \text{SO}_3\text{NHOSO}_3^{2-} + \text{HSO}_4^- \tag{4.17}$$

$$\text{HONHSO}_3^- + \text{HNO}_2(\text{aq}) \rightarrow \text{N}_2\text{O}(\text{aq}) + \text{HSO}_4^- + \text{H}_2\text{O}(\text{l}) \tag{4.18}$$

$$\text{SO}_2(\text{g}) \rightleftarrows \text{SO}_2(\text{aq}) \tag{4.1}$$

4.5 Model Calculations

$$SO_2(aq) + H_2O(l) \rightleftarrows HSO_3^- + H^+ \tag{4.2}$$

$$HSO_4^- \rightleftarrows H^+ + SO_4^{2-} \tag{4.26}$$

$$HNO_2(aq) \rightleftarrows H^+ + NO_2^- \tag{4.27}$$

$$HNO_3(aq) \rightleftarrows H^+ + NO_3^- \tag{4.28}$$

$$NO(g) \rightleftarrows NO(aq) \tag{4.29}$$

$$NO_2(g) \rightleftarrows NO_2(aq) \tag{4.30}$$

$$N_2O(g) \rightleftarrows N_2O(aq) \tag{4.31}$$

Equation (4.20) describes the temporal change of the partial pressure of NO_2 in the gas phase. NO_2 goes into solution through the concurrent reactions (4.21) to (4.23) and (4.19). Reactions (4.8) to (4.10), (4.16) to (4.19), and (4.21) to (4.25) run completely in the direction of products. The significance of each individual reaction depends on its kinetics and thus on its activation energy. Equations (4.1), (4.2), and (4.26) to (4.31) describe the equilibrium reactions which occur spontaneously and are to be considered at all times. The chemical reactions for the consumption of $NO_2(aq)$, represented by Eqs. (4.21) to (4.23) and (4.19) are particularly important. Reactions (4.21), (4.23), and (4.19) are frequently mentioned in literature (Sato et al., 1979; Lee and Schwartz, 1982; Littlejohn et al., 1993; Clifton et al., 1988; Schwartz and White, 1981,1983). Reaction (4.22) is postulated here for the formation of HADS directly from $NO_2(aq)$ and HSO_3^-. It has not been mentioned in literature before. It is postulated on the basis of experimental evidence as discussed here: In case HADS is formed exclusively through the path (4.23)-(4.8)-(4.9) the amount of HNO_2 needed for it must be provided by the reduction of NO_2 and as a consequence the tetravalent sulfur would be reduced to hexavalent sulfur, i.e. more SO_2 would be absorbed. The concentration of SO_2 in the gas phase after attaining a stable state in 6 h found experimentally is considerably larger than that calculated if reaction path (4.23)-(4.8)-(4.9) alone is assumed for HADS formation. The consideration of reaction (4.22) for the formation of a part of HADS directly from NO_2, which does not require the formation of hexavalent sulfur, ensures the observed lower absorption of SO_2.

Reactions (4.21) and (4.28) have already been discussed as net reaction (4.3). Equation (4.24) gives the formation of nitric acid and NO from nitrous acid in solution. Equation (4.25) shows the hydrolysis of nitrososulfonic acid to produce N_2O. The reactions (4.23), (4.26), and (4.27) form the net reaction (4.4), which has already been discussed. The reaction products N_2, N_2O, $NO_3^-/HNO_3HNO_3?$, HAODS, HAMS, and SO_4^{2-}/HSO_4^- are regarded as end products because they do not react further in the considered time period of 6 h (Candlin and Wilkins, 1961; Martin et al., 1981; Ellison and Eckert, 1984). The eventually possible oxidation reactions of NO, NO_2, HNO_2, HSO_3^- with O_2 produced in reaction (4.22) have been neglected as their concentrations are very low. The amount of nitrogen which could not be accounted for as NO_3^- or nitrogen-sulfur compounds was taken as N_2O and NO for the sake of simplicity and to keep the number of reactions to a minimum.

The gas phase reactions

$$NO_2(g) + SO_2(g) \rightarrow SO_3(g) + NO(g) \tag{4.32}$$

4 The Influence of NO_2 on Complex Phase and Reaction Equilibria

$$NO_2(g) + NO(g) + H_2O(g) \rightarrow 2\,HNO_2(g) \tag{4.33}$$

are principally possible (Sander and Seinfeld, 1976) but will not influence the results due to their low concentrations. They are, therefore, neglected.

On the basis of the reactions given above the following system of differential and algebraic equations may be formulated:

$$\frac{dp_{NO_2}}{dt} = \frac{dp_{NO_2(21)}}{dt} + \frac{dp_{NO_2(22)}}{dt} + \frac{dp_{NO_2(23)}}{dt} + \frac{dp_{NO_2(19)}}{dt} \tag{4.20}$$

with

$$\frac{dp_{NO_2(21)}}{dt} = -RT \frac{\hat{m}_{Liq}}{V - V_{Liq}} a \sqrt{\frac{4}{3} k_{21} D_{NO_2}} \left(\frac{K_{NO_2}}{p^0}\right)^{1.5} p_{NO_2}^{1.5} \tag{4.34}$$

$$\frac{dp_{NO_2(22)}}{dt} = -RT \frac{\hat{m}_{Liq}}{V - V_{Liq}} a \sqrt{\frac{4}{3} k_{22} m_{HSO_3^-}^{0.5} D_{NO_2}} \left(\frac{K_{NO_2}}{p^0}\right)^{1.5} p_{NO_2}^{1.5} \tag{4.35}$$

$$\frac{dp_{NO_2(23)}}{dt} = -RT \frac{\hat{m}_{Liq}}{V - V_{Liq}} a \sqrt{\frac{4}{3} k_{23} m_{HSO_3^-} D_{NO_2}} \left(\frac{K_{NO_2}}{p^0}\right)^{1.5} p_{NO_2}^{1.5} \tag{4.36}$$

$$\frac{dp_{NO_2(19)}}{dt} = -RT \frac{\hat{m}_{Liq}}{V - V_{Liq}} a \sqrt{k_{19} m_{NO} D_{NO_2}} \left(\frac{K_{NO_2}}{p^0}\right) p_{NO_2} \tag{4.37}$$

where $\hat{m}_{Liq}$ is the mass of liquid water, V is the total volume of the gas mixing vessel and V_{Liq} is the volume occupied by the aqueous solution.

$$r_{21} = k_{21} m_{NO_2}^2 \tag{4.38}$$
$$r_{22} = k_{22} m_{NO_2}^2 m_{HSO_3^-}^{0.5} \tag{4.39}$$
$$r_{23} = k_{23} m_{NO_2}^2 m_{HSO_3^-}^{0.5} \tag{4.40}$$
$$r_{19} = k_{19} m_{NO_2} m_{NO} \tag{4.41}$$
$$r_{24} = k_{24} \frac{m_{HNO_2}^4}{m_{NO}^2} \tag{4.42}$$
$$r_8 = k_8 m_{H^+}^{0.5} m_{N(III)} m_{S(IV)} \quad \text{(at 298.65 K)} \tag{4.43}$$
$$r_8 = k_8 m_{H^+} m_{N(III)} m_{S(IV)} \quad \text{(at 318.45 K)} \tag{4.44}$$
$$r_{25} = k_{25} m_H m_{NSS} \tag{4.45}$$
$$r_9 = k_9 m_{NSS} m_{HSO_3^-} \tag{4.46}$$
$$r_{10} = k_{10} m_{H^+} m_{HADS} \tag{4.47}$$
$$r_{16} = k_{16} m_{HADS} m_{N(III)} \tag{4.48}$$
$$r_{17} = k_{17} m_{HATS} m_{H^+} \tag{4.49}$$

4.5 Model Calculations

$$r_{18} = k_{18}\, m_{HAMS}\, m_{N(III)} \tag{4.50}$$

$$K_1 = K_{SO_2} = \frac{m_{SO_2}\, p^0}{m^*\, p_{SO_2}} \tag{4.51}$$

$$K_2 = \frac{\gamma^*_{HSO_3^-}\, \gamma^*_{H^+}\, m_{HSO_3^-}\, m_{H^+}}{m_{SO_2}\, m^*} \tag{4.52}$$

$$K_{26} = \frac{\gamma^*_{SO_4^{2-}}\, \gamma^*_{H^+}\, m_{SO_4^{2-}}\, m_{H^+}}{m_{HSO_4^-}\, m_{HSO_4^-}\, m^*} \tag{4.53}$$

$$K_{27} = \frac{\gamma^*_{NO_2^-}\, \gamma^*_{H^+}\, m_{NO_2^-}\, m_{H^+}}{m_{HNO_2}\, m^*} \tag{4.54}$$

$$K_{28} = \frac{\gamma^*_{NO_3^-}\, \gamma^*_{H^+}\, m_{NO_3^-}\, m_{H^+}}{m_{HNO_3}\, m^*} \tag{4.55}$$

$$K_{29} = K_{NO} = \frac{m_{NO}\, p^0}{m^*\, p_{NO}} \tag{4.56}$$

$$K_{30} = K_{NO_2} = \frac{m_{NO_2}\, p^0}{m^*\, p_{NO_2}} \tag{4.57}$$

$$K_{31} = K_{N_2O} = \frac{m_{N_2O}\, p^0}{m^*\, p_{N_2O}} \tag{4.58}$$

Equation (4.20) describes the total mass transfer of NO_2 into the liquid phase and is the sum obtained using Eqs. (4.34) to (4.37). The specific mass transfer surface area $a = 150$ m^2/m^3, as determined elsewhere (Petersen, 2000), was used. The values of diffusion coefficients, $D_{NO_2}(298.65\ K) = 1.51 \times 10^{-5}$ cm^2/s and $D_{NO_2}(318.45\ K) = 2.44 \times 10^{-5}$ cm^2/s were calculated from the correlation (Siddiqi and Lucas, 1986) using a molar volume of 31.7854 cm^3/mol at normal boiling point.

Whenever available the rate equations and the rate constants have been taken from the literature. These are listed in Table 4.1. The rate equations (4.41), (4.45), and (4.46) have been taken from the literature (Schwartz and White, 1981, 1983; Oblath et al., 1982). For Eqs. (4.48) and (4.50) the simple case of first order reaction for HADS and N(III) or HAMS and N(III) was assumed. The orders of reaction for HSO_3^-, 0.5 and 1 respectively in Eqs. (4.39) and (4.40), were obtained by fitting the observed concentration of $SO_2(g)$, $(m_{HADS} + m_{HAMS})$ and $m_{NO_3^-}$ for the simultaneous absorption of NO_2 and SO_2 in water at 298.65 K after 6 h. The equilibrium constants for the phase and chemical equilibrium for reactions (4.1), (4.2), (4.26) to (4.28) were calculated from the data given in Table 4.2 using the relation:

$$\begin{aligned}\ln K(T) = \ln K(T^0) &+ \frac{1}{R}\left[\left(\Delta c^0_{pR}(T^0) - \frac{1}{2}\frac{d\Delta c^0_{pR}}{dT}T^{0^2} - \Delta h^0_R(T^0)\right)\left(\frac{1}{T} - \frac{1}{T^0}\right)\right.\\ &\left. + \left(\Delta c^0_{pR}(T^0) - \frac{d\Delta c^0_{pR}}{dT}T^0\right)\ln\left(\frac{T}{T^0}\right) + \frac{1}{2}\frac{d\Delta c^0_{pR}}{dT}(T - T^0)\right]\end{aligned} \tag{4.59}$$

Table 4.1: Values of reaction rate constants from literature for some reactions.

Rate constant	298.65 K	318.45 K
k_{21}	7×10^{-7} [a]	$3 \times 10^{-7} \left(\frac{p^0}{K_{NO_2}}\right)^2$ [a]
k_{24}	9.4×10^{-7} [b]	–
k_8	142 [c]	17,431 [d]
k_{10}	1.75×10^{-2} [e]	0.1775 [f]
k_{17}	–	0.289 [g]

a) Schwartz and White, 1983; b) Abel and Schmid, 1928; c) Martin et al., 1981; d) Oblath et al., 1982; e) Petersen, 2000; f) Naiditch and Yost, 1941; g) Candlin and Wilkins, 1961.

Table 4.2: Thermodynamic data used for the calculation of equilibrium constants.

Equilibrium constant	Δg_R^0 kJ/mol	Δh_R^0 kJ/mol	Δc_p^0 J/(mol K)	$(d\Delta c_p/dT)^0$ J/(mol K^2)	Reference
K_1	−0.413	−26.35	155.1	1.65	Krissmann et al. (2000)
K_2	10.6	−17.8	−272	1.7	Goldberg and Parker (1985)
K_{26}	11.38	−21.93	−209	–	Wagman et al. (1982)
K_{27}	18.40	14.60	–	–	Wagman et al. (1982)
K_{28}	2.51	2.36	–	–	Aspen (1997)

Because the data for NO, N$_2$O, and NO$_2$ in molecularly dissolved state are not available the equilibrium constants K_{NO} and K_{N_2O} for Eqs. (4.29) and (4.31) respectively were calculated using the following empirical equations (CRC, 1991):

$$\ln K_{NO} = -58.8050 + \frac{82.342}{(T[K]/100)} + 22.8155 \ln\left(\frac{T[K]}{100}\right) \quad (4.60)$$

$$\ln K_{N_2O} = -56.743 + \frac{88.828}{(T[K]/100)} + 21.2531 \ln\left(\frac{T[K]}{100}\right) \quad (4.61)$$

The recommended value (Schwartz and White, 1983) of 0.01 was used for K_{NO_2} at 298.65 K. The nitrogen-sulfur compounds were assumed to be completely dissociated (Pasiuk-Bronikowska and Rudzinski, 1982). As the solutions are very dilute and the ionic strength is of the order of 10^{-3} mol/kg a modified Pitzer equation (Pitzer and Kim, 1974; Edwards et al., 1978) with only long range interactions is accurate enough for calculating the activity coefficient of any ionic species i.

$$\ln \gamma_i^* = -A_\phi(T) z_i^2 \left(\frac{\sqrt{I}}{1+1.2\sqrt{I}} + \frac{2}{1.2} \ln\left(1+1.2\sqrt{I}\right)\right) \quad (4.62)$$

4.5 Model Calculations

The ionic strength I is given by

$$I = \frac{1}{2} \sum_i m_i z_i^2 \qquad (4.63)$$

with m_i, the molality of species i and z_i its charge. The activity coefficients of all molecularly dissolved solutes and of water are set equal to unity. The temperature dependent Debye-Hückel parameter $A_\phi(T)$ is estimated from the relative dielectric constant ε_r, the density ρ_{H_2O} and the universal constants N_A, e, ε_0, and k via

$$A_\phi(T) = \frac{1}{3} \sqrt{2\pi N_A \rho_{H_2O}(T)} \left(\frac{e^2}{4\pi \varepsilon_0 \varepsilon_r(T) kT} \right)^{1.5} \qquad (4.64)$$

For the practical solution of the above system of equations it is useful to formulate the reaction rate equations and the equilibrium reactions in terms of the degree of reaction ξ^* written as

$$\xi^* = \frac{m_i - m_{i,0}}{v_i} \qquad (4.65)$$

where $m_{i,0}$ and m_i are the molalities of component i in the beginning ($t = 0$) and at any time t and v_i its stoichiometric number. The rate of reaction r_j for any reaction j is then

$$r_j = \frac{d\xi_j^*}{dt} \qquad (4.66)$$

and may be substituted in Eqs. (4.38) to (4.50). The molalities and the partial pressures of any component i may then be calculated with the help of the degree of reaction for all j reactions in which the component i participates via

$$m_i = m_{i,0} + \sum_j v_{ij} \xi_j^* \qquad (4.67)$$

$$p_i = p_{i,0} + \left(\sum_j v_{ij} \xi_j^* \right) \frac{\hat{m}_{Liq}}{V - V_{Liq}} RT \qquad (4.68)$$

The partial pressures and the molalities of different components may be written as:

$$p_{NO_2} = p_{NO_2,0} - \xi_{20}^+ - \xi_{30}^* \frac{\hat{m}_{Liq}}{V - V_{Liq}} RT \qquad (4.69)$$

$$p_{SO_2} = p_{SO_2,0} - \xi_1^* \frac{\hat{m}_{Liq}}{V - V_{Liq}} RT \qquad (4.70)$$

$$p_{NO} = p_{NO,0} - \left(\xi_{19}^* \frac{\hat{m}_{Film}}{\hat{m}_{Liq}} - 2\xi_{24}^* - 2\xi_{16}^* + \xi_{29}^* \right) \frac{\hat{m}_{Liq}}{V - V_{Liq}} RT \qquad (4.71)$$

$$p_{N_2O} = \left(0.5\xi_{22}^* \frac{\hat{m}_{Film}}{\hat{m}_{Liq}} + 0.5\xi_{25}^* + \xi_{18}^* - \xi_{31}^* \right) \frac{\hat{m}_{Liq}}{V - V_{Liq}} RT \qquad (4.72)$$

$$m_{NO_2} = \xi^*_{30} \tag{4.73}$$

$$m_{SO_2} = \xi^*_1 - \xi^*_2 \tag{4.74}$$

$$m_{NO} = \xi^*_{29} \tag{4.75}$$

$$m_{N_2O} = \xi^*_{31} \tag{4.76}$$

$$m_{HADS} = \xi^*_{22} \frac{\hat{m}_{Film}}{\hat{m}_{Liq}} + \xi^*_9 - \xi^*_{10} - \xi^*_{16} \tag{4.77}$$

$$m_{HAMS} = \xi^*_{10} - \xi^*_{18} \tag{4.78}$$

$$m_{NSS} = \xi^*_8 - \xi^*_{25} - \xi^*_9 \tag{4.79}$$

$$m_{HATS} = \xi^*_{16} - \xi^*_{17} \tag{4.80}$$

$$m_{HAODS} = \xi^*_{17} \tag{4.81}$$

$$m_{HNO_2} = \left(\xi^*_{21} + 2\xi^*_{23} + 2\xi^*_{19}\right)\frac{\hat{m}_{Film}}{\hat{m}_{Liq}} - 3\xi^*_{24} - \xi^*_8 - 2\xi^*_{16} - \xi^*_{18} - \xi^*_{27} \tag{4.82}$$

$$m_{HNO_3} = \xi^*_{21} \frac{\hat{m}_{Film}}{\hat{m}_{Liq}} + \xi^*_{24} - \xi^*_{28} \tag{4.83}$$

$$m_{HSO_3^-} = -\left(2\xi^*_{22} + \xi^*_{23}\right)\frac{\hat{m}_{Film}}{\hat{m}_{Liq}} - \xi^*_8 - \xi^*_9 - \xi^*_{16} + \xi^*_2 \tag{4.84}$$

$$m_{HSO_4^-} = \xi^*_{23}\frac{\hat{m}_{Film}}{\hat{m}_{Liq}} + \xi^*_{25} + \xi^*_{10} + \xi^*_{17} + \xi^*_{18} - \xi^*_{26} \tag{4.85}$$

$$m_{SO_4^{2-}} = \xi^*_{26} \tag{4.86}$$

$$m_{NO_2^-} = \xi^*_{27} \tag{4.87}$$

$$m_{NO_3^-} = \xi^*_{28} \tag{4.88}$$

$$m_{H^+} = \xi^*_2 + \xi^*_{26} + \xi^*_{27} + \xi^*_{28} \tag{4.89}$$

It is to be pointed out here that the mass transfer of NO_2 according to Eq. (4.20) has been treated formally as a reaction rate equation where the degree of reaction ξ^+_{20} has the unit Pa. Also the degrees of reaction ξ^*_{21}, ξ^*_{22}, ξ^*_{23}, and ξ^*_{19} have been multiplied with $\hat{m}_{Film}/\hat{m}_{Liq}$ to take into account that the relevant reactions for the mass transfer of NO_2 run completely in the liquid boundary film and the products are mixed completely with the bulk liquid. The mass of the liquid boundary film $\hat{m}_{Film}$ at any time may be calculated from the mass balance of NO_2 around the film:

$$\frac{\xi^*_{20}(V - V_{Liq})}{RT} - \hat{m}_{Film}\left(2\xi^*_{21} + 2\xi^*_{22} + 2\xi^*_{23} + \xi^*_{19}\right) = 0 \tag{4.90}$$

All other reactions are assumed to take place in the bulk of the solution. This gives a set of 21 equations for the calculation of 21 unknown ξ^*_j whereby the boundary condition for the differential equations is that $\xi^*_j(t=0) = 0$.

4.6 Results and Discussion

The rate constants k_{22}, k_{23}, and the ratio of the rate constants k_{25} and k_9 have been obtained by fitting the measured concentrations of $SO_2(g)$, $m_{NO_3^-}$, $(m_{HADS} + m_{HAMS})$ after a mixing period of 6 h by the method of least squares. As reactions (4.25) and (4.9) are two fast concurrent parallel reactions (Oblath et al., 1982) for NSS the absolute values of rate constants k_{25} and k_9 are not important; they must be chosen large enough to avoid the accumulation of NSS. The reactions (4.16) to (4.18) have not been considered at 298.65 K as the experimental results show that these are relevant only at higher temperatures. Reaction (4.19) which involves NO has also been excluded for the model calculations in this system. Although the formation of NO is formally taken into account by considering the reaction (4.24), it is much slower as compared to the concurrent reaction (4.8) that in fact through this only a small amount of NO (<10 mg/m^3) is formed. The following set of constants was obtained for the best agreement with the observed values of c_{SO_2}, $m_{NO_3^-}$, $(m_{HADS} + m_{HAMS})$.

$k_{22} = 2.0 \times 10^9$ kg$^{1.5}$/(mol$^{1.5} \cdot$ s)

$k_{23} = 8.5 \times 10^9$ kg^2/(mol$^2 \cdot$ s)

$k_9/k_{25} = 7 \times 10^9/1 \times 10^9 = 7$

$k_{19} = k_{16} = k_{17} = k_{18} = 0$

The results of model calculations for the concentration of SO_2 in the gas phase and the molalities of NO_3^- and (HADS+HAMS) are shown as drawn lines in Figures 4.2 and 4.3. The average mean deviation is ca. 2% for c_{SO_2}, $m_{NO_3^-}$, ca. 7% for $m_{NO_3^-}$, and ca. 10% for $(m_{HADS} + m_{HAMS})$. This agreement may be regarded as very good for such a complex system.

To test if this model can also predict the time dependence of the system (kinetics) some additional measurements have been performed at definite time intervals by analyzing the gas phase with a spectrophotometer and the liquid phase with an ion chromatograph. As the time needed for the ion chromatographic measurements is nearly 15 min, a 20 min time interval was chosen for the kinetic measurements. The duration of each experiment was restricted to 100 min to avoid any appreciable change in the liquid volume due to sampling. The concentrations of NO_2 and SO_2 in the gas phase and the molalities of NO_3^-, HADS, and HAMS in the liquid phase were measured. The measured and the calculated time dependence of the concentrations of $NO_2(g)$, $SO_2(g)$, NO_3^-, HADS, and HAMS are shown in Figures 4.5 and 4.6 for a typical initial concentration of SO_2 and NO_2. It may be seen from Figure 4.5 that the mass transfer of NO_2 which as a slow process determines the time dependence of the system, is predicted quite satisfactorily. It is also observed from Figures 4.5 and 4.6 that the agreement between the calculated values and experimental values after 20 min for all species is generally poorer as compared to that afterwards. A possible reason for this is the assumption that the tetravalent sulfur ($SO_2(g)$, $SO_2(aq)$, and HSO_3^-) attains equilibrium spontaneously (at $t = 0$) is in reality not fulfilled.

The modeling of the system at temperatures other than 298.15 K is not easy. The difficulty arises due to the lack of standard thermodynamic and kinetic data at these temperatures. An effort is made here to model the system at 318.45 K. It has been discussed in the

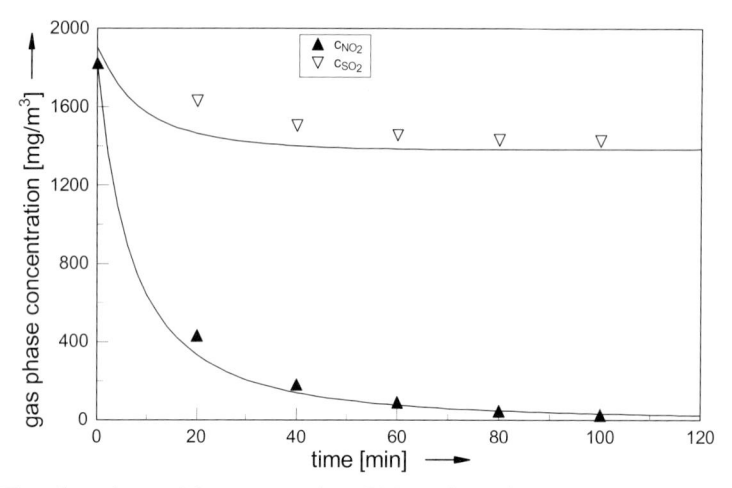

Figure 4.5: Time dependence of the concentration of SO_2 and NO_2 in the gas phase for a typical SO_2 + NO_2 gas mixture ($c_{SO_2,ini}$ = 8000 mg/m$_N^3$ and $c_{NO_2,ini}$ = 2000 mg/m$_N^3$) at 298.65 K.

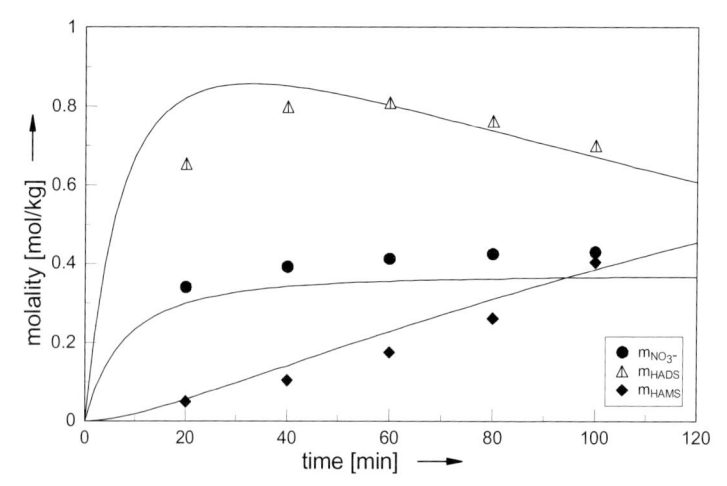

Figure 4.6: Time dependence of the liquid phase concentrations for a typical gas mixture ($c_{SO_2,ini}$ = 8000 mg/m$_N^3$ and $c_{NO_2,ini}$ = 2000 mg/m$_N^3$) at 298.65 K.

experimenttal section that additional reactions (4.16) to (4.18) have to be considered to describe the system at higher temperatures. Also at higher temperatures an appreciable amount of NO will be formed according to reaction (4.16) and so the reaction (4.19) for the simultaneous absorption of NO and NO_2 in water is also relevant to the SO_2 + NO_2 + N_2 + H_2O system.

The modeling at 318.45 K is done by considering reactions (4.16) to (4.19) in addition to those considered for 298.65 K. The value of $K_{NO_2} = 3 \times 10^{-3}$ at 318.45 K has been selected to reproduce the absorption of NO_2 according to Eq. (4.20). Using this value and the

4.6 Results and Discussion

temperature dependence of k_{21}^* from literature (Schwartz and White 1983) a value of $k_{21}^* = 3 \times 10^{-7}$ kg/(mol s Pa^2), the value of $k_{21} = k_{21}^*/(K_{NO_2}/p^0)^2 = 3.33 \times 10^8$ kg/(mol s) has been calculated. Reaction (4.24) is a very slow reaction and a change in k_{24} of the order of ten will not effect the results. The rate equation and the rate constant at 298.65 K are, therefore, used for 318.45 K too. Similarly for k_{27} and k_{28} the values for 298.65 K are used. This is legitimate because HNO$_3$ is completely dissociated at 298.65 K as well as at 318.45 K; also HNO$_2$(aq) and NO$_2^-$ are intermediate products and only the sum of these species is considered in the relevant rate equations. The remaining rate constants have been obtained by fitting the experimental results for SO$_2$(g), NO$_3^-$, and HAMS after a mixing period of 6.5 h. The constants are:

$k_{22} = 1.0 \times 10^{10}$ kg$^{1.5}$/(mol$^{1.5}$ s)

$k_{23} = 1.5 \times 10^{11}$ kg^2/(mol^2 s)

$k_{19} = 2.2 \times 10^9$ kg/(mol s)

$k_9/k_{25} = 7 \times 10^9/1 \times 10^9 = 7$

$k_{16} = 15$ kg/(mol s)

$k_{18} = 350$ kg/(mol s)

Some typical model calculations are compared with experimental results in Figures 4.7 and 4.8. Figure 4.7 shows the concentration of SO$_2$ in the gas phase after 2.5 h. It is seen that the model calculations reproduce the experimental values quite good (mean average deviation = 2.5%). The mean average deviation between the calculated and experimental values of the molality of NO$_3^-$, shown in Figure 4.8, is found to be 10%. The mean average

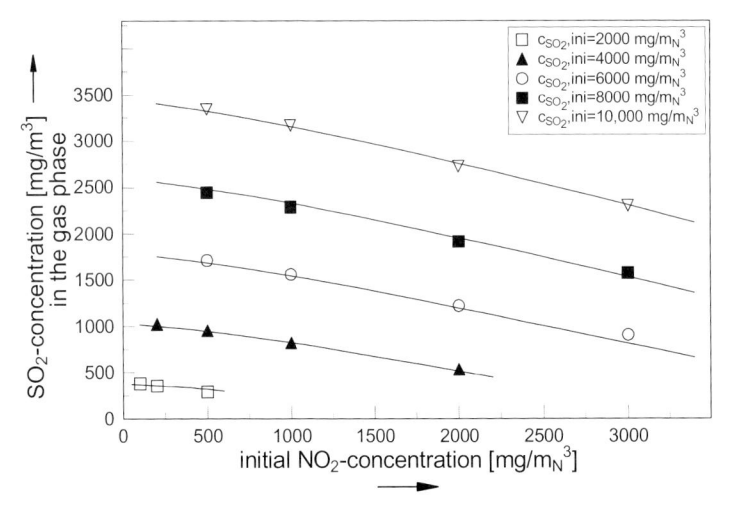

Figure 4.7: Measured and calculated SO$_2$-concentration in the gas phase as a function of initial NO$_2$-concentration at 318.45 K.

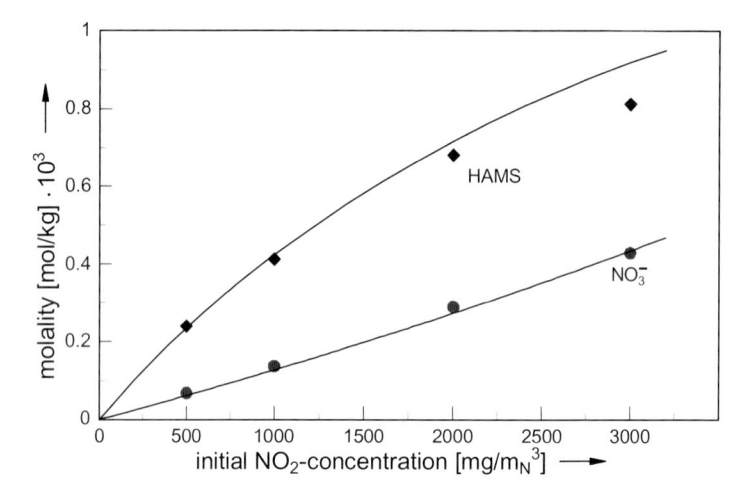

Figure 4.8: Measured and calculated molality of NO_3^- and of HAMS in the liquid phase as a function of initial NO_2-concentration at 318.45 K ($c_{SO_2,ini}$ = 10,000 mg/m$_N^3$).

deviation between the experimental and the calculated values of m_{HAMS} is 13%. This may be seen as satisfactory considering the approximations made in obtaining the basic thermodynamic and kinetic data at 318.45 K.

References

Abel, E.; Schmid, H. (1928): Kinetics of Nitrous Acid III. Kinetics of Nitrous Acid Decomposition. Z. Phys. Chem. 134, 279–300.

Aspen (1997): Aspen Plus Data Bank, AQUEOUS, Version 9.3.

Brauer, G. (1975): Handbuch der Präparativen Anorganischen Chemie. 1. Band. 3. Auflage, Enke Verlag, Stuttgart.

Candlin, J. P.; Wilkins, R. G. (1961): Sulphur-Nitrogen Compounds. Part II. The Hydrolysis of Hydroxylaminetrisulfonate and Hydroxylamine-NO-disulphonate Ions in Perchloric Acid. J. Chem. Soc., 3625–3633.

Chang, S. G.; Littlejohn, D.; Lin, N. H. (1982): Kinetics of Reactions in a Wet Flue Gas Simultaneous Desulfurization and Denitrification System. Flue Gas Desulfurization. ACS Symp. Ser. 188, 127–152.

Clifton, C.; Altstein, N.; Huie, R. E. (1988): Rate Constant for the Reaction of NO_2 with Sulfur (IV) over the pH Range 5.3–13. Environ. Sci. Technol. 22, 586–589.

CRC (1991): CRC Handbook of Chemistry and Physics. 72nd Ed., CRC Press, Boca Raton.

Edwards, T. J.; Maurer, J.; Newman, J.; Prausnitz, J. M. (1978): Vapor-Liquid Equilibria in Multicomponent Aqueous Solutions of Volatile Weak Electrolytes. AIChE J. 24, 966–976.

Ellison, T. K.; Eckert, C. A. (1984): The Oxidation of Aqueous SO_2. 4. The Influence of Nitrogen Dioxide at Low pH. J. Phys. Chem. 88, 2335–2339.

Goldberg, R. N.; Parker, V. B. (1985): Thermodynamics of Solution of SO$_2$(g) in Water and of Aqueous Sulfur Dioxide Solutions. J. Res. NBS 90, 341–358.

Gutberlet, H.; Finkler, S.; Pätsch, B.; van Eldik, R.; Prinsloo, F. (1996): Bildung von Schwefel- und Schwefel-Stickstoff-Verbindungen in Rauchgasentschwefelungsanlagen und ihr Einfluß auf die Oxidationskinetik von Sulfit. VGB Kraftwerkstechnik 76, 139–146.

Krissmann, J.; Siddiqi, M. A.; Lucas, K. (1997): Absorption of Sulfur Dioxide in Dilute Aqueous Solutions of Sulfuric and Hydrochloric Acid. Fluid Phase Equilibria 141, 221–233.

Krissmann, J.; Siddiqi, M. A.; Peters-Gerth, P.; Ripke, M.; Lucas, K. (1998): A Study of the Thermodynamic Behavior of Mercury in a Wet Flue Gas Cleaning Process. Ind. Eng. Chem. Res. 37, 3288–3294.

Krissmann, J.; Siddiqi, M. A.; Lucas, K. (2000): Improved Thermochemical Data for Computation of Phase and Chemical Equilibria in Flue-Gas/Water Systems. Fluid Phase Equilibria 169, 223–236.

Lee, Y. N.; Schwartz, S. E. (1982): Kinetics of Oxidation of Aqueous Sulfur (IV) by Nitrogen Dioxide. Fourth International Conference on Precipitation Scavenging, Dry Deposition, and Resuspension, St. Monica, California, New York, Ed. Pruppacher et al.

Littlejohn, D.; Wang, Y.; Chang, S.-G. (1993): Oxidation of Aqueous Sulfite Ion by Nitrogen Dioxide. Environ. Sci. Technol. 27, 2162–2167.

Martin, L. R.; Damschen, D. E.; Judeikis, H. S. (1981): The Reactions of Nitrogen Oxides with SO$_2$ in Aqueous Aerosols, Atmosph. Environ. 15, 191–195.

Naiditch, S.; Yost, D. M. (1941): The Rate and Mechanism of the Hydrolysis of Hydroxylamine Disulfonate Ion. J. Am. Chem. Soc. 63, 2123–2127.

Oblath, S. B.; Markowitz, S. S.; Novakov, T.; Chang, S. G. (1982): Kinetics of the Initial Reaction of Nitrite Ion in Bisulfite Solutions. J. Phys. Chem. 86, 4853–4857.

Pasiuk-Bronikowska, W.; Rudzinski, K. J. (1982): Complex Sulphite-Nitrite Reaction. Part I. Synthetic Review of Chemical-Kinetic Data. Chem. Eng. Commun. 18, 287–303.

Petersen, J. (2000): Untersuchungen zum Einfluss von Stickoxiden auf das physikochemische Verhalten von Stoffsystemen der nassen Rauchgasreinigung. VDI-Fortschritt-Bericht, Reihe 3, Verfahrenstechnik, No. 658, VDI Publishers, Düsseldorf.

Pitzer, K. S.; Kim, J. J. (1974): Thermodynamics of Electrolytes. IV. Activity and Osmotic Coefficients for Mixed Electrolytes. J. Am. Chem. Soc. 96, 5701–5706.

Raschig, F. (1923): Darstellung des hydroxylamin-iso-disulfonsauren Kaliums. Ber. Dt. Chem. Ges. 56, 206–208.

Rollefson, G. K.; Oldershaw, C. F. (1932): The Reduction of Nitrites to Hydroxylamine by Sulfites. J. Am. Chem. Soc. 54, 977–979.

Sander, S. P.; Seinfeld, J. H. (1976): Chemical Kinetics of Homogenous Atmospheric Oxidation of Sulfur Dioxide. Environ. Sci. Tech. 10, 1114–1123.

Sato, T.; Matani, S.; Okabe, T. (1979): The Oxidation of Sodium Sulfite with Nitrogen Dioxide. With Special Reference to Analytical Methods for Nitrogen-Sulfur-Compounds produced in the Reaction System. Nippon Kagaku Kaishi 7, 869–878.

Schwartz, S. E.; White, W. H. (1981): Solubility Equilibria of the Nitrogen Oxides and Oxyacids in Dilute Aqueous Solution. Advances in Environmental Science and Engineering, Vol. 4, 1–45, Gordon and Breach Science Publishers, New York.

Schwartz, S. E.; White, W. H. (1983): Kinetics of Reactive Dissolution of Nitrogen Oxides into Aqueous Solution. Adv. Environ. Sci. Tech. 12, 1–116.

Siddiqi, M. A.; Lucas, K. (1986): Correlations for Prediction of Diffusion in Liquids. Can. J. Chem. Eng. 64, 839–843.

Siddiqi, M. A.; Krissmann, J.; Peters-Gerth, P.; Luckas, M.; Lucas, K. (1996): Spectrophotometric Measurement of the Vapour-Liquid Equilibria of (Sulphur Dioxide + Water). J. Chem. Thermodyn. 28, 685–700.

Siddiqi, M. A.; Krissmann, J.; Lucas, K. (1997a): A New Fibre-optic Based Technique for the Spectrophotometric Measurement of Phase and Chemical Equilibria in Aqueous Solutions of Volatile Weak Electrolytes. Fluid Phase Equilibria 136, 185–195.

Siddiqi, M. A.; Krissmann, J.; Lucas, K. (1997b): A Novel Fibre-optic Based Technique for the Spectrophotometric Measurement of Phase and Chemical Equilibrium of (Sulphur Dioxide + Water) at Elevated Temperatures. J. Chem. Thermodyn. 29, 395–400.

Sisler, H.; Audrieth, L. F. (1938): Potassium Nitrolosulfonate. J. Am. Chem. Soc. 60, 1947–1948.

Takeuchi, H.; Ando, M.; Kizawa, N. (1977): Absorption of Nitrogen Oxides in Aqueous Sodium Sulfite and Bisulfite Solutions. Ind. Eng. Chem. Process. Des. Dev. 16, 303–308.

Wagman, D. D.; Evans, W. H.; Parker, V. B.; Schumm, R. H.; Halow, L.; Bailey, S. M.; Churney, K. L.; Nuttals, R. L. (1982): The NBS Tables of Chemical Thermodynamic Properties. J. Phys. Chem. Ref. Data 11, Suppl. 2.

Nomenclature

A_φ	Debye-Hückel coefficient
aq	aqueous
c	molarity or concentration in the gas phase
Δc_{pR}^0	change of standard molar heat capacity due to reaction at p^0
$(d\Delta c_{pR}^0/dT)$	derivative of Δc_{pR}^0 respect to temperature at p^0, T^0
D	diffusion coefficient
e	elementary charge (1.602×10^{-19} C)
g	gas
Δh_R^0	molar enthalpy of reaction at p^0
I	ionic strength on molality scale
K	K-value, equilibrium constant
k	rate constant, Boltzmann's constant
	liquid
m	molality
$\hat{m}$	mass
M	molecular mass
N_A	Avogadro's number (= 6.022×10^{23})
p	pressure, partial pressure
r	rate of reaction
R	molar gas constant
S(IV)	tetravalent sulfur
S(VI)	hexavalent sulfur
t	time
T	thermodynamic temperature
x	mole fraction
z	number of charges

Greek letters

ε_0	dielectric constant
ε_r	relative dielectric constant
γ	activity coefficient

Nomenclature

λ	wave length
v	stoichiometric number
ρ	density
ξ	extent of reaction (moles)
ξ^*	molal extent of reaction
ξ^+	extent of reaction (expressed in pressure units)

Superscripts

*	standard state of the hypothetical ideal aqueous solution of unit molality
0	standard state at standard pressure ($p^0 = 100$ kPa)

Subscripts

Film	film
i, j	species i, j or reaction i, j
Liq	bulk liquid
N	normal state ($T_N = 273.15$ K; $p_N = 101.3$ kPa)
R	rReaction
tot	totally dissolved
0	at time $t = 0$

5 An Infrared Spectroscopic Investigation on the Species Distribution in the System $NH_3 + CO_2 + H_2O$

Ute Lichtfers and Bernd Rumpf[1*]

5.1 Abstract

True species concentrations in the system $NH_3 + CO_2 + H_2O$ were determined by infrared spectroscopy in the temperature range from 313 to 393 K and overall molalities of ammonia and carbon dioxide up to 12 and 7 moles per kilogram of water. The experimental results for the true species concentrations were compared to predictions from a model solely fitted to data on vapor-liquid and vapor-liquid-solid equilibria. Especially for ionic species, significant deviations between experimental and predicted species concentrations were found. Therefore, the model was revised by incorporating measured species concentrations in the parameter determinations. The revised model is capable to describe the measured speciation as well as vapor-liquid and vapor-liquid- solid equilibria. The predictive power of the model was further tested by comparing experimental and predicted caloric effects in this complex, chemically reactive system.

5.2 Introduction

Aqueous solutions containing ammonia and sour gases like carbon dioxide, sulfur dioxide, hydrogen sulfide, or hydrogen cyanide have to be processed in many technical applications. Typical examples are the cleaning of raw gases from power stations, the production of fertilizers or applications in the field of environmental protection. Designing separation equipment requires reliable models to describe vapor-liquid equilibria (VLE), vapor-liquid-solid equilibria (VLSE) as well as caloric effects. Often, also information on kinetic effects, i.e. reaction kinetics as well as mass transfer kinetics is required.

---

[1] Technische Universität Kaiserslautern, Fachbereich Maschinenbau und Verfahrenstechnik, Lehrstuhl für Technische Thermodynamik, Erwin-Schrödinger-Straße, 67653 Kaiserslautern.
[*] Present adress: BASF AG, Ludwigshafen, Chemical Research and Engineering, GCT/C-L54c, D-67056 Ludwigshafen.

Due to chemical reactions in the liquid phase, modeling the thermodynamic properties of such solutions is an extremely difficult task. Therefore, reliable experimental information at least on the phase equilibrium is required to develop and improve models to describe the properties of such mixtures. However, although reliable models to describe VLE and VLSE in aqueous systems containing ammonia and sour gases exist (cf. Göppert and Maurer, 1988; Müller et al., 1988; Kurz et al., 1995; Rumpf et al., 1993), no reliable experimental data on the true species distributions in such systems are available.

Therefore, a new apparatus for the in-situ determination of species concentrations in aqueous mixtures containing ammonia and sour gases was developed. A special spectroscopic technique – attenuated total reflectance spectroscopy – is used to obtain spectra in these highly absorbant aqueous solutions. A method to resolve the complex spectra thus yielding true species concentrations was developed and tested. The apparatus was used to determine the true species distribution in the system $NH_3 + CO_2 + H_2O$ at temperatures from 313 to 393 K and overall molalities of ammonia and carbon dioxide up to 12 and 7 mol/kg, respectively. Results for the speciation were compared to predictions of a thermodynamic model originally developed by Edwards et al. (1978) with interaction parameters determined by Kurz et al. (1995) from fits to VLE and VLSE-data. Although the model is capable to reliably describe phase equilibria in that system, deviations between the new experimental results for the true species concentrations and model predictions were found. Therefore, the model was revised by including the data for the true speciation in the liquid phase in the parameter determination. The revised model is able to reliably describe both the true speciation as well as phase equilibria. The power of the model was further tested by predicting caloric effects in this complex, chemically reactive system.

5.3 Phase Equilibria and Quantitative Infrared Spectroscopy in the System $NH_3 + CO_2 + H_2O$

Figure 5.1 shows a scheme of the phase and chemical equilibria in the system $NH_3 + CO_2 + H_2O$. Due to chemical reactions in the liquid phase, ammonia and carbon dioxide are not only dissolved in molecular, volatile form, but also in ionic form. But as only the molecular dissolved gases are in equilibrium with the vapor phase, the phase equilibrium is strongly affected by the chemical reactions.

In the concentration range of interest here, five chemical reactions are considered in the liquid phase: The protolysis of ammonia (I) and carbon dioxide (II), the dissociation of the bicarbonate ion (III), the formation of carbamate ions (IV), and the autoprotolysis of water (V). The condition for chemical equilibrium of a reaction R in the liquid phase is

$$K_R^{(m)}(T) = \prod_i a_i^{\nu_{i,R}} \tag{5.1}$$

where a_i is the activity of a component i and $\nu_{i,R}$ is the stoichiometric number of a component i in a reaction R.

5 An Infrared Spectroscopic Investigation on the Species Distribution

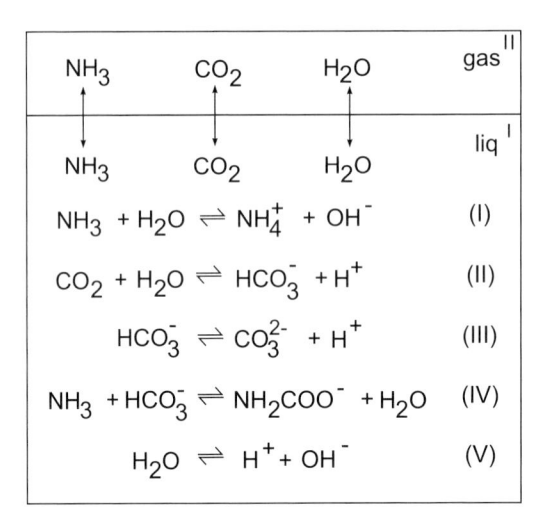

```
                                                     ‖
         NH₃           CO₂          H₂O         gas
          ↕             ↕            ↕
                                                     ‖ liq
         NH₃           CO₂          H₂O

         NH₃ + H₂O ⇌ NH₄⁺ + OH⁻                 (I)

         CO₂ + H₂O ⇌ HCO₃⁻ + H⁺                 (II)

               HCO₃⁻ ⇌ CO₃²⁻ + H⁺               (III)

         NH₃ + HCO₃⁻ ⇌ NH₂COO⁻ + H₂O            (IV)

               H₂O ⇌ H⁺ + OH⁻                   (V)
```

Figure 5.1: VLE and chemical reactions in the system $NH_3 + CO_2 + H_2O$.

In conventional phase equilibrium measurements, only the overall amounts of the dissolved gases per kilogram of water can be determined (cf. for example Göppert and Maurer, 1988; Müller et al., 1988; Rumpf et al., 1993; Kurz et al., 1995). Therefore, balance equations for the overall number of moles of a component i in the liquid phase have to be applied:

$$n_i = \bar{n}_i + \sum_R \nu_{i,R} \cdot \xi_R \tag{5.2}$$

where $\bar{n}_i$ is the overall number of moles of a component i and ξ_R is the extent of a chemical reaction R.

For a given temperature and given overall molalities of ammonia and carbon dioxide, from Eqs. (5.1) and (5.2) the speciation in the liquid phase can be calculated provided that the temperature dependent equilibrium constants $K_I^{(m)}$ to $K_V^{(m)}$ as well as a model for the activity coefficients in the liquid phase are known. The total pressure and the composition of the vapor phase is then calculated by applying extended Raoult's law for water and extended Henry's law for ammonia and carbon dioxide:

$$p \cdot y_w \cdot \varphi_w'' = p_w^s \cdot \varphi_w^s \cdot \exp\left(\frac{v_w^s \cdot (p - p_w^s)}{R \cdot T}\right) \cdot a_w \tag{5.3}$$

$$p \cdot y_i \cdot \varphi_i'' = H_{i,w}^{(m)}(T, p_w^s) \cdot \exp\left(\frac{v_{i,w}^\infty \cdot (p - p_w^s)}{R \cdot T}\right) \cdot m_i \cdot \gamma_i^{(m)} \qquad i = NH_3, CO_2 \tag{5.4}$$

Activity coefficients in the liquid phase are calculated using the Pitzer (1973) equation for the excess Gibbs energy of an aqueous electrolyte solution

5.3 Phase Equilibria and Quantitative Infrared Spectroscopy

$$\frac{G^E}{R \cdot T \cdot n_w \cdot M_w} = f_1(I_m) + \sum_{i \neq w}\sum_{j \neq w} m_i \cdot m_j \cdot \left(\beta_{i,j}^{(0)} + \beta_{i,j}^{(1)} \cdot f_2(I_m)\right)$$

$$+ \sum_{i \neq w}\sum_{j \neq w}\sum_{k \neq w} m_i \cdot m_j \cdot m_k \cdot \tau_{i,j,k} \tag{5.5}$$

That equation requires binary $\left(\beta_{i,j}^{(0)}, \beta_{i,j}^{(1)}\right)$ and ternary ($\tau_{i,j,k}$) interaction parameters which are usually assumed to be symmetric:

$$\beta_{i,j}^{(o)} = \beta_{j,i}^{(o)} \qquad o = 0,1 \tag{5.6}$$

$$\tau_{i,j,k} = \tau_{j,k,i} = \tau_{k,i,j} = \tau_{j,i,k} = \tau_{k,j,i} = \tau_{i,k,j} \tag{5.7}$$

Kurz et al. fitted interaction parameters to experimental data for the total and partial pressures above $NH_3 + CO_2 + H_2O$ in the temperature range from 313 to 473 K, overall molalities of ammonia up to about 26 mol/kg and total pressures up to about 10 MPa. The temperature dependent equilibrium constants $K_1^{(m)}$ to $K_V^{(m)}$ were taken from Bieling et al. (1989). Further details of the model (i.e. calculation of the remaining model parameters like dielectric constant of pure water, Henry's constants of ammonia and carbon dioxide, fugacity coefficients etc.) can be taken from Kurz et al. (1995).

In Figure 5.2, total pressures above $NH_3 + CO_2 + H_2O$ at 333 K as calculated from the model by Kurz et al. are compared to experimental data. Furthermore, Figure 5.3 shows the predicted species distribution in the liquid phase. Adding carbon dioxide to an ammoniacal solution at first results in a decrease in the total pressure as volatile ammonia and carbon dioxide are converted into ionic, nonvolatile form (cf. Fig. 5.3). After running through a minimum, the total pressure increases steeply. In that range, nearly no molecular ammonia is present in the liquid phase, whereas significant amounts of molecular dissolved carbon dioxide can be found.

However, although the model is able to describe VLE- and VLSE-data in the system $NH_3 + CO_2 + H_2O$ reliably, the predictions for the speciation in the liquid phase can not be tested as long as no experimental data are available.

In principle, only in-situ methods can be applied to determine species concentrations in that system as reactions I to V are reversible. Conventional infrared spectroscopic methods cannot be applied due to the large absorbance of the solvent water. Therefore, in the present work a special spectroscopic technique – attenuated total reflectance spectroscopy – had to be applied. The principles of that method are shown in Figure 5.4.

A cylindric crystal of zinc selenide is mounted in a thermostated high-pressure cell. The crystal is immersed in the aqueous solution which is pumped through the cell. The infrared beam enters the crystal and is totally reflected. At the interface between solid and liquid, the beam is partially absorbed in the liquid. The beam leaves the crystal and is recorded using a conventional infrared detector. The effective path length

$$d = \frac{n_R \cdot \lambda}{2 \cdot \pi \cdot n_1 \cdot \left[\sin^2 \alpha - \left(\frac{n_2}{n_1}\right)^2\right]^{0.5}} \tag{5.8}$$

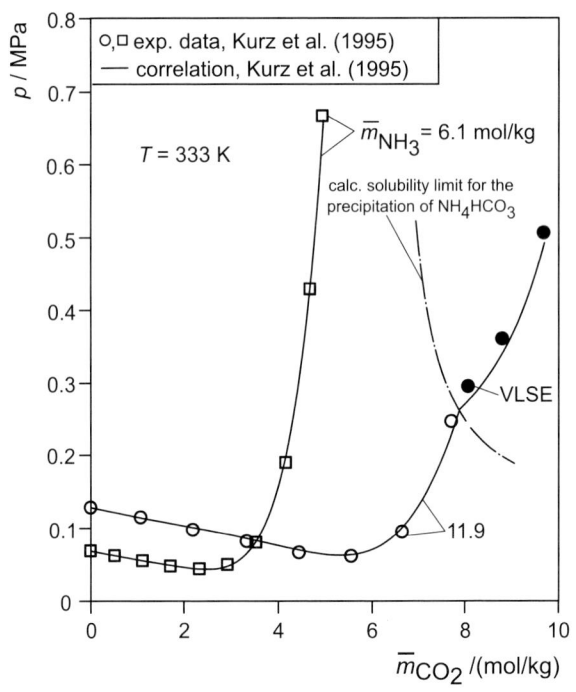

Figure 5.2: Total pressure above $NH_3 + CO_2 + H_2O$.

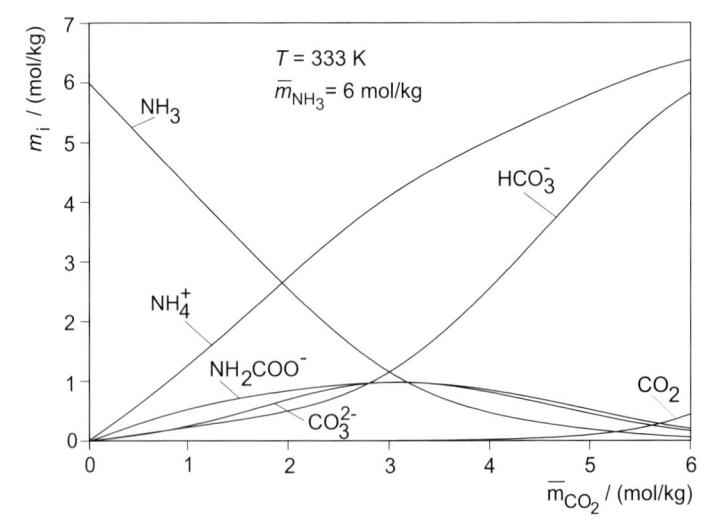

Figure 5.3: Predicted species distribution in the system $NH_3 + CO_2 + H_2O$.

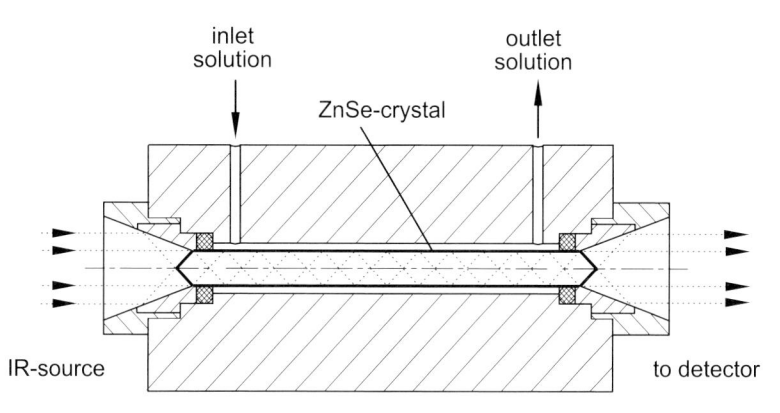

5.3 Phase Equilibria and Quantitative Infrared Spectroscopy

Figure 5.4: Principle of attenuated total reflectance spectroscopy.

where n_1 and n_2 is the refractive index of the crystal and the sample and n_R is the number of reflections, is typically in the order of about 2 μm. Therefore, highly absorbant – i.e. aqueous – mixtures can be investigated by that method. The dependence between absorbance $E_i(\tilde{v})$ at a wave number $\tilde{v}$ and the molar concentrations of a species i is given by Lambert-Beer's law similar to the case of transmission spectroscopy:

$$E_i = \varepsilon_i \cdot c_i \cdot d \tag{5.9}$$

where ε_i is the molar extinction coefficient of a species i and d is the effective path length.

In the system $NH_3 + CO_2 + H_2O$, nine species exist in the liquid phase. Assuming that the concentrations of H^+ and OH^- ions – which are typically of the order of 10^{-8} mol/kg – can be neglected, in principle seven concentrations have to be determined. However, for given overall molalities of ammonia and carbon dioxide, Eq. (5.2) can be used to reduce the number of unknown species concentrations:

$$c_{CO_3^{2-}} = c_{CO_2} - \bar{c}_{CO_2} + c_{NH_4^+} \tag{5.10}$$

$$c_{NH_3} = \bar{c}_{NH_3} + c_{HCO_3^-} - 2 \cdot (\bar{c}_{CO_2} - c_{CO_2}) \tag{5.11}$$

As can be seen from those equations, it is sufficient to determine three concentrations, e.g. the molarities of carbon dioxide, ammonium- and bicarbonate ions. The unknown molarity of carbamate ions can then be calculated from Eq. (5.2)

$$c_{NH_2COO^-} = \bar{c}_{NH_3} - c_{NH_3} - c_{NH_4^+} \tag{5.12}$$

The overall molarities of ammonia and carbon dioxide required in Eqs. (5.10) to (5.12) can be calculated from the density ρ of the solution and the overall molalities of those gases:

$$\bar{c}_i = \frac{\bar{m}_i \cdot \rho}{1 + \sum_j \bar{m}_j \cdot M_j} \qquad i = NH_3, \ CO_2 \tag{5.13}$$

5 An Infrared Spectroscopic Investigation on the Species Distribution

Finally, molarities can be converted to molalities using

$$m_i = \frac{c_i}{M_w \cdot c_w} \tag{5.14}$$

where c_w is the true molarity of water which is calculated from a mass balance and the density of the solution:

$$c_w = \frac{\rho - \sum_{i \neq w} c_i \cdot M_i}{M_w} \tag{5.15}$$

5.4 Experimental

5.4.1 Apparatus

The experimental equipment and procedure is described in detail by Lichtfers (2000), therefore only a few essentials are given here.

Figure 5.5 shows a scheme of the apparatus developed to determine true species concentrations in aqueous solutions containing ammonia and sour gases. In an experiment, a known amount of water is filled into a thermostated, evacuated vessel. A known amount of ammonia is added. Next, known amounts of carbon dioxide are added stepwise. The vessel is pressurized to avoid the formation of a gas phase. The solution is pumped through a high

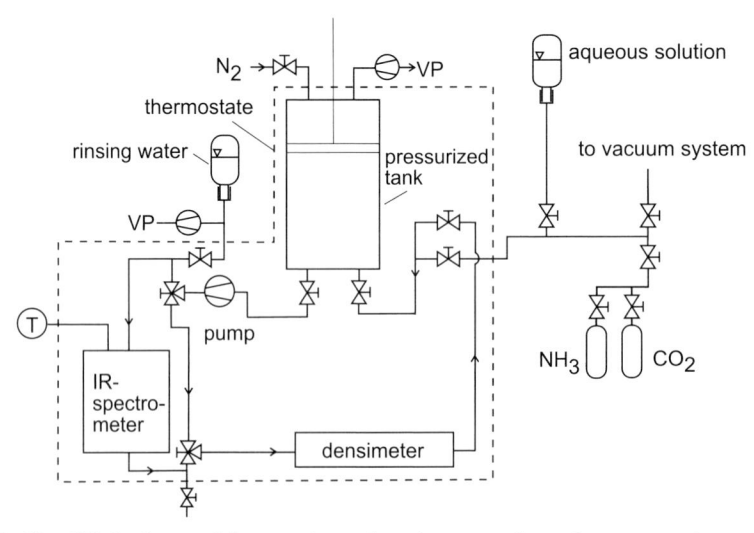

Figure 5.5: Simplified scheme of the experimental equipment to determine true species concentrations in aqueous systems containing ammonia and sour gases.

5.4   Experimental

pressure attenuated total reflectance cell which is mounted in an infrared spectrometer (Nicolet, Impact 400, resolution 2 cm^{-1}). To determine the density of the solution, a densimeter (Paar, DMA 602 HP) is used. Temperature in the ATR-cell was measured by platinum resistance thermometry. After equilibration, the spectrum, the density of the solution as well as the temperature are measured.

Due to the highly aggressive solutions, different ATR-cells had to be used. For temperatures up to 353 K, tunnel cells equipped with a zinc selenide crystal (cf. Fig. 5.4) with 11 or 6 reflections were used. At higher temperatures, a single reflection cell equipped with a diamond crystal (Graseby Specac, Kent, England) was used (for details cf. Lichtfers, 2000).

5.4.2   Materials

NH_3 ($\geq$ 99.999 mole%) and CO_2 ($\geq$ 99.995 mole%) were purchased from Messer-Griesheim, Ludwigshafen, Germany, and used without further purification. The salts used for the calibration measurements were degassed and dried under vacuum. Deionized water was distilled and further degassed under vacuum.

5.4.3   Results

In the ternary system $NH_3 + CO_2 + H_2O$, measurements were performed in the temperature range from 313 to 393 K. The overall molality of ammonia ranged up to 12 mol/kg. Depending on the temperature and the maximum allowable pressure in the apparatus, maximum overall molalities of carbon dioxide ranged up to about 7 mol/kg. Details of the experimental investigation are given in Table 5.1.

As an example, in Figure 5.6 spectra in the ternary system $NH_3 + CO_2 + H_2O$ at 333 K are shown. The identification of the various peaks to each species which is also shown in that figure is based on calibration measurements in binary systems (see below).

Starting from pure water, about 4.4 mol/kg of ammonia were added.

Adding ammonia to pure water results in a peak at 1110 cm^{-1} which is caused by a symmetric deformation (δ_{sy}) of molecular dissolved ammonia. The addition of carbon dioxide results in complex spectra which are caused by the reaction products in the ternary system $NH_3 + CO_2 + H_2O$. As can be seen from Figure 5.6, with increasing overall molality of carbon dioxide the peak caused by molecular ammonia decreases, whereas the peak at 1445 cm^{-1} caused by ammonium ions increases. Furthermore, peaks caused by bicarbonate- (at 1005, 1300, 1360, 1400 cm^{-1}) and carbonate ions (at 1375 and 1420 cm^{-1}) increase. At high overall amounts of the sour gas, molecular dissolved carbon dioxide causes a peak at about 2340 cm^{-1}.

Besides the peaks caused by NH_3, CO_2, NH_4^+, HCO_3^-, and CO_3^{2-} which can be identified in binary, non reacting mixtures, a further band is observed at about 1545 cm^{-1} which must be caused by carbamate ions.

5 An Infrared Spectroscopic Investigation on the Species Distribution

Table 5.1: Series of infrared spectroscopic measurements in the ammonia + carbon dioxide + water system.

T/K	Type of crystal	$\bar{m}_{NH_3}$/(mol/kg)	$\bar{m}_{CO_2}$/(mol/kg)	No. of measurements
313	ZnSe (11 refl.)	1.99	0–2.13	5
		3.04	1.18–3.01	5
		6.18	0–4.00	5
		12.01	0–5.97	7
313	diamond (1 refl.)	6.23	0–3.94	6
		12.18	0–6.02	7
333	ZnSe (11 refl.)	1.88	0–1.95	5
		3.25	0–3.09	6
		4.44	0–4.03	6
353	ZnSe (6 refl.)	4.08	0–3.35	6
		6.13	0–4.52	6
373	diamond (1 refl.)	3.00	0–2.22	6
		6.56	0–3.92	6
		11.98	0–6.96	8
393	diamond (1 refl.)	3.01	0–1.72	6
		6.2	0–2.51	5
		12.39	0–5.34	6

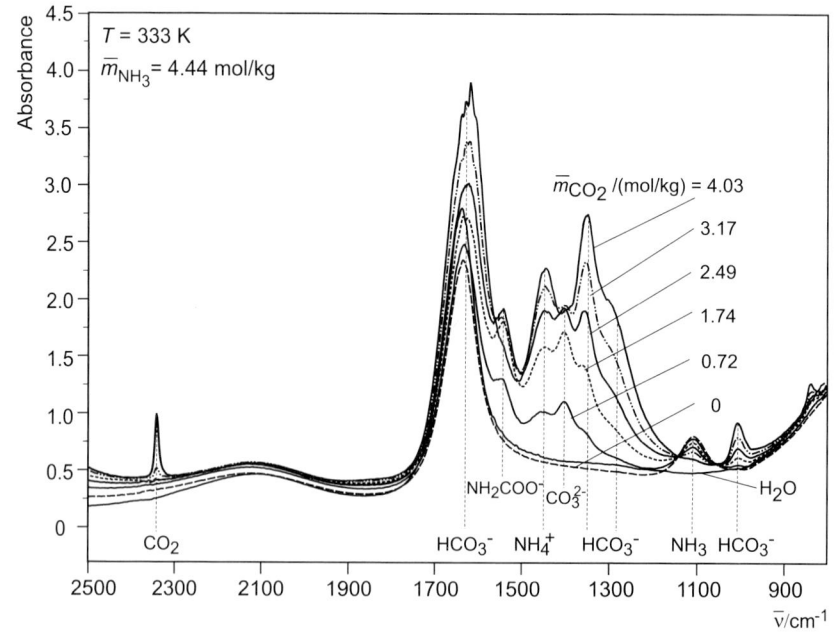

Figure 5.6: Spectra in the system $NH_3 + CO_2 + H_2O$.

5.4.4 Calibration Measurements

The determination of the species concentrations requires information on the peak size as well as the dependence of the peak height or area on the true concentration of each species. This dependence was determined for each temperature by calibration measurements in the binary systems $NH_3 + H_2O$, $CO_2 + H_2O$, $NH_4Cl + H_2O$, $KHCO_3 + H_2O$, and $K_2CO_3 + H_2O$. Aqueous solutions containing carbamate salts were not investigated as carbamate salts react to the same products as present in the ternary system $NH_3 + CO_2 + H_2O$. The calibration measurements are described in detail by Lichtfers (2000), therefore only a few examples are given here.

In Figure 5.7 spectra in the binary system $NH_3 + H_2O$ at 333 K and different molalities of ammonia are shown. At 1640 cm^{-1}, a peak caused by the H–O–H deformation (δ_d) is observed. Adding ammonia to pure water results in a second peak at 1110 cm^{-1} caused by the symmetric deformation (δ_{sy}) of dissolved ammonia. By further increasing the molality of ammonia, the peak at 1110 cm^{-1} increases whereas the peak at 1640 cm^{-1} decreases due to the decreasing molarity of water.

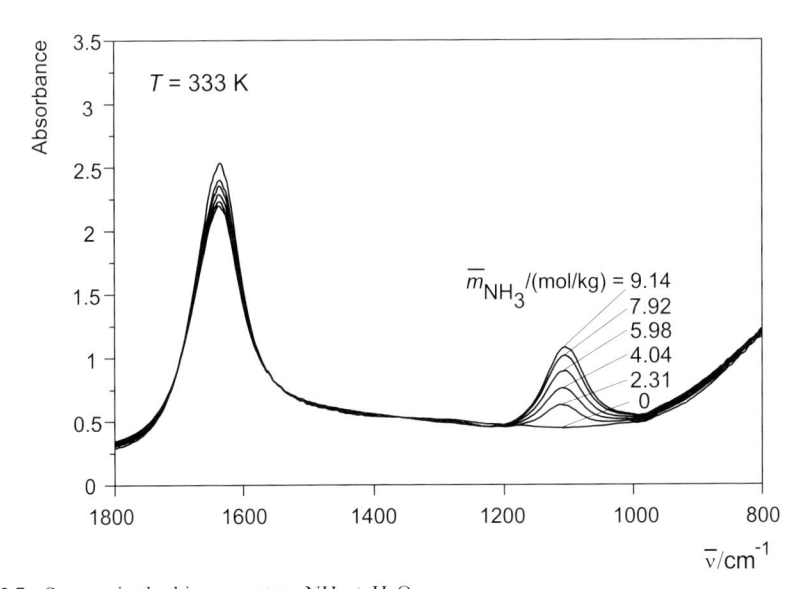

Figure 5.7: Spectra in the binary system $NH_3 + H_2O$.

More complex spectra are observed in the binary system $KHCO_3 + H_2O$ (cf. Fig. 5.8). Whereas the potassium ion is inactive in infrared, bicarbonate ions absorb at 1005, 1300, and 1360 cm^{-1}. At high concentrations of the dissolved salt, a shoulder at 1400 cm^{-1} appears. Furthermore, the peak at 1640 cm^{-1} increases with increasing molarity of potassium bicarbonate although the molarity of water decreases. Thus, bicarbonate ions obviously also absorb at about 1640 cm^{-1}. This is in agreement with bands observed in solid state spectra of bicarbonate salts (Davis and Oliver, 1972).

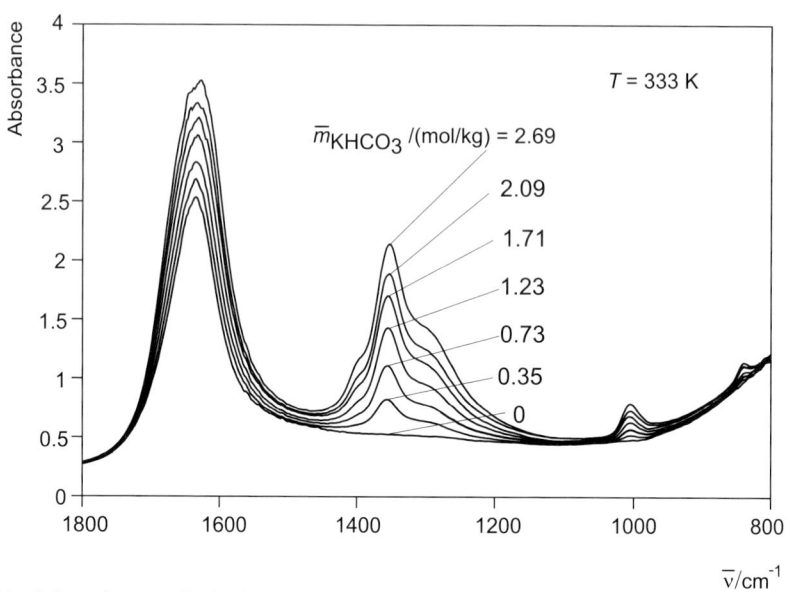

Figure 5.8: Selected spectra in the binary system KHCO$_3$ + H$_2$O.

5.4.5 Evaluation of Spectra

5.4.5.1 Spectra Taken in Calibration Measurements

The evaluation of spectra is repeated for each isothermal series of measurements, i.e. calibration measurements as well as measurements in the ternary system NH$_3$ + CO$_2$ + H$_2$O.

To evaluate the calibration measurements, for each absorbance, parameters E_{max}, $\bar{v}_{max}$, b, and c in the Gauss-Lorentz function

$$E(\bar{v}) = \frac{E_{max}}{1 + b^2 \cdot (\bar{v} - \bar{v}_{max})^2} \cdot \exp\left(-c^2 \cdot (\bar{v} - \bar{v}_{max})^2\right) \tag{5.16}$$

were determined by minimizing the difference between experimental and calculated absorbances.

In the first step, the spectrum of pure water was evaluated. Detailed calculations showed that six functions are required for a good representation of that spectrum. Next, spectra in the binary systems were evaluated. While keeping parameters b and c from the spectrum of pure water as well as the base line constant, additional Gauss-Lorentz functions describing the absorbances of the dissolved component were fitted to the spectra. A single additional band was fitted in each of the binary systems NH$_3$ + H$_2$O, CO$_2$ + H$_2$O, and NH$_4$Cl + H$_2$O. In the systems KHCO$_3$ + H$_2$O and K$_2$CO$_3$ + H$_2$O, more complex spectra are observed. Detailed calculations showed that the absorbances of the bicarbonate ion can be described by fitting five Gauss-Lorentz functions whereas the absorbances of the carbonate ion require two Gauss-Lorentz functions.

5.4 Experimental

Table 5.2 summarizes the wave numbers for NH_3, CO_2, H_2O, NH_4^+, CO_3^{2-}, and HCO_3^--ions as determined from the spectra in the binary systems. Furthermore, that table also contains numbers as taken from the literature. A good agreement between the results of this work and the data taken from the literature is observed.

Table 5.2: Wave numbers at peak maximum as determined from spectra in the binary systems and modes of motion.

Species	$\bar{\nu}_{max}/cm^{-1}$	Type	$\bar{\nu}_{max}^{Lit}/cm^{-1}$	Source
H_2O	1640	H–O–H deformation δ_d	1640	Nakamoto (1986)
NH_3	1110	N–H deformation δ_{sy}	1060	
CO_2	2342	asym. stretching ν_{as}	2349	
NH_4^+	1445	N–H deformation δ_d	1430	
CO_3^{2-}	1375	asym. C–O stretching ν_{as}	1380	Davis, Oliver (1972)
	1420	asym. C–O stretching ν_{as}	1436	
HCO_3^-	1005	C–OH stretching ν_{as}	1000	Davis, Oliver (1972)
	1300	C–OH deformation δ_d	1300	
	1360	sym. C–O stretching ν_{sy}	1355	
	(1400)	–	–	–
	1650	asym. C–O stretching ν_{as}	1620	
NH_2COO^-	1100*	C–N stretching	1120	Hisatsune (1984)
	1400*	sym. C–O stretching ν_{sy}	1400	
	1545*	asym. C–O stretching ν_{as}	1545	

* as evaluated from spectra in the system NH_3–CO_2–H_2O

For the series of measurements at 333 K, in Figure 5.9 the maximum absorbances as determined from the calibration measurements are plotted vs. the molarity of each species. Except for the peak at about 1360 cm^{-1} caused by bicarbonate ions, a linear dependence between absorbance and molarity is observed. Therefore, the isothermal results $E_{i,max} = f(c_i)$ were correlated by the equation

$$E_{i,max} = d_i + e_i \cdot c_i + f_i \cdot c_i^2 \tag{5.17}$$

where d_i to f_i are adjustable constants and c_i is the molarity of component i. Except for the bicarbonate band at about 1360 cm^{-1}, parameter f_i was set to zero.

5.4.5.2 Spectra in the System $NH_3 + CO_2 + H_2O$

In Figure 5.10, a spectrum in the ternary system $NH_3 + CO_2 + H_2O$ at 333 K and overall molalities of ammonia and carbon dioxide of 4.4 and 2.5 mol/kg is shown. For that overall composition, no molecular carbon dioxide could be detected, therefore the spectrum is only shown at wave numbers up to 1800 cm^{-1}.

Besides the experimental spectrum shown as a full line, the broken lines show the calculated contributions from the various peaks caused by the species present in the liquid phase. The summation of these contributions results in the dotted line.

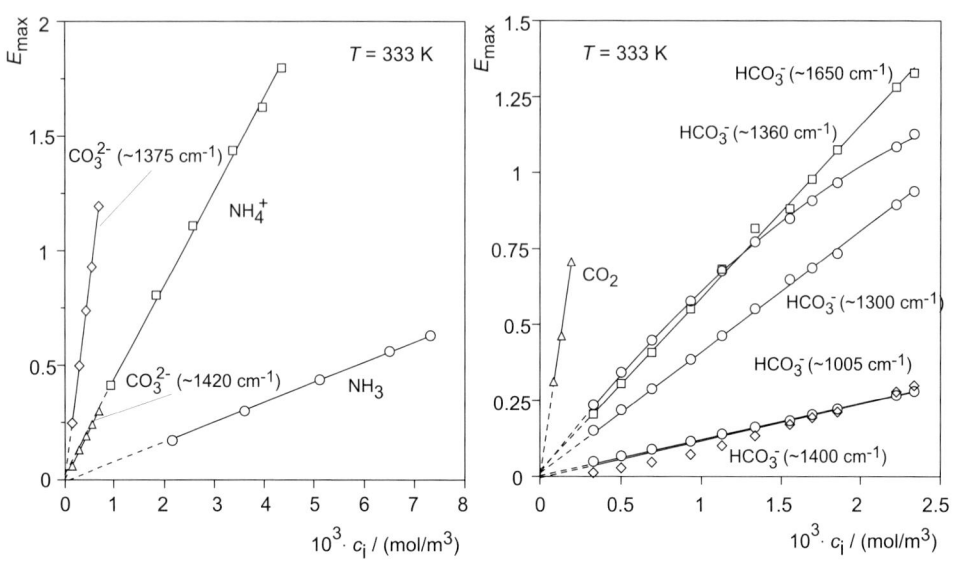

Figure 5.9: Results of the calibration measurements at 333 K.

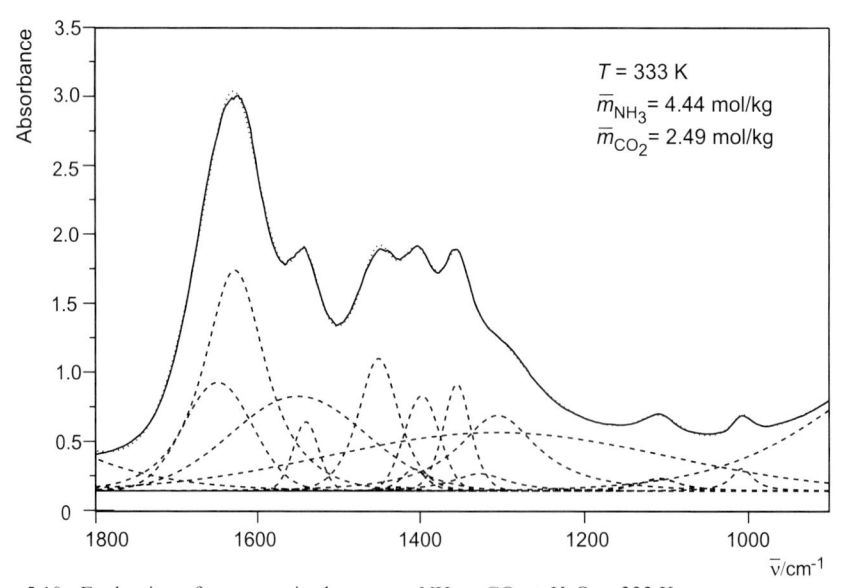

Figure 5.10: Evaluation of spectrum in the system $NH_3 + CO_2 + H_2O$ at 333 K.

To evaluate the ternary spectra, wave numbers at peak maximum as well as parameters describing the shape of each band (cf. Eq. (5.16)) were preset as determined in the calibration measurements. To further reduce the number of unknowns, Eqs. (5.10) and (5.11) together with the results from the calibration measurements (cf. Eq. (5.17)) were used.

5.4 Experimental

In the first step, the concentration of carbon dioxide was determined by evaluating the separated peak caused by molecular dissolved carbon dioxide in the range from 2200 to 2500 cm^{-1}. Secondly, the peak at 1005 cm^{-1} caused by bicarbonate ions was used to determine the concentration of that ion. Next, the concentrations of carbon dioxide and bicarbonate ions were used to calculate the true molarity of ammonia from Eq. (5.11). Together with the results from the calibration measurements, the bands caused by bicarbonate ions and molecular dissolved ammonia were preset. Finally, the remaining absorbances as identified from the calibration measurements were fitted to the ternary spectrum. Detailed investigations (cf. Lichtfers, 2000) showed that additional bands caused by carbamate ions had to be taken into account for a good description of the ternary spectra. The resulting wave numbers at peak maximum are compared to results from solid state spectra of NH$_2$COONH$_4$ (Hisatsune, 1984) in Table 5.2. Three additional peak maxima were found at wave numbers of 1100, 1400, and 1545 cm^{-1} whereas in the solid state spectra the wave numbers are 1120, 1400, and 1545 cm^{-1} (Hisatsune, 1984).

The concentration of carbonate- and carbamate ions were then calculated from Eqs. (5.10) and (5.12) (for a detailed description of the method cf. Lichtfers, 2000).

As can be seen from Figure 5.10, although the spectrum in the ternary system results from a complex superposition of the various peaks, a good agreement between experimental and calculated absorbances is observed. Similar results were obtained for the other temperatures and overall molalities investigated.

As a further test of the evaluation method, the absorbances E_{max} caused by the carbamate ions were plotted vs. the concentration of that ion as calculated from Eq. (5.12). The results are shown in Figure 5.11. At a wave number of 1100 cm^{-1}, a nearly linear depen-

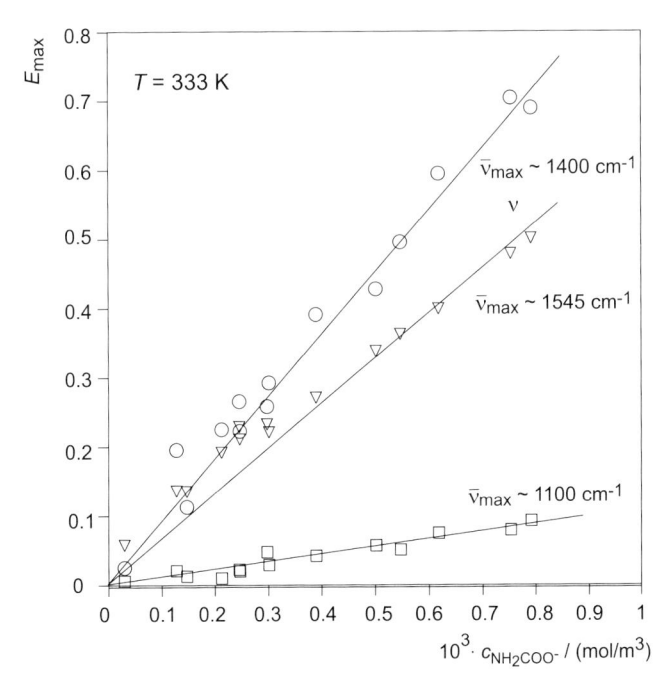

Figure 5.11: Carbamate absorbance vs. true concentration of carbamate ions at 333 K.

dence is observed. At the other wave numbers, a somewhat larger scattering is observed at molarities of the carbamate ion below about 0.4 mol/l. However, at those wave numbers the ternary spectra result from a complex superposition of several bands (cf. Fig. 5.10) which especially at low molarities of the carbamate ion are difficult to separate precisely.

5.5 Results

In Figure 5.12, as an example the results for the true species concentrations at 333 K and an overall molality of ammonia of 4.4 mol/kg are plotted vs. the overall molality of carbon dioxide in the liquid phase. Besides the experimental results, Figure 5.12 also shows the predicted species distribution from the model of Kurz et al. (1995). Furthermore, the results of the correlation (cf. below) are shown.

Adding carbon dioxide to an aqueous ammoniacal solution results in a decrease in the true concentration of ammonia. The amounts of ammonium- and bicarbonate ions increase, whereas carbamate- and carbonate ions run through a maximum. Only at high overall amounts of carbon dioxide, molecular carbon dioxide is present in the liquid phase.

Using model parameters from Kurz et al. which were solely determined from VLE- and VLSE measurements, a good agreement between measured and predicted species concentrations for molecular dissolved ammonia and carbon dioxide is observed. Only at higher

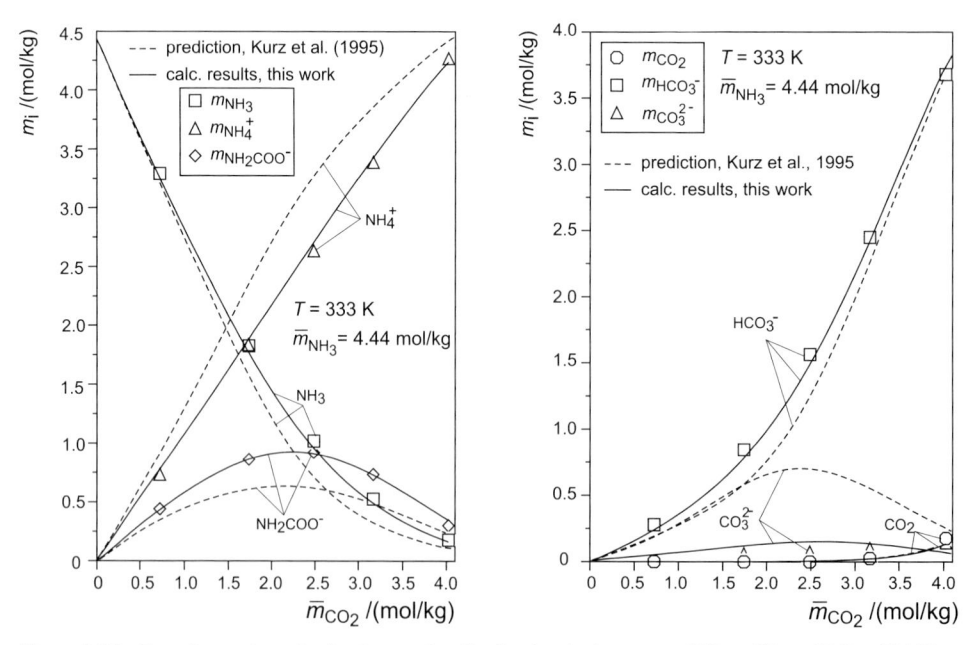

Figure 5.12: Experimental results for the species distribution in the system $NH_3 + CO_2 + H_2O$ at 333 K.

overall amounts of carbon dioxide, predicted true molalities of ammonia are slightly, but systematiccally smaller than the experimental results. However, as the model parameters by Kurz et al. were determined from fits to the partial and total pressures above $NH_3 + CO_2 + H_2O$, those deviations are not surprising as in that region the partial pressure of ammonia is rather small.

For the ionic species, although the principle dependencies are well predicted by the model, large and systematic deviations are observed. The carbamate concentrations are systematically smaller than the experimental results, whereas predicted carbonate concentrations are too large. Especially at higher overall amounts of carbon dioxide, larger deviations are also observed for the concentrations of ammonium ions.

A similar behavior – i.e. good agreement for the concentrations of molecular dissolved ammonia and carbon dioxide, but systematic deviations for ionic species – is also observed for the other isotherms (cf. Lichtfers, 2000). Therefore, the model was revised by including measured true species distributions into the parameter determination procedure.

5.6 Modeling

In the first step, the experimental data for the true speciation were used to determine numbers for the equilibrium constant of the carbamate reaction. At each overall molality of ammonia and carbon dioxide, measured species concentrations were used to calculate numbers for a pseudo-equilibrium constant $K_{IV}^{(m,*)}$

$$K_{IV}^{(m,*)} = \frac{m_{NH_2COO^-}}{m_{NH_3} \cdot m_{HCO_3^-}} \cdot \frac{\gamma_{NH_2COO^-}^{(m)} \cdot a_W}{\gamma_{NH_3}^{(m)} \cdot \gamma_{HCO_3^-}^{(m)}} \tag{5.18}$$

Activity coefficients as well as the activity of water were approximated using a modified Debye-Hückel term as given by Pitzer (1973). The true equilibrium constant was then determined by an extrapolation procedure

$$\ln K_{IV}^{(m)}(T) = \lim_{(\bar{m}_{CO_2}+\bar{m}_{NH_3}) \to 0} \ln K_{IV}^{(m,*)}(T, \bar{m}_{CO_2}, \bar{m}_{NH_3}) \tag{5.19}$$

For each temperature, numbers for the natural logarithm of the pseudo equilibrium constant were plotted vs. the sum of the overall molalities of ammonia and carbon dioxide and extrapolated to zero. As an example, Figure 5.13 shows the extrapolation procedure at 333 K resulting in a number for $K_{IV}^{(m)}$ of $0.682^{\pm 0.048}$.

For each series of isothermal measurements, the resulting numbers for $K_{IV}^{(m)}$ are given in Table 5.3 together with the estimated uncertainties. The results were correlated by the equation

$$\ln K_{IV}^{(m)}(T) = -4.8518 + \frac{1507.39}{T/K} \tag{5.20}$$

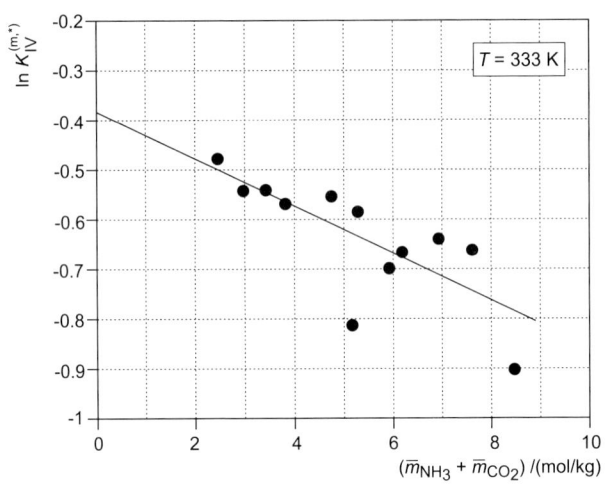

5 An Infrared Spectroscopic Investigation on the Species Distribution

Figure 5.13: Determination of $K_{IV}^{(m)}$ at 333 K.

Table 5.3: Numbers for $K_{IV}^{(m)}$ as determined from the spectroscopic measurements and estimated uncertainties.

T/K	$K_{IV}^{(m)}$
313	0.915 ± 0.056
333	0.682 ± 0.048
353	0.536 ± 0.059
373	0.491 ± 0.063
393	0.342 ± 0.038

In Figure 5.14, besides the experimental results from this work numbers for $K_{IV}^{(m)}$ as taken from the literature are plotted vs. temperature. The older experimental data of Faurholt (1921), van Krevelen et al. (1949), Marion and Dutt (1974), and Christensson et al. (1978) are about twice or three times larger than our data. The data of Szarawara et al. (1973) which were taken at lower temperatures confirm the trend of our new results. The recent data of Perez-Salado (1997) based on an electrochemical method are about a factor five larger than the results of this work. Bieling et al. (1989) determined the temperature dependency of $K_{IV}^{(m)}$ by fits to VLE-data in the system $NH_3 + CO_2 + H_2O$ at temperatures from 333 to 473 K. Their results deviate only slightly, but systematically from the experimental results of this work. For example at 333 K, the procedure applied by Bieling et al. results in a number for $K_{IV}^{(m)}$ of 0.596 whereas our method yields $0.682^{\pm 0.048}$.

In the next step, interaction parameters for the Pitzer equation were determined from fits to VLE-data in the system $NH_3 + CO_2 + H_2O$. The temperature dependence of $K_I^{(m)}$ was determined by correlating the data of Hitch and Messmer (1976), Bates and Pinching (1950), Wright et al. (1961), Noyes (1907), and Quist and Marshall (1968). Equilibrium constants for reactions II, III, and V were taken from Patterson et al. (1982, 1984) and Edwards et al. (1978), respectively (cf. Table 5.4).

5.6 Modeling

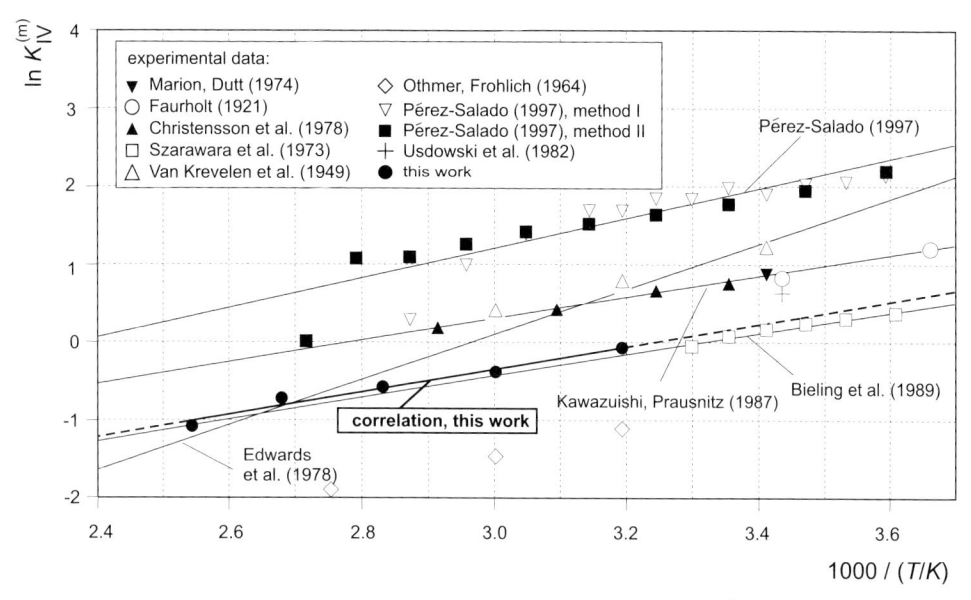

Figure 5.14: Comparison between selected data for the equilibrium constant $K_{IV}^{(m)}$ of the carbamate reaction.

Table 5.4: Equilibrium constants for reactions (I) to (V) and solubility product for the formation of NH$_4$HCO$_3$(s).

$$\ln K_R^{(m)} = A_R + \frac{B_R}{T/K} + C_R \cdot \ln(T/K) + D_R \cdot (T/K) + \frac{E_R}{(T/K)^2}$$

R	A_R	B_R	C_R	D_R	E_R
I	−686.514	33,110.5	110.718	−0.14596	−2.034 × 10^6
II	−1203.01	68,359.6	188.444	−0.2064	−4.713 × 10^6
III	175.360	−7230.6	−30.651	0.01315	−3.728 × 10^5
IV	−4.8518	1507.39	–	–	–
V	140.932	−13,445.9	−22.477	–	–
NH$_4$HCO$_3$(s)	7.4767	−2094.92	–	–	–

Interaction parameters for the binary system NH$_3$ + H$_2$O were taken from Rumpf et al. (1997). Due to the very small true molalities of H$^+$ and OH$^-$-ions, all interaction parameters between those ions and other species were set to zero. Furthermore, as molecular dissolved ammonia and carbon dioxide are simultaneously present only in very small amounts, all binary and ternary interaction parameters between those gases were set to zero:

$$\beta_{NH_3,CO_2}^{(0)} = \beta_{NH_3,CO_2}^{(1)} = 0 \tag{5.21}$$

$$\tau_{NH_3,NH_3,CO_2} = \tau_{NH_3,CO_2,CO_2} = 0 \tag{5.22}$$

Binary interaction parameters $\beta_{G,j}^{(1)}$ between molecular dissolved ammonia or carbon dioxide and ionic species were set to zero as the influence of ionic strength on the second virial coefficient for interactions between a neutral and a charged species is assumed to be very small (cf. Pitzer, 1991).

When a single gas G is dissolved in an aqueous solution of a single strong electrolyte $M_{\nu_M} X_{\nu_X}$, due to the condition of electroneutrality the influence of the separate ions M and X on the solubility of that gas cannot be separated. Therefore, it is common practice to define the following binary and ternary virial coefficients describing the influence of a single salt MX on the solubility of a single dissolved gas G (cf. for example Rumpf and Maurer, 1993):

$$B_{G,MX}^{(0)} = \nu_M \cdot \beta_{G,M}^{(0)} + \nu_X \cdot \beta_{G,X}^{(0)} \qquad (5.23)$$

$$\Gamma_{G,M,X} = \nu_M^2 \cdot \tau_{G,M,M} + 2 \cdot \nu_M \cdot \nu_X \cdot \tau_{G,M,X} + \nu_X^2 \cdot \tau_{G,X,X} \qquad (5.24)$$

$$\Gamma_{G,G,MX} = \nu_M \cdot \tau_{G,G,M} + \nu_X \cdot \tau_{G,G,X} \qquad (5.25)$$

As can be seen from those equations, in $B_{G,MX}^{(0)}$ and $\Gamma_{G,G,MX}$ one can arbitrarily set one of the parameters $\beta_{G,M}^{(0)}$ or $\beta_{G,X}^{(0)}$ and $\tau_{G,G,M}$ or $\tau_{G,G,X}$ to zero. Furthermore, in Eq. (5.24) two of the three parameters $\tau_{G,M,M}$, $\tau_{G,M,X}$ and $\tau_{G,X,X}$ can be arbitrarily set to zero. In the present work, $\beta_{G,NH_4^+}^{(0)}$, τ_{G,G,NH_4^+}, τ_{G,NH_4^+,NH_4^+}, and $\tau_{G,X,X}$ (where X = NH_2COO^-, HCO_3^-, or CO_3^{2-}) were set to zero.

However, in the present case ammonia and carbon dioxide are not only dissolved in a single electrolyte solution. Therefore, in principle the influence of ternary parameters $\tau_{G,X,Y}$ describing the combined effect of anions X and Y on the solubility of a gas G has to be considered. But as that influence is assumed to be very small, those parameters were set to zero.

Besides the parameters describing the influence of strong electrolytes on the solubility of a single gas in water, in the present case there exist some more parameters ($\tau_{G_1,G_2,M}$, $\tau_{G_1,G_2,X}$) as in principle two gases are dissolved in an ionic liquid. But as molecular ammonia and carbon dioxide together are present in the liquid phase only in very small amounts (cf. Fig. 5.12), those parameters were set to zero.

The remaining parameters describe interactions between dissolved ions. Similar to the case when a single strong electrolyte is dissolved in water, those parameters can be written as sums of neutral combinations of cations M and anions X (for details cf. Pitzer, 1973). For that type of interaction parameters, following Pitzer (1973) all parameters between ions carrying the same sign of charge were set to zero.

A sensitivity study revealed that only those parameters given in Table 5.5 are sufficient for a good description of the phase equilibria as well as the true speciation. The temperature dependence of the interaction parameters was approximated using

$$f(T) = A + \frac{B}{T/K} + C \cdot (T/K) \qquad (5.26)$$

The final set of interaction parameters is given in Table 5.5. Using those interaction parameters, the VLSE-data of Kurz et al. (1995) were used to determine numbers for the solubility constant of NH_4HCO_3. The resulting temperature dependence is given in Table 5.4.

The set of model parameters as given in Tables 5.4 and 5.5 describes VLE and VLSE in the ternary system in the temperature range from 313 to 473 K at overall molalities of

5.6 Modeling

Table 5.5: Interaction parameters in the ammonia + carbon dioxide + water system ($f(T) = \beta_{i,j}^{(0)}$ or $\beta_{i,j}^{(1)}$ or $\tau_{i,j,k}$).

$$f(T) = A + \frac{B}{T/K} + C \cdot (T/K) + D \cdot \ln(T/K)$$

Parameter	A	B	C	D	Source
$\beta_{NH_3,NH_3}^{(0)}$	−0.0197	9.864	–	–	Rumpf et al., 1997
τ_{NH_3,NH_3,NH_3}	5.539×10^{-3}	−0.179	–	-8.61×10^{-4}	
$\beta_{CO_2,HCO_3^-}^{(0)}$	−10.788	1880.32	0.0156	–	
$\beta_{NH_3,HCO_3^-}^{(0)}$	−0.022	–	–	–	
$\beta_{NH_3,CO_3^{2-}}^{(0)}$	0.372	−80.615	–	–	
$\beta_{NH_3,NH_2COO^-}^{(0)}$	0.041	–	–	–	
$\beta_{NH_4^+,HCO_3^-}^{(0)}$	0.047	−12.831	–	–	
$\beta_{NH_4^+,CO_3^{2-}}^{(0)}$	−0.356	107.752	–	–	this work
$\beta_{NH_4^+,NH_2COO^-}^{(0)}$	0.032	3.562	–	–	
$\beta_{NH_4^+,CO_3^{2-}}^{(1)}$	−4.207	1765.138	–	–	
$\tau_{CO_2,NH_4^+,HCO_3^-}$	−0.0024	–	–	–	
$\tau_{NH_3,NH_4^+,NH_2COO^-}$	−0.0007	–	–	–	
$\tau_{NH_4^+,NH_4^+,NH_2COO^-}$	−0.0011	–	–	–	

ammonia and carbon dioxide up to about 26 and 12 mol/kg as well as the true speciation in the temperature- and concentration range as determined in this work mostly within the experimental uncertainties.

As an example, Figure 5.15 shows the partial pressures of ammonia and carbon dioxide at 333 K as measured by Kurz et al. (1995) together with the results of the correlation. Generally, including the true speciation in the liquid phase into the parameter determination procedure results in a description of the phase equilibrium with an accuracy comparable to the work of Kurz et al. (for detailed comparisons cf. Lichtfers, 2000).

Bieling et al. (1995), Kurz et al. (1996), and Kamps et al. (2000) also investigated the effects of strong electrolytes on the simultaneous solubility of ammonia and carbon dioxide. As it was shown by those authors, the complex effects caused by salts like NaCl or Na_2SO_4 can be predicted quantitatively by the model using interaction parameters for the ternary system $NH_3 + CO_2 + H_2O$ as determined by Kurz et al. (1995). The new model parameters from this work were thus also used to test the quality of the predictions for the quaternary systems. Results using the new model parameters generally are of the same quality as in our previous work (for details cf. Lichtfers, 2000).

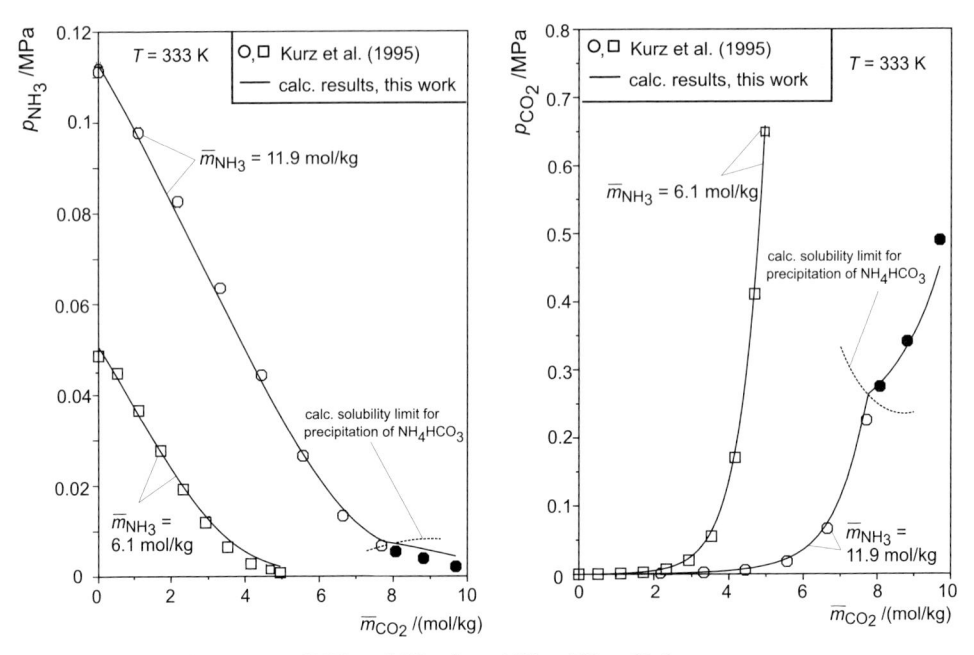

Figure 5.15: Partial pressures of NH_3 and CO_2 above $NH_3 + CO_2 + H_2O$.

5.7 Prediction of Caloric Effects

Weyrich (1997) measured enthalpies upon dilution of aqueous solutions containing ammonia and carbon dioxide with water at temperatures from 313 to 373 K and overall molalities of ammonia and carbon dioxide up to 12 and 6 mol/kg, respectively. A Calvet-type batch calorimeter equipped with special mixing cells in a differential arrangement was used. The mixing cells were divided into two parts by a thin Teflon foil. The upper part of a cell was filled with a known amount of an aqueous solution containing ammonia and carbon dioxide of known overall molalities, the lower part with pure water. The cells were pressurized to avoid the formation of a gas phase. After reaching thermal equilibrium, the Teflon foil was cut, thus the mixing process was started. Any caloric effects from that mixing process result in a heat transfer which was measured by a thermocouple block inside the calorimeter (for details cf. Weyrich, 1997).

The enthalpy change upon mixing can be expressed as

$$\Delta H_{\text{Dil}} = \sum_i n_i^{(2)} \cdot h_i^{(2)} - \sum_i n_i^{(u)} \cdot h_i^{(u)} - n_w^{(l)} \cdot h_{w,1}^{(l)} \qquad (5.27)$$

where $n_i^{(2)}$ is the true number of moles of a species i after mixing, $n_i^{(u)}$ is the true number of moles of a component i in the upper chamber before mixing and $n_w^{(l)}$ is the number of moles

5.7 Prediction of Caloric Effects

of water filled into the lower part of the chamber. Rearranging that equation into contributions from chemical reactions and excess effects yields

$$\Delta H_{Dil} = \underbrace{\sum_R \left(\xi_R^{(2)} - \xi_R^{(u)}\right) \cdot \Delta_R h^{Ref}(T)}_{\Delta H_1} +$$

$$\underbrace{\sum_R \left(\xi_R^{(2)} - \xi_R^{(u)}\right) \cdot \sum_i \nu_{i,R} \cdot h_i^E\left(T, m_j^{(2)}\right)}_{\Delta H_2} + \qquad (5.28)$$

$$\underbrace{\sum_i n_i^{(u)} \cdot \left(h_i^E\left(T, m_j^{(2)}\right) - h_i^E\left(T, m_j^{(u)}\right)\right)}_{\Delta H_3} + \underbrace{n_w^{(l)} \cdot h_w^E\left(T, m_j^{(2)}\right)}_{\Delta H_4}$$

As can be seen from that equation, the enthalpy change upon isothermal mixing can be interpreted as a shift in the chemical equilibria in the reference state (first term), excess effects on the reaction enthalpies (second term) as well as differences in the excess enthalpies resulting from the mixing process (third and fourth term).

Reaction enthalpies in the reference state were calculated from the temperature dependence of the equilibrium constants. Excess enthalpies $h_i^E(T,m_j)$ were calculated from the temperature dependent excess Gibbs energy Eq. (5.5). Details of the calculation procedure can be taken from Lichtfers (2000) and Weyrich (1997).

In Figure 5.16, the experimental data from Weyrich at 333 K and overall molalities of ammonia in the upper chamber of 6 and 12 mol/kg are plotted vs. the overall molality of carbon dioxide before mixing. The predictions using model parameters from Kurz et al. (1995) and from this work are also shown.

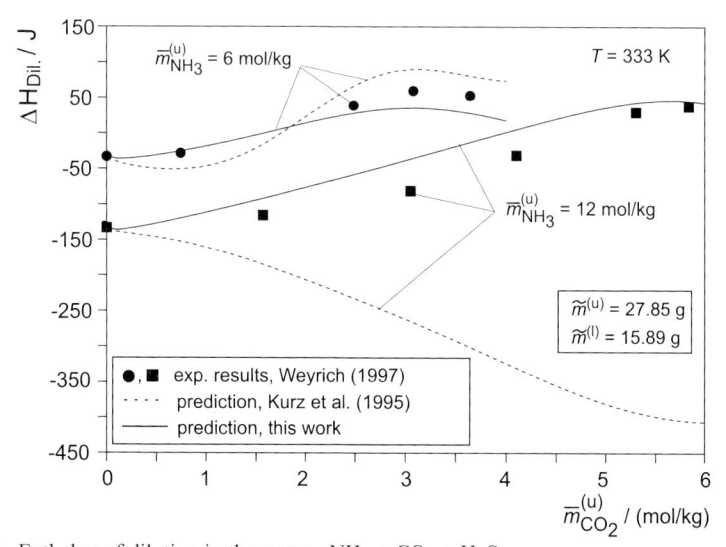

Figure 5.16: Enthalpy of dilution in the system $NH_3 + CO_2 + H_2O$.

As it was expected, upon diluting the system $NH_3 + H_2O$ with water exothermal mixing is observed. In the ternary system, that effect is reduced with increasing amount of carbon dioxide in the upper chamber of the mixing cell. At higher overall molalities of carbon dioxide, endothermal mixing is observed.

For the lower overall molality of ammonia, there is a good agreement between the experimental data and the prediction using model parameters by Kurz et al. However, for the series of measurements at $\bar{m}_{NH_3} = 12$ mol/kg, that model does not predict the change from exothermal to endothermal mixing. For example at $\bar{m}_{CO_2} = 5.3$ mol/kg, the model predicts an enthalpy change of -426 J whereas the experimental value is $+42.8$ J.

A much better agreement is observed using the model of the present work. Especially at $\bar{m}_{NH_3} = 12$ mol/kg, the model accurately predicts the change from exo- to endothermal mixing.

In Table 5.6, contributions to the enthalpy change in Eq. (5.28) as predicted by the models of Kurz et al. and this work are compared. As it was expected, the major contribution to the enthalpy change upon mixing comes from the shift in the chemical reactions (first term in Eq. (5.28)). However, for accurate predictions excess enthalpies cannot be neglected. The main difference between the prediction using the model by Kurz et al. and the model of the present work stems from contributions of the shift in the chemical reactions. As the reaction enthalpies are comparably large, small deviations in the prediction of the reaction numbers ξ_R before and after mixing cause large deviations in the predicted enthalpy changes.

Table 5.6: Comparison of contributions to the enthalpy of mixing at 333 K, $\bar{m}_{NH_3}^{(u)} = 11.9$ mol/kg and $\bar{m}_{CO_2}^{(u)} = 5.3$ mol/kg.

	ΔH_1/J	ΔH_2/J	ΔH_{3+4}/J	ΔH_{Dil}/J
Prediction, Kurz et al. (1995)	−504.9	165.1	−86.3	−426.1
Prediction, this work	217.7	−57.9	−117.1	42.7
Exp. result, Weyrich (1997)	−	−	−	42.81

Rumpf et al. (1998) also measured enthalpies upon partial evaporation of aqueous solutions containing ammonia and carbon dioxide. Those data were also used to successfully test the predictive capability of the model. Details of that test can be found in Lichtfers (2000).

5.8 Conclusion

An apparatus using an infrared spectroscopic in-situ technique to measure true species concentrations in aqueous systems containing ammonia and sour gases like carbon dioxide was developed. The apparatus was used to obtain spectra in the system $NH_3 + CO_2 + H_2O$ over wide ranges of temperature and composition. A method to resolve the complex spectra in that system was developed and tested. Thus, for the first time the true speciation in the system $NH_3 + CO_2 + H_2O$ was determined. Experimental results were compared to predictions of a model solely fitted to VLE- and VLSE-data. Using that model, predictions for the molecular dissolved gases showed a fair to good agreement with the experimental data, however larger deviations were observed for the ionic species. Therefore, the model was revised in two steps: First, numbers for the carbamate equilibrium constant were determined by an extrapolation procedure, at second, interaction parameters were determined from fits to VLE-data and measured species distributions. Both type of data are described well by the revised model. The thermodynamic power of the model was successfully tested by comparing measured and predicted enthalpy changes upon dilution in this complex, chemically reactive system.

Acknowledgements

The authors due thanks to Prof. Maurer for many helpful discussions and for providing the infrared spectrometer.

References

Bates, R. G.; Pinching, G. D. (1950): Dissociation constant of aqueous ammonia at 0 to 50° from e-m-f. studies of the ammonium salt of a weak acid. J. Am. Chem. Soc., 72, 1393–1396.

Bieling, V.; Rumpf, B.; Strepp, F.; Maurer, G. (1989): An evolutionary optimization method for modeling the solubility of ammonia and carbon dioxide in aqueous solutions. Fluid Phase Equilibria, 53, 251–259.

Bieling, V.; Kurz, F.; Rumpf, B.; Maurer, G. (1995): Simultaneous solubility of ammonia and carbon dioxide in aqueous solutions of sodium sulfate in the temperature range from 313 to 393 K and pressures up to 3 MPa. Ind. Eng. Chem. Res., 34, 1449–1460.

Christensson, F.; Koefoed, H. C. S.; Petersen A. C.; Rasmussen, K. (1978): Equilibrium constants in the ammonium carbonate-carbaminate system. The acid dissociation constant of carbamic acid. Acta Chem. Scand, A32(1), 15–17.

Davis, A. R.; Oliver, B. G. (1972): The species present in the CO_2-H_2O system. J. Sol. Chem., 4, 329–339.

Edwards, T. J.; Maurer, G.; Newman, J.; Prausnitz, J. M. (1978): Vapor-liquid equilibria in multicomponent aqueous solutions of volatile weak electrolytes. AIChE J., 24, 966–976.

Faurholt, C. (1921): Über die Prozesse $NH_2COONH_4 + H_2O \rightleftarrows (NH_4)_2CO_2$ und $CO_2 + H_2O \rightleftarrows H_2CO_3$. Z. Anorg. Allg. Chem., 120, 85–102.

Göppert, U.; Maurer, G. (1988): Vapor-liquid equilibria in aqueous solutions of ammonia and carbon dioxide at temperatures between 333 and 393 K and pressures up to 7 Mpa. Fluid Phase Equilibria, 41, 153–158.

Hisatsune, I. C. (1984): Low-temperature infrared study of ammonium carbamate formation. Can. J. Chem., 62, 945–948.

Hitch, B. F.; Messmer, R. E. (1976): The ionization of aqueous ammonia to 300 °C in KCl media. J. Sol. Chem., 5, 667–680.

Kawazuishi, K.; Prausnitz, J. M. (1987): Correlation of vapor liquid equilibria for the system ammonia-carbon dioxide-water. Ind. Eng. Chem. Res., 26, 1482–1485.

Kurz, F.; Rumpf, B.; Maurer, G. (1995): Vapor-liquid-solid phase equilibria in the system NH_3-CO_2-H_2O from around 310 to 470 K: New experimental data and modeling. Fluid Phase Equilibria, 104, 261–275.

Kurz, F.; Rumpf, B.; Sing, R.; Maurer, G. (1995): VLE and VLSE in the system ammonia-carbon dioxide-sodium chloride-water at temperatures from 313 K to 393 K and pressures up to 3 MPa. Ind. Eng. Chem. Res., 35, 3795–3802.

Kurz, F.; Rumpf, B.; Maurer, G. (1996): Simultaneous solubility of ammonia and carbon dioxide in aqueous solutions of ammonium sulfate and (ammonium sulfate + sodium sulfate) at temperatures from 313 K to 393 K and pressures up to 10 MPa. J. Chem. Thermodynamics, 28, 497–520.

Lichtfers, U. (2000): Spektroskopische Untersuchungen zur Ermittlung von Speziesverteilungen im System Ammoniak-Kohlendioxid-Wasser. Dissertation, Universität Kaiserslautern.

Marion, G. M.; Dutt, G. R. (1974): Ion association in the ammonia-carbon dioxide-water system. Soil. Sci. Soc. Amer. Proc., 38, 889–891.

Müller, G.; Bender, E.; Maurer, G. (1988): Das Dampf-Flüssigkeits-Gleichgewicht des ternären Systems Ammoniak-Kohlendioxid-Wasser bei hohen Wassergehalten im Bereich zwischen 373 und 473 K. Ber. Bunsen-Ges. Phys. Chem., 92, 148–160.

Nakamoto, K. (1986): Infrared and Raman spectra of inorganic and coordination compounds. 4th ed., John Wiley and Sons, Inc.: New York, Part III.

Noyes, A. A. (1907): The electrical conductivity of aqueous solutions. Carnegie Institution of Washington, Publication 63: Washington DC.

Othmer, D. F.; Frohlich, G. J. (1964): Carbon dioxide and ammonia in aqueous ammonium nitrate solutions. Vapor-liquid equilibria and gas separation by fractional absorption. Ind. Eng. Chem. Proc. Des. Dev., 3, 270–279.

Patterson, C. S.; Slocum, G. H.; Busey, R. H.; Messmer, R. E. (1982): Carbonate equilibria in hydrothermal systems: First ionization of carbonic acid in NaCl media to 300 °C. Geochim. Cosmochim. Acta, 46, 1653–1663.

Patterson, C. S.; Busey, R. H.; Messmer, R. E. (1984): Second ionization of carbonic acid in NaCl media to 250 °C. J. Sol. Chem., 13, 647–661.

Pérez-Salado Kamps, Á. (1997): Bestimmung chemischer Gleichgewichtskonstanten von Dissoziationsreaktionen in wässrigen Lösungen aus Messungen der elektromotorischen Kraft. Dissertation, Universität Kaiserslautern.

Pérez-Salado Kamps, Á.; Sing, R.; Rumpf, B.; Maurer, G. (2000): On the influence of single salts of NH_4Cl, NH_4NO_3 and $NaNO_3$ on the simultaneous solubility of ammonia and carbon dioxide in water. J. Chem. Eng. Data, 45, 796–809.

Pitzer, K. S. (1973): Thermodynamics of Electrolytes. I. Theoretical basis and general equations. J. Phys. Chem., 77, 268–277.

Pitzer, K. S. (1991): Activity coefficients in electrolyte solutions. 2nd ed., CRC Press, Boca Raton.

Quist, A. S.; Marshall, W. L. (1968): Ionization equilibria in ammonia-water solutions to 700 ° and to 4000 bars of pressure. J. Phys. Chem., 72, 3122.

Rumpf, B.; Maurer, G. (1993): An experimental and theoretical investigation on the solubility of carbon dioxide in aqueous solutions of strong electrolytes. Ber. Bunsen-Ges. Phys. Chem., 97, 85–97.

Rumpf, B.; Weyrich, F.; Maurer, G. (1998): Enthalpy changes upon partial evaporation of aqueous solutions containing ammonia and carbon dioxide. Ind. Eng. Chem. Res., 37, 2983–2995.

Rumpf, B.; Weyrich, F.; Maurer, G. (1993): Simultaneous solubility of ammonia and sulfur dioxide in water at temperatures from 313.15 K to 373.15 K and pressures up to 2.2 MPa. Fluid Phase Equilibria, 83, 253–260.

Rumpf, B.; Weyrich, F.; Maurer, G. (1997): Enthalpy of dilution in aqueous systems of single solutes ammonia, sodium sulfate and ammonium sulfate: Experimental results and modeling. Thermochimica Acta, 303, 77–91.

Szarawara, J.; Lugowska, M.; Zbigniew, B. (1973): Badania hydrolizy karbaminianu amonowego. Przemysl. Chemiczny, 52, 817–821.

Usdowski, H. E.; Menschel, G.; Hoefs, J. (1982): Kinetically controlled partitioning and isotopic equilibrium of ^{13}C and ^{12}C in the system CO_2-NH_3-H_2O. Z. Phys. Chem. Neue Folge, 130, 13–21.

van Krevelen, D. W.; Hoftijzer, P. J.; Huntjes, F. J. (1949): Composition and vapor pressures of aqueous solutions of ammonia, carbon dioxide and hydrogen sulphide. Rec. Trav. Chim., 68, 191–216.

Weyrich, F. (1997): Untersuchungen zum kalorischen Verhalten von Gemischen aus Ammoniak, Kohlendioxid, starken Elektrolyten und Wasser. Dissertation, Universität Kaiserslautern.

Wright, J. M.; Lindsay, W. T. Jr.; Truga, T. R. (1961): The behaviour of electrolytic solutions at elevated temperatures as derived from conductance measurements, Bettis Atomic Power Laboratory Report, WAPD-TM-204.

Nomenclature

A, B, C	constants in Eq. (5.26)
a_i	activity of component i
b, c	parameters in Gauss-Lorentz function
$B_{G,MX}^{(0)}$	combination of binary interaction parameters for interactions between gas G and salt MX
c_i	true molarity of species i
$\bar{c}_i$	overall molarity of component i
d	effective path length
$d_i \ldots f_i$	constants in Eq. (5.17)
E	absorbance
f_1, f_2	functions in Pitzer's equation
G^E	excess Gibbs energy
H	enthalpy
h_i	partial molar enthalpy of component i
$H_{i,w}^{(m)}$	Henry's constant for the solubility of component i in pure water (on molality scale)
I_m	ionic strength (on molality scale)
$K_R^{(m)}$	equilibrium constant for reaction R (on molality scale)
$K_{IV}^{(m,*)}$	pseudo equilibrium constant for the carbamate reaction (on molality scale)
M_i	molar mass of component i
$\bar{m}_i$	overall molality of component i
m_i	true molality of component i
$\bar{n}_i$	overall number of moles of component i
n_i	true number of moles of component i

117

5 An Infrared Spectroscopic Investigation on the Species Distribution

n_1, n_2	refractive index of crystal and sample
n_R	number of reflections
p	total pressure
p_i	partial pressure of component i
R	universal gas constant
T	absolute temperature
$v_{i,w}^\infty$	partial molar volume of component i at infinite dilution in water
v_w^s	molar volume of pure water
x_w	mole fraction of water
y_i	mole fraction of component i in gaseous phase
z_M, z_X	number of charges of cation M and anion X

Greek letters

α	angle
ε_i	molar extinction coefficient of species i
$\Delta_R h^{Ref}$	reaction enthalpy in the reference state $\left(\Delta_R h^{Ref} = \sum_i v_{i,R} \cdot h_i^{Ref}\right)$
$\gamma_i^{(m)}$	activity coefficient of component i (on molality scale)
φ_i''	fugacity coefficient of component i
φ_w^s	fugacity coefficient of saturated liquid water
λ	wave length
$\bar{\nu}$	wave number
$\bar{\nu}_{max}$	wave number at peak maximum
$v_{i,R}$	stoichiometric number of component i in reaction R
v_M, v_X	number of cations and anions in salt MX
ρ	mass density of aqueous phase
ξ_R	reaction number for reaction R
$\Delta\xi_R$	difference in reaction numbers ξ_R before and after mixing
$\beta_{i,j}^{(o)}$	binary interaction parameters in Pitzer's equation (o = 0, 1)
$\Gamma_{G,M,X}$	combination of ternary parameters for interactions between gas G and salt MX
$\Gamma_{G_1,G_2,MX}$	combination of ternary parameters for interactions between gases G_1, G_2, and salt MX
$\tau_{i,j,k}$	ternary interaction parameter in Pitzer's equation

Superscripts

E	excess
l	lower cell
s	saturation
u	upper cell
∞	infinite dilution
(2)	after mixing
'	liquid phase
"	gaseous phase

Subscripts

G	gas G
i, j, k	component i, j, k
M	cation M
m	on molality scale
R	reaction R
Ref	reference
w	water
X	anion X
1, 2	state 1, 2
Dil	dilution

II Associating Mixtures

6 VLLE for Mixtures of Water and Alcohols: Measurements and Correlations

Frank Gremer, Gerhard Herres, and Dieter Gorenflo [1*]

6.1 Abstract

The ternary systems of water with 2-ethyl-1-hexanol and 1-butanol or ethanol, respectively, important in industrial applications, have been chosen for investigating the vapour-liquid equilibrium (VLE) of systems with highly associating components and liquid separation (VLLE). A static equilibrium apparatus with two vibrating tube densimeters (made of glass) for measuring the densities of vapour and liquid developed by Rott (1990) has been modified and tested up to 30 bar and 200 °C (the densities only up to 150 °C). Vapour and liquid are circulated through the densimeters and the equilibrium cell by two micro metering pumps. All parts of the apparatus containing the test fluid are thermostatted at phase equilibrium temperature. Samples of vapour and liquid(s) are analysed by gas chromatography (GC-MSD).

The VLE and VLLE of the ternary systems water + 1-butanol + 2-ethyl-1-hexanol (2EH), ethanol + water + 1-butanol, and ethanol + water + 2-ethyl-1-hexanol and of the pertaining binaries has been investigated between 100 and 200 °C. Parallel to the measurements, the VLE and VLLE have been calculated by cubic equations of state (CEOS) of Trebble-Bishnoi-Salim type (TBS). The deviations between the measurements and calculations for the saturation pressure and the densities and compositions of vapour and liquid at equilibrium do not exceed a few percent. Association terms incorporated in the CEOS do not reduce the deviations significantly, if the small number of fitting parameters is maintained.

6.2 Introduction

The knowledge and prediction of vapour-liquid and liquid-liquid equilibria (VLLE) for mixtures of water with alcohols over a large range of temperature and pressure is important for industry. These associating components form highly non-ideal mixtures, many of them with

---
[1] Universität Paderborn, Wärme- und Kältetechnik, Warburger Str. 100, 33098 Paderborn.
[*] Author to whom correspondence should be addressed.

miscibility gaps. For mixtures of that kind, measurement and prediction of VLLE are difficult. Consequently only very few experimental data and no precise prediction methods have been found for higher pressures and temperatures, even in the case of ethanol + water, the most frequently investigated binary system of this group.

We present experimental investigations of binary and ternary systems formed by water, ethanol, 1-butanol, and 2-ethyl-1-hexanol (2EH) as components. The saturation pressure p_s, temperature T_s, and composition of the vapour (y) and liquid (x) phases (or x_I and x_{II}) have been measured between 100 and 200 °C, while the experimental temperature range for the measurement of the densities (liquid ρ' and vapour ρ'') is restricted to 150 °C.

Among various cubic equations of state (CEOS), the Trebble-Bishnoi-Salim-equation (1991) has been found to be apt best, and it predicts the VLLE for most of these associating components and their mixtures fairly well, even without special extension for association. After incorporating association terms in the CEOS, the deviations between measured and calculated data are not reduced significantly. This should mainly be due to the condition that the small number of fitting parameters applied has not been increased in order to demonstrate the effect of the association terms without further modifications.

6.3 Experimental

The experimental set up used for the measurements is shown in Figure 6.1. It contains a cylindrical equilibrium cell (internal volume: 600 cm^3) placed in a temperature controlled chamber. A magnetic stirrer helps to achieve phase equilibrium, together with two micro metering pumps which circulate vapour and liquid through the cell and through the two vibrating tube densimeters made of glass (type Paar DMA 602 HT, modified by us) which are connected to the cell for measuring the densities of the vapour and liquid at temperatures up

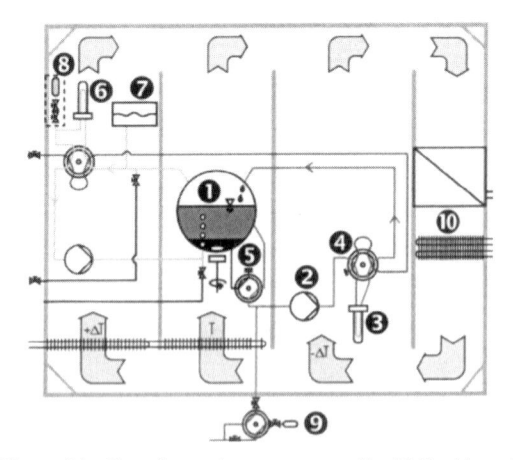

❶ equilibrium cell with magnetic stirrer
❷ micro metering pump (liquid circulation)
❸ vibrating tube densimeter (liquid)
❹ multi way valve – liquid densimeter
❺ liquid phase switch
❻ vibrating tube densimeter (vapour)
❼ differential pressure gauge
❽ vapour sample
❾ liquid sample
❿ air cooler and controlled heater

Figure 6.1: Experimental setup: vapour-liquid-liquid equilibrium cell in thermostatted chamber.

6.3 Experimental

to 150 °C. For measurements at higher temperatures, the whole densimeter unit can easily be withdrawn from the temperature controlled chamber.

The pressure is determined by a differential pressure gauge inside the chamber, in combination with absolute pressure gauges outside. The vapour densimeter, the differential pressure gauge, and the container for the vapour samples are placed in a separated air duct, which is slightly superheated over the equilibrium temperature ($T = T_s + \Delta T$), and the liquid densimeter is thermostatted at slightly lower temperature ($T = T_s - \Delta T$), see Figure 6.2. The temperatures in the vapour and liquid phases are measured with 100 Ω platinum resistance ther-

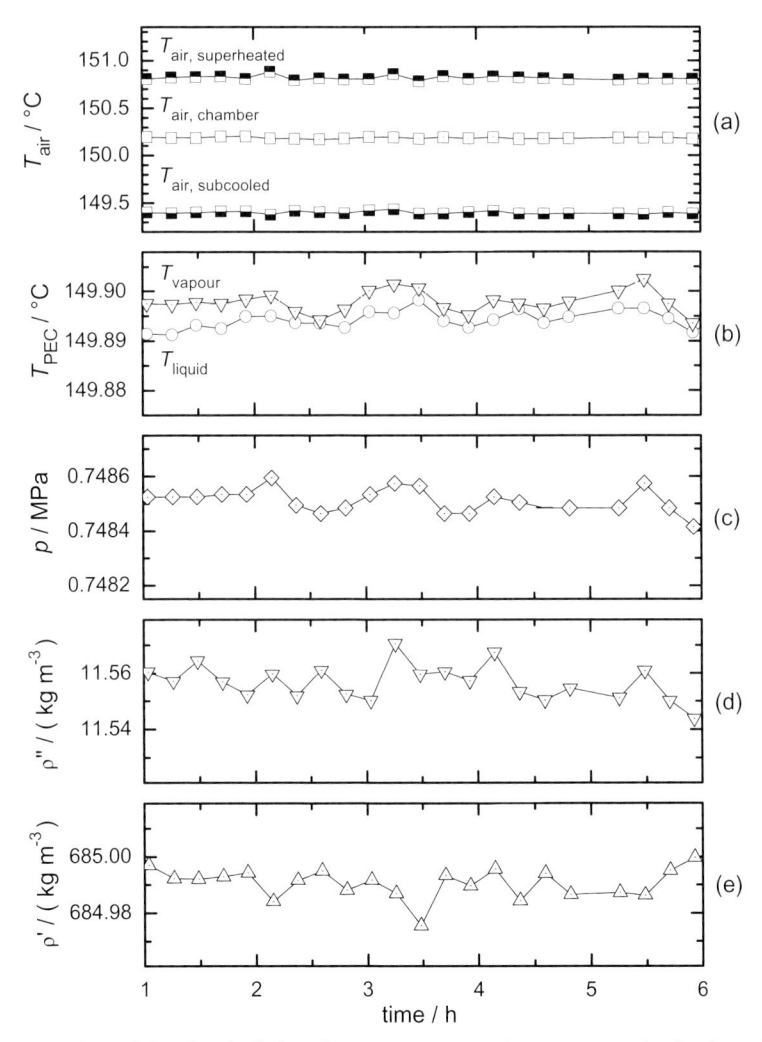

Figure 6.2: Stability of the signals during the measurements: a) temperatures in the three ducts of the chamber, b) temperatures of vapour and liquid in the cell, (c,d,e) saturation pressure and densities of vapour and liquid.

mometers. The vapour sample container (100 cm^3) can be closed from outside the apparatus. Approximately 0.2 g of the liquid phase is sucked in the container for the liquid samples.

The experimental limits of error for temperature T_s or pressure p_s are less than $T_s \pm 10$ mK or $p_s \pm 50$ Pa (or: $10^{-4} p_s$, if $> \pm 50$ Pa), respectively. As most of the samples separate in two liquid phases at ambient temperature, they have to be diluted in dried tetrahydrofuran (THF) before being analysed (four times) with a gas-chromatograph with mass-selective detector (GC-MSD). The accuracy is estimated to be ± 0.5 mol%. As a result of the uncertainty in the determination of the composition, the errors in the density measurements are considered to be within ± 0.1% or ± 1% for the liquid or vapour, respectively. Deviations up to ± 75 g/m^3 may occur at densities below 2.5 kg/m^3, due to breakdowns of the metering pump at these low densities and pressures.

The whole apparatus is a modified version of the apparatus used by Rott (1990) and Köster (1996) for measurements with mixtures of new refrigerants. A more detailed description of the modified experimental set up and procedure has been given by Gremer (2001).

In a comparison with literature data for the pure components water, ethanol, 1-butanol, toluene, and pyridine, the accuracy of the p, v, T-measurements was proved within the experimental limits of error given above. The sampling and the analytical methods were tested with mixtures of water + 1-butanol at atmospheric pressure (Fig. 6.3), and with ethanol + water at 120 and 150 °C (Fig. 6.4). In both cases the own measurements agree well with the reliable literature data, cf. the partially closed circles in Figure 6.4.

All substances (except THF) were obtained from Merck, Germany, and were used without further purification (purity better than 99.9% for ethanol, 1-butanol, and water or

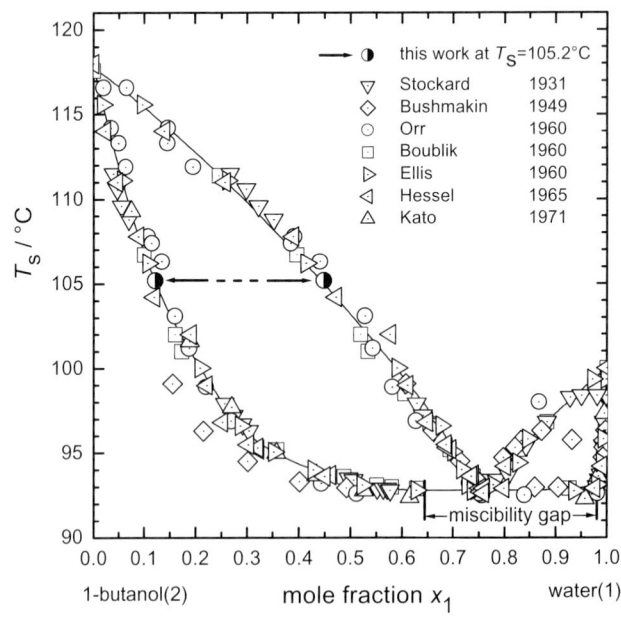

Figure 6.3: Comparison with literature data: temperature T_s against composition of liquid and vapour x_1, y_1 at equilibrium for water(1) + 1-butanol(2) at 1.013 bar. Literature data from Gmehling et al. (1977, 1981).

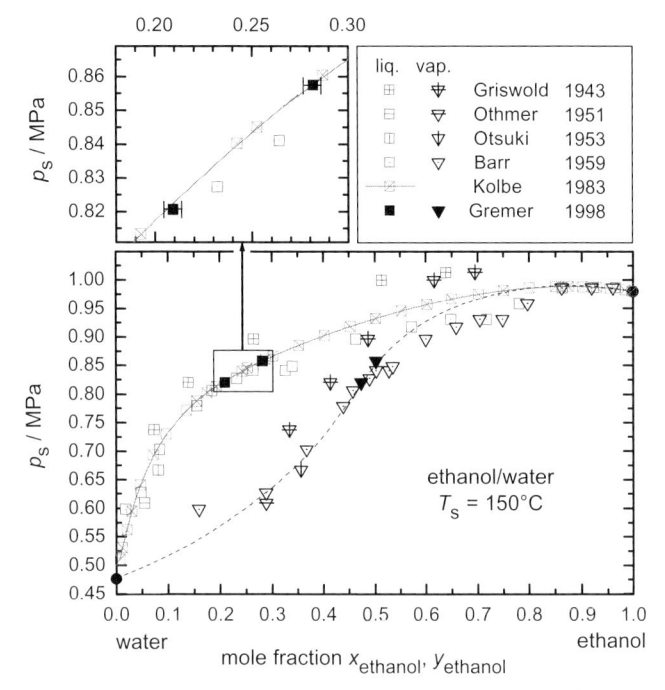

Figure 6.4: Comparison with literature data: pressure p_S against composition of liquid and vapour x_1, y_1 at $T_s = 150\,°C$ for ethanol(1) + water(2).

99.8 % for 2-ethyl-1-hexanol, respectively). THF was dried in a circulation apparatus over sodium. The purity of all the components was checked with the GC-MSD.

6.4 Experimental Results

First measurements were carried out with the ternary system water + 1-butanol + 2EH and the pertaining binaries. For water + 1-butanol, the liquid-liquid equilibrium and the vapour-liquid equilibrium *at atmospheric pressure* are well known (as shown in Fig. 6.3). At higher pressures and temperatures above 130 °C, however, only VLE-data of one source (Hessel and Geiseler, 1965) could be found. These authors postulate heterogeneous azeotropy even at 10 bar and temperatures above 170 °C. On the contrary, our own measurements confirm the upper critical temperature from literature near 125 °C (Sørensen and Arlt, 1979) and show homogeneous azeotropy at higher temperatures (Fig. 6.5, top), with a range of homoazeo-heterozeotropic behaviour between approximately 115 and 125 °C. The phase equilibrium was calculated with the CEOS of Trebble, Bishnoi, and Salim (TBS, 1991), see the solid lines in Figure 6.5, top. For fitting the TBS equation, critical data and three saturation

6 VLLE for Mixtures of Water and Alcohols: Measurements and Correlations

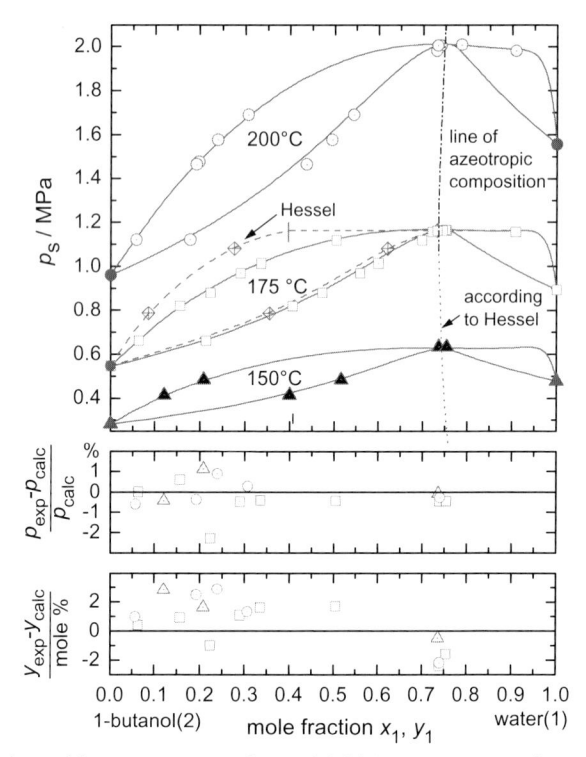

Figure 6.5: Comparison with measurements of Hessel (1965): pressure p_s against composition of liquid and vapour x_1, y_1 at equilibrium for water(1) + 1-butanol(2) at 150, 175, and 200 °C. p_s of pure water or 1-butanol from literature; lines calculated with TBS (top). Relative deviations of p_s (middle) and absolute deviation of y (bottom) from TBS for the own measurements.

pressures of each pure component were taken from literature and an asymmetrical quadratic mixing rule (Stryjek and Vera, 1986 a,b,c) with six binary interaction parameters was used for this highly non-ideal system. The azeotropic composition according to the new measurements between 150 and 200 °C (dotted or dot-dashed line) agree well with the data of Hessel below 175 °C (dotted).

The relative deviations for the pressure p_s – calculated at given temperature T and mole fraction x_1 of the liquid – are within $\pm 1\%$ (two exceptions), and the absolute deviations of the vapour phase mole fraction y_1 are within ± 2.5 mol% (lower diagrams). The deviations are mainly caused by inadequate description of the p, v, T-behaviour by the CEOS and by errors resulting from sampling and GC-analysing the liquid (in the case of p_s) and the vapour (in the case of y).

No VLE or LLE data above 100 °C could be found in literature for the binary system water + 2EH. This system is partially miscible in the whole experimental temperature range and over a great part of composition, and it forms heterogeneous azeotropes. The new measurements extend both, the description of the solubility limit (Fig. 6.6) and the vapour-liquid data up to 200 °C (Fig. 6.7). The heterogeneous azeotropes are described fairly well by the TBS-CEOS, viz. the three intersections of the solid lines and the corresponding symbols in

6.4 Experimental Results

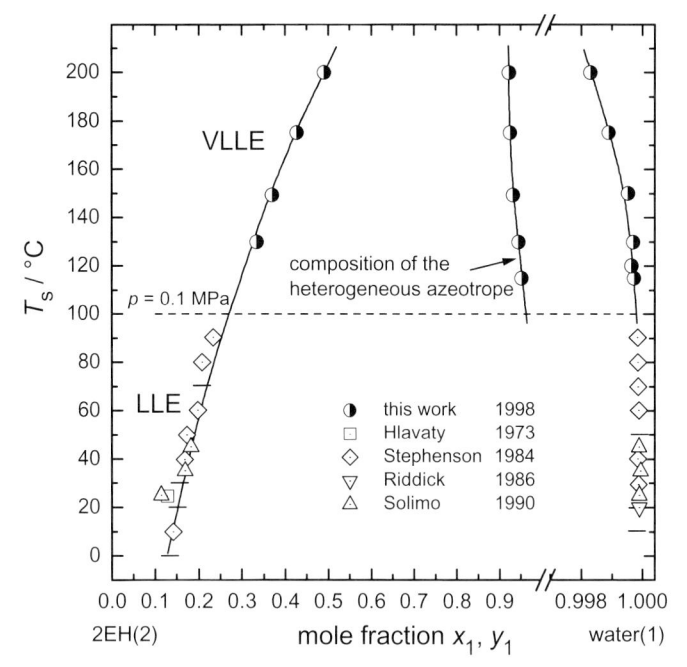

Figure 6.6: Comparison with literature at low pressures and own measurements at higher pressures: temperature T_s against composition of liquid and vapour x_1, y_1 at equilibrium for water(1) + 2EH(2) up to 200 °C.

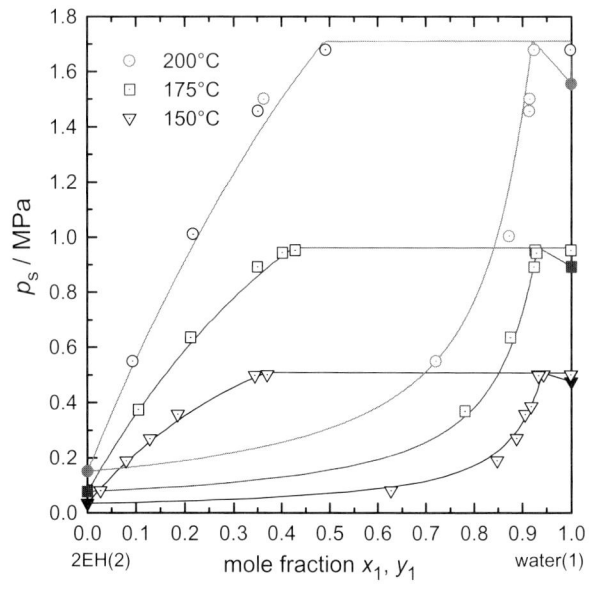

Figure 6.7: Pressure p_s against composition of liquid and vapour x_1, y_1 at equilibrium for water(1) + 2EH(2) at 150, 175, and 200 °C.

6 VLLE for Mixtures of Water and Alcohols: Measurements and Correlations

Figure 6.7. The TBS underestimates, however, the pressure to more than 5% in the homogeneous region close to the miscibility gap on the side with high content of alcohol.

The phase behaviour of binaries with two alcohols, like 1-butanol + 2EH, ethanol + 1-butanol, and ethanol + 2EH is less complex (without liquid separation). These systems could be calculated with only four binary interaction parameters in the TBS equation, and the results are quite good, as can be seen from Figures 6.8 and 6.9 for 1-butanol + 2EH and ethanol + 1-butanol as examples. One of the strong points of the TBS-CEOS is the comparatively good prediction of the liquid density, even without the frequently applied volume translation for v_c. Without any fitting to density data (except for ρ_c of the pure components)

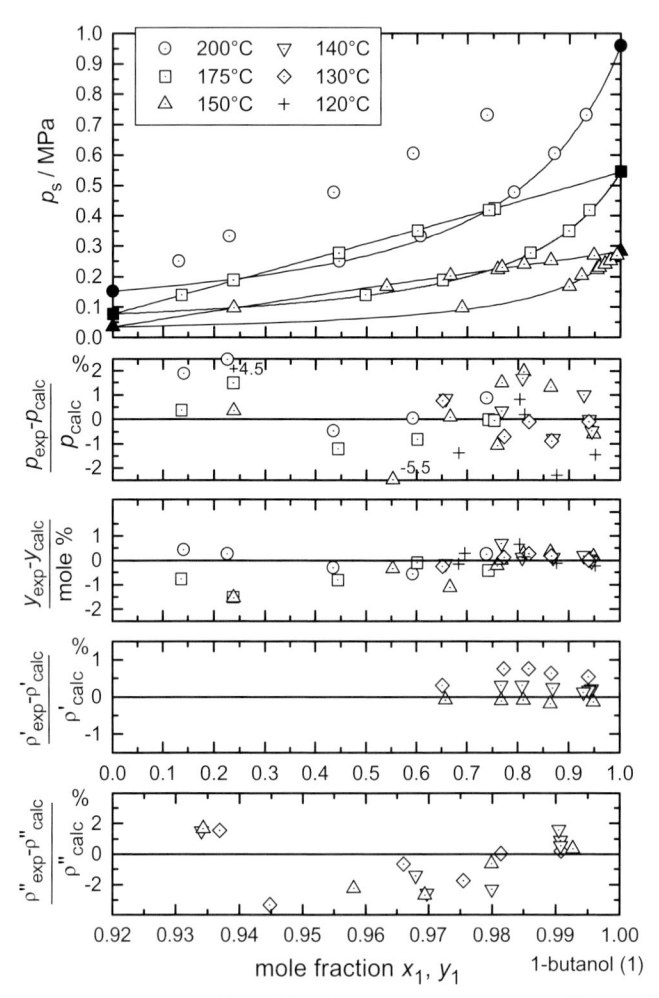

Figure 6.8: Pressure p_s against composition of liquid and vapour x_1, y_1 at equilibrium for 1-butanol(1) + 2EH(2). Deviations of calculated pressure, vapour-phase mole fraction and liquid and vapour densities.

130

6.4 Experimental Results

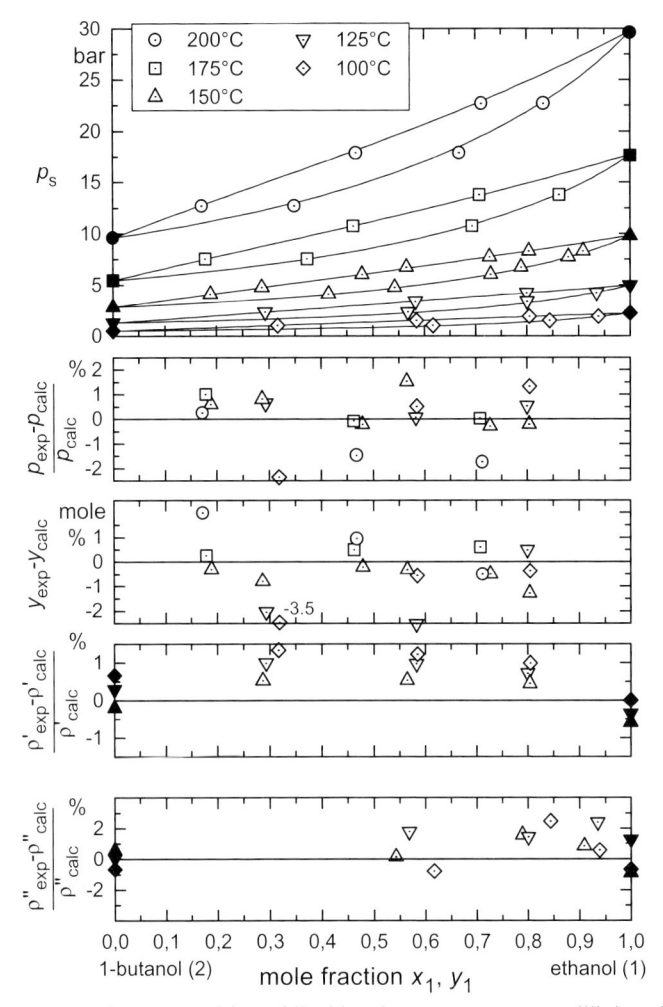

Figure 6.9: Pressure p_s against composition of liquid and vapour x_1, y_1 at equilibrium for ethanol(1) + 1-butanol(2). Deviations of calculated pressure, vapour-phase mole fraction and liquid and vapour densities.

the liquid density deviations for 1-butanol + 2EH are less than $\pm 1\%$ and not more than $\pm 5\%$ for ethanol + 1-butanol (Figs. 6.9 and 6.10, lower diagrams).

The ternary systems water + 1-butanol + 2EH, ethanol + water + 1-butanol, and ethanol + water + 2EH were calculated without fitting to ternary data, using only the binary interaction parameters. In order to test the results, approximately 15 data points of p, T, x, y (and ρ', ρ'' for part of them) have been measured for each of the ternaries. The *average* absolute error of the pressure p_s between measurements with water + 1-butanol + 2EH and this comparatively simple modelling is less than 2%. In Figure 6.10, the six ternary data points (open symbols) measured at $T_s = 175\,°C$ for this system (two of them with liquid separation) have been plotted as pressure p_s over the composition x of the liquid, together with the per-

131

6   VLLE for Mixtures of Water and Alcohols: Measurements and Correlations

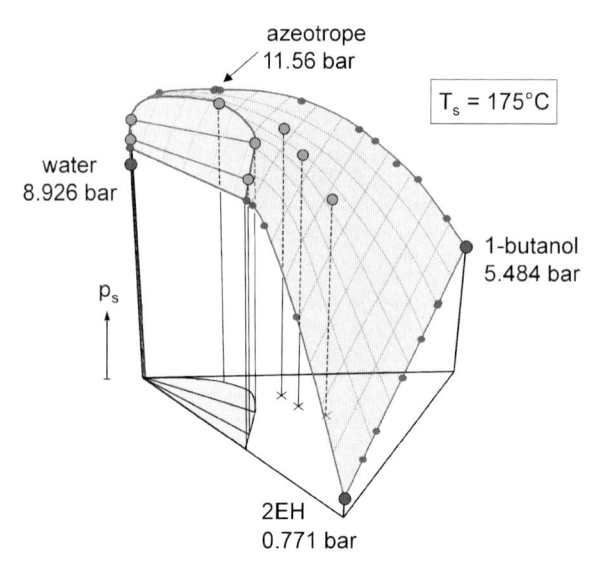

Figure 6.10: Pressure p_s over composition x of liquid at equilibrium for water + 1-butanol + 2EH at 175 °C.

taining measurements of the three binaries (closed symbols). For systems with *wide* miscibility gap, deviations of individual data points or within certain narrow parts of the VLLE, however, may be significant. This means that CEOS are suitable only within limits.

For testing improvements by incorporating association terms in the CEOS, the calculation method proposed by Anderko and coworkers (cf. e.g. Anderko, 1989 a,b,c, 1992; Anderko and Malanowski, 1989) have been applied. The method is based on the assumption that the compressibility factor $Z = Z^{ph} + Z^{ch} - 1$ can be separated into a *physical term* Z^{ph} without association and a *chemical term* Z^{ch} which models the associative behaviour of the molecules. In order to obtain a *direct* impression of the effect that is caused by adding the association term, the small, *total* number of fitting parameters has been kept constant in both cases, and also the pure components were modelled by CEOS. Under these conditions it came out that the different approach did not result in a substantial improvement, see Gremer (2001). It is likely, that the small number of fitting parameters initiated by the original CEOS without association term was too small for the more complicated structure of the modified CEOS to lead to an improvement (for more details see Gremer, 2001).

6.5   Conclusions

For three ternary systems of water and alcohols and for the pertaining binaries, the vapour-liquid equilibrium (and densities) have been measured between 100 and 200 °C. The experimental data can be described by a cubic equation of state of the Trebble-Bishnoi and Salim type within acceptable limits of error (of a few percent), even for azeotropic compositions. The limits of solubility occuring with most of the binaries and ternaries, can be predicted at least in a qualitative manner.

For systems with *wide* miscibility gap, such as water + 2-ethyl-1-hexanol, however, the cubic equations of state are suitable only within limits. This also holds for modified versions with association terms on the lines of Anderko being incorporated. If the *same* small number of fitting parameters is maintained, the deviations are not reduced significantly by these modifications.

References

Anderko A. (1989a): Calculation of vapor-liquid equilibria at elevated pressures by means of an equation of state incorporating association. Chem. Eng. Science, 44, 713–725.
Anderko A. (1989b): A simple equation of state incorporating association. Fluid Phase Equil., 45, 39–67.
Anderko A. (1989c): Extension of the AEOS model to systems containing any number of associating and inert components. Fluid Phase Equil., 50, 21–52.
Anderko A. (1992): Modeling phase equilibria using an equation of state incorporating association. Fluid Phase Equil., 75, 89–103.
Anderko A.; Malanowski A. (1989): Calculation of solid-liquid, liquid-liquid and vapor-liquid equilibria by means of an equation of state incorporating association. Fluid Phase Equil., 48, 223–241.
Barr-David F.; Dodge B. F. (1959): Vapor-liquid equilibrium at high pressure – the system ethanol-water and 2-propanol-water. J. Chem. Eng. Data, 4, 107–121.
Gmehling J.; Onken U. (1977 and 1981): Vapor-liquid equilibrium data collection, aqueous-organic systems (supplement 1). chemistry data series, Vol. 1, part 1a, Dechema, D-60486 Frankfurt, Theodor-Heuss-Allee 25.
Gremer F. (1998): Data of Fig. 6.4, presented at 15th ECTP 1999, Würzburg; see also:
Gremer F.; Herres G.; Gorenflo D. (2002): Vapor-liquid and liquid-liquid equilibria for mixtures of water and alcohols: Measurements and correlations. High Temperatures – High Pressures, 34, 355–362.
Gremer F. (2001): Phasengleichgewicht & Wärmeübergang beim Sieden von Mehrstoffsystemen aus Wasser und Alkoholen. PhD thesis, Universität-GH Paderborn.
Grisworld J.; Haney J. D.; Klein V. A. (1943): Ethanol-water system, vapor-liquid properties at high pressures. Ind. Eng. Chem. 35, 701–704.
Hessel D.; Geiseler G. (1965): Über die Druckabhängigkeit des heteroazeotropen Systems *n*-Butanol/Wasser. Z. Phys. Chem., 229, 199–209.
Hlavaty K.; Linek J. (1973): Liquid-liquid equilibria in four ternary acetic acid organic solvent-water systems at 24.6 degrees C. Collection of Czechoslovak Chemical Communications, 38, 374–378.
Köster R.; Herres G.; Buschmeier M.; Gorenflo D. (1996): Thermodynamic properties of binary and ternary mixtures of the refrigerants R32, R125, R134a and R143a. High Temp. High Press., 29, 25–31.
Kolbe B. (1983): Beschreibung der thermodynamischen Eigenschaften flüssiger Mischungen über einen größeren Temperatur- und Druckbereich am Beispiel des Systems Ethanol/Wasser. PhD thesis, Universität Dortmund.
Kolbe B.; Gmehling J. (1985): Thermodynamic properties of ethanol + water. I. Vapour-liquid equilibria measurements from 90 to 150 °C by the static method. Fluid Phase Equil., 23, 213–226.
Othmer D. F.; Möller W. P.; Englund S. W.; Christopher R. G. (1951): Composition of vapors from boiling binary solutions, recirculation-type still and equilibria under pressure for ethyl-alcohol-water system. Ind. Eng. Chem., 43, 707–711.

Otsuki H.; Williams F. C. (1953): Effect of pressure on vapor-liquid equilibria for the system ethyl alcohol-water. Chem. Eng. Progr. Symp. Ser., 49, 55–67.

Riddick J. A.; Bunger W. B.; Sakano T. K. (1986): Organic Solvents, 4th ed., edited by A. Weisenberger. Wiley and Sons, New York, Chichester, Brisbane, Toronto and Singapore.

Rott W. (1990): Zum Wärmeübergang und Phasengleichgewicht siedender R22/R114-Kältemittelgemische in einem großen Druckbereich. PhD thesis, Universität-GH Paderborn.

Salim P. H.; Trebble M. A. (1991): A modified Trebble-Bishnoi equation of state: thermodynamic consistency revisited. Fluid Phase Equil., 65, 59–71.

Solimo H. N. (1990): Liquid-liquid equilibria for the water + ethanol + 2-ethyl-1-hexanol ternary-system at several temperatures. Canadian Journal of Chemistry – Revue Canadienne De Chimie, 68, 1532–1536.

Sørensen J. M.; Arlt W. (1979): Liquid-liquid equilibrium data collection, binary systems. chemistry data series, Vol. 5, part 1, Dechema, Frankfurt/Main.

Stephenson R.; Stuart J.; Tabak M. (1984): Mutual solubility of water and aliphatic-alcohols. Journal of Chemical and Engineering Data, 29, 287–290.

Stryjek R.; Vera J. H. (1986a): PRSV: an improved Peng-Robinson equation of state for pure compounds and mixtures. Can. J. Chem. Eng., 64, 323–333.

Stryjek R.; Vera J. H. (1986b): PRSV: an improved Peng-Robinson equation of state with new mixing rules for strongly non ideal mixtures. Can. J. Chem. Eng., 64, 334–340.

Stryjek R.; Vera J. H. (1986c): A cubic equation of state for accurate vapor-liquid equilibria calculations. Can. J. Chem. Eng., 64, 820–826.

Nomenclature

p	pressure/bar
p_s	saturation pressure/bar
T	temperature/°C
T_s	saturation temperature/°C
v	specific volume/m^3/kg
v_c	volume at critical state/m^3/kg
x	mole fraction of liquid phase/%
y	mole fraction of vapour phase/%
Z	compressibility factor
Z^{ph}	physical term of compressibility factor
Z^{ch}	chemical term of compressibility factor
2EH	2-ethyl-1-hexanol
TBS	cubic equation of state of Trebble-Bishnoi-Salim type
LLE	liquid-liquid equilibrium
VLLE	vapour-liquid-liquid equilibrium

Greek letters

ρ'	saturation density of liquid/kg/m^3
ρ''	saturation density of vapour/kg/m^3
ρ_c	density at the critical state/kg/m^3

7 Phase Equilibria in Ternary Systems Containing Phenols, Hydrocarbons, and Water

Jürgen Schmelzer[1]*, Klaus Taubert, Antje Martin, René Meinhardt, and Jochen Kempe

7.1 Abstract

Liquid-liquid equilibrium data in ternary systems ((phenol or cresols) + (toluene or ethylbenzene or heptane or octane) + water) and some binary systems of these components were determined in the temperature range between 298 and 333 K. Additionally, vapor-liquid equilibrium data in ternary systems ((phenol or cresols) + (toluene or octane) + water) and in some binary systems of these components were measured at 333 and 363 or 353 K respectively. The experimental data in the binary systems were correlated using the Gibbs excess energy models NRTL and UNIQUAC as well as the generalized equation of state for nonspherical and associating mixtures by Elliott et al. (ESD-EOS). Infrared spectrometric studies were made, to test the possibility of getting association parameters (equilibrium constant, enthalpy of association). The models were used to predict liquid-liquid (-liquid) equilibria and vapor-liquid equilibria in the ternary systems. Such phase equilibria were also predicted using several versions of the group-contribution method UNIFAC. The quality of the predictions of the phase equilibrium behavior for all investigated ternary systems increases from UNIFAC to NRTL to UNIQUAC to ESD-EOS.

7.2 Introduction

The purpose of the present work was to search for methods for predicting the complex phase equilibrium behavior (liquid-liquid and vapor-liquid) of ternary systems containing (phenol or cresols), hydrocarbons and water. As phenols and water are associating substances, a method with chemical contributions had to be applied. The method by Elliott et al. (1990) is a generalized equation of state for mixtures of nonspherical and associating components.

---
[1] Hochschule für Technik und Wirtschaft Dresden, Fachbereich Maschinenbau/Verfahrenstechnik, Friedrich-List-Platz 1, 01069 Dresden.
* Author to whom correspondence should be addressed.

This equation (ESD-EOS) was tested by the authors, who developed the method, for aqueous systems. Therefore we tried to use this equation for calculations in our systems.

In addition, it should be tested to get association parameters (equilibrium constant, enthalpy on association) from infrared spectrometric studies.

The calculations using the ESD-EOS were compared with results obtained from the three Gibbs excess energy models: the group-contribution method UNIFAC (Fredenslund et al., 1975), the NRTL equation (Renon and Prausnitz, 1968), and the UNIQUAC model (Abrams and Prausnitz, 1975). As there were only a few data available in the literature, liquid-liquid equilibrium data in ternary systems (phenol or a cresol) + (toluene or ethylbenzene or heptane or octane) + water and some of the binary systems were determined in the temperature range between 298 and 333 K. Additionally the vapor-liquid equilibrium of ternary systems (phenol or a cresol) + (toluene or octane) + water and in some of binary systems were measured at 333 and 363 (or 353) K.

7.3 Experimental

7.3.1 Liquid-Liquid Equilibrium

Liquid-liquid equilibrium (LLE) data in 12 ternary systems (phenol or a cresols) + (toluene or ethylbenzene or heptane or octane) + water and in correlating binary systems were determined in a temperature range between 298 and 333 K by photometric turbidity-titration. Figure 7.1 shows a simplified scheme of the apparatus.

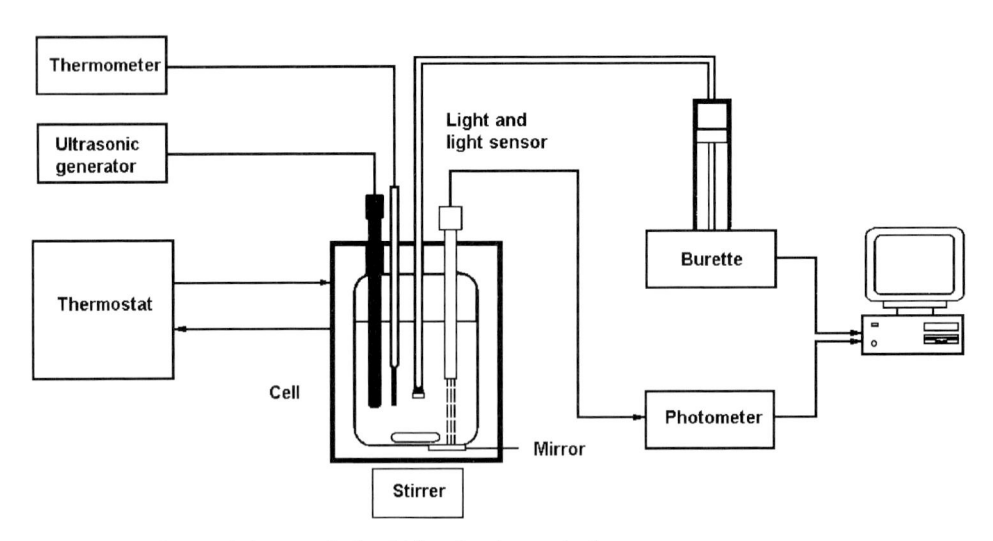

Figure 7.1: Scheme of photometrical turbidity-titration method.

7.3 Experimental

The core component is a glass cell containing a magnetic stirrer in conjunction with a temperature regulating system. One or two components are filled into this cell using syringes. The masses of the components are determined by weighing. The addition of the third component (in most cases water) takes place drop by drop by means of an automatic burette (DMS-Titrino 716, Deutsche Metrohm, Filderstadt). The solubility limit is obtained through measurement of the light transmission using the photometer 662 (Deutsche Metrohm, Filderstadt). The adjustment of equilibrium was accelerated by dispersing the water drops using an ultrasonic wave generator (UW 2070, BANDELIN Electronic, Berlin). The equilibrium compositions of the liquid-phases in the binary and ternary systems are then determined with a precision of ± 0.003 mol/mol. Liquid-liquid-liquid equilibrium data (LLLE) in systems containing alkanes were obtained in a temperature regulated glass cell and the liquid phases were analyzed by gas chromatography. The equilibrium compositions of the liquid-phases were determined with a precision of ± 0.005 mol/mol.

Table 7.1 gives an overview on the experiments. The experimental data will be published elsewhere (Schmelzer et al., 2004).

Table 7.1: Overview of the experimental investigations of liquid-liquid equilibria and vapor-liquid equilibria in ternary systems.

System	LLE T/K	VLE T/K
toluene + phenol + water	298+313	333+363
toluene + 2-cresol + water	298|323	333+363
toluene + 3-cresol + water	298+323	333+363
toluene + 4-cresol + water	298+323	333+363
ethylbenzene + phenol + water	298+323	
ethylbenzene + 3-cresol + water	298+323	
heptane + phenol + water	313+333	
heptane + 3-cresol + water	298+323	
octane + 2-cresol + water	298+323	333+353
octane + 3-cresol + water	298+323	333+363
octane + 4-cresol + water	298+323	

For some selected ternary systems the experimental results are shown together with calculations in Figures 7.4 and 7.6.

7.3.2 Vapor-Liquid Equilibrium

The vapor-liquid equilibrium (VLE) of six ternary systems (phenol or a cresol) + (toluene or octane) + water and in some binary systems were measured at 333 and 363 (or 353) K cf. Table 7.1. The experimental vapor-liquid equilibrium data were determined via an all-glass circulation still modified from the Röck and Sieg (1955) type (Fig. 7.2). This modified circulation still is the core component of the equipment which allows both isobaric and isother-

7 Phase Equilibria in Ternary Systems Containing Phenols, Hydrocarbons, and Water

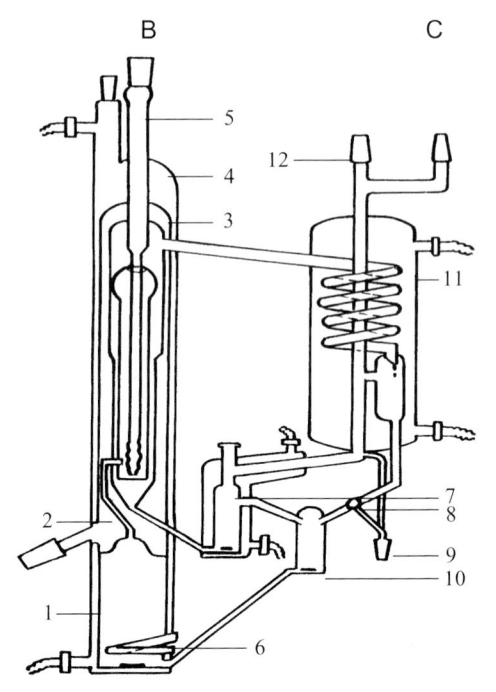

Figure 7.2: Scheme of modified Röck and Sieg typ circulation still. B: boiling part, C: condensation part, 1: boiling flask, 2: Cottrell pump, 3: vacuum jacket, 4: silicon oil jacket, 5: thermometer well, 6: tube for heater, 7: liquid sampler, 8: sampling cock, 9: to sampling tube, 10: mixing chamber, 11: cooler, 12: to controlled underpressure.

mal measurements by coupling of two automatic control loops in a process computer and a pulse modeled control of two magnetic valves.

The temperature is controlled by means of a computer system and measured by means of a calibrated platinum resistance thermometer connected to a precision temperature module (EUROTRONICS, Leipzig) with a precision of ± 0.03 K. A high accuracy pressure transducer (MKS Baratron, type 690) and a high accuracy signal conditioner (MKS 270D, MKS Instruments Deutschland, München) were used to determine the pressure in the apparatus directly with a precision of ± 40 Pa. Two magnetic valves (MV20, Fischer Technology, Bonn) were used to adjust the pressure or, indirectly, the temperature (for details cf. Schmelzer et al., 1998).

The equilibrium compositions of the liquid and vapor phases in the binary systems are determined from analyzes by a vibration tube densimeter (DMA 58, Anton Paar, Austria) with a precision of ± 0.002 mol/mol, in systems containing water by investigation of the water content by means of Karl-Fischer titration with a precision of ± 0.005 mol/mol, and in ternary systems through combination of both methods. In Figures 7.3 and 7.5 the experimental VLE data of some selected binary systems are shown. The experimental data will be published elsewhere (Schmelzer et al., 2004).

7.3.3 Infrared Spectroscopic Investigations

Geisler and Seidel (1977) showed that in associating liquid mixtures up to three associates can be quantitatively analyzed by infrared spectroscopic investigations (based on the change of the monomer bands). Extended interpretation of the IR absorption spectra allows for a deeper insight into association in complex systems (cf. Aspiron, 1996). However such interpretation of spectroscopic data might be rather difficult due for example to the influence of temperature on the extinction coefficient and on the position of band (e.g. Becker, 1961). Furthermore there are some limits in the mathematical analysis of the spectra (cf. Quadri and Shurvell, 1995). One of our objectives was to examine these problems in the (phenol + toluene + octane) system as well as to try to solve these problems by using newer methods, especially chemometrical procedures. The evaluation consisting of curve-fitting of spectra, assignment of peaks, and the determination of extinction coefficients.

The analysis was carried out through derivative spectra (first and second derivation) and multivariate statistical methods for the Principal Component Analysis (PCA) (cf. Hobert and Kempe, 1993) and factor analysis. QUANT (Perkin-Elmer) and STATGRAPHICS software was used.

Eight peaks were found in the region of the O–H-ground vibration as well as in the region of the first overtone of the phenol + toluene system. The peak number was much higher than would be expected from a scheme of a few association equilibria.

The peaks in that system could not be sufficiently assigned to defined associated and solvated complexes. Proceeding on the assumption that the empirical methods for the equilibrium analysis via IR spectra in this multiple associating and strong solvating component system have their limits, this insufficiency could only be overcome by using detailed structure analytical methods.

7.3.4 Materials

The investigated substances (see Table 7.1), purum grade materials of stated purity >98%, were fractionally distilled at reduced pressure and dried over Na_2SO_4 (phenol and cresols) or over molecular sieves type 4A (hydrocarbons). GLC purity of all investigated substances was between 99.1% (2-cresol) and 99.9% (3-cresol). The measured physical properties (vapor pressure, density, refractive index) were in sound agreement with published data.

7.4 Theory

7.4.1 UNIFAC Group-Contribution Method

The phase equilibrium of binary and ternary systems was predicted by applying several versions of UNIFAC: the original UNIFAC equation (Fredenslund et al., 1975) with the VLE parameter matrix of Hansen et al. (1991) and the LLE parameter matrix of Magnussen et al. (1981) and the UNIFAC-Oldenburg method by Gmehling et al. (1993).

7.4.2 NRTL and UNIQUAC Gibbs Excess Energy Models

The experimental results for the liquid-liquid and vapor-liquid equilibrium of the binary systems were correlated simultaneously using both the NRTL (Renon and Prausnitz, 1968) and the UNIQUAC (Abrams and Prausnitz, 1975) models. The binary interaction parameters were assumed to depend on temperature through:

NRTL: $\quad (g_{ij} - g_{jj})/R = C_{ij}^C + C_{ij}^T (T - 273.15)$ (7.1)

UNIQUAC: $\quad (u_{ij} - u_{jj})/R = C_{ij}^C + C_{ij}^T (T - 273.15)$ (7.2)

The parameters were obtained by minimizing the following objective function (Renon et al., 1971):

$$Q = \pi_2 \Sigma_i (100/P_{exp.})_i^2 (P_{calc.} - P_{exp.})_i^2 + \pi_3 \Sigma_i (100)_i^2 (x'_{1,calc.} - x'_{1,exp.})_i^2 +$$
$$+ \pi_3 \Sigma_i (100)_i^2 (x''_{1,calc.} - x''_{1,exp.})_i^2 \quad (7.3)$$

The binary NRTL and UNIQUAC parameters and the experimental data base are given in Table 7.2. These parameters were used for predictions of LLE (LLLE) and VLE data in the ternary systems.

7.4.3 ESD Equation of State (ESD-EOS)

The ESD-EOS (Elliott et al., 1990; Suresh and Elliott, 1991, 1992; Puhala and Elliott, 1993) combines an equation of state with an association model. The parameters of the self-association were estimated from a data base of vapor pressures and volumes of pure components. The cross-association parameters were calculated from parameters for self-association. For hydrocarbons and water, the pure component parameters from Elliott were adopted, parameters for phenol and cresols were determined. The pure component parameters are given in Table 7.3.

The binary ESD-EOS parameters k_{ij} were also assumed to depend on temperature:

$$k_{ij} = k_{ij}^C + k_{ij}^T (T - 273.15) \quad (7.4)$$

Numerical values were obtained by minimizing the objective function (Eq. (7.3)). The results are given in Table 7.2.

Table 7.2: Binary model parameters.

System	ESD k_{12}^C	ESD k_{12}^T	UNIQUAC C_{12}^C	UNIQUAC C_{21}^C	UNIQUAC C_{12}^T	UNIQUAC C_{21}^T	NRTL C_{12}^C	NRTL C_{21}^C	NRTL C_{12}^T	NRTL C_{21}^T	α_{12}	Ref. of exp. data[a]
toluene + phenol	−0.010265	0.00001061	369.57	−146.32	−1.6642	0.9573	857.14	−308.41	−4.3775	2.8430	0.2	1, 3
toluene + 2-cresol	−0.027326	0.00032133	372.85	−191.68	−2.7563	1.6261	942.67	−416.16	−5.9518	2.9683	0.2	1
toluene + 3-cresol	−0.011280	0.00011565	352.72	−149.49	−2.4085	1.3615	909.79	−303.03	−5.1134	2.2854	0.2	1, 3
toluene + 4-cresol	−0.008899	−0.00002098	108.66	4.74	0.9399	−0.8258	443.83	−50.11	1.6730	−1.6607	0.2	1, 3
ethylbenzene + phenol	−0.018370	−0.00000118	348.50	−119.64	−0.0556	−0.0582	681.76	−152.96	−0.4982	0.3624	0.2	1, 3
ethylbenzene + 3-cresol	−0.014787	0.00011872	339.75	−145.69	−0.0714	−0.0993	521.39	−39.50	−0.2917	0.0357	0.47	4
octane + 2-cresol	−0.027326	0.00056014	286.41	−57.92	0.0066	−0.2871	707.78	579.79	−2.8088	−1.4006	0.5056	1
octane + 3-cresol	−0.020558	0.00043516	519.83	−266.85	−2.7093	2.3092	606.70	−244.16	−1.2432	7.3141	0.4	1
toluene + water	0.071595	0.00059730	915.94	495.16	−4.0177	3.0247	1328.17	2027.73	−5.1206	5.9040	0.2	1
ethylbenzene + water	0.087057	−0.00015011	1125.46	322.16	−6.8689	1.4581	1555.04	2352.28	−9.6980	5.1204	0.2	2
octane + water	0.133486	0.00087780	1334.19	411.53	−4.4426	1.7854	1505.53	2271.75	−7.3150	8.3903	0.15	1
phenol + water	0.047965	0.00014692	−216.70	467.34	−0.9202	0.6408	−557.19	1821.63	1.5755	−5.0783	0.2	1, 2, 3
2-cresol + water	0.047370	0.00045420	−181.64	569.14	0.1756	−1.4436	439.33	1470.53	−3.1878	1.8177	0.4	1
3-cresol + water	0.059124	0.00027514	74.19	41.34	1.7006	−1.5048	−253.30	1397.66	−0.9899	1.7107	0.2	1, 2
4-cresol + water	0.050225	0.00017819	−87.22	243.42	−0.7445	0.6966	−307.39	1502.87	−1.8663	4.4922	0.2	1, 2

a) 1: this work, 2: Arlt et al. (1979), 3: Gmehling et al. (1977), 4: UNIFAC (Hansen et al., 1991).

7 Phase Equilibria in Ternary Systems Containing Phenols, Hydrocarbons, and Water

Table 7.3: Pure component parameters of ESD-EOS.

Component	T_c (K)	c	ε_{Disp}/k (K)	v^* (cm^3/mol)	K_{AB}/v^*	ε_{HB}/RT_c
octane	568.83	2.4842	285.211	54.157		
toluene	591.79	1.9707	332.752	36.227		
ethylbenzene	617.17	2.1223	333.658	42.242		
phenol	694.20	1.6503	415.407	34.391	0.02936	3.0470
2-cresol	697.60	1.1983	607.630	42.618	0.00098	4.3269
3-cresol	705.80	1.6317	444.525	41.836	0.01201	3.3574
4-cresol	704.60	2.5650	348.487	37.532	0.00064	3.5243
water	647.29	1.0053	427.254	9.411	0.10000	4.0000

7.5 Results and Discussions

7.5.1 Prediction Using UNIFAC

Applying the UNIFAC methods for predicting the phase equilibrium in those systems which were also investigated experimentally lead to unsatisfactory results. For example, the deviation between experimental results and predictions for vapor pressure were often greater than 20%, especially in systems with limited mutual solubility. The original UNIFAC equation with parameters by Magnussen et al. (1981) results in a wrong influence of temperature on the miscibility gap in (phenol or cresol) + water systems (cf. Figs. 7.3 and 7.4). In the ter-

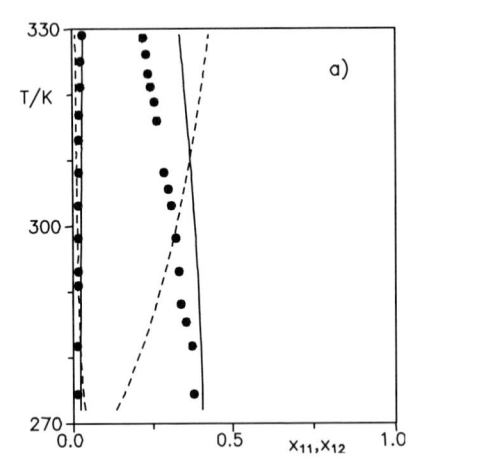

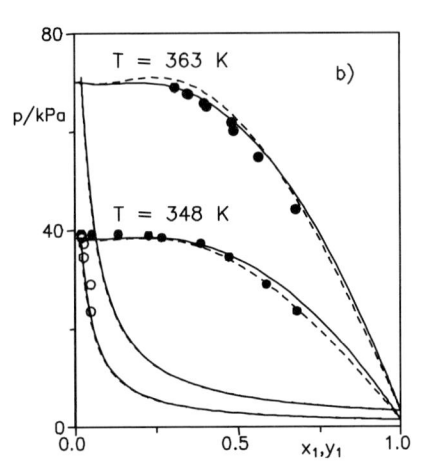

Figure 7.3: Experimental results and predictions (UNIFAC) for phase equilibrium in the binary system phenol(1) + water(2). ●●●●, ○○○○ exp. data (Arlt et al., 1979; Gmehling et al., 1977; this work), —— modified UNIFAC, - - - original UNIFAC (a: LLE-, b: VLE-parameters).

7.5 Results and Discussions

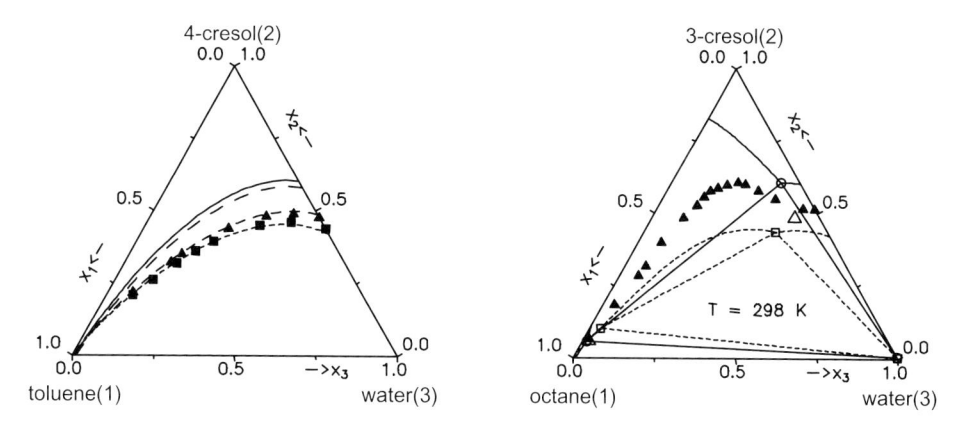

Figure 7.4: Experimental results and predictions (UNIFAC) for liquid-liquid equilibrium of ternary systems. exp. data (this work): ▲▲▲▲ 298 K, ■■■■ 323 K, △△△ LLLE; modified UNIFAC: —— 298 K, — — 323 K, -○-○-○ LLLE; original UNIFAC (LLE-parameters) - - - - 298 K, – – – 323 K, -□-□-□ LLLE.

nary systems with three liquid phases (alkane + cresol + water; cf. Fig. 7.4), that version of the UNIFAC method gives good results at 298 K, but the prediction overestimates the homogeneous region. At 323 K the predictions underestimate that region. The modification of UNIFAC gives predictions for the miscibility gap which are too large for the ternary systems. It also falsely predicts a liquid-liquid miscibility gap in binary systems (octane + cresols).

7.5.2 Correlation of Binary Data with NRTL, UNIQUAC, and ESD-EOS

The NRTL and UNIQUAC equations with two linear temperature-dependent interaction parameters give a somewhat better description of binary experimental data than the ESD-EOS (which has only a single binary temperature-dependent parameter). Some selected results are shown in Figure 7.5.

7.5.3 Prediction for Ternary Systems with NRTL, UNIQUAC, and ESD-EOS

The predictions of the phase equilibrium data using ESD-EOS in ternary systems are better or of the same quality, as results obtained by means of the best of the two activity coefficients equations. Figure 7.6 shows the results of LLE predictions in some ternary systems. The quality of the LLE (LLLE) predictions, for all investigated ternary systems considered, increases from NRTL to UNIQUAC to ESD-EOS. Thus, for the total of all systems (isotherms), best agreement between experimental and predicted LLE data was given by ESD-EOS in more than 50% of the cases. In the systems phenol + toluene or ethylbenzene +

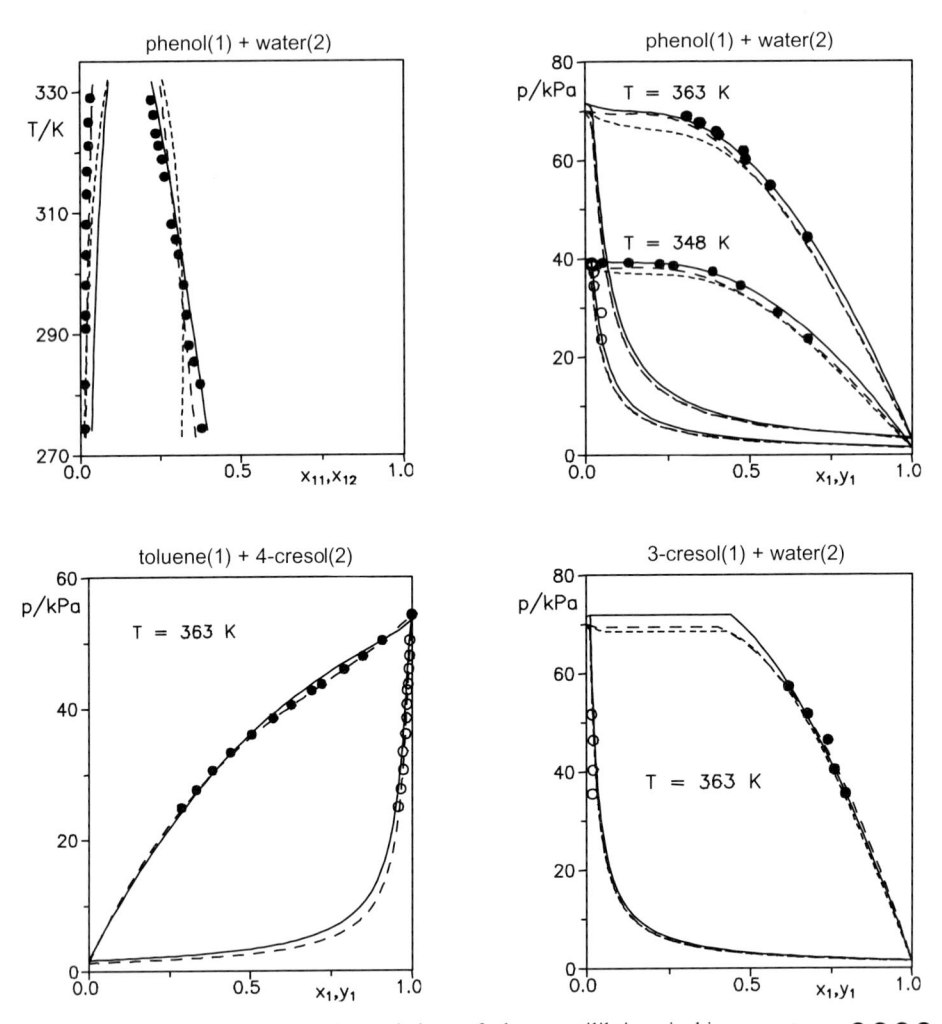

7 Phase Equilibria in Ternary Systems Containing Phenols, Hydrocarbons, and Water

Figure 7.5: Experimental results and correlations of phase equilibrium in binary systems. ●●●●, ○○○○ exp. data (phenol + water: Arlt et al., 1979; Gmehling et al., 1977; this work; other systems: this work), —— ESD-EOS, – – – UNIQUAC, - - - - NRTL.

water the NRTL equation wrongly predicts three liquid phases. The experimental three phase region of the octane + 2-cresol + water system is predicted qualitatively well by all models, but the NRTL equation gives a homogeneous region which is by far too small, and the UNIQUAC method predicts a miscibility gap in the binary octane + 2-cresol system at 298 K. For the ternary system octane + 3-cresol + water all models give a wrong influence of temperature for the two-phase and three-phase regions. The models give only two liquid phases (whereas there are three) at 298 K. At 323 K the UNIQUAC as well as the NRTL equation predicts three liquid phases. The ESD-EOS gives three liquid phases at 328 K.

7.5 Results and Discussions

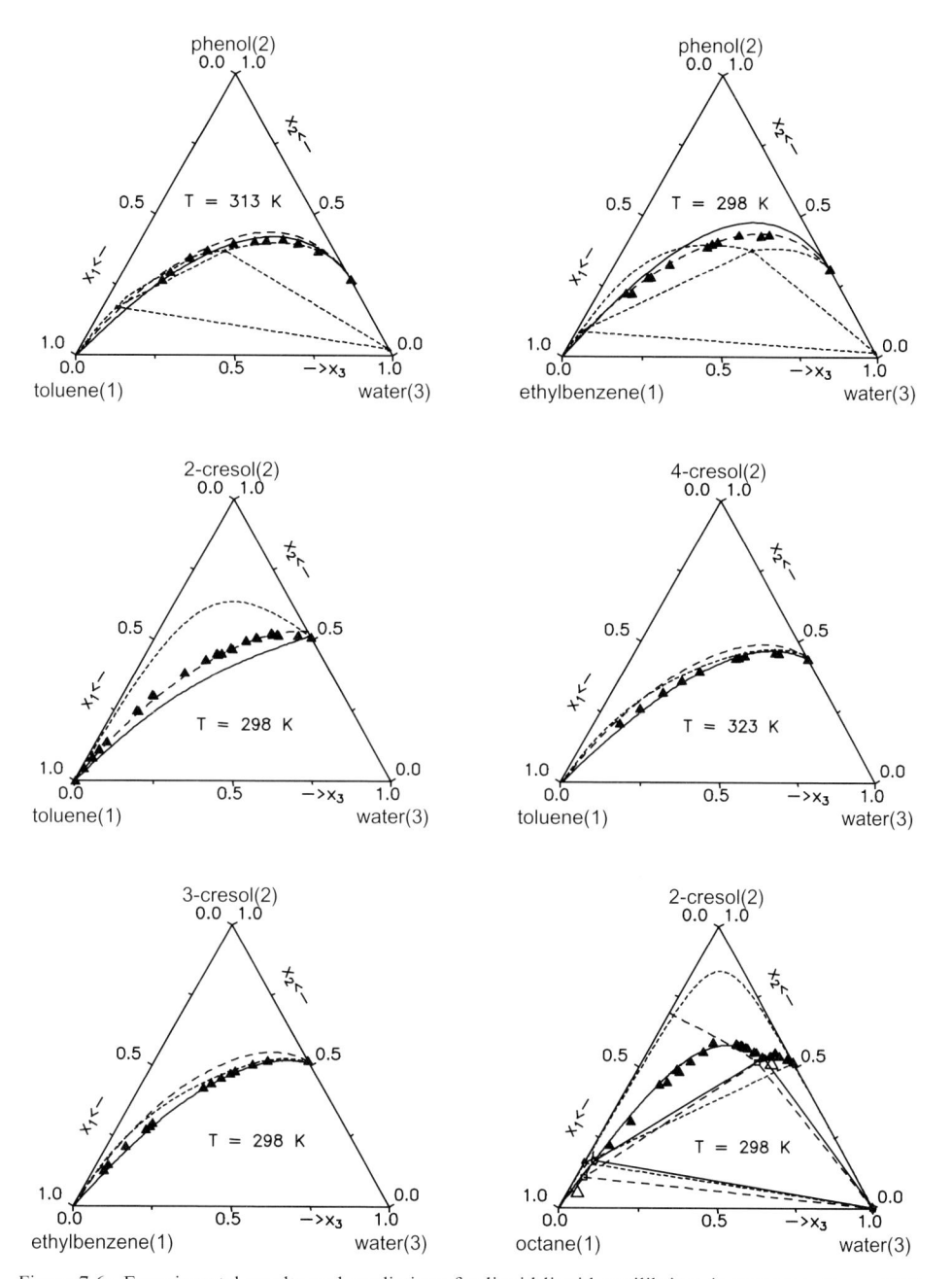

Figure 7.6: Experimental results and predictions for liquid-liquid equilibrium in ternary systems. LLE: ▲▲▲▲ exp. data (this work), —— ESD-EOS, – – – UNIQUAC, - - - - NRTL, LLLE: △△△ exp. data (this work), –○–○–○ ESD-EOS, -□-□-□ UNIQUAC, -◇-◇-◇ NRTL.

145

The deviations between experimental results and predictions for the vapor pressures and composition of the vapor phase are summarized in Table 7.4. The average deviations are defined as

$$\Delta p / \% = 100 \, \Sigma_i \, (|p_{calc} - p_{exp.}|/p_{exp.})_i / n \tag{7.5}$$

$$\Delta y / \text{mol}\% = 100 \, \Sigma_i \, [\Sigma_j |y_{calc.} - y_{exp.}|_j / 3]_i / n \tag{7.6}$$

For the total set of data ranges Δp is 4.8% (ESD-EOS), 5.2% (UNIQUAC), and 9.5% (NRTL) and Δy is 2.4 mol% (ESD-EOS), 2.9 mol% (UNIQUAC), and 3.8 mol% (NRTL). In three (of six investigated) ternary systems, the best agreement is achieved by the ESD-EOS. Δp and Δy are below 4% and 2 mol% for four toluene + phenol/cresol + water systems. The predictions with UNIQUAC are much worse for one of the ternary systems and the predictions with NRTL are much worse in two of the ternary systems (always compared with the two other methods). In general, the predictions for the octane + cresol + water systems are worse than those for the toluene + (phenol or cresol) + water systems. The deviations in the vapor pressure are illustrated in Figure 7.7 for two systems.

Table 7.4: Deviations between experimental results and predictions for VLE.

System	Number of datapoints p	y	ESD-EOS Δp %	Δy mol%	UNIQUAC Δp %	Δy mol%	NRTL Δp %	Δy mol%
toluene + phenol + water	22	22	3.8	1.4	3.2	1.1	5.1	1.9
toluene + 2-cresol + water	28	28	3.7	1.2	2.5	0.6	14.2	3.3
toluene + 3-cresol + water	22	12	2.2	2.4	8.7	3.9	3.5	3.0
toluene + 4-cresol + water	21	21	2.8	2.9	4.0	3.2	3.3	4.2
octane + 2-cresol + water	23	21	7.6	2.0	5.2	3.0	18.8	5.3
octane + 3-cresol + water	29	27	7.9	4.4	7.6	6.3	9.8	4.7
total	145	131	4.8	2.4	5.2	2.9	9.5	3.8

7.6 Conclusions

The investigated systems show complex phase equilibrium behavior. Thus, in ternary systems of cresol + alkane + water, three liquid phases are observed. The group-contribution method UNIFAC (used in several modifications) gives very poor predictions for the phase equilibria in such systems with limited mutual solubility. The NRTL and UNIQUAC equations are able to describe relatively well the binary LLE and VLE data simultaneously using linearly temperature-dependent interaction parameters, but in some cases the predictions for LLE (LLLE) and VLE are qualitatively incorrect. The ESD-EOS, a method using an equation of state in conjunction with chemical theory gives more reliable predictions of the com-

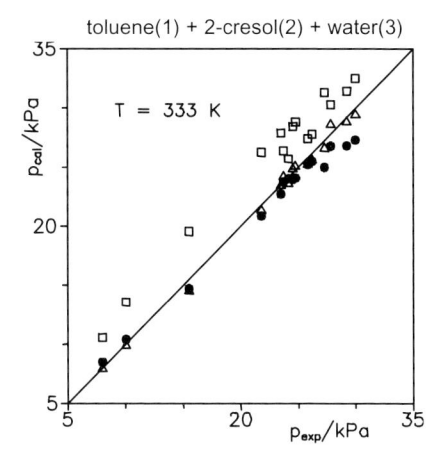

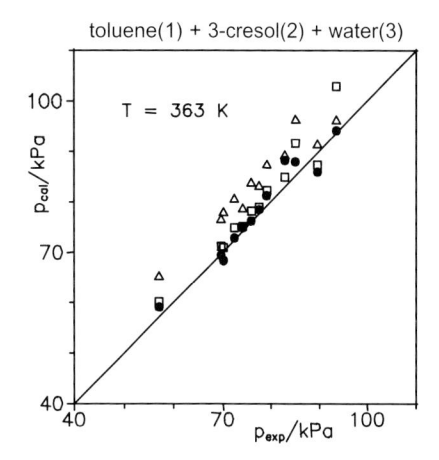

Figure 7.7: Comparison between experimental results (this work) and predictions for the vapor pressure of ternary systems. ●●●● ESD-EOS, △△△△ UNIQUAC, ☐☐☐☐ NRTL.

plex phase behavior in ternary systems. The infrared spectrometric studies in this multiple associating and strong solvating component system were not suitable to determine association parameters (equilibrium constant, enthalpy on association).

Acknowledgements

The support of undergraduate students Jörg Anderson, Antje Drews, Diana Martin, and Annett Precht is gratefully acknowledged.

References

Abrams, D. S.; Prausnitz, J. M. (1975): Statistical thermodynamics of liquid mixtures: a new expression for the excess Gibbs energy of partly or completely miscible systems. Am. Inst. Chem. Eng. J. 21, 116–128.

Arlt, W.; Macedo, M. E. A.; Rasmussen, P.; Sørensen, J. M. (1979 ff.): DECHEMA Chemistry Data Series, Vol. V. Liquid-liquid equilibrium data collection, DECHEMA, Frankfurt/M.

Aspiron, N. (1996): Anwendung der Spektroskopie in thermodynamischen Untersuchungen assoziierender Lösungen. Dissertation, Universität Kaiserslautern.

Becker, E. D. (1961): Infrared studies of hydrogen bonding in alcohol-base systems. Spectrochimica Acta 17, 436–447.

Elliott, J. R. Jr.; Suresh, S. J.; Donohue, M. D. (1990): A simple equation of state for nonspherical and associating molecules. Ind. Eng. Chem. Res. 29, 1476–1485.

Fredenslund, A.; Jones, R. L.; Prausnitz, J. M. (1975): Group-contribution estimation of activity coefficients in non-ideal liquid mixtures. Am. Inst. Chem. Eng. J. 21, 1086–1099.

Geisler, G.; Seidel, H. (1977): Die Wasserstoffbrückenbindung. Akademie-Verlag, Berlin.

Gmehling, J. et al. (1977 ff.): DECHEMA Chemistry Data Series, Vol. I. Vapor-liquid equilibrium data collection. DECHEMA, Frankfurt/M.

Gmehling, J.; Li, J.; Schiller, M. (1993): A modified UNIFAC model. 2. Present parameter matrix and results for different thermodynamic properties. Ind. Eng. Chem. Res. 32, 178–193.

Hansen, H. K.; Rasmussen, P.; Fredenslund, A.; Schiller, M.; Gmehling, J. (1991): Vapor-liquid equilibria by UNIFAC group-contribution. 5. Revision and Extension. Ind. Eng. Chem. Res. 30, 2352–2355.

Hobert, H.; Kempe, J. (1993): Charakterisierung von Braunkohlen durch chemometrische Hauptkomponentenanalyse der Infrarot-Spektren. Erdöl – Erdgas – Kohle 109, 426–430.

Magnussen, T.; Rasmussen, P.; Fredenslund, A. (1981): UNIFAC parameter table for prediction of liquid-liquid equilibria. Ind. Eng. Chem. Process Des. Dev. 20, 331–339.

Puhala, A. S.; Elliott, J. R. Jr. (1993): Correlation and prediction of binary vapor-liquid equilibrium in systems containing gases, hydrocarbons, alcohols and water. Ind. Eng. Chem. Res. 32, 3174–3179.

Quadri, S. M.; Shurvell, H. F. (1995): An infrared spectroscopic study of complex formation between meta-cresol and propionitrile. Spectrochimica Acta A 51, 1355–1365.

Renon, H.; Prausnitz, J. M. (1968): Local compositions in thermodynamic excess function for liquid mixtures. Am. Inst. Chem. Eng. J. 14, 135–144.

Renon, H.; Asselineau, L.; Cohen, G.; Raimbault, G. (1971): Calcul sur ordinateur des equilibres liquide-vapeur et liquide-liquide. Editions Technip, Paris.

Röck, H.; Sieg, L. (1955): Messungen von Verdampfungsgleichgewichten mit einer modernisierten Umlaufapparatur. Z. phys. Chem. (Frankfurt/M.) 3, 355–364.

Schmelzer, J.; Niederbröker, H.; Lerchner, J. (1998): Computergesteuerte Anlage zur dynamischen Messung von Flüssigkeit-Dampf-Gleichgewichtsdaten. Chem. Technik 50, 17–19.

Schmelzer, J.; Taubert, K.; Martin, A.; Meinhardt, R. (2003): ELDATA: The International Electronic Journal of Physico-Chemical Data (in preparation).

Suresh, S. J.; Elliott, J. R. Jr. (1991): Applications of a generalized equation of state for associating mixtures. Ind. Eng. Chem. Res. 30, 524–532.

Suresh, S. J.; Elliott, J. R. Jr. (1992): Multiphase equilibrium analysis via a generalized equation of state for associating mixtures. Ind. Eng. Chem. Res. 31, 2783–2794.

Nomenclature

c	shape factor for the repulsive term (ESD-EOS)
C_{ij}^C, C_{ij}^T	temperature coefficients of the interaction parameters
g_{ij}, g_{jj}	interaction parameter (NRTL)
k	molar distribution coefficient (ESD-EOS)
K_{AB}	measure of bonding volume (ESD-EOS)
k_{ij}	binary interaction parameter (EOS)
k_{ij}^C, k_{ij}^T	temperature coefficients of binary interaction parameter
n	number of data points
p	vapor pressure

Nomenclature

Δp	relative average deviation of pressure
Q	objective function
R	gas constant
T	temperature
T_C	critical temperature
u_{ij}, u_{jj}	interaction parameter (UNIQUAC)
v^*	characteristic size parameter (ESD-EOS)
x, x', x''	mole fraction in liquid phases
x_{ij}	mole fraction of component i in the liquid phase j
y	mole fraction in vapor phase
Δy	average deviation of mole fraction in vapor phases

Greek letters

α	non-randomness parameter (NRTL)
ε_{Disp}	dispersions term of potential energy well depth (ESD-EOS)
ε_{HB}	potential energy well depth of hydrogen bonding (ESD-EOS)
π	weighting coefficient in the objective function

8 Influence of Additives on Hydrophobic Association in Polynary Aqueous Mixtures. An NMR Relaxation and Self-Diffusion Study

Manfred Holz[1]* and Manghaiko Mayele[2]

8.1 Abstract

Aqueous mixtures were studied by application of NMR as a spectroscopic method. In particular a special kind of association was of interest, namely "hydrophobic association" which can play an important role when solutes with nonpolar molecular groups are dissolved in water. The association tendency of the organic compound was monitored using the NMR *A* parameter, obtained via the measurement of *inter*molecular ^{1}H relaxation rates and of self-diffusion coefficients, allowing the detection of changes in local concentration ratios. There were two main aims of this work. First we wanted to observe the influence of urea on hydrophobic association, where *tert*-butanol was the organic component. Second we wanted to demonstrate that our NMR method is suited for laboratory studies aiming to the later control of technically important reactions and as example the effect of different additives on the self-association of ethylene glycol in aqueous mixtures was detected. In addition we also studied the influence of urea on the translational dynamics of *tert*-butanol and water. We found that urea does not act as a "structure breaker" on water and with respect to the association it can have two opposing effects. Low concentrations of urea stabilize the association of *tert*-butanol, but high concentrations generate de-association. The weak self-association of ethylene glycol can be increased by the addition of certain salts, but not by different organic components. Self-diffusion coefficients, densities, and *A* parameters (from which ratios of local concentrations can be derived) are listed in a number of tables.

8.2 Introduction

Thermodynamic properties of non-ideal multi-component mixtures are determined by intermolecular interactions between different species and/or between species of the same kind in

---

1 Universität Karlsruhe (TH), Institut für Physikalische Chemie, Kaiserstr. 12, 76128 Karlsruhe.
2 NRC-Biotechnology Research Institute, Montreal, QC Canada H4P 2R2.
* Author to whom correspondence should be addressed.

8.2 Introduction

the mixture. Accordingly, for a better understanding and thus a more realistic modeling of those complex mixed systems, any specific information about the behavior of single components, as e.g. their local enrichment in the molecular vicinity of distinct particles due to attractive interactions, is of extraordinary value. If for example spectroscopic methods can be developed which deliver on a microscopic level such information for selected components, they could represent an ideal supplementation of the "macroscopic" methods which are usually applied in the field of applied thermodynamics.

Among the possible spectroscopic techniques, nuclear magnetic resonance (NMR) offers unique and very tempting advantages for the investigation of multi-component mixtures, namely (i) its selectivity, allowing the non-ambiguous assignment of observed signals to distinct species, and (ii) the great variety of NMR methods and parameters allows often a targeted adaptation of the experiment to the problem of interest, be it a structural or a dynamical problem. The "chemical shift", as the most familiar NMR parameter, is in general only sensitive to *intra*molecular properties, but ^{1}H-shifts of OH-, SH-, and NH- hydrogen are also sensitive to *inter*molecular hydrogen bonding and can thus be successfully used for the study of molecular association originating from H-bonds (see e.g. Tkadlecova et al.,1999; Luo et al., 2001). It is surprising that despite the unique power of this method for the investigation of H-bonded systems, as demonstrated in the physical chemical literature, it has not yet found an extended use in the area of chemical engineering. Of course, if in liquid mixtures components play a role, which can *not* form H-bonds, as e.g. nonpolar components, the chemical shift is a less suited tool for the study of intermolecular interactions. However, there is another, commonly not so well-known NMR-parameter which reflects sensitively all kinds of intermolecular interactions and this is the *inter*molecular magnetic dipole-dipole (D-D) relaxation rate $(1/T_1)_{inter}$, which occurs owing to interactions between strong nuclear dipole moments, e.g. of ^{1}H or ^{19}F (McConnell, 1987) or between an electron magnetic dipole moment and a nuclear magnetic dipole. In our laboratory, many years ago the so-called association-parameter (A-parameter) procedure has been developed (see e.g. Hertz et al., 1976; Holz, 1995), which makes use of this intermolecular D-D relaxation and which gives information about *relative* changes of local concentrations of distinct mixture components in the vicinity of the same and/or other components. The power of this method has been previously demonstrated by application to mainly binary liquid mixtures (Koch et al., 1983; Cebe et al., 1984) and electrolyte solutions (Hertz and Müller, 1989; Detscher et al., 1995; Holz, 1995).

In the present work the A-parameter concept was applied to more complex liquid mixtures than in previous studies, namely to ternary aqueous mixtures. The background for the choice of aqueous mixtures is the fact that those mixtures play an outstanding role in many fields of science and technology, as for example in life sciences, earth sciences, pure and applied chemistry, in many basic chemical industrial processes and in biotechnology. Owing to some unique physical properties of liquid water, aqueous liquid mixtures show peculiarities, which obviously make the modeling of aqueous mixtures an exceedingly difficult task. Very important among these peculiarities are the so-called "hydrophobic phenomena", namely "hydrophobic hydration" and "hydrophobic association" which have wide implications and which occur when nonpolar solutes or solutes with nonpolar molecular groups are dissolved in water (see e.g. Ben-Naim, 1980; Blokzijl and Engberts, 1993). Since the occurrence of hydrophobic association is restricted to aqueous systems, it has obviously not been taken into explicit consideration when general association models for technical applications were developed. On the other hand, hydrophobic hydration strongly influences dynamic quantities in aqueous mixtures, e.g. typical viscosity maxima and diffusion minima occur in the water rich region in aqueous mixtures of small organic compounds as low alcohols. Hydrophobic association plays a dominating role in molecular aggregation phenomena, e.g. in micelle or membrane formation and also often in the conformation of large biomolecules (see e.g. Blokzijl and Engberts, 1993). Moreover, since hydrophobic association is an entropy driven phenomenon (Ben-Naim, 1980), it shows a temperature dependence, which is opposite to "normal" association and accordingly it might cause a strange thermodynamic behavior of those aqueous mixtures with components having nonpolar groups. Consequently, a well-founded model of those aqueous solutions must regard hydrophobic association, at least if there are spectroscopic hints that this special kind of association is present in a mixture of interest. Therefore in this connection it is a first task for a complementary spectroscopic method to make those important hints available.

In our laboratory, there is a long standing general interest in hydrophobic effects (Hertz and Zeidler, 1964; Hertz and Leiter, 1982; Haselmeier et al., 1995) and thus we knew that also in aqueous mixtures with comparatively simple components hydrophobic effects must be taken into consideration and hence we thought it is worthwhile to investigate hydrophobic association also in connection with fluid mixtures of technical interest. In particular, we wanted to study the influence of certain additives on the hydrophobic *self*-association of organic components with apolar groups. There were two reasons for that, namely first we wanted to contribute to the microscopic understanding of the change of hydrophobic association or de-association by an additive and second we wanted to investigate if our *A*-parameter method might be suited for laboratory studies aiming to the later control of technically important reactions. For the first aim we studied the ternary mixture: water (component 1) + *tert*-butanol (t-BuOH, component 2) + urea (component 3), where the additive urea is an important denaturating agent in biochemistry and biotechnology. For the second aim we investigated the mixture water (1) + ethylene glycol (2) + additive (3) where the system water + ethylene glycol (EG) plays a role as a starting mixture used in industrial reaction processes. This mixture is, as previous experimental results show, an almost ideal mixture and the question was if it is possible to generate a controlled change of this ideality by applying suited additives which enhance the local EG concentration in the surrounding of EG.

8.3 Methods and Experimental

In this section we will briefly describe the theoretical background of the *A*-parameter procedure and we will discuss the physical meaning of the resulting quantity A_{ij}. Then the NMR techniques for the measurement of the required experimental quantities and the experimental set-up will be depicted. (For more detailed considerations we will refer to original works.)

8.3.1 Theoretical Background of the A-Parameter Procedure

Nuclear magnetic relaxation is the process of approach of a nuclear spin system to its thermal equilibrium state after a disturbance e.g. by a 90°- or 180°-radio-frequency pulse. The spin-lattice relaxation by dipole-dipole interaction of a given spin-½ nucleus is caused by the strength and fluctuation of local small magnetic fields caused by the nuclear magnetic dipole of neighboring spins residing on the same molecule (*intra*molecular D-D interaction) or on another molecule in the close vicinity (*inter*molecular D-D interaction). The fluctuation of these small magnetic fields originates in the intermolecular case from the random thermal motion of molecules relative to each other, which can be expressed by a mean self-diffusion coefficient D_{av}. The distance dependence of the strength of magnetic dipole fields is well known from classical physics and thus the intermolecular relaxation contains information about intermolecular distances and can be used for probing those distances. The intermolecular spin-lattice relaxation rate $(1/T_1)_{inter}$ for ^{1}H-^{1}H interaction is given (in the so-called "extreme narrowing limit") by (Müller and Hertz, 1996):

$$\left(\frac{1}{T_1}\right)_{inter} = \frac{1}{2} \frac{\gamma^4 \hbar^2 c'}{a^4 D_{av}} \int_a^\infty \left(\frac{a}{r}\right)^6 g_{ij}(r) 4\pi r^2 \, dr \tag{8.1}$$

where γ is the gyromagnetic ratio of the proton, $\hbar$ has its usual meaning, c' is the nuclear spin number density given in spins/cm^3, a is the closest distance of approach between the interacting nuclei, D_{av} the mean self-diffusion coefficient is given by $D_{av} = \frac{1}{2}(D_i + D_j)$ and $g_{ij}(r)$ is the atom-atom pair distribution function of the atoms carrying the interacting nuclei and residing on molecules i and j, where r is their separation distance. If we divide $(1/T_1)_{inter}$ by c' and by $(1/D_{av})$ we normalize the intermolecular relaxation rate with respect to the spin concentration and to the translational dynamics and we end up with a quantity, namely A_{ij}, which contains only constants and an integral over a pair correlation function multiplied with a "cut-off function" $(a/r)^6$:

$$A_{ij} = \left(\frac{1}{T_1}\right)_{inter} \frac{D_{av}}{c'} = \frac{1}{2} \frac{\gamma^4 \hbar^2}{a^4} \int_a^\infty \left(\frac{a}{r}\right)^6 g_{ij}(r) 4\pi r^2 \, dr \tag{8.2}$$

Thus the A-parameter is a measure for the affinity between molecules i and j. The "cut-off function" has a physical background, namely the distance dependence of the D-D interaction, resulting in an increased sensitivity of the A-parameter for short distances. As shown elsewhere (Müller and Hertz, 1996) A_{ij} is directly proportional to the "reduced first co-ordination number" $n_c^* = n_c / c'$, where n_c is the first co-ordination number, and consequently A_{ij} is proportional to the ratio of the local concentration of component j in the vicinity of component i to the bulk concentration of component j, c_{local}/c_{bulk}. It should be emphasized here that A_{ij} is related to a quantity which has been successfully used for the investigation of liquid mixtures, namely the Kirkwood-Buff integral (KBI) G_{ij}, which also contains an integral over a pair correlation function, however without a cut-off function (Matteoli and Mansoori, 1990; Holz, 1995; Sacco et al., 1996).

In the present work we were mostly dealing with self-association of the organic compound (component denoted 2) and thus the quantity of main interest is A_{22}. In this case D_{av} is

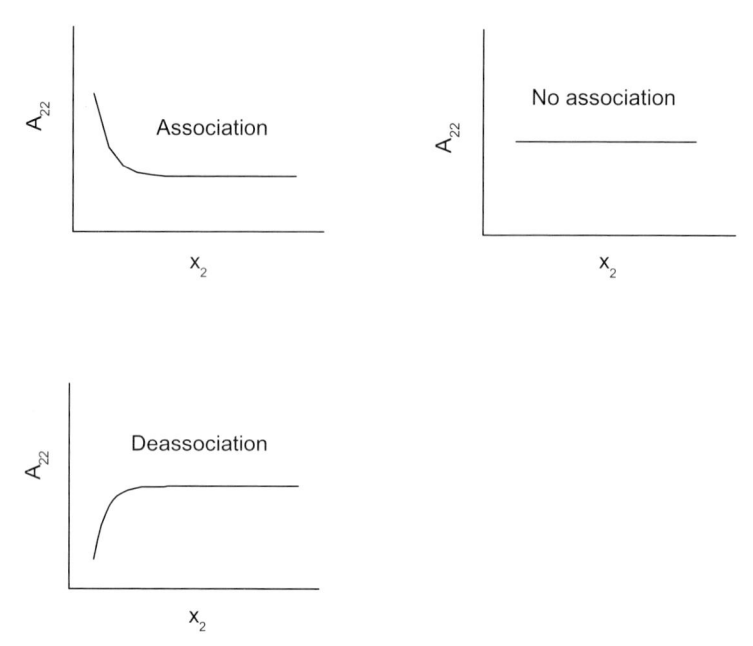

Figure 8.1: Behavior of the A_{22} parameter for three different cases. x_2 is the mole fraction of component denoted 2 in the aqueous mixture.

simply the self-diffusion coefficient of component 2 (D_2). The evaluation of the *absolute* values of A_{22} is difficult since the absolute value of a is not exactly known and since we obtain only a weighted integral value of an atom-atom pair correlation function. Therefore in the A-parameter procedure so far always the relative change of A_{ij}, upon a change of an experimental parameter as temperature or most important upon a change of concentration, is considered and evaluated.

If we take for example a given H-atom as reference and if under the process of dilution the H-atoms on neighbor molecules are attracted towards the reference atom, then $g_{22}(r)$ gets sharper, it contracts in the region of small r-values. As a consequence the reduced first co-ordination number and thus A_{22} increases as the component 2 is diluted. This effect reflects the association tendency, in the present case a self-association tendency. If on the other hand, during dilution the distance of nearest neighbor atoms to the reference atom increases, then A_{22} decreases; this situation we call de-association. For an ideal mixture $g_{22}(r)$ is independent of the concentration and so A_{22} remains constant over the entire composition range (Fig. 8.1).

8.3.2 Experimental Determination of the A-Parameter

8.3.2.1 Measurement of the Intermolecular Relaxation Rate

The determination of A_{22} requires the separation of the intermolecular contribution $(1/T_1)_{\text{inter}}$ from the total ^{1}H relaxation rate, given by

8.3 Methods and Experimental

$$\left(\frac{1}{T_1}\right)_{tot} = \left(\frac{1}{T_1}\right)_{intra} + \left(\frac{1}{T_1}\right)_{inter} \quad (8.3)$$

where $(1/T_1)_{tot}$, the total spin-lattice relaxation rate, is the first quantity directly measured in the experiment. For the experimental determination of the spin-lattice relaxation rate a $180° - \tau - 90°$ standard rf-pulse sequence (see e.g. Farrar and Becker, 1979) was used, where τ is the evolution time of the spin system, the time between the two radio-frequency (rf) pulses.

The needed intermolecular relaxation rate $(1/T_1)_{inter}$ for the determination of A_{22} is separated from $(1/T_1)_{tot}$ by performing isotopic dilution experiments (see e.g. Zeidler, 1965). In these experiments for a given mixture composition, the normal (protonated) form of the component of interest is replaced gradually by the full deuterated form and the ^1H relaxation rates of the remaining rest of the protonated form are every time measured. The composition of the other components of the mixture is thereby kept constant. The concentration of the component whose relaxation rate is measured is the sum of the concentrations of protonated and deuterated form, where the latter does not contribute to the ^1H-^1H D-D relaxation. By plotting the measured relaxation rates as function of the percentage of the protonated form and extrapolating to percentage zero one obtains the intramolecular relaxation rate $(1/T_1)_{intra}$ as axis intercept. The knowledge of $(1/T_1)_{intra}$ and of the total relaxation rate of the 100% protonated sample allows the determination of the quantity of interest, the intermolecular relaxation rate $(1/T_1)_{inter}$, via Eq. (8.3).

8.3.2.2 PFG NMR for the Determination of Self-Diffusion Coefficients

In this work the "Pulsed Field Gradient Spin-Echo" (PFGSE) method (Stejskal and Tanner, 1965; Price, 2000) was used for the determination of the self-diffusion coefficients. With the PFGSE technique the magnetic field gradient is not constantly applied, but switched on only twice for a short time. For the echo amplitude $M(2\tau)$ at the time 2τ the following relationship results:

$$M(2\tau) = M_0 \cdot \exp(-2\tau/T_2) \cdot \exp\left(-D\gamma^2\delta^2G^2(\Delta - \delta/3)\right) \quad (8.4)$$

M_0 is the signal amplitude at $\tau = 0$, G the amplitude of the gradient pulses, δ their length, and Δ the time interval between the two gradient pulses. D is the self-diffusion coefficient of the particle which carries the observed nuclear spin. With constant Δ we obtained the self-diffusion coefficient by variation of δ at constant G. The self-diffusion coefficient of the interesting species was measured thereby relative to a suitable reference sample with known self-diffusion coefficient (Mayele et al., 1999; Mayele and Holz, 2000; Holz et al., 2000).

8.3.2.3 Measurement of the Spin Density

The nuclear spin number density c_i' of the methyl protons was calculated from the density of the mixture using the following relation in which n_p represents the number of hydrogen atoms per molecule:

$$c'_i = \frac{x_i}{\sum_i x_i M_i} \cdot \rho \cdot n_p \cdot N_A \tag{8.5}$$

N_A is the Avogadro number and ρ the mixture density. M_i and x_i, respectively, are the molar mass and the mole fraction of the component i in the mixture.

8.3.3 Experimental

8.3.3.1 Apparatus

Measurements of spin-lattice relaxation rates and self-diffusion coefficients were mainly performed at 90 MHz ^{1}H resonance frequency using a Bruker SXP 4–100 spectrometer system, combined with an ASPECT 2000 computer, and coupled with a laboratory-made pulsed field gradient unit. The ASPECT 2000 computer permits, apart from data acquisition and pulse sequence control, also signal accumulation and mathematical data manipulations to improve signal-to-noise ratio. Further, it performs the Fourier transformation (FT) of the signal acquired in the time domain into the frequency domain signal. When better resolution and sensitivity were required, we used a Bruker AMX 300 spectrometer system operating at 300 MHz ^{1}H resonance frequency.

The temperature control in the sample head was carried out with a commercial liquid thermostat. The temperature was adjusted before the measurement using an external thermocouple with digital display and checked regularly. The temperature instability was less than 0.5 K. The relative experimental error limit was less than 2% for T_1 and D measurements.

The densities of the mixtures were measured with an Anton Paar digital densimeter whose temperature fluctuation was less than 0.005 K. The calibration was done before each series of measurements via determination of the oscillation period of pure water and air. The density of air was calculated from the ambient air pressure (Weast, 1979), for the density of pure water we used literature values (Kell, 1975). The absolute experimental error for density measurements was of the order 10^{-5} g cm^{-3}.

8.3.3.2 Sample Preparation

The sample preparation was made by weighing. As concentration scale we used the mole fraction of the organic component in aqueous mixtures and for salts the aqua-molality scale $\bar{m}$ i.e. the number of moles of salt in 55.5 moles of solvent. For consistency in the comparison to previously measured systems we expressed the concentration of the organic component as $x'_2 = x_2/(x_1 + x_2)$ where x_1 and x_2 are the mole fractions of D$_2$O and organic substance, respectively. The weighed mixtures were filled in Duran sample tubes of 10 mm outside diameter. In order to remove paramagnetic oxygen from the samples the freeze-pump-thaw cycle was applied at least five times for each sample. For this the liquids were frozen and the sample tubes degassed up to a residual pressure of approximately 10^{-6} bar.

8.3.4 Materials

The chemicals used were: H_2O (doubly distilled), D_2O 99.9 atom% D, *tert*-butanol-d$_1$ 99 atom% D, *tert*-butanol-d$_{10}$ 99 atom% D, urea-d$_4$ 98+ atom% D, ethylene glycol-d$_2$ 98 atom% D, ethylene glycol-d$_6$ 98 atom% D, *n*-butanol-d$_{10}$ 98 atom% D, dioxan-d$_8$ 99 atom% D, CsCl suprapur, and Na$_2$SO$_4$ suprapur. Except H_2O (from our laboratory), urea (from Aldrich), and salts (from Merck), all other chemicals were purchased from Cambridge Isotope Laboratories (CIL) and used as supplied. Urea and salts were dried under vacuum before use at 110 and 140 °C, respectively.

8.4 Results and Discussion

8.4.1 The Binary System: Water + *tert*-Butanol

In preceding works (see e.g. Sacco et al., 1996, 1997, 1998) on the association behavior of small alcohols in water it was found that the hydrophobic association tendency is more pronounced in water rich mixtures and increases with increasing temperature as theoretically expected for an entropy driven interaction (Ben-Naim, 1977, Sacco et al., 1997). In the present work we extend the study to the interesting case of *tert*-butanol. This alcohol, with its three CH_3 groups, should show the most pronounced hydrophobic effects among the alcohols studied so far by the A_{22} parameter method.

Measurements were performed at 10, 25, and 40 °C. Experimental results used for the determination of the A_{22} parameter are given in Table 8.1 along with the resulting A_{22} para-

Table 8.1: Experimental results for the D_2O (1) + t-BuOD-d$_1$ (2) system at 10, 25, and 40 °C: mixture density ρ, diffusion coefficient D of t-BuOD-d$_1$, and association parameter A_{22}.

x_2/ (mol mol^{-1})	10 °C ρ/ (g cm^{-3})	$10^9 D$/ (m^2 s^{-1})	$10^{39} A_{22}$/ (m^5 s^{-2})	25 °C ρ/ (g cm^{-3})	$10^9 D$/ (m^2 s^{-1})	$10^{39} A_{22}$/ (m^5 s^{-2})	40 °C ρ/ (g cm^{-3})	$10^9 D$/ (m^2 s^{-1})	$10^{39} A_{22}$/ (m^5 s^{-2})
0.010	1.09906	0.36	1.49	1.09370	0.56	2.14	1.08450	1.03	3.55
0.013	1.09630	0.34	1.93	1.09232	0.54	2.49	1.08334	0.97	3.42
0.017	1.09339	0.31	2.20	1.08850	0.51	2.67	1.07918	0.91	3.40
0.020	1.09048	0.29	2.28	1.08452	0.49	2.72	1.07509	0.86	3.30
0.030	1.08390	0.24	2.25	1.07628	0.43	2.62	1.06660	0.75	2.91
0.040	1.07727	0.20	1.98	1.06870	0.39	2.49	1.05733	0.67	2.74
0.050	1.07162	0.18	1.86	1.06055	0.35	2.33	1.04873	0.62	2.48
0.070	1.05782	0.14	1.53	1.04400	0.30	2.06	1.03020	0.54	2.20
0.100	1.03414	0.12	1.36	1.01902	0.25	1.81	1.00527	0.49	1.89
0.199	0.97247	0.10	1.19	0.95349	0.22	1.59	0.93908	0.43	1.68
0.300	0.92778	0.10	1.17	0.91487	0.23	1.56	0.89388	0.43	1.66
0.499	0.87821	0.10	1.03	0.85576	0.24	1.50	0.84090	0.43	1.63

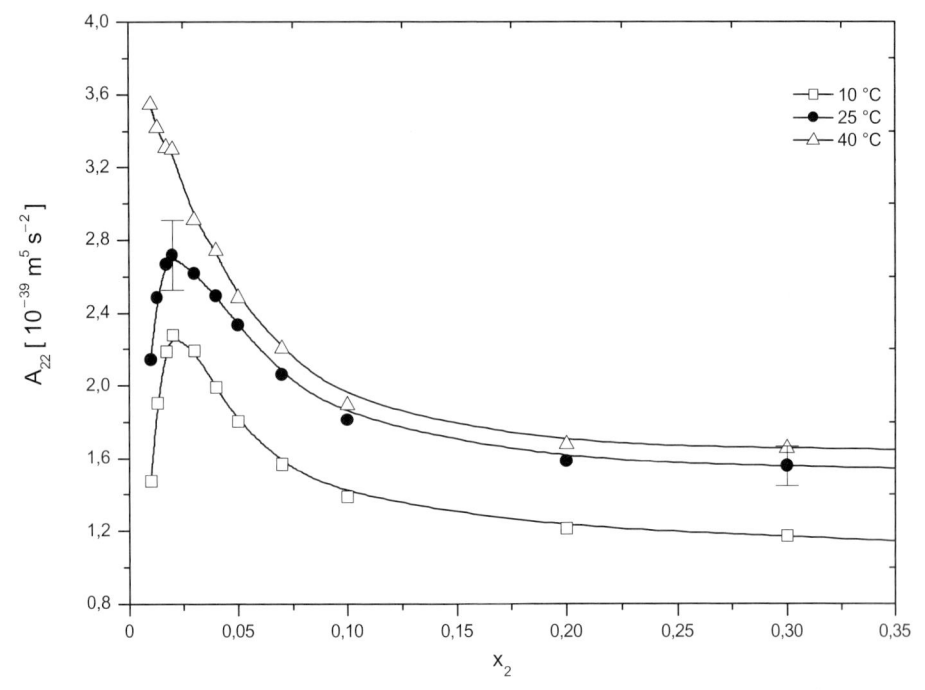

Figure 8.2: A_{22} parameter with respect to the association of *tert*-butanol-OD in aqueous mixtures at 10, 25, and 40 °C as a function of the mole fraction of the alcohol (x_2) in the region of low alcohol concentrations.

meters. The composition and temperature dependence of the A_{22} parameter in the interesting range of low alcohol concentrations is shown in Figure 8.2.

We first consider the curve at 10 °C. A_{22} reaches a maximum value at a t-BuOD concentration of about 2 to 3 mol%. This means that the local concentration of t-BuOD in the vicinity of t-BuOD at this concentration is about a factor of 2 higher than in pure t-butanol (taking the A_{22} value in pure t-butanol as equal to A_{22} at $x_2 = 0.5$, which is a good approximation, as experience shows). Further *dilution* of t-BuOD leads to a decrease of A_{22} which means that the hydrophobically associated t-BuOD molecules tend to dissociate again. In this composition range there are obviously enough water molecules available to build up a "hydrophobic hydration sphere" for the individual t-BuOD molecules. This result means that at 10 °C the mole fraction $x_2 = 0.02$ to 0.03 represents a transition composition at which the t-BuOD aqueous mixtures pass from solutions characterized by hydrophobic association (on the high concentration side) to solutions in which the hydrophobically hydrated single t-BuOD molecules dominate (on the low concentration side).

Increasing the temperature results in an increase in hydrophobic association as expected for an entropy controlled process from theory and from previous results (Lüdemann et al., 1996; Sacco et al., 1997). This can be seen by examining the $A_{22}(x_2)$ curves at higher temperatures in Figure 8.2. Further we recognize that the change from 10 to 25 °C shifts the maximum position of A_{22} to a slightly lower t-BuOD concentration and at 40 °C the maximum has disappeared. This observed shift of the maximum in the $A_{22}(x_2)$ function with increasing temperature shows that as the temperature is raised, the hydrophobic self-association becomes more and more favored and occurs already at lower t-BuOD concentrations. It is however most likely that the disappearance of the maximum at 40 °C is only apparent, that means, the inflection point in the $A_{22}(x_2)$ function is shifted to those small t-BuOD concentrations where accurate measurements can no more be obtained with the current experimental set-up.

The occurrence of structurally different regions in the composition range of aqueous solutions of some small organic compounds, with transition concentrations shifting to higher dilution of the organic component as temperature increases, has also been recently reported from studies based on the third derivatives of the free enthalpy $G(p,T,n_i)$ (Koga, 1996) and from the investigation of the magnetic susceptibility of aqueous *tert*-butanol (Mizuno et al., 1999). In addition, critical clathrate concentrations for t-BuOH found at concentrations of about 2 mol% from compressibility and IR studies (D'Angelo et al., 1994) suggest the existence of a region below this concentration threshold where t-BuOH molecules are essentially hydrated as single molecules.

As a further support for the conclusion that our results reflect a general phenomenon, we mention some further previous results: A maximum in the concentration dependence of the association tendency of small alcohols in water has been reported from KBI studies by Matteoli and Lepori (1984) on aqueous t-BuOH and by Ben-Naim (1977) on aqueous ethanol solutions. Furthermore, by comparing the pressure dependence of the intra-diffusion coefficients of pure water and of water in 2.5 mol% water-alcohol solutions, a maximum has been found in the $D(p)$ function (Harris and Newitt, 1998). This maximum was in case of alcohols with large apolar groups such as t-BuOH higher than in the case of relatively small alcohols and shifted to higher pressures than in pure water and the effect has been ascribed to hydrophobic interaction. Thus, the changeover from hydrophobic self-association regime to hydrophobic hydration regime is obviously a general feature of aqueous solutions of organic compounds with apolar groups in the water rich region and should be taken into account in modeling procedures.

8.4.2 The Ternary System: Water + *tert*-Butanol + Urea

On the basis of the accurate determination of the A_{22} parameter for the binary mixture of t-BuOD and water, presented in the preceding section, we could extend our studies, under the same conditions as we had for the binary system, to ternary solutions by adding urea.

The addition of a third component as additive is a frequently applied method for the modification of the aggregation behavior of particles in liquid media. However, in this way polynary liquid mixtures arise and hence the understanding of processes on a molecular level becomes very difficult. This difficulty even increases with increasing complexity of the molecules involved in the system as e.g. in biochemical systems. For example the influence of urea as denaturant of protein conformations in aqueous solutions, an effect of high scientific and technical interest (Myers, 1992), is not yet satisfactorily understood (Vanzi et al., 1998). Some of the numerous attempts made to elucidate the denaturing molecular mechanism of urea and the controversies they raised have recently been briefly surveyed elsewhere (Mayele and Holz, 2000). In the present work we started from the point that basic investigations of the urea effect on the hydrophobic association of small molecules, such as alcohols, can help in the understanding of the denaturation phenomenon. Previous NMR studies on aqueous solutions of propanol (Sacco et al., 1996) and ethanol (Sacco et al., 1997) revealed that the presence of relatively small quantities of urea enhanced the hydrophobic self-association, a fact which was unexpected within the generally accepted biochemical and biophysical point of view. However, these results corroborated previous findings from KBI investigations of the effect of urea on the self-association of propanol in water (Matteoli and Lepori, 1984, 1990). In thermodynamic studies on the effect of various solutes on the strength of hydrophobic interaction in aqueous solutions of methane and ethane it was concluded that urea concentrations up to 7 m enhance the hydrophobic interaction (Ben-Naim and Yaacobi, 1974). Owing to the greater hydrophobicity of t-BuOH compared with lower alcohols and alkanes mentioned above, one should expect more pronounced effects in its aqueous solutions. Furthermore, for the investigation of a possible transition from stabilization to destabilization we varied systematically the urea concentrations (see Section 8.4.2.2) from low values up to those values reported in the literature as typical at which protein denaturation occurs (Nandi and Robinson, 1984; Schiffer et al., 1995; Dötsch et al., 1995).

Two sets of measurements were performed: i) we determined the A_{22} parameter in the whole t-BuOD concentration range at all three temperatures, keeping the urea concentration constant at mole fractions $x_3 = 0.05$ and 0.12, respectively, and ii) at some chosen fixed concentrations of t-BuOD ($x_2' = 0.02$, 0.04, and 0.2) we varied the urea concentration from $x_3 = 0$ to 0.15, but at 25 °C only.

We point out here that we have a further source of information in the scope of the denaturing mechanism since we could also examine the self-diffusion results for *tert*-butanol, obtained in the *A* parameter procedure, along with results from *additional* water self-diffusion measurements.

8.4.2.1 Urea Effect on the Translational Dynamics of *tert*-Butanol and Water

Here, as in previous NMR studies (Endom et al., 1967; Hertz, 1973, 1986), the strength of increase or decrease of the self-diffusion coefficient of the solvent upon the addition of small amounts of a solute is taken as a measure of a "structure-breaking" or "structure-making" effect of the solute, respectively. (However, we should keep in mind that in this way the behavior of a dynamic quantity serves as a measure for structural changes!). In Figures 8.3a and 8.3b the composition dependence of the self-diffusion coefficient D of *tert*-butanol at different urea concentrations and temperatures is given.

Experimental data are listed in Tables 8.2 and 8.3.

The self-diffusion coefficients were measured at 10, 25, and 40 °C when 5 and 12 mol% urea were added to the aqueous *tert*-butanol solutions. Detailed discussion of these results can be found in our recent publication (Mayele and Holz, 2000) where it is shown that urea increases (acting as "structure-breaker") or decreases (acting as "structure-maker") the translational mobility of the *component tert*-butanol depending on the composition and/or the temperature range considered. Thus identifying urea in an aqueous system globally as a structure-breaker or structure-maker without those considerations appears in the light of the present work to be inappropriate.

Moreover, it should be emphasized that we observed here solely the molecular mobility behavior of *one* component at relatively low concentration in a ternary system and this

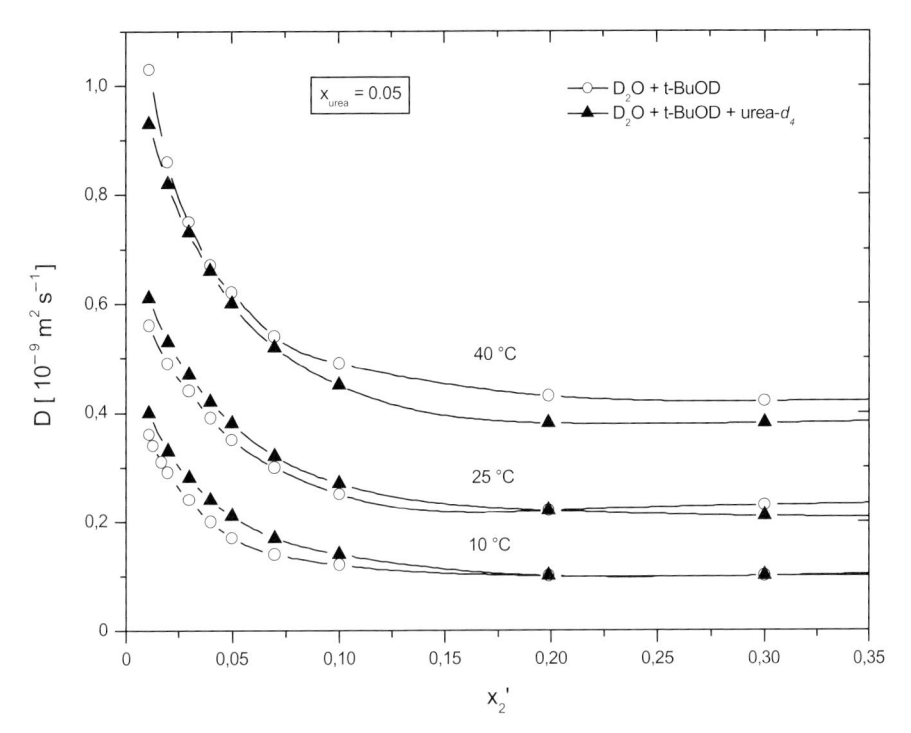

Figure 8.3a: Comparison of the self-diffusion coefficients, D, of *tert*-butanol-OD in D_2O + t-BuOD + urea-d_4 ($x_3 = 0.05$) and D_2O + t-BuOD mixtures at different temperatures as function of $x_2' = x_2 / (x_1 + x_2)$.

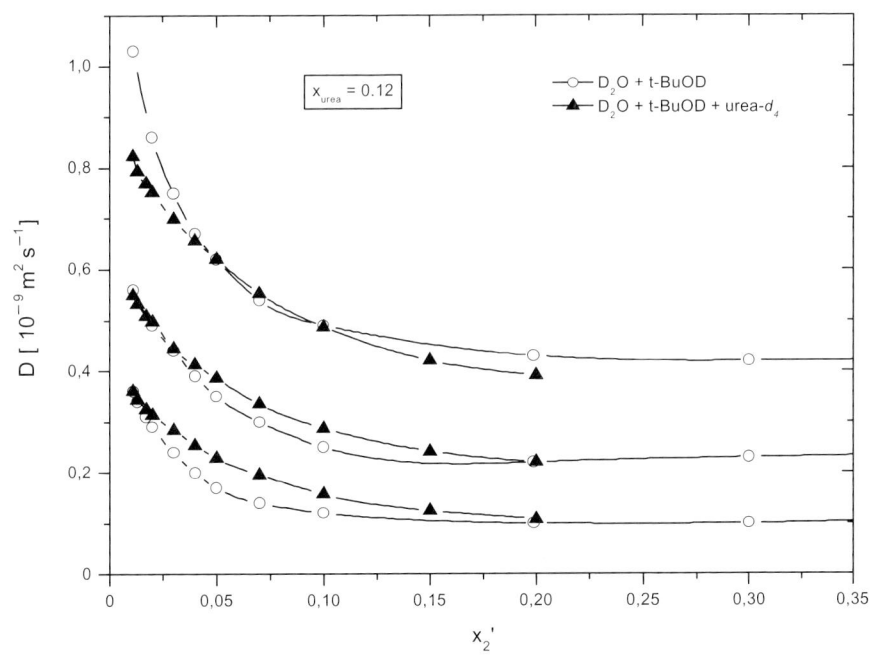

Figure 8.3b: As Fig. 8.3a, but with $x_3 = 0.12$.

Table 8.2: Experimental results for the ternary mixtures D$_2$O + t-BuOD-d$_1$ + urea-d$_4$ (x_3 = 0.05) at 10, 25, and 40 °C: mixture density ρ, diffusion coefficient D of t-BuOD-d$_1$, and association parameter A_{22}. $x'_2 = x_2/(x_1 + x_2)$.

	10 °C			25 °C			40 °C		
x'_2	ρ/ (g cm^{-3})	$10^9 D$/ (m^2 s^{-1})	$10^{39} A_{22}$/ (m^5 s^{-2})	ρ/ (g cm^{-3})	$10^9 D$/ (m^2 s^{-1})	$10^{39} A_{22}$/ (m^5 s^{-2})	ρ/ (g cm^{-3})	$10^9 D$/ (m^2 s^{-1})	$10^{39} A_{22}$/ (m^5 s^{-2})
0.010	1.14031	0.40	1.70	1.13107	0.61	4.22	1.12047	0.93	4.48
0.013	1.13967	0.37	2.56	–	–	–	–	–	–
0.017	1.13591	0.34	2.94	–	–	–	–	–	–
0.020	1.13225	0.33	3.04	1.12125	0.53	3.60	1.11085	0.82	3.94
0.030	1.12291	0.28	2.76	1.11194	0.47	3.26	1.10032	0.73	3.41
0.040	1.11571	0.24	2.43	1.10308	0.42	2.89	1.09098	0.66	3.13
0.050	1.10965	0.21	2.19	1.09467	0.38	2.62	1.08227	0.60	2.82
0.070	1.09271	0.17	1.83	1.07552	0.32	2.22	1.06224	0.52	2.40
0.100	1.06895	0.14	1.57	1.04904	0.27	1.89	1.03470	0.45	2.05
0.199	1.00106	0.10	1.19	0.97999	0.22	1.55	0.96832	0.38	1.60
0.300	0.95643	0.10	1.21	0.93712	0.21	1.46	0.92238	0.38	1.49
0.499	0.89509	0.09	1.03	0.87908	0.21	1.39	0.85974	0.39	1.38

Table 8.3: As Table 8.2 but x_3 = 0.12.

	10 °C			25 °C			40 °C		
x'_2	ρ/ (g cm^{-3})	$10^9 D$/ (m^2 s^{-1})	$10^{39} A_{22}$/ (m^5 s^{-2})	ρ/ (g cm^{-3})	$10^9 D$/ (m^2 s^{-1})	$10^{39} A_{22}$/ (m^5 s^{-2})	ρ/ (g cm^{-3})	$10^9 D$/ (m^2 s^{-1})	$10^{39} A_{22}$/ (m^5 s^{-2})
0.010	1.18817	0.36	0.82	1.17694	0.55	1.27	1.16476	0.82	0.77
0.013	1.18450	0.34	1.35	1.17327	0.53	1.72	1.16118	0.79	1.44
0.017	1.18081	0.32	1.66	1.16903	0.51	1.97	1.15706	0.77	1.94
0.020	1.17774	0.31	1.89	1.16636	0.50	2.12	1.15427	0.75	2.16
0.030	1.16981	0.28	2.07	1.15800	0.44	2.22	1.14718	0.70	2.39
0.040	1.16018	0.25	2.01	1.14887	0.41	2.19	1.13650	0.66	2.46
0.050	1.15235	0.23	1.87	1.14036	0.39	2.11	1.12785	0.62	2.41
0.070	1.13653	0.20	1.64	1.12374	0.34	1.93	1.11054	0.55	2.16
0.100	1.11245	0.16	1.32	1.09857	0.29	1.70	1.08720	0.49	1.81
0.150	1.07560	0.13	1.19	1.06104	0.24	1.56	1.05083	0.42	1.68
0.200	1.04578	0.11	1.12	1.03005	0.22	1.49	1.01807	0.39	1.58

behavior might be related to only a *local* structural change in the complex mixture. At this point we did not know whether the water molecules experience in the ternary mixture the same dynamic changes as induced by urea on *tert*-butanol. This question then has led us to measurements of the self-diffusion coefficients of water in these mixtures. (For an extended insight into the measurement procedure and a brief discussion of previous published works we refer to our recent paper (Mayele and Holz, 2000).) The results obtained at 25 °C are presented in Figure 8.4 and in Table 8.4.

Figure 8.4 reveals that the addition of urea both to pure water ($x'_{\text{t-BuOD}} = 0$, most upper curve) and to aqueous t-BuOD solutions (lower curves) does *not* increase the diffusion coef-

8.4 Results and Discussion

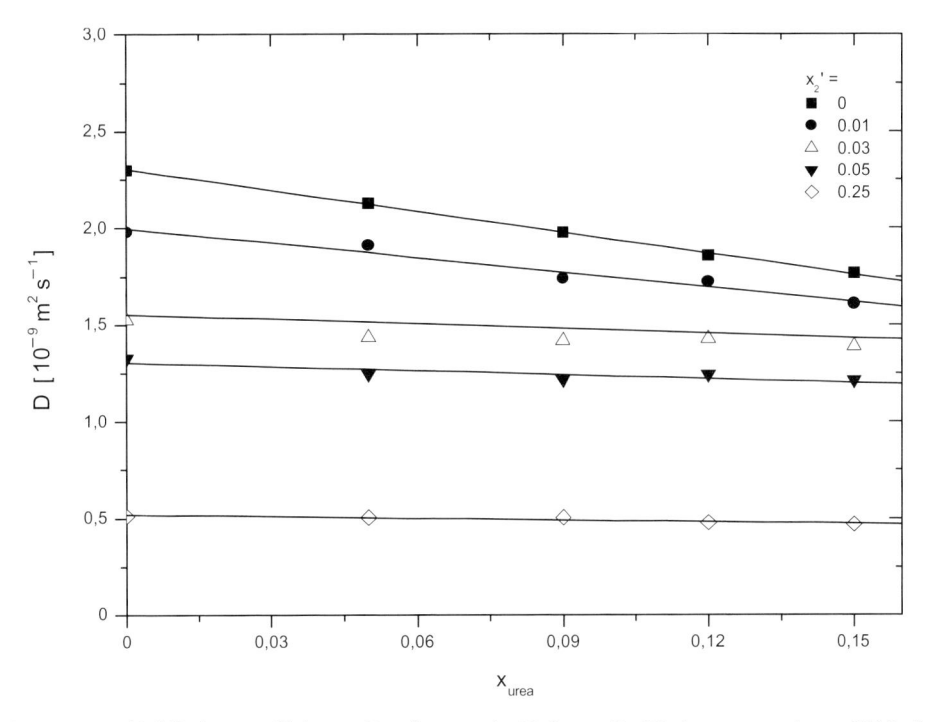

Figure 8.4: Self-diffusion coefficients, D, of water in H$_2$O + t-BuOD-d$_{10}$ + urea-d$_4$ at 25 °C for t-BuOD-d$_{10}$ concentrations $x'_2 = 0$, 0.01, 0.03, 0.05, and 0.25 as function of the urea concentration.

Table 8.4: Self-diffusion coefficients D of water in the ternary mixtures H$_2$O + t-BuOD-d$_{10}$ + urea-d$_4$ at 25 °C. D is given in $[10^{-9}$ m^2 s$^{-1}]$.

x_{urea}	$x'_{\text{t-BuOD}} = 0$	$x'_{\text{t-BuOD}} = 0.01$	$x'_{\text{t-BuOD}} = 0.03$	$x'_{\text{t-BuOD}} = 0.05$	$x'_{\text{t-BuOD}} = 0.25$
0	2.30	1.98	1.52	1.33	0.51
0.05	2.13	1.91	1.44	1.25	0.51
0.09	1.98	1.74	1.42	1.22	0.50
0.12	1.86	1.72	1.42	1.24	0.48
0.15	1.77	1.61	1.39	1.21	0.47

ficients of water molecules. We even find a decrease of the water diffusion coefficients with increasing urea concentration for the binary and ternary mixtures. Thus the results in Figure 8.4 mean that in the *binary* system as well as in the ternary system there is no hint for a structure-breaking effect of urea with respect to the bulk water.

In order to derive more quantitative information about the translational dynamics of water in the *direct neighborhood* of a *tert*-butanol molecule we further evaluate our results using the so-called B coefficient concept (Endom et al., 1967; Hertz, 1973), here applied with the self-diffusion B_D coefficient (Hertz, 1986; Sacco et al., 1989). In brief, the B_D coefficient is determined via the concentration dependence of the water self-diffusion coefficient in diluted aqueous solutions by application of the following expression (Hertz, 1986; Holz, 1996):

$$\frac{1/D}{1/D_0} = 1 + B_D \bar{m} \tag{8.6}$$

where D_0 is the self-diffusion coefficients of water in absence of the solute of interest and D is the self-diffusion coefficient of water in presence of this solute, whose concentration, expressed in the aqua-molality scale, is varied. Here we varied each time the t-BuOD concentration at a fixed urea concentration in the concentration range from $x_3 = 0$ to $x_3 = 0.15$. B_D is then derived from the slope of the function $D_0/D(\bar{m})$ in the limit $\bar{m} \to 0$. The sign of the B_D coefficient gives the direction, his absolute value gives the strength of the dynamic change induced by the solute in its vicinity. If we relate these *local* changes in the translational mobility to solvent structure changes, a positive sign of B_D reflects a local solvent structure enhancement and a negative sign a weakening of the solvent structure in the solvation sphere of the solute particle. By measurement of the B_D coefficient for t-BuOD at different urea concentrations, we derive information about the translational mobility of the local water in the hydration sphere of t-BuOD compared to the bulk water in presence of urea at the given concentrations. In Figure 8.5 the B_D coefficients are shown which are gained from the experimental results in Figure 8.4. (Numerical values obtained are summarized in Table 8.5.)

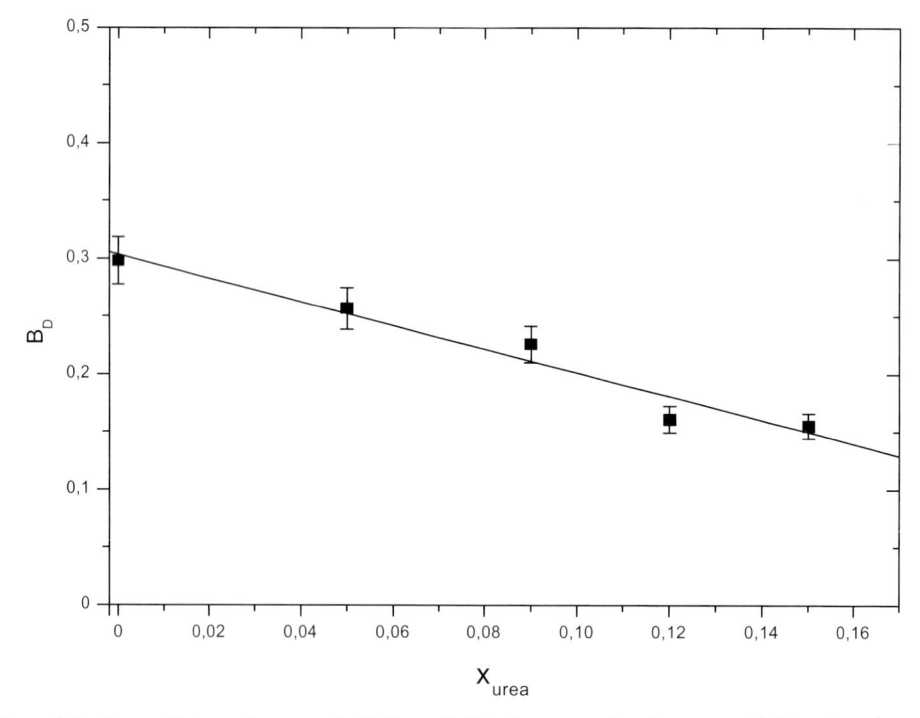

Figure 8.5: B_D coefficients for water in H_2O + t-BuOD-d_{10} + urea-d_4 mixtures at 25 °C as function of urea concentration.

8.4 Results and Discussion

Table 8.5: B_D coefficients at fixed urea concentrations in the system H_2O + t-BuOD-d_{10} + urea-d_4 and derived ratios of self-diffusion coefficients of bulk water and of water in the hydration sphere of t-BuOD. $D_{H_2O}^{hydrat}$ has been determined as described in the text.

x_{urea}	B_D	$D_{H_2O}^{bulk}/D_{H_2O}^{hydrat}$	$10^9 D_{H_2O}^{hydrat}/(m^2 \, s^{-1})$
0	0.298	1.66	1.38
0.05	0.257	1.57	1.37
0.09	0.226	1.50	1.35
0.12	0.161	1.36	1.33
0.15	0.155	1.34	1.32

Figure 8.5 shows that the B_D coefficient for t-BuOD is positive in the binary mixture ($x_3 = 0$) and also in the ternary mixtures over the whole urea concentration range examined. The first result reflects the well-known fact that the mobility of water is slowed down in hydrophobic hydration spheres (Hertz and Zeidler, 1964), a fact which is related to an enhanced structure of water near hydrophobic solutes (Frank and Evans, 1945; Franks, 1978). However, interesting in Figure 8.5 is the decrease of the absolute B_D values with increasing urea concentration, which means that under the influence of urea the diffusivity of water in the hydrophobic hydration sphere of t-BuOD is less reduced relative to the corresponding aqueous bulk phase than it is in the binary system. Consequently, we find an influence of urea on the strength of the hydrophobic slowing down effect, although, as we can see, urea concentrations up to 15 mol% are not enough to invert the sign of the B_D coefficient.

In order to gain more quantitative information about the translational dynamics of water in the hydrophobic hydration sphere from the NMR B_D coefficient we apply a "two-state model" with fast exchange (Hertz, 1973). This model permits the evaluation of the self-diffusion coefficient of hydration water $D_{H_2O}^{hydrat}$ relatively to that of bulk water $D_{H_2O}^{bulk}$ using the following expression:

$$\frac{1}{D_{H_2O}} = (1-x) \frac{1}{D_{H_2O}^{bulk}} + x \frac{1}{D_{H_2O}^{hydrat}} \tag{8.7}$$

where D_{H_2O} is the measured average self-diffusion coefficient and x is the mole fraction of water located in the hydration sphere. Taking into account that the number of water molecules in the hydration sphere is given by the product of the hydration number n_h and the number of the solute molecules, the following expression has been derived (Endom et al., 1967; Hertz, 1986; Holz, 1996):

$$\frac{D_{H_2O}^{bulk}}{D_{H_2O}^{hydrat}} = 1 + \frac{55.5}{n_h} \cdot B_D \tag{8.8}$$

If n_h is known, from the B_D coefficients and using Eq. (8.8) the ratio of the diffusion coefficient of bulk water to the diffusion coefficient of hydration water is available. Molecular dynamics simulations (Tanaka et al., 1984) and X-rays small angle scattering studies (Nishikawa and Iijima, 1990) yielded a hydration number for t-BuOD of approximately

$n_h = 25$. Assuming that n_h is independent of urea concentration, we obtain the ratio $D_{H_2O}^{bulk}/D_{H_2O}^{hydrat}$ as function of urea concentration and the results are presented in Table 8.5. It can be realized that for *tert*-butanol in water ($x_3 = 0$), $D_{H_2O}^{bulk}/D_{H_2O}^{hydrat} = 1.66$ is obtained which means that in the binary mixture the translational mobility of water in the hydration sphere of *tert*-butanol is reduced by about 40% relative to that of bulk water. (For more details see: Mayele and Holz, 2000.) Summarizing we can state: If there is really a structure-breaking effect by urea, as discussed in the literature (Wetlaufer et al., 1964; Frank and Franks, 1968; Castronuovo et al., 1996), then in the light of our above results, we conclude that it could only be a *local* effect in the hydrophobic hydration sphere. Such an interpretation would then support the "direct mechanism" in denaturing processes by urea. For the check of the validity of this interpretation further experiments are required, namely the study of the A_{23} parameter which would show if there is an affinity of urea towards the CH$_3$-groups of *tert*-butanol or not. These experiments are in progress in our laboratory.

8.4.2.2 Influence of Urea on *tert*-Butanol Self-Association

First we will examine the influence of urea on the hydrophobic self-association behavior of t-BuOD in water as monitored through the A_{22} parameter at 10, 25, and 40 °C (Fig. 8.6). Experimental results are summarized in Tables 8.2 and 8.3.

For the discussion of the results we will use as reference the concentration dependence of A_{22} in the binary water + t-BuOD system (open squares in Fig. 8.6) where A_{22} in pure *tert*-butanol is assumed to be equal to A_{22} at $x_2 = 0.5$, since in the alcohol rich region A_{22} remains constant.

As first ternary system we consider the t-BuOD aqueous solution with 5 mol% urea (solid circles in Fig. 8.6). We recognize at all measured temperatures an increase of the A_{22} parameter e. g. of about 30% at $x_2 = 0.02$ for 10 and 25 °C, that is the same percentage increase of the local butanol concentration relative to the binary system, i.e. urea promotes the self-association of *tert*-butanol. At 10 °C (Fig. 8.6a), like in the binary system, the A_{22} parameter passes also in the ternary system through a maximum at $x_2' \approx 0.02$. At 25 (Fig. 8.6b) and 40 °C (Fig. 8.6c) the A_{22} parameter increases continuously with increasing dilution and no maximum is observed. We emphasize that here, as in the binary system, the maximum could well have moved to smaller (no more observable) t-BuOD concentrations. It can also be seen that in the water-rich region at 40 °C the *absolute* difference to the binary system (e. g. ca. 20% at $x_2 = 0.02$) is smaller than at 25 °C which means that with a gradual raise of the temperature the association promoting effect of urea will gradually become smaller.

However, when we added 12 mol% urea (open triangles in Fig. 8.6) in the water-rich region a clear *de-association* of the *tert*-butanol molecules occurred, since with decreasing x_2', the A_{22} parameter, after having passed through a weak maximum, strongly decreases. A further remarkable fact is that in the ternary system with the high urea content only a weak temperature dependence is observable, which indicates that the urea influence cancels the temperature influence.

Summarizing the results of Figure 8.6, we can state that the urea effect on the "hydrophobic binding" depends strongly on the urea concentration. Relatively large amounts (ca. 7 molal, corresponding to 12 mol% in water-rich region) of urea are needed for a destabilization of the hydrophobic association of *tert*-butanol. This finding can be compared with

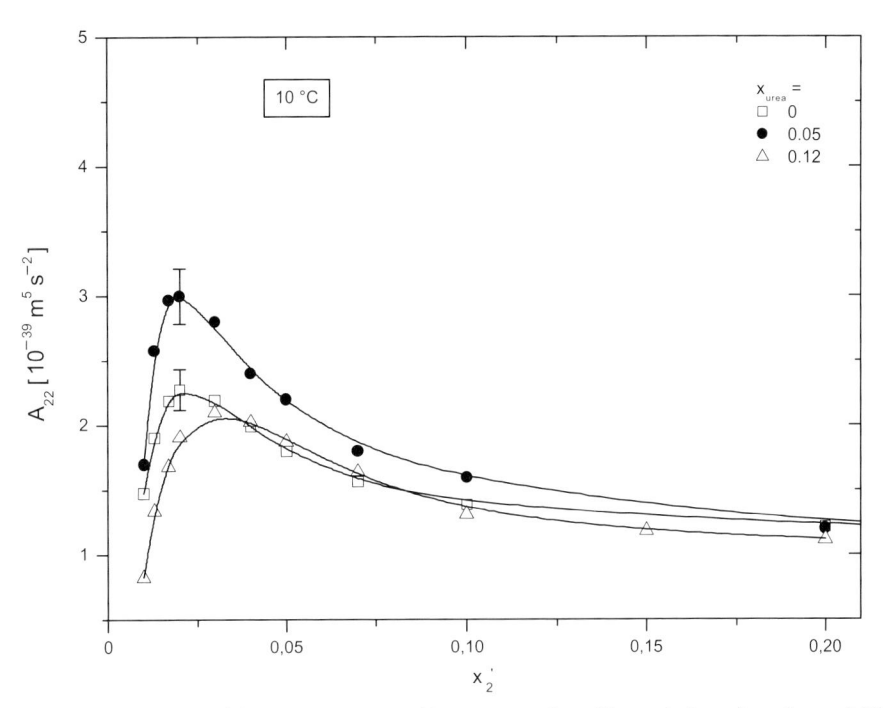

Figure 8.6a: Comparison of the A_{22} parameter with respect to the self-association of *tert*-butanol-OD in D$_2$O + t-BuOD + urea-d$_4$ (x_3 = 0.05 and 0.12) and D$_2$O + t-BuOD mixtures at 10 °C as function of $x'_2 = x_2 / (x_1 + x_2)$.

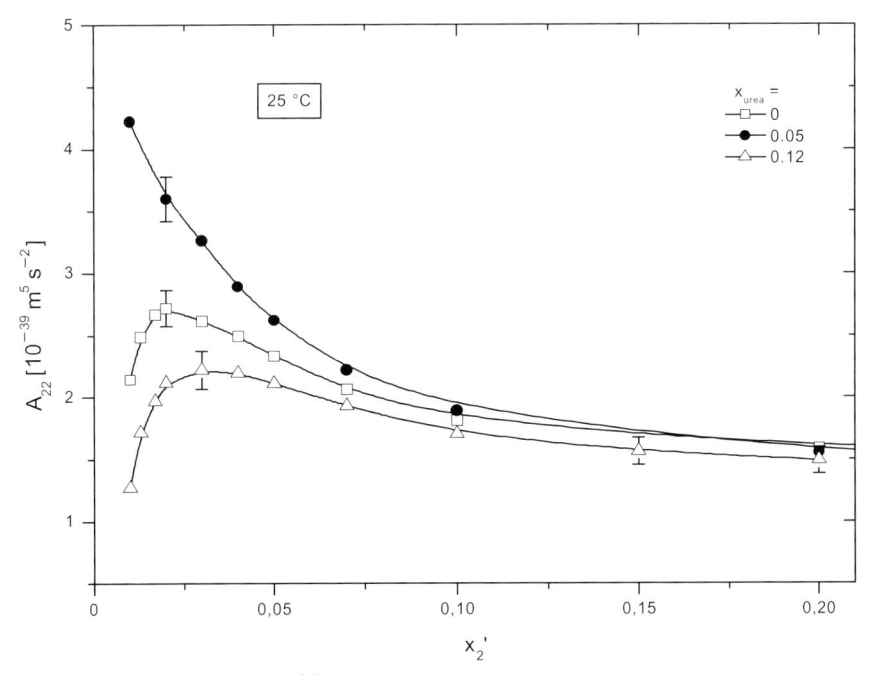

Figure 8.6b: As Fig. 8.6a but at 25 °C.

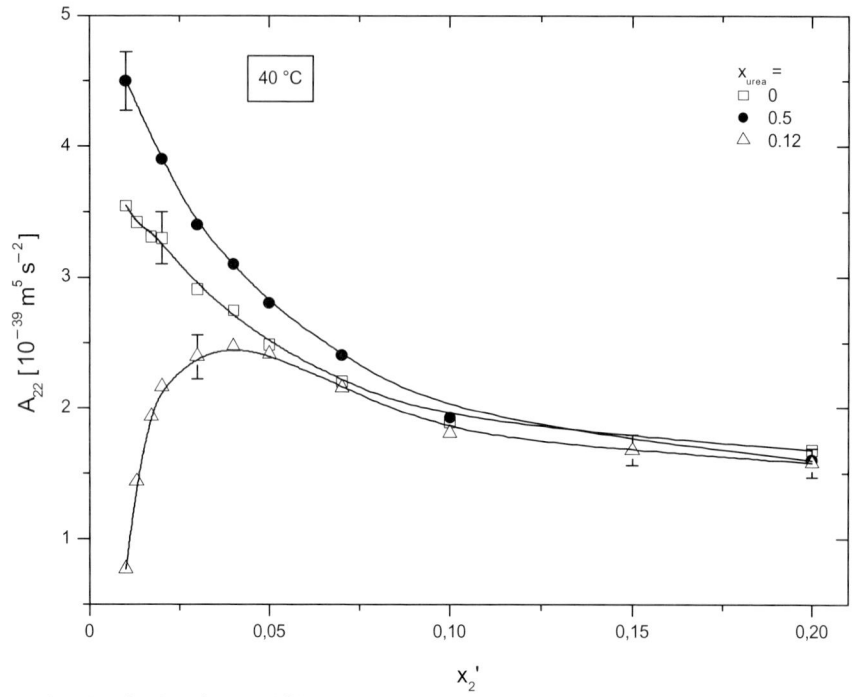

8 Influence of Additives on Hydrophobic Association in Polynary Aqueous Mixtures

Figure 8.6c: As Fig. 8.6a but at 40 °C.

studies on biomolecules, where e.g. unfolding simulations on proteins in presence of urea (Tirado-Rives et al., 1997; Thirumalai et al., 1997) as well as experimental examinations of the urea effect on peptides (Nandi and Robinson, 1984) led to the same conclusion since in these investigations 6 to 8 m of urea were required to achieve protein denaturation.

For further examinations with regard to the mixture composition and concentration dependence of the urea effect, we selected three fixed t-BuOD concentrations in the water-rich region, namely $x_2' = 0.02$, 0.04, and 0.2 and varied the urea concentration up to 15 mol% at 25 °C. The results are summarized in Figure 8.7 and in Table 8.6.

Figure 8.7 reveals that the A_{22}-values and hence the association tendency, shows for the lowest t-BuOD concentration $x_2' = 0.02$ a pronounced maximum at $x_3 = 0.035$. At the double concentration of t-BuOD a weaker maximum appears at the double mole fraction of urea $x_3 = 0.07$ and finally the maximum disappears at the highest alcohol concentration. The fact that we need for the double concentration of hydrophobic groups the double concentration of urea, at which a maximum stabilization of the hydrophobic interaction occurs, might support the presumption that a direct relation exists between the total number of water molecules in hydrophobic hydration spheres and the number of urea molecules at the stabilization maximum. Taking a hydration number $n_h = 25$ (Tanaka et al., 1984) and the above figures at the stabilization maximum we arrive at a ratio of about 12:1 for the number of water molecules in hydrophobic hydration spheres to the number of urea molecules.

8.4 Results and Discussion

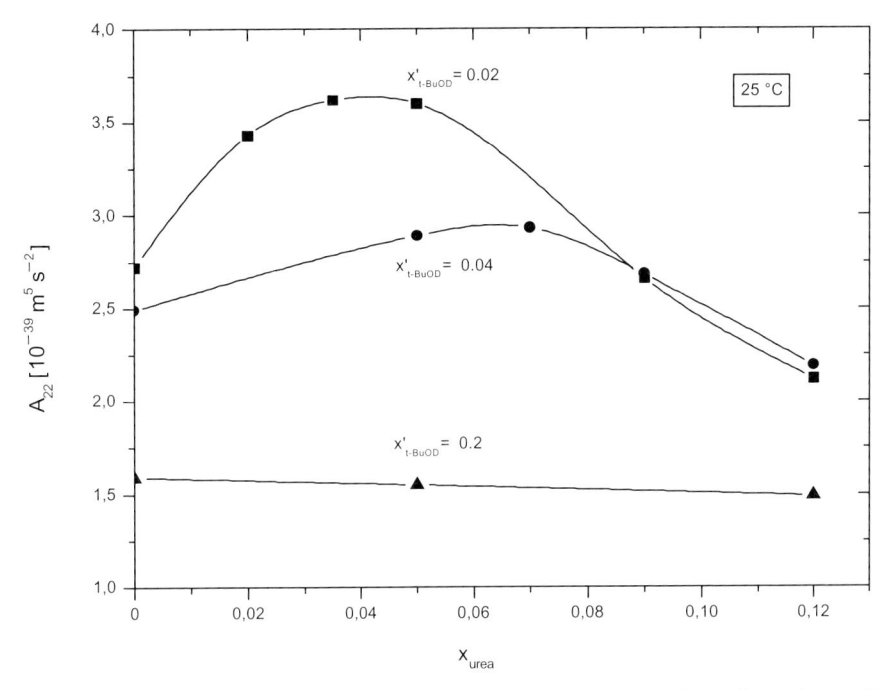

Figure 8.7: A_{22} parameter with respect to the association of *tert*-butanol-OD in D_2O + t-BuOD ($x'_2 = 0.02$, 0.04, and 0.2) + urea-d_4 mixtures at 25 °C as function of the urea concentration.

Table 8.6: Experimental results used for the determination of the urea concentration dependence of A_{22} parameters in the D_2O + t-BuOD-d_1 ($x'_2 = 0.02$, 0.04, and 0.2) + urea-d_4 system at 25 °C: mixture density ρ, diffusion coefficient D of t-BuOD-d_1, and association parameter A_{22}.

| x_{urea} | $x'_2 = 0.02$ |||| $x'_2 = 0.04$ ||| $x'_2 = 0.2$ |||
	ρ / (g cm^{-3})	$10^9 D$ / (m^2 s^{-1})	$10^{39} A_{22}$ / (m^5 s^{-2})	ρ / (g cm^{-3})	$10^9 D$ / (m^2 s^{-1})	$10^{39} A_{22}$ / (m^5 s^{-2})	ρ / (g cm^{-3})	$10^9 D$ / (m^2 s^{-1})	$10^{39} A_{22}$ / (m^5 s^{-2})
0	1.08452	0.49	2.72	1.06870	0.39	2.49	0.95349	0.22	1.59
0.02	1.10041	0.52$_2$	3.43	–	–	–	–	–	–
0.035	1.11160	0.53$_2$	3.62	–	–	–	–	–	–
0.05	1.12125	0.53	3.60	1.10308	0.42	2.89	0.97999	0.22	1.55
0.07	–	–	–	1.11766	0.41$_7$	2.93	–	–	–
0.09	1.14814	0.49$_9$	2.66	1.13045	0.41	2.63	–	–	–
0.12	1.16636	0.50	2.12	1.14887	0.41	2.19	1.03005	0.22	1.49

8.4.3 Water + Ethylene Glycol Systems

As already mentioned in the introduction, aqueous mixtures of ethylene glycol (EG) are of great technical importance as solvents in numerous industrial processes. Since the reactions which can be carried out in these mixtures are determined by intermolecular interactions between the molecules of the components (Engberts, 1979), any modification of water-water, water-EG, and EG-EG interactions can lead to specific solvation conditions which are favorable for a given chemical reaction. Thus appropriate modifications may serve for a reaction control.

Some few experimental results from Kirkwood Buff integral investigations on the intermolecular water-water, water-EG as well as EG-EG interactions in the binary system water + EG can be found in the literature (Marcus, 1990; Cheng et al., 1993; Matteoli, 1997). The KB integrals G_{ij} relate macroscopic thermodynamic properties to the integral over the pair distribution function of selected components of the mixture:

$$G_{ij} \equiv \int_0^\infty [g_{ij}(r) - 1] 4\pi r^2 \, dr = f(\gamma_i, \bar{v}_i, \kappa_T) \quad (8.9)$$

The function $g_{ij}(r)$ is the radial (center-center) pair distribution function between the components i and j. γ_i, and κ_T are the activity coefficient of the component i, the partial molecular volume of the component i in the mixture, and the isotherm compressibility of the mixture, respectively. In the investigations cited above it was stated that the concentration dependency of all G_{ij}'s in aqueous solutions of diols behaves differently from aqueous solutions of monofunctional alcohols. In the first case the functions $G_{ij}(x_i)$ reflect almost no correlation between the different components of the mixture indicating rather an uniform distribution. The hydrophobic part of the EG molecule ($-CH_2CH_2-$) shows neither a tendency to self-association nor to hydrophobic hydration. Therefore the intermolecular interactions in the system water + ethylene glycol seem to be determined for the major part by water-hydroxyl group interactions (dipolar H-bond interactions). These findings are in agreement with KBI investigations in the ternary systems water + ethylene glycol + acetonitrile and water + ethylene glycol + ethanol (Matteoli and Lepori, 1995). Here no preferential solvation of acetonitrile or ethanol by water or ethylene glycol was observed, i.e. the mixture water + ethylene glycol behaves like an ideal mixture. In this part of the work, first this ideal mixture behavior should be checked by means of our intermolecular magnetic dipole-dipole relaxation method in the binary system. Afterwards it should be examined whether third components can be found, which can reduce the dominance of the polar water-EG interactions in favor of the hydrophobic EG-EG self-association.

8.4.3.1   The Binary System: Water + Ethylene Glycol

The investigation of the system $DOCH_2CH_2OD + D_2O$ was performed using the methylene proton resonance. Measurements were made at 25 and 40 °C. Figure 8.8 (Table 8.7) shows the concentration dependence of the self-diffusion coefficient of ethylene glycol in water at 25 and 40 °C, which decrease with increasing concentration of ethylene glycol.

The measured values for D_2 (at 25 °C) in the present work, after correction for the dynamic isotope effect (Holz et al., 1996), are in good agreement with those from other authors (Ambrosone et al., 1997). These authors measured also the self-diffusion coefficient of water in this binary system and found that it also decreases with increasing concentration of ethylene glycol. This behavior was interpreted as the result of preferential formation of

8.4 Results and Discussion

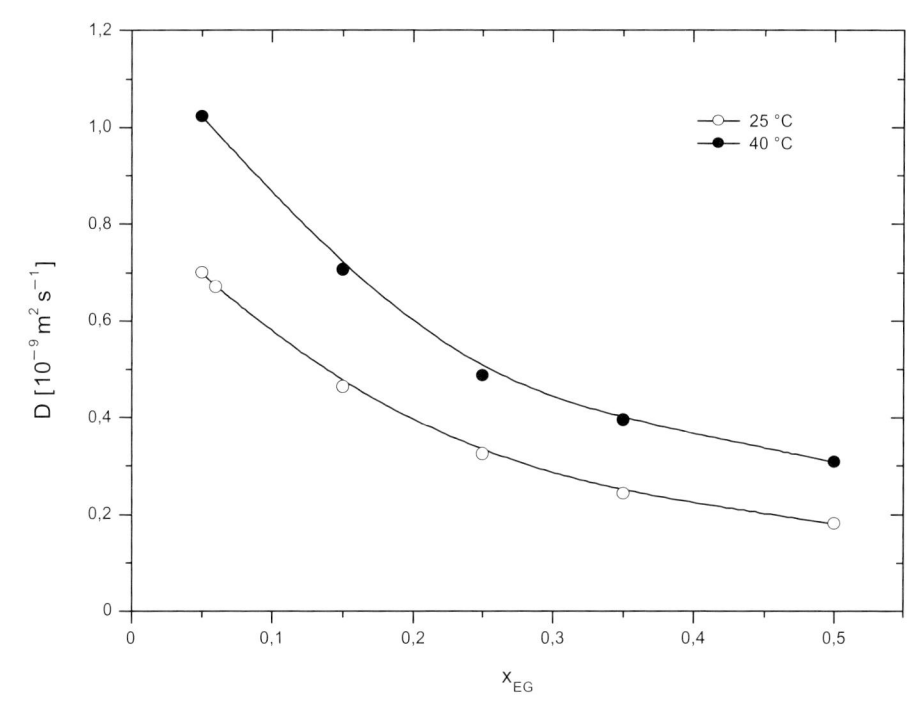

Figure 8.8: Self-diffusion coefficients of EG-d$_2$ in D$_2$O + EG-d$_2$ mixtures at 25 and 40 °C as function of the mole fraction x_{EG} of EG-d$_2$.

Table 8.7: Experimental results for the D$_2$O (1) + EG-d$_2$ (2) system at 25 and 40 °C: mixture density ρ, diffusion coefficient D of EG-d$_2$, and association parameter A_{22}.

x_2	25 °C			40 °C		
	ρ/(g cm^{-3})	$10^9 D$/(m^2 s^{-1})	$10^{39} A_{22}$/(m^5 s^{-2})	ρ/(g cm^{-3})	$10^9 D$/(m^2 s^{-1})	$10^{39} A_{22}$/(m^5 s^{-2})
0.05	1.11455	0.71	2.34	1.10355	1.02	2.74
0.06	1.11640	0.66	2.41	–	–	–
0.15	1.12910	0.47	2.41	1.11465	0.71	2.57
0.25	1.13786	0.32	2.37	1.12537	0.49	2.43
0.35	1.14259	0.25	2.32	1.13214	0.40	2.44
0.50	1.14524	0.18	2.25	1.13434	0.31	2.47

hydrogen bonds between ethylene glycol and water and the same conclusion has been made from calorimetric measurements of Ohta et al. (1998).

In Figure 8.9 the concentration and temperature dependence of A_{22} is given representing the self-association tendency of the CH$_2$ groups of ethylene glycol in water (cf. also Table 8.7).

The curve at 25 °C does *not* show, within the measuring error of 7%, a sizable tendency of the alcohol towards self-association, a result which confirms those from KBI inves-

171

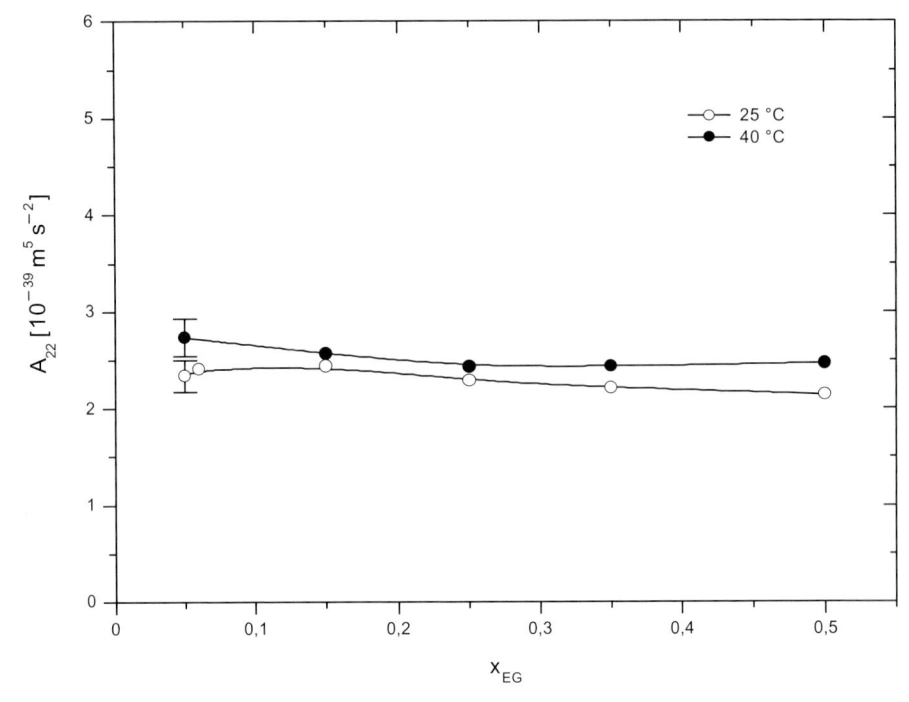

Figure 8.9: A_{22} parameter with respect to the association of EG-d$_2$ in D$_2$O + EG-d$_2$ mixtures at 25 and 40 °C as function of the mole fraction x_{EG} of EG-d$_2$.

tigations mentioned above. The behavior of the A_{22} parameter observed after transition from 25 to 40 °C gives a further support in this direction. There is no substantial modification in the concentration dependency of the A_{22} parameter at the higher temperature, i.e. no self-association tendency. The 40 °C-curve is only shifted to slightly higher A_{22} values however the shape remains unchanged within the margins of error. This curve shift could be related to an overall reduction of the distance of closest approach between EG molecules but is obviously not a result of an enhancement of local EG-EG hydrophobic association.

8.4.3.2 Ternary Systems: Water + Ethylene Glycol + Additive

The Additive Dioxane

It is well-known that pure ethylene glycol is one of the most structured liquids. The reason for that is the formation of strong three-dimensional networks of hydrogen bonds, which lead to a kind of polymerization of ethylene glycol (Corradini et al., 1994). It was already pointed out above that this low-molecular diol forms an ideal mixture with water. Contrary to EG, in pure dioxane the interactions are weak. This molecule does not contain H-bond donor sites. It can therefore behave only as H-bond acceptor. Due to its relatively small di-

8.4 Results and Discussion

pole moment (µ = 0.45 Debye at 25 °C) only weak dipolar interactions can exist in pure dioxane. In connection with the present work the reason for the selection of dioxane as third component lies in the fact that in the binary EG + dioxane system, dioxane acts as "structure breaker" with respect to the ethylene glycol network since it replaces EG-EG H-bonds by EG-dioxane H-bonds (Corradini et al., 1994). For the same reason one could expect that in the system ethylene glycol + water, dioxane could disturb the water-water and EG-water H-bonds and thus change the interaction conditions within the mixture. In order to test this hypothesis we measured the A_{22} parameter in the ternary system water + EG + dioxane. The binary system water + dioxane was regarded as modified solvent for ethylene glycol and the water + dioxane solvent was made from a mixture of 5 mol% full deuterated dioxane in deuterium oxide. The concentrations of ethylene glycol were 5, 15, and 25 mol% of EG in water + dioxane solution. The comparison of the obtained A_{22} parameter with that of the binary system (Fig. 8.10 and Table 8.8) does not, within the error limits, uncover a substantial modification of the self-association degree of ethylene glycol, i.e. dioxane does *not* cause a promotion of the EG-EG interactions in water. Dioxane is thus unsuitable for the promotion of self-association of small quantities of EG in aqueous environment.

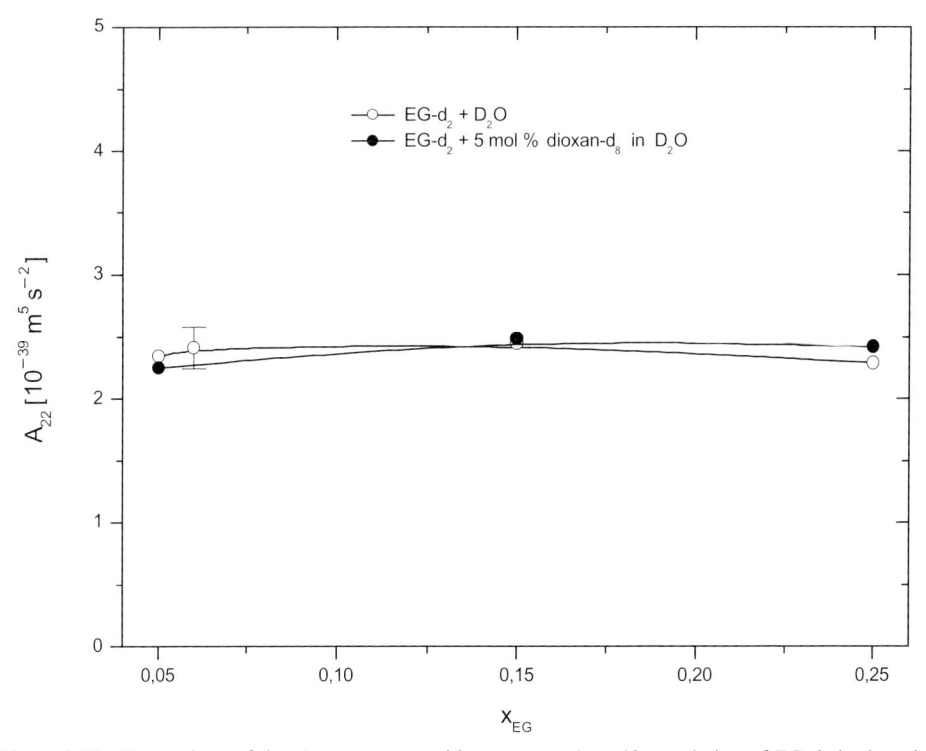

Figure 8.10: Comparison of the A_{22} parameter with respect to the self-association of EG-d$_2$ in the mixtures EG-d$_2$ + 5 mol% dioxane-d$_8$ in D$_2$O and EG-d$_2$ + D$_2$O at 25 °C as function of the mole fraction x_{EG} of EG-d$_2$.

8 Influence of Additives on Hydrophobic Association in Polynary Aqueous Mixtures

Table 8.8: Experimental results for the systems: 5 mol% dioxan-d$_8$ in D$_2$O (1) + EG-d$_2$ (2) and 1.8 mol% n-butanol-d$_{10}$ in D$_2$O (1) + EG-d$_2$ (2) at 25 °C. All other details as in Table 8.7.

x_2	Dioxane as additive			n-Butanol as additive		
	ρ/(g cm^{-3})	$10^9 D$/(m^2 s^{-1})	$10^{39} A_{22}$/(m^5 s^{-2})	ρ/(g cm^{-3})	$10^9 D$/(m^2 s^{-1})	$10^{39} A_{22}$/(m^5 s^{-2})
0.05	1.12646	0.62	2.25	1.10630	0.61	2.23
0.10	–	–	–	1.11511	0.51	2.48
0.15	1.13567	0.44	2.48	1.12186	0.42	2.46
0.25	1.14100	0.32	2.42	–	–	–

The Additive n-Butanol

In Kirkwood-Buff integral investigations it was found that ethanol did not markedly alter water-ethylene glycol and ethylene glycol-ethylene glycol interactions in the water-rich region of interest (Matteoli and Lepori, 1995). By using n-butanol, an alcohol with one additional methylene group as third component we wanted to check whether the total number of hydrophobic groups in the mixture can play a role. For this purpose we prepared solutions contain-

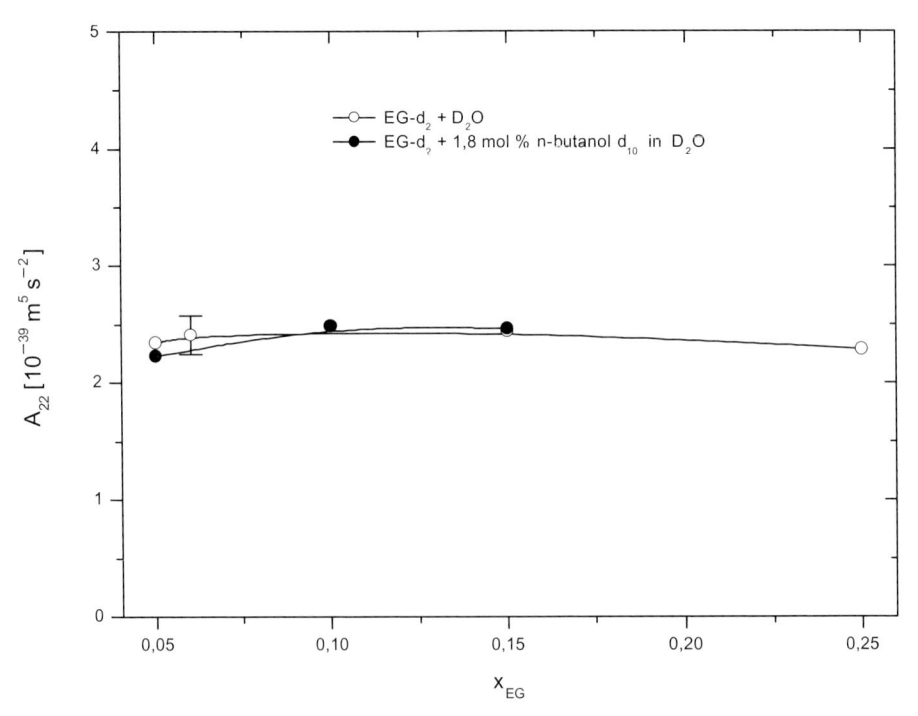

Figure 8.11: Comparison of the A_{22} parameter with respect to the self-association of EG-d$_2$ in the mixtures EG-d$_2$ + 1.8 mol% n-butanol-d$_{10}$ in D$_2$O and EG-d$_2$ + D$_2$O at 25 °C as function of the mole fraction x_{EG} of EG-d$_2$.

ing EG with concentrations of 5, 10, and 15 mol% in modified water (1.8 mol% *n*-butanol in water). The obtained A_{22} parameters are given in Table 8.8 and shown in Figure 8.11.

One can recognize, similar as with dioxane, that practically no modification of the self-association behavior of EG takes place. Thus additional CH_2-groups in alcohols used as additives do not lead to an enhancement of the hydrophobic self-association of EG in water.

Salts as Additives

In this investigation the solvent water was modified by addition of a salt and again the influence of this modification on the self-association behavior of ethylene glycol was measured. In electrolyte solutions ion-dipole interactions are important in addition to van der Waals forces and H-bonds. Some salts can cause a self-association of organic components or even phase separation by combination of these types of interaction. The self-association or the formation of two phases is related to the so-called salting-out effect. This effect decreases for cations and anions from the right to the left in the following series (see e. g. Collins et al., 1985):

$$Li^+ < Na^+ < K^+ = Cs^+$$

$$ClO_4^- < I^- < Br^- < Ac^- < SO_4^{2-}$$

which correspond to the empirical "lyotropic" or "Hofmeister series". Due to their position in the lyotropic series, the salts CsCl and Na_2SO_4, that means the association promoting cation Cs^+ and the association promoting anion SO_4^{2-} were selected here. The modification of the solvent water through the addition of CsCl and Na_2SO_4 was in each case made via preparation of solutions of 1 mole salt in 55.5 moles of D_2O. The measured A_{22} parameters are compared with the A_{22} parameters of the binary system (without salt) and it turns out that the addition of CsCl (cf. Fig. 8.12 and Table 8.9) increases clearly the degree of self-association of ethylene glycol within the measured concentration range. Additionally the A_{22} parameter passes through a maximum at approximately 10 mol% of ethylene glycol, which reminds of the maximum in the system water + *tert*-butanol. The salt Na_2SO_4 (cf. Fig. 8.12 and Table 8.9) causes likewise an increase of the EG-EG interaction. With this salt the effect is however more important at relatively high EG concentrations (compared with CsCl). Thus for the first time, third components which promote the self-association of EG in water were found. The local concentration increase is of the order of about 25% at 25 °C relatively to the binary system at $x_2 = 0.1$.

The effect of temperature on the self-association of ethylene glycol caused by addition of CsCl was also examined. For this purpose the A_{22} parameter was measured in the ternary system $D_2O + DOCH_2CH_2OD + CsCl$ at 40 °C. The aim was thereby to check whether the observed increase of the self-association degree of EG in this system is actually due to the hydrophobic interaction. If that is the case, then the hydrophobic association, being an entropy driven phenomenon, should be favored by the increase of the temperature. This is confirmed by the result presented in Figure 8.13 (cf. Table 8.9). The raise of the A_{22} parameter in the region of high dilution (below 10 mol% EG) is unambiguous. Similarly as in the system water + *tert*-butanol, it seems that the maximum at 25 °C is shifted towards regions of higher dilution of EG with increasing temperature.

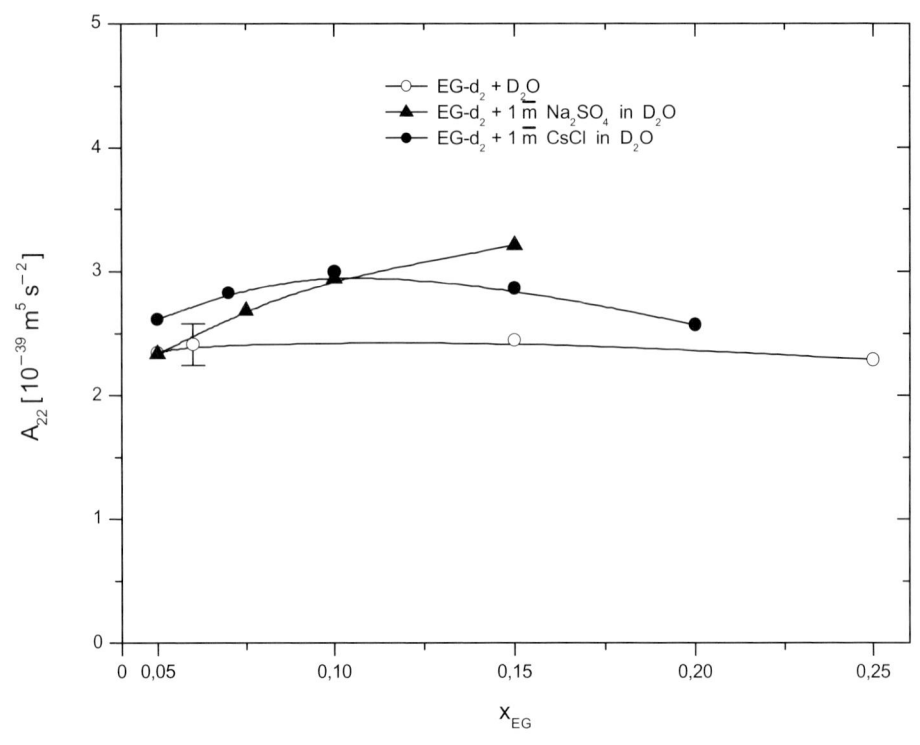

Figure 8.12: Comparison of the A_{22} parameter with respect to the self-association of EG-d$_2$ in the mixtures EG-d$_2$ + 1 mole CsCl in 55.5 mole D$_2$O and EG-d$_2$ + 1 mole Na$_2$SO$_4$ in 55.5 mole D$_2$O and EG-d$_2$ + D$_2$O at 25 °C as function of the mole fraction x_{EG} of EG-d$_2$.

Table 8.9: Experimental results for the systems: 1 $\bar{m}$ CsCl in D$_2$O (1) + EG-d$_2$ (2) at 25 and 40 °C and 1 $\bar{m}$ Na$_2$SO$_4$ in D$_2$O (1) + EG-d$_2$ (2) at 25 °C. All other details as in Table 8.7.

x_2	CsCl as additive (25 °C) ρ/ (g cm^{-3})	$10^9 D$/ (m^2 s^{-1})	$10^{39} A_{22}$/ (m^5 s^{-2})	CsCl as additive (40 °C) ρ/ (g cm^{-3})	$10^9 D$/ (m^2 s^{-1})	$10^{39} A_{22}$/ (m^5 s^{-2})	Na$_2$SO$_4$ as additive (25 °C) ρ/ (g cm^{-3})	$10^9 D$/ (m^2 s^{-1})	$10^{39} A_{22}$/ (m^5 s^{-2})
0.050	1.21575	0.74	2.61	1.21037	1.06	3.41	1.21089	0.52	2.34
0.070	1.21371	0.67	2.82	1.20999	0.98	3.21	–	–	–
0.075	–	–	–	–	–	–	1.20801	0.45	2.69
0.101	1.21055	0.57	3.00	1.20291	0.85	3.00	1.20548	0.41	2.94
0.015	1.20534	0.46	2.86	1.19624	0.70	2.75	1.20018	0.34	3.21
0.020	1.20039	0.39	2.57	1.19142	0.58	2.53	–	–	–

8.5 Conclusions

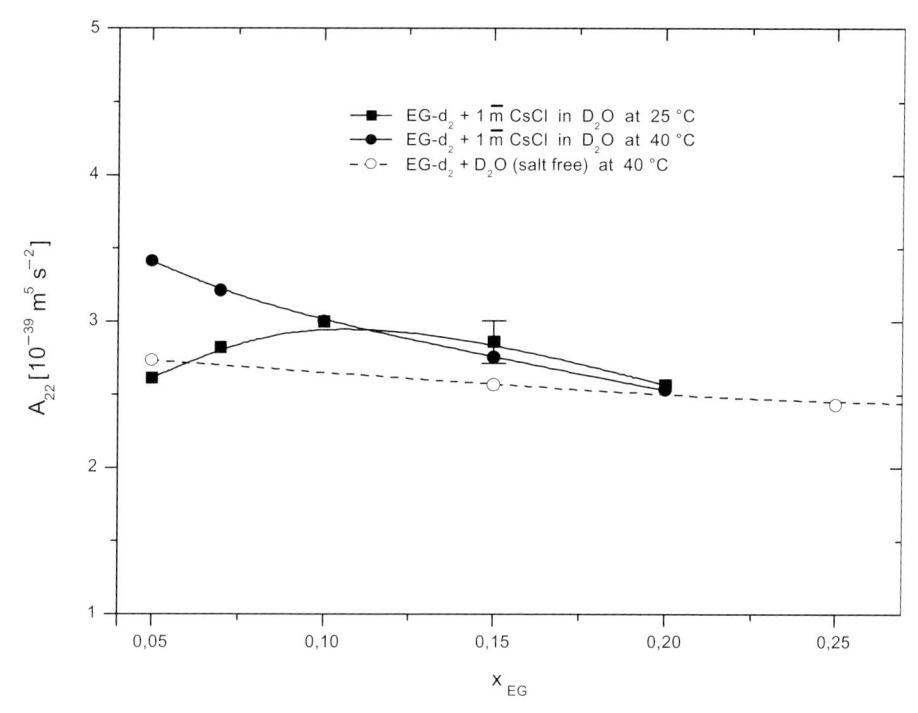

Figure 8.13: Comparison of the A_{22} parameter with respect to the self-association of EG-d$_2$ in the mixtures EG-d$_2$ + 1 mole CsCl in 55.5 mole D$_2$O at 25 and 40 °C and EG-d$_2$ + D$_2$O (salt-free) at 40 °C as function of the mole fraction x_{EG} of EG-d$_2$.

8.5 Conclusions

The application of NMR spectroscopic techniques to the investigation of complex multi-component mixtures demonstrated an unique advantage, namely the possibility to study the behavior of distinct mixture components by observation in a selective and unambiguous way. Thus, microscopic information on single component translational motion (via the self-diffusion coefficient) in the mixture and on local concentrations of single components at a given composition relative to another composition (via the A parameter) could be gained. We could show that in the binary system water (1) + *tert*-butanol (2) at 25 °C at a mole fraction $x_2 \approx 0.02-0.03$ a distinct transition between two characteristically different regions of the mixture occurs, namely from hydrophobically associated *tert*-butanol to hydrophobically hydrated *tert*-butanol at lower x_2-values. This transition in the water rich region seems to be a general property of aqueous mixtures with hydrophobic solutes and has to be taken into account when those mixtures are modeled. Further we found, as theoretically expected, an increasing association tendency with increasing temperature. This characteristic temperature dependence of the *A* parameter might serve in the future as an important source of information for the presence and dominance of hydrophobic association in aqueous mixtures of interest. In the ternary system water + *tert*-butanol + urea, it was shown that urea can speed up or slow down the self-diffusion of the *alcohol* depending on the mixture composition and/or temperature whereas the mean water translational dynamics is invariably slowed down. Therefore, a water structure-breaking mechanism of urea could not be confirmed.

Addition of relative small amounts of urea to diluted aqueous alcohol solutions stabilizes the hydrophobic interaction. Relatively high amounts of urea were necessary to cause a destabilization of the hydrophobic association. This finding suggests that the molecular mechanism which is responsible for the destabilization of hydrophobically associated small molecules by urea is the same as for the denaturation of large biomolecules by urea. Therefore, in the light of the present results, any proposed denaturation mechanism which insinuates that the need of high concentrations of urea is connected with the extraordinary great size of biomolecules, should be reconsidered.

Our *A* parameter measurements in the binary system water + EG did not show a tendency of EG towards self-association. The binary system behaved at ambient temperature like an ideal mixture. Rise in temperature, addition of an alcohol (*n*-butanol) as well as an ether (dioxane) could not induce EG-EG self-association. However a clear modification of the water-water and water-EG interactions in favor of the self-association of EG could be obtained by addition of the salts CsCl and Na_2SO_4.

All the results of the present work clearly demonstrate that microdynamical processes and weak association phenomena occurring in complex multi-component systems can be selectively monitored by NMR. The observation of the relative change of *A*-parameter as a function of the mixture composition allowed the determination of local concentration changes of a distinct mixture component and thus offers also a new way for the test of mixing rules applied in modeling of complex liquid mixtures.

References

Ambrosone, L.; D'Errico, G.; Sartorio, R.; Costantino, L. (1997): Dynamic properties of aqueous solutions of ethylene glycol oligomers as measured by the pulsed gradient spin-echo NMR technique at 25 °C. J. Chem. Soc., Faraday Trans. 93, 3961.
Ben-Naim, A. (1977): Inversion of the Kirkwood-Buff theory of solutions: application to the water-ethanol system. J. Chem. Phys. 67, 4884.
Ben-Naim, A. (1980): Hydrophobic interactions. Plenum, New York.
Ben-Naim, A.; Yaacobi, M. (1974): Effects of solutes on the strength of hydrophobic interaction and its temperature dependence. J. Phys. Chem. 7, 170.
Blokzijl, W.; Engberts, J.B.F.N. (1993): Hydrophobic effects: opinion and fact. Angew. Chem., Int. Ed. Engl. 105, 1610–1648.
Castronuovo, G.; D'Isanto, G.; Elia, V.; Velleca, F. (1996): Role of cosolvent in hydrophobic interactions. Calorimetric studies in alcanols in concentrated aqueous solutions of urea at 298 K. J. Chem. Soc., Faraday Trans. 92, 3087.

References

Cebe, E.; Kaltenmeier, D.; Hertz, H.G. (1984): A study concerning self-association in mixtures of cyclohexane, benzene and DMSO with CCl_4 by the nuclear magnetic relaxation method. Z. Phys. Chem. Neue Folge 140, 181–189.

Cheng, Y.; Pagé, M.; Jolicoeur, C. (1993): Comparative study of hydrophobic effects in water/alcohol and water/ethylene glycol, water/ethanolamine, water/ethylenediamine, and water/2-methoxyethanol systems. J. Phys. Chem. 97, 7359.

Collins, K.D.; Washabaugh, M.W. (1985): The Hofmeister effect and the behaviour of water at interfaces. Quart. Rev. Biophysics 18, 323.

Corradini, F.; Marchetti, A.; Tagliazucchi, M.; Tassi, L.; Tosi, G. (1994): Thermodynamic analysis of viscosity data of ethane-1,2-diol + 1,4-dioxane binary mixtures. Aust. J. Chem. 47, 1117.

D'Angelo, M.; Onori, G.; Santucci, A. (1994): Self-association of monohydric alcohols in water: compressibility and infrared absorption measurements. J. Chem. Phys. 100, 3107.

Detscher, E.; Hertz, H.G.; Holz, M.; Mao, X.A. (1995): Ion-ion de-association and association of strong electrolytes at high dilution from nuclear magnetic relaxation measurements. Z. Naturforsch. 50a, 487–501.

Dötsch, V.; Wider, G.; Siegal, G., Wüthrich, K. (1995): Salt-stabilized globular protein structure in 7 M aqueous urea solution. FEBS Letters 372, 288.

Endom, L.; Hertz, H.G.; Thül, B.; Zeidler, M.D. (1967): Microdynamic model of electrolyte solutions as derived from nuclear magnetic relaxation and self-diffusion data. Ber. Bunsen-Ges. Phys. Chem. 71, 1008.

Engberts, J.B.F.N. (1979): in "Water: A comprehensive treatise", Vol. 6.

Farrar, T.C.; Becker, E.D. (1979): Pulse and Fourier Transform NMR. Academic Press, Orlando.

Frank, H. S.; Franks, F. (1968): Structural approach to the solvent power of water for hydrocarbons; urea as a structure breaker. J. Chem. Phys. 48, 4647.

Frank, H.S.; Evans, M.W. (1945): Free volume and entropy in condensed systems. III. Entropy in binary liquid mixtures; partial molal entropy in dilute solutions; structure and thermodynamics in aqueous electrolytes. J. Chem. Phys. 13, 507.

Franks, F. (1978) (Ed.): Water: A comprehensive treatise. Plenum, New York, Vol. 4, Chap. 1, p. 2 and p. 85 and references cited therein.

Harris, K.R.; Newitt, P.J. (1998): Diffusion and structure in dilute aqueous alcohol solutions: Evidence for the effects of large apolar solutes on water. J. Phys. Chem. B. 102(44), 8874.

Haselmeier, R.; Holz, M.; Marbach, W.; Weingärtner, H. (1995): Water dynamics near a dissolved noble gas. First evidence for a retardation effect. J. Phys. Chem. 99, 2243–2246.

Hertz, H.G. (1973): in "Water: A comprehensive treatise". Ed. F. Franks, Plenum, New York, Vol. 3, Chap. 7.

Hertz, H.G. (1986): in „The chemical physics of solvation. Part B. Spectroscopy of solvation". Eds. R.R. Dogonadze, E. Kálmán, A.A. Kornyshev, and J. Ulstrup, Elsevier, Amsterdam, Chap. 7.

Hertz, H.G.; Kwatra, B.; Tutsch, R. (1976): Search for A-B association in liquids by the nuclear magnetic relaxation method. The system dimethylformamide – benzene. Z. Phys. Chem. Neue Folge 103, 259–278.

Hertz, H.G.; Leiter, H. (1982): Hydrophobic interactions in aqueous mixtures of methanol, ethanol, acetonitrile and dimethyl formamide. Z. Phys. Chem. Neue Folge 133, 45–67.

Hertz, H.G.; Müller, K.J. (1989): Evidence from ^{19}F relaxation measurements that cations and anions in strong electrolyte solutions at high dilution are associated. Chemica Scripta 29, 277–290.

Hertz, H.G.; Zeidler, M.D. (1964): Determination of the hydration of nonpolar groups in aqueous solution by nuclear magnetic resonance relaxation time measurements. Ber. Bunsen-Ges. Physik. Chem. 68, 821.

Hertz, H.G.; Zeidler, M.D. (1964): Kernmagnetische Relaxationszeitmessungen zur Frage der Hydratation unpolarer Gruppen in wässriger Lösung. Ber. Bunsen-Ges. Phys. Chem. 68, 821–837.

Holz, M. (1995): Nuclear magnetic relaxation as a selective probe of solute-solvent and solute-solute interactions in multi-component mixtures. J. Mol. Liquids 67, 175–191.

Holz, M. (1996): in "Encyclopedia of nuclear magnetic resonance". Eds. D.M. Grant and R.K. Harris, J. Wiley and Sons, Chichester, 1857–1864.

Holz, M.; Heil; S.R.; Sacco, A. (2000): Temperature-dependent self-diffusion coefficients of water and six selected molecular liquids for calibration in accurate ^{1}H NMR PFG measurements. Phys. Chem. Chem. Phys. 2, 4740–4742.

Holz, M.; Mao, X.; Seiferling, D.; Sacco, A. (1996): Experimental study of dynamic isotope effects in molecular liquids. Detection of translation-rotation coupling. J. Chem. Phys. 104, 669–679.

Kell, G.S. (1975): Density of liquid water in the range 0–150 deg. and 0–1 kbar. J. Chem. Eng. Data 20, 97.

Koch, W.; Leiter, H.; Mal, S. (1983): A study of association in the system methanol-CCl$_4$ by the nuclear magnetic relaxation method. Z. Phys. Chem. Neue Folge 136, 89–99.

Koga, Y. (1996): Mixing schemes in aqueous solutions of non-electrolytes: A thermodynamic approach. J. Phys. Chem. 100, 5172.

Lüdemann, S.; Schreiber, H.; Abseher, R.; Steinhauser, O. (1996): The influence of temperature on pairwise hydrophobic interactions of methane-like particles: A molecular dynamics study of free energy. J. Chem. Phys. 104(1), 286–95.

Luo, W.C.; Lay, J.L.; Chen, J.S. (2001): NMR Study of hydrogen bonding association of 2-isopropylphenol in carbon tetrachloride, acetone, dimethyl sulphoxide and acetonitrile. Z. Phys. Chem. (München) 215, 1–12.

Marcus, Y. (1990): Preferential solvation in mixed solvents. Part 5. Binary mixtures of water and organic solvents. J. Chem. Soc., Faraday Trans. 86, 2215.

Matteoli, E. (1997): A study on Kirkwood-Buff integrals and preferential solvation in mixtures with small deviations from ideality and/or with size mismatch of components. Importance of a proper reference system. J. Phys. Chem. B 101, 9800.

Matteoli, E.; Lepori, L. (1984): Solute-solute interactions in water. II. An analysis through the Kirkwood-Buff integrals for 14 organic solutes. J. Chem. Phys. 80, 2856.

Matteoli, E.; Lepori, L. (1990): The ternary system water + 1-propanol + urea at 298.15 K. Activity coefficients, partial molal volumes and Kirkwood-Buff integrals. J. Mol. Liquids. 47, 89.

Matteoli, E.; Lepori, L. (1995): Kirkwood-Buff integrals and preferential solvation in ternary non-electrolyte mixtures. J. Chem. Soc., Faraday Trans. 91, 431.

Matteoli, E.; Mansoori, G.A. (1990): Fluctuation theory of mixtures. Taylor and Francis, New York.

Mayele, M.; Holz, M. (2000): NMR studies on hydrophobic interactions in solution. Part 5: Effect of urea on the hydrophobic self-association of *tert*-butanol in water at different temperatures. Phys. Chem. Chem. Phys. 2, 2429–2434.

Mayele, M.; Holz, M.; Sacco, A. (1999): NMR Studies on Hydrophobic Interactions in Solution. Part 4: Temperature and concentration dependence of hydrophobic self-association of *tert*-butanol in water. Phys. Chem. Chem. Phys. 1, 4615–4618.

McConnell, J. (1987): The theory of nuclear magnetic relaxation in liquids. Cambridge University Press, Cambridge.

Mizuno, K.; Ochi, T.; Shimada, S.; Nishimura, Y.; Maeda, S.; Koga, Y. (1999): Magnetic susceptibility of aqueous *tert*-butanol. Phys. Chem. Chem. Phys. 1, 133.

Müller, K.J.; Hertz, H.G. (1996): A-parameter as an indicator for water-water association in solutions of strong electrolytes. J. Phys. Chem. 100, 1256–1265.

Myers, D. (1992): "Surfactant science and technology", 2nd ed., VCH, New York.

Nandi, P.K.; Robinson, D.R. (1984): Effects of urea and guanidine hydrochloride on peptide and nonpolar groups. Biochemistry 23, 6661.

Nishikawa, K.; Iijima, T. (1990): Structural study of *tert*-butyl alcohol and water mixtures by x-ray diffraction. J. Phys. Chem. 94, 6227.

Ohta, A.; Takiue, T.; Ikeda, N.; Aratono, M. (1998): Calorimetric study of dilute aqueous solutions of ethylene glycol oligomers. J. Phys. Chem. B 102, 4809.

Price, W.S. (2000): Probing molecular dynamics in biochemical and chemical systems using pulsed field gradient NMR diffusion measurements. New Adv. Anal. Chem. P1/31–P1/72.

Sacco, A.; Carbonara, M.; Holz, M. (1989): Translational and rotational molecular motion in solutions of alkali-metal halides in dimethyl sulfoxide. J. Chem. Soc. Faraday Trans I, 85, 1257.

Sacco, A.; Asciolla, A.; Matteoli, E.; Holz, M. (1996): NMR study of the influence of urea on the self-association of propan-1-ol in water and comparison with Kirkwood-Buff integral results. J. Chem. Soc., Faraday Trans. 92, 35–40.

Sacco, A.; De Cillis, F.M.; Holz, M. (1998): NMR studies on hydrophobic interactions in solution. Part 3. Salt effects on the self-association of ethanol in water at two different temperatures. J. Chem. Soc., Faraday Trans. 94, 2089.

Sacco, A.; Holz, M. (1997): NMR studies on hydrophobic interactions in solution. Part 2. Temperature and urea effect on the self-association of ethanol in water. J. Chem. Soc., Faraday Trans. 93, 1101.

Schiffer, C.A.; Dötsch, V.; Wüthrich, K.; van Gunsteren, W.F. (1995): Exploring the role of the solvent in the denaturation of a protein: A molecular dynamics study of the DNA binding domain of the 434 repressor. Biochemistry 34, 15,057.

Stejskal, E.O.; Tanner, J.E. (1965): Spin diffusion measurements: Spin echoes in the presence of a time-dependent field gradient. J. Chem. Phys. 42, 288.

Tanaka, H.; Nakanishi, K.; Nishikawa, K. (1984): Clathrate-like structure of water around some nonelectrolytes in dilute solution as revealed by computer simulation and x-ray diffraction studies. J. Inclusion Phenom. 2, 119.

Thirumalai, D.; Klimov, D.K.; Woodson, S.W. (1997): Kinetic partitioning mechanism as a unifying theme in the folding of biomolecules. Theor. Chem. Acc. 96, 14.

Tirado-Rives, J.; Jorgensen, W.L.; Orozco, M.D. (1997): Molecular dynamics simulations of the unfolding of barnase in water and 8 M aqueous urea. Biochemistry 36, 7313.

Tkadlecova, M.; Dohnal, V.; Costas, M. (1999): ^{1}H NMR and thermodynamic study of self-association and complex formation equilibria by hydrogen bonding. Phys. Chem. Chem. Phys.1, 1479–1486.

Vanzi, F.; Madan, B.; Sharp, K. (1998): Effect of the protein denaturants urea and guanidinium on water structure: A structural and thermodynamic study. J. Am. Chem. Soc. 120, 10748.

Weast, R.C. (1979) (Ed.): CRC-Handbook of Chemistry and Physics. 59th ed., CRC-Press, Boca Raton.

Wetlaufer, D.B.; Malik, S.K.; Stoller, L.; Coffin, R.I. (1964): Nonpolar group participation in the denaturation of proteins by urea and guanidinium salts. Model compound studies. J. Am. Chem. Soc. 86, 508.

Zeidler, M.D. (1965): Determination of the time of reorientation, time of transition, and of the quadrupole coupling constants for certain organic liquids by measurements of the nuclear magnetic relaxation time. Ber. Bunsen-Ges. Phys. Chem. 69, 659.

Nomenclature

$1/T_1$	nuclear magnetic spin-lattice relaxation rate
$1/T_{1,\,inter}$	intermolecular spin-lattice relaxation rate
$1/T_{1,\,intra}$	intramolecular spin-lattice relaxation rate
$1/T_{1,\,tot}$	total spin-lattice relaxation rate
$1/T_2$	spin-spin relaxation rate
a	closest distance of approach between the interacting nuclei
A_{ij}	association parameter between components i and j
A_{22}	self-association parameter of an organic component (denoted as component 2) in an aqueous mixture
B_D	"B-coefficient" derived from the concentration dependence of the water self-diffusion coefficient in diluted aqueous solutions
c'	nuclear spin number density (spins/cm^3)

c_{bulk}	bulk concentration
c_{local}	local concentration
D, D_i, D_j	self-diffusion coefficient, self-diffusion coefficient of mixture component i, j; i, j = 1, 2, 3
D_0	self-diffusion coefficient of the reference solvent, usually of water
D_{av}	mean self-diffusion coefficient, $D_{av} = \frac{1}{2}(D_i + D_j)$
D-D	dipole – dipole
EG	ethylene glycol
FT	Fourier transform
$G(p,T,n_i)$	free enthalpy
G	amplitude of the magnetic field gradient pulse
G_{ij}	Kirkwood Buff integral (KBI)
$g_{ij}(r)$	atom-atom pair distribution function of the atoms carrying the interacting nuclei and residing on molecules i and j
$\hbar$	Planck constant $h/2$
m	molality: number of moles of solute per 1 kg of solvent
$\bar{m}$	aqua-molality: number of moles of solute per 55.5 moles of solvent
$M(\tau)$	macroscopic magnetization after a delay (evolution) time τ
M_0	equilibrium nuclear magnetization at $\tau = 0$
M_i	molar mass of the component i in the mixture
N_A	Avogadro number
n_c	first co-ordination number
n_c^*	reduced first co-ordination number, $= n_c/c'$
n_h	hydration number, number of water molecules in the first coordination sphere of a solute
n_i	number of moles of component i
n_p	number of hydrogen atoms per molecule
p	pressure
PFGSE	Pulsed Field Gradient Spin-Echo
r	separation distance between atoms carrying the interacting nuclei
rf	radio-frequency
T	temperature
$\bar{v}_i$	partial molecular volume of the component i in a mixture
x_i	mole fraction of component i

Greek letters

δ	duration of the magnetic field gradient pulse
Δ	time interval between two magnetic field gradient pulses
γ	gyromagnetic ratio
γ_i	activity coefficient of component i
κ_T	isotherm compressibility
ρ	density
τ	time delay between two radio frequency (rf) pulses

Nomenclature

Subscripts

0	delay time $\tau = 0$ for M_0, neat liquid for D_0
1, 2, 3, ...	components denoted 1, 2, 3, ... (1 = water, 2 = organic component, 3 = additive)
c	co-ordination
h	hydration
i, j	mixture component i, j (i, j = 1, 2, 3)
p	hydrogen (proton)

III Aqueous Solutions of Strong Electrolytes

9 Phase Equilibria of Aqueous Solutions Containing Volatile Electrolytes

Jürgen Zipprian[1*], Nils Elm, and Karlheinz Schaber

9.1 Abstract

A new technique has been developed for measuring the vapour-liquid equilibrium of aqueous solutions containing volatile electrolytes in highly diluted vapour phases (<300 ppm). The analysis of the vapour phase is performed by FT-IR spectroscopy. The apparatus has been validated by VLE measurements with pure water. In the presented work binary and ternary aqueous mixtures with the components H_2O, NH_3, HCl, HBr, and $CaCl_2$ have been investigated. The concentrations of the solutions were regulated to obtain electrolyte partial pressures between 0.15 and 30 Pa. The experiments have been carried out at temperatures between 30 and 70 °C. Equilibrium data in this range of temperature and concentration are important especially for the design of flue gas cleaning equipment. The measurements of the $HCl + H_2O$ system yield lower partial pressures of HCl in comparison with data taken from literature.

Thermodynamic models are presented to calculate the electrolyte partial pressures over their aqueous solutions.

9.2 Introduction

Phase equilibria of aqueous solutions containing volatile electrolytes in highly diluted vapour phases are of particular interest in gas purification and in environmental technology. For the design of industrial absorption processes with stringent limits concerning emissions of components like HCl or HBr accurate phase equilibria data are required, especially at concentrations of the volatile electrolytes lower than 300 ppm. Only for the $HCl + H_2O$ system sufficient data are available also at very low vapour phase concentrations of HCl. Up to

---
[1] Universität Karlsruhe (TH), Institut für Technische Thermodynamik und Kältetechnik, Engler-Bunte-Ring 21, 76128 Karlsruhe.
* Author to whom correspondence should be addressed.

now the data of Zeisberg (1925) which are published in Perry's Chemical Engineers' Handbook (1984) are applied commonly. But these data, i.e. the HCl partial pressures at very low concentrations are in contradiction with newer data of Fritz and Fuget (1956) who indicate lower HCl partial pressures. This is often the consequence of the extrapolation of experimental results at high concentrations with mathematical models. Partial pressures for the system HBr + H$_2$O are available for solutions with 48 to 60 mass% at 50 and 55 °C (Perry and Green, 1984). Thermodynamic properties of ammonia are reported over the entire concentration range, but there are no data available for very diluted solutions (Macriss et al., 1964). Only a few systems of more than one electrolyte like the ternary system HCl + HBr + H$_2$O have been investigated (Lutugina et al., 1981). In general most of all published experimental data are limited to high concentrations and temperatures.

In this work binary and ternary solutions of strong and weak volatile electrolytes at partial pressures of the electrolytes below 30 Pa in presence of an inert gas are investigated. In contrast to most of all investigations so far known, in this work the concentrations of the vapour phase are measured directly with FT-IR spectroscopy. For this purpose a new apparatus was developed. The parts in contact with the solution consist of glass or resistant plastics, so there is no limitation in concentration. The spectrometer is equipped with a heatable gas cell to avoid condensation. The 6.4 m path length of the cell enables the detection of substances in the range of a few ppm. In the present paper the apparatus and the experimental procedures are described. The measured partial pressures of the electrolytes are shown in diagrams, including literature data wherever possible.

9.3 Experimental

9.3.1 Apparatus

The experimental set-up consists of two main parts: the calibration system for the FT-IR spectrometer and the equilibrium cell. A schematic diagram is given in Figure 9.1. The two sections are separated by the valves V 2, V 3, V 4, and V 5.

The spectrometer has to be calibrated for the substances of interest. For this purpose, gas mixtures of nitrogen with variable concentrations of H$_2$O, HCl, HBr, and NH$_3$ are generated in the calibration system. The calibration system is placed in the air thermostat (7), whose temperature is held at 120 °C. Water or a solution containing an electrolyte is furthered by hose pump (5) from glass vessel (4) in the evaporator (8), where it is completely evaporated. The mass flow is measured by weighing the glass vessel. The vapour is mixed with a preheated nitrogen flow (9). The heater (11) regulates the calibration gas temperature to 120 °C before the gas stream enters the cell of the spectrometer. Behind the thermostat, water and the electrolyte are removed in a condenser (14) to measure the volume flux of N$_2$ in a soap film bubble flow meter (15). The mole fractions of the components in the gas mixture are calculated with the massflow of the solutions, its concentrations, and the volume flux of N$_2$. Finally, the partial pressures are calculated with respect to the total pressure in the gas cell.

9.3 Experimental

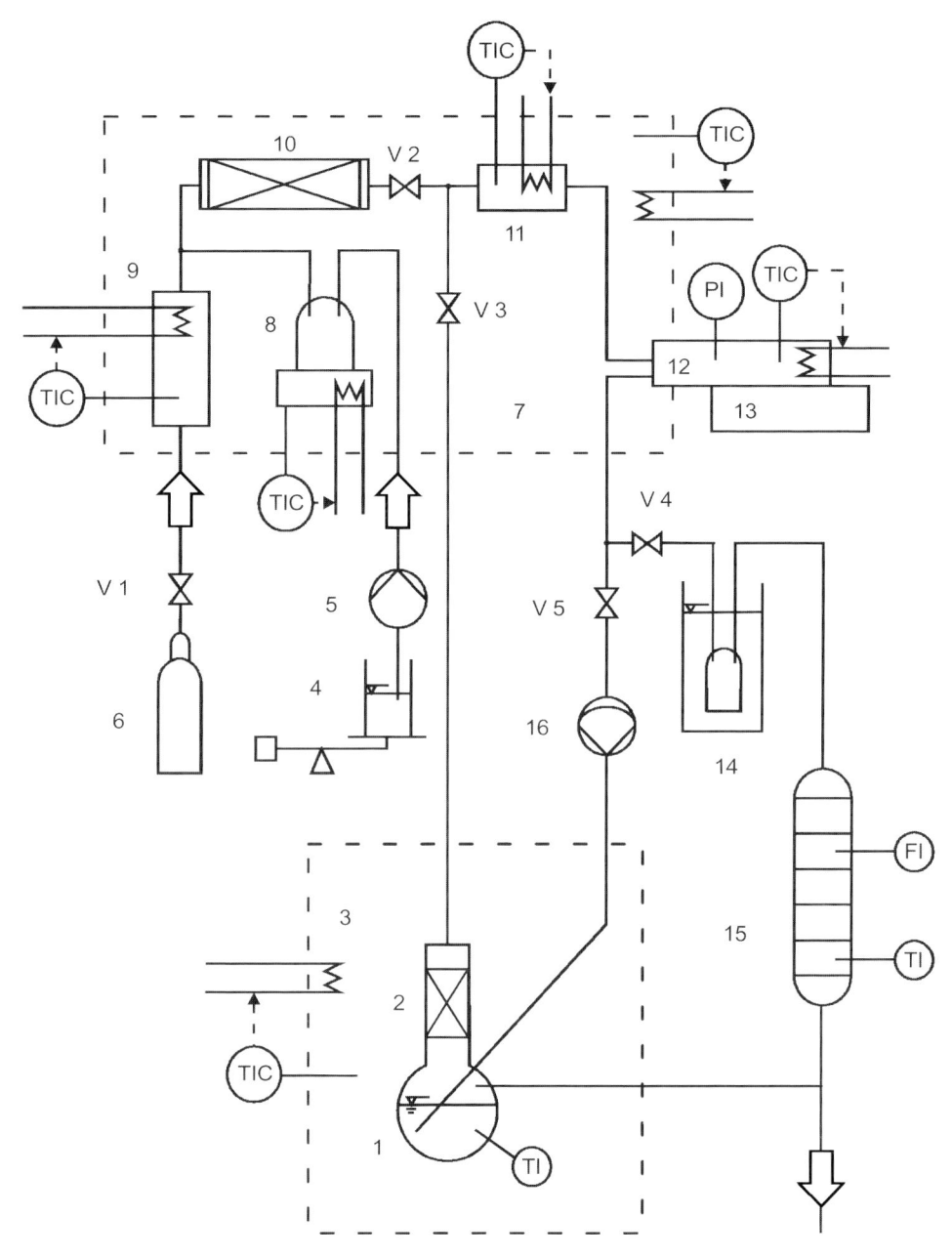

Figure 9.1: Experimental set-up: 1, glass equilibrium cell; 2, mist collector; 3, air thermostat; 4, high precision balance with glass vessel; 5, hose pump; 6, compressed N_2 bottle; 7, air thermostat; 8, evaporator; 9, N_2 heater; 10, mixer; 11, gas heater; 12, heatable gas cell; 13, FT-IR spectrometer; 14, condenser; 15, soap film bubble flow meter; 16, heatable diaphragm pump.

9 *Phase Equilibria of Aqueous Solutions Containing Volatile Electrolytes*

The equilibrium measurements are performed in the presence of air at atmospheric pressure. The electrolyte concentrations in the calibration solution as well as for the phase equilibria measurements are set by diluting and weighing commercial standard solutions. There is no need for further analysis. The mass fractions of HCl, HBr, or NH_3 in the liquid phase can be adjusted with an accuracy of 1.2% of the indicated value. About 1 dm^3 of the solution is filled in the glass equilibrium cell (1), which is placed in the air thermostat (3). This liquid quantity is high enough that the displacement of volatile electrolytic components into the gas phase during the equilibrium measurements at different temperatures can be neglected. Also the losses via a Teflon tube of 3 m length and an inner diameter of 3 mm, which maintains atmospheric pressure in the equilibrium cell, can be neglected. The temperature in the cell is measured by a platinum resistance thermometer with an accuracy of 0.05 °C.

The vapour over the solution is sucked by a heatable diaphragm pump (16). Drops are precipitated in a mist collector (2). The vapour is superheated in the heater (11). To avoid cold spots the Teflon tube between both thermostats lies in a copper pipe with big wall thickness, which is heated in the thermostat (7). For the same reason the inlet and the outlet of the heatable gas cell of the spectrometer are placed in the thermostat. After the pump the vapour is bubbled through the solution. The agitation is so efficient that a supplementary stirrer is not necessary.

The accuracy of the partial pressure measurements is about 3.5% for water. It is lower than 5.0% for electrolyte partial pressures below 3 Pa and 3.5% at higher values.

9.3.2 IR-Spectra and Calibration Curves

A spectrometer type MB 100 (BOMEM/Hartmann & Braun) has been used. Infrared spectrums can be measured within the wavenumber range from 750 to 6000 cm^{-1}. Figure 9.2 shows the spectrum of water at a partial pressure of 6.3 kPa in ambient air.

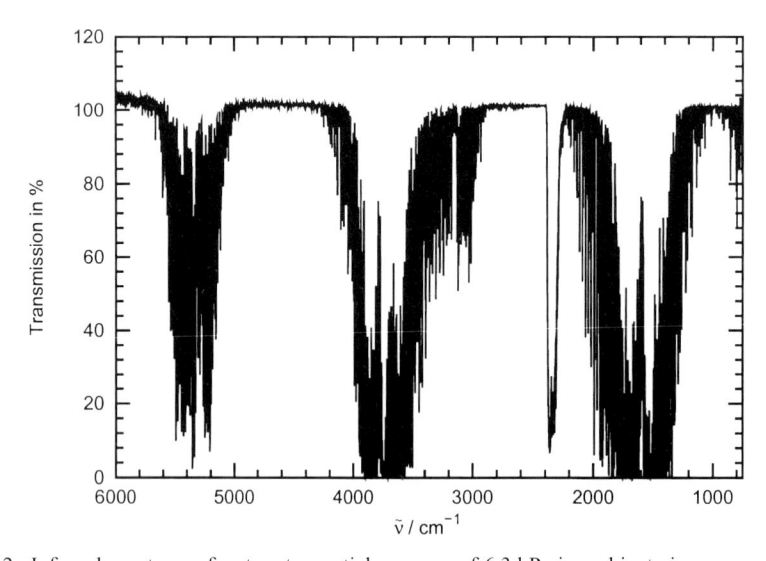

Figure 9.2: Infrared spectrum of water at a partial pressure of 6.3 kPa in ambient air.

9.3 Experimental

The peak between 2260 and 2393 cm^{-1} results from 300 ppm CO_2 in the air. The other peaks are due to the absorbency of water. Only little ranges of the entire spectrum can be used to measure the partial pressure of the volatile electrolytes because of the extensive water peaks.

The characteristic peaks for the substances analysed in this work are listed in Table 9.1 together with interference with other substances.

Table 9.1: Wavenumber ranges for the analysis of different substances and interferences of absorption peaks.

Spectral range	Wavenumber/cm^{-1}	Substances and interferences
range I	3007.2854 – 3018.3753	H_2O influenced by HCl
range II	2818.2753 – 2823.0970	HCl influenced by H_2O
range III	1133.0962 – 1138.8822	H_2O influenced by NH_3
range IV	957.10463 – 969.64101	NH_3 influenced by H_2O
range V	2468.2209 – 2472.5605	HBr

For HBr there is no interference with water. The influence of water can be taken into account by a simple mathematical method. This will be demonstrated for HCl. The procedure is likewise for NH_3.

The calibration curve is given by the peak area A_i of the substance as a function of the partial pressure p_i. For HCl it is necessary to determine the area of water in the range II as a function of the area in a different range without HCl absorbency, for example range III. Experiments have shown that this function is given by:

$$A_{H_2O, \text{range II}} = 0.03 \cdot A_{H_2O, \text{range III}} \qquad (9.1)$$

The absorbency of water in the range II is so small that the total area of the peak can be taken as the sum of the area caused by HCl and that caused by water. The calibration curve is given by the Eq. (9.2) which is shown in Figure 9.3.

$$A_{HCl} = A_{\text{range II}} - 0.03 \cdot A_{H_2O, \text{range III}} = f(p_{HCl}) \qquad (9.2)$$

Up to partial pressures of 4 Pa the function is linear according to Lambert-Beer's law. At higher partial pressures the slope decreases. For each substance analysed in this study the calibration curve has the same form. The curves are always separated in the linear part and the function for higher partial pressures.

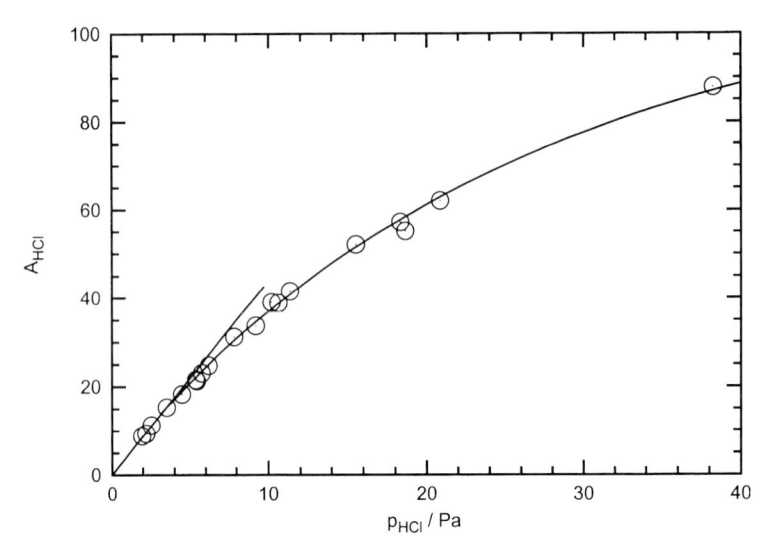

Figure 9.3: Calibration curve for HCl up to a partial pressure of 40 Pa.

9.4 Results and Discussion

The apparatus has been validated by measuring the partial pressure of pure water. The agreement between the results and the literature values is quite satisfying. Both the calibration system and the phase-equilibrium apparatus are shown to be convenient for the required experiments. The analysis of the experimental procedure shows that the liquid phase concentration can be adjusted with an accuracy better than 1.2% of the given mole or mass fractions. All concentrations are given in mass%.

9.4.1 System $NH_3 + H_2O$

The first system contains the volatile weak electrolyte NH_3. A weak electrolyte dissociates scarcely in water, that means molecules prevail against ionic species. The molecular form accounts for much greater partial pressures in comparison to strong electrolytes. Thus the mass fractions of NH_3 in the solutions analysed in the present work are extremely low compared to those of strong electrolytes. Because of the highly diluted liquid phase in the $NH_3 + H_2O$ system the measurements are more susceptible to failures in the performance of the experiments than the measurements with HCl or HBr. High accuracy for measurements of solutions of strong electrolytes can be expected, if the partial pressure of a weak electrolyte can be measured with high precision.

9.4 Results and Discussion

Solutions with ammonia concentrations between 9.9×10^{-4} and 0.01% were analysed (Table 9.2). The partial pressures over these solutions are in the range of those measured with the strong electrolytes.

Figure 9.4 shows the experimental results for NH_3 and calculations with a model from Edwards et al. (1975, 1978).

Table 9.2: Partial pressures of NH_3 and H_2O over aqueous solutions of NH_3 as a function of the concentration in mass% and the temperature.

% NH_3	$t/°C$	p_{NH_3}/Pa	p_{H_2O}/Pa
9.9×10^{-4}	36.4	1.46	6120
	45.6	2.08	9790
	54.8	2.83	15,400
	31.9	1.12	4740
	64.2	3.67	24,620
5.2×10^{-3}	31.8	6.68	4640
	41.0	9.85	7480
	36.5	8.04	6100
	45.7	11.74	9750
	54.8	16.87	15,160
	31.9	6.49	4690
	50.3	13.62	11,680
1.0×10^{-2}	41.0	19.73	7630
	30.4	11.85	4320
	36.5	15.57	6080
	41.1	19.29	7790

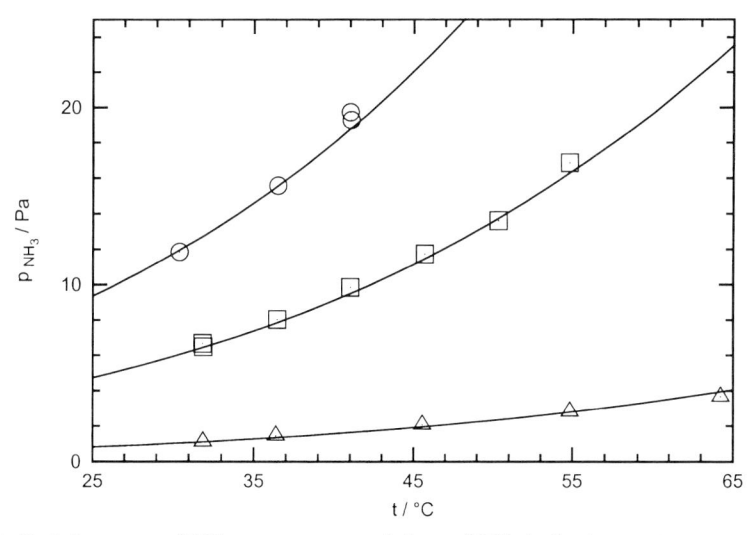

Figure 9.4: Partial pressure of NH_3 over aqueous solutions of NH_3 in the temperature range from 25 to 65 °C. ○ 1.0×10^{-2} mass%; □ 5.2×10^{-3} mass%; △ 9.9×10^{-4} mass%; —— correlation (Edwards et al., 1975, 1978).

9 Phase Equilibria of Aqueous Solutions Containing Volatile Electrolytes

The agreement between the own experimental data and the correlation of Edwards et al. is very good at each given temperature and concentration. The largest difference is 6.3% at 9.9×10^{-4} mass%. At small amounts of ionic species (less than 0.01 mass%) in the solution no decrease of the water partial pressure can be observed. The measured values are equal to the vapour pressure of pure water taken from literature (VDI, 1963) within the experimental accuracy of 3.5%.

Table 9.3: Partial pressures of HCl and H$_2$O over aqueous solutions of HCl as a function of the concentration in mass% and the temperature.

% HCl	$t/°C$	p_{HCl}/Pa	p_{H_2O}/Pa
4.5	41.1	0.15	7450
	45.8	0.26	9330
	50.4	0.25	11,760
	50.5	0.27	11,750
	55.1	0.34	14,700
	60.0	0.58	18,460
	64.6	0.78	22,860
	64.7	0.83	23,320
	69.7	1.14	27,400
6.1	31.8	0.15	4300
	41.1	0.37	7250
	50.2	0.81	11,330
	45.6	0.57	9160
	54.8	1.23	14,330
	64.2	2.36	22,420
7.5	31.5	0.42	4220
	36.5	0.53	5540
	45.8	1.11	8970
	55.4	2.17	14,120
	64.9	4.20	22,090
	69.8	6.10	25,320
9.0	41.1	1.56	6840
	50.4	2.95	10,760
	59.8	5.87	16,670
	69.3	12.33	25,660
	31.7	0.89	4090
11.9	41.3	6.58	6520
	50.5	13.27	10,310
	54.9	18.63	12,800
	59.7	27.32	16,030
	41.1	6.76	6510
	45.8	9.54	8230
	55.0	18.42	12,760
	31.8	3.08	3920
	36.5	4.60	5090
	41.1	6.69	6540
	31.5	3.12	3870
15.0	31.6	9.48	3410
	41.1	19.92	5890
	45.7	28.06	7400

9.4.2 System HCl + H₂O

The phase equilibrium of aqueous solutions of HCl was measured at six concentrations in the temperature range from 30 to 70 °C. Mass fractions between 4.5 and 15.0 % were chosen to obtain HCl partial pressures down to the spectrometer resolution limit of 0.15 Pa and up to 30 Pa. The results are listed in Table 9.3. Figure 9.5 shows the logarithm of the HCl partial pressure versus the reciprocal temperature.

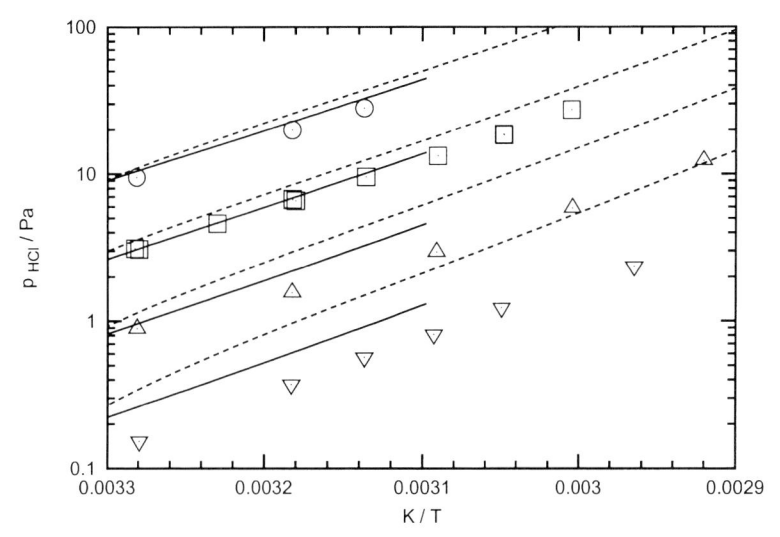

Figure 9.5: Partial pressure of HCl over aqueous solutions of HCl in the temperature range from 30 to 70 °C. ○ 15.0 mass%; □ 11.9 mass%; △ 9.0 mass%; ▽ 6.1 mass%; ——— , Fitz and Fuget (1956); - - - - - - , Zeisberg (1925).

In this type of diagram the saturation vapour pressure of a pure substance can be considered with good accuracy as a straight line. The partial pressure of HCl exhibits the same behaviour. The curves of constant mass fractions are nearly parallel lines. This effect is observed also for data taken from literature. That means these curves can easily be extrapolated with reference to the temperature by prolonging the lines.

In Figure 9.5 one can see that the slope of the curves is nearly the same for all concentrations. A simple linear empirical equation to estimate the measured partial pressure of HCl is:

$$\log_{10} p_{HCl} = \log_{10}\left[f(\xi_{HCl})\right] - 3263.7076 \cdot \left(\frac{1.0}{(T/K)} - \frac{1.0}{304.8}\right) \quad (9.3)$$

The function $f(\xi_{HCl})$ gives the measured partial pressure at 304.8 K:

$$f(\xi_{HCl}) = (a \cdot \xi_{HCl} + b \cdot \xi_{HCl}^2 + c \cdot \xi_{HCl}^3 + d \cdot \xi_{HCl}^4) \cdot 100 \quad (9.4)$$

9 Phase Equilibria of Aqueous Solutions Containing Volatile Electrolytes

The value of the coefficients a, b, c, d are listed in Table 9.4. Figure 9.6 shows the data points in comparison to the calculated values.

Table 9.4: Coefficients of Eq. (9.4) for HCl at 304.8 K and HBr at 314.7 K.

	a	b	c	d
HCl	−0.10254	4.98718	−75.74387	500.93369
HBr	−1.16963	1.60575	−6.57087	8.49902

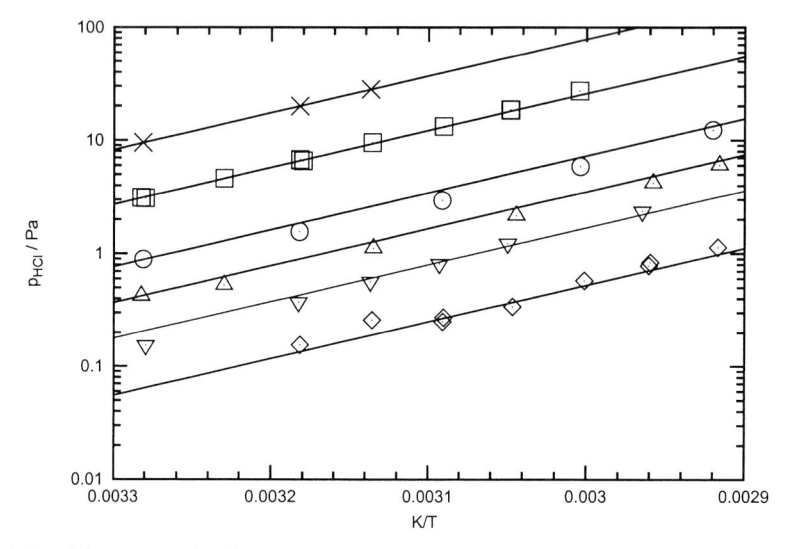

Figure 9.6: Partial pressure of HCl over aqueous solutions of HCl in the temperature range from 30 to 70 °C. X 15.0 mass%; □ 11.9 mass%; ○ 9.0 mass%; △ 7.5 mass%; ▽ 6.1 mass%; 4.5 mass%; —— calculated curves.

Some experiments were carried out two or three times to demonstrate the reproducibility of the results which is better than ±3.0% in pressure. The broken lines in Figure 9.5 show the data given by Zeisberg (1925). The solid curves represent data reported by Fritz and Fuget (1956).

In 1925 Zeisberg evaluated the data available at that time and compiled a table of partial vapour pressures of HCl and H$_2$O for the International Critical Tables (1928). Fritz and Fuget evaluated Zeisberg's estimations concerning the accuracy of his data as somewhat optimistic. They calculated the HCl partial pressures from electromotive force data of Harned and Ehlers (1932, 1933) and Åkerlöf and Teare (1937). Over the entire range of concentration Fritz and Fuget report lower values than Zeisberg. As an example, the partial pressure above a solution with 11.9% HCl at 41.1 °C measured in our work is 1.9% lower than the value of Fritz and Fuget. For all other concentrations our measurements are lower than the data reported in the literature. Each spectrum gives the partial pressure of the volatile electrolyte and that of water. The vapour pressure is about a thousand times higher than that of

HCl. For these quantities all authors give nearly the same values. The maximal difference of all reported data is about 2%. The agreement between the data of this work and the literature data is also within this range as shown in Table 9.3.

9.4.3 System HBr + H$_2$O

The absorption of IR radiation by HBr is much lower than the absorption by HCl. The area of the peak in the range V is about a quarter of that of the peak in the range II at the same partial pressures of HBr and HCl. This leads to a linear calibration function up to 20 Pa. Three different solutions with concentrations between 32.0 and 39.0% have been analysed in the temperature range from 40 to 65 °C. The results are given in Table 9.5. The partial pressure of HBr over its aqueous solution is shown in Figure 9.7. The maximum pressure is 19 Pa at 60.2 °C.

Table 9.5: Partial pressures of HBr and H$_2$O over aqueous solutions of HBr as a function of the concentration in mass% and the temperature.

% HBr	$t/°C$	p_{HBr}/Pa	p_{H_2O}/Pa
32.0	41.6	0.83	5280
	50.9	1.42	8470
	46.1	1.06	6750
	55.4	1.88	10,570
	60.1	2.64	13,150
	64.8	3.49	16,250
36.0	41.5	2.12	4560
	50.7	4.31	7340
	60.2	7.99	11,360
	46.1	3.09	5860
	55.5	5.89	9190
	65.0	11.02	14,020
39.0	41.2	5.41	4010
	50.2	10.94	6330
	60.2	19.23	10,090

As it was seen for HCl the logarithm of the HBr partial pressure is also a linear function of the reciprocal temperature. The modified Eqs. (9.3) and (9.4) can be used to calculate the HBr partial pressures. The values of the coefficients of HBr are listed in Table 9.4. The slope of the curves is taken to be (−3227.4414).

It is necessary to calculate the molality of the electrolyte from the mass fraction to compare the partial pressure of HCl and HBr. The molar mass of HBr is about twice of that of HCl. Therefore the molality of HBr has approximately the same value as the molality of HCl, if the mass fraction is doubled. It is obvious that the partial pressure of HBr is much lower than that of HCl given the same molality and temperature.

9 Phase Equilibria of Aqueous Solutions Containing Volatile Electrolytes

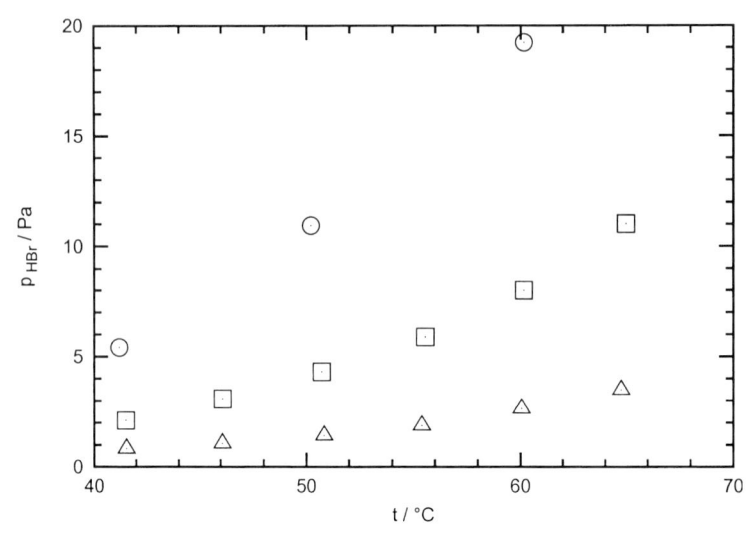

Figure 9.7: Partial pressure of HBr over aqueous solutions of HBr in the temperature range from 40 to 70 °C. ○ 39.0 mass%; □ 36.0 mass%; △ 32.0 mass%.

The high amounts of ionic species in the solution cause an important reduction of the water partial pressure as shown in Table 9.5.

9.4.4 System HCl + CaCl$_2$ + H$_2$O

This ternary system contains the volatile electrolyte HCl and the non volatile salt CaCl$_2$. Three solutions with 10% CaCl$_2$ and 1.0, 3.0, and 5.0% HCl, respectively, and a solution with 5.0% of both electrolytes were analysed. Figure 9.8 shows the measured partial pressures of HCl above the four different solutions. The experimental values are listed in Table 9.6.

In comparison to the binary system HCl + H$_2$O lower mass fractions of HCl were chosen to obtain HCl partial pressures from 0.16 to 20.7 Pa. The two electrolytes are sharing the common ion Cl$^{(-)}$. The partial pressure of HCl is increased by the salt. The recombination of dissociated HCl molecules is intensified by the higher concentration of the common ion. Over the entire temperature range the HCl partial pressure above the solution containing 9.9% CaCl$_2$ and 3.0% HCl is nearly the same as that above the solution with 5.0% of both electrolytes. In that case the increasing of the partial pressure caused by the salt compensates the decreasing by the lower HCl mass fraction.

Figure 9.8 shows the logarithm of the HCl partial pressure versus the reciprocal temperature in the unit of Kelvin. In this diagram the partial pressure is a linear function as it was seen for the binary systems HCl + H$_2$O. Furthermore the slope of the curves is the same for the binary and for the ternary systems. The presence of high amounts of CaCl$_2$ do not influence the shape of the curve. This fact can be used for extrapolating data points.

9.4 Results and Discussion

Table 9.6: Partial pressures of HCl and H$_2$O over aqueous solutions of HCl and CaCl$_2$ as a function of the concentration in mass% and the temperature.

% HCl	% CaCl$_2$	$t/°C$	p_{HCl}/Pa	p_{H_2O}/Pa
1.0	10.1	41.2	0.16	7200
		50.4	0.20	11,340
		59.8	0.38	17,540
		69.3	0.80	27,750
3.0	9.9	31.7	0.40	4200
		41.1	0.79	6960
		50.5	1.65	11,030
		59.9	3.00	17,110
		69.3	5.57	26,880
5.0	10.0	41.0	2.57	6530
		50.3	5.23	10,360
		59.7	10.72	16,060
		69.1	20.66	24,890
		31.6	1.40	3910
5.0	5.0	41.2	0.86	7000
		50.2	1.57	10,950
		59.7	2.79	16,930
		69.1	5.44	25,970
		31.7	0.41	4230

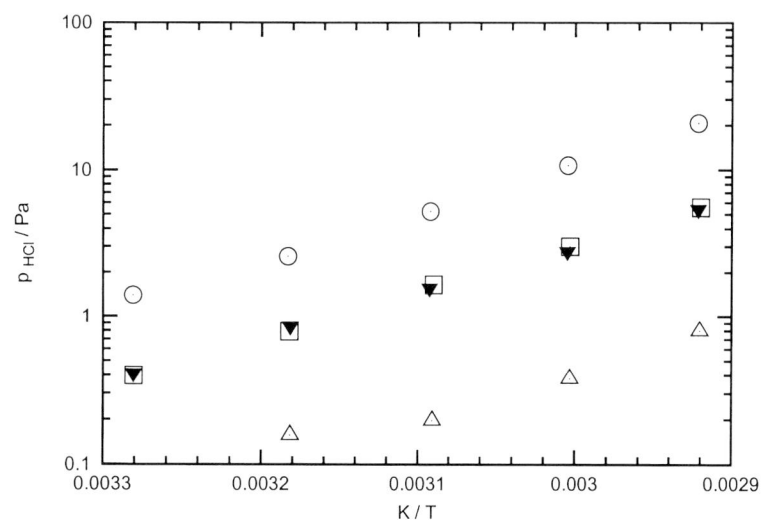

Figure 9.8: Partial pressure of HCl over aqueous solutions of HCl and CaCl$_2$ in the temperature range from 30 to 70 °C. ○ 5.0 mass% HCl, 10.0 mass% CaCl$_2$; □ 3.0 mass% HCl, 9.9 mass% CaCl$_2$; △ 1.0 mass% HCl, 10.1 mass% CaCl$_2$; ▼ 5.0 mass% HCl, 5.0 mass% CaCl$_2$.

9.4.5 System HCl + HBr + H$_2$O

The second ternary system which has been investigated contains the two volatile electrolytes HCl and HBr. Equilibrium data were measured for six different solutions with 28.0, 32.0, and 36.0% HBr. The HCl mass fractions were regulated to values between 0.2 and 1.0%. All experimental data are listed in Tables 9.7 and 9.8.

The most important relations are demonstrated in two diagrams. Figure 9.9 shows the partial pressure of HCl above three solutions containing 32.0% HBr and 0.20, 0.40, and 1.0% HCl, respectively, as a function of temperature. The two electrolytes are sharing the common ion H$^{(+)}$. This effects an increase of the partial pressures as it was seen for the system HCl + CaCl$_2$ + H$_2$O. Because of the high HBr concentration, very low HCl mass fractions are sufficient to obtain HCl partial pressures up to 26 Pa.

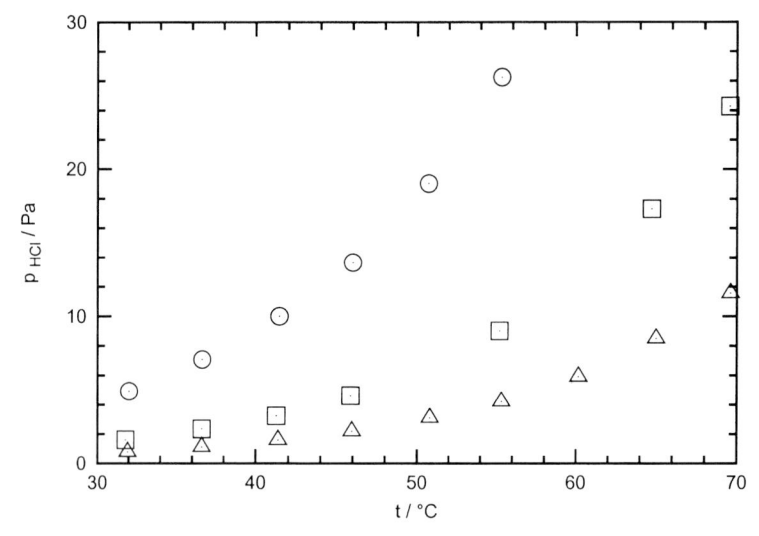

Figure 9.9: Partial pressure of HCl over aqueous solutions of HCl and HBr in the temperature range from 30 to 70 °C. ○ 1.0 mass% HCl, 32.0 mass% HBr; □ 0.40 mass% HCl, 32.0 mass% HBr; △ 0.20 mass% HCl, 32.0 mass% HBr.

In Figure 9.10 the corresponding HBr partial pressures are marked in together with values of the solutions containing 28% HBr. At the same temperature and mass fraction of HCl the HBr partial pressure above the solution with 32.0% HBr is more than two times higher than that above the solution with 28% HBr. In this diagram the effect of HCl on the partial pressure of HBr is demonstrated very clearly. The augmentation of the HCl mass fraction from 0.2 to 1.0% causes an increase of the HBr partial pressure of nearly 50%. As it was seen before the logarithm of the HCl partial pressure is a linear function of the reciprocal temperature. Even in the case of very low HCl concentrations of the other electrolyte the slope of the curves is nearly the same as for the binary system HCl + H$_2$O.

9.4 Results and Discussion

Table 9.7: Partial pressures of HCl, HBr, and H$_2$O over aqueous solutions of HCl and HBr as a function of the concentration in mass% and the temperature.

% HCl	% HBr	$t/°C$	p_{HCl}/Pa	p_{HBr}/Pa	p_{H_2O}/Pa
0.24	27.7	45.8	0.98	0.81	7310
		50.5	1.37	0.72	9210
		55.2	1.93	0.86	11,500
		60.0	2.68	1.02	14,230
		64.8	3.85	1.28	17,760
		69.5	5.36	1.69	21,820
1.0	28.0	46.0	5.68	0.76	7020
		50.5	7.97	0.89	8740
		55.3	11.31	1.30	10,990
		59.9	15.51	1.46	13,500
		64.7	22.33	1.99	16,900
		69.4	32.16	2.54	20,750
0.20	32.0	36.6	1.11	0.66	4010
		46.0	2.20	1.24	6620
		55.3	4.21	2.01	10,320
		41.4	1.59	0.82	5220
		65.0	8.50	3.95	16,010
		69.6	11.57	5.17	19,710
		31.9	0.78	0.73	3110
		50.8	3.10	1.70	8330
		60.2	5.91	2.73	12,770

Table 9.8: Partial pressures of HCl, HBr, and H$_2$O over aqueous solutions of HCl and HBr as a function of the concentration in mass% and the temperature.

% HCl	% HBr	$t/°C$	p_{HCl}/Pa	p_{HBr}/Pa	p_{H_2O}/Pa
0.40	32.0	36.6	2.35	0.89	3940
		45.9	4.63	1.38	6510
		55.2	9.02	2.24	10,170
		31.8	1.61	0.51	3090
		41.3	3.24	0.95	5070
		64.7	17.31	3.90	15,740
		69.6	24.28	5.38	19,530
1.0	32.0	36.6	7.08	0.93	3810
		46.0	13.66	1.72	6260
		55.3	26.26	3.03	9790
		32.0	4.93	0.71	3040
		41.5	10.01	1.23	4980
		50.7	19.03	2.15	7950
0.20	36.0	31.9	2.14	1.23	2780
		41.2	4.10	2.48	4460
		50.6	7.96	4.96	7210
		60.0	14.80	8.92	11,180
		69.6	27.57	16.32	17,030

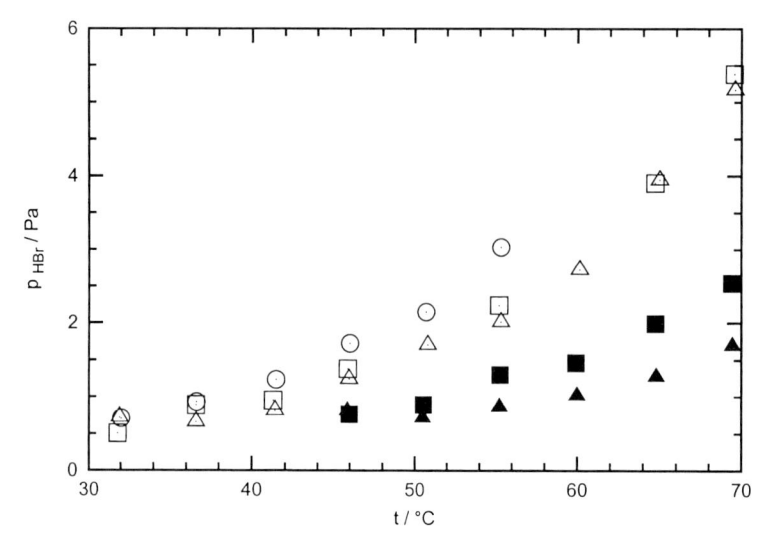

Figure 9.10: Partial pressure of HBr over aqueous solutions of HCl and HBr in the temperature range from 30 to 70 °C. ○ 1.0 mass% HCl, 32.0 mass% HBr; □ 0.40 mass% HCl, 32.0 mass% HBr; △ 0.20 mass% HCl, 32.0 mass% HBr; ■ 1.0 mass% HCl, 28.0 mass% HBr; ▲ 0.24 mass% HCl, 27.7 mass% HBr.

9.5 Theory

Phase equilibria of aqueous solutions of HCl are calculated with assumption of total dissociation (Elm, 2000). The partial pressures of HCl and H$_2$O are calculated separately as a function of temperature and mass fraction.

The partial pressure of HCl is calculated with a summarized equation of Henry's law and the dissociation constant (Elm, 2000).

$$p_{HCl} = m_{HCl}^2 \left(\gamma_{HCl}^r\right)^2 H_{HCl}^* \tag{9.5}$$

The rational activity coefficient is obtained with an equation from Pitzer (1973, 1974). In Eq. (9.6) parameters describing the molecule-molecule interaction of HCl were not taken into account.

$$\ln \gamma_{HCl}^r = f^\gamma + m_{HCl} B_{HCl}^\gamma + \frac{3}{2} m_{HCl}^2 C_{HCl}^\Phi + \frac{4}{3} m_{HCl}^3 D_{HCl} \tag{9.6}$$

Henry's constant from Eq. (9.5) is fitted to the semiempirical Eq. (9.7).

$$\ln H_{HCl}^* = H_1 + H_2 T + \frac{H_3}{T} + H_4 \ln T \tag{9.7}$$

f^γ includes the Debye-Hückel constant A_Φ which is calculated with the equation reported by Bradley and Pitzer (1979).

$$f^\gamma = A_\Phi \left[\frac{\sqrt{I}}{1 + 1.2\sqrt{I}} + \frac{2}{1.2} \ln\left(1 + 1.2\sqrt{I}\right) \right] \tag{9.8}$$

I is the ionic strength. The partial pressure of water is obtained with the generalized Raoult's law.

$$p_{H_2O} = a_{H_2O} \, p^s_{H_2O}(T) \tag{9.9}$$

$p^s_{H_2O}(T)$ is the vapour pressure of pure water.
The activity of water is calculated with the osmotic coefficient.

$$\ln a_{H_2O} = -\frac{18.015}{1000} \, 2 m_{HCl} \, \Phi \tag{9.10}$$

Equation (9.10) is just simplified for aqueous HCl solutions.
The osmotic coefficient is obtained with a simplified equation from Pitzer (1973, 1974).

$$\Phi = 1 + f^\Phi + m_{HCl} B^\Phi_{HCl} + m^2_{HCl} C^\Phi_{HCl} \tag{9.11}$$

$$f^\Phi = -A_\Phi \frac{\sqrt{I}}{1 + 1.2\sqrt{I}} \tag{9.12}$$

Values for the coefficients B^γ_{HCl}, B^Φ_{HCl}, and C^Φ_{HCl} are described by Holmes et al. (1987). The Henry coefficient and D_{HCl} are fitted to measured partial pressures of HCl.

9.6 Results and Discussion

9.6.1 Partial Pressure of HCl

Figure 9.11 shows the calculated partial pressures and the experimental results for HCl.
The reproduction of the own experimental data with the correlation of the Henry coefficient H^*_{HCl} and D_{HCl} is not sufficient. Surely it can be improved by adding terms on Eq. (9.6). But that would have a wrong calculation of the activity coefficient.

9 Phase Equilibria of Aqueous Solutions Containing Volatile Electrolytes

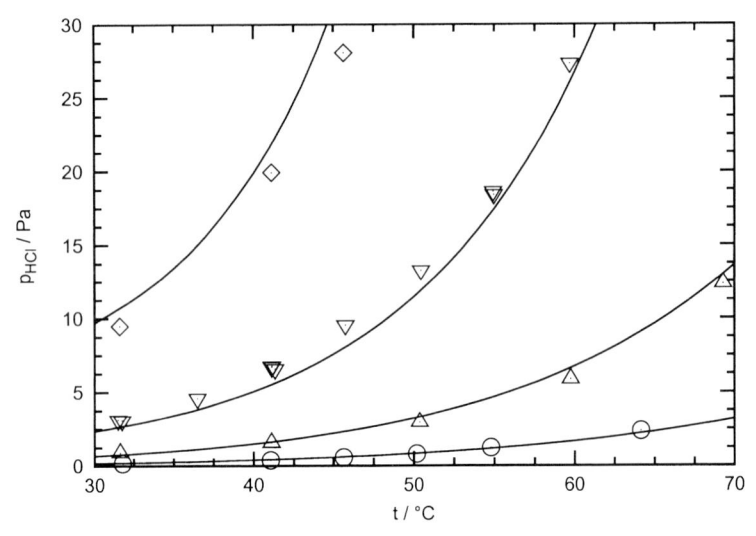

Figure 9.11: Partial pressure of HCl over aqueous solutions of HCl in the temperature range from 30 to 70 °C. ◇ 15.0 mass%; ▽ 11.9 mass%; △ 9.0 mass%; ○ 6.1 mass%; ——— correlations.

9.6.2 Partial Pressure of H$_2$O

Figure 9.12 shows the calculated and measured partial pressures of H$_2$O as function of the temperature.

Over the entire range of temperature and concentration the reproduction is very good. The largest difference is 4.6% at 9.0 mass%. This shows that the values of Holmes et al. (1987) are correct.

9.7 Conclusions

Experiments with pure water, binary and ternary solutions containing strong electrolytes and measurements with the system NH$_3$ + H$_2$O have demonstrated the operatability of the new apparatus. The FT-IR spectroscopy is an appropriate method for analysing vapour phases containing components like HCl and HBr in high dilution even in the presence of large amounts of water. Gaseous substances can be analysed in a concentration range down to a few ppm. Measurements of the phase equilibria of the systems NH$_3$ + H$_2$O, HCl + H$_2$O, HBr + H$_2$O, HCl + CaCl$_2$ + H$_2$O, and HCl + HBr + H$_2$O were carried out. In the literature different values are reported for the partial pressure of HCl above its aqueous solution. The experiments of this work commonly gave lower results, except for solutions with 12% HCl and maximum temperatures of 41 °C. The measured partial pressures of water are in good

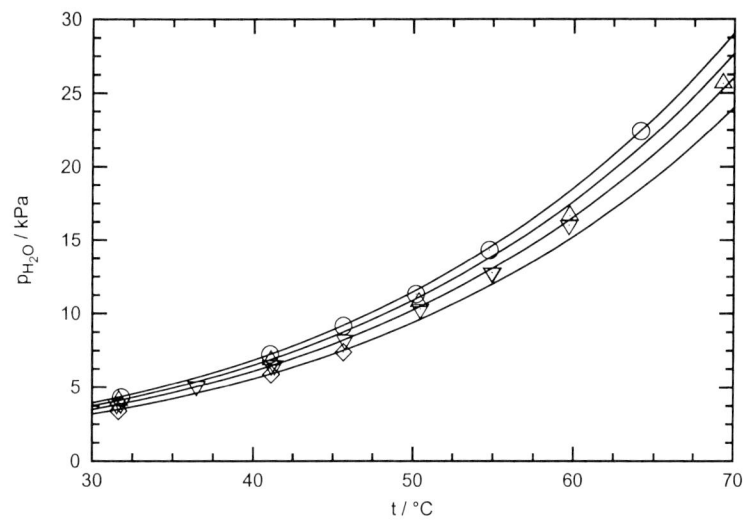

Figure 9.12: Partial pressure of H₂O over aqueous solutions of HCl in the temperature range from 30 to 70 °C. ◇ 15.0 mass%; ▽ 11.9 mass%; △ 9.0 mass%; ○ 6.1 mass%; —— correlations.

agreement with the literature data. At the same molality and temperature the partial pressures of HBr are smaller than the HCl values. The results for the HBr system could not be compared with literature because no literature data are available in the concentration range of our experiments. Ternary systems with electrolytes sharing a common ion were analysed. In that case the partial pressure of an electrolyte is increased by the other. The logarithmic representation of the HCl and HBr partial pressure is a linear function of the reciprocal temperature. The slope of the straight line is nearly the same for all investigated systems.

The theoretical description of the partial pressure from HCl above aqueous HCl solutions with the model from Pitzer is not sufficient. Only the prediction of the partial pressure of H₂O with this method is successful.

References

Åkerlöf, G.; Teare, J. W. (1937): Thermodynamics of Concentrated Aqueous Solutions of Hydrochloric Acid. J. Am. Chem. Soc. 59, 1855–1868.

Bradley, D. J.; Pitzer, K. S. (1979): Thermodynamics of Electrolytes. 12. Dielectric Properties of Water and Debye-Hückel Parameters to 350 °C and 1 kbar. J. Phys. Chem. 83, 1599–1603.

Edwards, T. J.; Newman, J.; Prausnitz, J. M. (1975): Thermodynamics of Aqueous Solutions Containing Volatile Weak Electrolytes. AIChE J. 21, 248–259.

Edwards, T. J.; Maurer, G.; Newman, J.; Prausnitz, J. M. (1978): Vapor-Liquid Equilibria in Multicomponent Aqueous Solutions of Volatile Weak Electrolytes. AIChE J. 24, 966–976.

Elm, N. P. (2000): Dampf-Flüssig-Gleichgewichte starker Elektrolytmischungen mit flüchtigen Komponenten. Dissertation, Universität Karlsruhe.

Fritz, J. J.; Fuget C. R. (1956): Vapour Pressure of Aqueous Hydrogen Chloride Solutions, 0° to 50°C. J. Chem. Eng. Data 1, 10–12.
Harned, H. S.; Ehlers, R. W. (1932): The Dissociation Constant of Acetic Acid from 0 to 35° Centigrade. J. Am. Chem. Soc. 54, 1350–1357.
Harned, H. S.; Ehlers, R. W. (1933): The Thermodynamics of Aqueous Hydrochloric Acid Solutions from Electromotive Force Measurements. J. Am. Chem. Soc. 55, 2179–2193.
Holmes, H. F.; Busey, R. H.; Simonson, J. M.; Mesmer, R. E.; Archer, D. G.; Wood, R. H. (1987): The Enthalpie of Dilution of HCl(aq) to 648 K and 40 Mpa. J. Chem. Thermodynamics 19, 863–890.
International Critical Tables (1928). Vol. III. McGraw-Hill Book Company. New York.
Lutugina, N. V.; Reshetova, L. I.; Syrovarova, T. K. (1981): Deposited Doc., VINITI, 3562-81.
Macriss, R. A.; Eakin, B. E.; Ellington, R. T.; Huebler, J. (1964): Physical and Thermodynamic Properties of Ammonia – Water Mixtures. Research Bulletin No. 34. Inst. Of Gas Technology. Chicago.
Perry, R. H.; Green, D. W. (1984): Perry's Chemical Engineers' Handbook. 6th Edition. McGraw-Hill Book Company. New York.
Pitzer, K. S. (1973): Thermodynamics of Electrolytes. I. Theoretical Basis and General Equations. J. Phys. Chem. 77, 268–277.
Pitzer, K. S.; Kim, J. J. (1974): Thermodynamics of Electrolytes. IV. Activity and Osmotic Coefficients for Mixed Electrolytes. J. Am. Cem. Soc. 96, 5701–5707.
VDI – Wasserdampftafeln (1963). Oldenbourg. München.
Zeisberg, F. C. (1925): Partial Vapor Pressures of Aqueous HCl Solutions. Chemical and Metallurgical Engineering 32, 326–327.

Nomenclature

$A_{i,j}$	relative peak area of component i in range j
A_Φ	Debye-Hückel constant
a_{HCl}	activity of HCl
a	coefficient
B^Φ_{HCl}	interaction coefficient of HCl
B^γ_{HCl}	interaction coefficient of HCl
b	coefficient
C^Φ_{HCl}	interaction coefficient of HCl
c	coefficient
D_{HCl}	interaction coefficient of HCl
d	coefficient
$f(p_{HCl})$	function of p_{HCl}
$f(\xi_{HCl})$	function of ξ_{HCl}
f^γ	function
f^Φ	function
H^*_{HCl}	Henry constant of HCl
H_1	Henry coefficient
H_2	Henry coefficient
H_3	Henry coefficient
H_4	Henry coefficient

Nomenclature

I	ionic strength on molality scale
m_{HCl}	molality of HCl
p_i	partial pressure of component i
$p_{H_2O}^s$	vapour pressure of pure water
t	temperature (Celsius)
T	temperature (Kelvin)

Greek letters

γ_{HCl}^r	rational activity coefficient of HCl
$\tilde{\nu}$	wave number
ξ_i	mass percent of component i
Φ	osmotic coefficient

10 Experimental Determination and Prediction of Phase Equilibria in Systems Containing Strong Electrolytes

Magnus Topphoff, Christian Rose, Jörn Kiepe, and Jürgen Gmehling [1]*

10.1 Abstract

The LIQUAC model is widely used for the prediction of the vapor-liquid equilibrium (VLE) behavior, osmotic coefficients, and mean ionic activity coefficients of systems containing strong electrolytes. It consists of a Debye-Hückel term, the UNIQUAC term, and the osmotic virial equation for the middle-range contribution. One objective of this work is to provide a model based on the group contribution concept that can be used to predict phase equilibria in mixed solvent electrolyte systems. Therefore, in the LIQUAC model the UNIQUAC equation has been substituted by the original UNIFAC equation and group interaction parameters have been introduced into the middle-range term. With the help of a large database consisting of literature data and own measurements, 129 binary group interaction parameters for 8 solvent groups, 13 cations, and 8 anions have been fitted. The model parameters have been used to calculate reliably the VLE behavior, osmotic coefficients, and mean ionic activity coefficients for a large number of systems. Additionally the salt solubilities of several strong electrolytes in water can be predicted using the mentioned thermodynamic models. For the prediction of solid-liquid equilibria (SLE) of aqueous salt solutions the solubility product K_{sp} and the mean ionic activity coefficient ($\gamma_\pm$) has to be known. K_{sp} can be calculated from tabulated thermodynamic data and $\gamma_\pm$ with the help of the model developed. To extend the applicability of the model to predict phase equilibria in the presence of non-condensable gases and strong electrolytes, the LIFAC model is combined with the PSRK (*predictive Soave Redlich Kwong*) model, whereby the published group interaction parameters of the PSRK and LIFAC model are used directly. In the experimental part several VLE, activity coefficients at infinite dilution, and mean activity coefficients of salts have been determined using ebulliometry, headspace gas chromatography, the dilutor technique, and an apparatus for the determination of the electromotive force (EMF) of a cell.

---

[1] Universität Oldenburg, Technische Chemie (FB9), 26111 Oldenburg. e-mail: gmehling@tech.chem.uni-oldenburg.de
* Author to whom correspondence should be addressed.

10.2 Introduction

The reliable knowledge of the phase equilibria of electrolyte systems is essential for the design and simulation of different chemical processes including wastewater treatment, extractive distillation, extractive and gas antisolvent salt crystallization of salts, petroleum and natural gas exploitation, petroleum refining, coal gasification, environmental protection, formation of gas hydrates, and various absorption processes. The quality of the description of the phase equilibrium behavior for electrolyte systems strongly depends on the thermodynamic model (g^E-model, equation of state (EOS), EOS + g^E mixing rules) and the quality of the parameters used. In order to obtain reliable model parameters for predictive methods, a large database should be applied. The Dortmund Data Bank (DDB) developed in our research group formed the basis for various thermodynamic models. It is a steadily growing database, which today contains more than 2600 VLE and 3100 salt solubility data sets for electrolyte systems besides other data. But for a large number of electrolyte systems no VLE data are available.

In 1994, Li et al. proposed a g^E-model for single and mixed solvent electrolyte systems (Li et al., 1994), in which the contribution of the short-range interactions to the excess Gibbs energy is described by the UNIQUAC model, therefore the model was named LIQUAC. One aim of this study was the introduction of the "solution of groups concept" into the LIQUAC model. Therefore, in the LIQUAC model the UNIQUAC equation has been substituted by the original UNIFAC equation and new group interaction parameters have been introduced into the middle-range term. It has been shown that both models can be used to predict reliably vapor-liquid equilibria (VLE) for single solvent and mixed solvent/salt systems as well as osmotic coefficients and mean activity coefficients for water/salt systems (Li et al., 1994; Polka et al., 1994). To ensure a large range of applicability additional interaction parameters between the non-charge groups and ions have been fitted with the help of the Dortmund Data Bank (Gmehling, 1985, 1991). With the help of the mentioned thermodynamic models also salt solubilities of several strong electrolyte systems are predicted. For the prediction of solid-liquid equilibria (SLE) of aqueous salt solutions only the solubility product K_{sp} and the mean ionic activity coefficient ($\gamma_\pm$) has to be known. K_{sp} can be calculated from tabulated thermodynamic data and the required $\gamma_\pm$ with the help of the proposed models.

Another purpose of our research project was the further development of the PSRK model for the prediction of gas solubilities in electrolyte systems. Since phase equilibria of electrolytes with supercritical gases cannot be predicted with the help of g^E-models an equation of state (EOS) approach had to be used. Therefore in the proposed PSRK model, the LIFAC model is used to predict the excess Gibbs energy required in the g^E mixing rule, whereby the published group interaction parameters of the PSRK and LIFAC model can be used directly. The missing group interaction parameters between CO_2, N_2, CH_4 and the ionic groups Na^+, NH_4^+, Ca^{2+}, Cl^- were fitted using the experimental gas solubilities for electrolyte systems stored in the Dortmund Data Bank. Using these parameters the solubilities of CO_2, N_2, and CH_4 in aqueous electrolyte solutions were predicted in a large temperature and pressure range.

The objective of the experimental part of this work was to study systematically the effect of electrolytes on the vapor-liquid equilibrium behavior of several binary solvent systems at different temperatures resp. pressures and different constant salt concentrations, for which no data were available in the literature. These data are essential for the determination of reliable model parameters for the presented g^E- or group contribution model. 282 isothermal resp. isobaric data sets in a temperature range from 30–70 °C (66.6–101.33 kPa) were measured. Many of the obtained results were published already (Yan et al., 1997, 1998a–c, 2001; Topphoff et al., 2001). Also the activity coefficients at infinite dilution of various solutes in solvent/salt mixtures were analyzed by the dilutor technique. Activity coefficients for six different solutes at four temperatures (30–60 °C) in three different solvent/salt mixtures were determined.

To get reliable information also about the temperature dependence of the activity coefficients of salts an apparatus for the determination of the electromotive force (EMF) of a cell was build up. To verify the apparatus first measurements for aqueous 1:1 electrolyte systems for which experimental data at 25 °C are available were carried out. Additionally measurements at temperatures up to 60 °C and for ternary systems containing two different salts were performed.

10.3 Group Contribution Model for the Prediction of Activity Coefficients in Systems Containing Strong Electrolytes

The model developed in this work consists of three terms for the excess Gibbs energy: (1) a Debye-Hückel term which represents the long-range interactions, (2) a virial term which accounts for the middle-range interactions caused by the ion-dipole effects, (3) a UNIFAC term (Fredenslund et al., 1977) which represents the short-range interactions.

This model is different from other group contribution models (Kikic et al., 1991; Achard et al., 1994), focusing mainly on the charge-dipole and charge-induced dipole interactions between solvent groups and ions. Therefore, a virial term is introduced into the expression for the excess Gibbs energy. In the UNIFAC term the ion-solvent and ion-ion interactions are not taken into account because these parameters show only a minor influence on the predicted activity coefficients of the solvents. The new model has the advantage that only a relative small number of parameters had to be fitted, without diminishing the precision of the prediction. The required group-interaction parameters between solvent groups in the UNIFAC term were directly taken from literature (Hansen et al., 1991).

10.3.1 Thermodynamic Framework

As a result of the dissociation of salts into ions, the solvent-solvent, ion-solvent, and ion-ion interactions in electrolyte systems determine the thermodynamic properties of electrolyte systems. In this work, various types of interactions have been taken into account, the excess Gibbs energy has been calculated as the sum of three contributions:

10.3 Group Contribution Model for the Prediction of Activity Coefficients

$$G^E = G^E_{LR} + G^E_{MR} + G^E_{SR} \tag{10.1}$$

The term G^E_{LR} represents the long-range (LR) interaction contribution caused by the Coulomb electrostatic forces. Corresponding activity coefficients of the solvent s and the ion j can be expressed using the extended Debye-Hückel theory (Li et al., 1994):

$$\ln \gamma^{LR}_S = [2AM_S d/(b^3 d_S)][1 + bI^{1/2} - (1+bI^{1/2})^{-1} - 2\ln(1+bI^{1/2})] \tag{10.2}$$

$$\ln \gamma^{LR}_S = -z_j^2 A I^{1/2}/(1+bI^{1/2}) \tag{10.3}$$

where M_S is the molecular weight of solvent s, I is the ionic strength, d_S is the molar density of solvent s calculated from the DIPPR Tables (1984), d is the molar density of the solvent mixture and A and b are the Debye-Hückel parameters. The dependence of the Debye-Hückel parameters on the absolute temperature T, the density d and the dielectric constant D of the mixed solvent is given by:

$$A = 1.327757 \times 10^5 d^{1/2}/(DT)^{3/2} \tag{10.4}$$

$$b = 6.35969 \, d^{1/2}/(DT)^{1/2} \tag{10.5}$$

The G^E_{LR} term takes into account the electrostatic interactions between the charged species, but the physical validity of the equation is limited to the very dilute region. The purpose of this term is mainly to provide the true limiting law at infinite dilution.

The G^E_{MR} term is the contribution of the indirect effects of the charge interactions (such as the charge-dipole interactions and the charge-induced dipole interactions) to the excess Gibbs energy, which are called middle-range (MR) interactions. For a solution containing n_k moles of solvent group k ($k = 1, 2, \ldots, s$) and n_j moles of ion j ($j = 1, 2, \ldots,$ ion), G^E_{MR} is expressed as described by Li et al. (1994):

$$G^E_{MR}/RT = (n_{kg})^{-1} \left[\sum_k \sum_{\text{ion}} B_{k,\text{ion}}(I) n_k n_{\text{ion}} + \sum_c \sum_a B_{c,a}(I) n_c n_a \right] \tag{10.6}$$

where n_{kg} denotes n kg solvent and B is the interaction coefficient. c covers all cations, while a covers all anions. However, in this work the second virial coefficient $B_{i,j}$ is the interaction coefficient between the species i and j. The relation of ion-ion interaction parameter $B_{c,a}$ and ion-solvent group interaction parameter $B_{k,\text{ion}}$ to the ionic strength are described by Li et al. (1994):

$$B_{c,a} = b_{c,a} + c_{c,a} \exp(-I^{1/2} + 0.13 I) \tag{10.7}$$

$$B_{k,\text{ion}} = b_{k,\text{ion}} + c_{k,\text{ion}} \exp(-1.2 I^{1/2} + 0.13 I) \tag{10.8}$$

where $b_{i,j}$ and $c_{i,j}$ are the middle-range interaction parameters between species i and j ($b_{i,j} = b_{j,i}$, $c_{i,j} = c_{j,i}$, $c_{i,i} = b_{i,i} = 0$, and $c_{j,j} = b_{j,j} = 0$), where i and j denote ions or solvent groups. In Eqs. (10.7) and (10.8), the constants were determined using a number of reliable experimental data for single solvent and mixed solvent electrolyte systems.

By differentiating Eq. (10.6) with respect to the number of moles of solvent group k and ions, one obtains:

$$\ln \gamma_k^{MR} = \sum_{ion} B_{k,ion}(I) m_{ion} - \frac{M_k}{M} \sum_{k'} \sum_i [B_{k',ion}(I) + IB'_{k',ion}(I)] x'_{k'} m_{ion}$$

$$- M_k \sum_c \sum_a [B_{c,a}(I) + IB'_{c,a}(I)] m_c m_a \quad \text{with } k' = 1, 2, \ldots, n_k \quad (10.9)$$

$$\ln \gamma_c^{*MR} = \frac{1}{M} \sum_k B_{c,k}(I) x'_k + \frac{z_j^2}{2M} \sum_k \sum_j B'_{k,ion}(I) x'_k m_{ion} + \sum_a B_{c,a}(I) m_a$$

$$+ \frac{z_a^2}{2} \sum_{c'} \sum_a B'_{c',a}(I) m_{c'} m_a - \frac{B_{c,s}(I)}{M_s} \quad \text{with } c' = 1, 2, \ldots, n_c \quad (10.10)$$

$$\ln \gamma_a^{*MR} = \frac{1}{M} \sum_k B_{a,k}(I) x'_k + \frac{z_j^2}{2M} \sum_k \sum_j B'_{k,ion}(I) x'_k m_{ion} + \sum_c B_{c,a}(I) m_c$$

$$+ \frac{z_a^2}{2} \sum_c \sum_{a'} B'_{c,a'}(I) m_c m_{a'} - \frac{B_{a,s}(I)}{M_s} \quad \text{with } a' = 1, 2, \ldots, n_a \quad (10.11)$$

where x'_k is the salt-free mole fraction of solvent group k, M_k is the molecular weight of solvent group k, M_s is the molecular weight of the reference solvent s, M is the molecular weight of mixed solvent, $B'_{i,j}(I)$ is equal to $dB_{i,j}(I)/dI$ and $B_{i,j}/M_s$ is needed for the normalization of the activity coefficient of ions to infinite dilution in the reference solvent s.

$$M = \frac{\sum_k \sum_i v_k^{(i)} x'_i M_k}{\sum_k \sum_i v_k^{(i)} x'_i} \qquad x'_{k'} = \frac{\sum_i v_{k'}^{(i)} x'_i}{\sum_i \sum_k v_k^{(i)} x'_i} \quad (10.12)$$

From Eq. (10.9), the activity coefficient γ_S^{MR} of solvent s can be obtained using the following relation:

$$\ln \gamma_S^{MR} = \sum_k v_k^{(s)} \ln \gamma_k^{MR} \quad (10.13)$$

where $v_k^{(s)}$ is the number of groups of type k in the solvent s, and γ_k^{MR} is the activity coefficient of group k in the mixture for the middle-range contribution.

The G_{SR}^E term expresses the contribution of the short-range (SR) interactions to the excess Gibbs energy, and can be described using the UNIFAC group-contribution method published by Fredenslund et al. (1977). The group volume and surface area parameters R_k and Q_k for the solvents and the UNIFAC group interaction parameters between the solvent groups were directly taken from Hansen et al. (1991). The values of R_k and Q_k for the ions can be estimated from the ionic radii (Bondi, 1968). However, the very small values of the ionic radii of the cations lead to values between 0.1 and 0.5. Values of this magnitude signif-

10.3 Group Contribution Model for the Prediction of Activity Coefficients

icantly reduce the fitting capabilities of the short-range contribution. Therefore, the values of the volume and surface parameters for ions were treated as adjustable parameters. The values are listed in Table 10.1.

Table 10.1: Relative van der Waals group volume R_k and surface parameter Q_k.

	\multicolumn{12}{c}{Cations}
	Li^+	Na^+	K^+	NH_4^+	Ca^{2+}	Ba^{2+}	Mg^{2+}	Sr^{2+}	Co^{2+}	Ni^{2+}	Cu^{2+}	Zn^{2+}	Hg^{2+}
R_k	1.0	3.0	3.0	3.0	1.0	3.0	1.0	1.0	1.0	1.0	1.0	1.0	3.0
Q_k	1.0	3.0	3.0	3.0	1.0	3.0	1.0	1.0	1.0	1.0	1.0	1.0	3.0

	\multicolumn{8}{c}{Anions}
	F^-	Cl^-	Br^-	I^-	NO_3^-	CH_3COO^-	SCN^-	SO_4^{2-}
R_k	0.4195	0.9861	1.2331	1.6807	1.6400	2.0500	1.2325	1.9862
Q_k	0.5597	0.9917	1.1510	1.4118	1.6000	1.9000	1.1506	0.2332

The complete equation for the activity coefficient of the solvent s is

$$\ln \gamma_s = \ln \gamma_s^{LR} + \ln \gamma_s^{MR} + \ln \gamma_s^{SR} \tag{10.14}$$

where the concentration scale of the solvent s is mole fraction. The reference state is the pure solvent s, $x'_s = x_s \to 1$ and $\gamma_s \to 1$, at system temperature and pressure.

For the activity coefficient of an ion j the following equation is obtained:

$$\ln \gamma_j = \left(\ln \gamma_j^{LR} + \ln \gamma_j^{MR} + \ln \gamma_j^{SR} \right) - \ln \left(M_s / x'_s M_s + M_s \sum_j m_j \right) \tag{10.15}$$

where the concentration scale of the ion j is molality, the reference state $x'_j = x_j \to 1$, $I \to 0$, $\gamma_j \to 1$, and the standard state is the hypothetical ideal solution of unit molality at system temperature and pressure. The mean activity coefficient of a salt MX is calculated by

$$\ln \gamma_{MX(\pm)} = (1/\nu) \left[\nu_M \ln \gamma_M + \nu_X \ln \gamma_X \right] \tag{10.16}$$

where ν_M and ν_X represent the stoichiometric factors for the cation M and the anion X. $\nu = \nu_M + \nu_X$ is the stoichiometric number of moles of ions for one mole of salt MX. The terms $\ln \gamma_M$ and $\ln \gamma_X$ are calculated from Eqs. (10.10) and (10.11).

10.3.1.1 Estimation of Parameters

A computerized database was used to fit the model parameters for 10 solvents (water, methanol, ethanol, 1-propanol, 2-propanol, 1-butanol, acetone, ethyl acetate, dichloromethane, and THF which correspond to 8 structural groups: H_2O, CH_2, OH, CH_3OH, CH_2CO, $CCOO$, CCl_2, and CH_2O), 13 cations (lithium, sodium, potassium, calcium, barium, magnesium, strontium, copper, nickel, zinc, cobalt, mercury, and ammonium), and 8 anions (fluoride, chloride, bromide, iodide, nitrate, acetate, sulfate, and thiocyanate). The maximum salt concentration considered was 20 mol kg^{-1} for 1:1 salts and 22 mol kg^{-1} for 2:1 salts. Overall 306 binary and 264 ternary VLE data sets have been used to fit 129 binary interaction parameters. The VLE data types included are: isobaric T-x-y and T-x, isothermal P-x-y, osmotic coefficients (isothermal, x-Φ), and mean activity coefficients (isothermal, x-$\gamma_\pm$).

10.3.2 Results and Discussion for VLE Data

While fitting the interaction parameters it was found that the group interaction parameters $a_{i,j}$ between solvent groups and ions and between cations and anions in the UNIFAC equation are not very sensitive to the values of the objective function. This means that the quality of the estimations is not affected by changing the values of these parameters. Two sets of parameters ($a_{i,j} = 0$ and $a_{i,j} \neq 0$) for the UNIFAC energy interaction parameters between solvent groups and ions and between cations and anions were used to calculate the mean deviations between experimental and calculated values for 31 ternary systems. The mean absolute deviation over all 112 data sets is given in Table 10.2 for both procedures.

Table 10.2: Mean deviation between experimental and calculated data using different procedures.

		$a_{i,j} \neq 0$			$a_{i,j} = 0$	
	$\|\Delta y\|$	$\|\Delta T\|$[K]	$\|\Delta P\|$[kPa]	$\|\Delta y\|$	$\|\Delta T\|$[K]	$\|\Delta P\|$[kPa]
Mean deviation	0.016	0.710	0.34	0.017	0.90	0.37

The deviations obtained for both sets of parameters are almost identical. In order to reduce the number of parameters to be fitted, the parameters $a_{i,j}$ between solvent groups and ions and between cations and anions were set equal to zero and only the middle-range interaction parameters $b_{i,j}$ and $c_{i,j}$ in Eqs. (10.7) and (10.8) were fitted. The interaction parameters obtained can be found in the literature (Yan et al., 1999). The current status of the interaction parameters matrix is shown in Figure 10.1.

To check the reliability of the proposed VLE model for mixed-solvent electrolyte systems, a comparison with the electrolyte model based on the UNIFAC equation proposed by Kikic et al. (1991) was performed. The results of the comparison are given in Table 10.3. Mean absolute deviations for the vapor-phase mole fractions and the temperatures were calculated with the UNIFAC parameters of Kikic et al. (1991) and with parameters for the new VLE model obtained in this work. It can be seen that the deviations in vapor-phase mole fractions and temperature predicted by Kikic et al. (1991) are generally twice as large as the deviations obtained in this work, although the same number of parameters are used. Figures 10.2 and 10.3 show y-x and ϑ-x-y diagrams for selected ternary systems predicted with the UNIFAC model of Kikic et al. (1991) and the proposed model. Furthermore it can be recog-

10.3 Group Contribution Model for the Prediction of Activity Coefficients

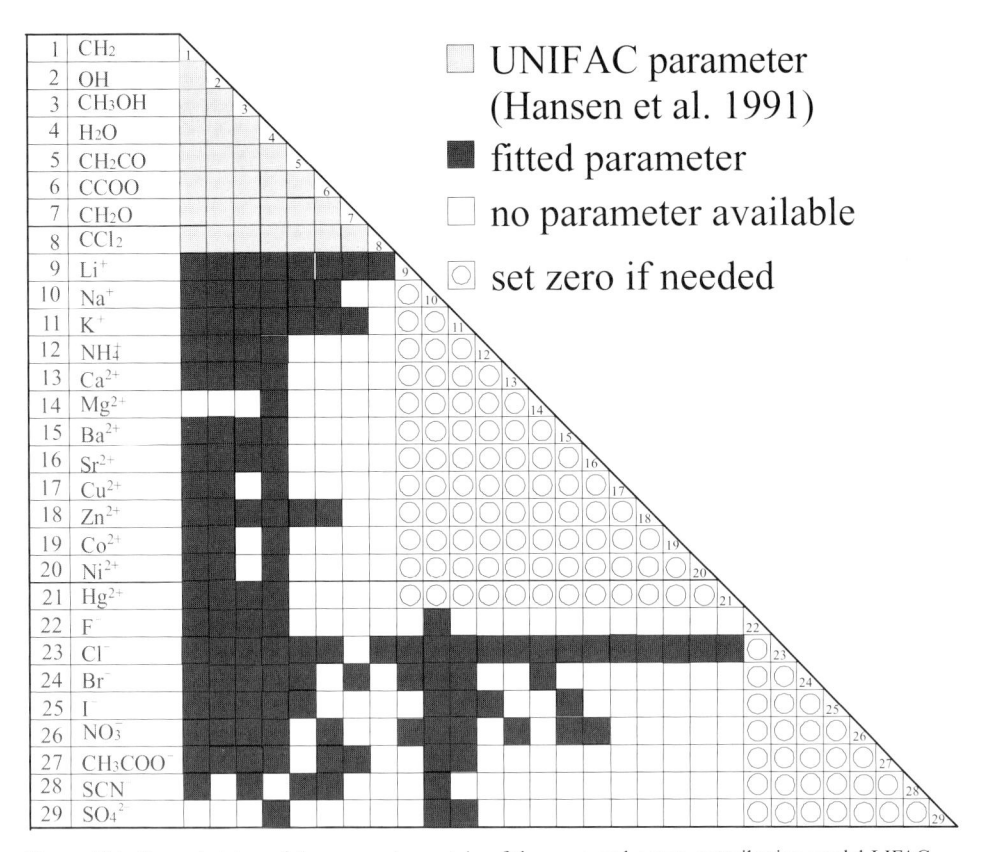

Figure 10.1: Present status of the parameter matrix of the proposed group contribution model LIFAC.

nized in Figures 10.2 and 10.3 that more reliable results are obtained with the proposed model. In both systems a typical salting out effect of the more volatile component occurs, which means that the vapor phase concentration of the alcohol rises with increasing salt concentration compared to the salt-free system. For the system methanol + water + lithium chloride the UNIFAC model of Kikic et al. (1991) predicts a salting in effect for methanol at low concentrations which is not in accordance with the experimental data.

The total mean absolute deviation of the osmotic coefficients is 0.052 at moderate and high salt concentrations up to 20 mol kg^{-1} solvent. The results for the mean ionic activity coefficients are even better than those for osmotic coefficients across the whole salt concentration range. In most cases the calculated results for the mean ionic activity coefficients are in good agreement with the experimental ones for moderate and high salt concentration. Figure 10.4 shows some results in which the salt molalities vary from 0 up to 20 and the values for the mean ionic activity coefficients between 0.2 to 500.

For 264 ternary VLE data sets the total mean absolute deviation for the vapor phase mole fraction is 0.019, for the temperature is 0.92 K, and for the pressure is 0.43 kPa. For the 306 binary VLE data sets the total mean absolute deviation for the mean osmotic

Table 10.3: A comparison of the results for ternary VLE using parameters from Kikic et al. (1991) and model parameters from this work.

			Kikic et al. (1991)		This work	
System[a]	Data points	Max. molality	$|\Delta y|$	$|\Delta T|/K$	$|\Delta y|$	$|\Delta T|/K$
methanol + water						
+NaBr	23	3.80	0.029	0.75	0.026	0.79
+NaOOCCH$_3$	21	1.35	0.034	1.12	0.009	0.20
+LiCl	24	3.76	0.048	1.72	0.017	0.53
+KCl	21	1.97	0.022	0.34	0.017	0.28
ethanol + water						
+NaF	12	0.84	0.007	0.42	0.009	0.45
+NaCl	32	6.03	0.024	1.30	0.011	0.34
+KI	20	6.43	0.029	1.03	0.016	1.07
+CuCl$_2$	177	5.88	0.046	1.23	0.018	0.95
+Sr(NO$_3$)$_2$	69	3.04	0.033	0.96	0.009	0.38
+KOOCCH$_3$	42	4.32	0.072	5.65	0.023	1.302
Ca(NO$_3$)$_2$	44	2.04	0.032	1.05	0.011	0.82
1-propanol + water						
+KCl	32	4.68	0.058	0.81	0.028	0.50
+NaCl	27	4.26	0.014	0.54	0.023	0.79
+LiCl	24	3.92	0.074	2.32	0.018	0.65
2-propanol + water						
+LiBr	10	0.60	0.040	0.51	0.030	0.36
acetone + water						
+KCl	28	2.36	0.067	3.62	0.023	1.05
Average	623		0.039	1.46	0.018	0.61

[a] data taken from the Dortmund Data Bank (DDB)

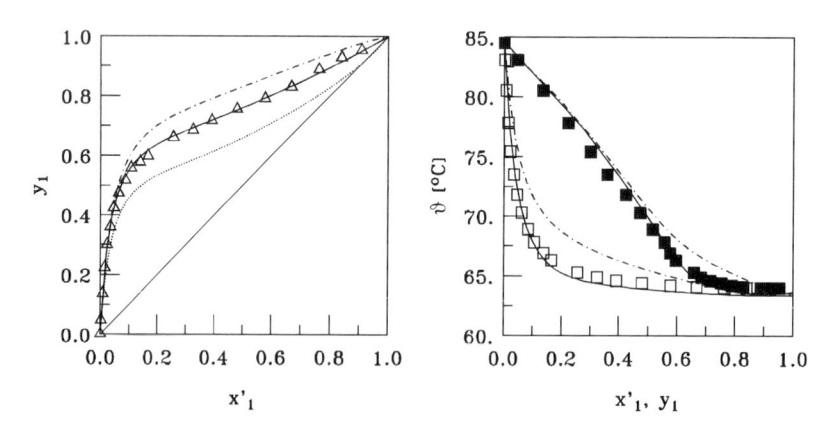

Figure 10.2: y–x and ϑ–y–x phase diagram for the system ethanol + water + calcium nitrate at 50.66 kPa; salt molality: 2.049 mol kg^{-1}. ($\triangle$, $\square$, $\blacksquare$) Experimental data (Polka and Gmehling, 1994); (——) prediction this work; (– –) Kikic et al. (1991); (······) salt-free system, UNIFAC model (Fredenslund et al., 1977).

10.3 Group Contribution Model for the Prediction of Activity Coefficients

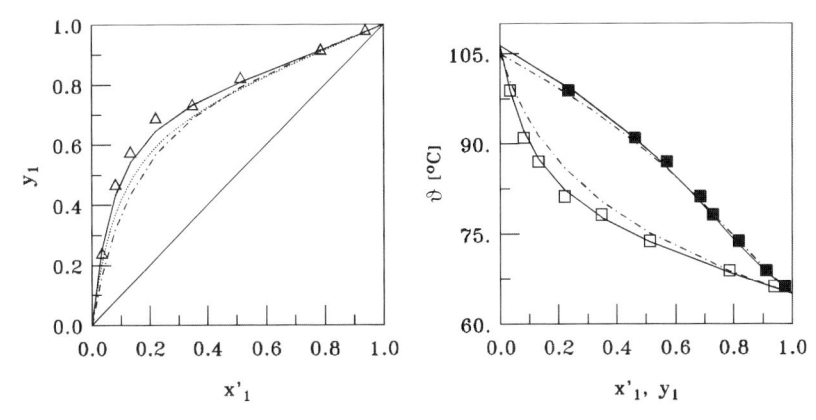

Figure 10.3: Predicted and experimental VLE for methanol + water + LiCl at 101.33 kPa; (———) prediction this work; (......) salt-free system, UNIFAC model (Fredenslund et al., 1977); ($\triangle, \square, \blacksquare$) experimental values ($x_{salt}$ = 0.004–0.061) from Boone et al. (1976).

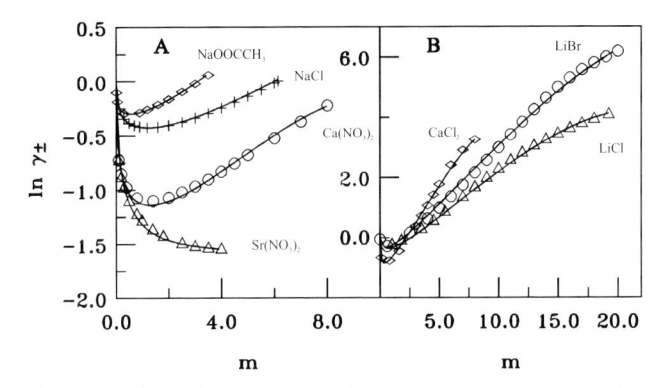

Figure 10.4: Experimental and predicted mean activity coefficients for aqueous electrolyte solutions at 25 °C. Salt molalities 0–20 mol kg^{-1}, $\gamma_{\pm}$ 0.2–500. (———) Prediction this work; ($\triangle, \bigcirc, +, \diamondsuit$) experimental values from (A) Na(OOCCH$_3$) and NaCl (Hamer and Wu, 1972); Ca(NO$_3$)$_2$ and Sr(NO$_3$)$_2$ (Stokes, 1948); (B) LiBr and LiCl (Hamer and Wu, 1972); CaCl$_2$ (Robinson and Stokes, 1965).

coefficients is 0.052, for the mean activity coefficients is 0.120, and for the pressure is 0.44 kPa. Compared with the group contribution model of Kikic et al. (1991), the proposed model allows a more reliable representation of VLE in mixed electrolyte systems.

10.3.3 Prediction of Solid-Liquid Equilibria

The study of phase equilibria in electrolyte systems and in particular the determination of salt solubilities is extremely important, either from a scientific or an industrial point of view. For the prediction of solid-liquid equilibria (SLE) of aqueous salt solutions the solubility product K_{sp}

and the mean ionic activity coefficient ($\gamma_\pm$) have to be known. K_{sp} can be calculated from tabulated thermodynamic properties (Wagman et al., 1982) and $\gamma_\pm$ using the proposed model. For the prediction of salt solubilities in strong electrolyte systems no additional parameters are introduced. For the prediction of SLE the parameters derived from VLE data are taken.

10.3.3.1 Theory

The solubility product at 25 °C can directly be predicted from the standard Gibbs energy of formation or, if all mean activity coefficients are known, from experimental solubility data.

The temperature dependence of the solubility product of a salt ($M_{v_+} X_{v_-} \cdot n \cdot H_2O$) can be obtained using the Gibbs-Helmholtz equation (Prausnitz et al., 1999):

$$\ln K_{sp}(T) = \ln K_{sp}^0(T_0) - \frac{\Delta h_R^0(T_0)}{R}\left(\frac{1}{T} - \frac{1}{T_0}\right) + \frac{1}{R}\int_{T_0}^{T}\left[\frac{\int_{T_0}^{T}\Delta c_{p,R}^0(T)\,dT}{T^2}\right]dT \quad (10.17)$$

with $\Delta h_R^0 = \left(v_+ \Delta h_{f\,M^{z+}}^0 + v_- \Delta h_{f\,X^{z-}}^0 + n \Delta h_{f\,H_2O}^0\right) - \Delta h_{f\,Salz}^0$

and $\Delta c_{p,R}^0 = \left(v_+ \Delta c_{p,M^{z+}}^0 + v_- \Delta c_{p,X^{z-}}^0 + n \Delta c_{p,H_2O}^0\right) - \Delta c_{p,Salz}^0$

For the numerical determination of the integral the temperature dependence of the molar heat capacities has to be known. In most cases only constant molar heat capacities at reference temperature are given. If the temperature difference between T and T_0 is moderate (approx. 50 K), Δc_p^0 can be assumed to be constant (Wagman et al., 1982), and Eq. (10.17) then becomes:

$$\ln K_{sp}(T) = -\frac{\Delta g_R^0(T_0)}{RT} - \frac{\Delta h_R^0(T_0)}{R}\left(\frac{1}{T} - \frac{1}{T_0}\right) + \frac{\Delta c_{p,R}^0(T_0)}{R}\left(\ln\frac{T}{T_0} + \frac{T_0}{T} - 1\right)$$

(10.18)

For the prediction of the saturation molality at given temperature we obtain:

$$m_\pm(T) = \sqrt[v]{\frac{K_{sp}(T)/Q\,a_{H_2O}^n}{\gamma_\pm}} \quad (10.19)$$

with $m_\pm^v = m_+^{v_+} m_-^{v_-}$, $Q = v_+^{v_+} v_-^{v_-}$, $v = v_+ v_-$, $m_{+(-)} = v_{+(-)} m_\pm$

10.3.3.2 Results and Discussion

The experimental solubility data are directly connected via the tabulated thermodynamic values with the mean activity coefficients of the investigated electrolytes. Therefore the used thermodynamic data can be verified by comparing predicted salt solubilities at 25 °C with experimental data and calculated mean activity coefficients at saturation. If the experimental

10.3 Group Contribution Model for the Prediction of Activity Coefficients

mean activity coefficients at saturation are predicted correctly the resulting errors in the salt solubilities must depend on the values for the standard Gibbs energy. Errors in the tabulated standard values directly diminish the precision of the prediction. In Table 10.4 experimental and predicted saturation molalities and the mean deviation between calculated and experimental mean activity coefficients at saturation at 25 °C in water are presented for several electrolytes. The experimental mean activity coefficients are taken from Hamer and Wu (1972), Robinson and Stokes (1965), Stokes (1948), and Rabie et al. (1999). If no data at saturation are available the deviation at the highest concentration is considered. If the difference between the highest concentration and saturation molality is more than 0.5 mol kg^{-1} no deviation is given.

Table 10.4: Calculated and experimental saturation molality at 25 °C in water.

System[a] water +	Saturation molality exp.	calc.	%	$\|\Delta\gamma_\pm\|$ %
KCl	4.905	4.825	1.66	1.51
KBr	5.770	5.768	0.03	0.98
LiCl · H$_2$O[b]	19.97	19.66	1.56	6.02
NaF	0.987	0.987	0.03	0.40
NaCl	6.139	6.013	2.07	1.66
NaBr · 2 H$_2$O[b]	9.120	8.998	1.34	0.37
NH$_4$Cl	7.459	8.029	7.64	1.80
KI	8.933	8.089	9.45	–
BaCl$_2$ · 2 H$_2$O[b]	1.776	1.765	0.64	0.27
MgCl$_2$ · 6 H$_2$O[b]	5.686	5.702	0.29	0.25
NiCl$_2$ · 6 H$_2$O[b]	5.06	5.15	1.82	2.34
Na$_2$SO$_4$ · 10 H$_2$O	1.955	1.912	2.22	2.28
CoCl$_2$ · 6 H$_2$O[b]	4.308	4.099	4.84	6.69
K$_2$SO$_4$	0.687	0.674	1.90	0.21

[a] data taken from the Dortmund Data Bank (DDB)
[b] no data for molar heat capacities ($\Delta c_{p,i}^0$) available

The predicted salt solubilities are in good agreement with the experimental data and only minor deviations between calculated and experimental mean activity coefficients are predicted.

The basis for modeling the saturation molalities as a function of temperature are Eqs. (10.18) and (10.19). Besides the standard Gibbs energy of formation of all species the standard heat of reaction Δh_R^0 and the difference of the molar heat capacities Δc_p^0 have to be known. Because Δc_p^0 is assumed to be constant in a small temperature range (approx. 80 °C) only systems in a comparable temperature interval (–13.4 and 110 °C) are investigated in this work. Systems for which no information about the molar heat capacity is available are not taken into account. In Table 10.5 predicted salt solubilities as a function of temperature are presented for a few electrolytes.

The mean deviation between experimental and predicted results varies between 1.3 and 11.7% for all investigated systems. Considering that the prediction is based on model parameters determined with the help of VLE data the results are in satisfactory agreement. In Figure 10.5 the obtained results for four selected systems are represented.

10 Determination of Phase Equilibria in Electrolyte Systems

Table 10.5: Predicted and experimental temperature dependent saturation molalities.

System[a] Water +	Data points	$\Delta/°C$	Molality max.	Saturation mol. $\|\Delta S\|$ abs.	%
KBr	51	−13.4–110.0	9.27	0.409	7.28
KCl	49	−11.0–109.6	7.9	0.204	4.237
NaCl	64	0.0–109.7	7.05	0.317	4.960
KI	13	0.0– 92.2	12.08	1.061	11.743
NaF	9	0.0– 35.0	0.985	0.012	1.259
Na$_2$SO$_4$	10	33.5–101.9	3.44	0.093	2.915
Na$_2$SO$_4$ · 10 H$_2$O	11	0.0– 32.0	3.38	0.045	3.880
NH$_4$Cl	21	−15.4– 75.0	11.4	0.503	5.523
K$_2$SO$_4$	38	0.4–101.9	1.39	0.024	3.390

[a] data taken from the Dortmund Data Bank (DDB)

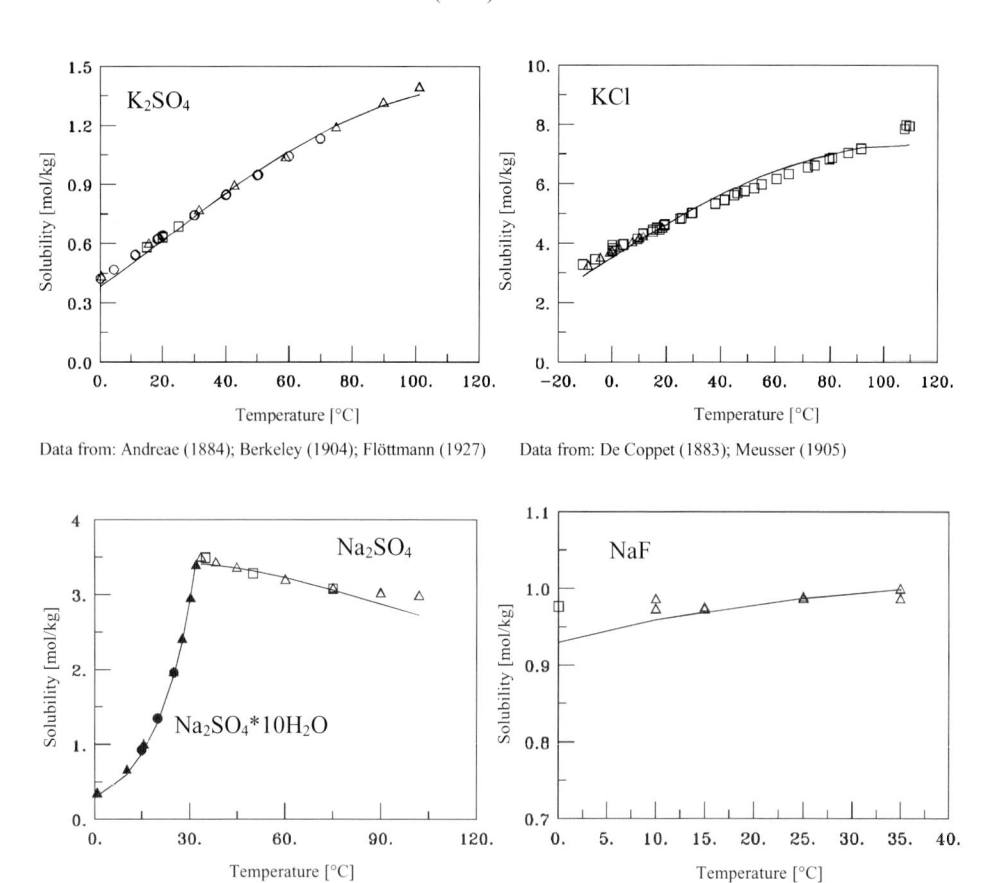

Data from: Andreae (1884); Berkeley (1904); Flöttmann (1927)

Data from: De Coppet (1883); Meusser (1905)

Daten von: Berkeley (1904); Flöttmann (1927); Caspari (1924)

Daten von: Foote (1930); Titova (1995)

Figure 10.5: Experimental and predicted solubilities for the aqueous electrolyte systems water + K$_2$SO$_4$, water + KCl, water + Na$_2$SO$_4$, and water + NaF as a function of temperature; (———) prediction this work.

10.3 Group Contribution Model for the Prediction of Activity Coefficients

Only minor deviations between experimental and predicted values are observed. At higher temperatures the difference between experimental and predicted data increase. This error occurs because the temperature dependence of the molar heat capacities is neglected. The assumption of constant molar heat capacities is only acceptable in a small temperature range (approx. 50 °C). In the systems water/Na_2SO_4 different solid phases occur. At temperatures between 0 and 32.4 °C the solid hydrate $Na_2SO_4 \cdot 10\ H_2O$ (Mirabilite) crystallizes. By increasing the temperature the hydrate becomes unstable in relation to the unhydrated Na_2SO_4 and at temperatures above 32.4 °C only solid Na_2SO_4 is present. At the invariant point at 32.4 °C the two solid phases Mirabilite and Na_2SO_4 are in the equilibrium with the aqueous solution.

The prediction of ternary systems is a little more complex. Figure 10.6 shows the system water + Na_2SO_4 + NaCl exhibiting an invariant point. These two salts have a common ion thereby simplifying the calculation. After calculating the solubility product K_{sp} from thermodynamic data or experimental saturation molalities in the single salt-solutions with the help of the proposed model, the solubility can be predicted. The molality of one of the non-common ions is fixed and therefore, the molality for the other one can be calculated. For the determination of the invariant points and the solubility curves a FORTRAN program has been written, which uses the Simplex-Nelder-Mead method (Nelder and Mead, 1965). The intersection of the two curves gives the composition at the invariant point where two solid phases are in equilibrium with the aqueous solution.

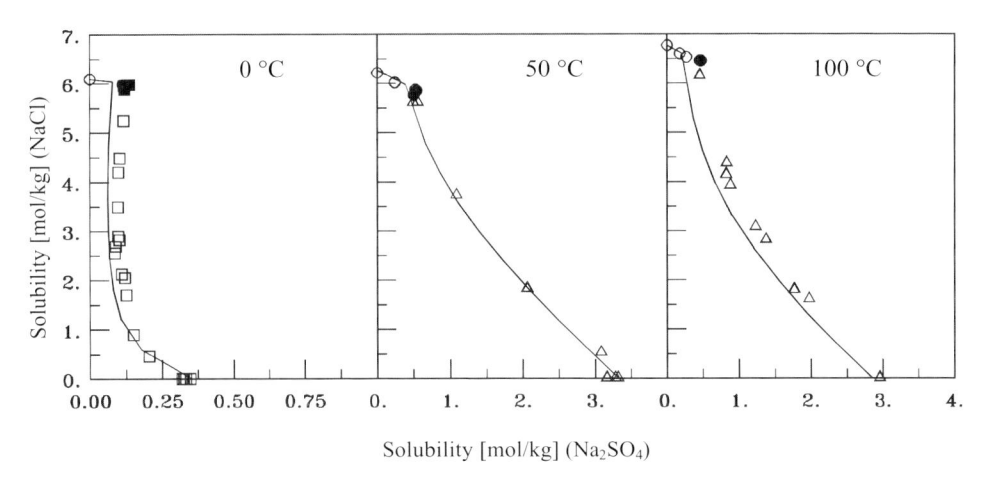

Figure 10.6: Experimental and predicted solubilities in the ternary mixture water + NaCl + Na_2SO_4 at several temperatures (at 50 °C, K_{sp} calculated from thermodynamic data) (experimental data from Silcock, 1979); (———) prediction this work. Solid phase: ○ NaCl, □ $Na_2SO_4 \cdot 10\ H_2O$, △ Na_2SO_4, invariant points: ■ NaCl and $Na_2SO_4 \cdot 10\ H_2O$, ● NaCl and Na_2SO_4.

10.3.4 Prediction of Gas Solubilities

The knowledge of gas solubilities in aqueous electrolyte systems is important for different industrial processes, especially for petroleum and natural gas exploitation, petroleum refining, coal gasification, environmental protection, gas antisolvent salt crystallization, formation of gas hydrates, and various absorption processes. For the design and optimization of these processes reliable information on the phase equilibrium behavior in a large temperature and pressure range is required. The predictive Soave-Redlich-Kwong (PSRK) group contribution equation of state proposed by Holderbaum and Gmehling (1991) is already widely used for the reliable prediction of vapor-liquid equilibria and gas solubilities of non-electrolyte systems in a large temperature and pressure range, regardless whether the systems are non-polar, polar, sub- or supercritical, symmetric, or highly asymmetric (Horstmann et al., 2000).

The prediction of phase equilibria containing non-condensable gases is not possible with classical g^E-models resp. group contribution methods, because their applicability is limited to subcritical compounds. Therefore our research concentrated on the further development of the PSRK model for the prediction of gas solubilities in electrolyte systems. In the proposed PSRK model, the group contribution method (LIFAC) is used to calculate the excess Gibbs energy required in the g^E mixing rule, where the published group interaction parameters of the PSRK and LIFAC model are used directly. The missing group interaction parameters between CO_2, N_2, CH_4 and the ionic groups Na^+, NH_4^+, Ca^{2+}, Cl^- are fitted using the experimental gas solubilities stored in the Dortmund Data Bank. Using these parameters the solubilities of CO_2, N_2, and CH_4 in aqueous electrolyte solutions can be predicted in a large temperature and pressure range.

10.3.4.1 Modified PSRK Group Contribution Method

The PSRK model based on the Soave-Redlich-Kwong (1972) equation for non-electrolyte systems can be written as

$$P = \frac{RT}{v-b} - \frac{a(T)}{v(v+b)} \tag{10.20}$$

The mixture b-parameter is calculated by the conventional linear mixing rule:

$$b = \sum_i x_i b_i \tag{10.21}$$

where is mole fraction of the component i in the liquid or vapor phase. b_i is the parameter of pure component i, and it is given by

$$b_i = 0.08664 \frac{RT_{c,i}}{P_{c,i}} \tag{10.22}$$

The attractive mixture parameter a is given by the modified Huron-Vidal excess Gibbs energy mixing rule (Michelsen, 1990). In Eq. (10.23) the PSRK mixing rule is described

$$a(T) = b \left[\frac{g_o^E}{A_1} + \sum_i x_i \frac{a_{ii}(T)}{b_i} + \frac{RT}{A_1} \sum_i x_i \ln \frac{b}{b_i} \right] \tag{10.23}$$

In the PSRK model A_1 has a value of -0.64663. The attractive parameter a_i of pure component i is obtained from

10.3 Group Contribution Model for the Prediction of Activity Coefficients

$$a_{ii}(T) = 0.42748 \frac{R^2 T_{c,i}^2}{P_{c,i}} \left[f(T_{r,i}) \right]^2 \tag{10.24}$$

In Eq. (10.25), g_0^E is the molar excess Gibbs energy at 1 atm:

$$g_0^E = RT \sum_i x_i \ln \gamma_i \tag{10.25}$$

where γ_i is the activity coefficient of component i which is calculated using the UNIFAC model. It should be noticed that in the UNIFAC expression constant group interaction parameters are used

$$\Psi_{mn} = \exp\left(-\frac{a_{mn}}{T}\right) \tag{10.26}$$

while in the PSRK model temperature dependent interaction parameters are applied, if necessary:

$$\Psi_{mn} = \exp\left(-\frac{\alpha_{mn} + \beta_{mn} T + \chi_{mn} T^2}{T}\right) \tag{10.27}$$

For systems with strong electrolytes, the equations given above can directly be used for the vapor phase assuming that the strong electrolytes are not present in the vapor phase. For the liquid phase the Eqs. (10.21) and (10.22) are changed to

$$b = \sum_i x_i' b_i \tag{10.28}$$

and

$$a(T) = b \left[\frac{RT}{A_1} \sum_i x_i' \ln \gamma_i' + \sum_i x_i' \frac{a_{ii}(T)}{b_i} + \frac{RT}{A_1} \sum_i x_i' \ln \frac{b}{b_i} \right] \tag{10.29}$$

where x_i' is the salt-free mole fraction of component i (gas or solvent) in the liquid phase, which allows that the a-, b- mixture parameters of the liquid phase can be calculated even though the critical properties (T_c, P_c) of the ions are not available. Different suggestions for the determination of the critical data T_c, P_c of the ions have been published (Dahl and Macedo, 1992). Finally, it has been found that the approximation proposed in this work seems to be reasonable because the salting effect on gas solubilities may be described by the activity coefficients of the gases and solvents in the liquid phase.

In Eq. (10.29) γ_i is the activity coefficient of component i (gas or solvent) in the liquid phase predicted by the LIFAC model. It should be pointed out that if the ionic strength I is equal to zero the LIFAC model reduces to the UNIFAC model, and the PSRK model extended for electrolyte systems proposed in this work is also reduced to the PSRK model used for non-electrolyte systems. This means that the large number of parameters already available for the UNIFAC and PSRK model can directly be used in the extended PSRK model.

For a detailed description of the method see Li et al. (2001).

10.3.4.2 Results and Discussion

The overall results for more than 60 data sets predicted by the PSRK model are summarized in Table 10.6, where

$$\frac{\Delta P}{P}(\%) = 100 \times \frac{|P_{exp} - P_{calc}|}{P_{exp}} \quad (10.30)$$

shows that in the salt concentration 0.1–6 m, temperatures between 298 and 523 K, and pressure range 1–667 bar the total mean relative deviation between experimental and predicted pressures is not more than 6.57%. These results confirm that the PSRK model can be used reliably to predict gas solubilities for aqueous electrolyte systems in a wide salt concentration, temperature, and pressure range. It should be pointed out that the above results include all data sets for the $CO_2 + H_2O + CaCl_2$ and $CH_4 + H_2O + NaCl$ systems although some of the experimental data points scatter significantly.

Table 10.6: Deviations between experimental and predicted pressures.

System	P/bar	T/K	m/mol · kg^{-1}	Data points	$\Delta P/P$ (%)
$CO_2 + H_2O + NaCl$	1–600	313–523	1.0–6.0	90	5.69
$CO_2 + H_2O + NH_4Cl$	1–96	313–433	3.0–6.0	71	3.01
$CO_2 + H_2O + CaCl_2$	1–667	298–394	0.1–5.3	148	9.89
$CH_4 + H_2O + NaCl$	14–616	303–398	0.5–4.0	46	10.60
$CH_4 + H_2O + CaCl_2$	8–606	298–398	0.2–1.35	48	5.37
$N_2 + H_2O + NaCl$	100–616	324–398	1.0	18	1.94
Mean deviation				421	6.57

The projection of gas solubilities in different water-gas-salt systems is shown in Figure 10.7. While Figure 10.7 a, b describe CO_2 solubilities at 393.15 K in aqueous NH_4Cl resp. NaCl solutions at different salt concentrations Figures 10.7 c, d show the temperature dependence of the CO_2 solubility in 4.0 resp. 6.0 m aqueous NaCl solution. From Figure 10.7 a, b it can be seen that quite good agreement between predicted and experimental results for gas solubilities are obtained by the PSRK model. The solubilities of CO_2 presented in Figure 10.7 c, d are strongly temperature dependent and the CO_2 solubility decreases with increasing temperature, which is almost predicted by PSRK. It can be seen that the calculated results are in good agreement with the experimental data even at high salt concentrations (e.g. 6.0 m).

10.4 Determination of Vapor-Liquid Equilibria for Systems Containing Strong Electrolytes

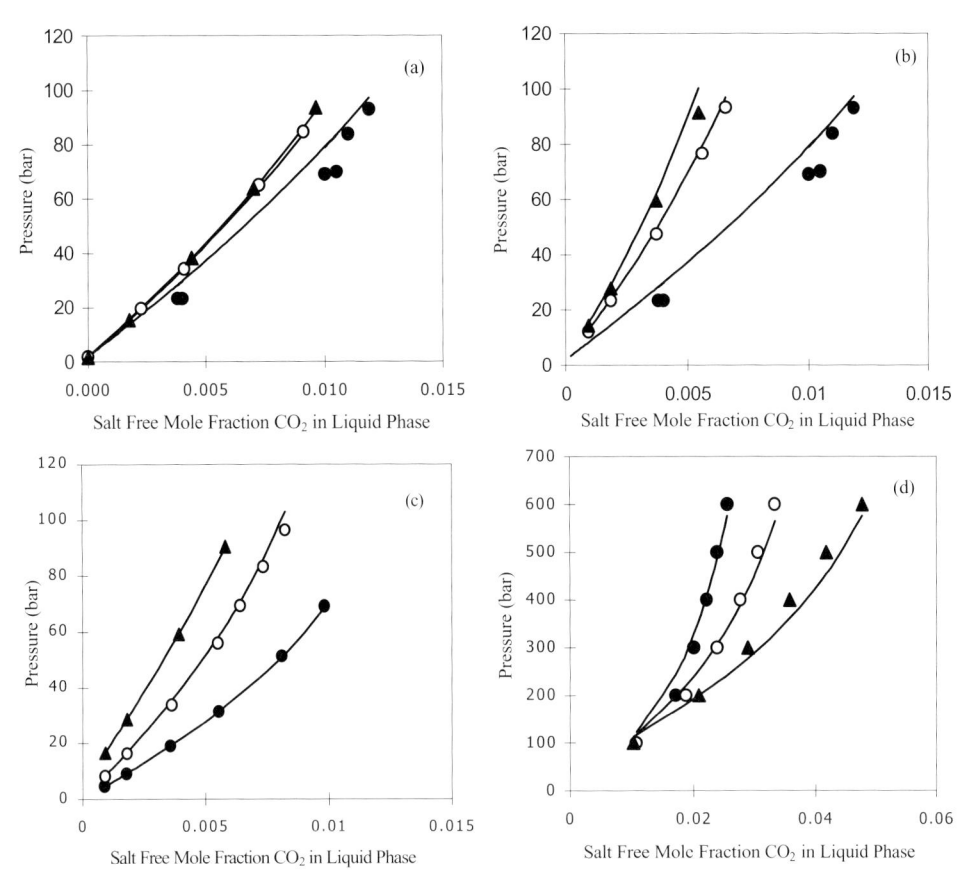

Figure 10.7: Experimental and predicted CO_2 solubilities in water + gas + salt systems. (a) Aqueous NH_4Cl solution at 393.15 K; (b) aqueous NaCl solution at 393.15 K; (c) 4.0 m aqueous NaCl solution; (d) 6.0 m aqueous NaCl solution. (—) PSRK model; (○, ▲) experimental values from (a) Rumpf et al., (1994 a) ($m = 4.0$, 6.0), (b) Rumpf et al. (1994 b) ($m = 4.0$, 6.0), (●) Prutton and Savage (1945) ($m = 0$); (●, ○, ▲) experimental values from (c) Rumpf et al. (1994 b) (313.15 K, 353.15 K, 433.15 K), (●, ○, ▲) experimental values from (d) Takenouchi and Kennedy (1965) (423.15 K, 473.15 K, 523.15 K).

10.4 Determination of Vapor-Liquid Equilibria for Systems Containing Strong Electrolytes

For the determination of vapor-liquid equilibria several experimental methods can be applied. In this work the ebulliometry and the headspace gas chromatography has been used to investigate several vapor-liquid equilibria. The objective of this work was the systematical study of the salt effect on ternary VLE for which no or obviously wrong data were available in the literature. These data have been used to fit new parameters for the LIQUAC (Li et al.,

1994) and the LIFAC model. Additionally, also the activity coefficients of solvents at infinite dilution and the mean ionic activity coefficients of salts are of most interest for the reliable knowledge of phase equilibria in the presence of strong electrolytes. Therefore measurements with the help of the dilutor technique resp. an apparatus for the determination of the electromotive force (EMF) of a cell were performed. A detailed description of the measurements is given in the following chapters.

10.4.1 Materials

The solvents were purified by distillation using a 1.0 m Vigreux column, dried, and stored over 3Å molecular sieves. The water content was determined using the Karl-Fischer method. Finally the water content of all liquid components was <100 ppm. The final purity was better than 99.7 wt. % for all solvents determined by GC using a TCD detector. The salts were dried at 90–120 °C in a vacuum oven until constant mass was reached.

10.4.2 Headspace Gas Chromatography

All mixtures were prepared directly by using a Sartorius analytical balance, the accuracy of which was 0.1 mg. For each experimental point approximately 8 cm^3 of sample solution was put into the sample vial (22 cm^3). After the sample vials were tightly closed by means of a special aluminum lid, with a washer and a Teflon disc, they were kept at the desired temperature in a thermostatic bath controlled within 0.1 °C. The measurements were started after the samples were thermostated for at least 12 h to ensure phase equilibrium.

For the determination of the vapor-phase composition, samples of the vapor phase were automatically withdrawn using a Perkin-Elmer F45 GLC vapor analyzer and analyzed by a F22 gas chromatograph with the help of a thermal conductivity detector and an integrator (Hewlett-Packard 3390A). For the separation in all cases a 1.2 m stainless steel column filled with Porapak Q 80/100 was used. The operating conditions were the following: injection temperature, 210 °C; oven temperature, 190 °C; detector temperature, 210 °C; carrier gas, Helium (purity 99.9%) and a flow rate of 0.41 cm^3 s^{-1}. More details of the experimental setup have already been described before by Weidlich and Gmehling (1985).

10.4.3 Ebulliometry

The modified Scott ebulliometer is similar to the one used by Dallinga et al. (1993). However some modifications have been carried out. A detailed description about this technique is given by Topphoff et al. (2001).

All liquid mixtures were prepared gravimetrically with a Sartorius analytical balance with an accuracy of ±0.1 mg. For each experimental point approx. 300 cm^3 of the mixture was filled into the ebulliometer. The pressure was set to 101.3 kPa, before the ebulliometer

10.4 Determination of Vapor-Liquid Equilibria for Systems Containing Strong Electrolytes

was heated and steady-state boiling condition was observed. The equilibrium temperature was measured with the help of a Hewlett-Packard quartz thermometer (model 2804 A) with an accuracy of ±0.001 °C.

10.4.3.1 Experimental Results

A list with all measured systems and the covered temperature and pressure range is given in Table 10.7 Altogether 282 isothermal resp. isobaric data sets in a temperature range from 30–70 °C (66.6–101.3 kPa) were investigated. The measurements were carried out at different constant salt concentrations from the salt-free system up to saturation. Many of the obtained results were already published (Yan et al., 1997, 1998 a–c, 2001; Topphoff et al., 2001). To validate the experimental methods measurements for the binary salt-free systems were carried out. The obtained results were compared with literature data. For all systems the results are in good agreement with the published data. All experimental data obtained were used for the determination of new interaction parameters for the LIQUAC resp. the LIFAC model.

Table 10.7: Investigated isothermal and isobaric ternary VLE systems.

System	ϑ/°C	P/kPa	Data sets
Headspace			
ethanol + THF + CaCl$_2$	40–60		17
ethanol + THF + NaI	40–60		21
CHCl$_3$ + ethanol + NaI	40–60		18
CHCl$_3$ + ethanol + Ca(NO$_3$)$_2$	40–60		17
ethyl acetate + ethanol + NaI	30–60		40
ethanol + 2-propanol + BaI$_2$	40–70		9
acetone + methanol + LiNO$_3$	30–70		12
2-propanol + water + KI	40–70		27
ethanol + 1-butanol + LiI	30–70		24
methanol + 2-propanol + KF	40–60		21
acetone + methanol + LiBr	40–70		15
acetone + THF + LiBr	30–50		12
Ebulliometry			
methyl acetate + methanol + LiNO$_3$		101.3	11
ethyl acetate + ethanol + LiNO$_3$		101.3	11
acetone + methanol + ZnCl$_2$		66.6–101.3	8
acetone + methanol + NaI		100.0	8
acetone + methanol + NaAc		101.3	5
CH$_2$Cl$_2$ + methanol + LiCl		101.3	6
Overall			282

In Figure 10.8 the experimental data for the systems methyl acetate + methanol + LiNO$_3$ are plotted as an x/y diagram on a salt-free basis. On the right hand side the salt-free systems together with literature data and calculated values (UNIQUAC parameters taken from Gmehling et al., 1977) are presented. The experimental data are in a good accordance

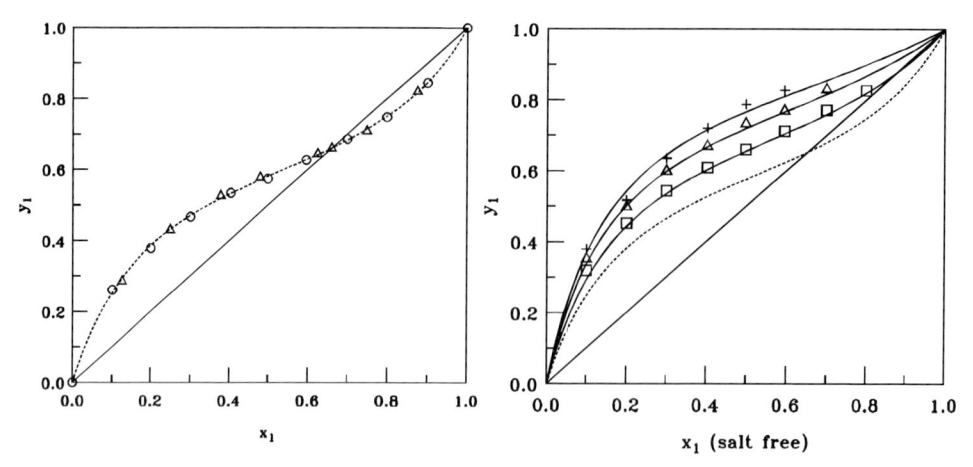

Figure 10.8: Vapor-liquid equilibrium diagram for the ternary system methyl acetate (1) + methanol (2) + LiNO$_3$ (3) at 101.33 kPa; *left figure*: salt-free system: (○) this work; (△) Vasileva et al. (1983); *right figure*: (□) $x_3 = 0.05$, (△) $x_3 = 0.10$, (+) $x_3 = 0.15$; continuous line calculated by the LIQUAC model; dashed line calculated by the UNIQUAC model ($x_3 = 0$).

with the published and predicted values. On the right hand side it can be seen that the addition of LiNO$_3$ to the systems methyl acetate + methanol increases the amount of the ester in the vapor phase. A salting out effect of the ester occurs and the azeotropic point shifts to higher ester concentrations with increasing salt concentration. Generally the model is able to present the experimental data with a high accuracy.

The present status of the LIQUAC parameter matrix is shown in Figure 10.9 Overall 221 binary interaction parameters between 15 solvents, 19 cations, and 11 anions are available. This provides a large applicability of the model.

10.4.4 Measurement of Activity Coefficients at Infinite Dilution in Electrolyte Systems Using the Dilutor Technique

10.4.4.1 Experimental and Measurement Procedure

A detailed description for the measurement of activity coefficients at infinite dilution using the dilutor technique is given by Krummen et al. (2000). For the measurement both cells are filled with approx. 80 cm^3 of the water/salt mixture. After weighing the cell, the highly diluted component (solute) is injected through the septum. It has to be noted that the solute concentration is within the diluted range, i.e. at the start of the measurement the mole fraction of the solute x_i is already smaller than 10^{-3}. The carrier gas flows through the capillaries in the equilibrium cell, whereby through the small diameters of the capillaries very small gas bubbles with a constant size are produced, which lead to a large surface area required for the mass transfer. This is an advantage over the assigned fritted glass tips, which

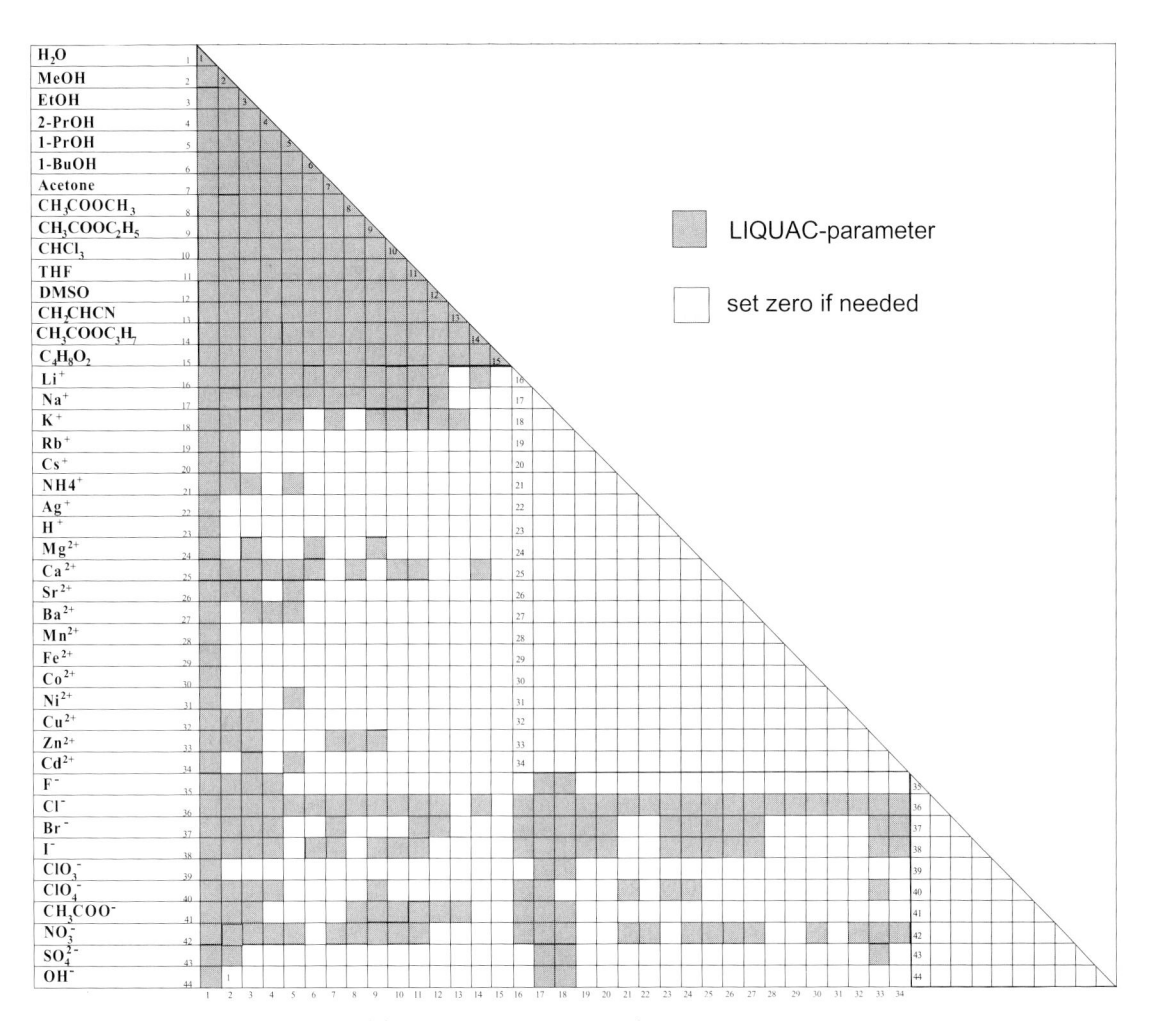

10.4 Determination of Vapor-Liquid Equilibria for Systems Containing Strong Electrolytes

Figure 10.9: Present status of the LIQUAC parameter matrix.

produce gas bubbles of strongly varying size. Supplementary, the retention time of the gas bubbles is increased by two baffles. During the measurement (duration approx. 1.5 h) the solute is stripped out of the cell with the carrier gas. The decrease of the solute concentration with time can then be used to determine γ^∞.

In order to avoid condensation effects, the carrier gas leaving the equilibrium cell passes a heated (at least 30 °C above the temperature in the cell) transfer line before it enters the gas-sampling valve. In periodic intervals the content of the sample loop ($V_{\text{sample loop}} = 400$ mm^3) is analyzed with the help of a gas chromatograph (United Technologies Packard; model 438A). It would be possible to examine several solutes at the same time, but then a column (e.g. Porapak P, mesh 80/100; 1/8″ diameter, length 0.66 m) with a satisfying separation of these components is required. The detection is carried out using a flame ionization detector (FID) and the signals are analyzed with the help of an integrator (Hewlett

Packard; HP 3390A). The obtained peak areas with the corresponding time are transferred to a computer for analyzing all the data with the help of a computer program.

10.4.4.2 Principle of the Applied Method

The principle of the dilutor method was described in detail by Leroi et al. (1977). Beyond that in this work a vapor phase correction (important for solutes with higher volatilities) introduced by Duhem and Vidal (1978) is considered.

In contrast to the papers mentioned, in this work a saturation cell is used, in order to guarantee a constant amount of solvent in the equilibrium cell. The same applies to solvent mixtures. In case of vapor-liquid equilibrium Eq. (10.32) can be used for the highly diluted component i:

$$x_i \gamma_i \varphi_i^s P_i^s = y_i \varphi_i^v P \qquad (10.31)$$

Whereby following assumptions are made:

- The solute is highly diluted ($x_i < 10^{-3}$).
- $\varphi_i^V \approx 1$, since helium is used as carrier gas:

$$x_i \gamma_i^\infty \varphi_i^s P_i^s = y_i P \qquad (10.32)$$

Since the investigated solvents show a negligible vapor pressure, γ_i^∞ can be calculated using Eq. (10.33).

$$\gamma_i^\infty = -\frac{n_{solv} RT}{\varphi_i^s P_i^s \left(\dfrac{\dot{F}_{in}}{a} + V_g\right)} \qquad (10.33)$$

with

$$\dot{F}_{in} = \dot{F}_{out}\left(1 - \frac{P_i^s}{P}\right) \quad \text{and} \quad a = \ln\frac{A_1}{A_0} \bigg/ t \qquad (10.34)$$

The slope a is determined by linear regression of the logarithmic peak area A_i vs. time t. For the analysis, only the change of the peak area with time and not the absolute peak area is necessary. Therefore no time-consuming calibration is required. The analysis is carried out using the program mentioned above which is based on Eq. (10.34). The saturation fugacity coefficient φ_i^s is calculated using second virial coefficients. The vapor pressure P_i^s can be determined by different vapor pressure equations (Antoine, Wagner, or DIPPR).

In addition, the vapor space volume V in the equilibrium cell can be determined using the density of the solvent (at system temperature) and the total mass. The necessary pure component properties, i.e. Antoine constants, critical data, and acentric factors, originate from the Dortmund Data Bank (DDB). These quantities needed for the calculation are afflicted with an error of $\pm 0.5\%$ (Weidlich and Gmehling, 1987). A detailed description of the derived equations and the measurement procedure is given by Krummen et al. (2000).

10.4 Determination of Vapor-Liquid Equilibria for Systems Containing Strong Electrolytes

Furthermore, for the calculation of γ^∞ the carrier gas flow rate F, the slope a, the absolute pressure P, and the temperature T in the equilibrium cell as well as the mass of the solvent for the calculation of the solvent's mole number n is required. The gas flow rate F represents the dominating source of error with max. 0.85% (i.e. ±0.2 s over a volume of 20 cm^3 of the soap bubble flow meter). The resulting relative standard error in γ^∞ is about ±2% (estimated by error propagation).

10.4.4.3 Experimental Results

The reliability of the apparatus used for the determination of limiting activity coefficients in pure solvents (non-electrolyte systems) has already been demonstrated (Gruber et al., 1999).

In this work activity coefficients at infinite dilution in aqueous systems with strong electrolyte have been determined. All obtained experimental data are listed in Table 10.8. These experimental data will be published in the near future.

Table 10.8: Experimental γ_i^∞ values for aqueous electrolyte systems.

Solvent + Solute	mass% salt	30 °C	40 °C	50 °C	60 °C
water + ethanol	NaCl				
	0	5.12	5.41	5.80	6.02
	2.85	5.60	6.07	6.51	7.01
	5.54	6.15	6.62	7.17	7.70
	13.86	8.75	9.40	10.08	10.7
	22.67	14.09	15.14	15.57	16.3
water + 2-propanol	0	9.72	10.8	11.7	12.8
	2.85	12.8	13.4	14.2	15.6
	5.54	13.2	14.7	16.5	18.3
	13.86	21.3	23.6	25.7	27.9
	22.67	44.0	47.0	48.4	51.1
water + 1-propanol	0	17.4	18.9	19.9	20.9
	2.85	21.4	22.4	23.0	23.8
	5.54	22.3	24.5	26.5	28.7
	13.86	35.7	37.6	39.1	40.7
	22.67	70.1	71.3	70.4	71.8
water + ethanol	KCl				
	5.54	6.40	6.82	6.81	7.18
	13.86	8.02	8.57	9.01	9.65
	22.67	11.1	11.7	12.4	13.2
water + 2-propanol	2.85	10.8	12.0	13.4	15.0
	5.54	13.2	13.9	14.6	16.0
	13.86	17.7	19.7	21.8	23.8
	22.67	29.7	31.7	34.1	36.7
water + 1-propanol	5.54	23.0	24.0	25.0	26.2
	13.86	29.7	32.7	35.2	37.3
	22.67	49.9	51.7	52.7	54.8
water + methanol	LiCl				
	0	2.20	2.27	2.29	2.40
	5.54	1.81	1.87	2.38	2.67

Table 10.8 (continued)

Solvent + Solute	mass% salt	30 °C	40 °C	50 °C	60 °C
	13.86	1.95	2.03	2.36	2.59
	22.67	1.91	2.09	2.78	2.92
water + ethanol	5.54	5.77	6.61	7.71	8.22
	13.86	7.95	8.34	8.93	9.85
	22.67	8.94	9.82	11.0	11.3
water + 2-propanol	5.54	12.3	13.7	16.1	17.4
	13.86	19.6	21.5	22.3	23.9
	22.67	29.5	30.1	31.1	32.7
water + 1-propanol	5.54	24.5	25.8	29.4	30.4
	13.86	37.4	39.8	40.2	40.8
	22.67	51.4	53.3	53.3	53.7
1-propanol + water	LiCl				
	0	3.60	3.35	3.14	2.97
	5.00	1.61	1.44	1.35	1.25
ethylene glycol + methanol	KoAc				
	0	1.37	1.31	1.28	1.27
	3.90	1.36	1.35	1.38	1.34
ethylene glycol + ethanol	0	2.16	2.07	2.01	1.99
	3.90	2.62	2.20	2.19	2.12
ethylene glycol + 2-propanol	0	3.09	2.95	2.88	2.82
	3.90	3.43	3.22	3.14	3.08
ethylene glycol + 1-propanol	0	3.63	3.27	3.10	3.05
	3.90	3.79	3.57	3.33	3.30

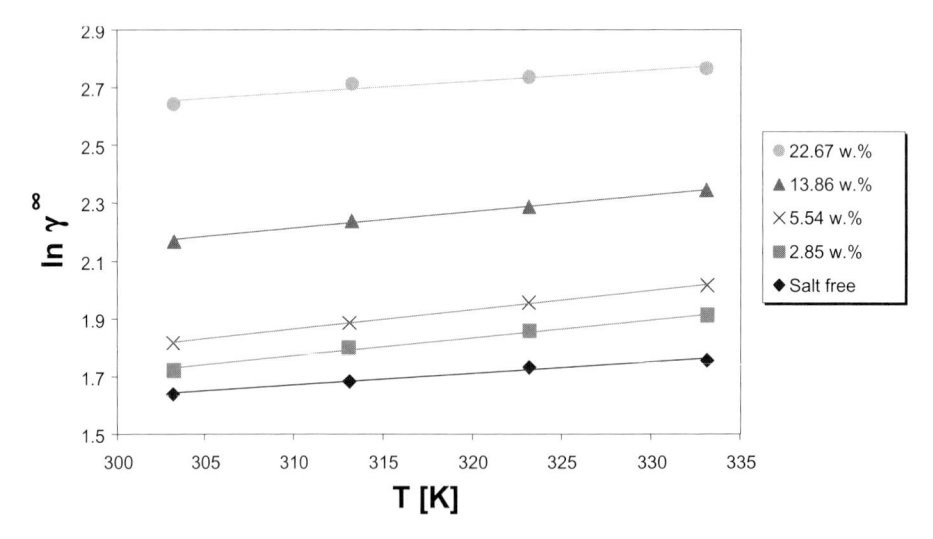

Figure 10.10: Experimental activity coefficients at infinite dilution for ethanol in a water + NaCl solution.

10.4 Determination of Vapor-Liquid Equilibria for Systems Containing Strong Electrolytes

In Figure 10.10 the determined experimental activity coefficients for ethanol in water/ NaCl-solution for different salt concentrations are shown. It can be seen that the activity coefficient increases with increasing salt concentration. A salting out effect of the less polar component ethanol is observed. This effect is typical for systems containing electrolytes and was observed for all systems investigated.

10.4.5 Measurement of the Mean Activity Coefficients of Ions by Determination of the Electromotive Force

Once the standard potential of the cell is known, the activities of the ions can be determined by measuring the electromotive force (EMF) of the cell at the salt concentration of interest. The potential difference between the two ion sensitive electrodes is measured with the help of an ion meter (model ORION 920A). A scheme of the used apparatus is given in Figure 10.11.

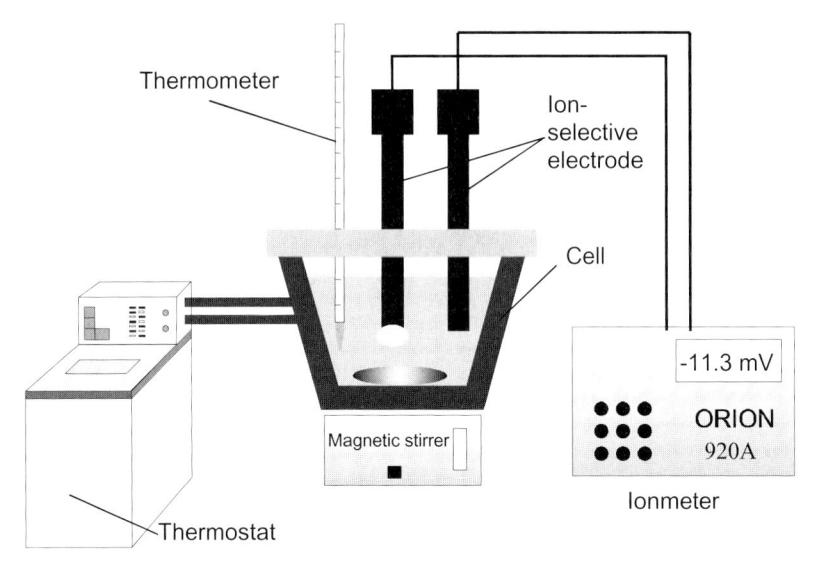

Figure 10.11: Experimental set-up for the determination of mean activity coefficients of ions by EMF measurement.

In contrast to other studies the measurements are not performed by determination of the potential difference between a standard reference electrode and an ion selective electrode. To prevent an extensive calibration the potential difference (ΔE) is directly measured between the cation and the anion selective electrode. 80 cm³ of sample solution are filled into the cell that is kept at constant temperature. The temperature is measured with the help of a mercury thermometer (accuracy ± 0.1 K).

After determining the standard potential (E^0) of the cell by a graphical method the mean ionic activity coefficient ($\gamma_\pm$) can be calculated by the Nernst equation. Starting from the Nernst equation for 1,1-electrolytes it can be written

$$E = E^0 + \frac{RT}{F} \ln\left[(\gamma_\pm)^2 \cdot m_\pm^2\right] \tag{10.35}$$

Replacing $\ln(\gamma_\pm)$ in Eq. (10.36) with the Debye-Hückel limiting law

$$\log \gamma_\pm = -A \cdot |z_+ z_-| \sqrt{m} \qquad I = m \text{ for monovalent ions} \tag{10.36}$$

with A = Debye-Hückel parameter and I = ionic strength, the expression can be rearranged for concentrations smaller than 0.1 mol kg^{-1}

$$E - 2\frac{RT}{F} \ln m = E^0 - 2\frac{RT}{F} \cdot A \cdot \ln 10 \cdot \sqrt{m} \tag{10.37}$$

The expression on the left hand side of Eq. (10.37) $[E - 2(RT/F) \ln m]$ is plotted against $\sqrt{m}$, and extrapolated to $m = 0$. The intersection at $\sqrt{m} = 0$ is the value of E^0. Once the standard potential of the cell is known, the mean activity coefficient of the ions can be calculated. For ternary systems consisting of two salts and one solvent the standard potential has been determined with the help of an iteration cycle based on the Pitzer equation, because a graphical determination is only possible for binary systems.

Until now the binary systems NaCl + water, KCl + water, and NaBr + water in the temperature range between 25 and 60 °C and at different salt concentrations have been measured to verify the apparatus. Additionally the ternary system NaBr + NaCl + water has been investigated at 25 °C. Figure 10.12 shows experimental data for the system NaCl + water in a temperature range between 25–60 °C. The values for the mean ionic activity coefficients decrease with the amount of salt and finally increase again, where raising the temperature generally leads to higher mean ionic activity coefficients especially at higher concentrations.

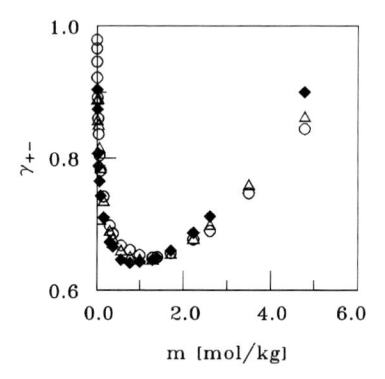

Figure 10.12: Experimental mean activity coefficients for NaCl + water solution at different temperatures: (○) 25 °C, (△) 40 °C, (◆) 60 °C.

10.5 Conclusion

In this project a reliable model for the prediction of phase equilibria in the presence of strong electrolytes has been developed. The addition of salt to a binary or ternary system has a significant effect on the relative volatility of the solvents. Furthermore, the added salts have a pronounced effect on the azeotropic composition. Overall 282 VLE data sets for 18 ternary systems in the presence of strong electrolyte systems were measured with the help of the headspace gas chromatography and ebulliometry. The covered temperature and pressure range was between 30–60 °C and 66.6–101.325 kPa. The results were used to fit new interaction parameters for the LIQUAC and the new developed group contribution model LIFAC. Additionally several activity coefficients at infinite dilution have been determined by the dilutor technique. The solutes methanol, ethanol, 1-propanol, 2-propanol, and water were investigated in the solvents water + (NaCl, KCl, or LiCl), 1-propanol + LiCl, and ethylene glycol + KoAc in a temperature range between 30–60 °C at five different salt concentrations. Also mean activity coefficients for ions have been measured by the determination of the electromotive force (EMF) of a cell. First measurements have been carried out to verify the reliability of the apparatus used; they were in good agreement with the literature data.

A new group contribution model for the description of phase equilibria in the presence of strong electrolytes has been proposed in this work. Based on the g^E-model LIQUAC the group contribution concept has been developed and overall 306 binary and 264 ternary VLE data sets from own experiments and from the literature have been used to fit 129 binary interaction parameters. For 264 ternary VLE data sets the total mean absolute deviation of the vapor phase mole fraction is 0.019, for the temperature it is 0.92 K, and for the pressure it is 0.43 kPa. For the 306 binary VLE data sets the total mean absolute deviation of the mean osmotic coefficients is 0.052, for the mean activity coefficients it is 0.120, and for the pressure it is 0.44 kPa. Compared with the group contribution model of Kikic et al. (1991), the proposed model allows a more reliable prediction of VLE in mixed electrolyte systems.

The model was also applied for the prediction of salt solubilities. The obtained results for the investigated systems were in good agreement with the experimental data. Because only thermodynamic data for systems containing water are available in the literature only aqueous salt systems were studied.

Since phase equilibria of electrolytes with supercritical gases cannot be predicted with the help of g^E-models an equation of state (EOS) approach has to be used. Therefore in the proposed PSRK model, the LIFAC model is used to predict the excess Gibbs energy required in the g^E mixing rule, whereby the published group interaction parameters of the PSRK and LIFAC model were used directly. The total mean relative deviation between experimental and predicted pressures is less than 6.57% for a salt concentration range of 0.1–6 m, temperatures between 298 and 523 K, and a pressure range of 1–667 bar. It should be pointed out that the above results include all data sets for the $CO_2 + H_2O + CaCl_2$ and $CH_4 + H_2O + NaCl$ system although some of the experimental data points scatter significantly. Concluding, these results confirm that the PSRK model can be reliably used to predict gas solubilities for aqueous electrolyte systems not only in a wide salt concentration but also for a large temperature and pressure range.

References

Achard, C.; Dussap, C. G.; Gros, J. B. (1994): Representation of Vapor-Liquid Equilibria in Water-Alcohol-Electrolyte Mixtures with a Modified UNIFAC Group-Contribution Method. Fluid Phase Equilibria 98, 71–89.

Andreae J. L. (1884): Die Löslichkeit fester Körper in Wasser bei verschiedenen Temperaturen. J. Prakt. Chem. 29, 456–477.

Berkeley, Earl of (1904): On some Physical Constants of Saturated Solutions. Philos. Trans. Roy. Soc. London 203, 189–215.

Bondi, A. (1968): Physical Properties of Molecular Crystals, Liquids and Glasses. Wiley, New York.

Boone, J. E.; Rousseau, R. W.; Schoenborn, E. M. (1976): The Correlation of Vapor-Liquid Equilibrium Data for Salt-Containing Systems. Adv. Chem. Ser. 155, 36–51.

Caspari, W. A. (1924): The System Sodium Carbonate-Sodium Sulphate-Water. J. Chem. Soc. 125, 2381–2387.

Dahl, S.; Macedo, E. A. (1992): The MHV2 Model: A UNIFAC-based Equation of State Model for Vapor-Liquid and Liquid-Liquid Equilibria of Mixtures with Strong Electrolytes. Ind. Eng. Chem. Res. 31, 1195–1201.

Dallinga, L.; Schiller, M.; Gmehling, J. (1993): Measurement of Activity Coefficients at Infinite Dilution Using Differential Ebulliometry and Non-Steady-State Gas-Liquid Chromatography. J. Chem. Eng. Data 38, 147–155.

De Coppet, L. C. (1883): Recherches sur la solubilité des chlorures, bromures et iodures de potassium et de sodium. Ann. Chim. Phys. (Paris) 30, 411–429.

DIPPR (1984): Tables of Physical and Thermodynamic Properties of Pure Compounds. AIChE, New York.

Dortmund Data Bank (DDB, www.ddbst.de) (2001).

Duhem, P.; Vidal, J. (1978): Extension of the Dilutor Method to Measurement of High Activity Coefficients at Infinite Dilution. Fluid Phase Equilib. 2, 231–235.

Flöttmann, Fr. (1927): Über Löslichkeitsgleichgewichte. Anal. Chem. 73, 1–39.

Foote, H. W.; Hickey, F. C. (1937): The Ternary Systems Barium Hydroxide and Water with Barium Chloride, Thiocyanate, Chlorate or Acetate at 25 °C. J. Am. Chem. Soc. 59, 648–650.

Fredenslund, A.; Gmehling, J.; Rasmussen, P. (1977): Vapor-Liquid Equilibria Using UNIFAC. Elsevier, Amsterdam, p. 380.

Gmehling, J. (1985): Dortmund Data Bank Basis for the Development of Prediction Methods. CODATA Bulletin 58, Pergamon, Oxford, 56–64.

Gmehling, J. (1991): Development of Thermodynamic Models with a View to the Synthesis and Design of Separation Processes. In: J. Gmehling (Ed.), Software Development in Chemistry 5. Springer-Verlag, Berlin, 1991, p. 1–14.

Gmehling, J.; Onken, U. (1977): Vapor-Liquid Equilibrium Data Collection. Chemistry Data Series 1, Part 2a, DECHEMA, Frankfurt.

Gruber, D.; Krummen, M.; Gmehling J. (1999): The Determination of Activity Coefficients at Infinite Dilution with the Help of the Dilutor Technique (Inert Gas Stripping). Chem. Eng. Technol. 22, 827.

Hamer, W. J.; Wu, Y. C. (1972): Osmotic Coefficients and Mean Activity Coefficients of Uni-Univalent Electrolytes in Water at 25 °C. J. Phys. Chem. Ref. Data 1, 1047–1099.

Hansen, H. K.; Rasmussen, P.; Fredenslund, A.; Schiller, M.; Gmehling, J. (1991): Vapor-Liquid Equilibria by UNIFAC Group Contribution. 5. Revision and Extension. Ind. Eng. Chem. Res. 30, 2352–2355.

Horstmann, S. (2000): Theoretische und experimentelle Untersuchungen zum Hochdruckphasengleichgewichtsverhalten fluider Stoffgemische für die Erweiterung der PSRK-Gruppenbeitragszustandsgleichung. Ph. D. Thesis, Shaker, Aachen.

References

Horstmann, S.; Fischer, K.; Gmehling J. (2000): PSRK Group Contribution Equation of State: Revision and Extension III. Fluid Phase Equilib. 167, 173–186.

Holderbaum, T.; Gmehling, J. (1991): PSRK: A Group Contribution Equation of State Based on UNIFAC. Fluid Phase Equilib. 70, 251–265.

Kikic, I.; Fermeglia, M.; Rasmussen, P. (1991): UNIFAC Prediction of Vapor-Liquid Equilibria in Mixed Solvent-Salt Systems. Chem. Eng. Sci. 46, 2775–2780.

Krummen, M., Gruber, D., Gmehling, J. (2000): Measurement of Activity Coefficients at Infinite Dilution in Solvent Mixtures Using the Dilutor Technique. Ind. Eng. Chem. Res. 39, 2114–2123.

Leroi, J.-C.; Masson, J. C.; Renon, H.; Fabries, J.-C.; Sannier, H. (1977): Accurate Measurement of Activity Coefficients at Infinite Dilution by Inert Gas Stripping and Gas Chromatography. Ind. Eng. Chem. Process Des. Dev. 16 Is. 1, 139–144.

Li, J.; Polka, H.-M. (1994): A g^E-Model for Single and Mixed Solvent Electrolyte Systems 1. Model and Results for Strong Electrolytes. Fluid Phase Equilib. 94, 89–114.

Li, J.; Topphoff, M.; Fischer, K.; Gmehling, J. (2001): Prediction of Gas Solubilities in Aqueous Electrolyte Systems Using the PSRK Model. Ind. Eng. Chem. Res. 40, 3703–3710.

Meranda, D.; Furter, W. F. (1972): Vapor-Liquid Equilibrium in Alcohol-Water Systems Containing Dissolved Halide Salts and Salt Mixtures. AIChE J. 18, 111–116.

Meusser, A. (1905): Zur Löslichkeit von Kaliumchlorid, -bromid, -jodid in Wasser. Z. Anorg. Allg. Chem. 44, 79–80.

Michelsen, M. L. (1990): A Modified Huron-Vidal Mixing Rule for Cubic Equations of State. Fluid Phase Equilib. 60, 213–219.

Nelder, J. A.; Mead, R. A. (1965): A Simplex Method for Function Minimization. Computer J. 7, 308–313.

Polka, M. H.; Gmehling, J. (1994): Effects of Calcium Nitrate on the Vapor-Liquid Equilibria of Ethanol + Water and 2-Propanol + Water. J. Chem. Eng. Data 39, 621–624.

Polka, H.-M.; Li, J.; Gmehling, J. (1994): A g^E-Model for Single and Mixed Solvent Electrolyte Systems 2. Results and Comparison with Other Models. Fluid Phase Equilib. 94, 115–127.

Prausnitz, M. J.; Lichtenthaler, R. N.; Gomes de Azevedo, E. (1999): Molecular Thermodynamics of Fluid-Phase Equilibria, 3rd Ed., Prentice-Hall, New Jersey.

Prutton, C. F.; Savage, R. L. (1945): The Solubility of Carbon Dioxide in Calcium Chloride-Water Solutions at 75, 100, 120°C and High Pressures. J. Am. Chem. Soc. 67, 1550–1554.

Rabie, H. R.; Wilczek-Vera, G.; Vera, J. H. (1999): Activities of Individual Ions From Infinite Dilution to Saturated Solutions. J. Solution Chem. 28, 885–913.

Robinson, R. A.; Stokes, R. H. (1965): Electrolyte Solutions. 2nd Ed., Butterworth, London.

Rumpf, B.; Nicolaisen, H.; Maurer, G. (1994a): Solubility of Carbon Dioxide in Aqueous Solutions of Ammonium Chloride, Temperatures from 313 K to 433 K and Pressures up to 10 MPa. Ber. Bunsen-Ges. Phys. Chem. 98, 1077–1081.

Rumpf, B.; Nicolaisen, H.; Öcal, C.; Maurer, G. (1994b): Solubility of Carbon Dioxide in Aqueous Solutions of Sodium Chloride Experimental Results and Correlation. J. Sol. Chem. 23, 431–448.

Schmitt, D. (1979): Der Einfluß von Salzen auf das Dampf-Flüssigkeitsgleichgewicht binärer Stoffgemische und die Rektifikation azeotroper Gemische bei der Zugabe von Salzen. Ph. D. Thesis, Karlsruhe University, FR Germany, p. 106.

Silcock, H. L. (1979): Solubilities of Inorganic and Organic Compounds. Monograph, Pergamon, Oxford.

Soave, G. (1972): Equilibrium Constants from a Modified Redlich-Kwong Equation of State. Chem. Eng. Sci. 4, 1197–1203.

Stokes, R. H. (1948): Thermodynamic Study of Bivalent Metal Halides in Aqueous Solution. Part XVII – Revision of Data for All 2:1 and 1:2 Electrolytes at 25 °C, and Discussion of Results. Trans. Faraday Soc. 44, 295–307.

Takenouchi, S.; Kennedy, G. C. (1965): The Solubility of Carbon Dioxide in NaCl Solutions at High Temperatures and Pressures. Am. J. Sci. 263, 445–454.

Titova, K. V.; Kolmakoa, E. I. (1995): Löslichkeit von Alkalimetall- und Erdkalimetallsulfaten in Wasserstoffperoxyden. Zh. Neorg. Khim. 40(3), 393–395.

Topphoff, M.; Kiepe, J.; Gmehling, J. (2001): Effects of Lithium Nitrate on the Vapor-Liquid Equilibria of Methyl Acetate + Methanol and Ethyl Acetate + Ethanol. J. Chem. Eng. Data 46, 1333–1337.

Vasileva, I. I.; Marinichev, A. N.; Susarev, M. P. (1983): Deposited Doc., 2400–83.

Vercher, E.; Munoz, R.; Martinez-Andreu, A. (1991): Isobaric Vapor-Liquid Equilibrium Data for Ethanol + Water + Potassium Acetate and Ethanol-Water-(Potassium Acetate/Sodium Acetate) Systems. J. Chem. Eng. Data 36, 274–277.

Wagman, D. D.; Evans, W. H.; Parker, V. B.; Schumm, R. H.; Halow, I.; Baily, S. M.; Churney, K. L.; Nuttall, R. L. (1982): National Bureau of Standards: The NBS Tables of Chemical Thermodynamic Properties. J. Phys. Chem. Ref. Data 11 (2).

Weidlich, U.; Gmehling, J. (1985): Extension of UNIFAC by Headspace Gas Chromatography. J. Chem. Eng. Data 30, 95–101.

Weidlich, U.; Gmehling, J. (1987): Measurement of γ^∞ Using Gas-Liquid Chromatography. 1. Results for the Stationary Phases n-Octacosane, 1-Docosanol, 10-Nonadecanone, and 1-Eicosene. J. Chem. Eng. Data 32, 138–142.

Yan, W.; Rose, C.; Gmehling, J. (1997): Isothermal Vapor-Liquid Equilibrium Data for the Ethanol + Ethyl Acetate + Sodium Iodide System at Five Temperatures. J. Chem. Eng. Data 42, 603–608.

Yan, W.; Rose, C.; Zhu, M.; Gmehling, J. (1998a): Isothermal Vapor-Liquid Equilibrium Data for the Acetone + Methanol + Lithium Nitrate System. J. Chem. Eng. Data 43, 482–485.

Yan, W.; Rose, C.; Zhu, M.; Gmehling, J. (1998b): Measurement and Correlation of Isothermal Vapor-Liquid Equilibrium Data for the System Acetone + Methanol + Lithium Bromide. J. Chem. Eng. Data 43, 585–589.

Yan, W.; Topphoff, M.; Zhu, M.; Gmehling, J. (1998c): Measurement and Correlation of Isobaric Vapor-Liquid Equilibrium Data for the System Acetone + Methanol + Zinc Chloride. J. Chem. Eng. Data 44, 314–318.

Yan, W.; Topphoff, M.; Rose, C.; Gmehling, J. (1999): Prediction of Vapor-Liquid Equilibria in Mixed-Solvent Electrolyte Systems Using the Group Contribution Concept. Fluid Phase Equilib. 162, 97–113.

Yan, W.; Topphoff, M.; Gmehling, J. (2001): Measurement and Correlation of Isothermal Vapor-Liquid Equilibrium Data for the System Ethanol + i-Propanol + Barium Iodide. J. Chem. Eng. Data, submitted.

Nomenclature

a_i	activity of component i
a_i	parameter for pure component i in equation of state
a_{ij}	interaction parameter of the UNIFAC equation
A	Debye-Hückel parameter
b	Debye-Hückel parameter
b_i	parameter for pure component in equation of state
b_{ij}	binary virial interaction parameter (middle range)
B	second virial coefficient
c_{ij}	binary virial interaction parameter (middle range)
c_p	molar heat capacity
d	density
E	potential difference of the cell
E^0	standard potential of the cell
D	dielectric constant

Nomenclature

F	Faraday's constant
F	flow
g	molar Gibbs Energy
G	absolute Gibbs Energy
I	ionic strength
h	molar enthalpy
K_{sp}	solubility product
m	molality/mol kg^{-1} solvent
m	mass
M	molecular weight
n_i	mole number of the component i
n_{kg}	mass solvent
P	pressure
P_i^s	saturation pressure of component i
q_i	relative van der Waals surface area of component i
Q	type of dataset
Q_k	relative van der Waals surface area of group k
R	universal gas constant
R_k	relative van der Waals volume of group k
T	absolute temperature
V	volume
v	molar volume
w_Q	weighting factor
x_i	mole fraction of component i in the liquid phase
x'_i	salt free mole fraction of component i in the liquid phase
y_i	mole fraction of component i in the vapor phase
z_j	charge number

Greek letters

α, β, χ	temperature dependent UNIFAC parameter
γ_i	activity coefficient of component i
$\gamma_\pm$	mean ionic activity coefficient
ϑ	temperature in degree Celsius
v_k	number of groups
v_+, v_-	stochiometric coefficient of the ions
Φ	molar osmotic coefficient
τ	UNIQUAC parameter
φ_i	fugacity coefficient
Ψ	Boltzmann factor
∞	value at infinite dilution

Subscripts

i, j, k	components of the system
C	critical data
k	group k

P	constant pressure
T	constant temperature
V	constant volume
+	cation
−	anion
±	mean value

Superscripts

0	standard state
calc	calculated quantity
exp	experimental quantity
E	excess property
s	saturated
LR	long range
MR	middle range
SR	short range
nt	number of dataset types
np	number of data points
*	normalization to infinite dilution

11 Ion Coordination and Thermodynamic Modeling of Molten Salt Hydrate Mixtures

Wolfgang Voigt[1], Kay Hettrich, and Dewen Zeng

11.1 Abstract

Water-anion coordination competition within the concentration range of molten hydrates was investigated by Raman and UV-Vis spectroscopy in the systems $MgCl_2$ + $CsCl$ + H_2O and $MgCl_2$ + $CuCl_2$ + H_2O. For the latter system also the solid-liquid equilibria at 383 K have been determined. Results are discussed in terms of formation of chlorocomplexes of magnesium and copper. Thermodynamic properties (activities, enthalpies of mixing and evaporation, solid-liquid phase equilibria) have been modeled by means of step-wise hydration and complexation reactions involving the entire concentration range from aqueous solutions to anhydrous molten salts.

11.2 Introduction

Thermodynamic properties of liquid water + salt systems are enormously changing going from dilute solutions toward molten salts. Characteristic changes in the concentration dependence of properties occur within certain intervals on the concentration scale, which are caused by changes in the predominant state of hydration as illustrated in Figure 11.1. The range of molten hydrates separates solutions from molten salts. It represents itself a transition range from prevailing water-water to cation-anion contact interactions located at molar water/salt ratios between 3–10. Understanding of the properties in this composition range in terms of changing coordination is crucial for the development of thermodynamic models valid within broad composition ranges.

Within our concept we define an ideal molten salt hydrate as an ionic liquid where all the water is coordinated at the strongest hydrating cation within its first coordination sphere preventing direct cation-anion contacts (Emons et al., 1986). In a number of crystalline hy-

---
[1] TU Bergakademie Freiberg, Institut für Anorganische Chemie, Leipziger Str. 29, 09596 Freiberg.

11 Ion Coordination and Thermodynamic Modeling of Molten Salt Hydrate Mixtures

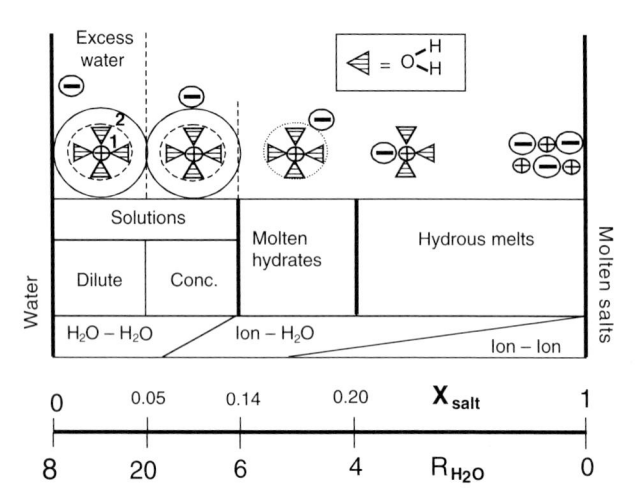

Figure 11.1: Characteristic concentration ranges from dilute solutions to molten salts.

drates like $MgCl_2 \cdot 6\,H_2O$, $CaCl_2 \cdot 6\,H_2O$, or $Mg(NO_3)_2 \cdot 6\,H_2O$ this structural situation is realized. In molten state most of the water will still be coordinated at the cations, which results in low water activities and enhanced enthalpies of evaporation compared with pure water (Braunstein, 1971). However, depending on coulombic and specific chemical interactions between cations, water, and anions in molten hydrates anions compete more or less successful with water for coordination sites at the cations forming ion pairs or anionic complex ions. The extent of this ligand exchange at the cations governs the concentration dependence of liquids properties. Mixtures of cations in molten hydrates give rise to a redistribution of water and anions corresponding to their different affinities. This can be accompanied by unusual behavior as for instance a salting-out of KCl by addition of water in the system $MgCl_2$ + KCl + H_2O (Voigt and Emons, 1995) or the absolute increase of vapor pressure by addition of anhydrous salts (Emons et al., 1987). In principle redistributions of ligands on cations can be modelled by applying the mass action law on simultaneous reaction equilibria. But contrary to complexation equilibria in dilute electrolyte solutions the environment of the complexes itself is subjected to a change from solution-like to molten salt-like liquid. In order to provide thermodynamic descriptions covering the range between aqueous solutions and molten salts, a suited model frame-work should be able to incorporate thermodynamic information from solutions and molten salts. Unfortunately, there are not many thermodynamic and spectroscopic data available within the concentration range of molten hydrates. The concept of the present work was to provide new experimental data characterizing the anion-water coordination competition in the system $MgCl_2$ + H_2O within the range of molten hydrates adding a chloride ion donor component (alkali metal chloride) or a chloride ion acceptor ($CuCl_2$). The data and structural information gained from this work will be used to develop a reaction-based thermodynamic model.

11.3 Experimental

11.3.1 Solid-Liquid Equilibria

Solid-liquid equilibria of the system $MgCl_2 + CuCl_2 + H_2O$ were determined at T = 383 K by an isothermal method. The apparatus is described in detail elsewere (Voigt et al., 1985). In brief, appropriate amounts of salts and water (25–30 g) are filled into titanium autoclaves lined by PTFE. For homogenisation the mixture is heated up to 443 K for 24 h. Then the mixture was equilibrated at 383 K for 72 h. Thereby the autoclaves rotated around its vertical axes with changing directions inside a metal block thermostat. Liquid and solid phases were separated at equilibrium temperature by filtration through a quartz filter plate inside the autoclave applying 2000 g centrifugal power. After cooling down to room temperature the phases were analyzed for magnesium (complexometrically), copper (gravimetrically as CuSCN), and chloride (argentometrically) content. Solid phase compositions were corrected for adhering mother liquid according to Schreinemakers method (Findlay, 1958). Solid phases were investigated additionally by X-ray powder diffraction (D5000, Siemens) and thermal analysis (DTA/TG with SEIKO instruments SII).

11.3.2 UV-Vis Spectroscopy

UV-Vis spectra were recorded in a wave length interval of 187 to 1021 nm using a one-beam spectrometer SPECORD S 100 (Analytik GmbH Jena, Germany). This spectrometer allowed to place a home-built furnace in the beam. Cuvettes of 0.1 to 10 mm path length made from SUPRASIL and closed by a TEFLON cup have been used. Salt + water mixtures were prepared mostly in the cuvettes itself by weighing appropriate amounts of salt hydrates, anhydrous salts and additions of water. Every sample was equilibrated between 30 and 60 min and three spectra were recorded before a new temperature was applied. Measurements were taken at increasing and decreasing temperature of the same sample. After all measurements the sample was checked for water loss by weight. Masses were converted to molar concentrations using the densities of the system $MgCl_2 + H_2O$ (Dietzel, 1959).

After subtraction of the solvent spectrum of pure magnesium chloride hydrate, the spectra of the absorbing solute species were deconvoluted by means of the program "Spectra Handler" (Griffith et al., 1999). A gaussian peak profile was assumed and known wave length of absorption maxima have been fixed. Base line and peak width were fitted by the least square method.

11.3.3 Raman Spectroscopy

The measurements in the system $CsCl-MgCl_2-H_2O$ were performed with a dispersive spectrometer equipped with argon laser, CCD detector, and a holographic notch-filter "Supernotch" (Kaiser Optical Systems Inc.). The excitation line at 514 nm was used with 600 mW power (at the sample). Molten hydrate mixtures were prepared in glass ampoules of approx. 150 mm length and 20 mm diameter closed by a pressure-tight Teflon cup. Spectra were treated using the program "Multifit" (Koß, 1998) including base line correction, the rayleigh wing, and peak profile fitting (Gaussian, Lorentz, Voigt). Relative intensities were calculated from peak areas relative to a band at 580 cm^{-1} arising from the glass wall. Samples containing $CuCl_2$ were measured in closed glass capillaries with a Raman microscope system T64000 (JOBIN YVON) using an argon laser beam of 3 mW at 514 nm.

11.3.4 Chemicals

$MgCl_2 \cdot 6\,H_2O$, $CuCl_2 \cdot 2\,H_2O$, and CsCl (reagent grade, Merck KGaA, Darmstadt) were used as starting materials for the salt mixtures. Anhydrous copper(II) chloride was prepared by drying the hydrate at 80 °C for several days. For mixtures with water contents less than 6 mol per mol $MgCl_2$ a larger amount of hexahydrate was dried to approx. tetrahydrate composition, which was analyzed and stored in a bottle inside of an exsiccator.

11.4 Results and Discussion

11.4.1 System $MgCl_2 + CuCl_2 + H_2O$

11.4.1.1 Solid-Liquid Equilibria

The numerical data of our results are summarized in Table 11.1. In Figure 11.2 the solubility isotherm is plotted as determined by us at 383 K together with data from literature at lower temperatures. At 383 K two double salt hydrates ($MgCl_2 \cdot CuCl_2 \cdot 6\,H_2O$ and $MgCl_2 \cdot 2\,CuCl_2 \cdot 6\,H_2O$) appear, which were not reported at 298 and 323 K. From the latter a crystal structure analysis has been performed (Bremer et al., 2002), which reveals dimeric $Cu_2Cl_6^{2-}$ species and octahedral $Mg(H_2O)_6^{2+}$ as the building units.

The structure of the other double salt could not be determined, because of a phase transition during cooling (Hettrich, 2002). From known structures of homologous compounds one can expect a structure built up from $CuCl_4^{2-}$ and $Mg(H_2O)_6^{2+}$ units in this case. Another striking feature of the solubility curve at 383 K is the deep minimum in $CuCl_2$ solubility, which typically occurs in systems with extensive complex ion formation. The occurence of anhydrous cupric chloride as a solid phase demonstrates the dehydrating effect of magnesium chloride relative to cupric chloride. The molar water/salt ratio R of the liquid phase reaches a minimum value of 4.0 in this region (Fig. 11.3). From Figure 11.3 it becomes also evident that the water content of the liquid phase is well within the range of molten hydrates.

11.4 Results and Discussion

Table 11.1: Experimental results for solid-liquid equilibria in the system $MgCl_2 + CuCl_2 + H_2O$ at 383 K.

Sample No.	Liquid phase composition mol/kg H_2O		Solid phase
	$MgCl_2$	$CuCl_2$	
Z1/1	8.693	1.271	$MgCl_2 \cdot 6H_2O$
Z1/2	8.840	1.309	$MgCl_2 \cdot 6H_2O$
Z1/3	8.672	1.145	$MgCl_2 \cdot 6H_2O$
Z1/4	8.349	1.223	$MgCl_2 \cdot 6H_2O$
Z2/2	8.927	1.560	$MgCl_2 \cdot 6H_2O$
Z2/4	9.175	1.849	$MgCl_2 \cdot 6H_2O$
Z3/1	8.859	2.551	$MgCl_2 \cdot 4H_2O$
Z3/3	8.981	2.523	$MgCl_2 \cdot 4H_2O$
Z3/4	9.048	2.815	$MgCl_2 \cdot 4H_2O + MgCuCl_4 \cdot 6H_2O$
Z4/1	8.194	3.549	$MgCuCl_4 \cdot 6H_2O$
Z4/3	8.202	3.217	$MgCuCl_4 \cdot 6H_2O$
Z6/1	8.942	2.694	$MgCl_2 \cdot 4H_2O + MgCuCl_4 \cdot 6H_2O$
Z6/3	9.164	2.789	$MgCl_2 \cdot 4H_2O + MgCuCl_4 \cdot 6H_2O$
Z11/2	6.580	6.510	$MgCu_2Cl_6 \cdot 6H_2O$
Z12/1	9.124	2.754	$MgCuCl_4 \cdot 6H_2O + MgCu_2Cl_6 \cdot 6H_2O$
Z12/2	5.927	8.124	$CuCl_2$
Z12/3	9.126	2.789	$MgCl_2 \cdot 4H_2O + MgCuCl_4 \cdot 6H_2O$
Z12/4	5.912	8.098	$CuCl_2$
Z13/1	0	9.608	$CuCl_2 \cdot 2H_2O$
Z13/3	0	9.622	$CuCl_2 \cdot 2H_2O$
Z13/2	1.743	7.500	$CuCl_2 \cdot 2H_2O$
Z13/4	1.733	7.487	$CuCl_2 \cdot 2H_2O$
Z14/1	8.856	2.681	$MgCl_2 \cdot 4H_2O + MgCuCl_4 \cdot 6H_2O$
Z14/2	8.801	2.618	$MgCl_2 \cdot 4H_2O + MgCuCl_4 \cdot 6H_2O$
Z14/3	8.839	2.627	$MgCl_2 \cdot 4H_2O + MgCuCl_4 \cdot 6H_2O$
Z14/4	8.809	2.626	$MgCl_2 \cdot 4H_2O + MgCuCl_4 \cdot 6H_2O$
Z16/2	6.040	7.460	$MgCu_2Cl_6 \cdot 6H_2O$
Z16/3	9.080	2.010	$MgCl_2 \cdot 4H_2O$
Z17/1	7.209	5.789	$MgCuCl_4 \cdot 6H_2O + MgCu_2Cl_6 \cdot 6H_2O$
Z17/2	3.026	6.900	$CuCl_2 \cdot 2H_2O$
Z17/3	7.112	5.765	$MgCuCl_4 \cdot 6H_2O + MgCu_2Cl_6 \cdot 6H_2O$
Z18/1	8.860	2.941	$MgCuCl_4 \cdot 6H_2O$
Z18/3	6.911	6.263	$MgCu_2Cl_6 \cdot 6H_2O$
Z18/4	4.461	6.976	$CuCl_2 \cdot 2H_2O$
Z19/1	5.601	8.183	$CuCl_2$
Z19/3	7.264	5.996	$MgCuCl_4 \cdot 6H_2O + MgCu_2Cl_6 \cdot 6H_2O$
Z20/1	8.455	3.191	$MgCuCl_4 \cdot 6H_2O$
Z20/2	7.697	4.357	$MgCuCl_4 \cdot 6H_2O$
Z20/3	8.330	3.088	$MgCuCl_4 \cdot 6H_2O$
Z20/4	7.717	4.302	$MgCuCl_4 \cdot 6H_2O$
Assarson (1950)	7.923	0	$MgCl_2 \cdot 6H_2O$
Dietzel, Serowy (1959)	8.252	0	$MgCl_2 \cdot 6H_2O$

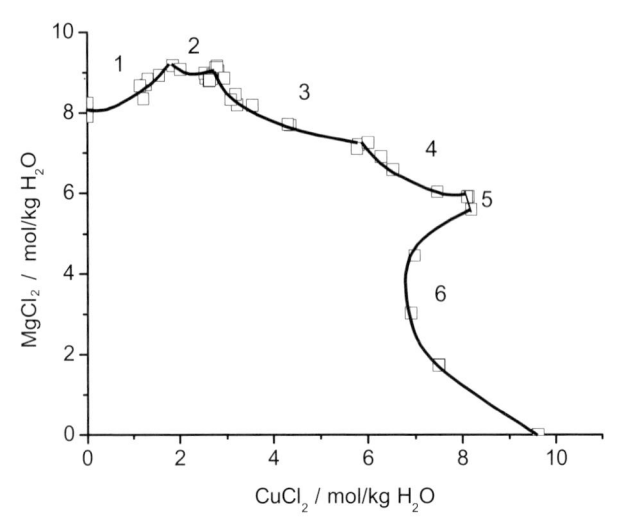

11   Ion Coordination and Thermodynamic Modeling of Molten Salt Hydrate Mixtures

Figure 11.2: Experimental results of solid-liquid equilibria in the system $CuCl_2 + MgCl_2 + H_2O$ at $T = 383$ K. Solid phases: 1: $MgCl_2 \cdot 6\,H_2O$; 2: $MgCl_2 \cdot 4\,H_2O$; 3: $CuCl_2 \cdot MgCl_2 \cdot 6\,H_2O$; 4: $2\,CuCl_2 \cdot MgCl_2 \cdot 6\,H_2O$; 5: $CuCl_2$; 6: $CuCl_2 \cdot 2\,H_2O$.

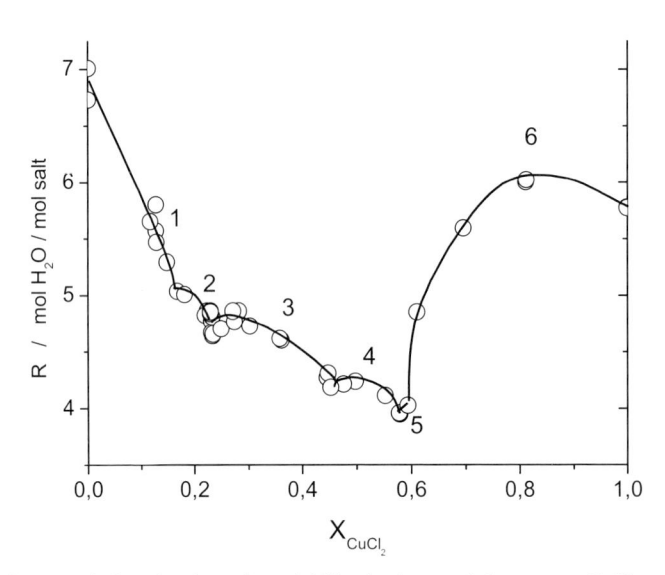

Figure 11.3: Molar water/salt ratio along the solubility isotherm of the system $CuCl_2 + MgCl_2 + H_2O$ at T = 383 K. Numbers denote solid phases as in Fig. 3.2.

246

11.4 Results and Discussion

11.4.1.2 UV-Vis Spectroscopy

Spectra of CuCl$_2$ solute in {MgCl$_2$ + R H$_2$O (R = 6, 8, 10)} were recorded in a temperature range of 303 to 443 K. Solute concentration varied between 7×10^{-4} to 3×10^{-2} mol/L. Bands of d–d transitions of aquo- and chlorocomlexes of Cu(II) are very broad and featureless extending from visible to the near infrared (Figgis and Hitchman, 2000). Therefore we focused on the charge-transfer (CT) spectra of Cu(II) exhibiting absorption maxima in a range of 200 to 500 nm. CT bands of Cu(II) completely coordinated by chloride ions are characterized by two overlapping bands with maxima at 270 to 280 nm and 375 to 390 nm with shoulders at the short and long wave length side in a range of 210 to 220 nm and 420 to 450 nm, respectively (Smith, 1976). Unfortunately, it is not possible to distinguish unambiguously between the different coordination geometries of Cu(II) (tetrahedral, tetragonal and its distorsions) from simple band assignments (Smith, 1976). Also chlorobridged dimeric or polymeric species possess similar spectral features (Smith, 1976).

Figure 11.4 shows this part of the electronic spectrum and its deconvolution into the four component bands mentioned above centered at 215, 280, 382, and 432 nm. The 380 nm band growths with decreasing water content as shown in Figure 11.5. Similar increase in intensity was observed at increasing temperature at a molar water/salt ratio R = 10 (Fig. 11.6). At lower water contents (R = 8, 6) the temperature dependence becomes small and reversed in direction (Fig. 11.7). Only the long wave length shoulder continues to grow with temperature.

Very recently measurements of UV-Vis-NIR spectra of Cu(II) in aqueous LiCl solutions up to 18 mol/kg H$_2$O have been published (Brugger et al., 2001). The authors stated the same trends in the spectral features of the CT bands in the region of 250 to 450 nm. The band at 389 nm was detectable at 2.5 mol/kg H$_2$O LiCl and grew to a molar absorbance of about 1600 L mol^{-1}cm^{-1} (read from diagram) at 18.1 mol/kg H$_2$O LiCl. This value can be compared with 1700 L mol^{-1}cm^{-1} at 384 nm found by us in the molten hexahydrate of MgCl$_2$ at 403 K. In both cases the molar ratio water/chloride ions is equal, however, the

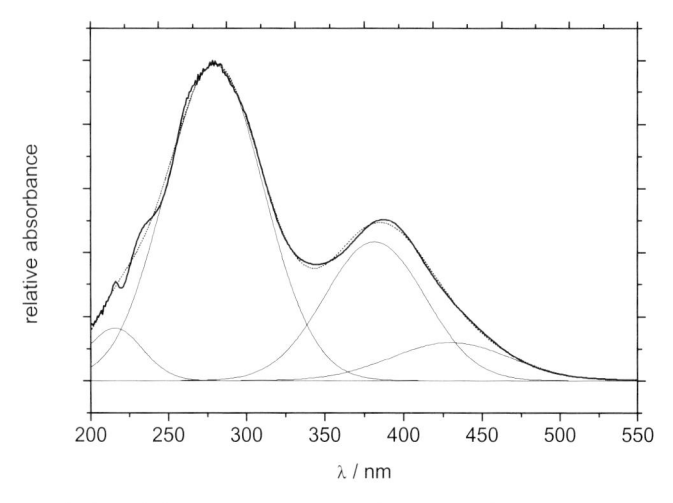

Figure 11.4: Deconvolution of the UV-Vis spectrum of Cu(II) in liquid (MgCl$_2$ + 8 H$_2$O) at 403 K. m_{Cu} = 0.00174 mol/L; d = 5 mm; —— exp. spectrum, ······ fitted spectrum.

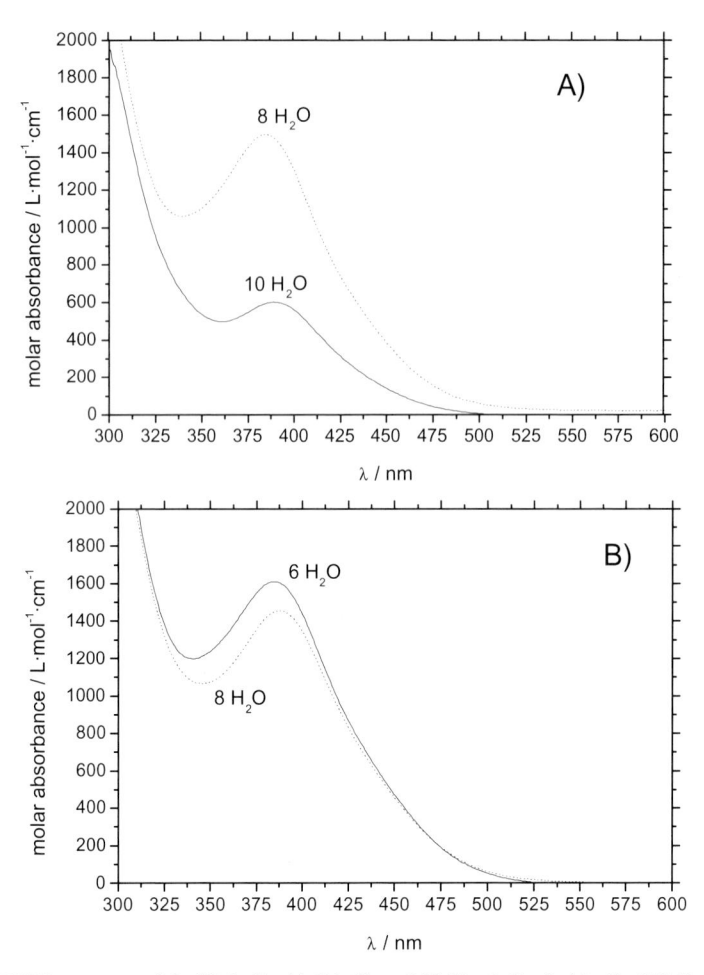

Figure 11.5: UV-Vis spectra of Cu(II) in liquid (MgCl$_2$ + R H$_2$O). a) R = 8, 10; T = 373 K; b) R = 6, 8; T = 413 K.

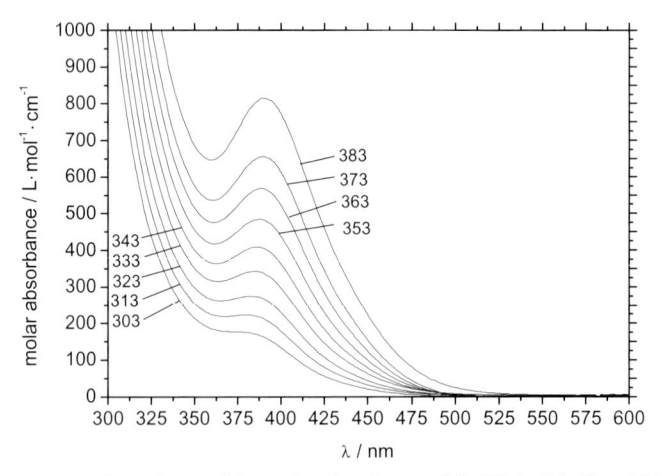

Figure 11.6: Temperature dependence of the molar absorbance of Cu(II) in (MgCl$_2$ + 10 H$_2$O).

11.4 Results and Discussion

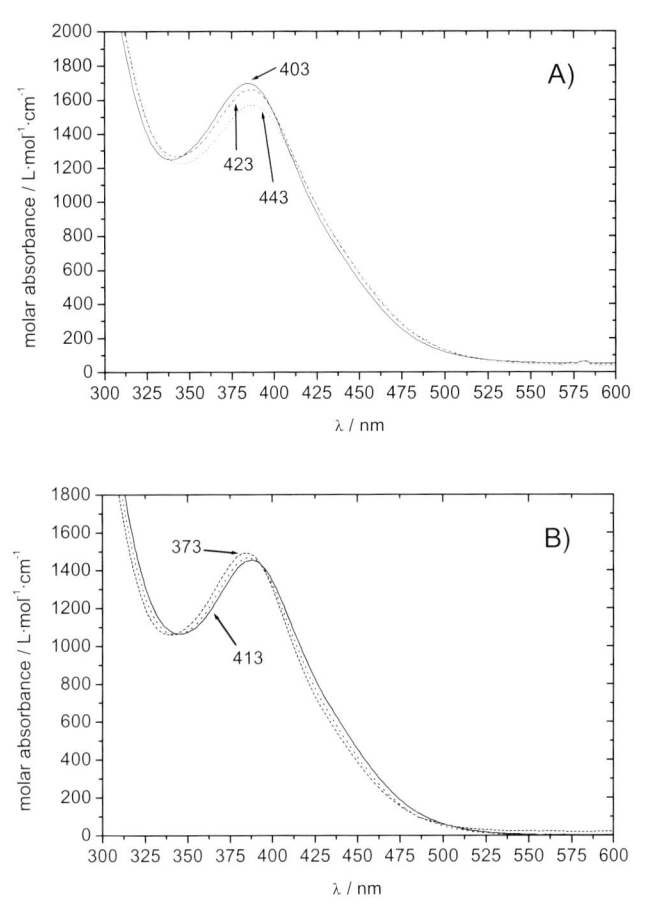

Figure 11.7: Temperature dependence of the molar absorbance of Cu(II) in the concentration region of molten hydrates. Numbers at lines denote temperatures. a) $MgCl_2 + 6\ H_2O$; b) $MgCl_2 + 8\ H_2O$.

ionic strength of molten hexahydrate (27.7 mol/kg H_2O) considerable exceeds that of the LiCl solution (18.1 mol/kg H_2O). Brugger et al. interpreted their spectra by a step-wise chlorocomplex formation and reported formation constants using Debye-Hückel based activity corrections. Principal factor analyses of the spectra indicated a minimum of five factors ($Cu^{2+}_{(aq)}$, $CuCl^{+}_{(aq)}$, $CuCl^{0}_{2(aq)}$, $CuCl^{-}_{3(aq)}$, $CuCl^{2-}_{4\ (aq)}$) necessary to describe the CT bands. For the highest LiCl concentrations the existence of a sixth species $CuCl^{3-}_{5}$ was suggested.

The main difference to the present results consists in the long wave length component at around 425 nm, which was not detected by Brugger et al. Von Barner et al. (1985) observed this long wave length component in spectra of Cu(II) in anhydrous melts of KCl + $AlCl_3$ and assigned it to a $CuCl^{-}_{3}$ species. Since this band developed at the lowest chloride ion activities (pCl = 0.1 on mole fraction base) and high concentrations of Cu(II) (approx. 0.1 mol/L) a dimeric nature of this species like $Cu_2Cl^{2-}_{6}$ becomes probable. In our case at much higher chloride activities and lower Cu(II) concentration the existence of chlorobridged species is quite unlikely. An explanation should rather be seen in the occurence of five- or sixfold coordinated chlorocomplexes. From our spectra the conclusion can be drawn that at $R \leq 8$ nearly all Cu(II) is completely coordinated by chloride ions. Between $8 < R < 10$ mixed aquo-chlorocomplexes of copper exist.

11.4.1.3 Raman Spectra

At $CuCl_2$ concentrations comparable with the content of $MgCl_2$ the high molar absorbance does not allow to obtain UV-Vis spectra. Therefore, we attempted to record Raman spectra of the almost black liquids enclosed in glass capillaries. In Figure 11.8 spectra of molten magnesium chloride hexahydrate with additions of 0.5, 1.0, and 2.0 mol $CuCl_2$ per mol hexahydrate are shown. A very broad asymmetric band with a maximum at 285 cm^{-1} dominates the spectra. For the mixture containing 2 mol $CuCl_2$ the band is remarkable broadend. Spectra quality did not allow further refinements.

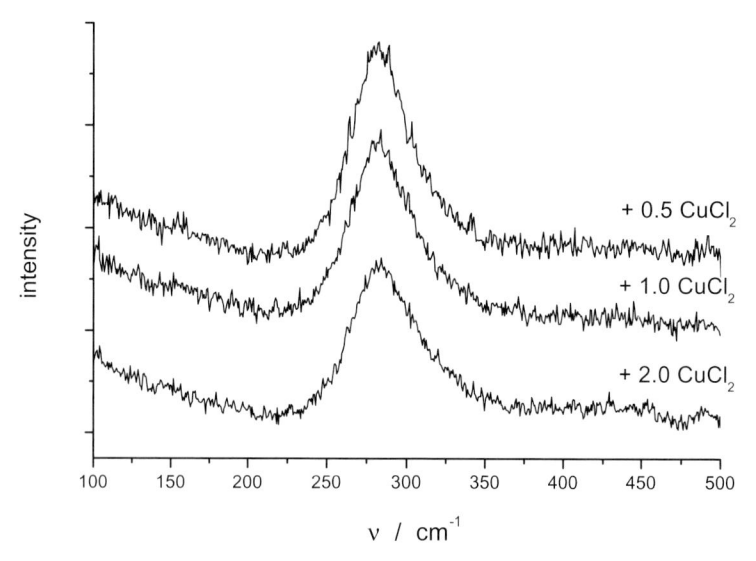

Figure 11.8: Raman spectra of molten mixtures $\{(MgCl_2 + 6 H_2O) + n\ CuCl_2\}$ at $T = 443$ K. $n = 0.5$, 1.0, 2.0.

The observed band covers a wave number range typically reported for Raman active Cu–Cl vibrations of solids like Cs_2CuCl_4 or NH_4CuCl_3 (Adams and Newton, 1971). Cu–O vibrations should appear at wave numbers from 350 to 410 cm^{-1} (Beattie et al., 1969). Thus our spectra give evidence that also at higher $CuCl_2$ contents copper ions are still coordinated by chloride ions and the water is bound to Mg^{2+} propably as $[Mg(H_2O)_6]^{2+}$. The latter should give rise to a band at approx. 330 to 350 cm^{-1} (see 3.2), which is obviously covered by the high wave number wing of the Cu–Cl band. At least for the $[(MgCl_2 + 6 H_2O) + 0.5\ CuCl_2]$ mixture (Fig. 11.9) the band is highly polarized, which gives an indication for a symmetric species like tetrahedral $CuCl_4^{2-}$.

11.4 Results and Discussion

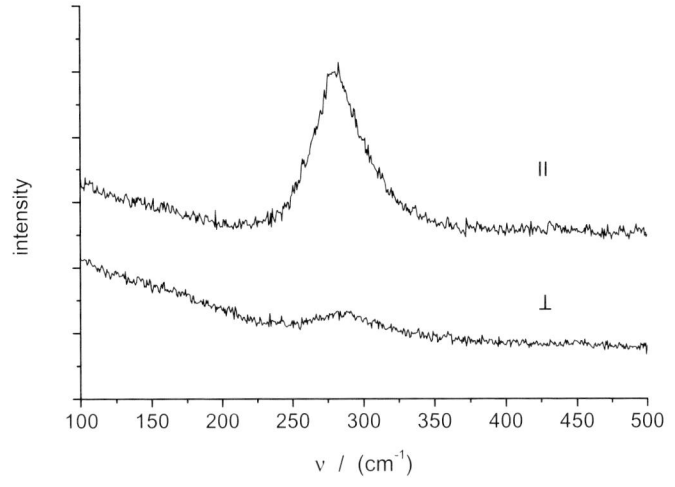

Figure 11.9: Raman spectrum of the mixture {(MgCl$_2$ + 6 H$_2$O) + 0.5 CuCl$_2$} in parallel and vertical orientation to polarization. T = 423 K.

11.4.2 MgCl$_2$ + CsCl + H$_2$O

Raman spectra were recorded for mixtures of the type [(MgCl$_2$ + R H$_2$O) + n CsCl] with R = 6 and 12.9 and n = 0, 0.5, 1.0 and 2.0 at temperatures between 403 and 473 K. Attention was focused on the detection of chlorocomplexes of magnesium. Therefore the interesting wave number range was limited to 100 to 500 cm^{-1}. Figure 11.10 gives an example of a recorded original spectrum and its deconvoluted bands. The band at 580 cm^{-1} arises from the

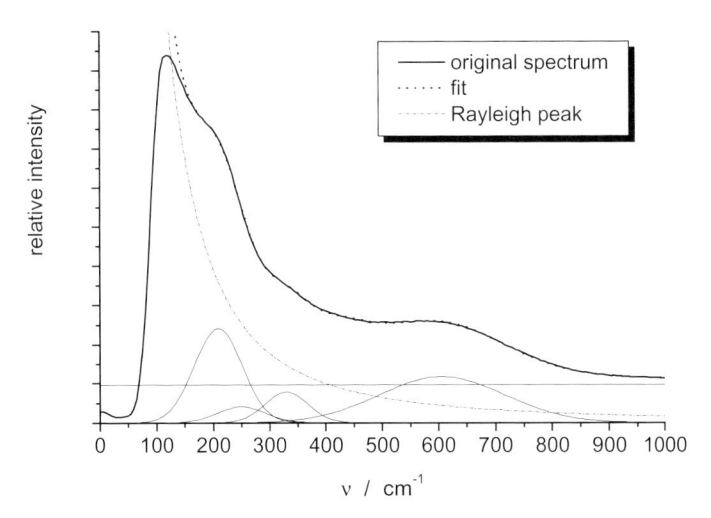

Figure 11.10: Deconvolution of the Raman spectrum of liquid {(MgCl$_2$ + 6 H$_2$O) + 0.5 CsCl} at 443 K.

glass ampoule and was used as intensity reference. The bands at 200, 250, and 335 cm^{-1} are correlated with structural entities of the liquid. The bands at 250 and 335 cm^{-1} were already observed earlier in the system MgCl$_2$ + KCl + H$_2$O (Voigt and Klaeboe, 1986), where the 250 cm^{-1} band was assigned to Mg–Cl vibrations of chlorocomplexes and the 335 cm^{-1} band to the Mg–O vibration of the aqua ion [Mg(H$_2$O)$_6$]$^{2+}$.

The origin of the 200 cm^{-1} component is not yet clear. Whereas the 250 cm^{-1} band disappears at higher water contents, the 200 cm^{-1} band remains relatively constant in height and width (Fig. 11.11). The enlarged intensity of the 335 cm^{-1} band at $R = 12.87$ supports its assignment to the magnesium hexaaqua ion.

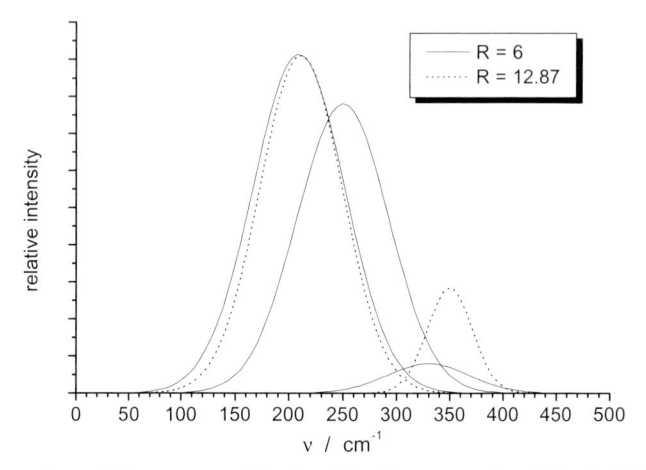

Figure 11.11: Comparison of the component bands of the Raman spectra of liquid {(MgCl$_2$ + 2 CsCl) + R H$_2$O} with $R = 6$ and 12.87.

In Figure 11.12 the dependence of the relative intensities of these three bands are plotted against the CsCl amount added to a magnesium chloride hexahydrate melt. As expected the intensity of the band arising from the aquacomplex (335 cm^{-1}) decreases where at the same time the corresponding chlorocomplex band (250 cm^{-1}) gains intensity. The behavior of the 205 cm^{-1} band parallels that of the 335 cm^{-1} band.

The degree of polarization of all bands is not high as can be seen from Figure 11.13. This points to strong distorsions of the coordination polyhedra of aqua- as well as chlorocomplexes at the enhanced temperatures.

Only the 250 cm^{-1} band shows a significant temperature dependence. Its intensity grows with temperature.

In conclusion the Raman investigation supports the view derived earlier from trends in thermodynamic properties (Voigt and Emons, 1995) that addition of akali metal chloride MCl (M = K, Rb, Cs) to molten hydrated magnesium chloride causes ligand exchange H$_2$O ↔ Cl$^-$ around the magnesium cations. The proximity of the 250 cm^{-1} band to the wave numbers reported for the MgCl$_4^{2-}$ complex in anhydrous molten mixtures MgCl$_2$ + (NaCl, KCl, RbCl, CsCl) can be an indication for the occurrence of tetrahedral chlorocomplexes in molten hydrates.

11.5 Thermodynamic Modeling

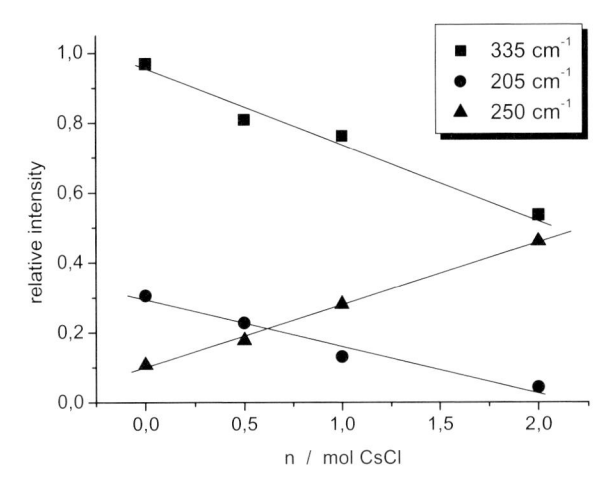

Figure 11.12: Relative intensities of the component Raman bands in liquid {(MgCl$_2$ + 6 H$_2$O) + n CsCl} in dependence on mole number n of added CsCl.

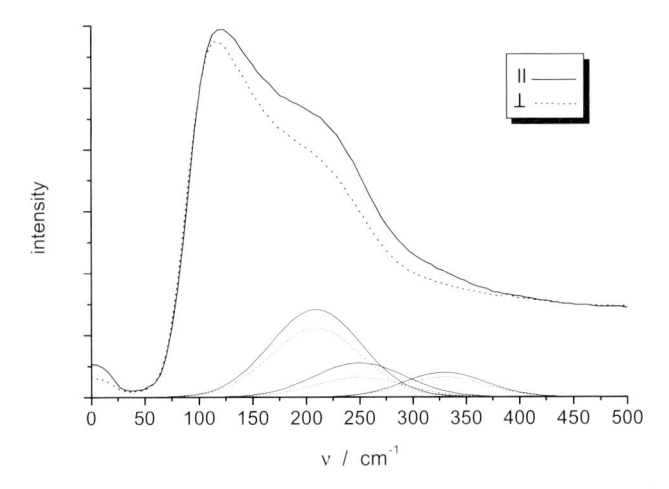

Figure 11.13: Polarization of Raman spectrum and component bands in liquid {(MgCl$_2$ + 6 H$_2$O) + 1.0 CsCl)} at T = 442 K.

11.5 Thermodynamic Modeling

11.5.1 Modeling Strategy

In general thermodynamic models of concentrated electrolyte solutions are composed of a long-range Debye-Hückel term and specific interaction parameters from virial equations approaches (Pitzer, 1991), local composition models (Chen and Evans, 1986; Pinsky and Gruber, 1994; Abovsky et al., 1998), or simple non-electrolyte mixing parameters like the van-Laar interaction coefficients (Geerlings et al., 1997; Pitzer, 1980). Sometimes step-wise ion hydration is explicitly introduced by appropriate equilibrium constants (Lu and Maurer, 1993; Lu et al., 1996; Schönert, 1990, 1994). Within the concentration range of molten hydrates the modified BET model (Stokes and Robinson, 1948) is successfully applied to describe water activities and salt activities (Voigt, 1993; Ally and Braunstein, 1993) in binary systems. In absence of strong chemical interactions activities ternary systems can be predicted using simple additivity rules (Sangster et al., 1987) or equations derived from a quasi-lattice statistic model (Sacchetto and Kodejs, 1982; Ally and Braunstein, 1998). Stokes and Robinson (1973) showed that the osmotic coefficients of a number of chemically quite different electrolytes can be described up to 30 mol/kg H_2O by means of a step-wise hydration model characterized by only two parameters k, K_o by means of Eq. (11.1)

$$K_n = K_o \cdot k^{(n-1)} \tag{11.1}$$

$$\Delta_R G_n^0 = \Delta_R G_o^0 - RT \cdot (n-1) \cdot \ln k$$

with K_n the equilibrium constant of the ion hydrate including n molecules of water. Correlation (Eq. (11.1)) is supported by mass-spectroscopic determinations of hydration equilibria in the gas phase (Conway, 1981; Peschke et al., 1998, 1999). Following this line reaction chains can be built-up for the description of all water-ligand exchange processes correlating the step-wise equilibrium constants by incremental parameters k_i, thereby reducing the number of independent constants needed. For the selection of relevant types of species as well as for the relative magnitude of the parameters extra-thermodynamic knowledge especially from spectroscopy should be used. In fine-tuning of the model composition dependent reaction quantities $\Delta_R G_i^0$ should be introduced as shown for more simple models earlier (Østvold, 1972; Barnes and Thompson, 1988). One problem represents the ideal mixing statistics going from aqueous solutions to molten salts. In molten salt mixtures formulations are based on cation and anion sub-lattice mixing, whereas in electrolyte solutions such charge-specific differentiation is not made. For molten salt hydrates it was suggested to put the water molecules into the anion lattice (Braunstein, 1971). We found that the chosen form of lattice statistics is less important (because of the compensating effect from reference state) than the effect of particle size. In a first version our model will therefore be formulated as a pure reaction model with correlated equilibrium constants and activities defined on a volume fraction species basis.

11.5.2 Computational Problems of Parameter Estimation

For calculations of chemical equilibria envolving hundreds of species appropriate codes are available for specific interaction models (ChemSage; EQ3/6). However, there exists no ready to use software for the reversed task, that is estimation of sets of equilibrium constants or Gibbs energies of formation of species from a variety of experimental data sets. After having tested several programs we decided to use the freely available non-linear optimization package LANCELOT (Conn et al., 1992). This package allows solving of large-scale non-linear problems including arbitrary constraints on the variables. The objective function was formulated in a way that model parameters and equilibrium mole numbers of species were estimated simultaneously during minimization. Thus the number of unknowns amounts to the number of model parameters plus the number of species multiplied by the number of experimental data. Typically between 100 and 3000 unknowns were handled. The available program option for automatic calculation of the first and second derivatives was not activated to ensure stable solutions. Instead variables were transformed into LANCELOT specific variable types including analytical first and second derivatives. For this time consuming procedure input scripts based on the UNIX tools "awk" and "cshell" were written, which generate the required input cards for LANCELOT automatically for every change of the models structure (type of species, correlations between equilibrium constants) or experimental data sets. Using the non-linear equation solver DONLP2 and GNUPLOT a graphical oriented simulation environment was organized to investigate the effect of chosen reactions and magnitude of model parameters on the calculated thermodynamic properties and its comparison with experimental data.

11.5.3 Model Development for the Systems $MgCl_2 + KCl + H_2O$ and $MgCl_2 + CuCl_2 + H_2O$

11.5.3.1 Binary Systems

$MgCl_2 + KCl$

For the molten mixtures of $MgCl_2$ and KCl enthalpy of mixing, activity data, and the solid-liquid phase diagram are available. Pure liquid $MgCl_2$ and KCl were taken as reference state, independent of its structural units. This means the corresponding enthalpies and entropies were set to zero:

$$H°(MgCl_2)(l) = H°(KCl)(l) = 0 \qquad (11.2)$$

$$S°(MgCl_2)(l) = S°(KCl)(l) = 0$$

Formation of the tetrachlorocomplex $MgCl_4^{2-}$ represents the dominant process on mixing of molten $MgCl_2$ and KCl (Brooker and Huang, 1980). However, models taken into account only this complex cannot reproduce the thermodynamic data in melts rich of $MgCl_2$ (Østvold, 1972). Therefore, a reaction chain of the tetrachlorocomplex with excess of $MgCl_2$ was formulated in this work, which is equivalent to a "solvation" effect.

$$2\ KCl + MgCl_2 \rightarrow K_2MgCl_4 \qquad H_o = -53.0\ kJ/mol \qquad (11.3)$$

$$K_2MgCl_4 + MgCl_2 \rightarrow K_2MgCl_4(MgCl_2) \qquad H_1 = -23.0\ kJ/mol \qquad (11.4)$$

$$K_2MgCl_4(MgCl_2) + MgCl_2 \rightarrow K_2MgCl_4(MgCl_2)_2 \qquad H_2 = -5.0\ kJ/mol \qquad (11.5)$$

$$K_2MgCl_4(MgCl_2)_2 + MgCl_2 \rightarrow K_2MgCl_4(MgCl_2)_3 \qquad H_3 = -5.0\ kJ/mol \qquad (11.6)$$

$$K_2MgCl_4(MgCl_2)_3 + MgCl_2 \rightarrow K_2MgCl_4(MgCl_2)_4 \qquad H_4 = -5.0 \text{ kJ/mol} \qquad (11.7)$$

Entropy changes were set to zero. The description of the mixing quantities and component activities is illustrated in Figures 11.14 and 11.15 for 1073 K.

Using the Gibbs energy of melting for KCl (Chou et al., 1992) and MgCl$_2$ (Janz et al., 1979) from literature as given below

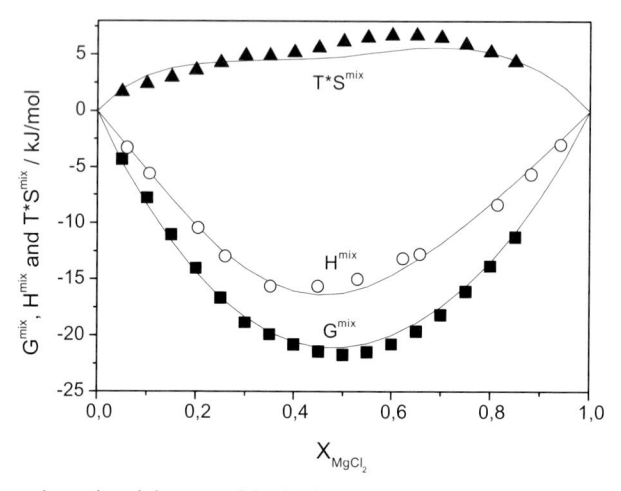

Figure 11.14: Thermodynamic mixing quantities in the system MgCl$_2$ + KCl at 1072 K. Symbols: exp. data taken from (Markov and Volkov, 1985); lines: model.

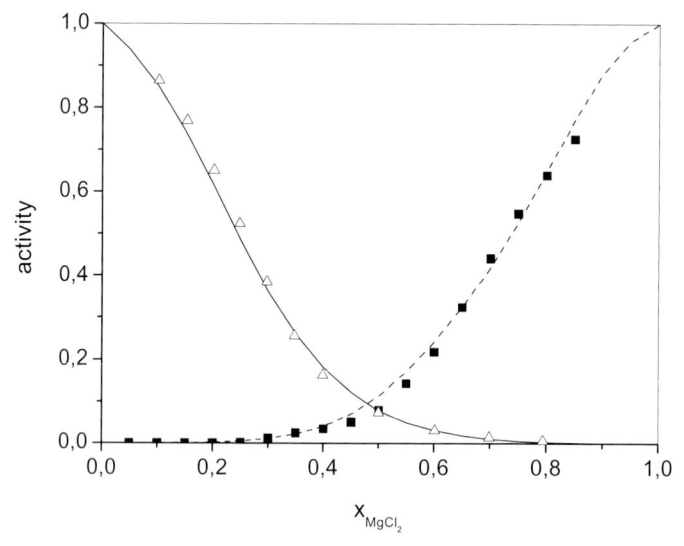

Figure 11.15: Component activities in the system MgCl$_2$ + KCl at 1073 K. Lines: model; Δ exp. data KCl (Østvold, 1972); exp. data MgCl$_2$ (Neil et al., 1965).

11.5 Thermodynamic Modeling

$$\Delta_{SL}G°(KCl) = -12{,}580 - 13.43 \cdot T + 0.03684 \cdot T^2 - 1.192 \times 10^{-5} \cdot T^3 \text{ J/mol} \quad (11.8)$$

$$\Delta_{SL}G°(MgCl_2) = 37{,}390 - 31.6 \cdot T - 6.394 \times 10^{-3} \cdot T^2 \text{ J/mol} \quad (11.9)$$

the liquidus lines have been calculated and compared with experimental data as shown in Figure 11.16. This calculation represents a prediction in respect to a temperature interval of more than 300 K. The reasonable agreement gives hope that exploitation of the model at even lower temperatures should be possible.

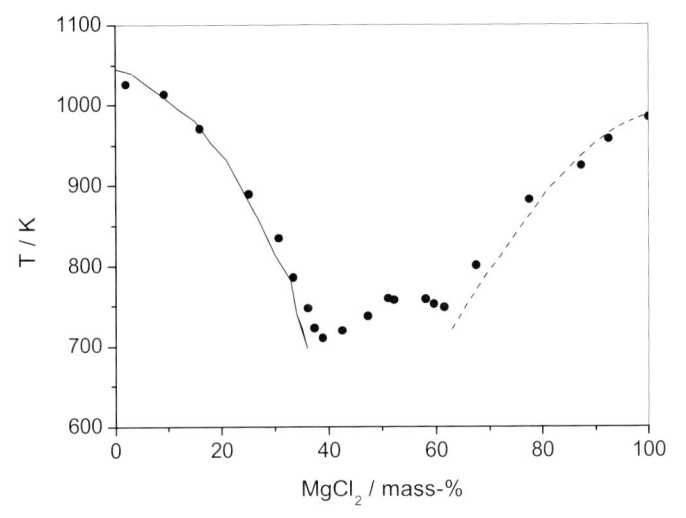

Figure 11.16: Calculated liquidus curves of MgCl$_2$ and KCl in the system MgCl$_2$ + KCl. Lines: model; points: exp. data (Menge, 1911).

KCl + H$_2$O

Again, pure liquid water is defined as a reference with standard formation functions set to zero at all temperatures. Only two hydrated species are assumed according to reactions Eqs. (11.10) and (11.11)

$$KCl + H_2O \rightarrow K(H_2O)^+ + Cl^- \qquad H_1 = 30.0 \text{ kJ/mol} \qquad S_1 = -20 \text{ J/mol K} \quad (11.10)$$

$$K(H_2O)^+ + H_2O \rightarrow K(H_2O)_2^+ \qquad H_2 = -41.4 \text{ kJ/mol} \qquad S_2 = -10 \text{ J/mol K} \quad (11.11)$$

The four parameters were determined from the solid-liquid equilibrium in the system KCl + H$_2$O. In Figure 11.17 experimental data of the crystallization curve of KCl from water are compared with the model. Agreement can be stated from the melting point of KCl down to about 350 K.

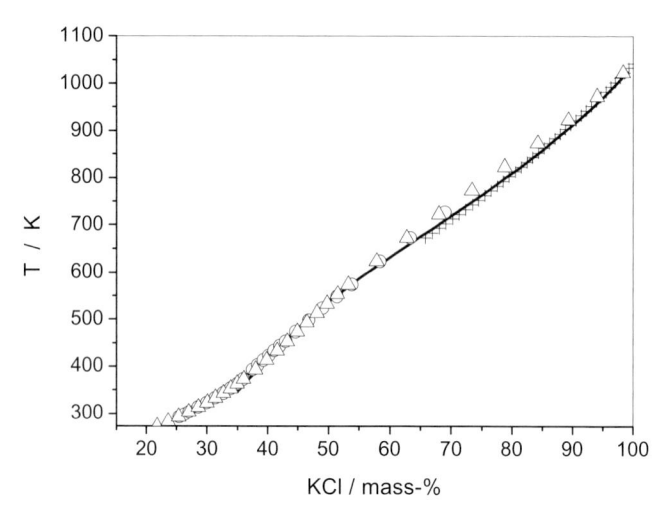

Figure 11.17: Solubility of KCl in water between room temperature and melting point of KCl. Line: model; symbols: exp. data (○ Linke and Seidell, 1965; △ Cohen-Adad and Lorimer, 1992; Chou et al., 1992).

MgCl$_2$ + H$_2$O

Two hydration chains were formulated for this strong hydrating salt. One chain for the hydration of the molecule MgCl$_2$ forming MgCl$_2$(H$_2$O), MgCl$_2$(H$_2$O)$_2$, ..., MgCl$_2$(H$_2$O)$_6$. Another chain represents a second hydration sphere of the cation formulated as [Mg(H$_2$O)$_6$]$^{2+}$, [Mg(H$_2$O)$_6$](H$_2$O)$^{2+}$, [Mg(H$_2$O)$_6$](H$_2$O)$_2^{2+}$, ..., [Mg(H$_2$O)$_6$](H$_2$O)$_5^{2+}$. Gibbs energy for both series is described by Eqs. (11.12) and (11.13)

$$G^0\{MgCl_2(H_2O)_i\} = i \cdot H_2 + i \cdot (i-1)/2 \cdot h_2 - (i \cdot S_2 + i \cdot (i-1)/2 \cdot s_2) \cdot T + HH \tag{11.12}$$

$$G^0\{Mg(H_2O)_{5+i}\} = i \cdot H_0 + i \cdot (i-1)/2 \cdot h_0 - (i \cdot S_0 + i \cdot (i-1)/2 \cdot s_0) \cdot T + HH \tag{11.13}$$

with i identifying the hydration step running from 1, ..., 6. The parameters given in Table 11.2 were determined from water activity data at 373 to 623 K (Wollny, 1983; Urusova and Valyashko, 1983) and water salt ratios 3 < R < 50. Figure 11.18 represents experimental data at three temperatures in comparison with the model. In the region of molten hydrates (low values of a_w and R) the experimental points coincide with the model curve. Systematic deviations arise at high water activities where the model needs further refinements.

The dependence of the enthalpy of evaporation on the molar water/salt ratio is plotted in Figure 11.19. As can be seen also a simplified model with only one reaction chain describes the data reasonable. The model was then applied to the solid-liquid equilibria in the binary system and the thermodynamic constants of the solid phases were derived (Table 11.3). Figure 11.20 shows the crystallization branches of magnesium chloride and its hydrates. The branch for anhydrous magnesium chloride was calculated using the Gibbs en-

Table 11.2: Hydration parameters of Eqs. (11.12) and (11.13) for the system MgCl$_2$ + H$_2$O.

Enthalpy parameter	Value kJ/mol	Entropy parameter	Value J/mol K
HH	−11.80		
H_2	−30.31	S_2	−21.8
h_2	0.0	s_2	−6.93
H_0	−103.44	S_0	−150.5
h_0	17.80	s_0	0.0

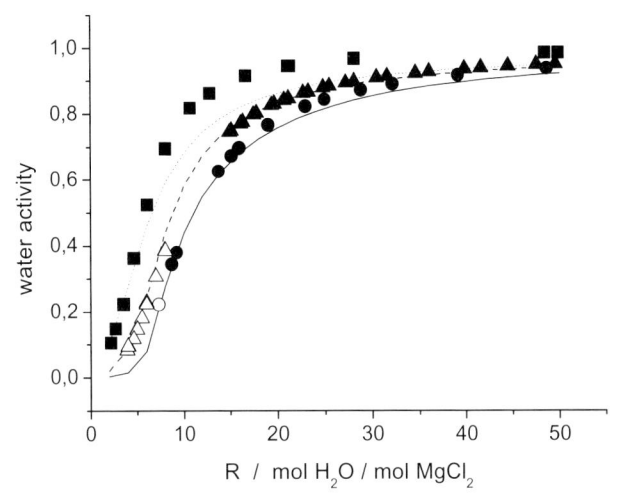

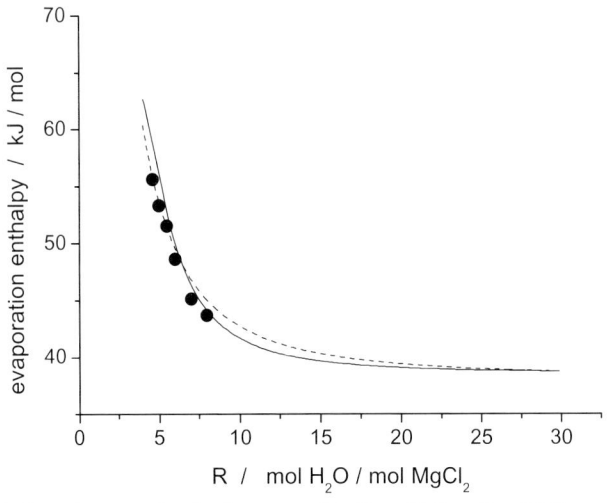

Figure 11.18: Water activity in the system MgCl$_2$ + H$_2$O in dependence of molar water/salt ratio R. Lines: model; ——— 373.15 K; – – – 473.15 K; ······ 623.15 K. Exp. Data: ● = 373 K, Fanghänel and Grjotheim (1990); ▲ = 473 K, Holmes and Mesmer (1996), Wollny (1983), Dittrich (1986); ■ = 623 K, Urusova and Valyashko (1983).

Figure 11.19: Enthalpy of evaporation in the system MgCl$_2$ + H$_2$O at T = 473 K. ——— model with two hydration chains; – – – model with one hydration chain; ● exp. data Wollny (1983).

ergy function of Eq. (11.9) and represents a prediction not considering eventual precipitation of basic magnesium chlorides due to hydrolysis at high temperatures. Below 350 K the model is not enough fine-tuned to describe the experimental data.

Table 11.3: Enthalpies and entropies of formation of solid hydrates from pure liquid components $MgCl_2$, KCl, H_2O.

Solid phase	H^0 kJ/mol	S^0 J/mol K
$MgCl_2 \cdot H_2O$	−72.11	−60.4
$MgCl_2 \cdot 2\,H_2O$	−103.92	−97.02
$MgCl_2 \cdot 4\,H_2O$	−166.99	−192.44
$MgCl_2 \cdot 6\,H_2O$	−224.05	−301.36
$KCl \cdot MgCl_2 \cdot 6\,H_2O$	−243.44	−300.0

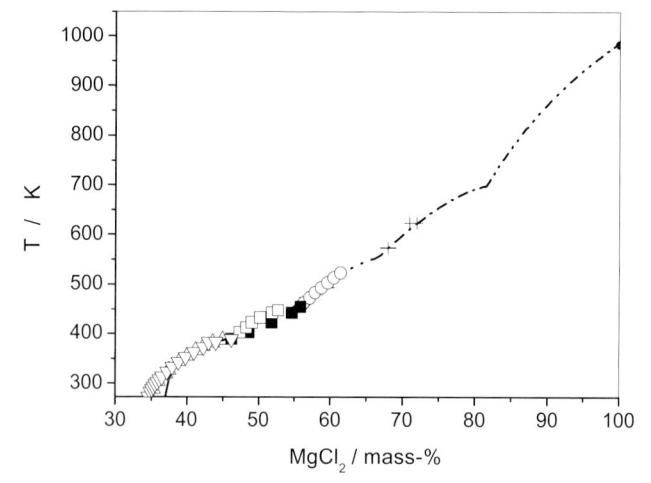

Figure 11.20: Liquidus curve of the system $MgCl_2 + H_2O$ between room temperature and the melting point of $MgCl_2$. Lines: model; symbols: exp. solubility data: $MgCl_2 \cdot 6\,H_2O$: ▽ Linke and Seidell (1965); △ Clynne and Potter (1979); $MgCl_2 \cdot 4H_2O$: ■ Linke and Seidell (1965); □ Fanghänel et al. (1987); $MgCl_2 \cdot 2\,H_2O$: ◇ Linke and Seidell (1965); ○ Fanghänel et al. (1987); $MgCl_2 \cdot H_2O$ (composition not proved): Urusova and Valyashko (1983).

11.5.3.2 Ternary Systems

$MgCl_2 + KCl + H_2O$

For this system solid-liquid equilibrium data (Fanghänel and Emons, 1988; Fanghänel and Emons, 1989) and water activities from vapor pressure measurements are available for $4.5 < R < 8$ and temperatures up to 523 K (Emons et al., 1987). These data were fitted using

11.5 Thermodynamic Modeling

the parameters of the binary subsystems and the enthalpy and entropy of formation of two additional species formed according to reactions (11.14) and (11.15) as given below

$$KCl + MgCl_2 + 3 H_2O \rightarrow KMgCl_3(H_2O)_3 \qquad (11.14)$$

$$H° = -12.72 \text{ kJ/mol} \quad S° = -108.1 \text{ J/mol K}$$

$$KCl + MgCl_2 + 4 H_2O \rightarrow KMgCl_3(H_2O)_4 \qquad (11.15)$$

$$H° = -15.75 \text{ kJ/mol} \quad S° = -158.0 \text{ J/mol K}$$

As can be seen from Figure 11.21 the model reproduces the solubility isotherms in the ternary system within a wide range of temperature and composition. Also the increase of water activity when anhydrous KCl is added to molten magnesium chloride hydrates (Emons et al., 1987) is reflected by our model. Thus, the thermodynamic behavior of the system is well described by suited choice of hydrated cations and chlorocomplexes of magnesium.

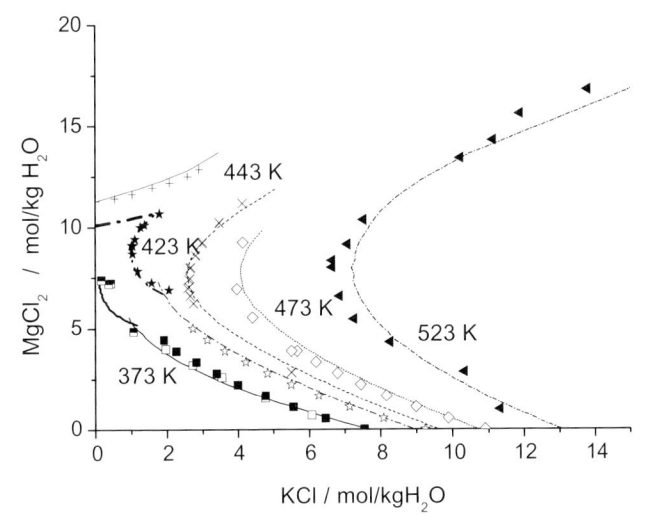

Figure 11.21: Selected solubility isotherms in the ternary system $MgCl_2$ + KCl + H_2O. Lines: model; symbols: exp. data from different sources. Solid phases: KCl, KCl · $MgCl_2$ · 6 H_2O, $MgCl_2$ · 6 H_2O, $MgCl_2$ · 4 H_2O.

$MgCl_2$ + $CuCl_2$ + H_2O

The previous investigations showed that Cu(II) behaves as a chloride ion acceptor in respect to $MgCl_2$. The predominant coordination number of Cu(II) is four. In cases of higher coordination numbers, bonding of the fifth or sixth ligand is much weaker. Therefore, we assumed the existence of complexes of the type $CuCl_n(H_2O)_{4-n}$ with n = 0, 1, 2, 3, 4. The Gibbs en-

11 Ion Coordination and Thermodynamic Modeling of Molten Salt Hydrate Mixtures

ergy of formation of the complexes (pure liquid CuCl$_2$ and H$_2$O as reference state) were correlated by Eq. (11.16).

$$G^0\{CuCl_n(H_2O)_{4-n}\} = n \cdot G_0 + n \cdot (n-1)/2 \cdot g \qquad (11.16)$$

with $G^0 = 3801.5$ J/mol and $g = -6960.4$ J/mol. The parameters were determined from fitting of the solid-liquid equilibria at 383 K. For all magnesium containing species the parameters given for the binary system MgCl$_2$ + H$_2$O were applied. In addition, the formation of a dimeric species Cu$_2$Cl$_6^{2-}$ ($G^0 = -72.6$ kJ/mol) had to be included in order to describe the solubility isotherm correctly. Values of the Gibbs energies of formation G^0 of the solids from our model are as follows:

CuCl$_2$	−24.28 kJ/mol	CuCl$_2 \cdot$ 2 H$_2$O	−33.12 kJ/mol
MgCl$_2 \cdot$ CuCl$_2 \cdot$ 6 H$_2$O	−16.97 kJ/mol	MgCl$_2 \cdot$ 2 CuCl$_2 \cdot$ 6 H$_2$O	−141.37 kJ/mol

The liquid species parameters and the Gibbs energies of the solid phases had to be estimated from a relatively small number of experimental data. To obtain chemical meaningful parameters during the fitting procedure the absolute and relative magnitudes of parameters were constrained by values or relations based on known tendencies in hydration and complexation thermodynamics. In this way the solubility isotherm at 383 K could be described correctly as shown in Figure 11.22.

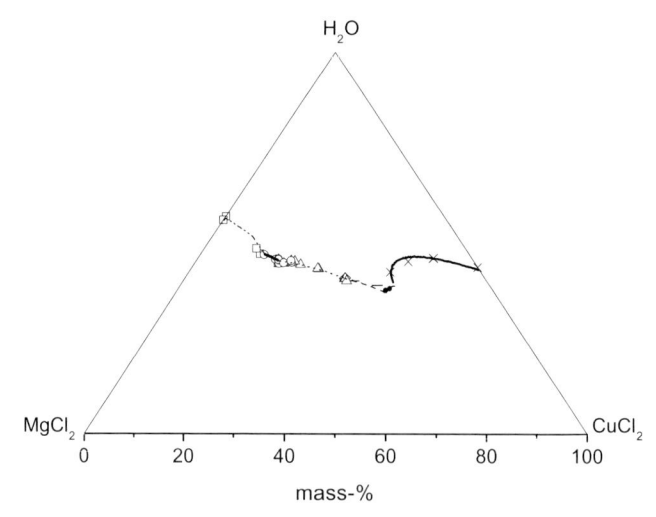

Figure 11.22: Description of the solubility isotherm in the system CuCl$_2$ + MgCl$_2$ + H$_2$O at $T = 483$ K by the reaction model. Lines: model; symbols: exp. data this work.

11.6 Conclusions

Molten salt hydrates define a region on the water-salt concentration scale, where the molar water/salt ratio R approaches the coordination number of the strong hydrating cations. The thermodynamic properties of mixtures of molten hydrates are first of all controlled by water-anion coordination competition at the strong hydrating cations. In the system $MgCl_2$ + H_2O the spherical chloride anion provides a structural simple situation to study the anion-water ligand exchange. Raman spectra showed the appearance of magnesium chlorocomplexes and the disappearance of magnesium hexaaqua ions when adding CsCl as a chloride donor. Contrary, $MgCl_2$ acts as a chloride ion donor in mixtures with $CuCl_2$ as shown by UV-Vis spectroscopy, Raman spectroscopy, and the structure of solid phases crystallizing from the molten hydrate mixtures. The distribution of water and anions between the coordination spheres of cations can be modeled by formulation of reaction chains, connecting the parameters of step-wise hydration and complexation and thereby reducing the number of model parameters.

At the examples of the systems $MgCl_2$ + KCl + H_2O and $MgCl_2$ + $CuCl_2$ + H_2O it was shown that such reaction models reproduce the thermodynamic properties sufficiently. Difficulties in describing activity data at lower temperatures and concentrations should be overcome in future by considering the dependence of the species formation quantities (G^0, H^0) on the chemical environment or introducing a few simple interaction coefficients.

Thermodynamic modeling with emphasis on species and their correlated formation quantities opens perspectives in predicting phase equilibria of aqueous electrolyte systems at extreme high concentrations and complex compositions.

Acknowledgements

The authors wish to thank Prof. Knoche and his group (Dr. Koß, Dipl.-Phys. Behmann) at RWTH Aachen as well as Prof. Papatheodorou (ICE/HT Patras, Greece) for providing the possibility to perform Raman measurements and to participate from the experience in spectra evaluation.

References

Abovsky, V.; Liu, Y.; Watanasiri, S. (1998): Representation of nonideality in concentrated electrolyte solutions using the Electrolyte NRTL model with concentration-dependent parameters. Fluid Phase Equil. 150–151, 277–286.

Adams, D. M.; Newton, D. C. (1971): Single-crystal Raman and far-infrared spectra of K$_2$CuCl$_4$(H$_2$O)$_2$ and related compounds. J. Chem. Soc. A, 3507–3510.

Ally, M.; Braunstein, J. (1993): BET model for calculating activities of salts and water, molar enthalpies, molar volumes and liquid-solid phase behavior in concentrated electrolyte solutions. Fluid Phase Equil. 87, 213–236.

Ally, M.; Braunstein, J. (1998): Statistical mechanics of multilayer adsorption: electrolyte and water activities in concentrated solutions. J. Chem. Thermodyn. 30, 49–58.

Assarson, G. (1950): Equilibria in aqueous systems containing K$^+$, Na$^+$, Ca^{2+}, Mg^{2+} and Cl$^-$. I. The ternary system Cacl$_2$-KCl-H$_2$O. J. Am. Chem. Soc. 72, 1433–1436.

Barnes, C.; Thompson, W. T. (1988): A determination of the thermodynamic properties of MgCl$_2$ in LiCl-containing melts from Emf measurements. Can. Met. Quarterly, 267–276.

Beattie, I. R.; Gilson, T. R.; Ozin, J. (1969): Single-crystal Raman spectroscopy of "square-planar" and "tetrahedral" CuCl$_4^{2-}$ Ions, of the ZnCl$_4^{2-}$ ion and of CuCl$_2$ · 2 H$_2$O. J. Chem. Soc. A, 534–541.

Braunstein, J. (1971): Ionic Interactions, Vol. 1, 180. Academic Press, New York, London.

Bremer, M.; Hettrich, K.; Voigt, W. (2002): paper submitted to Eur. J. Solid State Chem.

Brooker, M. H.; Huang, C. H. (1980): Raman spectroscopic studies of structural properties of solid and molten states of the magnesium chloride – alkali chloride system. Can. J. Chem. 58, 168–179.

Brugger, J.; McPhail, D. C.; Black, J.; Spiccia, L. (2001): Complexation of metal ions in brines: application of electronic spectroscopy in the study of the Cu(II)-LiCl-H$_2$O system between 25 and 90 °C. Geochim. Cosmochim. Acta 65, 2691–2708.

Chen, Ch.-Ch.; Evans, L. B. (1986): A local composition model for the excess Gibbs energy of aqueous electrolyte systems. AIChE J. 32, 444–454.

Chou, I.-M.; Sterner, S. M.; Pitzer, K. S. (1992): Phase relations in the system NaCl–KCl–H$_2$O: IV. Differential thermal analysis of the sylvite liquidus in the KCl–H$_2$O binary, the liquidus in the NaCl–KCl–H$_2$O ternary, and the solidus in the NaCl–KCl binary to 2 kb pressure, and a summary of experimental data for thermodynamic PTX analysis of solid-liquid equilibria at elevated P-T conditions. Geochim. Cosmochim. Acta 56, 2281–2293.

Clynne, M. A.; Potter, R. W. (1979): Solubility of some alkali and alkaline earth chlorides in water at moderate temperatures. J. Chem. Eng. Data 24, 338–340.

Cohen-Adad, R.; Lorimer, J. W. (1992): Solubility data series: Alkali metal and ammonium chlorides in water and heavy water (binary systems). Vol. 47. Pergamon Press, Oxford.

Conn, A. R.; Gould, N. I. M.; Toint, P. L. (1992): Lancelot – A fortran package for large-scale nonlinear optimization. Springer-Verlag, Berlin, Heidelberg, N.Y.

Conway, B. E. (1981): Ionic hydration in chemistry and biophysics. Elsevier scientific publishing company, Amsterdam, Oxford, New York.

Dietzel, H. (1959): Strukturprobleme im System MgCl$_2$–H$_2$O. Freib. Forschh. A132, 317–332.

Dietzel, H.; Serowy, F. (1959): Die Lösungsgleichwichte des Systems MgCl$_2$–H$_2$O zwischen 20 und 200 °C. Freib. Forschh. A132, 1–32.

Dittrich, A. (1986): Zum Einfluß von Alkalimetallchloriden auf den Dampfdruck von Magnesiumchloridhydratschmelzen. Dissertation, TU Bergakademie Freiberg.

Emons, H.-H.; Fanghänel, Th.; Naumann, R.; Voigt, W. (1986): Salzhydratschmelzen. Sitzungsber. Akad. Wiss. DDR, 3N, 5–35.

Emons, H.-H.; Dittrich, A.; Voigt, W. (1987): Thermodynamic properties of molten magnesium chloride hydrates in presence of potassium chloride. Proc. Joint Intern. Symp. On Molten Salts, Honululu, Vol 87–7, 111.

Fanghänel, Th.; Kravchuk, K.; Voigt, W.; Emons, H.-H. (1987): The binary system MgCl$_2$–H$_2$O at 130–250 °C. Z. anorg. allg. Chem. 547, 21–26.

Fanghänel, Th.; Emons, H.-H. (1988): Solid-liquid phase equilibria in the system KCl–MgCl$_2$–H$_2$O at elevated temperatures. II. Isotherms at 130, 140, and 150 °C. Z. anorg. allg. Chem. 562, 165–169.

Fanghänel, Th.; Emons, H.-H. (1989): Solid-liquid phase equilibria in the system KCl–MgCl$_2$–H$_2$O at elevated temperatures. II. Isotherms at 160, 170, and 180 °C. Z. anorg. allg. Chem. 576, 99–107.

References

Fanghänel, Th.; Grjotheim, K. (1990): Thermodynamics of aqueous reciprocal systems. III. Isopiestic determination of osmotic and activity coefficients of aqueous $MgCl_2$, $MgBr_2$, KCl, and KBr at 100.3 °C. Acta Chem. Scand. 44, 892–895.

Figgis, B. N.; Hitchman, M. A. (2000): Ligand field theory and its applications. Wiley-VCH, New York, Chichester, Weinheim.

Findlay, A. (1958): Die Phasenregel und ihre Anwendungen. VCH, Weinheim.

Geerlings, J.; Richter, J.; Rørmark, L.; Oye, H. A. (1997): Activity coefficients and solubilities of the system $(Ag/Cs)NO_3$–H_2O from vapour pressure measurements. Ber. Bunsen-Ges. Phys. Chem. 101, 1129.

Griffith, T. R.; Nerukh, D. A.; Eremenko, S. A. (1999): The application of theoretical models of complex shape to the fitting of experimental spectra having closely overlapping bands. Phys. Chem. Chem. Phys. 3199–3298.

Hettrich, K. (2002): Untersuchung der Koordinationskonkurrenz in Magnesiumchlorid-Hydratschmelzen mit spektroskopischen Methoden und Fest-Flüssig-Gleichgewichten. Dissertation, Technische Universität Bergakademie Freiberg.

Holmes, H. F.; Mesmer, R. E. (1996): Aqueous solutions of the alkaline-earth metal chlorides at elevated temperatures. Isopiestic molalities and thermodynamic properties. J. Chem. Thermodyn. 28, 1325–1358.

Janz, G. J.; Allen, C. B.; Bansal, N. P., Murphy, R. M.; Tomkins, R. P. T. (1979): NSRDS-NBS 61, Part II Molten Salts: Data on Single and Multi-component Systems. Washington D.C.

Koß, H.-J. (1998): Multifit, Lehrstuhl für Technische Thermodynamik RWTH Aachen, Germany.

Linke, W. F.; Seidell, A. (1965): Solubilities – Inorganic and metal-organic compounds. American Chemical Society, Washington, DC.

Lu, X.; Maurer, G. (1993): Model for describing activity coefficients in mixed electrolyte aqueous solutions. AIChE J. 39, 1527.

Lu, X.; Zhang, L.; Wang, Y.; Shi, J.; Maurer, G. (1996): Prediction of activity coefficients of electrolytes in aqueous solutions at high temperatures. Ind. Eng. Chem. Res. 35, 1777–1784.

Markov, B. F.; Volkov, S. V. (1985): Thermodynamic properties of molten salt systems (Russ.). Naukova Dumka, Kiev.

Menge, O. (1911): Die binären Systeme von $MgCl_2$ und $CaCl_2$ mit den Chloriden der Metalle K, Na, Ag, Pb, Cu, Zn, Sn und Cd. Z. Anorg. Chem. 72, 162

Neil, D. E.; Clark, H. M.; Wiswall, R. H. (1965): Thermodynamic properties of molten solutions of $MgCl_2$–KCl, $MgCl_2$–NaCl, and $MgCl_2$–KCl–NaCl. J. Chem. Eng. Data, 21–24.

Østvold, T. (1972): EMF measurements of the change in chemical potential of one component on mixing in fused binary alkali – alkaline earth halide systems. High Temp. Sci., 51–81.

Peschke, M.; Blades, A. T.; Kebarle, P. (1998): Hydration energies and entropies for Mg^{2+}, Ca^{2+}, Sr^{2+}, and Ba^{2+} from gas-phase ion-water molecule equilibria determinations. J. Phys. Chem. A, 9978–9985.

Peschke, M.; Blades, A. T.; Kebarle, P. (1999): Formation, acidity and charge reduction of the hydrates of doubly charged ions M^{2+} (Be^{2+}, Mg^{2+}, Ca^{2+}, Zn^{2+}), Intern. J. Mass Spectrosc., 685–699.

Pinsky, M. L.; Gruber, G. (1994): Phase equilibria in aqueous systems containing Na^+, K^+, Mg^{2+}, Cl^-, and SO_4^{2-} ions using the NRTL model, in "Thermophysical Properties for Industrial Process design". AIChE Symp. Series No. 298, Vol. 90.

Pitzer, K. S. (1980): Electrolytes. From dilute solutions to fused salts. J. Am. Chem. Soc. 102, 2902–2906.

Pitzer, K. S. (1991): Activity coefficients in electrolyte solutions. 2nd Ed., CRC Press, Boca Raton, Ann Arbor, Boston, London.

Sacchetto, G. A.; Kodejs, Z. (1982): Water solubility in molten salt mixtures. A theory for selective ionic hydration. Faraday Trans., 3519–3527.

Sangster, J. M.; Abraham, M.-C.; Abraham, M. (1987): Vapour pressure of aqueous solutions of $(Ag, Tl, Cd)NO_3$ at 98.5 °C. I. Brunauer-Emmett-Teller isotherms. Can. J. Chem. 56, 348–351.

Schönert, H. (1990): The Debye-Hückel theory for hydrated ions. I. Osmotic and activity coefficients of binary aqueous solutions of some 1:1 electrolytes at 25 °C. Ber. Bunsen-Ges. Phys. Chem. 94, 658–664.

Schönert, H. (1994): Debye-Hückel theory for hydrated ions. 6. Thermodynamic properties of aqueous solutions of 1:1 chlorides between 273 and 523 K. J. Phys. Chem. 98, 643–653.

Smith, D. W. (1976): Chlorocuprates(II). Coord. Chem. Rev. 21, 94–157.

Stokes, R. H.; Robinson, R. A. (1948): Ionic hydration and activity in electrolyte solutions. J. Am. Chem. Soc. 70, 1870–1878.

Stokes, R. H.; Robinson, R. A. (1973): Solvation equilibria in very concentrated electrolyte solutions. J. Solution Chem. 2, 173–191.

Urusova, M. A.; Valyashko, V. M. (1983): Solubility, vapor pressure and thermodynamic properties of the system $MgCl_2$–H_2O at 300–350 °C. Zh. Neorg. Khim., 1845–1849.

Voigt, W. (1993): Calculation of salt activities in molten salt hydrates applying the modified BET equation, I: Binary systems. Monatsh. Chem. 124, 839–848.

Voigt, W.; Emons, H.-H. (1995): Thermodynamic properties of the systems $MgCl_2$-ACl–H_2O (A = Li, ..., Cs) within the range of molten hydrates. 109–114, Proc. Intern. Symp. In Honour of H. Oye, Trondheim.

Voigt, W.; Fanghänel, Th.; Emons, H.-H. (1985): Zur Bestimmung von Fest-Flüssig-Gleichgewichten in hochkonzentrierten Salz-Wasser-Systemen bei Temperaturen bis 250 °C, Z. phys. Chem. (Leipzig) 3, 522–528.

Voigt, W.; Klaeboe, P. (1986): Raman spectroscopic evidence for cation-anion contact in molten hydrates of magnesium chloride. Acta Chem. Scand. A40, 354–357.

von Barner, J. H.; Brekke, P. B.; Bjerrum, N. J. (1985): Chloro complexes in molten salts. 10. Potentiometric and spectrophotometric study of the system KCl-$AlCl_3$-$CuCl_2$ at 300 °C. Inorg. Chem. 24, 2162–2168.

Wollny, F.-W. (1983): Zur Bestimmung der Dampfdrücke über Magnesiumchlorid-Hydratschmelzen. Dissertation, TU Bergakademie Freiberg.

Nomenclature

a	activity
G	Gibbs energy
g	model parameter
H	enthalpy
HH	model parameter
h	model parameter
K	equilibrium constant
k	model parameter
l	liquid
m	molarity
n	mole number
R	molar ratio water/salt
R	universal gas constant
S	entropy
T	absolute temperature
x	mole fraction

Greek letters

λ wave length
ν wave number

Superscripts

0   standard condition
mix   mixture property

Subscripts

n   number of reaction step
SL   solid-liquid
R   reaction property
w   water

12 Effects of Salts on Excess Enthalpies of Binary Liquid Mixtures

Peter Ulbig[1], Thorsten Friese, and Katrin Wagner

12.1 Abstract

The effect of salts (sodium and potassium chloride) on excess molar enthalpies of binary liquid solutions was measured at various temperatures by flow calorimetry. In order to achieve a high number of excess enthalpy isotherms (125 isotherms with 2772 data points were measured), the whole experimental set-up was automated. Therefore, a flow-mixing cell was developed. If the saturation concentration of the salts in the binary liquid mixture was unknown, gravimetrically based solubility measurements were carried out additionally. The knowledge of the saturation point is important to prevent the precipitation of solid salt inside the calorimetric cell causing a blockage of the flow path. For the investigation of the association behavior of 1,1-electrolytes, conductivities were determined. The association constants of the ions were calculated. Since the corresponding conductivity theory is volume-based in concentration, the influence of salt on the molar volume of the binary liquid mixtures was investigated, as well. Based on a group contribution model introduced by Achard and coworkers, an improved model was developed to describe both, activity coefficients and excess enthalpies of systems including salts.

12.2 Introduction

Aqueous electrolyte systems have been described excessively in the literature. During the last 15 years, however, electrolyte mixtures with organic or aqueous-organic solvents have raised more and more interest, as well. While salt-free systems have been described successfully by several models, there are only few theories for the description of salt-containing systems.

In the past, the thermodynamic consistency of g^E-models has successfully been improved using different kinds of thermodynamic data. Not only VLE-data but also excess en-

---

[1] Universität Dortmund, Lehrstuhl für Thermodynamik, Emil-Figge-Straße 76, 44227 Dortmund.

thalpies, activity coefficients at infinite dilution, etc. were used for fitting model parameters. Up to now, however, excess enthalpies of aqueous-organic systems including salts have not been regarded in the g^E-model development for electrolyte systems (Loehe and Donohue, 1997). Nevertheless, the importance of the effect of salts on excess enthalpy data has been recognized by several authors (Achard et al., 1994; Renon, 1996).

The aim of the project under consideration was the formation of a data base consisting of excess enthalpies of binary liquid systems including salts at various temperatures and salt concentrations. A combination of these excess enthalpy data with VLE-data for the same systems provides the basis for the development of a thermodynamically consistent g^E-model.

12.3 Experimental/Methods

12.3.1 Apparatus

A detailed description of the experimental devices was given by Friese (1999), Friese et al. (1998), and Wagner et al. (1998).

12.3.1.1 Calorimetric Measurements

For the calorimetric measurements, aquaeous salt solutions were prepared by mass from degassed water and dried salt. Excess enthalpies of the aqueous salt solution and an alcohol were measured with a flow micro calorimeter (LKB 2277, now: TAM 2277, Thermometric, Sweden). The measuring cylinder of this isoperibolic calorimeter operated according to the Tian-Calvet principle is located inside a thermostated water bath (see Fig. 12.1).

Two self-built computer-controlled syringe pumps were used to deliver the alcohol and the aqueous salt solution to a flow mixing measuring cell which is located inside the measuring cylinder of the calorimeter. Within the cell, the liquids are mixed and the heat effect is detected and recorded. In order to make use of the so-called twin measuring principle, the generated mixture flows through a second flow cell used as a reference cell in order to compensate influences on both cells (e.g. flow fraction) by subtracting the heat events measured in each cell from each other.

In order to prevent corrosion due to the chloride solutions, the measuring cells were made of Hastelloy C. Different mixing devices were investigated to ensure both, the creation of a homogeneous mixture and the detection of the entire heat effect within the cell. The following set-up proved as most efficient (see Fig. 12.2).

At the inlet of the measuring cell, the two liquids are mixed intensively by flowing through a sinter plate. To maximize the residence time in the measuring cell, the mixture flows down to the bottom of the cell through a thread in the core part of the cell before being transported back analogously through a thread in the peripheric part to the outlet at the top of the cell.

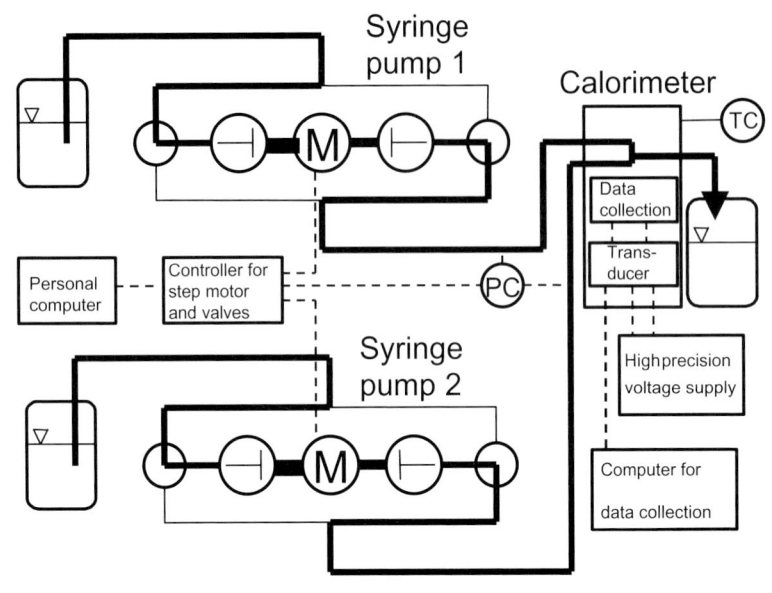

12 Effects of Salts on Excess Enthalpies of Binary Liquid Mixtures

Figure 12.1: Calorimeter and syringe pumps for excess enthalpy measurements.

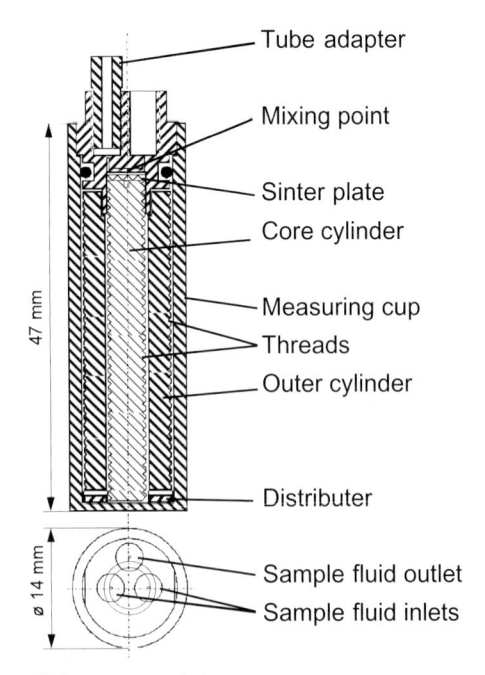

Figure 12.2: Flow mixing cell for excess enthalpy measurements.

The total volume flow rate of the mixture being pumped through the calorimeter is kept at a constant value of 0.08 ml min^{-1}. Starting with salt water, an excess enthalpy isotherm is measured by a stepwise increase of the alcohol fraction. The maximum alcohol fraction is limited by the solubility of the salt in the (alcohol + water) mixture. The concentration of the mixture is established with a standard deviation smaller than 0.8 mol%. The standard deviation of the excess enthalpy is estimated to be smaller than 2% (Friese et al., 1999). The heat flow released during the mixing of the salt solution and the alcohol is detected by Peltier elements and recorded by a computer program. Having obtained the last data point of the isotherm, the calorimetric signal is calibrated using the built-in electric resistance of the calorimeter. Before continuing with the next isotherm, the tubings and the measuring cell are flushed with distilled and degassed water.

12.3.1.2 Solubility Measurements

The solubility of sodium and potassium chloride in the binary aqueous-organic solvent mixtures was determined using a gravimetric method. The solvent mixture was prepared by mass on a Sartorius (Göttingen, Germany) LP1200S balance (accuracy: ± 1 mg, precision: ± 2 mg) with an overall accuracy of the concentration of 0.2 mg g^{-1}. With a Sartorius (Göttingen, Germany) MC210P balance (accuracy: ± 10 µg, precision: ± 20 µg), salt was weighed into special self-built glass cylinders having a glass sinter plate with a pore diameter of 10 to 16 µm at the bottom. The solvent mixture was added and the cylinders were closed on both ends with gas-tight lids. The glass cylinders were placed into an ultrasonic bath and transferred into a thermostated shaking water bath (GFL 1086, Burgwedel, Germany), see Figure 12.3.

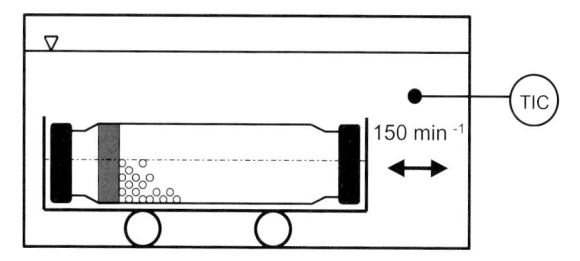

Figure 12.3: Shaking water bath with glass cylinder for solubility measurements.

Once solid-liquid equilibrium had been achieved, the saturated solution was separated from the remaining salt. The mass of dissolved salt and thus the concentration of the saturated solution was determined by differential weighing on the Sartorius MC210P balance with an overall concentration accuracy of ± 0.5 mg g^{-1}. The experimental procedure for the solubility measurements was verified by measuring the solubility of sodium chloride in the pure solvents methanol and ethanol as well as in mixtures of methanol and cyclohexane. A good agreement with literature data was achieved for these systems. The solubility data could be reproduced to better than 2% (Wagner et al., 1998).

12 Effects of Salts on Excess Enthalpies of Binary Liquid Mixtures

12.3.1.3 Conductivity Measurements

For the conductivity measurements, a conductivity meter (Radiometer CDM230, Copenhagen, Denmark) and a four-pole flow measuring cell with platinated platinum electrodes (Radiometer CDC865) were used (Friese, 1999). The measurements could be carried out without a polarisation of the electrodes. The small volume of the cell (0.6 ml) permits a rapid change of concentrations. The set-up of the conductivity measurements is illustrated in Figure 12.4.

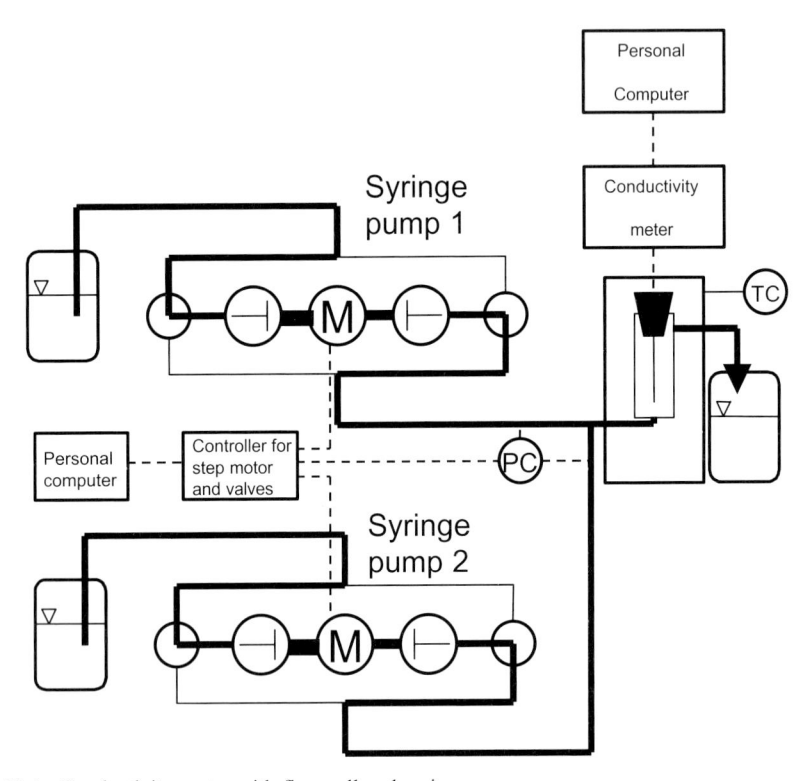

Figure 12.4: Conductivity meter with flow cell and syringe pumps.

The procedure for measuring the specific conductivities is similar to the measurement of the excess enthalpies. The aqueous salt solution and the alcohol are delivered by computer-controlled syringe pumps and are mixed in a tee-piece. The mixture is thermostated inside the water bath of the calorimeter and is pumped through the conductivity measuring cell which is located in the calorimeter water bath, as well. The conductivity meter is calibrated using standard calibration buffers. The measurement starts with an alcohol-free aqueous salt solution. The mole fraction of the alcohol is then increased stepwise. The volumetric flow rate for the mixture is set to a constant value of 0.5 ml min^{-1}. 30 min of measuring time with data points taken every 30 s proved to be sufficient for each concentration step. For the conductivity measurement itself, the pumps are stopped for 7 min in order to

12.3 Experimental/Methods

carry out the measurement under non-flow conditions and to receive a more stable signal. The conductivity measurements were reproduced by beginning the measurements with the pure alcohol and by decreasing the alcohol concentration stepwise. The conductivity was measured for highly diluted solutions of sodium and potassium chloride, i.e. for solutions with 0.005, 0.01, 0.05, 0.25, and 0.5 wt.% of salt in water. Additionally, measurements were carried out for the salt concentrations used in the excess enthalpy measurements.

12.3.1.4 Measurement of the Excess Volume

Excess volumes were measured using a thermostated flexural resonator (Paar Physica, DMA60/602, Graz, Austria) in combination with an autosampler (Paar Physica, SP3, Graz, Austria). A detailed description of the set-up was given by Geyer (2000).

For the measurements, the apparatus was calibrated using nitrogen and pure water. Pseudo-binary mixtures of an alcohol and an aqueous salt solution of a given concentration were prepared gravimetrically and filled into the 12 ml sample tubes of the autosampler. The temperature in the measuring cell was kept constant to within ± 0.005 K.

12.3.2 Materials

Sodium chloride and potassium chloride were supplied by Fluka (puriss. p.a., purity $\geq 99.5\%$). Prior to the measurements, the salts were dried under vacuum for at least 24 h at 513.15 K. Water was distilled and filtered four times (conductivity ≤ 1 µS cm^{-1}). The organic solvents, i.e. methanol (Roth, Roti Solv. HPLC, $\geq 99.9\%$), ethanol (Roth, Rotipuran, $\geq 99.8\%$), 1-propanol (Roth, Rotipuran, $\geq 99.5\%$), 2-propanol (Merck, gradient grade, $\geq 99.8\%$), cyclohexane (Fluka, for residue analysis, $\geq 99.8\%$), cyclohexanol (Fluka, puriss., $\geq 99.0\%$), and benzyl alcohol (Fluka, puriss. p.a., $\geq 99.0\%$) were used without further purification. They were dried with molecular sieve of 3 Å pore diameter supplied by Fluka (Dehydrat Fluka with indicator), which had been activated under vacuum at 513.15 K for at least 6 h. The remaining water content was determined by coulometric titration using a Karl-Fischer coulometer (Mettler-Toledo, Giessen, Germany; KF coulometer DL36). It was below 80 wt.ppm in all cases. Directly before the measurements, all pure liquids were degassed by ultrasonic treatment and with a water-jet vacuum pump.

12.3.3 Analysis

Excess enthalpy measurements were carried out by mixing a pure alcohol (component 1) with an aqueous salt solution (water (component 2) + salt (component 3)). The following nomenclature is used to specify mole fractions or weight fractions, respectively: In the system alcohol (1) + water (2) + salt (3), $w_{3,23}$ is, for example, the weight fraction of salt in the aqueous salt solution and $x_{1,123}$ is the mole fraction of alcohol in the entire mixture.

12 Effects of Salts on Excess Enthalpies of Binary Liquid Mixtures

The measured heat effect has to be corrected by a calibration factor, by a factor taking into account flow depending terms, temperature and friction effects, and by the offset of the baseline (Friese, 1999). The resulting value of the measurements is the excess enthalpy of the mixture (alcohol + salt water), i.e. h^E_{1+23}. The excess enthalpy of the mixture (alcohol + water + salt), h^E_{123}, is calculated from Eq. (12.1):

$$h^E_{123} = h^E_{1+23} + (1 - x_{1,123})\, h^E_{23} \tag{12.1}$$

Thus, the excess enthalpy, h^E_{23}, taking into account the heat effect occurring when mixing water with salt has to be known. This value is calculated according to Eq. (12.2) using the integral enthalpy of solution of the salt in water for the given salt concentration, $h^{int}_{solv,23}(x_{3,23})$ and the integral enthalpy of solution of the salt in water at infinite dilution, $h^{int,\infty}_{solv,23}$ (Friese et al., 1999):

$$h^E_{23} = x_{3,23} \left(h^{int}_{solv,23}(x_{3,23}) - h^{int,\infty}_{solv,23} \right) \tag{12.2}$$

The integral molar enthalpy of solution and the integral molar enthalpy of solution at infinite dilution are given in Joule per mole solute and were taken from Clarke and Glew (1985).

The conductivity measurements were analyzed using the theory of Barthel (1976) with an iterative calculation of the degree of dissociation, α. In the flow cell conductivity measurement procedure, water with a constant salt concentration is diluted stepwise by the addition of alcohol. To make use of the conductivity theory, the experimental data of an isotherm, $\sigma = f(x_{1,123})$ at constant $w_{3,23}$, have to be converted into specific conductivity data, $\sigma = f(w_{3,23})$ at constant alcohol concentration $x_{1,123}$ by interpolation. The equivalence conductivity, Λ, is thus given by

$$\Lambda(w_{3,23}, x_{1,123}) = \frac{\sigma(w_{3,23}, x_{1,123})}{c_{3,123}} \quad \text{for } T = \text{const. and } x_{1,123} = \text{const. with} \tag{12.3}$$

$$c_{3,123} = (1 - w_{1,123})\, w_{3,23}\, \frac{\rho_{123}}{M_3} \quad \text{and} \tag{12.4}$$

$$w_{1,123} = \frac{x_{1,123}\, M_1}{x_{1,123}\, M_1 + (1 - x_{1,123})\, M_{23}} \tag{12.5}$$

The excess volumes of the pseudo-binary mixtures of alcohol (1) and salt water (23) were calculated according to

$$v^E_{1+23} = v_{123} - x_{1,123}\, v_1 - (1 - x_{1,123})\, v_{23} \tag{12.7}$$

$$= x_{1,123}\, M_1 \left(\frac{1}{\rho_{123}} - \frac{1}{\rho_1} \right) + (1 - x_{1,123})\, M_{23} \left(\frac{1}{\rho_{123}} - \frac{1}{\rho_{23}} \right) \tag{12.8}$$

This procedure allows a comparison of the excess volumes of the liquid systems with and without salt, i.e. v^E_{1+23} and v^E_{12}.

12.4 Results and Discussion

12.4.1 Excess Enthalpy

125 excess enthalpy isotherms with 2772 data points were measured using the optimized and automated calorimetric procedure described above. Table 12.1 gives an overview of all measurements. Most of the results are published and have been discussed in detail elsewhere previously (Friese et al., 1998, 1999; Friese, 1999). Therefore, only a brief summary of the results is given here.

12.4.1.1 Effect of the Salt Concentration

Figure 12.5 shows that the effect of salt becomes more significant with higher salt concentrations. For the example given in Figure 12.5, an influence of salt on the excess enthalpy of (alcohol + water) mixtures can already be observed at low salt concentrations ($w_{3,23}$ = 1.00 wt.%). At high salt concentrations (here: 7.5 wt.% sodium chloride in water), however, the excess enthalpy is reduced by about one third compared to the salt-free system at 308.15 K.

12.4.1.2 Effect of Temperature

The shift to less exothermic excess enthalpies due to the effect of a salt as observed in Figure 12.5 cannot be generalized. Figure 12.6 displays excess enthalpy isotherms for the system (1-propanol + water + NaCl). For the salt-free system, excess enthalpies are predominantly negative at low temperatures (285.65 K). At 353.15 K, however, the excess enthalpy is positive over the entire concentration range. For a better understanding, the measured values in Figure 12.6 are connected by lines. Along with the change in sign of the excess enthalpy with increasing temperature, the shift to less exothermic excess enthalpies under the influence of salt decreases. For temperatures greater than 323.15 K, the isotherms for the salt-free system and for the system containing NaCl intersect. A shift to less endothermic excess enthalpies can thus be observed in relation to the salt-free isotherm if the alcohol mole fraction exceeds the point of intersection. With increasing temperature, the intersection is moved to lower alcohol mole fractions although the total salt concentration, $w_{3,123}$, rises with decreasing alcohol mole fractions. Regarding the 353.15 K isotherms, the behavior of the excess enthalpy in the alcohol mole fraction range from about 0.06 to 0.2 is hardly affected by the addition of salt even though the salt concentration in the mixture is relatively high.

12.4.1.3 Effect of Different Alcohols

In Figure 12.7, the influence of salt at a constant salt concentration of $w_{3,23}$ (KCl) = 5.00 wt.% in water and at constant temperature is displayed for different alcohols. The difference to the salt-free excess enthalpy isotherm due to the addition of salt is significantly smaller for mix-

Table 12.1: Excess enthalpy measurements.

T/K	Solvent (1)	Solvent (2)	Salt (3)	$w_{3,23}$/wt.%
285.65	methanol	water	NaCl	0, 1, 5, 10
	methanol	water	KCl	1, 5, 7.5
	ethanol	water	NaCl	0, 1, 3.962, 5, 5.976, 10
	ethanol	water	KCl	1, 5, 7.5
	1-propanol	water	NaCl	0, 1, 3, 5
	1-propanol	water	KCl	1, 3, 5
	2-propanol	water	NaCl	0, 2.5, 5.976, 7.5
	2-propanol	water	KCl	2.5, 5, 7.5
	2-propanol	water	(NaCl+KCl)	6.744
298.15	cyclohexane	methanol	NaCl	0, 0.2, 0.4, 0.8
	cyclohexanol	water	NaCl	0, 3.5
	benzylalcohol	water	NaCl	0, 5, 10
	methanol	water	NaCl	0, 1, 5, 10
	methanol	water	KCl	1, 5, 7.5
	ethanol	water	NaCl	0, 1, 5.53, 10, 10.45
	ethanol	water	KCl	1, 5, 7.5
	1-propanol	water	NaCl	0, 1, 3, 5
	1-propanol	water	KCl	1, 3, 5
	2-propanol	water	NaCl	0, 2.5, 5, 7.5
	2-propanol	water	KCl	2.5, 5, 7.5
308.15	methanol	water	NaCl	0, 1, 5, 10
	methanol	water	KCl	1, 5, 7.5
	methanol	water	(NaCl+KCl)	5
	ethanol	water	NaCl	0, 1, 5, 10
	ethanol	water	KCl	1, 5, 7.5
	1-propanol	water	NaCl	0, 1, 3, 5
	1-propanol	water	KCl	1, 3, 5
	2-propanol	water	NaCl	0, 2.5, 5, 7.5
	2-propanol	water	KCl	2.5, 5, 7.5
323.15	methanol	water	NaCl	0, 1, 5, 10
	methanol	water	KCl	1, 5, 7.5
	ethanol	water	NaCl	0, 1, 5, 10
	ethanol	water	KCl	1, 5, 7.5
	1-propanol	water	NaCl	0, 1, 3, 5
	1-propanol	water	KCl	1, 3, 5
	2-propanol	water	NaCl	0, 2.5, 5, 7.5
	2-propanol	water	KCl	2.5, 5, 7.5
338.15	ethanol	water	NaCl	0, 1, 5, 10
	ethanol	water	KCl	1, 5, 7.5
	1-propanol	water	NaCl	0, 1, 3, 5
	1-propanol	water	KCl	1, 3, 5
	2-propanol	water	NaCl	0, 2.5, 5, 7.5
	2-propanol	water	KCl	2.5, 5, 7.5
353.15	1-propanol	water	NaCl	0, 1, 3, 5

12.4 Results and Discussion

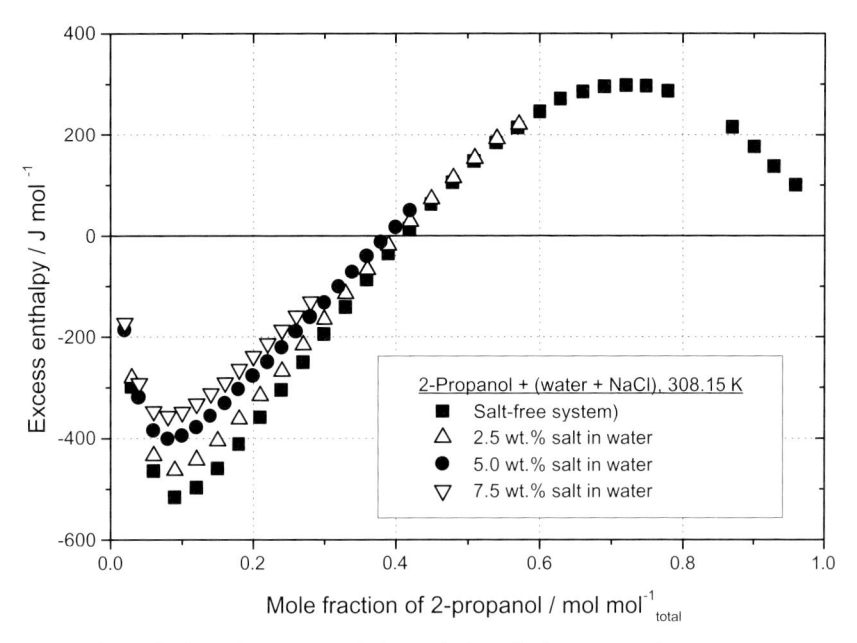

Figure 12.5: Effect of salt on the excess enthalpy, variation of salt concentration.

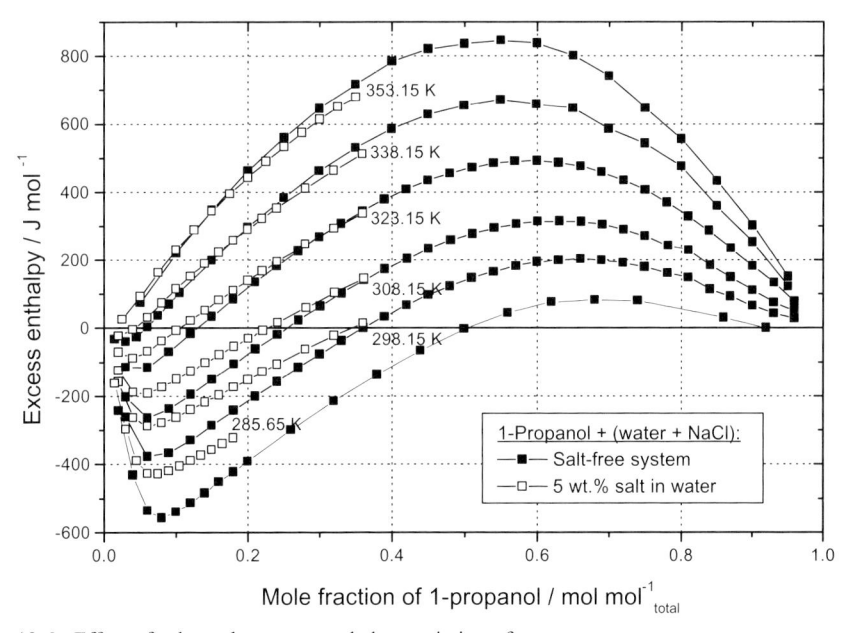

Figure 12.6: Effect of salt on the excess enthalpy, variation of temperature.

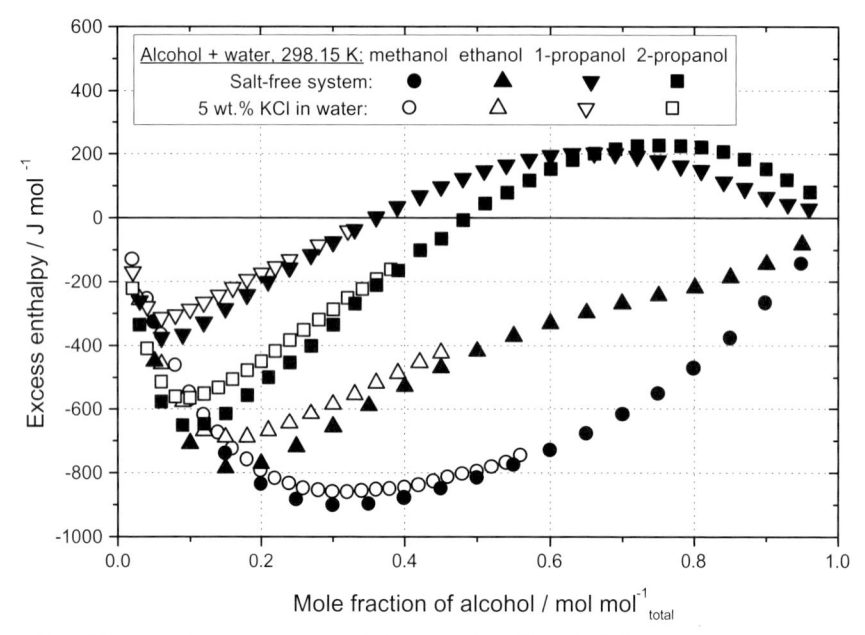

Figure 12.7: Effect of salt on the excess enthalpy, variation of the alcohol.

tures with methanol (relatively and absolutely) than for the other alcohols. For each system, the greatest effect can be observed at the minimum of the salt-free isotherm. While in mixtures of ethanol, 1-propanol or 2-propanol with water, the addition of KCl causes a reduction of the exothermic effect of 13 to 16% at the minimum of the isotherms, in mixtures with methanol only about 5% of reduction are achieved. Thus, the interaction between methanol and water is much less affected by the addition of salt than the interaction between other alcohols and water. The methanol molecule is smaller and more polar than the other alcohols. Therefore, it is easier for a methanol molecule to interact with water molecules in the solvation shell of an ion or to replace it.

12.4.1.4 Effect of Different Salts

The influence of different kinds of salt on the excess enthalpy of (alcohol + water) mixtures is shown in Figure 12.8. Both salts under investigation, sodium chloride and potassium chloride, are 1,1-electrolytes with chloride as the anion. Therefore, the differences in the influence of the salt on the excess enthalpy of the solvent mixture are likely to result from the type of cation. Generally speaking, for alkali halides, cations rather than anions are responsible for the observed shift in phase equilibria and in excess enthalpies because they are smaller and thus carry a higher surface charge. Provided that the salt is completely dissociated, the influence of sodium and potassium ions can be compared if the same number of moles of chloride ions is dissolved in the solvent mixture. The assumption of complete dissociation is at least valid at low mole fractions of the alcohol. In Figure 12.8, two concentrations

12.4 Results and Discussion

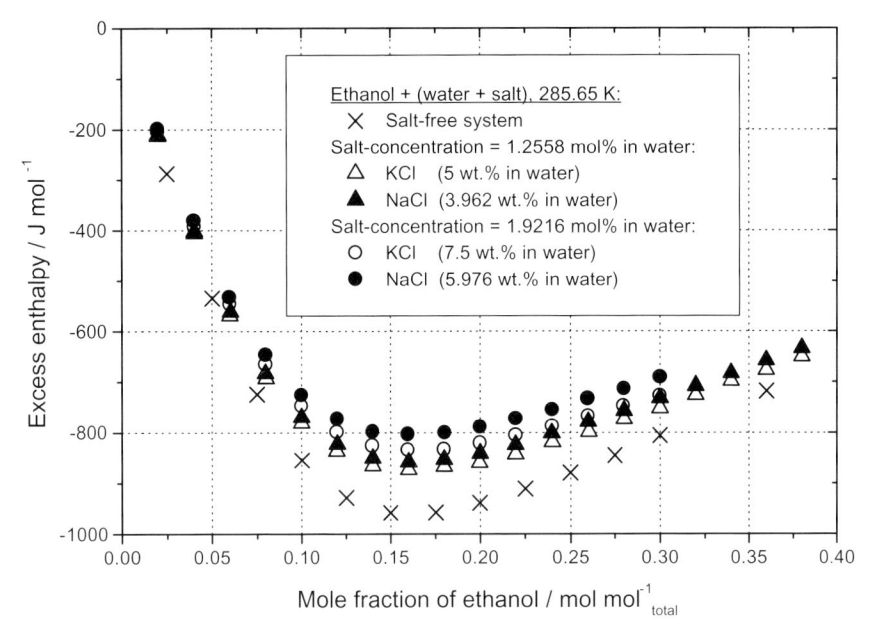

Figure 12.8: Effect of salt on the excess enthalpy, variation of the salt.

(1.2558 and 1.9216 mol% salt in water) were investigated for both, sodium and potassium chloride. The excess enthalpy isotherm of (ethanol + water) at 285.65 K is slightly more affected by sodium chloride than by potassium chloride. As mentioned above, this observation can be attributed to the size of the cation because the sodium ion is the smaller one with a higher surface charge.

12.4.2 Solubility

The measured solubilities of NaCl and KCl in several organic and aqueous-organic mixtures have been discussed in detail previously (Wagner et al., 1998; Friese, 1999). The solubility of the salts is related to the polarity of the solvents: the more polar the solvents, the higher the amount of salt which can be dissolved. The solubility of a salt within a miscibility gap of the binary liquid mixture can be regarded as a linear function of the solvent concentration.

12.4.3 Conductivity

Conductivity measurements were performed for (methanol + salt water) and (ethanol + salt water) systems. NaCl and KCl were used as salts. The results of the measurements were given by Friese (1999). Figure 12.9 shows a typical example for the results of the conductivity measurement.

12 Effects of Salts on Excess Enthalpies of Binary Liquid Mixtures

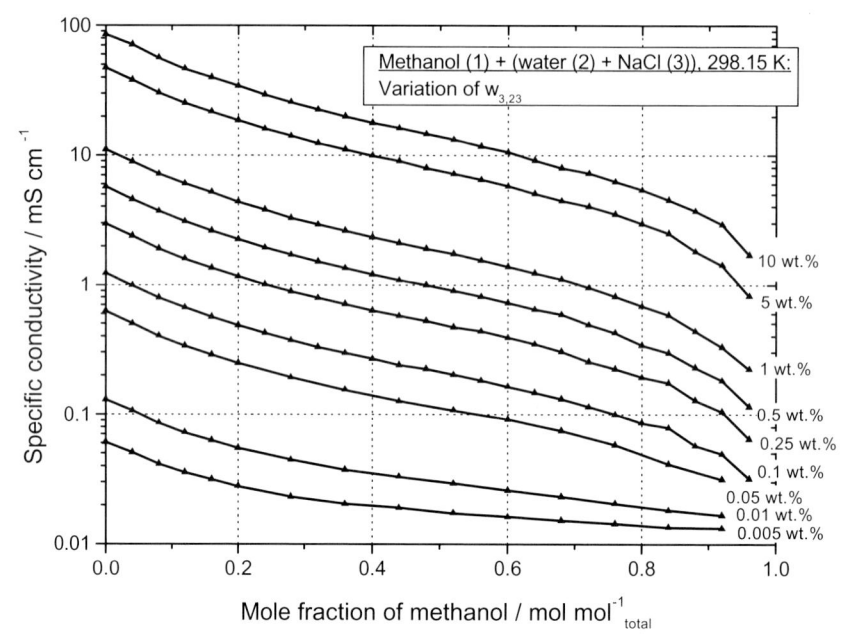

Figure 12.9: Effect of salt onto the specific conductivity.

12.4.4 Excess Volume

Density measurements were carried out for (methanol + water) and (ethanol + water) mixtures under the influence of NaCl or KCl at 298.15 and 308.15 K (Friese, 1999). The effect of the salts onto aqueous mixtures of either one of the alcohols is very similar: the higher the salt concentration, the smaller is the excess effect (see Fig. 12.10). The difference between the effect of NaCl and KCl, respectively, is within the experimental error.

For the analysis of the conductivity measurements, the excess volumes of the binary liquid mixtures should not be neglected because a linear combination of the reciprocal densities of the pure liquids leads to an uncertainty of up to 4% for the density of the mixture. Compared to this, the influence of the salt on the density calculation is small.

12.5 Theory

12.5.1 The HEACE Model

Based on the group contribution model UNIFAC modified by Achard et al. (1994), the so-called HEACE model (Excess Enthalpies h^E and Activity Coefficients of Electrolyte Solutions) was developed by Friese (1999). This model can be used to describe the effect of salts

12.5 Theory

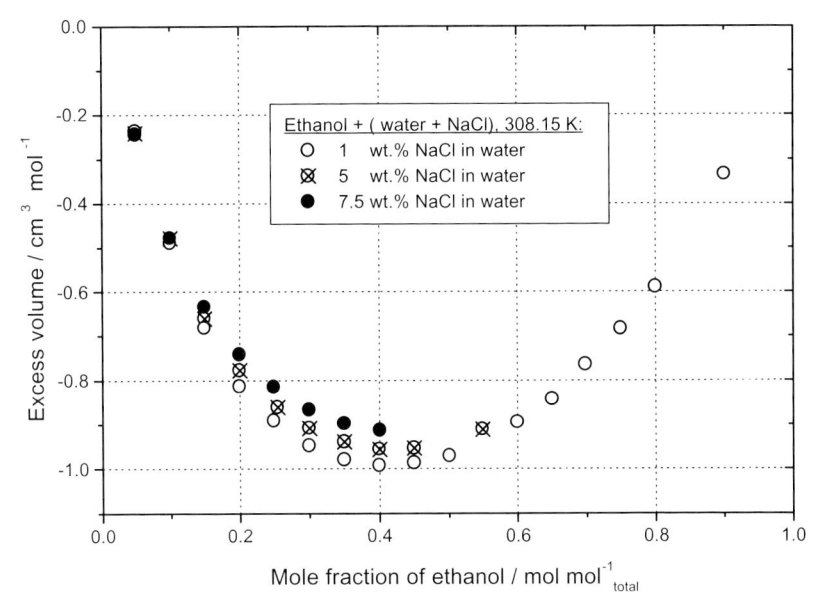

Figure 12.10: Effect of salt onto the excess volume.

onto both, excess enthalpies and vapor-liquid equilibria. The relevant modifications included in the HEACE model in comparison with the basic approach of Achard et al. are as follows:

- In highly concentrated electrolyte solutions, short-range interaction forces become dominant compared to the long-range Coulomb interactions. Therefore, the latter are neglected.

- Short-range contributions of the salt-free system are described with the modified UNIFAC model (Weidlich and Gmehling, 1987).

- An extended temperature function with two parameters (u_{mn_1}, u_{mn_2}) instead of only one parameter is introduced for describing the ion-solvent interaction, a_{mn}. This extension is legitimate because the fitting procedure of the parameters uses two kinds of thermodynamic information (VLE-data and h^E-data) instead of only one. The corresponding relationships are given by Eq. (12.9):

$$a_{mn} = u_{mn_1} - u_{nn} + u_{mn_2} T \quad \text{with } u_{mn_1} = u_{nm_1} \text{ and } u_{mn_2} = u_{nm_2} \tag{12.9}$$

- Achard et al. define the reference interaction parameter between equal species, u_{nn} (see Eq. (12.9)), to be constant. In the HEACE model, the same assumption is made for ions and water, respectively. For solvents, however, the following equation is used:

$$u_{solv,solv} = a_{solv,H_2O} - a_{H_2O,solv} + u_{H_2O,H_2O} \tag{12.10}$$

$a_{H_2O,solv} = a_{H_2O,solv}(T)$ and $a_{solv,H_2O} = a_{solv,H_2O}(T)$, respectively, represent the extended temperature dependence of the modified UNIFAC model by Weidlich and Gmehling (1987).

- The cations Na$^+$ and K$^+$ belong to the same main group, i.e. the same interaction parameters are valid for both of them. The different effects of these cations on the excess enthalpy and the VLE, respectively, are taken into account by different Van der Waals volumes and surfaces.

The solvation concept of Achard et al. remains without changes. This concept takes into account the hydration of ions by water molecules. A complete dissociation of the salts into ions is assumed. The hydration numbers, N_k^H, are taken from Achard et al. (1994) while the structural parameters, R_m and Q_m, and the interaction parameters, a_{mn_1}, a_{mn_2}, and a_{mn_3}, are taken from Schiller (1993). The conversion of the g^E-form of the HEACE model into its h^E-form results from the application of the Gibbs-Helmholtz equation (Eq. (12.11)):

$$h^E = -RT^2 \left(\sum_{i=1}^{4} x_i \frac{\partial \ln \gamma_i}{\partial T} - x_3 \bar{h}_3^{E,\infty} - x_4 \bar{h}_4^{E,\infty} \right) \qquad (12.11)$$

The reference states for components 3 and 4 in Eq. (12.11) are the corresponding ions at infinite dilution.

Details concerning the fitting procedure itself and the numerical algorithms used in the fitting procedure are given by Friese (1999). For a salt concentration of zero, the equations of the HEACE model become equal to the well known modified UNIFAC model (Weidlich and Gmehling, 1987). It was assured that the interaction parameters between ions and solvents describe only the influence of the salts and are not used to compensate deviations of the salt-free model in comparison to experimental salt-free data.

12.5.2 Conductivity Theory

The conductivity, Λ, of an electrolyte solution is a function of the electrolyte concentration, c, the limiting value of the equivalence conductivity, Λ^∞ ($c \to 0$), the distance parameter, R^b, being considered as an association border, the temperature of the solution, T, and characteristic data of the solvent such as the viscosity, η, or the relative permittivity, ε_r: $\Lambda = \Lambda(c, \Lambda^\infty, R^b, \varepsilon_r, \eta, T)$.

The undisturbed movement of the ions is slowed down by interactions between the ions. In extended conductivity theories, (cf. Barthel, 1976), this is taken into account by introducing a relaxation term, Λ_I, and the so-called electrophoretical term, Λ_{II}:

$$\Lambda = \Lambda^\infty - \Lambda_I(c, \Lambda^\infty, R^b, \varepsilon_r, T) - \Lambda_{II}(c, R^b, \varepsilon_r, \eta, T) \qquad (12.12)$$

In case of an association of the ions, the corresponding association equilibrium has to be considered. Thus, Barthel (1976) developed the following conductivity equation:

$$\frac{\Lambda}{\alpha} = \Lambda^\infty - S(\alpha c)^{1/2} + E\alpha \ln(\alpha c) + J_1(R_1)\alpha c + J_2(R_2)(\alpha c)^{3/2} \qquad (12.13)$$

For the analysis of the measured conductivity data of this investigation, the different parameters of Eq. (12.13) are calculated with the equation of Fuoss and Onsager in combination with an equation for the J_2-term by Justice (Barthel, 1976). The necessary equations were given by Friese (1999). The association constant for the association equilibrium can be formulated as follows:

$$K_A = \frac{1-\alpha}{\alpha^2 c} \frac{\gamma_{CA}}{\gamma_\pm^2} \tag{12.14}$$

The activity coefficient of the ion pairs, γ_{KA}, equals one in the ideally diluted region. The choice of an equation for the ionic activity coefficient depends on the concentration range. Equation (12.15) is used for the range of the association theory of Bjerrum:

$$\ln \gamma_\pm = -\frac{\kappa R_\gamma}{1+\kappa R_\gamma} \quad \text{with} \tag{12.15}$$

$$\kappa^2 = 16\pi R^b N_A 1000 \alpha c \tag{12.16}$$

Equations (12.12) to (12.16) have to be solved by iteration. For compatibility, it holds

$$R_1 = R_2 = R_\gamma = R^b \tag{12.17}$$

The distance parameter, R^b, represents the sum of the radii of all ions and the length of a solvent molecule. However, the behavior of electrolytic alcohol solutions can be better described by adding the length of a hydroxyl group (Barthel et al., 1983). Therefore the following equation is to be used:

$$R^b = \sum_C r_C + \sum_A r_A + l_{solv} + l_{OH} \tag{12.18}$$

The degree of dissociation, α, is finally obtained by an iteration procedure during which the three parameters, Λ^∞, K_A, and J_2, have to be calculated.

12.6 Results and Discussion

The model parameters of the HEACE model were fitted using both, VLE-data and h^E-data simultaneously. The VLE-data (204 data points) were taken from literature while 1550 h^E-data points were taken from this work. The uncertainty for the fitting of the VLE-data with the HEACE model is calculated to be 14% and thus is similar to the uncertainty of 15% determined using the model of Achard et al. For the excess enthalpy, an uncertainty of 10.5% can be achieved with the HEACE model (see example in Fig. 12.11) using the same set of parameters as for the VLE-calculation. With the h^E-expression of the model of Achard et al., however, the effect of salt on the excess enthalpy cannot be described correctly (uncertainty of 300%). Thus, the HEACE model reveals a better thermodynamic consistency with-

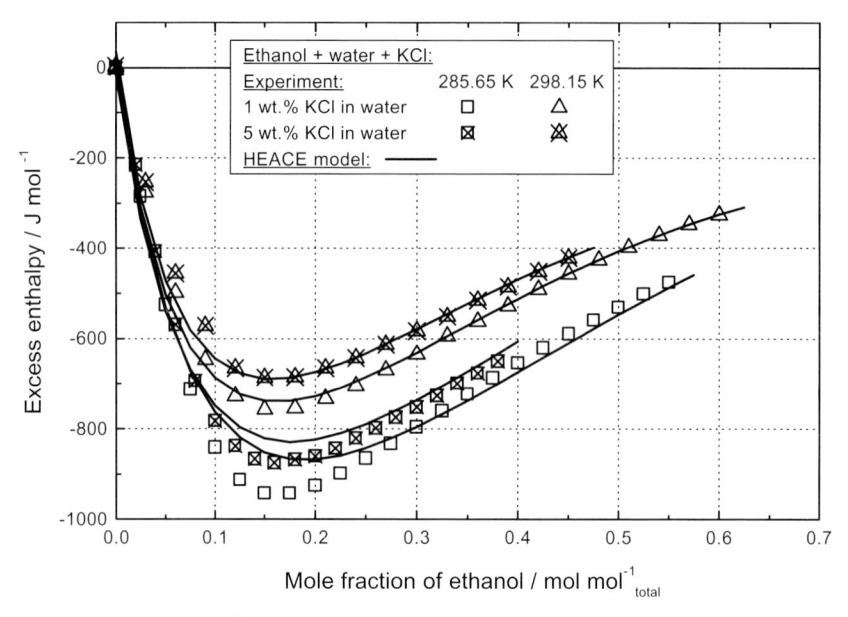

Figure 12.11: Description of h^E-data with the HEACE model.

out increasing the number of fitting parameters. The predictive capability of the HEACE model was tested with 850 h^E-data points which had not been used in the fitting procedure. For this prediction, an uncertainty of 13.5% was achieved. In the example given in Figure 12.12, only the data for 298.15 and 323.15 K were used in the fitting procedure, whereas a prediction was made for the data for 308.15 and 338.15 K.

Using the measured conductivity data and the conductivity theory of Barthel (1976), the equivalence conductivity at infinite dilution, the association constants, and the degrees of dissociation for the systems under investigation could be calculated as a function of the molar composition of the mixture (Friese, 1999). Although, the results for the equivalent conductivity data are in good agreement with literature (see Fig. 12.13), the calculation for the degree of dissociation is not satisfying (see Fig. 12.14).

It is reasonable that the degree of dissociation decreases with an increasing fraction of the less polar solvent (alcohol). Furthermore, lower degrees of dissociation at a higher salt concentration at the same mole fraction of alcohol were observed. This can be explained by the closer approach of the ions to each other approaching the solubility limit. The absolute values of the calculated degrees of dissociation, however, are much too low. Solutions at molarities of 0.008 or 0.04, respectively, should reveal complete dissociation, i.e. $\alpha = 1$, for alcohol molar fractions approaching zero. The reason for these unsatisfying results is the uncertainty of the conductivity measurements. The slight scattering of the equivalence conductivity data points has little effect on the extrapolation accuracy for the equivalence conductivity at infinite dilution but it is too high for an accurate calculation of association constants. Therefore, a conductivity measuring cell with a higher accuracy should be used.

12.6 Results and Discussion

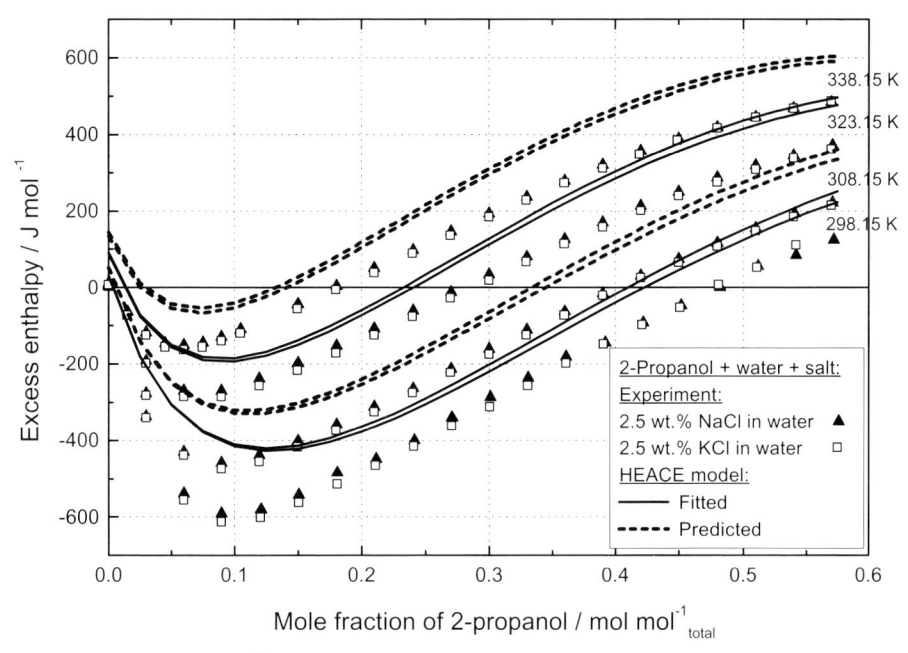

Figure 12.12: Prediction of h^E-data with the HEACE model.

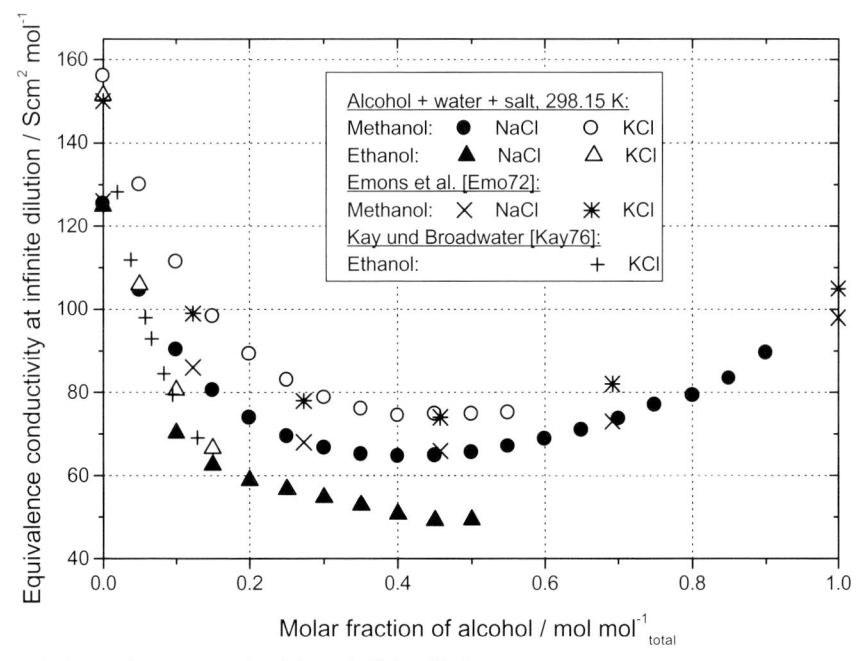

Figure 12.13: Equivalence conductivity at infinite dilution.

285

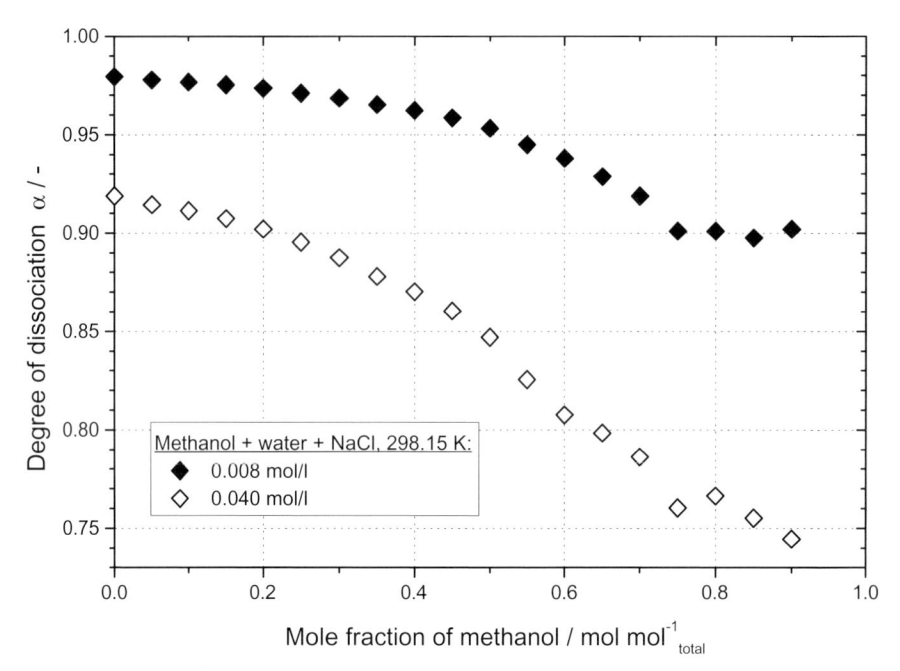

Figure 12.14: Degree of dissociation.

12.7 Conclusion

Up to now, only very few excess enthalpy data were available for binary liquid systems including salts. This data base was too small to be used for the fitting of parameters for group contribution models together with VLE-data of electrolyte systems. In this project, a data base for the excess enthalpies of (water + alcohol) systems including salts was created. 125 excess enthalpy isotherms with 2772 data points for different salt concentrations and temperatures were measured by flow calorimetry. Excess enthalpies were measured with an uncertainty of about 2%. In order to avoid a precipitation of the salt inside the calorimetric cell, the solubility of the salts in the binary liquid mixtures was measured gravimetrically.

With a newly developed group contribution model for (water + alcohol + salt) mixtures (HEACE – Excess Enthalpies h^E and Activity Coefficients of Electrolyte Solutions), group contribution parameters were fitted simultaneously to VLE-data for systems including salts which were taken from literature and to the excess enthalpy isotherms obtained within this project. Originally, it had been intended to introduce the degree of dissociation into the HEACE model in order to describe the association behavior of 1,1-electrolytes in binary aqueous mixtures. This was not realized, however, due to a too high uncertainty in the conductivity measurements.

With the HEACE model, VLE-data are represented with an average uncertainty of 14%. Excess enthalpy data are represented with an average uncertainty of 10.5%. The uncertainty for a prediction of the salts on the excess enthalpy amounts to 13.5%. The model uses the same set of parameters for VLE- and h^E-data with two parameters per main group interaction salt/solvent. It enables a thermodynamically consistent prediction of both, VLE-data and excess enthalpies for electrolyte systems.

Acknowledgements

The authors would like to thank Prof. Lichtenthaler (University of Heidelberg) for helpful discussions concerning the optimization of the calorimetric flow mixing cell, Prof. Holz (University of Karlsruhe) for interesting discussions concerning the association of ions in (alcohol + water) mixtures, and Prof. Gmehling and co-workers (University of Oldenburg) for the compilation of relevant VLE-data from the literature. The authors are especially thankful to Dr. Neueder (University of Regensburg) for the analysis of the conductivity data and for helpful discussions concerning this topic.

References

Achard, C.; Dussap, C. G.; Gros, J. B. (1994): Representation of Vapor-Liquid Equilibria in Water-Alcohol-Electrolyte Mixtures with a Modified UNIFAC Group-Contribution Method. Fluid Phase Equilibria 98, 71.
Barthel, J. (1976): Ionen in nichtwäßrigen Lösungen. Dr. Dietrich Steinkopff Verlag, Darmstadt.
Barthel, J.; Neueder, R.; Feuerlein, F.; Strasser, F.; Iberl, L. (1983): Conductance of Electrolytes in Ethanol Solutions from –45 °C to 25 °C. J. Sol. Chem. 12, 449.
Clarke, E. C. W.; Glew, D. N. (1985): Evaluation of the Thermodynamic Functions for Aqueous Sodium Chloride from Equilibrium and Calorimetric Measurements below 154 °C. J. Phys. Chem. Ref. Data 14, 489.
Emons, H.-H.; Ponsold, B.; Viehweger, U. (1972): Untersuchungen an Systemen aus Salzen und gemischten Lösungsmitteln. XIV. Leitfähigkeitsbestimmungen an Salz-Methanol-Wasser-Systemen. Wissenschaftl. Zeitschr. TH Merseburg 14, 2, 93.
Friese, T.; Schulz, S.; Ulbig, P.; Wagner, K. (1998): Effect of NaCl on the Excess Enthalpies of Binary Liquid Systems. Thermochim. Acta 310, 87–94.
Friese, T.; Ulbig, P.; Schulz, S.; Wagner, K. (1999): Effect of NaCl or KCl on the Excess Enthalpies of Alkanol+Water Mixtures at Various Temperatures and Salt Concentrations. J. Chem. Eng. Data 44, 701–714.
Friese, T. (1999): Experimentelle Untersuchung und Modellierung der Exzeßenthalpie flüssiger Systeme unter Salzeinfluß. PhD-Thesis, Universität Dortmund, Shaker-Verlag, Aachen.
Geyer, H. (2000): Entwicklung und Untersuchung von Gruppenbeitragsmethoden zur Vorhersage thermodynamischer Stoffgrößen unter Verwendung von Methoden der Computational Intelligence. PhD-Thesis, Universität Dortmund, Shaker-Verlag, Aachen.

Kay, R. L.; Broadwater, T. L. (1976): Solvent Structure in Aqueous Mixtures. III. Ionic Conductances in Ethanol-Water Mixtures at 10 and 25 °C. Journal of Solution Chemistry 5, 1, 57.

Loehe, J. R.; Donohue, M. D. (1997): Recent Advances in Modelling Thermodynamic Properties of Aqueous Strong Electrolyte Solutions. AIChE J. 43, 1, 180.

Renon, H. (1996): Models for Excess Properties of Electrolyte Solutions: Molecular Bases and Classification, Needs and Trends for New Developments. Fluid Phase Equilibria 116, 217.

Schiller, M. (1993): Vorausberechnung thermodynamischer Eigenschaften mit der Gruppenbeitragsmethode Modified UNIFAC Dortmund und Messung von Aktivitätskoeffizienten bei unendlicher Verdünnung. PhD-Thesis, Universität Oldenburg, Dissertations Druck Darmstadt, Darmstadt.

Wagner, K.; Friese, T.; Schulz, S.; Ulbig, P. (1998): Solubilities of Sodium Chloride in Organic and Aqueous-Organic Solvent Mixtures. J. Chem. Eng. Data 43, 5, 871–875.

Weidlich, U.; Gmehling, J. (1987): A Modified UNIFAC Model. Ind. Eng. Chem. Res. 26, 1372.

Nomenclature

a_{mn}	interaction parameter/K
c	molar concentration/(mol l^{-1})
$\bar{h}_i^E$	partial molar excess enthalpy/(J mol^{-1})
h^E	excess enthalpy/(J mol^{-1})
K_A	association constant
l	characteristic length/m
m	mass/kg
M	molar mass/(g mol^{-1})
N_A	Avogadro number
Q_m	relative van der Waals surface of group m
r	radius/m
R	universal gas constant
R	distance parameter/m
R_m	relative van der Waals volume of group m
T	temperature/K
u_{mn}	interaction parameter/K
v	molar volume/(cm^3 mol^{-1})
v^E	excess volume/(cm^3 mol^{-1})
w	weight fraction/(g g^{-1})
x	molar fraction/(mol mol^{-1})
S, E, J_1, J_2	coefficients of the conductivity equation

Greeks letters

α	degree of dissociation
γ_i	activity coefficient of component i
ρ	density/(kg m^{-3})
ε_r	relative permittivity constant

Nomenclature

η	dynamic viscosity/(Pa s)
Λ	equivalent conductivity/(S cm^2 mol^{-1})
σ	specific conductivity/(S cm^{-2})

Superscripts

∞	infinite dilution
b	border
E	excess property
H	hydration
int	integral property

Subscripts

$\pm$	averaged electrolytic property
12	mixture of components 1 and 2
1+23	mixture of component 1 with mixture of components 2 and 3
3,23	fraction of component 3 in mixture of components 2 and 3
1,123	fraction of component 1 in mixture of components 1, 2, and 3
A	anion
C	cation
CA	pair: cation – anion
OH	hydroxyl group
salt	salt
sat	saturated
solv	solvent
r	relative

13 Hydrate Equilibria in Aqueous Solutions Containing Inhibitors

Armin M. Rock and Lothar R. Oellrich [1]

13.1 Abstract

Three-phase (V-L$_W$-H) hydrate equilibria of pure gases and binary gas mixtures from methane, ethane and propane in pure water and in the presence of electrolyte and mixed organic-electrolyte inhibitors were treated experimentally and theoretically. For gas mixtures, the range of gas composition mainly addressed in this work was limited to methane-rich mixtures which have major significance with respect to natural gas hydrate formation behavior. Experimental investigations were performed employing the isochoric method in a temperature and pressure range of $-15 \leq t/°C \leq 25$ and $0 \leq p/\text{MPa} \leq 30$, respectively. Our modeling work was based on a modified statistical theory of van der Waals and Platteeuw for the solid hydrate phases. Fluid phases were described with a cubic equation-of-state based fugacity model. For liquid phase calculations, a modified Huron-Vidal (MHV1) mixing rule was combined with a new group contribution G^E-model proposed by Yan et al. which applies to mixed solvent-electrolyte solutions. A new set of hydrate phase cell potential parameters is presented which enables accurate and thermodynamically consistent representation of hydrate dissociation pressures and structural behavior for the here concerned pure gases and gas mixtures, including structural transitions. Applicability over an extended temperature range was achieved by improving the temperature dependent behavior of the hydrate model which is of particular interest for hydrate formation in the presence of inhibitors.

13.2 Introduction

Gas hydrates are assigned to the family of inclusion or clathrate compounds. They exhibit regular crystalline lattice structures formed by hydrogen-bonded water (host) molecules. These structures contain various cavities of specific shape and size. The host lattice receives

---

[1] Universität Karlsruhe (TH), Institut für Technische Thermodynamik und Kältetechnik, Engler-Bunte-Ring 21, 76128 Karlsruhe.

13.2 Introduction

its thermodynamic stability from partial cavity occupation with small non-polar or weakly polar guest substances, e.g. light hydrocarbon gases like methane.

Operating temperature and pressure conditions, typically encountered in natural gas production and storage, range within the thermodynamic stability region of gas hydrates, i.e. ambient temperature and elevated pressures. Under these conditions, the contact of a potentially hydrate forming gas with a free aqueous liquid phase most likely results in the formation of solid gas hydrates, for example in multiphase offshore production streams or with condensate produced through Joule-Thomson expansion of water-vapor saturated gas streams. Production lines can be plugged by deposition of hydrate particles causing operational hazards and production losses. Therefore, prevention of hydrate formation has been an ever present topic in the gas industry since hydrates have been identified as a major problem in gas processing in the early 1930s (Hammerschmidt, 1934). Among different hydrate prevention strategies, the use of widely available alcohols like methanol or glycols as inhibiting agents has gained great importance in practical application over the last decades.

The underlying inhibition mechanism is a depression of equilibrium hydrate formation temperature similar to the freezing point depression of ice, the effect since being referred to as thermodynamic inhibition. Solute electrolytes, like seawater, show the same effect.

In recent years, gas hydrates have attracted increasing scientific and public attention since vast natural gas hydrate deposits have been discovered, particularly in the seafloor at the continental margins (Kvenvolden, 1994), being regarded as a potential future energy resource.

Thermodynamic hydrate stability conditions in the presence of inhibitors need to be calculated accurately for the design of inhibition utilities having large effect on operational, economical and safety issues of natural gas recovery and production (Sloan, 1994). Furthermore, reliable predictive tools are required in order to assess the depth to which natural gas hydrate deposits may be expected in the ocean (Zatsepina and Buffett, 1998).

A comprehensive treatise, covering the current state-of-the-art of hydrate related research, was published by Sloan (1998). In addition, his work includes a large compilation of experimental simple and mixed gas hydrate equilibrium data.

State-of-the-art hydrate phase models are commonly based on the statistical thermodynamics approach derived by van der Waals and Platteeuw (1959). Utilizing this approach, Parrish and Prausnitz (1972) developed an algorithm to predict hydrate phase equilibrium conditions of gas mixtures in pure water, suitable for engineering purposes. Since then, modifications of the Parrish and Prausnitz method have been adopted in most hydrate related studies, including this one. Even though sophisticated, this method keeps to be semi-empirical since model parameters have to be adjusted to experimental hydrate equilibrium data. Therefore, the need for reliable and consistent experimental data will be particularly emphasized in this work.

Initially, it was the main objective of the present study to extend a computational method, developed in a previous work (Nixdorf, 1996; Nixdorf and Oellrich, 1997), for incorporating the effect of inhibitors. Necessarily, any thermodynamically consistent hydrate phase equilibrium model accounting for the influence of inhibitors should equally represent hydrate equilibria with pure water as a limiting case. However, when using published (Nixdorf, 1996; Sloan, 1998), self-consistent model parameter sets, we experienced partly large discrepancies between experimental and predicted data, particularly at high pressures (>15 MPa) and for methane-rich gas mixtures in certain ranges of composition. Thus, we had to focus on improving predictions for gas hydrate equilibria with pure water under the above conditions, in the first place.

In the present study, we maintained the basic features of the original procedure of Parrish and Prausnitz. Modifications, as discussed in the following, mainly concern the strategy for obtaining model parameters. This strategy includes both the determination of fitting parameters and the selection of appropriate experimental information required to obtain specific model parameters. The new parameters are required to enable thermodynamically consistent predictions for pure and mixed gas hydrate equilibria over an extended temperature range with respect to the depression of equilibrium temperature in the presence of inhibitors. We have further used an equation of state based fugacity model to account for gas and liquid phase non-idealties, required for multiple phase equilibrium calculations.

Our efforts were focused on three-phase (V-L_W-H) hydrate equilibria from pure gases and binary gas mixtures containing methane, ethane, and propane, exclusively. Since these mixtures show all specific phenomena with respect to hydrate phase equilibrium, they are of particular theoretical interest. In addition, the above gases are the key components determining the hydrate formation behavior of most natural gases, which makes them proper model gas components.

Reviewing openly available literature sources revealed that there is a lack of experimental data for gas mixtures containing methane, ethane, or propane in the presence of aqueous single and mixed electrolyte solutions. Information on systems containing electrolyte and non-electrolyte mixed inhibitors is almost entirely missing. Some data sets for simple gas hydrates of methane, ethane, and propane in the presence of single salt solutions exhibit large scattering and seem to be inconsistent. Therefore, we supported our theoretical work with additional experimental investigations in order to extend the data base and to prove existing data for reliability. Macroscopic hydrate equilibrium data were measured for pure methane, ethane and propane gas in sodium chloride solutions and for binary methane-propane and methane-ethane gas mixtures in the presence of aqueous solutions with single and mixed electrolytes, methanol (MeOH), diethylene glycol (DEG), and MeOH-electrolyte. Furthermore, we examined the corresponding hydrate equilibria with pure water in order to obtain consistent data sets needed for hydrate model parameter determination.

13.3 Fundamental Hydrate Structural and Phase Behavior

In contrast to hydrated ions, where ion-water bonding has a covalent character, water and gas molecules in hydrate phases interact via van der Waals forces only. As a consequence, there is no defined hydration number in gas hydrates, meaning that cavities are occupied statistically with a variable fraction of cages remaining empty under varying temperature and pressure conditions. Therefore, gas hydrates are also referred to as non-stoichiometric compounds.

Crystallographic information on the most common cubic hydrate structures I and II was provided by v. Stackelberg (1949), v. Stackelberg and Müller (1954), McMullan and Jeffrey (1965), and Mak and McMullan (1965) from X-ray diffraction studies and was reviewed

13.3 Fundamental Hydrate Structural and Phase Behavior

by Jeffrey (1984). The existence and crystallographic characteristics of a hexagonal structure H was reported from Ripmeester et al. (1987). The occurence of hydrate structures I and II is typical for natural gas components, i.e. the low n- or i-alkanes, the inorganic gases nitrogen, carbon dioxide, hydrogen sulfide and the rare gases argon, krypton, and xenon. Structure H hydrates are generally formed from gas mixtures containing larger cyclic or branched hydrocarbons like methyl-cyclohexane or 2,2-dimethylbutane occuring in gas condensates or crude oils.

Table 13.1 lists some important structure characteristics for the hydrate structures I and II which have been considered in the present study. The cavities are illustrated in Figure 13.1.

Table 13.1: Structure characteristics for structure I and structure II hydrates (Sloan, 1998).

Hydrate structure	I		II	
Water molecules/unit cell	46		136	
Cavity	Small	Large	Small	Large
Notation	5^{12}	$5^{12}6^2$	5^{12}	$5^{12}6^4$
Coordination number	20	24	20	28
Mean cavity radius/Å	3.95	4.33	3.91	4.73
Number of cavities/unit cell	2	6	16	8
Number of cavities i/number of water molecules v_i	1/23	3/23	2/17	1/17

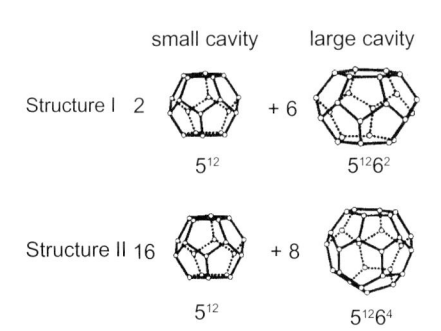

Figure 13.1: Hydrate cavities for structure I and structure II (Nixdorf, 1996).

Either structure is built of two polyhedral cavity types. A unit cell of structure I consists of two small cavities with twelve pentagonal faces (dodecahedron, notation: 5^{12}) and six large cavities with twelve pentagonal and two hexagonal faces (tetrakaidecahedron, $5^{12}6^2$). A unit cell of structure II consists of sixteen small cavities of type 5^{12} and eight large cavities with twelve pentagonal and four hexagonal faces (hexakaidecahedron, $5^{12}6^4$). The hydrate structure formed from a particular pure gas depends on the stability which can be provided to each specific cavity type and on the fractional number of either cavity type in a certain structure, determining its statistical weight. The statistical weight refers to the fact that there are three times more large cavities than small cavities in structure I, whereas in structure II there are two small cavities for each large cavity. The ratio of guest diameter (de-

293

13 Hydrate Equilibria in Aqueous Solutions Containing Inhibitors

termined from van der Waals volumes) to the accessible (free) cavity diameter is used to estimate whether a cavity can be stabilized by a certain guest molecule or not. To obtain the free cavity radii, the van der Waals radius of water (0.14 nm) is subtracted from the mean cavity radii given in Table 13.1. Guest-cavity diameter ratios were reported by Christiansen and Sloan (1994) and Sloan (1998) and are listed in Table 13.2 for the natural gas hydrate formers examined in this work.

Table 13.2: Guest-cavity diameter ratios [a] (Sloan, 1998).

Hydrate structure		I		II	
Cavity		Small	Large	Small	Large
Guest	Guest diameter/Å		Diameter ratio		
CH_4	4.36	**0.855**	**0.744**	0.868	0.655
C_2H_6	5.50	1.08	**0.939**	1.10	0.826
C_3H_8	6.28	1.23	1.07	1.25	**0.943**

a) Boldface numbers indicate cavities occupied from the pure gas.

It is assumed that values of approximately 0.70:<1 are required to provide sufficient stability to a certain cavity type. When the diameter ratio exceeds unity the guest molecule cannot enter into the cavity without lattice distortion. The numbers typed in boldface indicate the cavity type(s) occupied by a pure component.

Obviously, methane can stabilize both the large and the small cavities of structure I. Moreover, spectroscopic data from Sum et al. (1997) suggest that the large cavity of structure I is even better stabilized by methane than the small one. In structure II, methane gives less stability to the large cavities, compared with structure I. Considering the larger statistical weight of large cages in structure I, it follows that pure methane forms structure I hydrates. Ethane stabilizes the large cavities in either structure but forms structure I hydrates as a pure gas for statistical reasons. Pure propane forms structure II hydrates since it fits into the large ($5^{12}6^4$) cavity of structure II only.

For mixtures of hydrate formers, the structural behavior is generally influenced by gas composition, temperature, and pressure, respectively. Hendriks et al. (1996) concluded from theoretical considerations that structure II formation should be energetically most favorable, at least in certain ranges of gas composition, when both the small and the large cavities can simultaneously be stabilized by small and large mixture gas components. As an important result, matching closely experimental equilibrium data, they predicted an overall minimum of the Gibbs free energy for structure II formation from a ternary methane-ethane-water mixture in a concentration range between roughly 70 and 99 mol% of methane (on a water-free basis) at 280 K, even though the pure components form structure I. Most recently, the lower structure transition point was experimentally confirmed by spectroscopic results at about 72 mol% of methane and 274 K (Subramanian et al., 2000).

Similarly, a structure transition is inferred for methane-propane-water mixtures at low propane concentrations (Sloan, 1990a). Adding only 1 mol% propane to methane gas results in a sharp decrease of hydrate equilibrium pressure by 50% at 275 K compared with pure methane hydrate (Deaton and Frost, 1946). This marked concentration dependence of hydrate equilibrium pressure can be reasonably explained only with a change of hydrate structure from I for pure methane to structure II for the mixture caused by the strong affinity of the propane molecule to the large cavity of structure II. However, spectroscopic evidence for the structure transition has been missing until present, as well as macroscopic equilibrium data and hydrate phase coexistence points in the considered concentration range of less than 1 mol% of propane.

To compensate for this lack of information, three-phase (V-L_W-H) hydrate equilibria of binary methane-propane gas mixtures with pure water were examined in this work by varying the gas composition systematically in the range of low propane concentration.

Further peculiarities for mixed gas hydrates appear in terms of solid-vapor azeotropic phenomena. Local minima of hydrate equilibrium pressure as a function of binary gas composition were confirmed experimentally for methane-ethane (Holder and Grigoriou, 1980) and methane-propane gas mixtures (Thakore and Holder, 1987). Both, hydrate phase coexistence and azeotropic points represent stringent criteria to check for thermodynamic consistency of the hydrate model.

Fluid phase equilibria of hydrocarbon-water mixtures are strongly asymmetric due to different intermolecular forces ruling the pure component phase behavior, namely dispersion forces between non-polar hydrocarbon molecules and polar association (hydrogen-bonding) between water dipoles. Thus, in the mixture interactions between like species are energetically favored over unlike species. This effect is commonly referred to as hydrophobic interaction, the most striking consequence of which is a local structuring (clustering) within the condensed phases, in particular the formation of clathrate structures for low hydrocarbon-liquid water mixtures and the transition to high order, crystalline states at elevated pressures. The corresponding macroscopic phase behavior is outlined by the large discrepancy of normal boiling points of the pure components, the strictly limited mutual liquid phase solubility and, finally, the occurrence of solid hydrate phases at high pressures. Detailed analysis of the global phase behavior encountered in hydrate forming systems was presented by Sloan (1990b) and Harmens and Sloan (1990). A simplified, schematic (pressure, temperature) phase diagram for a (mixed) hydrocarbon-water hydrate forming system in a constant composition cross-section is given in Figure 13.2.

V denotes the vapour phase, L_W and L_{HC} the water-rich and the hydrocarbon-rich liquid phase, respectively. The solid phases are labelled with I for ice and H for any hydrate phase. According to Gibbs' phase rule, invariable states are represented by the triple-point T of water in the mixture and the quadruple-points Q_i where four phases are in equilibrium with each other. Three- and four phase equilibrium lines, where only one intensive variable can be changed independently, separate areas of two and three different stable phases, respectively.

Regarding an isobaric cooling process $A \rightarrow B$, the water dew line is crossed first. In the two phase (V-L_W) region the hydrocarbon vapor is saturated with water. When the three-phase (V-L_W-H) equilibrium line is reached, temperature remains constant until all liquid water or hydrocarbon vapor is consumed through hydrate formation, depending on which component is present in excess. If liquid water is the excess phase, an additional solid ice phase will be formed at low temperatures. Three-phase hydrate equilibrium lines expand from the first quadruple-point Q_1 where hydrate, ice, water-rich liquid and hydrocarbon-rich vapor are in equilibrium. No second quadruple-point is found when the hydrocarbon mixture is supercritical (supc). In contrast, when the hydrocarbon vapor-liquid phase envelope is in-

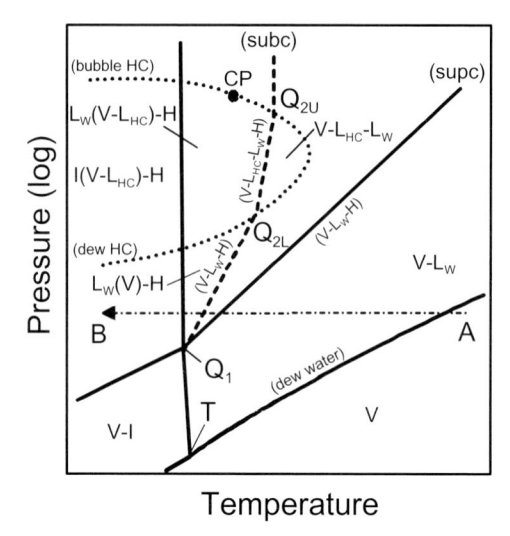

Figure 13.2: Schematic phase behavior of a hydrocarbon-water hydrate forming system in a constant composition pressure-temperature cross section.

tersected (subc), a second lower quadruple-point Q_{2L} is encountered on the hydrocarbon dew line (dew HC) and a second upper quadruple-point Q_{2U} on the bubble line or the retrograde dew line.

A four-phase hydrate equilibrium line (V-L$_{HC}$-L$_W$-H) spreads between the lower and upper second quadruple-points. For a single condensable hydrocarbon component, the four phase hydrate equilibrium line reduces to a single second quadruple-point on the vapor pressure curve of the hydrocarbon.

13.4 Experimental

13.4.1 Apparatus

A schematic of the experimental set-up used in the present work for hydrate equilibrium measurements is illustrated in Figure 13.3.

The core part is a high-pressure autoclave consisting of a 1000 ml stainless steel vessel designed for pressures up to 30 MPa and an outer pressure less sleeve used for autoclave temperature control. Temperature is set by a cryo-thermostat with a water-glycol mixture as heat-/cold carrier circulating through the outer sleeve. The thermostat was digitally controlled by a computer, thus we were able to run well defined temperature programs. A temperature range of 243 K $\leq T \leq$ 303 K was covered with a constancy of ± 0.01 K. We used a high performance magnetic stirrer with a stirrer speed between 250 and 1800 rpm to obtain

13.4 Experimental

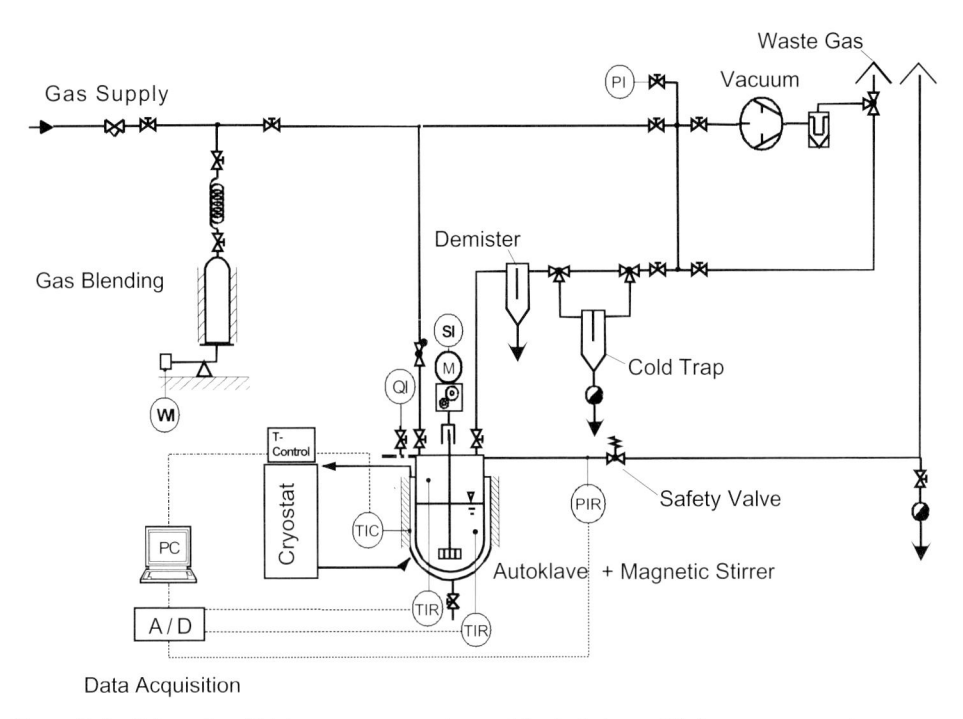

Figure 13.3: Schematic of high pressure apparatus used for hydrate equilibrium measurements.

intensive gas-liquid contact. Tremendous agitation and creation of large interfacial area is needed to overcome liquid phase metastability in the region of thermodynamic hydrate stability and to reduce heat and mass transfer resistances which limit the macroscopic rate of hydrate formation and dissociation. Temperature sensors were located at different axial positions within the autoclave volume and in the outer sleeve to obtain gas and liquid phase temperatures and for temperature control, respectively. The stirrer speed was chosen such that gas and liquid temperatures were balanced completely within the range of experimental accuracy and uniform cooling and heating was ensured throughout the experimental run. We deployed 100 Ω-platinum resistance thermometers (Philips Thermocoax) housed in a thin stainless steel shell which were fixed in the autoclave lid with pressuretight locks. Thermometers were calibrated according to ITS 90 against a normalized 25 Ω-platinum resistance thermometer (Rosemount Scientific).

An accuracy of ± 0.01 K was achieved for a single temperature measurement. The systems absolute pressure was measured with an electronic, piezoresistive pressure transmitter (Rosemount, model 3051TA) which was fed by a high precision 24 V direct current source. The pressure transmitter provided a linear 4–20 mA output signal in the operating span between 0 and 30 MPa which was transformed by a precision 500 Ω-resistance to a 2–10 V signal. By calibration the guaranteed absolute accuracy of 0.075% of the operating span (22.5 kPa) was improved to ± 10 kPa. All analogue signals were sampled by a multi-channel digital multimeter (PREMA, Model 5017SC) and digitally transformed with a resolution of 24 bit for computerized data acquisition and recording.

Aqueous solutions, examined in this work, were produced gravimetrically with an absolute accuracy of ±0.01 g at a total weight of 1000 g. We used deionized water with a residual conductivity of less than 1 μS/cm, salts with a certified purity of 99.5%, and methanol or diethylene glycol with a purity of 99.8%. Pure gases (99.95% purity) and gas mixtures were purchased from Messer-Griesheim (Germany) and were used as received, except for methane-propane gas mixtures with less than 1 mol% of propane which were produced gravimetrically in our institute by blending the pure components.

Gas compositions were determined with a Hewlett-Packard gas chromatographic system (HP6890) to an absolute accuracy of ±0.05 mol%. Before hydrate equilibrium measurements were performed, the apparatus was tested thoroughly. Functionality and reliability were proven over the entire temperature and pressure range of interest by measuring vapor-pressures of pure ethane and propane and simple hydrate equilibria of pure methane and comparison with reference data available in the literature.

13.4.2 Procedure

In preparing a hydrate equilibrium measurement, the autoclave was thoroughly cleaned with deionized water and completely dried under vacuum before a new aqueous solution was filled in. The autoclave was then charged with liquid to half volume and was evacuated to remove solute gases. Subsequently, the tubing and the autoclave were flushed with the examined gas (mixture) and evacuated in turn for several times before the autoclave was finally pressurized. To determine hydrate equilibrium data, we employed the isochoric (constant volume) method in this study. The result obtained from the isochoric method is a loop-like pressure-temperature trace as shown in Figure 13.4 for the example of a binary methane-ethane gas mixture in the presence of an aqueous methanol solution.

The loop exhibits characteristic sections which were described and interpreted on a microscopic scale by Christiansen and Sloan (1994). The sections are discussed below:

- *Start*: Gas-liquid equilibrium is adjusted outside the hydrate stability region. Labile, short-lived clusters are formed by dissolution of gas in water due to hydrophobic effects.

- *Cooling*: The mixture is rapidly ($\approx$ –10 K/h) cooled down into a metastable gas-liquid region whereby the cooling trace follows real p,V,T-behavior. In the liquid phase clusters agglomerate to more long-lived, but yet unstable, structures by ejecting some clustered water molecules to the disorderly surroundings and sharing cluster faces. Cluster agglomeration is attributed to hydrophobic attraction of non-polar molecules and the natural tendency to minimize the absolute solution excess entropy (Franks, 1975). Agglomerates are in a dynamic equilibrium with each other or with labile clusters, meaning that they can decompose or grow until their size exceeds a critical radius and stable nuclei are formed at a sufficiently low temperature. The period of primary nucleation is also referred to as induction period, since no macroscopic effects of nucleation can be observed in this period.

- *Hydrate formation*: Macroscopic hydrate growth can be detected by spontaneous, rapid pressure reduction since gas molecules are more densely packed in the hydrate phase than in the gas phase. An increase of liquid temperature may be observed for a short

13.4 Experimental

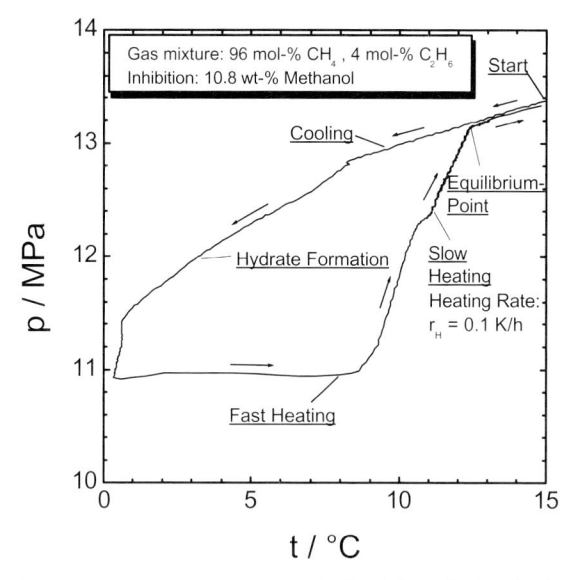

Figure 13.4: Characteristic pressure-temperature loop obtained from the isochoric procedure. Example: binary methane-ethane (4 mol%) gas mixture and 10.8 mass% methanol inhibition.

time due to the exothermic phase transition. The subcooling is the isobaric temperature difference between the initial hydrate formation temperature and the corresponding hydrate equilibrium temperature. The extent of agitation strongly affects the subcooling required to form enough critical nuclei which initiate macroscopic crystal growth. However, for comparable operating conditions, the molecular size and composition of hydrate formers will also have large effect on the required subcooling. Additionally, to our experience, there is an influence of liquid composition. Most surprisingly, we qualitatively found that electrolytes in certain ranges of ionic strength might even promote hydrate formation in the sense of less subcooling required. This might be explained by some local order imposed on the liquid phase by ionic hydration.

- *Fast Heating* – After a certain amount of hydrate was formed, the system was reheated fast ($\approx$ 10 K/h) to approach the hydrate stability border. The pressure remained nearly constant in this section in most cases.

- *Slow Heating* – Upon reaching the three-phase (V-L$_W$-H) equilibrium line, hydrate dissociation causes a sharp increase of pressure due to the release of enclathrated gas molecules into the gas phase. We then adjusted a static state at sufficient distance to the cooling trace. From that point the system temperature was automatically increased by computer control with rates between 0.03 and 0.2 K/h such that static states were obtained for each temperature step.

- *Equilibrium*: Dissociation of the last hydrate crystal causes the heating trace to change its slope discontinuously (break point) and rejoin the initial cooling trace. The break point on the heating trace, representing the final hydrate stability condition, was taken as equilibrium point and was assigned to the original gas composition since we chose the gas volume large enough to ensure that gas composition would not change notably through solubility effects. The break point pressure and temperature conditions were evaluated from the raw data by regressing data in linear regimes neighbored on both sides of the break point and intersecting the regressed lines. The total error assigned to an equilibrium point was conservatively estimated accounting for single measurement accuracy, statistical deviations of temperature and pressure and regression parameter confidence intervals, and was evaluated to ± 0.1 K and ± 25 kPa, respectively.

At the final hydrate stability point, heat and mass transfer resistances, which affect the dissociation rate of extended hydrate particles, should disappear for the hydrate phase volume tending to zero. However, at the particle surface there is an additional intrinsic resistance against hydrate dissociation, as was lately determined by Clarke and Bishnoi (2000). Thus, the break point coordinates might be dependent on the heating rate. In order to exclude this source of error from our results, we systematically examined the effect of heating rate for all mixtures considered in our work. For each mixture experiments were repeated on the highest pressure level with gradually decreased heating rates. For a complete experimental row we then chose the largest heating rate suitable for providing reproducable results. In most cases appropriate heating rates were 0.1 K/h.

For few mixtures we found that hydrate dissociation was obstructed in the presence of electrolytes and we had to reduce heating rates to 0.05 or 0.03 K/h.

13.5 Theory

13.5.1 Thermodynamic Framework

Our modeling work dealt with three-phase equilibria of systems containing a hydrocarbon-rich gas phase, a water-rich liquid phase and a solid hydrate phase (V-L$_W$-H). Besides thermal and mechanical equilibrium ($T^G = T^L = T^H$, $p^G = p^L = p^H$), the Gibbs free energy minimum condition ($dG = 0$) yields the mathematical relations for chemical equilibrium, here in terms of the iso-fugacity criterion

$$f_j^G = f_j^L = f_j^H \tag{13.1a}$$

where f_j denotes the fugacity of component j in the mixture and superscripts G, L, and H indicate the gas, liquid, or hydrate phase. We assumed that the presence of inhibitors be limited to the liquid phase, in particular the vapor fugacities of water and organic inhibitors were neglected due to their low vapor pressures at ambient temperature, compared with the absolute system pressure. Thus, the general Eqs. (13.1a) can be simplified to

$$f_j^G = f_j^L = f_j^H; \quad f_W^L = f_W^H \tag{13.1b}$$

with index j covering all gaseous species and W for water.

13.5 Theory

Setting the system temperature and the initial gas and liquid composition, Eqs. (13.1b) were solved iteratively for the incipient hydrate stability pressure. We performed calculations for both structure I and structure II formation in order to predict which hydrate structure is thermodynamically stable. Corresponding to an absolute minimum of the Gibbs free energy, the lower pressure obtained was taken as the equilibrium pressure. Liquid composition changes following gas dissolution were accounted for by isothermal and isobaric gas-liquid flash calculations preceding each pressure iteration step. The calculational procedures for obtaining the fugacities in the different phases are introduced below.

Solid Hydrate Phase

The fugacity of water f_W^H in a hydrate phase is obtained from Eq. (13.2):

$$f_W^H = f_W^\beta \exp\left\{-\frac{\Delta \mu_W^{\beta-H}}{RT}\right\} \tag{13.2}$$

where $\Delta \mu_W^{\beta-H}$ ($\Delta \mu_W^{\beta-H} = \mu_W^\beta - \mu_W^H$) is the chemical potential difference of water between the hydrate phase and a hypothetical β-modification which represents a new reference state characterizing the properties of a thermodynamically unstable empty hydrate lattice. Van der Waals and Platteeuw (1959) utilized statistical thermodynamics to derive $\Delta \mu_W^{\beta-H}$ from the statistical distribution of guest molecules in the host lattice regarding cavities as guest-host interaction sites and making the following basic assumptions:

- The cavities shall not be distorted by guest molecules. Thus, the lattice energy does not depend on cavity occupation.
- Cavities may be occupied by a single guest only.
- Cavities are occupied independently which implies that guest-guest interactions are neglected and the guest potential energy does not depend on whether neighbored cavities are occupied or not.
- Classical statistics are valid.

With the above assumptions the chemical potential difference of water in the hydrate phase $\Delta \mu_W^{\beta-H}$ is represented by Eq. (13.3)

$$-\frac{\Delta \mu_W^{\beta-H}}{RT} = \sum_i v_i \ln\left(1 - \sum_j \Theta_{ij}\right) \tag{13.3}$$

v_i denotes the number of cavities of type i per water molecule in the hydrate phase and Θ_{ij} is the fractional occupancy of cavity type i with guest molecules of type j and is defined by the ratio of the number N_{ij} of guest molecules j enclathrated in cavities i to the total number N_i of cavities i: $\Theta_{ij} = N_{ij}/N_i = N_{ij}/v_i N_W^H$.

Including the partial iso-fugacity criterion $f_j^G = f_j^H$, the fractional occupancy is expressed in terms of a generalized Langmuir sorption isotherm according to Eq. (13.4).

$$\Theta_{ij} = \frac{C_{ij}(T)f_j^G}{1 + \sum_j C_{ij}(T)f_j^G} \tag{13.4}$$

The temperature dependent Langmuir-coefficients C_{ij} are determined by the intermolecular interactions between enclathrated guest molecule j and the surrounding water molecules of cavitiy i and can be regarded as an appropriate measure for the occupied cavity's stability. When it is assumed that enclathrated molecules possess the same translatorial, rotational, and vibrational degrees of freedom as in the ideal gas state the Langmuir-coefficients can directly be related to molecular properties by the guest's molecular partition function. Van der Waals and Platteeuw (1959) simplified the real radial distribution of water molecules and treated cavities as ideally spherical. They further accounted for interactions with directly coordinated water molecules only and used the Lennard-Jones-Devonshire cell theory to obtain a spherically symmetric (isotropic) and smoothed cavity potential which allows the calculation of Langmuir-coefficients by Eq. (13.5).

$$C_{ij} = \frac{4\pi}{kT} \int_0^{R^f} \exp\left\{-\frac{w_{ij}(r)}{kT}\right\} r^2 dr \tag{13.5}$$

The integral of Eq. (13.5) represents the residual part of the guest's molecular partition function and $w_{ij}(r)$ is the spherical, smoothed cavity interaction potential between guest and water molecules at the radial distance r from the cavity center. The upper limit R^f is the free cavity radius accessible for the guest body center.

Parrish and Prausnitz (1972) used a different cavity potential which was derived by McKoy and Sinanoglu (1963). In their approach the cavity potential was based on the Kihara (spherical core) pair potential which gave better results for non-spherical guest molecules. The Kihara (spherical core) pair potential $\Gamma(r)$ is given by Eq. (13.6)

$$\Gamma(r) = \begin{cases} \infty, r \leq 2a \\ 4\varepsilon\left\{\left(\frac{\sigma}{r-2a}\right)^{12} - \left(\frac{\sigma}{r-2a}\right)^6\right\}, r > 2a \end{cases} \tag{13.6}$$

where $2a$ is the hard core diameter, ε is the characteristic energy and σ is the core to core distance at zero potential ($\Gamma = 0$). The collision diameter is given by $\sigma + 2a$ and r is the body center to center distance.

Summing over all binary guest-water interactions of the cell's first coordination shell and setting the hard core diameter of water equal zero, McKoy and Sinanoglu obtained an isotropic, smooth cavity potential in the following analytical form:

$$w_{ij}(r) = \begin{cases} \infty, r \geq R_i - a_j \\ 2z_i\varepsilon_j\left\{\frac{\sigma_j^{12}}{R_i^{11}r}\left(\delta_{ij}^{(10)} + \frac{a_j}{R_i}\delta_{ij}^{(11)}\right) - \frac{\sigma_j^6}{R_i^5 r}\left(\delta_{ij}^{(4)} + \frac{a_j}{R_i}\delta_{ij}^{(5)}\right)\right\}, r < R_i - a_j \end{cases}$$

$$\tag{13.7}$$

13.5 Theory

$$\delta_{ij}^{(N)} = \frac{1}{N}\left\{\left(1 - \frac{r}{R_i} - \frac{a_j}{R_i}\right)^{-N} - \left(1 + \frac{r}{R_i} - \frac{a_j}{R_i}\right)^{-N}\right\}, N = 10, 11, 4, 5$$

Molecular parameters σ, ε, and a in Eq. (13.7) have the same physical meaning as in the pair potential, however, σ and ε are mixture parameters characterizing the binary interaction between guest and water, whereas the core radius a is specific for each guest component. Information on the radial distribution of water molecules is included in the cell potential of Eq. (13.7) by means of the radius R_i and the coordination number z_i of the first coordination shell of cavity type i which are known from crystallographic data given in Table 13.1.

In this work, we used effective cell potential parameters, specific for each guest-cavity combination, which were calculated with the following classical cross coefficient combination rules:

$$\sigma_{ij} = \frac{(\sigma_j^G + \sigma_W)}{2}(1 - k_{ij}^\sigma), \quad \varepsilon_{ij} = \sqrt{\varepsilon_j^G \varepsilon_W}(1 - k_{ij}^\varepsilon), \quad a_j = a_j^G \tag{13.8}$$

σ_j^G, ε_j^G, and a_j^G are the Kihara (spherical core) parameters for the pure guest components j which were determined for the individual gases from low pressure viscosity and second virial coefficient data by Tee et al. (1966). Parameters σ_W and ε_W for water in the hydrate phase were obtained from Holder and John (1983). In their work, potential parameters for water were fitted to hydrate equilibrium data from mixtures containing small, spherical guest molecules like methane, argon, and krypton which satisfy the restrictions of the original van der Waals and Platteeuw model. However, these parameters cannot be used to represent thermophysical properties of pure water since the Kihara potential is not adequate for describing the strongly directed intermolecular forces between water molecules. In accordance with the assumptions of McKoy and Sinanoglu, the hard core radius of water a_W was set to zero.

We introduced cavity interaction parameters k_{ij}^σ and k_{ij}^ε which are specific for each guest-cavity combination to account for cell potential asymmetries caused by the real guest molecule and cavity asphericity. Interaction parameters for σ and ε were fitted to experimental hydrate equilibrium data of pure and mixed gas hydrates.

No correction was needed for parameter a since predictions are much more sensitive to σ and ε than to a.

Finally, to calculate the absolute value of the hydrate phase fugacity of water f_W^H we need the fugacity of water in its reference state f_W^β, the hypothetical empty lattice. f_W^β is related to the fugacity of pure liquid water f_{oW}^L at system temperature and pressure by Eq. (13.9):

$$f_W^\beta = f_{oW}^L \exp\left\{\frac{\Delta\mu_W^{\beta-L}}{RT}\right\} \tag{13.9}$$

$\Delta\mu_W^{\beta-L}$ is the chemical potential difference of water between the pure liquid state L and the reference state β and was calculated with a classical thermodynamics correlation (Eq. (13.10)) following a suggestion of Holder et al. (1988):

$$\frac{\Delta\mu_W^{\beta-L}(T,p)}{RT} = \frac{\Delta\mu_{W,0}^{\beta-L}(T_0,p_0)}{RT_0} - \int_{T_0}^{T}\frac{\Delta h_W^{\beta-L}(T)}{RT^2}dT + \int_{p_0}^{p}\frac{\Delta v_{W,0}^{\beta-L}}{RT}dp$$

(13.10)

$$\Delta h_W^{\beta-L}(T) = \Delta h_{W,0}^{\beta-\alpha} - \Delta h_{W,0}^{L-\alpha} + \int_{T_0}^{T}\left(\Delta c_{p,W,0}^{\beta-L} + d(T-T_0)\right)dT, \quad \Delta v_{W,0}^{\beta-L} = \Delta v_{W,0}^{\beta-\alpha} - \Delta v_{W,0}^{L-\alpha}$$

$\Delta h_{W,0}^{\beta-\alpha}$ and $\Delta v_{W,0}^{\beta-\alpha}$ are the differences of molar enthalpy and volume between the β-modification and ice (α) at reference temperature and pressure ($T_0 = 273.15$ K, $p_0 = 0$ Pa). $\Delta h_{W,0}^{L-\alpha}$ and $\Delta v_{W,0}^{L-\alpha}$ are the molar enthalpy and volume differences of fusion. Tables 13.3 and 13.4 list all Kihara potential parameters and thermodynamic parameters used in this work.

Table 13.3: Kihara potential parameters used in Eq. (13.8).

	$2a^G$ in Å (10^{-10} m)	σ^G in Å	$(\varepsilon/k)^G$ in K
CH_4 a)	0.7667	2.7383	232.20
C_2H_6 a)	1.1302	2.8468	425.32
C_3H_8 a)	1.3003	3.2187	493.71
Water b)	$a_W = 0$ Å	$\sigma_W = 3.5644$ Å	$(\varepsilon/k)_W = 102.13$ K

a) Tee et al. (1966), b) Holder and John (1983)

Table 13.4: Thermodynamic parameters used in Eq. (13.10).

		Structure I	Structure II
$\Delta\mu_{W,0}^{\beta-L}$ a)	in J/mol	1297	937
$\Delta h_{W,0}^{\beta-\alpha}$ a)	in J/mol	1389	1025
$\Delta h_{W,0}^{L-\alpha}$ b)	in J/mol	6008	6008
$\Delta c_{p,W,0}^{\beta-L}$ c)	in J/(mol K)	−38.12	−38.12
d c)	in J/(mol K^2)	0.141	0.141
$\Delta v_{W,0}^{\beta-\alpha}$ c)	in cm^3/mol	3.0	3.4
$\Delta v_{W,0}^{L-\alpha}$ c)	in cm^3/mol	−1.6	−1.6

a) Dharmawardhana et al. (1980), b) Nixdorf (1996), c) Sloan (1998)

Fluid Phases

We used an equation of state based model to calculate the gas and liquid phase fugacities $f_j^G, f_j^L,$ and f_W^L with

13.5 Theory

$$f_j^G = \tilde{y}_j \, \varphi_j^G \, p, \quad f_j^L = \tilde{x}_j \, \varphi_j^L \, p, \quad f_W^L = \tilde{x}_W \, \varphi_W^L \, p \qquad (13.11)$$

$\tilde{y}, \tilde{x}$ are the mole-fractions and φ^G, φ^L are the fugacity coefficients of the gas and the liquid phase and p is the system pressure. Fugacity coefficients were calculated with a two parameter cubic equation of state in the form proposed by Soave (SRK-EOS) (Soave, 1972):

$$p = \frac{RT}{v-b} - \frac{a(T)}{v(v+b)} \qquad (13.12)$$

with the van der Waals volume b and the temperature dependent attraction parameter a.

From Eq. (13.12) the fugacity coefficient of a component k in the mixture becomes

$$\ln \varphi_k = \left(\frac{b_k'}{b_{mix}}\right)(\zeta_{mix} - 1) - \ln(\zeta_{mix} - B_{mix}) + \Lambda \alpha_k'; \quad B_{mix} = \frac{b_{mix}\, p}{RT}$$

with
$$\qquad (13.13)$$

$$b_k' = \left.\frac{\partial(n b_{mix})}{\partial n_k}\right|_{T,p,n_{l \neq k}}, \quad \alpha_{mix} = \frac{a_{mix}}{b_{mix}\, RT}, \quad \alpha_k' = \left.\frac{\partial(n \alpha_{mix})}{\partial n_k}\right|_{T,p,n_{l \neq k}}, \quad \Lambda = \ln\left(\frac{\zeta_{mix}}{\zeta_{mix} + B_{mix}}\right)$$

with the mixture compressibility factor $\zeta_{mix} = p v_{mix}/RT$ and the energy parameter α. n is the total number of moles in the mixture and n_k is the number of moles of component k. The mixture volume parameter b_{mix} is obtained from a linear mixing rule in the simplest form (Eq. (13.14)) where b_{ol} is the volume parameter of the pure component l. For the gas phase $\tilde{x}$ has to be replaced by $\tilde{y}$, accordingly.

$$b_{mix} = \sum_l \tilde{x}_l \, b_{ol} \qquad (13.14)$$

The mixture energy parameter α_{mix} and attraction parameter a_{mix} were determined differently for the gas and the liquid phase. For the gas phase containing low hydrocarbons only, we used a classical quadratic mixing rule

$$a_{mix} = \sum_l \sum_m \tilde{y}_l \tilde{y}_m a_{lm}, \quad a_{lm} = \sqrt{a_{ol}\, a_{om}}\,(1 - k_{lm}) \qquad (13.15)$$

with binary interaction parameters k_{lm} which were obtained from Knapp et al. (1982).

For the liquid phase containing strongly polar solvents, ionic species, and solute gases we employed a first order approximation of the modified Huron-Vidal-mixing rule (MHV1) (Huron and Vidal, 1979; Michelsen, 1990) as proposed by Holderbaum and Gmehling (1991) in their PSRK-method.

$$\alpha_{mix} = \frac{1}{\Lambda_0}\left[\frac{\bar{G}_0^E}{RT} + \sum_l \tilde{x}_l \ln\left(\frac{b_{mix}}{b_{ol}}\right)\right] + \sum_l \tilde{x}_l \, \alpha_{ol}; \quad \Lambda_0 = -0.64663 \qquad (13.16)$$

Equation (13.16) links the equation of state energy parameter α with the molar Gibbs excess free energy $\bar{G}_0^E$ of the fluid mixture at atmospheric pressure and constant fluid density. The mixing rule (Eq. (13.16)) can be combined with any $\bar{G}_0^E$-model valid for the liquid state at atmospheric pressure, like UNIQUAC or UNIFAC, in order to account for the strongly non-ideal fluid phase mixing effects.

The calculational procedures for obtaining the SRK-EOS pure component parameters b_{ol} and a_{ol} for polar and non-polar components were discussed in detail by Holderbaum (1991). Solute electrolyte species were incorporated in the equation of state formulation with a procedure proposed by Zuo and Guo (1991).

The molar Gibbs excess free energy $\bar{G}_0^E$, introduced in the mixing rule of Eq. (13.16) to account for the strongly non-ideal mixing effects in the solvent-electrolyte mixture, was modeled according to a suggestion from Yan et al. (1999). In their work, $\bar{G}_0^E$ was composed from additive contributions of different interaction mechanisms:

$$\bar{G}_0^E = \bar{G}_0^{E,LR} + \bar{G}_0^{E,MR} + \bar{G}_0^{E,SR} \tag{13.17}$$

$\bar{G}_0^{E,LR}$ accounts for the long range (LR), electrostatic interactions, the second $\bar{G}_0^{E,MR}$ term represents all middle range (MR) contributions from charge-dipole and charge-induced dipole interactions between solvents and ionic species and between cations and anions, and the short range (SR) term $\bar{G}_0^{E,SR}$ was modelled with the UNIFAC group contribution method (Fredenslund et al., 1977; Hansen et al., 1991). We further used the extensions to the UNIFAC interaction parameter matrix which were introduced by Gmehling et al. (1997) for the prediction of gas solubilities with the PSRK-method. All gas-ion interaction parameters, which were not available so far, were arbitrarily set to zero. This approximation seemed to be reasonable since it can be assumed that ionic species are directly surrounded by polar solvents preferably.

13.5.2 Estimation of Hydrate Cavity Interaction Parameters

Cavity interaction parameters k_{ij}^σ and k_{ij}^ε, introduced for obtaining effective Kihara potential parameters σ_{ij} and ε_{ij}, were adjusted to experimental hydrate equilibrium data by minimizing the following objective function OF with the number N of experimental data points:

$$OF = \sum_{n=1}^{N} WF_n \left(\frac{\Delta \mu_{W,n}^{\beta-L} - RT_{\exp,n} \ln a_{W,n} - \Delta \mu_{W,n}^{\beta-H}(k_{ij}^\sigma, k_{ij}^\varepsilon)}{RT_{\exp,n}} \right)^2 \tag{13.18}$$

We used the chemical potential difference of water between the liquid and the hydrate phase which corresponds to a direct minimization of the system Gibbs free energy. WF is a weighting factor accounting for the standard deviations of the chemical potential differences with the experimental uncertainties of temperature, pressure, and gas and liquid composition. The activity of water in the liquid phase a_W was incorporated in the objective function since our fitting data base embraced systems containing inhibitors in order to extend the experimentally covered temperature range. However, for parameter estimation, the only inhibitor considered was sodium chloride because accurate reference data and correlations of the osmotic coefficient of aqueous sodium chloride solutions are available over a large range of temperature, pressure, and ionic strength. In this work, we used an expression presented by

13.5 Theory

Pitzer et al. (1984) which also includes the complete temperature and pressure dependencies of $\ln a_W$.

Gas solubility was neglected in the fitting procedure since we chose equilibrium points with dissociation pressures lower than 10 MPa only where hydrocarbon liquid mole fractions have no significance concerning the calculation of the water activity. Table 13.5 lists the complete set of cavity interaction parameters determined in this work together with the corresponding experimental data sources.

Table 13.5: Cavity interaction parameters [a].

	Structure I Small cavity k^σ	k^ε	Large cavity k^σ	k^ε	Structure II Small cavity k^σ	k^ε	Large cavity k^σ	k^ε
CH_4	−0.017	−0.021	−0.059	0.050	0.030	0.040	−0.030	0.079
	Data base A				**Data base B**			
C_2H_6			−0.080	0.050			−0.179	0.165
	Data base C				**Data base D**			
C_3H_8							−0.088	0.129
					Data base E			

Data base A: pure methane; Nixdorf (1996) (pure water), this work (TW) (pure water; 15 mass% NaCl), de Roo et al. (1983) (11.7; 17.1; 21.5; 24.1 mass% NaCl)

Data base B: methane-propane; pure water – Deaton and Frost (1946) (4.8; 11.7 mol% propane); Nixdorf (1996) (2.93 mol% propane); TW (2.42; 4.86 mol% propane); 10 mass% NaCl – TW (2.46; 4.86 mol% propane)

Data base C: pure ethane; pure water – Roberts et al. (1940); Deaton and Frost (1946); Reamer et al. (1952); Galloway et al. (1970); Holder and Grigoriou (1980); Holder and Hand (1982); Avlonitis (1988); Nixdorf (1996); 20 mass% NaCl – Tohidi et al. (1993); Englezos and Bishnoi (1991); TW

Data base D: methane-ethane-propane; pure water – Nixdorf (1996) (12.55 mol% ethane/2.93 mol% propane; 3.98 mol% ethane/1.0 mol% propane)

Data base E: pure propane; pure water – Miller and Strong (1946); Deaton and Frost (1946); Reamer et al. (1952); Robinson and Mehta (1971); Verma (1974); Kubota (1984); Thakore and Holder (1987); Patil (1987); Nixdorf (1996); 15 mass% NaCl – TW

a) Our cavity interaction parameter matrix is subject to a continuous evolutionary process. Kindly address to the corresponding author for currently available parameters of further hydrate formers.

As far as possible, parameters were adjusted to equilibrium data of pure gas hydrates in pure water and in aqueous sodium chloride solutions (methane SI, ethane SI, and propane SII). Structure II parameters of methane were fitted to data of binary methane-propane gas mixtures with pure water and sodium chloride solutions. Finally, structure II parameters of ethane were adjusted to hydrate equilibrium data of ternary methane-ethane-propane gas mixtures which contained sufficient propane so that structure II formation could safely be assumed.

13.6 Results and Discussion

In this section, experimental hydrate dissociation pressures obtained in this study and literature data for pure and mixed gas hydrates in pure water and in the presence of inhibitors are being compared with calculated results which are based on the new cell potential parameters presented in the previous section. In our fitting procedure pure gas hydrate data were used primarily, only structure II parameters for methane and ethane were adjusted to mixed gas hydrate data since pure methane and ethane form structure I under normal conditions. Therefore, the experimental and theoretical results for pure gas hydrates will be discussed first.

Figure 13.5 displays the experimental data and calculations for structure I formation of pure methane gas in pure water and sodium chloride solutions. We should point out that for parameter adjustment only data points below 10 MPa were used. For predictions over the entire range of pressure, temperature, and salt concentration a different liquid phase model was used (as discussed in Section 13.5.1) accounting for the influence of gas solubility. The calculations agree well with the experimental data which are predicted within 3% of relative pressure deviation.

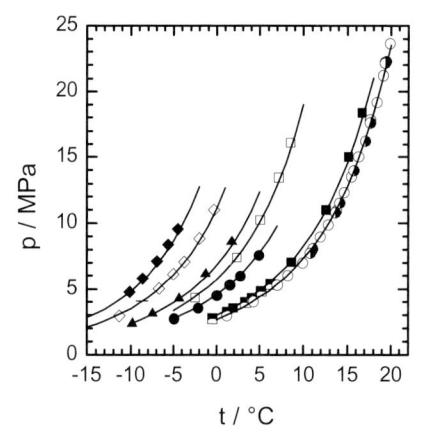

Figure 13.5: Experimental and predicted hydrate dissociation pressures vs. temperature for methane gas in pure water and sodium chloride solutions. Nixdorf (1996), ○ pure water; this work (TW), ◑ pure water, ■ 3 mass% NaCl, □ 15 mass% NaCl; Dholabhai et al. (1991), ◩ 3 mass% NaCl; de Roo et al. (1983), ● 11.7 mass% NaCl, ▲ 17.1 mass% NaCl, ◇ 21.5 mass% NaCl, ◆ 24.1 mass% NaCl; — calculated.

The results for three-phase (V-L$_W$-H) equilibria of pure ethane hydrates in pure water and in a 20 mass% sodium chloride solution are shown in Figure 13.6. The overall predictive accuracy of our model for pure ethane hydrates corresponds to the one for pure methane hydrates. However, it has to be noted that the absolute pressure deviations are much less, in the range of 10 to 30 kPa, due to the significantly lower absolute dissociation pressures.

Three-phase (V-L$_W$-H) SII-hydrate equilibrium data and calculations for pure propane gas in pure water and in the presence of various sodium chloride solutions are given in Figure 13.7. It can be seen that the experimental literature data for different sodium chloride so-

13.6 Results and Discussion

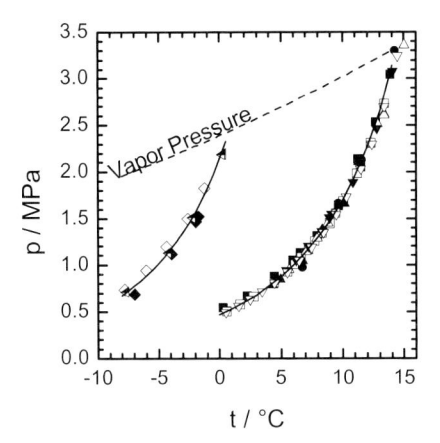

Figure 13.6: Experimental and predicted hydrate dissociation pressures vs. temperature for ethane gas in pure water and in a 20 wt.% sodium chloride solution. Pure water – ■ Roberts et al. (1940); ☐ Deaton and Frost (1946); ● Reamer et al. (1952); ○ Galloway et al. (1970); ▲ Holder and Grigoriou (1980); △ Holder and Hand (1982); ▼ Avlonitis (1988); ▽ Nixdorf (1996); 20 mass% NaCl – ◆ Tohidi et al. (1993); ◇ Englezos and Bishnoi (1991); ◄ this work; — calculated; -- vapor pressure pure ethane.

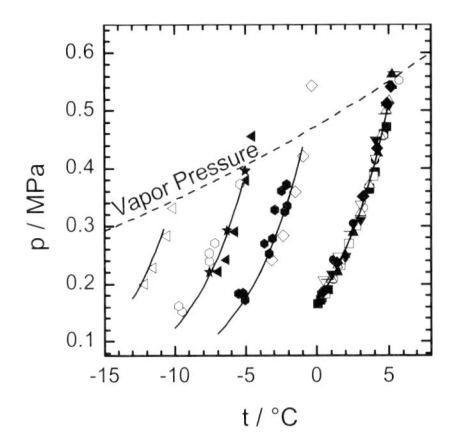

Figure 13.7: Experimental and predicted hydrate dissociation pressures vs. temperature for propane gas in pure water and sodium chloride solutions. Pure water – ■ Miller and Strong (1946); ☐ Deaton and Frost (1946); ● Reamer et al. (1952); ○ Robinson and Mehta (1971); ▲ Verma (1974); △ Kubota et al. (1984); ▼ Thakore and Holder (1987); ▽ Patil (1987); ◆ Nixdorf (1996); 10 mass% NaCl – ◇ Tohidi et al. (1993); ● Bishnoi and Dholabhai (1993); 15 mass% NaCl – ◄ Tohidi et al. (1993); ○ Bishnoi and Dholabhai (1993); ★ this work; 20 mass% NaCl – ◁ Tohidi et al. (1993); — calculated; -- vapor pressure pure propane.

lutions exhibit substantial scattering, even though we used only the most recent literature sources available. For 10 and 15 mass% sodium chloride solutions, the data of Tohidi et al. (1991) and Bishnoi and Dholabhai (1993) show a discrepancy of 1 to 2 K at constant pressure. This discrepancy allows no definite determination of Kihara potential parameters due to the strong temperature dependence of the Langmuir coefficient. Therefore, it was necessary to check the reliability of the existing data by own measurements. Our own data for a 15 mass% sodium chloride solution and the literature data for propane hydrates in pure water are described accurately by our model with a mean relative pressure deviation of 2%. Gas solubility has no effect on the predicted hydrate dissociation pressures in the considered pressure range. It is interesting to note that our predictions for a 10 mass% sodium chloride solution agree well with the results from Bishnoi and Dholabhai (1993) which were obtained by slowly changing the system temperature (referred to as procedure 2 in the original literature), similar to our procedure. They further used an isothermal procedure (referred to as procedure 1) which resulted in higher dissociation pressures.

Mixed gas hydrate equilibrium data for binary methane-propane mixtures in pure water were obtained in this work for propane mole-fractions between 0.2 and 5%. In Figure 13.8a, the experimental and predicted hydrate dissociation pressures are plotted vs. temperature for each constant composition. By observing a discontinuous change of slope in the slow heating regime of the corresponding experimental loop (Fig. 13.4), already below the final hydrate stability point, a change of hydrate structure was directly confirmed for gas mixtures with 0.24 and 0.37 mol% propane, even without spectroscopic information. For increasing temperature, the hydrate structure changes from structure II to structure I between 15 and 16 °C for 0.24 mol% and between 17 and 18 °C for 0.37 mol% propane. Our model predicts the transition between 14 and 15 °C and 17 and 18 °C, respectively. For 0.24 mol% propane, the hydrate dissociation pressure is overestimated by approximately 10% in the structure II regime, in the structure I regime deviations are less than 1%. It should be considered, however, that the relative pressure deviation of 10% can easily be explained with the experimental concentration error of ±0.05 mol% which is large compared with the absolute propane mole-fraction. Hydrate dissociation pressures of the other mixtures examined are represented by our model with a mean relative deviation of 2.5% and a

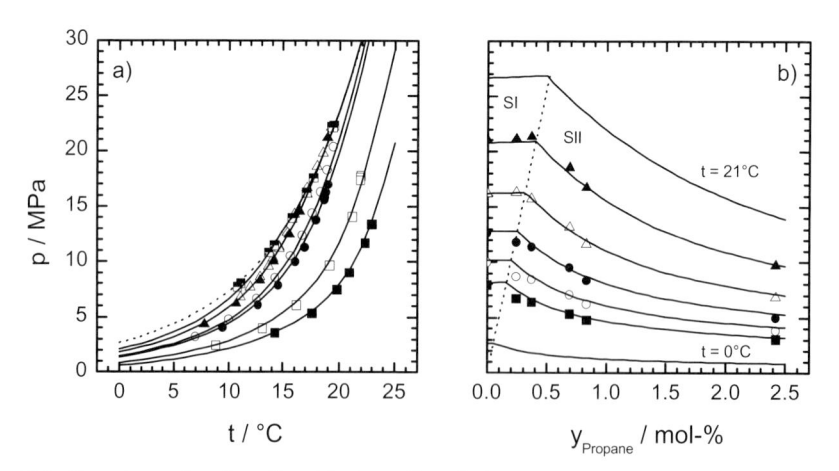

Figure 13.8: Experimental and predicted hydrate dissociation pressures vs. temperature a) and (water-free) gas composition b) for binary methane-propane gas mixtures in pure water. Experimental data obtained in this work. a) (■ 0 mol% propane; △ 0.24 mol%; ▲ 0.37 mol%; ○ 0.69 mol%; ● 0.83 mol%; □ 2.42 mol%; ■ 4.86 mol%; — calculated methane-propane; ··· calculated pure methane.
b) Smoothed data; ■ t = 11°C; ○ 13°C; ● 15°C; △ 17°C; ▲ 19°C; ··· SI-SII phase coexistence; — calculated methane-propane.

13.6 Results and Discussion

maximum deviation of 6.5%. The dependence of hydrate dissociation pressure on gas composition is given in Figure 13.8b where isothermal data sets are plotted versus propane mole-fraction.

The dotted line in Figure 13.8b represents the geometric locus of structure II and structure I hydrate phase coexistence. The transition pressure depends on gas composition and sharply increases with rising propane concentration. The isothermal data sets were obtained by properly regressing the original data given in Figure 13.8a. In the structure I region, hydrate dissociation pressures slightly increase, compared with pure methane hydrates, at constant temperature and for rising propane concentration, since propane cannot contribute to the stabilization of structure I cavities. Beyond the transition point, the hydrate dissociation pressure drops markedly with increasing propane concentration which can be attributed to the strong affinity of the propane molecule to the large cavity of structure II. It can be observed (and predicted) that the concentration dependence of the hydrate dissociation pressure becomes stronger for higher temperatures and gets weaker for increasing propane concentration.

Even though the range of propane concentration, which was experimentally covered in this work, was limited to low propane mole-fractions, our new potential parameter set applies well to the entire concentration range, which can be seen from Figure 13.9. The solid-vapor azeotropic behavior which was investigated by Thakore and Holder (1987) is accurately represented by our model, even though these data were not used for parameter adjustment. The dotted line represents the composition of the hydrate formers in the hydrate phase on a water-free basis. One can note that propane is enriched in the hydrate phase by a factor of 20 to 30 in the range of low propane concentration which offers an interesting, theoretical prospective for hydrate applications in gas separation and conditioning.

For methane-propane gas mixtures, the effect of electrolyte inhibitors was examined for gas compositions with 2.2 to 2.5 mol% propane (Fig. 13.10a) and 4.86 mol% propane (Fig. 13.10b).

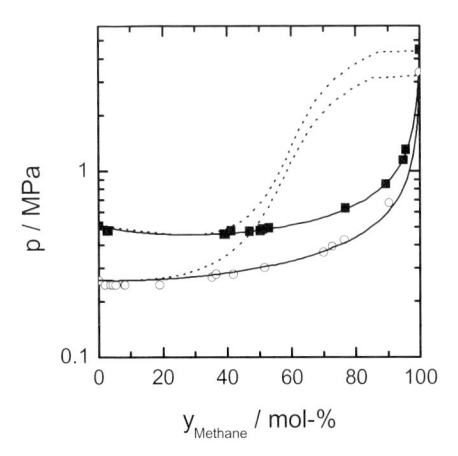

Figure 13.9: Solid-vapor azeotropic behavior in methane-propane-water systems. Minima of hydrate dissociation pressures vs. (water-free) gas composition. Experimental data from Thakore and Holder (1987); ■ $t = 5°C$; ○ $2°C$; — calculated (water-free gas composition); ··· calculated (water-free hydrate composition).

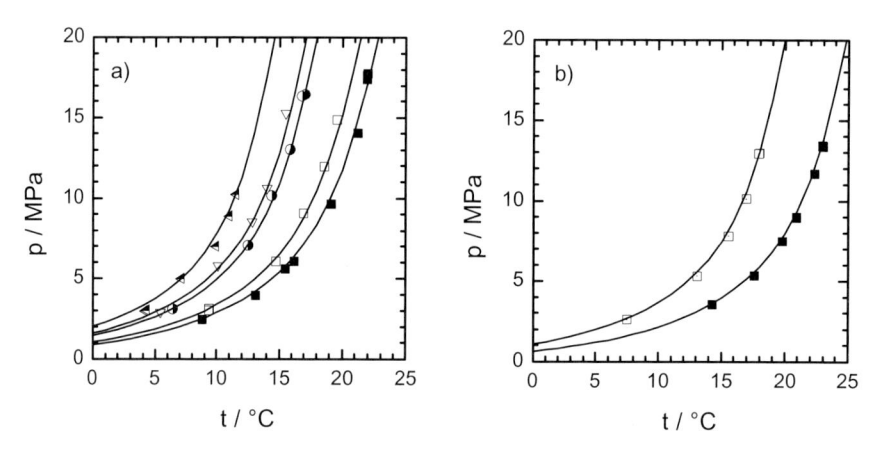

Figure 13.10: Experimental and predicted hydrate dissociation pressures vs. temperature for binary methane-propane gas mixtures in aqueous solutions containing electrolyte inhibitors. Experimental data obtained in this work. a) Methane-propane gas mixtures with 2.2÷2.5 mol% propane; ■ pure water; □ 3 mass% NaCl; ◐ 10 mass% NaCl; ◀ 15 mass% NaCl; ▽ 12.33 mass% CaCl$_2$; — calculated. b) Methane-propane gas mixture with 4.86 mol% propane; ■ pure water; ⊡ 10 mass% NaCl; — calculated.

The predictive accuracy for the mixtures considered in Figure 13.10 agrees well with our results for salt-free systems, except for two data points (15 mass% NaCl; Fig. 13.10a) where the relative pressure deviation exceeded 10%.

Figure 13.11 depicts our experimental and calculated results obtained for hydrate formation from methane-ethane gas mixtures (a) in pure water and (b) in the presence of electrolyte inhibitors.

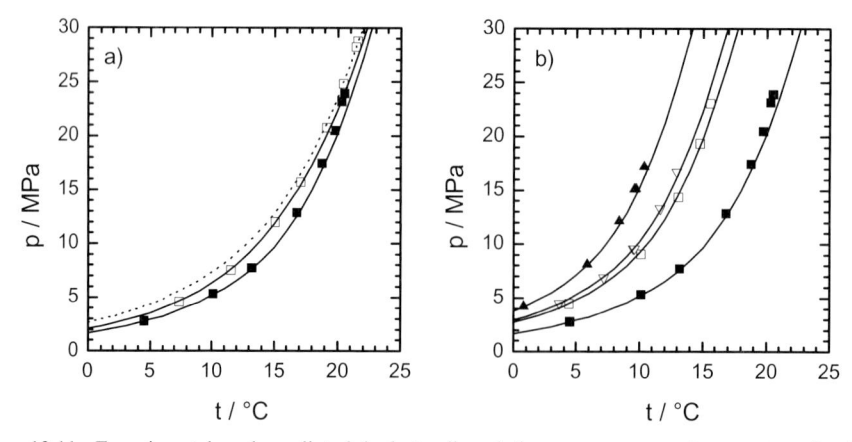

Figure 13.11: Experimental and predicted hydrate dissociation pressures vs. temperature for binary methane-ethane gas mixtures in pure water a) and in the presence of electrolyte inhibitors b). Experimental data obtained in this work. a) ■ 4.02 mol% ethane; □ 1.97 mol% ethane; — calculated; ··· calculated pure methane. b) 4.02 mol% ethane; ■ pure water; □ 10 mass% NaCl; ▲ 15 mass% NaCl; ▽ 5 mass% NaCl/6.89 mass% CaCl$_2$; — calculated.

13.6 Results and Discussion

For the gas mixture containing 1.97 mol% ethane and pure water, a structure transition from structure II at lower temperatures to structure I at higher temperatures is predicted between 15 and 16 °C whilst the macroscopic dissociation pressures are predicted precisely with maximum relative deviations of 5%. Compared with pure methane (structure I) hydrate formation, dissociation pressures are lower for methane-ethane mixtures because ethane provides higher stability to the large cage of structure I than methane. In contrast to methane-propane mixtures, the structure transition is smooth, in this case, since ethane can stabilize the large cages in either structure. Thus, a clear macroscopic observation of structure change was difficult for methane-ethane gas mixtures, and further spectroscopic evidence is desirable. Similar overall predictive accuracy was achieved for a mixture containning 4.02 mol% ethane for which structure II formation was predicted for temperatures lower than 24 °C. However, our experimental data suggest a structure change at approximately 21 °C when compared with calculated structure I dissociation pressures. Accordingly, the predicted (structure II) equilibrium pressures are too low by about 8% in the region of inferred structure change. The latter gas mixture was used to form hydrates in the presence of single and mixed electrolyte inhibitors. Structure II formation was predicted for all data points shown in Figure 13.11b, and experimental dissociation pressures were represented accurately by our model. It is worth mentioning that, along with the depression of formation temperature, the predicted structure transition point shifts to lower temperatures and higher pressures with increasing inhibitor mass-fraction.

Hydrate dissociation pressures for a methane-propane gas mixture with 2.46 mol% propane in the presence of different aqueous solutions containing methanol and mixed methanol-electrolyte inhibitors are shown in Figure 13.12. The inhibitor mass-fractions were chosen such that the water liquid mole-fractions were almost identical for all solutions examined ($\tilde{x}_W \cong 0.94$) corresponding to a 10 mass% sodium chloride solution and presumed total dissociation of electrolyte species. Obviously, these solutions were dilute enough so

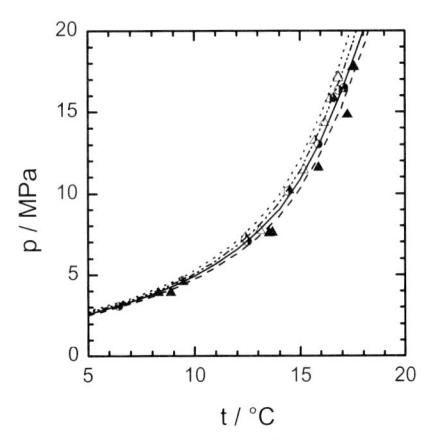

Figure 13.12: Experimental and predicted hydrate dissociation pressures vs. temperature for a binary methane-propane gas mixture in different aqueous solutions containing electrolyte and alcoholic inhibitors with comparable water mole fraction. Experimental data obtained in this work. Methane-propane gas mixture with 2.46 mol% propane; (exp ●/calc —) 10 mass% NaCl; (▲/--) 10.85 mass% MeOH; (△/---) 5 mass% NaCl/5.44 mass% MeOH; (▲/-----) 6.17 mass% CaCl$_2$/5.44 mass% MeOH.

that the nature of inhibitor had only minor effect on the experimental results and the predicted formation temperature depression. For methanol inhibition, the formation temperature depression is predicted slightly smaller, and for mixed methanol-electrolyte inhibition, slightly larger compared with a sodium chloride solution of equal total solute mole-fraction, which agrees with our experimental results. With maximum relative pressure deviations of less than 10%, for the methanol solution, the predictive accuracy of our model is still satisfactory in the examined range of organic inhibitor mole-fraction. Similar results were obtained for methanol and mixed methanol-electrolyte inhibition of methane-ethane hydrate formation with 4.02 mol% ethane (Fig. 13.13).

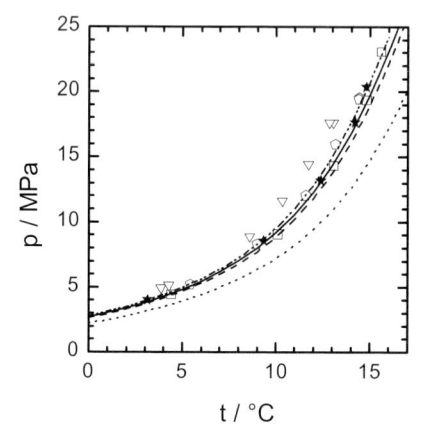

Figure 13.13: Experimental and predicted hydrate dissociation pressures vs. temperature for a binary methane-ethane gas mixture in different aqueous solutions containing electrolyte and organic inhibitors with comparable water mole fraction. Experimental data obtained in this work. Methane-ethane gas mixture with 4.02 mol% ethane; (exp □/calc —) 10 mass% NaCl; (★/--) 10.86 mass% MeOH; (○/---) 5 mass% NaCl/5.43 mass% MeOH, (▽/-----) 28.73 mass% DEG.

For the latter gas mixture, we examined the inhibiting effect of diethylene glycol (DEG), additionally. As in the aqueous methanol solutions, the chosen DEG mass-fraction corresponded to a mole-fraction of 0.06. As can be seen from Figure 13.13, DEG was the most effective of the here addressed inhibitors producing the largest formation temperature depression on the basis of equal total inhibitor mole-fractions. However, using available VLE UNIFAC group interaction parameters, the liquid phase model was not capable of representing the DEG-results even qualitatively, revealing its limits of applicability.

13.7 Conclusions

The main objective of our modeling efforts was an improved mathematical description of water fugacity in the solid hydrate phases which forms the backbone for all multiple phase equilibrium calculations in hydrate forming systems. The new set of cell potential parameters for methane, ethane, and propane gas, enclathrated in the cavities of hydrate structures I and II, is the result of evaluating reliable literature data and supplementary own consistent data of hydrate dissociation pressures for pure and mixed gas hydrates. Special attention was paid to specific hydrate phase effects, namely structural transitions and solid-vapor azeotropic phenomena, which impose stringent requirements to hydrate phase modeling and which are now represented reasonably by means of our new parameter set.

It is a primary requisite for a thermodynamically consistent modeling of hydrate formation in pure water and in inhibitor solutions that the strongly exponential temperature dependencies contained in the hydrate model are described correctly since for a given gas composition formation temperature can be depressed considerably in the presence of inhibitors. In this work, we decided to accept a published parameter set for thermodynamic reference hydrate properties as *correct* and focused our modifications on the temperature dependent Langmuir coefficients by adjusting cell potential parameters to experimental hydrate equilibrium data. Increasing the model flexibility, we introduced cavity interaction parameters, specific for each guest-cavity combination. For the purpose of significantly determining hydrate model parameters, we extended the temperature range, covered by the experimental data base, to lower temperatures by including mixtures containing concentrated sodium chloride solutions. In the fitting procedure, the activity of water was calculated by means of accurate reference data for the osmotic coefficient of water in sodium chloride solutions, available in the literature. The parameter set obtained in this way gave the best overall performance in calculating hydrate dissociation pressures of inhibitor-free systems, at the same time. Typical relative deviations between experimental and calculated hydrate dissociation pressures were 2–5% for the inhibitor-free mixtures considered in this work, only in a few cases deviations exceeded 10%. Structure transitions were predicted for methane-propane and methane-ethane gas mixtures, in accordance with our macroscopic equilibrium data and with the latest spectroscopic results reported in the literature.

On the basis of reliable hydrate phase calculations, the applicability of liquid phase models, proposed in the literature to account for the influence of electrolyte and polar organic components, can be examined. To this end, we used a group contribution Gibbs excess free energy (G^E) model presented by Yan et al. (1999) which was especially developed for treating mixed solvent-electrolyte systems. Their approach also covers the boundary cases of single solvent-electrolyte systems and salt-free polar solvent mixtures by incorporating the original UNIFAC concept, and seemed to be the most useful for our purposes. For calculating the fugacity of water in the mixed liquid phase under high pressure conditions and to account for the effect of gas solubility, we combined the G^E-model and the simple cubic equation of state from Soave-Redlich-Kwong by means of a modified Huron-Vidal (MHV1) mixing rule.

For aqueous single and mixed electrolyte (NaCl - NaCl/$CaCl_2$) solutions, the model gave excellent results over the entire range of salt concentration up to an ionic strength near 6 mol/kg water. It was assumed that the strong electrolytes, used in this study, were completely dissociated in the aqueous liquid. Hydrate dissociation pressures were predicted without loss of accuracy compared with the inhibitor-free systems. We further examined the hydrate formation behavior of gas mixtures in the presence of aqueous methanol and methanol-electrolyte solutions at constant total inhibitor mole-fraction. Our experimental and predicted results for these mixtures revealed that the chosen total inhibitor mole-fraction (≈ 0.06) was so dilute that the specific nature of inhibitor had almost neglectable effect on the formation temperature depression whilst predicted dissociation pressures were in satisfactory agreement with the experimental data. Therefore, in order to finally judge the predictive capabilities of the deployed G^E-model with respect to methanol inhibition, the hydrate formation behavior in concentrated methanol solutions has yet to be analyzed which will be subject of subsequent work.

The inhibiting effect of diethylene glycol (DEG) was examined, additionally. Our experimental results clearly showed, that on the basis of a comparable inhibitor mole-fraction, DEG suppressed the hydrate formation temperature most effectively, compared with the other inhibitors considered in this work. However, it turned out that the model was not capable of properly describing the inhibiting effect since the formation temperature depression was estimated too small by some 3–4 K. This result can be assumed to be a consequence of the temperature dependence of the UNIFAC group interaction parameters.

UNIFAC interaction parameters, readily available in the literature, were adjusted to cover the temperature range of interest for vapor-liquid equilibria, i.e. saturated liquids at their bubble points. In our application, however, strongly subcooled liquids have to be treated and reasonable extrapolations into the temperature range of our concern appear to be hardly possible. Therefore, in our subsequent theoretical work, alternative G^E-approaches shall be considered for data correlation with interaction parameters to be tuned especially for the temperature range addressed by gas hydrate formation.

Acknowledgement

The friendly support of Dr. K. Fischer, University of Oldenburg, Germany, who provided UNIFAC interaction parameters used in this study, is gratefully acknowledged.

References

Avlonitis D. (1988): Multiphase Equilibria in Oil-Water Hydrate Forming Systems. M.Sc. Thesis, Heriot-Watt University, Edinburgh, UK.
Bishnoi P.R.; Dholabhai P.D. (1993): Experimental Study on Propane Hydrate Equilibrium Conditions in Aqueous Electrolyte Solutions. Fluid Phase Equilibria 83, 455–462.
Christiansen R.L.; Sloan E.D. (1994): Mechanisms and Kinetics of Hydrate Formation. In: NGH 1994, 283–305.

Clarke M.; Bishnoi P.R. (2000): Determination of the Intrinsic Rate of Gas Hydrate Decomposition Using Particle Size Analysis. In: NGH 2000, 556–563.

de Roo J.L.; Peters C.J.; Lichtenthaler R.N.; Diepen G.A.M. (1983): Occurrence of Methane Hydrate in Saturated and Unsaturated Solutions of Sodium Chloride in Water in Dependence of Temperature and Pressure. AIChE J. 29, 651–657.

Deaton W.M.; Frost E.M. (1946): Gas Hydrates and Their Relation to the Operation of Natural-Gas Pipelines. Monograph 8, United States Dep. of the Interior, Bureau of Mines. Washington D.C., USA.

Dharmawardhana P.B; Parrish W.R.; Sloan E.D (1980): Experimental Thermodynamic Parameters for the Prediction of Natural Gas Hydrate Dissociation Conditions. Ind. Eng. Chem. Fund. 19, 410–414.

Dholabhai P.D.; Englezos P.; Kalogerakis N.; Bishnoi P.R. (1991): Equilibrium Conditions for Methane Hydrate Formation in Aqueous Mixed Electrolyte Solutions. Can. J. Chem. Eng. 69, 800–805.

Englezos P.; Bishnoi P.R. (1991): Experimental Study on the Equilibrium Ethane Hydrate Formation Conditions in Aqueous Electrolyte Solutions. Ind. Eng. Chem. Res. 30, No. 7, 1655–1659.

Franks F. (1975): The Hydrophobic Interaction. Water: A Comprehensive Treatise, Vol. 4. Franks F. (ed.), Plenum Press, New York, USA.

Fredenslund A.; Gmehling J.; Rasmussen P. (1977): Vapor-Liquid Equilibria Using UNIFAC. Elsevier, Amsterdam, The Netherlands.

Galloway T.J.; Ruska W.; Chappelear P.S.; Kobayashi R. (1970): Experimental Measurement of Hydrate Numbers for Methane and Ethane and Comparison with Theoretical Values. Ind. Eng. Chem. Fund. 9, No. 2, 237–243.

Gmehling J.; Li J.; Fischer K. (1997): Further Development of the PSRK model for the Prediction of Gas Solubilities and Vapor-Liquid Equilibria at Low and High Pressures II. Fluid Phase Equilibria 141, 113–127.

Hammerschmidt E.G. (1934): Formation of Gas Hydrates in Natural Gas Transmission Lines. Ind. Eng. Chem. 26, No. 8, 851–855.

Hansen H.K.; Rasmussen P.; Fredenslund A.; Schiller M.; Gmehling J. (1991): Vapor-Liquid Equilibria by UNIFAC Group Contribution. 5th Revision and Extension. Ind. Eng. Chem. Res. 30, 2352–2355.

Harmens A.; Sloan E.D. (1990): The Phase Behavior of the Propane-Water System: A Review. Can. J. Chem. Eng. 68, No. 1, 151–158.

Hendriks E.M.; Edmonds B.; Moorwood R.A.S.; Szczepanski R. (1996): Hydrate Structure Stability in Simple and Mixed Hydrates. Fluid Phase Equilibria 117, 193–200.

Holder G.D.; Grigoriou G.C. (1980): Hydrate Dissociation Pressures of Methane-Ethane-Water: Existence of a Locus of Minimum Pressures. J. Chem. Thermodynamics 1980, 1093–1104.

Holder G.D.; Hand J.H. (1982): Multiple-Phase Equilibria in Hydrates from Methane, Ethane, Propane and Water Mixtures. AIChE J. 28, No. 3, 440–447.

Holder G.D.; John V.T. (1983): Thermodynamics of Multicomponent Hydrate Forming Mixtures. Fluid Phase Equilibria 14, 353–361.

Holder G.D.; Zetts S.P.; Pradhan N. (1988): Phase Behavior in Systems Containing Clathrate Hydrates. Reviews in Chemical Engineering 5, No. 1, 1–70.

Holderbaum T. (1991): Die Vorausberechnung von Dampf-Flüssig-Gleichgewichten mit einer Gruppenbeitragszustandsgleichung. Fortschrittsberichte VDI, Reihe 3, Nr. 243, VDI-Verlag, Düsseldorf, Germany.

Holderbaum T.; Gmehling J. (1991): PSRK: A Group Contribution Equation of State Based on UNIFAC. Fluid Phase Equilibria 70, 251–265.

Huron M.-J.; Vidal J. (1979): New Mixing Rules in Simple Equations of State for Representing Vapor-Liquid Equilibria of Strongly Non-Ideal Mixtures. Fluid Phase Equilibria 3, 255–271.

Jeffrey G.A. (1984): Hydrate Inclusion Compounds. Inclusion Compounds, Vol. 1, Academic Press, London, UK, 135–190.

Knapp H.; Döring R.; Oellrich L.; Plöcker U.; Prausnitz J.M. (1982): Vapor-Liquid Equilibria for Mixtures of Low Boiling Substances. DECHEMA Chemistry Data Series, VI, DECHEMA, Frankfurt, Germany.

Kubota H.; Shimizu K.; Tanaka Y.; Makita T. (1984): Cited in Sloan (1998). J. Chem. Eng. Japan 17, 423.

Kvenvolden K.A. (1994): Natural Gas Hydrate Occurrence and Issues. In: NGH 1994, 232–246.
Mak T.C.W.; McMullan R.K. (1965): Cited in Sloan (1998). J. Chem. Phys. 42, No. 8, 2732.
McKoy V.; Sinanoglu O. (1963): Dissociation Pressures of Some Gas Hydrates. J. Chem. Phys. 38, 2946–2956.
McMullan R.K.; Jeffrey G.A. (1965): Polyhedral Clathrate Hydrates IX: Structure of Ethylene Oxide Hydrate. J. Chem. Phys. 42, No. 8, 2725–2732.
Michelsen M.L. (1990): A Modified Huron-Vidal Mixing Rule for Cubic Equations of State. Fluid Phase Equilibria 60, 213–219.
Miller B.; Strong E.R. (1946): Cited in Sloan (1998). Am. Gas. Assn. Monthly 28, No. 2, 63.
NGH1994 (1994): Natural Gas Hydrates. Proc. 1st Int. Conf., New York 1993. Sloan E.D., Happel J., Hnatow M.A. (eds.), Annals of the New York Academy of Sciences, 715, New York, USA.
NGH2000 (2000): Gas Hydrates: Challenges for the Future. Proc. 3rd Int. Conf., Salt Lake City 1999. Holder G.D., Bishnoi P.R. (eds.), Annals of the New York Academy of Sciences, 912, New York, USA.
Nixdorf J. (1996): Experimentelle und theoretische Untersuchung der Hydratbildung von Erdgasen unter Betriebsbedingungen. Doctoral Thesis, University of Karlsruhe, Germany.
Nixdorf J.; Oellrich L.R. (1997): Experimental Determination of Hydrate Equilibrium Conditions for Pure Gases, Binary and Ternary Mixtures and Natural Gases. Fluid Phase Equilibria 139, 325–333.
Parrish W.R.; Prausnitz J.M. (1972): Dissociation Pressures of Gas Hydrates Formed by Gas Mixtures. Ind. Eng. Chem. Proc. Des. Dev. 11, No. 1, 26–35.
Patil S.L. (1987): Measurements of Multiphase Gas Hydrates Phase Equilibria: Effect of Inhibitors and Heavier Hydrocarbon Components. M.Sc. Thesis, University of Alaska, USA.
Pitzer K.S.; Peiper J.C.; Busey R.H. (1984): Thermodynamic Properties of Aqueous Sodium Chloride Solutions. J. Phys. Chem. Ref. Data 13, No. 1, 1–102.
Reamer H.H.; Selleck F.T.; Sage B.H. (1952): Cited in Sloan (1998). J. Petrol. Techn. 4, No. 8, 197.
Ripmeester J.A.; Tse J.S.; Ratcliffe C.I.; Powell B.M. (1987): A New Clathrate Hydrate Structure. Nature 325, 135–136.
Roberts O.L.; Brownscombe E.R.; Howe L.S. (1940): Constitution Diagrams and Composition of Methane and Ethane Hydrates. Oil and Gas J. 39, No. 30, 37–43.
Robinson D.B.; Mehta B.R. (1971): Cited in Sloan (1998). J. Can. Petr. Tech. 10, 33.
Sloan E.D. (1990a): Natural Gas Hydrate Phase Equilibria and Kinetics: Understanding the State-of-the-Art. Revue de l'Institut Francais du Petrole 45, No. 2, 246–265.
Sloan E.D. (1990b): Clathrate Hydrates of Natural Gases, 1st ed., Marcel Dekker Inc., New York, USA.
Sloan E.D (1994): Conference Overview. In: NGH 1994, 1–23.
Sloan E.D. (1998): Clathrate Hydrates of Natural Gases. 2nd ed., Marcel Dekker Inc., New York, USA.
Soave G. (1972): Equilibrium Constants from a Modified Redlich Kwong Equation of State. Chem. Eng. Sci. 27, No. 6, 1197–1203.
Subramanian S.; Kini R.A.; Dec S.F.; Sloan E.D. (2000): Structural Transition Studies in Methane-Ethane Hydrates Using Raman and NMR. In: NGH 2000, 873–886.
Sum A.K.; Burruss R.C.; Sloan E.D. (1997): Measurement of Clathrate Hydrates via Raman Spectroscopy. J. Phys. Chem. 101, 7371–7377.
Tee L.S.; Gotoh S.; Stewart W.E. (1966): Molecular Parameters for Normal Fluids. Ind. Eng. Chem. Fund. 5, No. 3, 363–367.
Thakore J.L.; Holder G.D. (1987): Solid-Vapor Azeotropes in Hydrate Forming Systems. Ind. Eng. Chem. Res. 26, 462–469.
Tohidi B.; Burgass R.; Danesh A.; Todd A. (1993): Cited in Sloan (1998). SPE26701, Proc. SPE Offshore Europ. Conf., 255, Aberdeen, UK.
van der Waals J.H.; Platteeuw J.C. (1959): Clathrate Solutions. Advances in Chemical Physics, Vol. II. Interscience Publishers Inc., New York, USA, 1–57.
Verma V.K. (1974): Gas Hydrates from Liquid Hydrocarbon-Water Systems. Ph.D. Thesis, University of Michigan, Univ. Microfilms No. 75–10,324, Ann Arbor, MI, USA.
von Stackelberg M. (1949): Feste Gashydrate. Naturwissenschaften 36, No. 11, 327–359.
von Stackelberg M.; Müller H.R. (1954): Feste Gashydrate II. Zeitschrift für Elektrochemie 58, No. 1, 25.

Yan W.; Topphoff M.; Rose C.; Gmehling J. (1999): Prediction of Vapor-Liquid Equilibria in Mixed Solvent Electrolyte Systems Using the Group Contribution Concept. Fluid Phase Equilibria 162, 97–113.

Zatsepina O.Y.; Buffet B.A. (1998): Thermodynamic Conditions for the Stability of Gas Hydrate in the Seafloor. J. Geophys. Res. 103, No. B10, 24.127–24.139.

Zuo Y.-X.; Guo T.-M. (1991): Extension of the Patel-Teja Equation of State to the Prediction of the Solubility of Natural Gas in Formation Water. Chem. Eng. Sci. 46, No. 12, 3251–3258.

Nomenclature

a_j	Kihara core radius of hydrate former j
$a(T)$	EOS-attraction parameter
b	EOS-volume parameter
B	EOS-parameter
$C_{ij}(T)$	Langmuir coefficient
c_p	molar isobaric heat capacity
f_k	fugacity of component k
G	Gibbs free energy
$\bar{G}$	molar Gibbs free energy
h	molar enthalpy
$k_{ij}^{\sigma\varepsilon}$	guest-cavity interaction parameters
k_{lm}	binary interaction parameter (gas phase)
N	absolute number
n	number of moles
p	pressure
r	radial distance
R_i	mean radius of cavity i
T	absolute temperature
t	Celsius temperature
v	molar volume
w	cell potential
$\tilde{x}$	liquid mole-fraction
$\tilde{y}$	gas mole-fraction
z_i	coordination number of cavity i

Greek letters

α	EOS-energy parameter
$\Gamma(r)$	isotropic pair potential
ε	Kihara characteristic energy
ζ	compressibility factor
Θ_{ij}	fractional cavity occupancy
Λ	EOS-parameter

μ_k	chemical potential of component k
v_i	structure parameter: number of cavities i/number of water molecules
σ	Kihara core-to-core distance at zero potential
φ_k	fugacity coefficient of component k

Superscripts

G, L, H	gas phase, liquid phase, hydrate phase
E	excess property
α	ice
β	hypothetical empty hydrate lattice

Subscripts

i	cavity type i
j	hydrate former j
k	component k
l, m	summation indices
W	water
calc	calculated
exp	experimental
mix	mixture property
o	pure component property
0	reference state

Physical Constants

universal gas constant	$R = 8.31441$ J/(mol K)
Boltzmann constant	$k = 1.38066 \times 10^{-23}$ J/K

IV Phase Equilibrium of Aqueous Two-Phase Systems

14 Experimental and Theoretical Investigations on the Precipitation of Polyelectrolytes from Aqueous Solutions by Neutral Polymers

Thomas Grünfelder and Gerd Maurer [1]

14.1 Abstract

Adding a neutral hydrophilic polymer to an aqueous solution of a polyelectrolyte at moderate polymer concentrations often causes a liquid-liquid phase split, whereas at very high polymer concentrations one of the polymers might precipitate. This phenomenon has been investigated in the work presented here at 25 and 50 °C. Experimental results for the liquid-liquid equilibrium in systems polyelectrolyte + neutral polymer + water are reported and compared to calculations. The phase equilibrium was calculated applying an extension of the VERS model for the Gibbs excess energy of aqueous solutions. Some of the new experimental data were used to parameterize the model. Some other experimental data were used to test the model's ability for predicting the influence of the molecular weight of both polyelectrolyte and uncharged polymer on the liquid-liquid equilibrium.

14.2 Introduction

Aqueous solutions of polyelectrolytes play an important role in numerous industrial applications, e.g. in coating processes (paper industry, pharmaceutical industry), as flocculents (waste water treatment, mining), and for the controlled release of active ingredients in pharmaceutical applications. For this reason, aqueous solutions of polyelectrolytes have been the focus of various studies in the last decades.

When two neutral, hydrophilic but nevertheless incompatible polymers are dissolved simultaneously in water, a phase split (resulting in an aqueous two-phase system) is often obtained even at low polymer concentrations. This phenomena has been used in several applications, e.g. in downstream processing in biotechnology, as e.g. proteins partition unevenly between the coexisting phases.

---
1 Technische Universität Kaiserslautern, Fachbereich Maschinenbau und Verfahrenstechnik, Lehrstuhl für Technische Thermodynamik, Erwin-Schrödinger-Straße, 67653 Kaiserslautern.

When e.g. two polyelectrolytes with opposite charges on the polymer backbone are simultaneously dissolved in water, insoluble gel-like polymer complexes are formed. This behavior is used e.g. to manufacture membranes, for the precipitation of residues in waste water treatment processes as well as in some medical and pharmaceutical applications.

The phase behavior observed when a neutral polymer and a polyelectrolyte are simultaneously dissolved in water has attained only little attention (cf. Perrau et al., 1989; Yu and De Swaan Arons, 1994). When the polymers used in such experiments are incompatible, an aqueous two-phase is observed at low and intermediate polymer concentrations whereas at higher concentrations also a solid phase might precipitate from the solution. However, precipitation usually appears only at rather high polymer concentrations i.e. in very viscous solutions which are difficult to handle. The phase equilibrium in aqueous systems of a neutral hydrophilic polymer and a polyelectrolyte is studied in experimental work at 25 and 50 °C as well as in phase equilibrium calculations (at 25 °C only). Six polyelectrolytes were used together with two neutral polymers in these studies. One particular interest of the present work was on the influence of the molecular weight of the (eight) polymers on such phase equilibria. The experimental work aimed to provide a data base for developing and testing Gibbs excess energy models for describing such phase equilibria. The experimental work comprised for each system the determination of the phase boundary (which was approximated by the cloud point curve) and for most systems also the determination of the composition of the coexisting phases. As analytical procedures for all components are either not available or rarely give reliable results, only in a few cases a complete analysis of both coexisting phases was performed. In most experiments, the compositions of the coexisting phases were determined by combining the experimental results for the phase boundary on the one side with experimental results for the water content of a phase (which was determined by freeze drying) on the other side. That method was tested by checking if the feed point and the composition of coexisting phases are on the same mass balance line and, for some selected systems, by comparison with experimental results from a more complete analysis of the liquid phases. In those experiments the concentration of a second component in a liquid phase was also measured and used in a mass balance test.

In the systems examined, usually a liquid-liquid equilibrium of two aqueous phases is observed even at overall polymer concentrations as low as 10 to 15 mass%. Based on the experimental database, the phase behavior in such systems was modeled using an extension of the VERS-model of Großmann et al. (1995) for the Gibbs excess energy of such aqueous solutions. The VERS model is a group contribution method derived from Pitzer's osmotic virial equation for aqueous solutions of strong electrolytes (Pitzer, 1973). The modification of the VERS-model is mainly achieved by considering an incomplete dissociation of the electrolyte groups of the polyelectrolyte backbone. That dissociation is treated as a chemical equilibrium reaction in an aqueous solution of polymer groups. For that purpose the polymers are split into groups, the chemical reaction equilibrium is considered in a "real" mixture of water and polymer groups and then the electrolyte monomers are linked to polymer chains again. Any other influence of the degree of dissociation (e.g. changes of the conformation of the polyelectrolyte) on the properties of the aqueous solutions is neglected. Further assumptions and approximations had to be made to reduce the number of interaction parameters, which had to be fitted to experimental data. It was aimed to get as many as possible of those parameters from experimental data for the properties of aqueous solutions of the single polymers.

14.3 Experimental/Methods

14.3.1 Materials

There are some requirement for polymers used in the experiments. At first, the solubility in water should be sufficiently high. At second, the polymers should be well characterized. While there are many polymers which fulfill the first condition, it is more difficult to comply with the second one. Usually, natural polymers or polymers processed from natural substances (e.g. acacia, gelatin, methyl cellulose) are more difficult to characterize than synthetic polymers. Therefore, only synthetic polymers were used. For very similar reasons only homopolymers and no copolymers were selected. Polymers with a very narrow molecular weight distribution would have been desirable. However, such polymers are usually very expensive and therefore, only commercially available polymers – with a probably broad and in most cases unknown molecular weight distribution – were used. Tables 14.1 and 14.2 give a summary of the polyelectrolytes and the neutral polymers used in the present work. Both tables give the full names and the average molecular masses as well as abbreviations (for both the name and the molecular mass). Only for one polymer (PDADMACl) the repeat unit is a cation, whereas for the other five polyelectrolytes it is an anion. The anionic polyelectrolytes were mostly sodium salts with a molecular mass of between 2000 and 70,000. In most cases two samples of a single polyelectrolyte were used. The molecular mass of the samples differed by a factor of about 2 to 3. For the cationic polyelectrolyte no information about the polymer mass was available. Ethylene glycol and vinyl pyrrolidone were the repeat units of the neutral polymers. The molecular mass varied between 6000 and 103,000 (PEG/PEO) and 4000 and 18,000 (PVP). The polymers were supplied either as powder or in aqueous solution. Solid material was vacuum-dried in an exsiccator (approx. 10^{-2}–10^{-3} mbar) over silica gel at room temperature for approximately four weeks. Tests revealed that after this period of time the mass of the polymer remained unchanged, i.e. it can be considered as "dry". Between the experiments, the polymers were stored under these conditions.

Tables 14.1: Neutral polymers.

Polymer	Abbreviation	Molecular mass	$v^{(NP)}$
poly(ethylene glycol) [a]	PEG 6	6230 [b]	141
	PEG 34	34,400 [b]	781
poly(ethylene oxide) [a]	PEO 53	52,900 [c]	1201
	PEO 103	103,000 [c]	2338
poly(vinyl pyrrolidone)	PVP 4	3882 [d]	35
	PVP 18	17,749 [d]	160

[a] PEO is the more common name for PEG of high molecular mass
[b] M_n, from gel permeation chromatography (Tintinger, 1995)
[c] M_n, from manufacturer's information
[d] M_n, from gel permeation chromatography (Kany, 1998)

Tables 14.2: Polyelectrolytes.

Polyelectrolyte	Abbreviation	Molecular mass [a]	$v^{(PE)}$
poly(acrylic acid sodium salt)	NaPA 5 NaPA 15	5000 15,000	53 159
poly(acrylic acid ammonium salt)	NH$_4$PA 5 NH$_4$PA 10 NH$_4$PA 20	5000 10,000 20,000	56 111 222
poly(methacrylic acid sodium salt)	NaPMA 6 NaPMA 15	6500 15,000	56 139
poly(styrene sulfonic acid sodium salt)	NaPSS 70	70,000	330
poly(ethylene sulfonic acid sodium salt)	NaPES 2 NaPES 10	2000 10,000	15 77
poly(diallyl dimethyl-ammonium chloride)	PDADMACl lo MW PDADMACl hi MW	low high	– –

[a] from manufacturer; no further specification

The polymer content of the aqueous solutions was measured by freeze drying and assuming that all the remaining solid found is polymer.

14.3.2 Experimental Procedures

Experimental investigations of the phase equilibrium consisted of two steps. At first, the boundary between the homogeneous phase and the two-phase region was determined by turbidity measurements. At second, feed solutions (in two-phase regions) were prepared, equilibrated, separated into the coexisting phases and samples were taken for analyzing the composition. Details of the experimental procedures are given in the following sections.

14.3.2.1 Turbidity Measurements

The procedure applied in a turbidity experiment is shown in Figure 14.1. An aqueous solution of one polymer (in Fig. 14.1, a neutral polymer is taken as an example) was prepared in a centrifuge glass tube by either dissolving a known amount of polymer in a known amount of water or by diluting a known amount of an aqueous solution of a polymer with an also known amount of water. The composition of that aqueous solution is marked by B_1 in Figure 14.1. To this solution, another, similarly prepared aqueous solution of the other polymer (composition indicated by C in Fig. 14.1) is added in small steps. For each step the amount of the aqueous mixture in the glass tube is known. Between two additions of the polymer solution C, the sample is stirred magnetically and kept at constant temperature until equilibrium is achieved. Each step results in a change of composition, starting at B_1 and continuing along the mass balance line $\overline{B_1C}$. After equilibration the mixture is visually checked for

14.3 Experimental/Methods

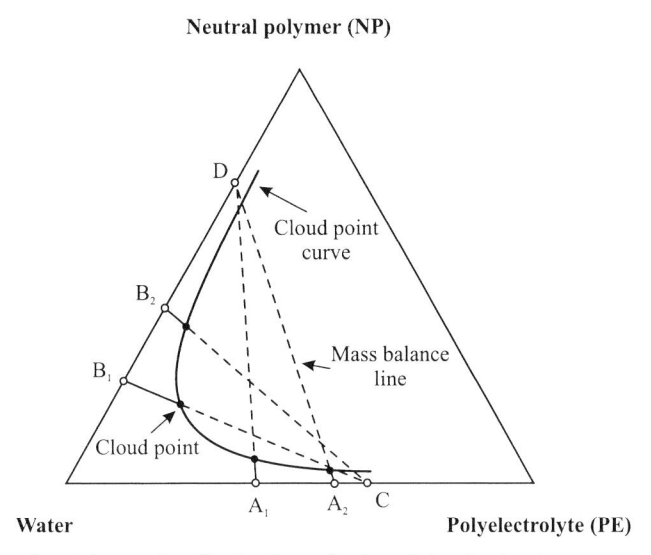

Figure 14.1: Experimental procedure for the determination of the cloud points.

turbidity. Turbidity is used as an indicator for the appearance of a second (either liquid or solid) phase. The further addition of solution C is stopped as soon as turbidity is confirmed.

In a typical experiment turbidity is already observed immediately after each addition of the second solution. However, as long as the equilibrated aqueous solution in the tube is still in a homogeneous region it will disappear during equilibration. Typically, turbidity disappears within a couple of minutes. When turbidity did not disappear within about 30 min, it was assumed that the mixture was in the two-phase region. Nevertheless equilibration was continued for two days and the visual check for turbidity was verified. From the known masses before and after such a step which resulted in a remaining turbidity and the also known compositions in B_1 and C, a point on the cloud point curve is determined. Repeating this procedure with different solutions B_2, B_3, etc., as well as adding solution D to solutions A_1, A_2, etc., results in a series of cloud points, between which the cloud point curve is interpolated. However, particularly at high polymer concentrations, the liquid phases are rather viscous and the magnetic field might not be sufficient to move the stirrer. In such cases, the experiment was stopped and no data taken. In most polymer systems investigated here, the viscosity of the aqueous polymer solutions limited the experimentally accessible concentration range to about 30 to 50 mass% polymer. The experimental uncertainty due to errors in weighing and due to the finite volume of solution added stepwise, results in absolute uncertainties for the concentration of a polymer on the cloud point curve of less than 1 mass% (cf. Grünfelder, 2001).

There are two aspects which have to be discussed further: The relation between the cloud point curve and the binodal curve and the influence of the polymer molecular weight distribution on those curves. Usually a cloud point curve does not coincide with the boundary between the one-phase and the two-phase regions, i.e. the binodal curve, but lies in between the spinodal and the binodal curves. However, there are several reasons for the assumption that in the experiments of the present work both curves coincide at least within experimental uncertainty. When for example the mass balance line $\overline{B_1C}$ in Figure 14.1 represents the mixing procedure, in equilibrium the second (polyelectrolyte-rich) phase is very close to point C and the formation of a metastable phase is not probably to occur. Furthermore, in those examples where the coexisting phases were completely analyzed, it could be shown that deviations between the binodal curve and the cloud point curve are smaller than the aforementioned experimental uncertainties. The second question (i.e. that on the influence of polymer polydispersity) is yet unanswered. All polymer samples were considered to behave like monodisperse solutions. From earlier work (cf. Kany, 1998) it is known that the polydispersity factor (which is the ratio of mass averaged to number averaged polymer molecular mass) of PEG is close to unity, whereas no information is available for all other polymers investigated here.

14.3.2.2 Compositions of Coexisting Phases

For each system three to five feed solutions inside the two phase region were prepared. A composition was chosen to provide two phases of approximately equal volume. After equilibration at constant temperature, the turbid solutions were separated into the coexisting phases in a centrifuge. Samples of both phases were weighed into a plate and diluted with water before they were further analyzed. In order to determine the composition of a ternary mixture, the concentrations of at least two components need to be known. The concentration of water in an aqueous solution of two polymers (i.e. non-volatile solutes) was determined by freeze drying. In some experiments atomic absorption spectroscopy was used to determine the sodium concentration (and thus the concentration of the polyelectrolyte). However, in most cases the results for the water concentration were combined with the experimental data for the cloud point curve to determine the compositions of the coexisting phases.

Freeze Drying

In order to determine the concentration of water, a sample of the aqueous polymer phase was filled into a flat container, weighed, and diluted with water. Dilution reduces the depression of the freezing point as well as it expands the average distance between polymer chains resulting in a porous dried product without trapped water. The diluted samples were frozen, cooled by liquid nitrogen to $-190\,°C$ and broken into small pieces to increase the surface. After that preparation, the sample was dried in a freeze dryer (Amsco/Finn Aqua, Type Lyovac GT2) at a pressure of about 51 Pa (corresponding to a sublimation temperature of water of $-30\,°C$). After freeze drying, the dried samples were kept for another two to three days in vacuum (ca. 10^{-2}–10^{-3} mbar) and over silica gel in an exsiccator at room temperature before they were weighed. From the mass change caused by freeze drying, the water concentration in the aqueous phase was calculated. The absolute experimental error for the water concentration is approximately 1.1 mass%, whereas it is less than 0.1 mass% for the concentrations of both polymers.

Atomic Absorption Spectroscopy

After estimating the polyelectrolyte concentration from the cloud point curve the aqueous sample was diluted with water in several steps to a final sodium concentration of 0.1 to 1.5 ppm mass. The sodium concentration of these diluted samples was then measured by atomic absorption spectroscopy (instrument: Varian, Type SpektrAA 100 with MK IV Burner and SpectrAA Software Version 2.10) using sodium chloride samples for calibration. The concentration of the sodium containing polyelectrolyte is then calculated assuming that each monomer unit carries one sodium ion. As expected, the presence of uncharged polymer does not affect the result of the analysis. The reproducibility of the experimental results for the polymer concentration is approximately 10%. Therefore, that method was only applied to determine the (small) polyelectrolyte concentration in the aqueous phase enriched by the neutral polymer. In such phases the polyelectrolyte concentration was typically between 1 and 5 mass%; the absolute uncertainty of that concentration is therefore typically 0.1 to 0.5 mass%.

Polymer Concentrations

The composition of an aqueous solution of two nonvolatile solutes can be determined from the binodal line when the water concentration is known. As mentioned above, the binodal line was approximated by the cloud point curve. The accuracy of this method can be verified by applying a mass balance test (using the known composition of the feed point) or by measuring the concentration of one of the polymers in each of the coexisting phases. The mass balance for water (concentration of water in the feed and in the coexisting phases) was combined with the results for the cloud point curve to give the concentration of each polymer in both coexisting phases. The water mass balance was also combined with the experimenttal results for the polyelectrolyte concentration in the neutral polymer-rich liquid phase to give all concentrations in the coexisting phases. Comparisons between results from both methods revealed neither a systematic deviation nor deviations beyond the experimental uncertainty of the turbidity experiments (cf. Grünfelder, 2001). This means that the cloud point curve is a good approximation of the binodal curve. Consequently the method to determine the composition of coexisting phases (from the cloud point and freeze drying data alone) provides good experimental data.

14.4 Results and Discussion

14.4.1 Overview

Cloud points were determined experimentally for 92 systems polyelectrolyte + neutral polymer + water at 25 and 50 °C. For about two thirds of those systems the compositions of coexisting phases were also determined (cf. Tables 14.3 and 14.4).

Tables 14.3: Systems experimentally examined at 25 °C: +, composition of coexisting phases and cloud point curve determined; o, only cloud point curve determined.

t = 25 °C	PEG 6	PEG 34	PEO 53	PEO 103	PVP 4	PVP 18
NaPA 5	+	+	+	+	+	+
NaPA 15	+	+	+	+	+	+
NH$_4$PA 5	o	o			o	o
NH$_4$PA 10	o					o
NH$_4$PA 20	o	o			o	o
NaPMA 6	+	+			+	+
NaPMA 15	+	+			+	+
NaPSS 70	o					
NaPES 2	o	+	+	+	+	
NaPES 10	+	+	+	+	+	+
PDADMACl lo MW	o					
PDADMACl hi MW	o	o				

Tables 14.4: Systems experimentally examined at 50 °C: +, composition of coexisting phases and cloud point curve determined; o, only cloud point curve determined.

t = 50 °C	PEG 6	PEG 34	PEO 53	PEO 103	PVP 4	PVP 18
NaPA 5	+	+	+	+	+	+
NaPA 15	+	+	+	+	+	+
NH$_4$PA 5	o	o			o	o
NH$_4$PA 10						
NH$_4$PA 20	o	o			o	
NaPMA 6	+	+			+	+
NaPMA 15	+	+			+	+
NaPSS 70	o	o				o
NaPES 2	+	+	+	+	+	+
NaPES 10	+	+	+	+	+	+
PDADMACl loMW						
PDADMACl hiMW	o	o				

14.4.2 Cloud Points

Figure 14.2 shows a typical example for the results of the experimental investigations. The cloud points of the system NaPMA 15 + PVP 18 + water at 50 °C are marked by circles. By interpolation the cloud point curve is established. Measurements could not be performed at concentrations beyond 50 mass% PVP due to the high viscosity of such solutions. NaPMA 15 was supplied in aqueous solution (30 mass% polymer in water) and used as supplied or after dilution with water. The cloud point experiments were therefore restricted to NAPMA-concentrations below 30 mass%. Even at a relatively low total polymer concentration (i.e. around 10 mass%) a phase split into two liquid phases is observed. The polyelec-

14.4 Results and Discussion

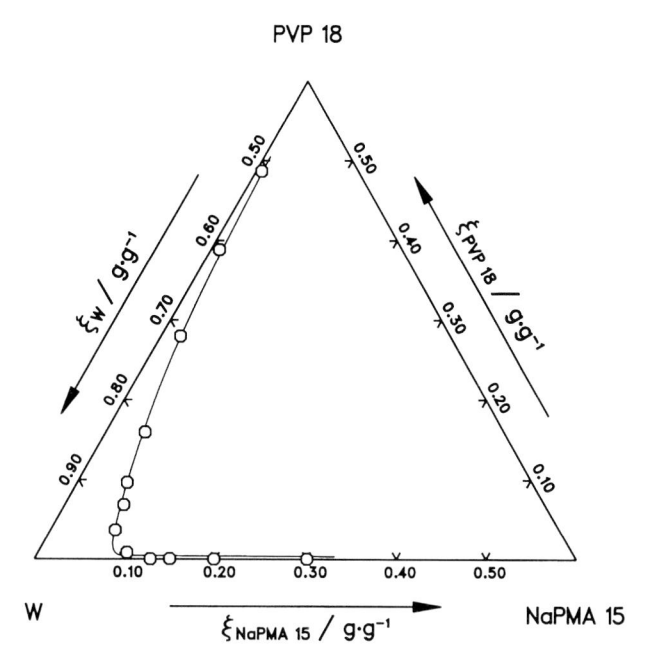

Figure 14.2: Cloud points of NaPMA 15 + PVP 18 + water at 50 °C.

trolyte-rich phase ("polyelectrolyte phase") contains only very little neutral polymer, whereas the neutral polymer-rich phase ("neutral polymer phase") contains a considerable amount of the polyelectrolyte. Therefore, the cloud point curve is not symmetric. In almost all systems examined, the cloud point curve shows a similar behavior. A different behavior was found only in the system NaPSS 70 + PVP 18 + water at 50 °C (where a closed loop miscibility gap is found) and in several systems with the neutral polymer PEG 6 (where at high concentrations PEG precipitation – i.e. a solid phase – was observed). The detailed experimental results are available elsewhere (Grünfelder, 2001).

14.4.3 Composition of Coexisting Phases

Figure 14.3 shows a typical example for the experimental results for the compositions of the coexisting liquid phases. The feed points are given together with the experimental results for the coexisting phases. The total polymer mass concentrations in both phases are rather different, e.g. an aqueous polyelectrolyte-rich phase of about 20 mass% polymer coexists with a neutral polymer-rich aqueous phase of about 30 mass% polymer. Therefore, in Figure 14.3 the tie lines are relatively "steep".

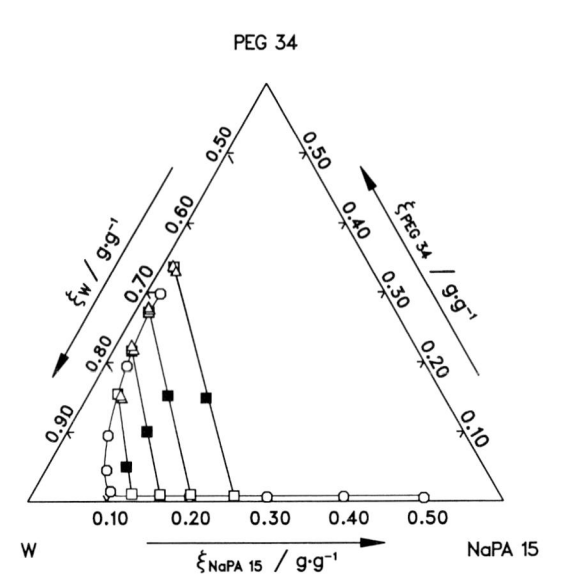

14 Precipitation of Polyelectrolytes from Aqueous Solutions by Neutral Polymers

Figure 14.3: Composition of coexisting phases of NaPMA 15 + PEG 34 + water at 25 °C: ■, feed point; ○, cloud point; coexisting phases: □, from cloud point curve and freeze drying; △, from atomic absorption spectroscopy and freeze drying.

14.4.4 Influence of Molecular Weight on the Phase Behavior

Figure 14.4 shows an example for the influence of the polymer molecular mass on the liquid-liquid two-phase equilibrium. As it was to be expected, increasing the molecular mass (for example that of PEG from 6000 to 34,000 in the system NaPES 10 + PEG + water) results in an enlargement of the miscibility gap. In most systems investigated here, the slope of the tie lines is changed very little when the molecular mass of one of the polymers is varied.

14.4.5 Influence of Temperature on the Phase Behavior

Figure 14.5 shows a typical example for the influence of a temperature change from 25 to 50 °C on the liquid-liquid equilibrium. The miscibility gap is increased, but the change is mostly rather small, often within the experimental uncertainty. However, there is usually a remarkable change of the slope of the tie lines. In all examined systems, the slope of the tie lines increases with rising temperature. For the same aqueous feed in a two-phase region the mass fraction of water in the polyelectrolyte phase increases whereas that in the neutral polymer phase decreases when the temperature rises. Only a few systems show deviations from that rule. For example, there is no miscibility gap for NaPSS 70 + PVP 18 + water in the experimentally accessible concentration range (i.e. total polymer concentrations below 50 mass%) at 25 °C, whereas there is a closed miscibility loop at 50 °C.

14.4 Results and Discussion

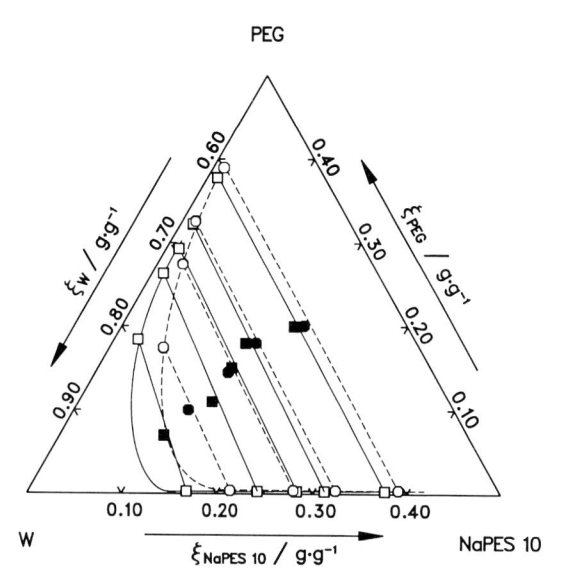

Figure 14.4: Influence of molecular weight on the liquid-liquid equilibrium of NaPES 10 + PEG + water at 25 °C: ■, feed point; ○, cloud point; □, coexisting phases.

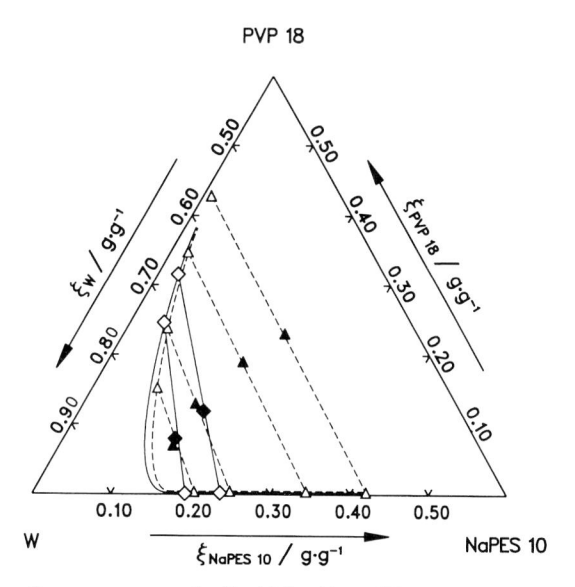

Figure 14.5: Influence of temperature on the liquid-liquid equilibrium of NaPA 15 + PVP 18 + water: feed points and coexisting phases at 25 °C (▲, △) and at 50 °C (◆, ◇).

14.5 Modeling

14.5.1 Introduction to VERS-PE

The liquid-liquid equilibrium observed in aqueous solutions of two neutral polymers as well as in solutions of a neutral polymer and a low molecular salt can be correlated by several methods. One of these methods is the VERS model, which is a modification of the osmotic virial equation (Großmann et al., 1995), where relative surface fractions are used to express the solute concentration. However, the properties of an aqueous polyelectrolyte solution differ remarkably from that of an aqueous solution of a neutral polymer. This is due to charge density effects which result in phenomena such as counterion condensation etc. (cf. Lifson and Katchalsky, 1954; Manning, 1969; Oosawa, 1971). Therefore, it was decided to modify the VERS-model in order to take the dissociation equilibrium as well as electrostatic interactions between counter ions and the polymeric chain into account. Due to its focus on systems containing polyelectrolytes (PE), this modification is referred to as VERS-PE model. The Gibbs excess energy is considered with a normalization according to Raoult's law for the solvent (i.e. water) and according to Henry's law on molality scale (where the solvent again is pure water) for the polymers. A polyelectrolyte is considered to be undissociated in the reference state. The influence of pressure on the chemical potential is neglected:

$$\mu_i = \left(\frac{\partial G}{\partial n_i}\right)_{T,p,n_{j \neq i}} \approx \mu_i(T, n_j) \tag{14.1}$$

The difference between the chemical potential of a solute in the aqueous solution, μ_i, and that in the reference state, μ_i^{ref},

$$\mu_i^{ref} = \mu_i(T, \bar{m}_i = 1 \text{ (amount of substance)}, \bar{m}_i = 0 \text{ in pure water (interactions)}) \tag{14.2}$$

is split into four contributions:

$$\mu_i - \mu_i^{ref} = \Delta_{12}\mu_i + \Delta_{23}\mu_i + \Delta_{34}\mu_i + \Delta_{45}\mu_i \tag{14.3}$$

The first contribution describes the change from the reference state to an ideal dilution, still neglecting any dissociation:

$$\Delta_{12}\mu_i = RT \ln \bar{m}_i \tag{14.4}$$

where $\bar{m}_i$ is the molality of solute i in the aqueous phase. The second contribution contains all effects resulting from the splitting of all polymers into monomers. It considers contributions from e.g. combinatorial and free volume terms. The third term includes the effects caused by the dissociation of the electrolyte monomers as well as from differences in the interactions (in comparison to the reference state, where the monomers are surrounded by solvent molecules only). The forth contribution comes from the polymerization of the monomers to form the polymer. For the sake of simplicity, it is assumed that the second and forth

14.5 Modeling

term have the same number, but different signs, so they disappear in the final expression for the chemical potential of a solute:

$$\Delta_{23} \mu_i + \Delta_{45} \mu_i = 0 \tag{14.5}$$

For a neutral polymer ($i \equiv NP$) of groups A, the third contribution is:

$$\Delta_{34} \mu_{NP} = \mu_{NP,4} - \mu_{NP,3} = v^{(NP)} RT \ln \Gamma_{A,4} \tag{14.6}$$

where $v^{(NP)}$ and Γ_A are the number of neutral monomer groups in NP and the activity coefficient of groups A, respectively. For a polyelectrolyte ($i \equiv PE$) of groups B, which dissociate according to

$$B \rightleftarrows C + D \tag{R14.1}$$

the third contribution is:

$$\Delta_{34} \mu_{PE} = \mu_{PE,4} - \mu_{PE,3}$$
$$= v_{B,4} \mu_{B,4} + v_{C,4} \mu_{C,4} + v_{D,4} \mu_{D,4}$$
$$- v^{(PE)} \left[\mu_B^{ref} + RT \ln \left(v^{(PE)} \bar{m}_{PE} \right) \right] \tag{14.7}$$

where $v^{(PE)}$ is the number of monomer groups per polyelectrolyte (PE) chain and $v_{g,4}$ (g = B, C, D) is the number of groups B, C, and D resulting from $v^{(PE)}$ monomers in chemical equilibrium. For example, when B represents the non-dissociated monomer and the charge numbers of group C and counterion D differ only by the sign, then: $v^{(PE)} = v_{B,4} + v_{C,4}$ and $v_{C,4} = v_{D,4}$. The chemical potential of a group g (g = B, C, D) in state 4 is:

$$\mu_{g,4} = \mu_{g,4}^{ref} + RT \ln \left(m_g \Gamma_{g,4} \right) \tag{14.8}$$

Introducing the degree of dissociation α

$$\alpha = \frac{v_{C,4}}{v^{(PE)}} \tag{14.9}$$

gives, as $v_{B,4} = (1 - \alpha) v^{(PE)}$ and $v_{C,4} = v_{D,4} = \alpha v^{(PE)}$:

$$\Delta_{34} \mu_{PE} = v^{(PE)} \left[(-\mu_{B,4} + \mu_{C,4} + \mu_{D,4}) + \mu_{B,4} - \mu_B^{ref} - RT \ln \left(v^{(PE)} \bar{m}_{PE} \right) \right] \tag{14.10}$$

Taking into account that in chemical equilibrium

$$\mu_{B,4} = \mu_{C,4} + \mu_{D,4} \tag{14.11}$$

and introducing Eq. (14.8) (for g ≡ B), Eq. (14.10) becomes:

$$\Delta_{34}\,\mu_{PE} = v^{(PE)} RT \ln\left[(1-\alpha)\,\Gamma_{B,4}\right] \quad (14.12)$$

Adopting the main outline of the VERS model gives (cf. Grünfelder, 2001) for the activity coefficient of a neutral group, e.g. group A:

$$\ln \Gamma_A = 2 \cdot \frac{1000}{M_W} \cdot \frac{q_A}{q_W} \cdot \sum_{\text{all groups L}} \frac{\vartheta_L}{\vartheta_W} a_{AL} \quad (14.13)$$

The activity of the neutral polymer (NP) of groups A therefore is:

$$\ln a_{NP} = \ln \bar{m}_{NP} + 2 \cdot v^{(NP)} \cdot \frac{1000}{M_W} \cdot \frac{q_A}{q_W} \cdot \sum_{\text{all groups L}} \frac{\vartheta_L}{\vartheta_W} a_{AL} \quad (14.14)$$

and that of the polyelectrolyte is:

$$\ln a_{PE} = \ln \bar{m}_{PE} + v^{(PE)} \left[\ln(1-\alpha) + 2\left(\frac{1000}{M_W}\right) \frac{q_B}{q_W} \sum_{\text{all groups L}} \frac{\vartheta_L}{\vartheta_W} a_{BL}\right] \quad (14.15)$$

where the relative surface fractions are given by:

$$\frac{\vartheta_A}{\vartheta_W} = \frac{M_W}{1000} \bar{m}_{NP}\, v^{(NP)} \frac{q_A}{q_W} \quad (14.16\,a)$$

$$\frac{\vartheta_B}{\vartheta_W} = \frac{M_W}{1000} \bar{m}_{PE}\, v^{(PE)} (1-\alpha) \frac{q_B}{q_W} \quad (14.16\,b)$$

$$\frac{\vartheta_C}{\vartheta_W} = \frac{M_W}{1000} \bar{m}_{PE}\, v^{(PE)} \alpha \frac{q_C}{q_W} \quad (14.16\,c)$$

and

$$\frac{\vartheta_D}{\vartheta_W} = \frac{M_W}{1000} \bar{m}_{PE}\, v^{(PE)} \alpha \frac{q_D}{q_W} \quad (14.16\,d)$$

q_I and $a_{JK} = a_{KJ}$ are the surface parameter of group I and binary interaction parameters between solute groups K and J, respectively.

The degree of dissociation α is calculated from the dissociation equilibrium for Eq. (R14.1):

$$K(T) = \frac{m_C\, m_D}{m_B} \cdot \frac{\Gamma_C \Gamma_D}{\Gamma_B} = \frac{\alpha^2}{1-\alpha} v^{(PE)} \bar{m}_{PE} \frac{\Gamma_C \Gamma_D}{\Gamma_B} \quad (14.17)$$

The activity coefficient of the uncharged groups (i.e. A and B) is given by Eq. (14.13) for group A (and for groups B by replacing subscript A by B in that equation) while the activity coefficients of the charged species (i.e. J is either C or D) is given by:

14.5 Modeling

$$\ln \Gamma_J = \left[2 \cdot \frac{1000}{M_W} \cdot \frac{q_J}{q_W} \cdot \sum_{\text{all groups L}} \frac{\vartheta_L}{\vartheta_W} a_{JL} \right]$$

$$- \frac{1}{2 \, \alpha v^{(PE)}} A_\varphi \left[\frac{\sqrt{I_m}}{1 + b\sqrt{I_m}} + \frac{2}{b} \ln\left(1 + b\sqrt{I_m}\right) \right] \quad (14.18)$$

A_φ is the Debye-Hückel parameter for water. I_m is the ionic strength in the unpolymerized state on molality scale:

$$I_m = \frac{1}{2} \sum_{\text{all groups L}} m_L z_L^2 \quad (14.19)$$

where z_L is the number of electrical charges on group L and $b = 1.2$ mol/kg.

The activity of water follows from Eqs. (14.14) and (14.15) by applying the Gibbs-Duhem equation (cf. Grünfelder, 2001):

$$\ln a_W = -\frac{M_W}{1000} (\bar{m}_{NP} + \bar{m}_{PE}) - v_A^{(NP)} \frac{q_A}{q_W} \frac{\vartheta_A}{\vartheta_W} a_{AA} \bar{m}_{NP}$$

$$- \frac{M_W}{1000} \bar{m}_{NP} \left(\mu_{NP} (\bar{m}_{PE}, \bar{m}_{NP}) - \mu_{NP} (\bar{m}_{PE} = 0, \bar{m}_{NP}) \right)$$

$$- \frac{M_W}{1000} \int_{\bar{m}_{PE}=0}^{\bar{m}_{PE}} \bar{m}_{PE} \left(\frac{\partial \ln \gamma_{PE}}{\partial \bar{m}_{PE}} \right) d\bar{m}_{PE} \quad (14.20)$$

The integration in Eq. (14.20) is performed numerically.

For calculating the liquid-liquid equilibrium in a system neutral polymer + polyelectrolyte + water (for given temperature and total feed) the following properties must be known:

- Number of groups of both polymers ($v^{(NP)}$ and $v^{(PE)}$), which were determined from the average molecular mass of the polymers and their monomers. That information is given in Tables 14.1 and 14.2.

- Surface parameters q_i for water as well for groups A, B, and C and the counter ion D. The surface parameter for water as well as for groups A, B and C were calculated according to Bondi (1968). The surface parameter of sodium was taken from Tintinger (1995). All surface parameters are given in Table 14.5.

- The Debye-Hückel constant for water: A_φ (298.15 K) = 0.39147 (Tintinger, 1995).

- Binary parameters $a_{JK} = a_{KJ}$ for interactions between all solute species and the dissociation constant of the polyelectrolyte groups $K(T)$, which were estimated as described in the following section.

Tables 14.5: Surface parameters.

Substance	Group	q
water	W	1.40
PEG, PEO	EG	1.320
PVP	VP	3.120
sodium	Na^+	1.4
NaPA	NaA	3.084
	A^-	1.684
NaPMA	NaMA	3.704
	MA^-	2.304
NaPES	NaES	4.208
	ES^-	2.808

14.5.2 Determination of Parameters of the VERS-PE Model

There are four different groups in a system NaPE+NP+water: A (groups from the neutral polymer), B and C (monomer units of the polyelectrolyte in neutral and charged form), and sodium ions D (counter ions of charged polyelectrolyte monomer units, D in Eq. (R14.1)). Therefore, there are 10 binary interaction parameters. As common in Pitzer's method, interaction parameters between ionic groups carrying the same charge are neglected ($a_{CC} = a_{DD} = 0$). Furthermore, it is assumed that binary parameters represent van der Waals interactions which are not influenced by the charge of a group. That assumption reduces the number of adjustable parameters further, as $a_{CB} = a_{BB}$ and $a_{CA} = a_{BA}$. To further reduce the number of parameters, parameters for interactions between the sodium ion and all other groups were also neglected: $a_{AD} = a_{BD} = a_{CD} = 0$. Thus there are only three unknown binary parameters: a_{AA} (for interactions between groups of the neutral polymer), a_{BB} (for interactions between polyelectrolyte groups), and a_{AB} (for interactions between groups of the neutral polymer and those of the polyelectrolyte). Parameter a_{AA} was fitted to the activity of water in aqueous solutions of the neutral polymer. Figure 14.6 shows a typical example for PEG. The activity of water in an aqueous solution of PEG (as calculated from an osmotic virial equation with virial coefficients reported by Hasse et al., 1995) can be represented with the VERS-PE model with $0.008 \leq a_{AA} \leq 0.010$ for both PEG6 and 34 (water activity data for PEO 53 and 103 are not available). Similar results were obtained for PVP 4 and 18 (cf. Grünfelder, 2001). There are two model parameters (interaction parameter a_{BB} and dissociation constant K) which can be determined similarly from experimental results for the activity of water in aqueous solutions of a single polyelectrolyte (cf. Lammertz, 2004). However, it turned out that the quality of the experimental data does not allow to find a unique set of parameters a_{BB} and K. Both binary interaction parameter (a_{AA} and a_{BB}) proved to be very important for a good description of the liquid-liquid equilibrium in the ternary system PE + NP + water. Therefore, both parameters were adjusted also taking liquid-liquid equilibrium data into account. But in all cases, it was assumed that the molecular mass of the polyelectrolyte has no influence neither on a_{AA} and a_{BB} nor on K. In principle, when a_{AA}, a_{BB}, and K are determined from binary data, a_{AB} is the only parameter which was to be adjusted to the ternary liquid-liquid equilibrium. The experimental results for the liquid-liquid equili-

14.5 Modeling

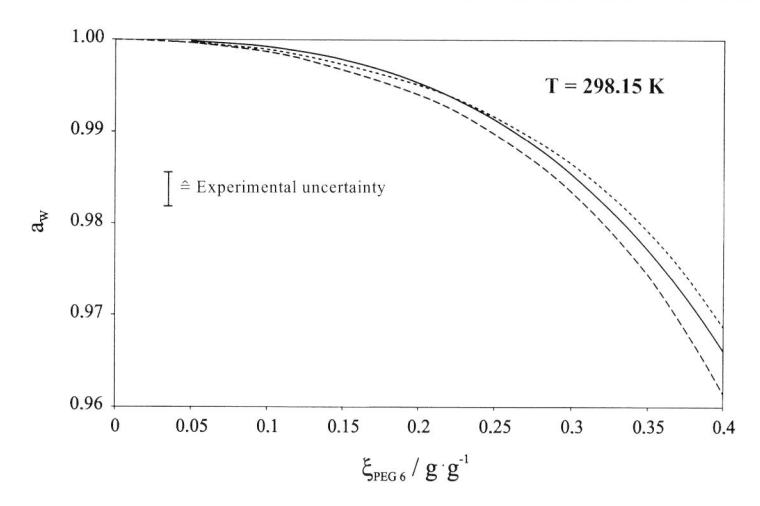

Figure 14.6: Activity of water in aqueous solutions of PEG 6: ———, from Hasse et al. (1995); correlation of the present work with $a_{AA} = 0.008$ (······) and $a_{AA} = 0.010$ (– – – – –).

brium in aqueous solutions of a polyelectrolyte (NaPA, NaPMA, and NaPES) and a neutral polymer (PEG/PEO and PVP) at 25 °C were correlated, but only data for polymers with low molecular mass (e.g. PEG 6, NaPMA 6, PVP 4) were taken into account (cf. Table 14.6). The interaction parameters are given in Table 14.7. With those parameters the influence of the molecular mass of the polymers on the liquid-liquid equilibrium was predicted for systems with polymers of higher molecular mass (cf. Table 14.6).

Tables 14.6: Modeling of liquid-liquid equilibrium: C, correlation using the VERS-PE-model; P, prediction (of the influence of molecular mass).

t = 25 °C	PEG 6	PEG 34	PEO 53	PEO 103	PVP 4	PVP 18
NaPA 5	C	P	P	P		
NaPA 15	P	P	P	P		
NaPMA 6	C	P			C	P
NaPMA 15	P	P			P	P
NaPES 2						
NaPES 10	C	P	P	P		

Tables 14.7: Parameters for the model VERS-PE used to describe the liquid-liquid equilibrium at 25 °C.

t = 25 °C	a_{BB}	a_{AB}	a_{AA}	K
NaPA/(PEG or PEO)	0.026	0.010	0.0195	0.01
NaPMA/(PEG or PEO)	0.018	0.010	0.020	0.07
NaPMA/PVP	0.018	0.007	0.018	0.07
NaPES/(PEG or PEO)	0.0035	0.010	0.015	0.10

14.5.3 Results of Correlations/Predictions with the VERS-PE Model

A typical example for a comparison between experimental results for the liquid-liquid equilibrium and the correlation with the VERS-PE model is shown in Figure 14.7. The correlation for NaPMA 6 + PEG 6 + water at 25 °C shows a reasonable agreement with the experimental data, i.e. when the feed is given and the polymer concentrations in coexisting phases

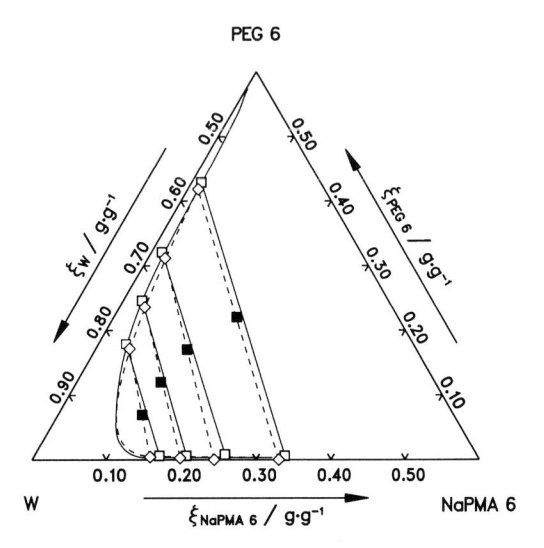

Figure 14.7: Liquid-liquid equilibrium of NaPMA 6 + PEG 6 + water at 25 °C: ■, feed point; coexisting phases: □, experimental results; ◇, correlation using VERS-PE.

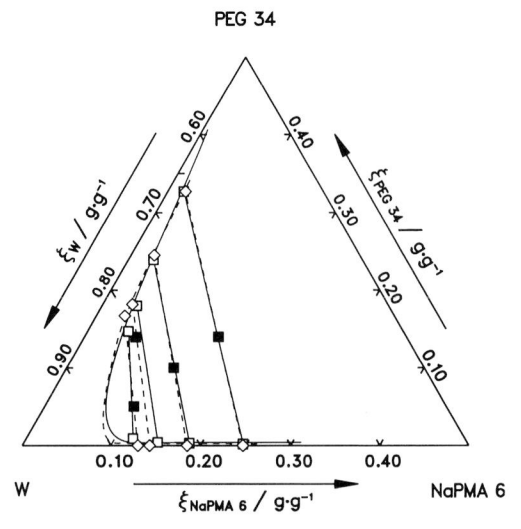

Figure 14.8: Liquid-liquid equilibrium of NaPMA 6 + PEG 34 + water at 25 °C. ■, feed point; coexisting phases: □, experimental results; ◇, prediction using VERS-PE.

are compared, the deviations of calculated results from the experimental data are in most cases within the experimental uncertainty and rarely exceed that experimental uncertainty by a factor of two. The largest deviations are always observed near the critical point. A similar statement holds when predictions for the influence of the molecular mass of the polymers on the liquid-liquid equilibrium are compared to experimental data. Figure 14.8 shows a typical example. Model parameters were adjusted for NaPMA 6 + PEG 6 + water and predictions were performed for the system, where the molecular mass of PEG was increased from 6000 to 34,000. Prediction of the effect of molecular weight was successful in all examined systems at least in a qualitative way. In some systems the deviations between experimental results for the polymer concentrations in the coexisting phases and the predictions were even smaller than the experimental uncertainty (Grünfelder, 2001).

14.6 Conclusions

The phase behavior of 45 ternary systems polyelectrolyte + neutral polymer + water was investigated experimentally at 25 and 50 °C. The polyelectrolytes were predominantly anionic polymers – sodium as well as ammonium salts of polymers of acrylic acid, methacrylic acid, styrene sulfonic acid, and ethylene sulfonic acid. The range of molecular mass varied between about 2000 and 70,000. PEG/PEO and PVP were the only neutral polymers, their molecular mass was between 4000 and 103,000. In nearly all systems only liquid-liquid phase equilibrium was observed, as solid-liquid equilibria occur only at rather high polymer concentrations where the aqueous solutions are too viscous to achieve equilibrium in a reasonable period of time. Polymer molecular weight as well as temperature has only a small influence on the liquid-liquid equilibrium. Modeling of such phase equilibria was achieved by extending the VERS model to electrolyte systems. That extension accounts for the partial dissociation of the polyelectrolyte groups. The experimental results can be correlated reliably, i.e. deviations to the experimental results for the polymer concentrations are typically not larger than the experimental uncertainty. The model allows for the prediction of the (small) influence of the molecular mass of a polymer on the phase equilibrium. Such predictions are always quantitatively correct, in some cases even agree with the experimental data within the range of experimental uncertainty.

Acknowledgements

The authors gratefully acknowledge assistance to the experimental work by undergraduate students V. Lopes Aguiar, L. Storbeck, P. Jockers, S. Wenz, M. Mahr von Staszewski, and K. Lengemann.

References

Bondi, A. (1968): Physical Properties of Molecular Crystals, Liquids, and Gases. John Wiley & Sons Inc., New York.

Großmann, C.; Tintinger, R.; Zhu, J.; Maurer, G. (1995): Aqueous Two-Phase Systems of Poly(ethylene glycol) and Dextran – Experimental Results and Modeling of Thermodynamic Properties. Fluid Phase Equilibria 106, p. 111–138.

Grünfelder, T. (2001): Experimentelle und theoretische Untersuchungen zum Flüssig-Flüssig-Gleichgewicht in ternären Systemen Polyelektrolyt/ungeladenes Polymer/Wasser. PhD Thesis, Universität Kaiserslautern, Germany.

Hasse, H.; Kany, H.-P.; Tintinger, R.; Maurer, G. (1995): Osmotic Virial Coefficients of Aqueous Poly(ethylene glycol) from Laserlight Scattering and Isopiestic Measurements. Macromolecules 28, p. 3540–3552.

Kany, H.-P. (1998): Thermodynamische Eigenschaften wässriger Polymer Lösungen. PhD Thesis, Universität Kaiserslautern, Germany.

Lammertz, S. (2004): Thermodynamische Eigenschaften wässriger Polyelektrolytlösungen, PhD Thesis (in preparation), Technische Universität Kaiserslautern, Germany.

Lifson, S.; Katchalsky, A. (1954): The Electrostatic Free Energy of Polyelectrolyte Solutions. II. Fully Stretched Macromolecules. J. Polym. Sci. 13, p. 43–55.

Manning, G. S. (1969): Limiting Laws and Counterion Condensation in PolyelectrolyteSolutions. I. Colligative Properties. J.Chem. Phys. 51 (3), p. 924–933.

Perrau, M. B.; Iliopoulos, I.; Audebert, R. (1989): Phase Separation of Polyelectrolyte Nonionic Polymer Systems in Aqueous Solution – Effects of Salt and Charge Density. Polymer 30, p. 2112–2117.

Pitzer, K. S.(1973): Thermodynamics of electrolytes. I. Theoretical Basis and General Equation. J. Phys. Chem. 77, p. 268–277.

Oosawa, F. (1971): Polyelectrolytes. Marcel Dekker Inc., New York.

Tintinger, R. (1995): Thermodynamische Eigenschaften ausgewählter wässriger Zwei-Phasen-Systeme. PhD Thesis, Universität Kaiserslautern, Germany.

Yu, M.; De Swaan Arons, J. D. (1994): Phase Behavior of Aqueous Solutions of Neutral and Charged Polymer Mixtures. Polymer 35, p. 3499–3502.

Nomenclature

a	activity
a_{JK}	binary parameter for interactions between groups J and K
A	group A
A_φ	Debye-Hückel parameter for water
b	parameter (= 1.2 mol/kg)
B	neutral group B
C	charged group C (by dissociation of B)
D	counter ion dissociated from B
G	Gibbs energy
I_m	ionic strength on (group) molality scale
K	chemical reaction equilibrium constant

Nomenclature

m	molality (of groups)
$\bar{m}$	molality
M	molecular mass
M_n	number averaged molecular mass
n	mole number
NaPES	sodium salt of poly(ethylene sulfonic acid) e.g. NaPES 10 = NaPES of average molecular mass of 10,000
NaPMA	sodium salt of poly(methacrylic acid) e.g. NaPMA 15 = NaPMA of average molecular mass of 15,000
NaPSS	sodium salt of poly(styrene sulfonic acid) e.g. NaPSS 70 = NaPSS of average molecular mass of 70,000
NP	neutral polymer
p	pressure
PE	polyelectrolyte
PEG	poly(ethylene glycol), e.g. PEG 6 = PEG of average molecular mass of 6000
PEO	poly(ethylene oxide) e.g. PEO 103 = PEO of average molecular mass of 103,000
PVP	poly(vinyl pyrrolidone) e.g. PVP 18 = PVP of average molecular mass of 18,000
q	surface parameter
R	gas constant
T	temperature
VERS	virial equation with relative surfaces
z	charge number

Greek letters

α	degree of dissociation
Δ	difference
Γ	group activity coefficient
ϑ	surface fraction
γ	activity coefficient
μ	chemical potential
ν	number of groups
ξ	mass fraction

Subscripts

A	group A
B	neutral group B
C	charged group C (by dissociation of B)
D	counter ion dissociated from B
g	group
i	component
j	component
J	group
K	group
L	group

NP   neutral polymer
PE   polyelectrolyte
W   water

Superscripts

(NP)   neutral polymer
(PE)   polyelectrolyte
ref   reference state

15 Experimental and Theoretical Studies on Partitioning of Native and Unfolded Enzymes in Aqueous Two-Phase Systems

Maria-Regina Kula[1] and Christian Rämsch

15.1 Abstract

The phase diagrams of the quarternary mixtures PEG + sulfate + urea + water as well as PEG + dextran + urea + water were measured for various molecular weight fractions of the polymers employed. High concentrations of urea (5–6 M) are compatible with these aqueous two-phase systems. This allowed to partition T 4 lysozyme and variants of the enzyme in the folded and unfolded state and to study the influence of protein conformation and charge on the partition coefficient of proteins. The influence of protein conformation on partition – reflecting available surface area for interaction with the solvent – is higher than charge differences. Sulfate stabilizes protein conformation considerably as shown following the fluorescence of tryptophan residues. At room temperature and neutral pH values complete unfolding of T 4 lysozyme can only be reached in PEG + dextran two-phase systems. Proteins can be unfolded, however, by a combination of temperature, pH, and urea and extracted also in PEG + sulfate systems.

15.2 Introduction

Liquid-liquid extraction is a well-known unit operation in the chemical industry. Compared to other separation technologies extraction is considered to have special advantages handling labile substances or when distillation is impossible for economic reasons or product properties. In the biotechnological industry solvent extraction is employed e.g. for the isolation of antibiotics from fermentation broth. The commonly used solvents, however, can not be used to extract biologically active proteins because of the polar and macromolecular structure of proteins. For protein extraction aqueous two-phase systems can be exploited with advantage. Such systems are based on polymer/polymer interaction, polymer/salt interaction, or thermo

---

[1] Heinrich-Heine-Universität Düsseldorf, Institut für Enzymtechnologie, 52426 Jülich.

separation and have been intensively studied over the last decades (Albertsson, 1986; Walter et al., 1985; Hustedt et al., 1985; Kula, 1979; Kula and Selber, 1999; Zaslavski, 1995). Over the last 20 years the production of biologically active proteins also has dramatically changed applying molecular genetic techniques and the number of recombinant proteins for pharmaceutical and diagnostic use is growing rapidly (Walsh, 2000). In parallel with this development the demand for protein-separation technology is increasing as the isolation and purification of a pharma protein contributes up to 80% to the production cost. During the cultivation of recombinant *Escherichia coli* to high expression levels the target protein is often found in an aggregated, inactive form as „inclusion bodies". These inclusion bodies have to be solubilized with high concentrations of urea or guanidinium hydrochloride and then refolded to the biologically active, native conformation. Hart et al. (1994, 1995) described a process for the isolation of insulin-like growth factor I from recombinant *Escherichia coli* utilizing an aqueous phase system composed of PEG and Na_2SO_4 in the presence of urea. The process is very remarkable as the product was extracted from whole culture broth in 90% yield and 97% purity. This approach has many advantages if denatured proteins have to be recovered and processed. Since the knowledge of aqueous phase systems containing urea in high concentration was extremely limited, we investigated the quaternary phase diagrams of PEG + sodium sulfate + urea + water mixtures and PEG + dextran + urea + water mixtures, we studied the conformation of T 4-lysozyme variants in such phase systems and the influence of protein conformation on the partition coefficient. T 4 lysozyme is a well-studied protein (Matthews, 1995) and was chosen as a model. We obtained several recombinant strains of T 4-lysozyme mutants from Prof. Brian Matthews (Univ. of Oregon) with known 3 D structure and different stability and charge for our investigations. Some parts of the results of these studies have been published previously (Rämsch et al., 1999, 2000; Rämsch, 2000).

15.3 Material and Methods

15.3.1 Chemicals and Lysozyme Mutants

The chemicals employed were commercial products of the highest purity available. The recombinant *Escherichia coli* strains carrying T 4 lysozyme or mutants thereof were cultivated and the proteins purified as described by Dao-Pin et al. (1991) to a purity of 95% as judged by SDS PAGE. The cystein content of the samples was determined according to Ellmann (1959) to verify the presence (two residues) or absence of the amino acid in the different constructs (see Table 15.1). All variants used had activity similar to the wild type enzyme with the exception of L 99 A* and WT 4 which exhibited 60 and 40% of wild type activity, respectively.

Table 15.1: Properties of T 4-lysozyme variants (from Dao-Pin et al., 1991, and Matthews, 1995). Single letter abbreviations are used to denote amino acid replacements, + indicates that detailed structural information is available.

MUTANT code substitutions	$T_m/°C$	Activity	X-Ray structure
WT	67.2	100	+
WT* C49T/C97A	65.8	103	+
L 99 A* C49T/C97A/L99A	51.6	60	+
V 149 T* C49T/C97A/V149T	57.1	100	+
WT 2 K16E/K135E	65.6	142	−
WT 3 K16E/K135E/K147E	64.7	93	−
WT 4 K16E/R119E/K135E/K147E	63.2	40	+

15.3.2 Determination of Phase Diagrams

Predetermined amounts of the components were weight and made up to 10.0 g systems on a tabletop balance. The components were dissolved and mixed using an overhead shaker, incubated, and separated at $25 \pm 0.5\,°C$ by gravity for 8 h. Top and bottom phase were sampled carefully and analyzed for PEG + dextran or sulfate and urea as described in detail by Rämsch et al. (1999), using appropriate HPLC methods. Each phase diagram was defined by 8–12 tie lines measured at constant urea concentration. All concentrations are given in mass%.

15.3.3 Protein Conformation

Protein conformation was determined by fluorescence spectroscopy making use of the fact that the emission maximum of tryptophan depends on the solvent exposure of tryptophan side chains in the protein. The emission maximum is shifted from 332 nm in the native to 354 nm in the unfolded conformation. A luminescence spectrometer LS 50 B from Perkin Elmer (Überlingen, Germany) was employed in the studies at constant temperature (25 °C). The excitation was set to 280 nm, a slit width of 7.5 nm was used.

15.3.4 Partition Coefficients

As in solvent extraction the partition coefficient of a compound in an aqueous two-phase system is defined as

$$k = \frac{m'}{m''} \approx \frac{c'}{c''} \tag{15.1}$$

and was expressed as ratio of concentration in the top and bottom phase. Lysozyme was determined by UV spectroscopy at 280 nm using the known specific adsorption coefficient for lysozyme of $oD_{280} = 1.28$ m AU/mg (Tsugita, 1971). In some experiments the enzymatic activity of WT* was assayed according to Tsugita (1971) and the partition coefficient expressed as the ratio of activity in the top and bottom phase.

15.4 Results and Discussion

15.4.1 Phase Diagrams of Quaternary Systems

Phase diagrams of PEG + Na$_2$SO$_4$ systems and PEG + dextran systems containing urea were determined experimentally. As a typical example Figure 15.1 illustrates the results for PEG 3000 + dextran T 500 and 30% urea in 10 mM potassium phosphate buffer, pH 7. In Figure 15.2 the binodals are presented in PEG 6000 + dextran systems at various urea con-

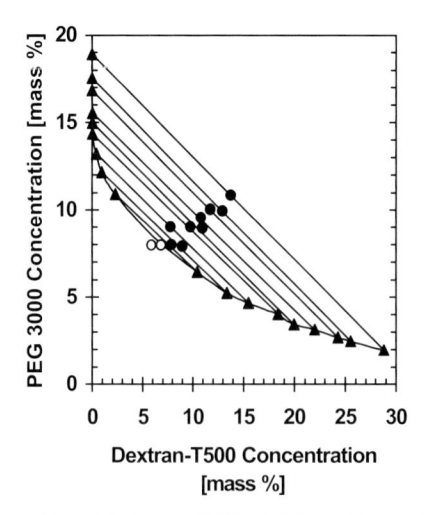

Figure 15.1: Phase equlibria PEG 3000 + dextran T 500 + 30 mass% urea 25 °C, pH 7 (● feed concentration, ▲ top/bottom phase, ○ one phase).

15.4 Results and Discussion

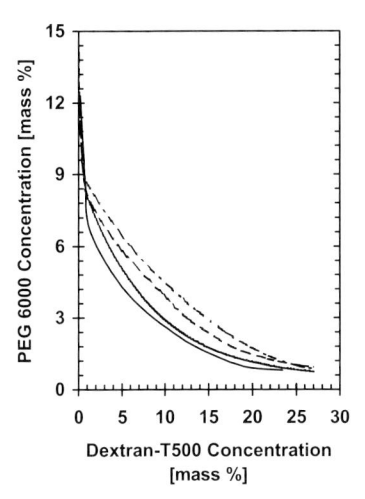

Figure 15.2: Binodials PEG 6000 + dextran T 500 + urea systems, 10 mM KPi pH 7, 25 °C (— 0% (Forciniti et al., 1990); from bottom to top: 20 mass%; 30 mass%; 35 mass% urea).

centrations illustrating that the presence of urea increases the mutual solubility of the polymers and decreases the miscibility gap. Phase separation is observed up to 35% (~6 M) urea in PEG + dextran systems. Lowering the chain length of PEG shifts the binodal away from the origin (Rämsch et al., 1999). Phase diagrams could be described quite well using the VERS model (Brenneisen et al., 2002; Großmann et al., 1997 and 1998; Tintinger et al., 1997a,b). The partition coefficient of urea was determined in PEG 6000 + dextran T 500 systems as a function of the tie line length. As shown in Figure 15.3 k_{urea} is close to one increasing very slightly with the tie line length, similar trends as described above were found

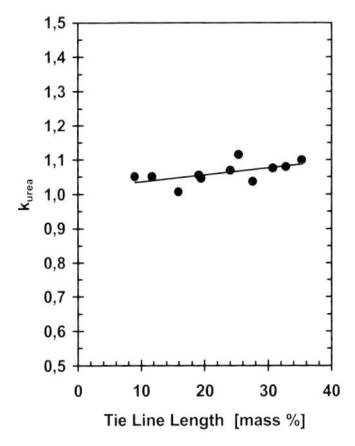

Figure 15.3: Urea partitioning in PEG 6000 + dextran T 500 + 20 mass% urea system, 10 mM Kpi, pH 7, 25 °C.

in PEG + Na$_2$SO$_4$ systems containing urea (Rämsch et al., 1999). Phase separation in PEG + Na$_2$SO$_4$ systems was observed up to 30% (~5 M) urea. In PEG + salt as well as in PEG + dextran systems solubility of urea in the complex fluids limits the urea concentration that can be utilized to unfold proteins.

15.4.2 Stability of T 4-Lysozyme Variants

Table 15.1 summarizes the different T 4-lysozyme variants investigated. These mutants were studied first in aqueous buffer containing increasing urea concentrations following the unfolding of the proteins by fluorescence emission. The results are presented in Figures 15.4a, b and show the typical behavior of monomeric proteins (Creighton, 1993). At an emission maximum of ca. 332 nm tryptophan residues are buried inside the protein structure while at a maximum of about 354 nm the side chain of tryptophans are fully exposed to the solvent after the protein unfolds. With the exception of variant WT 2 the unfolding by urea – taken as the midpoint of the transition curves (T_m) presented in Figure 15.4 – corresponds to the trend in temperature stability.

Since the maximal urea concentration in the aqueous phase systems is 5–6 M, it appears possible to unfold all mutants fully in aqueous phase systems except WT 2.

To examine the influence of the phase forming components on protein unfolding the maximum of the fluorescence emission spectra of the variants was determined in the presence of PEG 6000 and dextran T 500 at concentrations typically found in the top or the bottom phase of a corresponding biphasic system. The results are illustrated with the variant V 149 T* in Figure 15.5. The presence of the polymers PEG and/or dextran stabilizes protein structure but only weakly in contrast to sulfate salts, which strongly stabilize protein

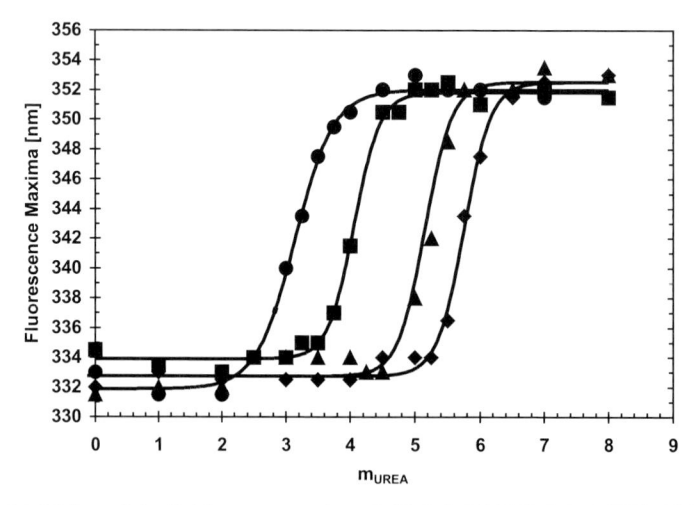

Figure 15.4a: Unfolding of the T 4-lysozyme variants L 99 A (●), V 149 T (■), WT* (▲) and WT (◆) in aqueous urea solution, 10 mM potassium phosphate at 25 °C, pH 7 observed by fluorescence emission spectra.

15.4 Results and Discussion

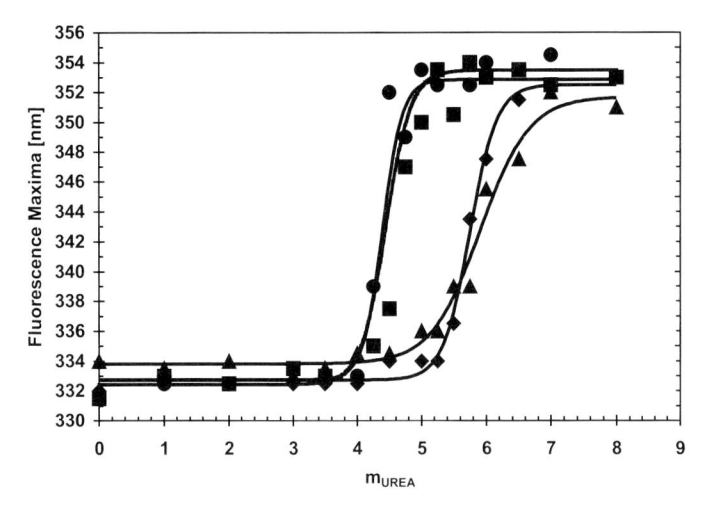

Figure 15.4b: Unfolding of the T4-lysozyme variants WT 4 (■), WT 3 (●), WT 2 (▲), and WT (◆) in aqueous urea solution, 10 mM potassium phosphate at 25 °C, pH 7 observed by fluorescence emission spectra.

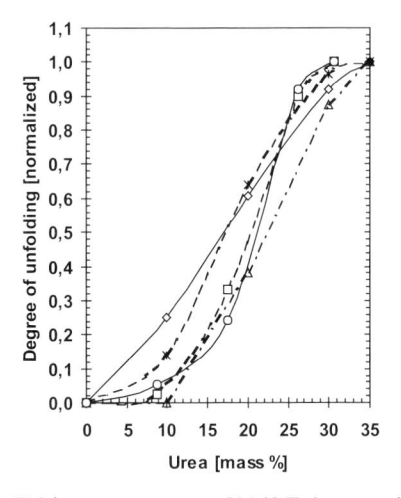

Figure 15.5: Stability of the T 4-lysozyme mutant V 149 T in a typical phase environment of a PEG 6000 + dextran T 500 + 10 mM potassium phosphate system at 25 °C, pH 7 as judged by fluorescence emission spectra (◇ 0% PEG, 0% DEX; × 20% PEG, 0% DEX; △ 9% PEG, 0.5% DEX; □ 2% PEG, 15% DEX; ○ 0% PEG, 20% DEX).

structure against unfolding by urea (Rämsch et al. 1999, 2000). Similar results as shown in Figure 15.5 were obtained for all variants, no change was found in the relative order of the variants as seen in Figures 15.4 a,b. These results confirm that unfolding of proteins by urea can be achieved at neutral pH values and ambient temperatures only in PEG + dextran systems. In PEG + Na_2SO_4 systems extreme pH values as well as increased temperature are necessary (see below) to destabilize the protein conformation in the presence of sulfate.

15.4.3 Partition of T 4-Lysozyme Variants

The partition of the most unstable variants L 99 A* and V 149 T* in comparison to the wild type enzyme and WT* was discussed in detail by Rämsch et al. (2000). The partition coefficient increases dramatically upon unfolding of the protein chain, which appears to be a combined effect of the increase in total surface area as expected from the equation derived by Brønsted (1931) and of the increase in the hydrophobicity of the surface during unfolding. The results are remarkably different for the partition of mutants with different charges, however. Figures 15.6 and 15.7 show partition coefficients of charge variants in PEG + Na_2SO_4

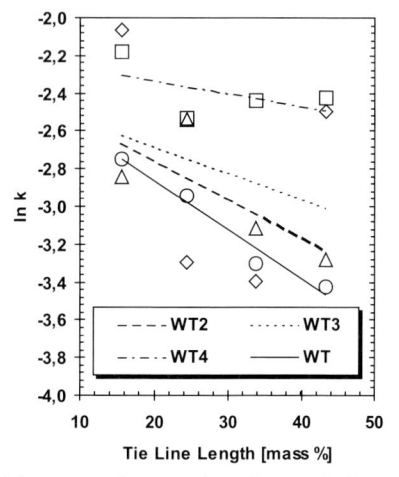

Figure 15.6: Partitioning of T 4-lysozyme charge variants in a PEG 6000 + Na_2SO_4 system, pH 7, 25 °C (□ WT 4, ◇ WT 3, △ WT 2, ○ WT).

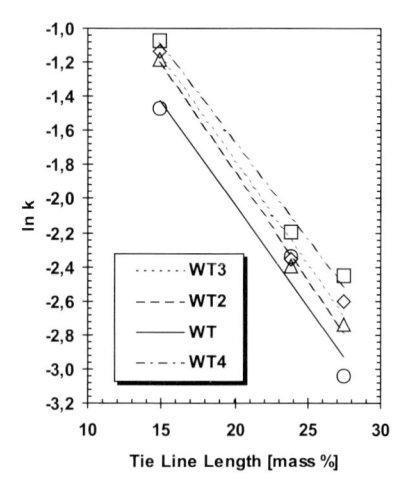

Figure 15.7: Partitioning of T4-lysozyme charge variants in a PEG 6000 + Na_2SO_4 + urea 30 mass% system, pH 7, 25 °C (□ WT4, ◇ WT3, △ WT2, ○ WT).

15.4 Results and Discussion

systems in the presence/absence of (30 mass%) urea. Due to the stabilizing effect of the sulfate ions the proteins should be mostly in the native conformation and the observed differences possibly reflect changes in the hydrophobicity of the systems. The results are in agreement with studies of Albertsson (1986), which indicate that the slope of ln k should decrease with charge and increase with the tie line length. In the charge variants lysine's (K) and and arginine (R) are replaced by glutamate (E) resulting in a decreasing charge from 9 e of the wild type enzyme at neutral pH to 5 e (WT 2), 3 e (WT 3), and 1 e (WT 4), respectively, which was controlled by titration experiments. Figures 15.8 and 15.9 show the partition of

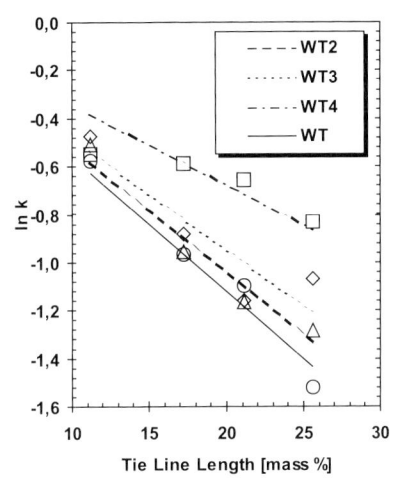

Figure 15.8: Partitioning of T4-lysozyme charge variants in a PEG 6000 + dextran T 500 system, 10 mM KPi, pH 7, 25 °C (□ WT4, ◇ WT3, △ WT2, ○ WT).

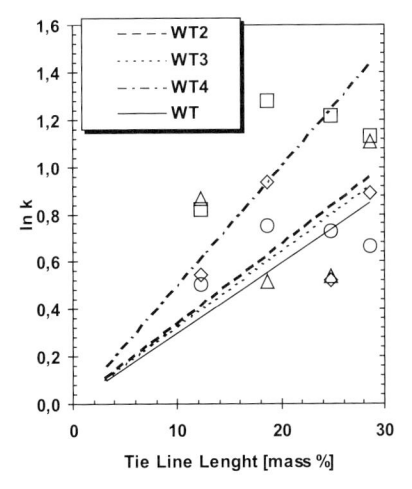

Figure 15.9: Partitioning of T 4-lysozyme charge variants in a PEG 6000 + dextran T 500 + 35 mass% urea system, 10 mM KPi, pH 7, 25 °C (□ WT4, ◇ WT3, △ WT2, ○ WT).

15 Partitioning of Native and Unfolded Enzymes in Aqueous Two-Phase Systems

these variants in PEG 6000 + dextran T 500 systems in the presence and absence of (35 mass%) urea. From the results shown in Figures 15.4 and 15.5 it can be expected, that the wild type enzyme and at least variants WT 4 and WT 3 should be unfolded in the urea containing system. As a result the partition coefficient is above one and the slope is reversed. The figures also illustrate that the effect of protein charge on partition is much less pronounced than changes in conformation. From the values for WT and WT 4 the electrostatic potential $\Delta\phi$ may be accessed according to equation

$$\Delta\Phi(\text{TLL}) = -\frac{RT}{F}\left(\frac{\ln k_{\text{WT}}(\text{TLL}) - \ln k_{\text{WT4}}(\text{TLL})}{Z_{\text{WT}} - Z_{\text{WT4}}}\right) \tag{15.2}$$

Figure 15.10 illustrates that the addition of urea to PEG 6000 + dextran T 500 systems alters the potential between the phases only slightly despite the large influence of urea on the binodial of the systems. From these results it can be concluded that even slight changes of protein conformation with reaction conditions will influence the observed partition in aqueous two-phase system, which is difficult to access and control and has not been considered previously in attempts to model the partition coefficient of a protein.

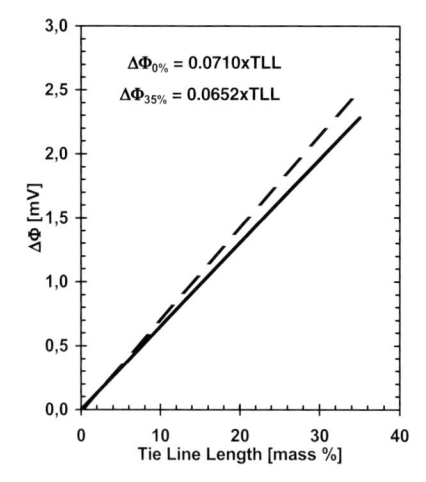

Figure 15.10: Potential difference calculated from the protein partition data of T 4 lysozym WT and WT 4 in PEG 6000 + dextran T 500 + urea systems, 10 mM KPi, pH 7, 25 °C (- - - - 0 mass%, ——— 35 mass% urea).

15.4.4 T 4-Lysozyme Unfolding and Refolding in PEG + Na₂SO₄ Systems

As already discussed proteins are stabilized in PEG + sulfate systems containing urea at neutral pH values by the sulfate ion and even the mutant L 99 A* exhibits a low partition coefficient of about one. Altering the pH, in addition to the presence of urea, however, results in changes in protein conformation as illustrated in Figure 15.11. It can be seen that the partition coefficient increases strongly at pH values <5 and >9 concomitant with

15.4 Results and Discussion

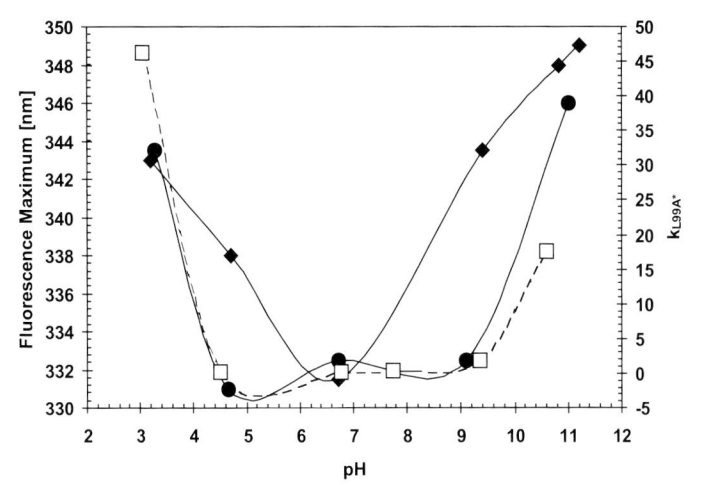

Figure 15.11: Relation between partition coefficient of T 4-lysozyme variant L 99 A* (□) and the change of conformation through pH-change (fluorescence maximum ◆ top-phase and ● bottom-phase).

changes of the emission maximum of the protein measured under these conditions. The fluorescence emission spectra were measured in the separated phases and again show the strong effect of the salt predominately present in the bottom phase on protein stabilization. Even at pH 3 and 9.5 complete unfolding of the rather unstable mutant L 99 A* was not obtained as judged by the emission maxima. Experiments with WT* were carried out to test, whether in principle the protein can be unfolded and extracted at extreme pH values and then be renatured. The variant WT* was partitioned at rather high concentrations in a PEG 6000 + Na_2SO_4 system in the presence of 20 mass% urea at pH 10.5. By activity assays a partition coefficient of 0.3 was found up to 30 °C which compares with k <0.1 at pH 7 at 25 °C. At temperatures >35 °C activity decreased and fluorescence measurements showed complete unfolding of WT*. Concomitantly k increased to values >10. The top phase was removed after phase separation at 40 °C and diluted at 40 °C in prewarmed 6 M urea, 2.5 mM DTT, 10 mM potassium phosphate buffer pH 7.0 in various proportions. The dilutions were cooled slowly to 25 °C over a period of 3.5 h, then activity was measured. At high dilutions (1:7 to 1:11) 78–90% lysozyme activity was recovered. Re-partitioning in the original PEG + sulfate system (pH 7) yielded a k value <0.1. In the experiment 74% of the activity introduced could be recovered. The results show that proteins can be unfolded, extracted, and renatured in urea containing PEG + Na_2SO_4 system employing high pH values and increased temperatures for the destabilization of the protein conformation resulting in a dramatic change in k from <0.1 to >10. These findings can explain the surprising results obtained by Hart et al. (1994), who solubilized insulin-like growth factor I from inclusion bodies mainly formed in the periplasm using whole broth of recombinant *Escherichia coli* at pH 10 and 2 M urea and extracted 90% of the solubilized non-native protein in 97% purity in a PEG 8000 + Na_2SO_4 system at 22 °C. Apparently the fine tuning of the opposing influence of sulfate, pH, and urea concentration on protein conformation allows highly selective extractions in PEG + Na_2SO_4 systems.

15.5 Conclusion

Conformation was demonstrated to be an important parameter of the partition of proteins in aqueous two-phase systems which has been neglected so far. Protein conformation depends on solution properties e.g. pH, and salt composition, and is difficult to access. But unfolding of a protein can be achieved by a combination of high urea concentration, high pH values, and increased temperature and may be employed to isolate aggregated recombinant proteins by extraction under suitable conditions.

Acknowledgements

We thank Prof. B. Matthews (U. Oregon) for the gift of the mutant strains, Prof. G. Maurer, Dr. J. Brenneisen, and Dr. J. Thömmes for fruitful discussions and Dipl. Ing. L. Kleinelangehorst and Dipl. Ing. E. Knieps for expert help.

References

Albertsson, P.Å. (1986): Partition of cell particles and macromolecules. 3rd ed., John Wiley & Sons, New York.
Brenneisen, J.; Rämsch, C.; Thömmes, J.; Kula, M.-R.; Maurer, G. (2002): Aqueous two-phase systems containing urea: Modeling the partitioning of T4-lysozyme variants with VERS (Virial Equation with Relative Surface fractions). Manuscript in preparation.
Brønsted, J.N. (1931): Molekülgröße und Phasenverteilung. Z. Phys. Chem. 1931, Abt. A (Bodenstein-Festband), 257–266.
Creighton, T.E. (1993): Proteins: Structures and molecular properties. 2nd ed., W.H. Freeman and Company, New York.
Dao-Pin, S.; Söderlind, E.; Baase, W.A.; Wozniak, J.A.; Sauer, U.; Matthews, B.W. (1991): Cumulative site-directed charge-change replacements in bacteriophage T 4 lysozyme suggest that long-range electrostatic interactions contribute little to protein stability. J. Mol. Biol. 221: 873–887.
Ellmann, G.L. (1959): Modification and quantitative detection of sulfhydryl groups. Arch. Biochem. Biophys. 82: 70–77.
Forciniti, D.; Hall, C.K.; Kula, M.-R. (1990): Interfacial tension of poly(ethylene glycol)-dextran-water systems: Influence of temperature and polymer molecular weight. J. Biotechnol. 16: 279–296.
Großmann, C.; Tintinger, R.; Zhu, J.; Maurer, G. (1997): Partitioning of some amino acids and low molecular peptides in aqueous two-phase systems of poly (ethylene glycol) and dipotassium hydrogen phosphate. Fluid Phase Equilibria 137: 209–228.
Großmann, C.; Tintinger, R.; Zhu, J.; Maurer, G. (1998): Partitioning of low molecular combination peptides in aqueous two-phase systems of poly(ethylene glycol) and dextran in the presence of small amounts of K_2HPO_4/KH_2PO_4 buffer at 293 K: experimental results and predictions. Biotechnol. Bioeng. 60: 699–711.

Hart, R.A.; Lester, P.M.; Reifsnyder, D.H.; Ogez, J.R.; Builder, S.E. (1994): Large scale, in situ isolation of periplasmic IGF-I from *E. coli*. Bio/Techn. 12: 1113–1117.

Hart, R.A.; Ogez, J.R.; Builder, S.E. (1995): Use of multifactorial analysis to develop aqueous two-phase systems for isolation of non-native IGF-I. Bioseparation 5: 113–121.

Hustedt, H.; Kroner, K.H.; Kula, M.-R. (1985): Applications of phase partitioning in biotechnology, in: "Partitioning in aqueous two-phase systems: Theory, methods, uses and applications to biotechnology". H. Walter, D.E. Brooks, D. Fisher (eds.), Academic Press, New York: 529–587.

Kula, M.-R. (1979): Extraction and purification of enzymes using aqueous two-phase systems. Appl. Biochem. Bioeng. 2: 71–95.

Kula, M.-R.; Selber, K. (1999): Protein purification by aqueous liquid-liquid extraction, in: "Encyclopedia of bioprocess technology". M.C. Flickinger and S.W. Drew (eds.), John Wiley & Sons, New York: 2179–2191.

Matthews, B.W. (1995): Studies on protein stability with T4-lysozyme. Adv. Prot. Chem. 46: 249–278.

Rämsch, C.; Kleinelanghorst, L.B.; Knieps, E.A.; Thömmes, J.; Kula, M.-R. (1999): Aqueous two-phase systems containing urea: Influence on phase separation and stabilisation of protein-conformation by phase components. Biotechnol. Progr. 15: 493–499.

Rämsch, C.; Kleinelanghorst, L.B.; Knieps, E.A.; Thömmes, J.; Kula, M.-R. (2000): Aqueous two-phase systems containing urea: Influence of protein structure on protein partitioning. Biotech. Bioeng. 69: 83–90.

Rämsch, C. (2000): Untersuchung zur Verteilung von nativen und entfalteten Enzymen in wässrigen Zweiphasensystemen. Shaker Verlag, Aachen.

Tsugita, A. (1971): Phage lysozymes and other lytic enzymes, in: The enzymes. P.D. Boyer (ed.), 3rd ed., Vol. 5, Academic Press, New York: 343–411.

Tintinger, R.; Zhu, J.; Großmann, C.; Maurer, G. (1997a): Partitioning of some combination di- and tripeptides in aqueous two-phase systems of poly(ethylene glycol) and dipotassium hydrogen phosphate at 293 K: experimental results and prediction. Ber. Bunsen-Ges. Phys. Chem. 101: 687–697.

Tintinger, R.; Zhu, J.; Großmann, C.; Maurer, G. (1997b): Partitioning of some amino acids and low molecular mass peptides in aqueous two-phase systems of poly(ethylene glycol) and dextran in the presence of small amounts of dipotassium hydrogen phosphate/potassium dihydrogen phosphate-buffer at 293 K: experimental results and correlation. J. Chem. Eng. Data 42: 975–984.

Walsh, G. (2000): Biopharmaceutical benchmarks. Nature/Biotechnology 18: 831–833.

Walter, H.; Brooks, D.E.; Fisher, D. (1985): Partitioning in aqueous two-phase systems. Academic Press, New York.

Zaslavski, B.Y. (1995): Aqueous two-phase partitioning: Physical chemistry and bioanalytical applications. Marcel Dekker Inc., New York.

Nomenclature

c	concentration
F	Faraday constant
k	partition coefficient
m	molality
PEG	polyethylene glycol
R	universal gas constant
T	absolute temperature
TLL	tie line length [mass%]

T_m midpoint of transition during the unfolding of protein
z number of charges

Greek letters

$\Delta\phi$ electrostatic potential difference

Superscripts

I top phase
II bottom phase

16 Experimental and Theoretical Investigations on the Partitioning of Proteins in Aqueous Two-Phase Systems

Jochen Brenneisen and Gerd Maurer [1]

16.1 Abstract

Liquid-liquid extraction using aqueous two-phase systems is a suitable process for the recovery/purification of proteins. The partitioning of proteins in such systems is often described assuming an "electrical potential difference" between the coexisting liquid phases. Recently a method has been described for predicting that potential difference from a model for the Gibbs excess energy of the liquid phases. When the protein partitioning is dominated by the effects summarized in the "electrical potential difference" the net charge of the protein is the only additional property required to predict protein partitioning to an aqueous two-phase system. That procedure is tested by comparing new experimental results for the partitioning of several proteins to aqueous two-phase systems of poly(ethylene glycol) (PEG) and either K_2HPO_4 or dextran (DEX) with model predictions.

16.2 Introduction

Proteins are usually produced by fermentation in aqueous solutions. Downstream processing must take into account the special properties of proteins as, e.g., the observation that most proteins are only stable in aqueous media within narrow limits of parameters such as temperature, pH, and ionic strength. Therefore, extraction by an organic solvent is rarely an appropriate process to concentrate and purify proteins. However, extraction using aqueous two-phase systems allows to adjust the extraction conditions in such a way that product loss by protein denaturation can be minimized. Aqueous two-phase systems are observed when two hydrophilic, but incompatible solutes are simultaneously dissolved in water. Typical examples for such solutes are high molecular weight poly(ethylene glycol) (PEG) on one side and dextran (DEX) or an inorganic salt (e.g. Na_2SO_4 or K_2HPO_4) on the other side. Although

---

[1] Technische Universität Kaiserslautern, Fachbereich Maschinenbau und Verfahrenstechnik, Lehrstuhl für Technische Thermodynamik, Erwin-Schrödinger-Straße, 67653 Kaiserslautern.

the water concentration in such mixture might reach about 80 mass%, a phase split into two water-rich liquid phases is observed where each of the hydrophilic solutes prefers one of the coexisting phases. In aqueous systems of PEG and DEX both pH and ionic strength can be adjusted by addition of buffers and neutral salts to minimize denaturation of the proteins.

As proteins are polyelectrolyte macromolecules the partitioning of a protein to an aqueous two-phase system depends on its net charge as well as on the pH-difference of the coexisting phases. The pH as well as the pH-difference depends on the phase forming systems as well as on the addition of buffers and salts (cf. Albertsson, 1986; Walter et al., 1985). It is common practice to correlate the partitioning of a protein to an aqueous two-phase systems by applying the concept of an "electrical potential difference" $\Delta\Phi$ (cf. Albertsson, 1986):

$$\ln K = \ln K_0 + F z \Delta\Phi / RT \tag{16.1}$$

where K is the partition coefficient of the protein (concentration in top phase/concentration in bottom phase), K_0 is the partition coefficient in the absence of an electrical potential difference, F is Faraday's constant, z is the number of net charges of the protein, $\Delta\Phi$ is the electrical potential difference between the coexisting phases, R is the universal gas constant, and T is the thermodynamic temperature. When either K_o is close to unity or $\Delta\Phi$ is in the order of a few millivolts and the protein carries a sufficiently high net charge, the partitioning of a protein to an aqueous two-phase system is governed by the second term on the right hand side of Eq. (16.1). Methods to determine the net charge of a protein are available, but, until recently there was no method available neither to measure nor to calculate the "electrical potential difference" $\Delta\Phi$. Großmann and Maurer (1995) presented an explanation for that property and derived a method to calculate $\Delta\Phi$ from an expression for the Gibbs energy of the aqueous solution. It was shown that the "electrical potential difference" results from the assumption that each of the coexisting phases is electrically neutral, i.e. that not all ionic species can partition independently between the coexisting phases (as the condition of electroneutrality of each of the coexisting single phases has to be obeyed).

16.3 Theory

16.3.1 Model

The compositions of the coexisting phases in a salt-containing aqueous two-phase system can be calculated by minimizing the Gibbs energy of the system under the constraint that both phases have to be electrically neutral (Großmann and Maurer, 1995). When the phase boundary allows all species to partition between both coexisting phases (characterized by superscripts ' and ") the conditions for phase equilibrium are:

$$T' = T'' \tag{16.2}$$

$$p' = p'' \tag{16.3}$$

16.3 Theory

$$\mu'_i - \mu''_i = \frac{z_i}{z_k}\left(\mu'_k - \mu''_k\right) \tag{16.4}$$

where μ'_i is the chemical potential of species i which results from the Gibbs energy G through

$$\mu_i = \left(\frac{\partial G}{\partial n_i}\right)_{T,p,n_{j \neq i}} \tag{16.5}$$

and subscript k stands for an arbitrarily selected ionic species. Using the same reference state for species in both phases the difference $\mu'_k - \mu''_k$ can be replaced by introducing the activities a'_k and a''_k:

$$\mu'_k - \mu''_k = RT \ln\left(\frac{a'_k}{a''_k}\right) \tag{16.6}$$

resulting in:

$$\mu'_i - \mu''_i = \frac{z_i}{z_k} RT \ln\left(\frac{a'_k}{a''_k}\right) \tag{16.7}$$

The right hand side can be rearranged by introducing the concept of an electrical potential difference $\Delta\Phi = \Phi' - \Phi''$:

$$RT \ln\left(\frac{a'_k}{a''_k}\right) = -\Delta\Phi F z_k \tag{16.8}$$

resulting in

$$\mu'_i - \mu''_i = -z_i F \Delta\Phi \quad (i = 1, 2, N; i \neq k) \tag{16.9}$$

where N is the number of partitioning species.

As from Eq. (16.6)

$$\mu'_i - \mu''_i = RT \ln\left(\frac{a'_i}{a''_i}\right) \tag{16.10}$$

the condition of phase equilibrium (cf. Eq. (16.4)) becomes:

$$\ln\left(\frac{a''_i}{a'_i}\right) = \frac{z_i F}{RT} \Delta\Phi \tag{16.11}$$

When the activity of a partitioning solute i is expressed by the product of molality m_i and activity coefficient $\gamma_i^{(m)}$

$$a_i = m_i \gamma_i^{(m)} \tag{16.12}$$

and the partition coefficient of species i is defined as

361

$$K_i = m_i''/m_i' \tag{16.13}$$

that partition coefficient becomes:

$$\ln K_i = \ln\left(\gamma_i^{(m)'}/\gamma_i^{(m)''}\right) + z_i F\Delta\Phi/(RT) \tag{16.14}$$

which agrees with Eq. (16.1) as

$$\ln K_{o,i} = \ln\left(\gamma_i^{(m)'}/\gamma_i^{(m)''}\right) \tag{16.15}$$

At high dilution of partitioning species i its activity coefficient γ_i can be replaced by the infinite dilution activity:

$$\gamma_i^\infty = \lim_{m_i \to 0} \gamma_i^{(m)} \tag{16.16}$$

resulting in

$$\ln K_{o,i}^\infty = \ln\left(\gamma_i'^{,\infty}/\gamma_i''^{,\infty}\right) \tag{16.17}$$

where $\gamma_i'^{,\infty}$ and $\gamma_i''^{,\infty}$ are the activity coefficients of partitioning species i at infinite dilution in the coexisting phases ' and ". Similarly, the potential difference is approximated by the limiting value, i. e. the potential difference in the phase forming aqueous two-phase system:

$$\Delta\Phi \to \Delta\Phi^{(o)} = \lim_{\text{all add.solutes} \to 0} \Delta\Phi = \frac{RT}{F z_k} \lim_{\text{all add.solutes} \to 0} \ln\left(\frac{a_k''}{a_k'}\right) \tag{16.18}$$

$K_{o,i}^\infty$ is a property of the coexisting phases as well as of the partitioning solute i whereas $\Delta\Phi^{(o)}$ only depends on the aqueous two-phase system. The partitioning of a protein in an aqueous two-phase system at high dilution is then given by:

$$\ln K_i = \ln K_{o,i}^\infty + z_i F\Delta\Phi^{(0)}/(RT) \tag{16.19}$$

The partition coefficient of a protein in an aqueous two-phase system can therefore be calculated from a model for the Gibbs energy of the aqueous solutions and the charge number of that protein in the coexisting phases. The pH of the aqueous solutions is one of the most important parameters for that charge number. In the present work it is assumed that the number of electric charges on a protein molecule in an aqueous solution at a fixed temperature depends only on the pH of that solution. That pH is usually adjusted by small amounts of a buffer. The net charge of a protein as a function of the pH is usually determined experimentally.

16.3.2 Gibbs Energy of the Aqueous Solutions

The Gibbs energy of an aqueous solution is calculated using the VERS model (*Virial Equation with Relative Surface Fractions*) (Großmann, 1994) for the excess Gibbs energy. The VERS-Model is a semi-empirical G^E-equation based on Pitzer's equation for aqueous electrolyte solutions (e.g. Pitzer, 1991). The excess is defined on an asymmetric normalization: for water the reference state is the pure liquid and for the solute it is a hypothetical liquid (one molal solution in pure water; interactions as at infinite dilution in pure water). As usual the influence of pressure on G^E is neglected.

$$\frac{G^E 1000}{n_w RT M_w} = f_1(I) + \left(\frac{1000}{M_w}\right)^2 \sum_{i \neq w} \sum_{j \neq w} \frac{\Theta_i \Theta_j}{\Theta_w \Theta_w} A_{ij} + \left(\frac{1000}{M_w}\right)^3 \sum_{i \neq w} \sum_{j \neq w} \sum_{k \neq w} \frac{\Theta_i \Theta_j \Theta_k}{\Theta_w \Theta_w \Theta_w} B_{ijk}$$

(16.20)

Θ_i is the relative surface fractions of component i

$$\Theta_i = \frac{m_i q_i}{\sum_j m_j q_j} \qquad (16.21)$$

m_i and q_i are molality and surface parameter of component i. $f_1(I)$ is a modified Debye-Hückel term which depends on ionic strength I

$$I = \frac{1}{2} \sum_i m_i z_i^2 \qquad (16.22)$$

$$f_1(I) = -A_\phi \frac{4I}{b} \ln\left(1 + b\sqrt{I}\right) \qquad (16.23)$$

A_ϕ is the Debye-Hückel constant of water and b is a constant ($b = 1.2$). A_{ij} and B_{ijk} are binary and ternary interaction parameters, respectively. Similar to Pitzer's equation for aqueous solutions of strong electrolytes, binary interaction parameter A_{ij} is split into two terms. Only the first term depends on ionic strength I:

$$A_{ij} = A_{ij}^{(0)} + A_{ij}^{(1)} \frac{2}{\alpha^2 I}\left[1 - \left(1 + \alpha\sqrt{I}\right)\exp\left(-\alpha\sqrt{I}\right)\right] \qquad (16.24)$$

The constant α ($\alpha = 2.0$) was adopted from Pitzer (1973).
All interaction parameters are symmetric: $A_{ij}^{(v)} = A_{ji}^{(v)}$ ($v = 0, 1$) and $B_{ijk} = B_{ikj} = B_{jki} = B_{jik} = B_{kij} = B_{kji}$. The excess Gibbs energy model requires interaction parameters which were either taken from the literature or fitted to experimental results (cf. Section 16.5).

16.3.3 Chemical Reactions

It is assumed that all salts dissociate completely in water. When phosphates are present the following chemical equilibrium reactions are considered:

$$HPO_4^{2-} + H_2O \rightleftharpoons PO_4^{3-} + H_3O^+ \qquad (16.25)$$

$$H_2PO_4^- + H_2O \rightleftharpoons HPO_4^{2-} + H_3O^+ \qquad (16.26)$$

$$H_3PO_4 + H_2O \rightleftharpoons H_2PO_4^- + H_3O^+ \qquad (16.27)$$

Furthermore the dissociation of water is taken into account:

$$2 H_2O \rightleftharpoons H_3O^+ + OH^- \qquad (16.28)$$

The chemical reaction equilibrium constants (as pK-numbers) of reactions (16.25) to (16.28) were taken from Rumpf and Maurer (1994). They are given in Table 16.1.

Table 16.1: pK-values for reactions 16.25 to 16.28 at 298.15 K (from Rumpf and Maurer, 1994).

Reaction	pK
16.25	11.960
16.26	7.201
16.27	2.131
16.28	13.998

16.4 Experimental/Methods

Several types of experiments were performed. The experiments aimed to provide not only experimental data for the partitioning of some proteins in some aqueous two-phase systems, but also a data base required for the determination of model parameters. Only the essentials of those experiments are given here. Details of the experimental work are available elsewhere (Brenneisen, 2001).

16.4.1 Liquid-Liquid Equilibrium Measurements

The partitioning of small amounts ($\approx$ 1 mg/g) of eight single proteins to aqueous two-phase systems of PEG 6000 + K$_2$HPO$_4$ and PEG 6000 + DEX T500 was determined experimentally at 298 K. In aqueous two-phase systems of PEG 6000 + K$_2$HPO$_4$ neither the pH nor the ionic strength was adjusted, whereas those properties were adjusted in aqueous solutions of PEG 6000 + DEX T500. The pH was adjusted to pH $\approx$ 4.6 (by potassium acetate/acetic acid), to pH $\approx$ 7 (by K$_2$HPO$_4$/KH$_2$PO$_4$), and to pH $\approx$ 10 (by butylamine/hydrochloric acid) – for details cf. Brenneisen (2001). Potassium chloride was used to adjust the ionic strength in aqueous two-phase systems of PEG 6000 + DEX T500 to about 200 and 500 mmolal.

16.4 Experimental/Methods

Aqueous two-phase systems were prepared by mixing stock solutions of the single substances. The mixtures were equilibrated in centrifugal-tubes in a thermostatted water bath over night. The coexisting phases were careful removed with a pipette for analysis. Both phases were analyzed. Table 16.2 gives a summary of the analytical procedures as well as uncertainties of the experimental results for the concentration of the dissolved species.

Table 16.2: Analytical methods and experimental uncertainties.

Substance	Analytical method	Relative experimental uncertainty/%
PEG	spectroscopy	3.5
DEX	polarimetry	2
phosphate		
(phase forming compound)	titration	2
phosphate (buffer)	titration	6
KCl	spectroscopy	4
acetate (buffer)	biochemical test kid	<10
amine/HCl (buffer)	spectroscopy	5
protein	spectrophotometry	5

16.4.2 Experiments on Subsystems

The model for describing the partitioning of proteins to aqueous two-phase systems requires parameters for interactions between dissolved species. Some of those parameters can be determined from independent experimental data, e.g. from experimental data for the activity of water in aqueous solutions and for the pH in aqueous solutions of salt mixtures. Furthermore, as follows from Eq. (16.19) predictions for the partition coefficient of a protein require the number of electric net charges on a protein molecule.

16.4.2.1 Isopiestic Measurements

The isopiestic method is particularly suited for the determination of the activity of a single solvent in mixtures with nonvolatile solutes. Isopiestic measurements were performed on several subsystems e.g. aqueous solutions of K_2HPO_4 and KH_2PO_4 and aqueous solutions of PEG and KCl at 298.15 K. The uncertainty of the experimental results for the activity of water in such mixtures is $\Delta a_w = \pm 0.001$.

16.4.2.2 Measurement of pH

The pH of aqueous solutions of the single salts K_2HPO_4 and KH_2PO_4 and aqueous solutions of mixtures of those salts was determined at 298.15 K as a function of the salt concentration with an uncertainty of $\Delta pH = \pm 0.1$. The results were used for the estimation of parameters of the Gibbs excess energy model.

16.4.2.3 Determination of the Protein Net-Charge

The net charge of a protein in an aqueous solution was determined by combining two experimental techniques: acid/base titration and isoelectric focussing (IEF). The change of the protein's net charge was determined in a titration experiment. An aqueous solution of a known amount of a protein (which was stabilized by small amounts of NaCl) was titrated using aqueous solutions of NaOH and HCl, respectively. The procedures applied followed recommendations by Tanford and Wagner (1954). Isoelectric focussing was applied to determine the isoelectric point pI, i.e. that pH-value where the protein's net charge is zero. The IEF experiments were carried out in a horizontal electrophoresis cell (type Bio-Phoreses Horizontal Electrophoresis Cell) with a supply unit (Power Pac 3000, Bio-Rad Laboratories, Hercules, California USA) and ready to use gels (Servalyt Precotes, Serva, Heidelberg, Germany) with a pH range from 3 to 10. The experiments were carried out according to manufacturer's specifications. The uncertainty of the new experimental results is estimated to $\Delta pI = \pm 0.3$.

16.4.3 Materials

Poly(ethylene glycol) (PEG) 6000 number average molecular weight $M_n = 6230$ (lot #664762) was supplied from Clariant GmbH, Germany. Dextran (DEX) T500 number average molecular weight $M_n = 179{,}350$ was purchased from Pharmacia, Uppsala, Sweden. All salts were bought from various suppliers in analytical grade. All salts and polymers were dried before use. HPLC- water was degassed by vacuum distillation before it was used in all experiments. All proteins were purchased either from Sigma (Deisenhofen, Germany) or from Boehringer (Mannheim, Germany), cf. Table 16.3.

Table 16.3: Proteins: suppliers and isoelectric points at 298.15 K.

Protein	Supplier etc.	pI This work	pI Literature
bovine serum albumin	Sigma A7030 (lot 16H03601)	4.4	4.5 (Manon and Zydney, 1998)
chicken serum albumin	Sigma A2512 (lot 37H7010)	4.8	–
cytochrome c	Sigma C2037 (lot 68H7001)	10.2	10.09 (Theorell and Åkesson, 1941); 10.25 (Marini et al., 1980)
β-galactosidase	Boehringer (567779)	4.5	–
glucosidase	Sigma G0395 (lot 102H4009)	4.4	–

Table 16.3 (continued)

Protein	Supplier etc.	pI This work	pI Literature
catalase	Boehringer (106828)	5.4 – 5.8	5.7 (Malamud and Drysdale, 1978)
lysozyme	Sigma L2879 (lot 15H790)	>10.7	11.1 (Tanford and Wagner, 1954)
myoglobin	Sigma M0630 (lot 48H7021)	7.4	6.5 – 7.8 (Righetti and Cravaggio, 1976)

16.5 Experimental Results/Modelling

16.5.1 Parameters for the Gibbs Excess Energy Model

The expression for the Gibbs excess energy (cf. Eqs. (16.20) and (16.24)) requires the Debye-Hückel parameter A_ϕ for water (at 298.15 K: $A_\phi = 0.391466$), pure component surface parameters (q_i) and binary ($A_{ij}^{(0)}$, $A_{ij}^{(1)}$) as well as ternary (B_{ijk}) interaction parameters.

The number for the surface parameter of water ($q_w = 1.40$) was also assigned to any low molecular ion as well as to all buffer species. PEG 6000 was considered to consist of 139 PEG middle groups (–CH$_2$–O–CH$_2$–) and two end groups (HO–CH$_2$–) with different surface parameters: $q_{PEG,middle} = 1.320$ and $q_{PEG,end} = 1.740$ (Brenneisen, 2001) resulting in $q_{PEG6000} = 187$. DEX T500 ($M_n = 179,350$) is assumed to consist of 1106 dextran ring groups of surface parameter $q_{DEX,ring} = 5.760$, resulting in $q_{DEX\ T500} = 6371$. The most important interaction parameters were fitted to experimental results, whereas less important interaction parameters were neglected, i.e. set to zero. Mainly parameters for interactions between the phase forming components were considered. Parameters for interactions between further solute species were either adopted from the literature, estimated or neglected. All non-zero interaction parameters are given Table 16.4 together with information on the underlying experimental data.

Table 16.4: Interaction parameters at 298.15 K.

Species		$A_{m,k}^{(0)}$	$A_{m,k}^{(1)}$	$B_{m,m,k}$	Source
m	k		Neutral polymers		
PEG	PEG	0.008350	–	–	Großmann et al. (1995)
PEG	DEX	0.004638	–	–	Tintinger (1999)
DEX	DEX	0.001524	–	–	Großmann et al. (1995)
m	k		Potassium phosphates		
K$^+$	H$_2$PO$_4^-$	–0.04867	–0.1330	–	Brenneisen (2001)

Table 16.4 (continued)

Species		$A_{m,k}^{(0)}$	$A_{m,k}^{(1)}$	$B_{m,m,k}$	Source
K$^+$	HPO$_4^{2-}$	0.00903	1.8547	–	Brenneisen (2001)
K$^+$	PO$_4^{3-}$	0.5510	9.6913	–	Brenneisen (2001)
m	k	\multicolumn{3}{c}{Sodium, potassium, and hydrogen chlorides}	
Na$^+$	Cl$^-$	0.7650	0.2664	0.0004233	Pitzer (1991)
K$^+$	Cl$^-$	0.04808	0.21875	–0.00026267	Pitzer (1991)
H$^+$	Cl$^-$	0.20332	–0.01688	–0.00121	Kim, Frederick (1988)
m	k	\multicolumn{3}{c}{PEG/potassium phosphates}	
PEG	K$^+$	0.04487	–	–	Brenneisen (2001)
PEG	HPO$_4^{2-}$	0.06627	–	–	Brenneisen (2001)
PEG	PO$_4^{3-}$	0.0000069	–	–	Brenneisen (2001)
m	k	\multicolumn{3}{c}{Polymer/chloride}	
PEG	Cl$^-$	0.01855	–	–	Brenneisen (2001)
DEX	Cl$^-$	–0.008145	–	–	Brenneisen (2001)
m	k	\multicolumn{3}{c}{Acetate buffer}	
K$^+$	Ac$^-$	0.154632	0.338016	–0.01613	Yan, Han (1994)
K$^+$	Cl$^-$	0.046523	0.223016	–0.00014233	Yan, Han (1994)*
HAc	HAc	–0.032	–	–	Kirsch et al. (1997)
PEG	HAc	0.01237	–	–	Brenneisen (2001)
m	k	\multicolumn{3}{c}{Amine buffer}	
PEG	butyl-amine	–0.06402	–	–	Brenneisen (2001)

* only used in systems with (acetate buffer + KCl)

For example, only a single parameter was considered for interactions between PEG/PEG ($A_{PEG/PEG}^{(0)}$), DEX/DEX ($A_{DEX/DEX}^{(0)}$) and PEG/DEX ($A_{PEG/DEX}^{(0)}$). Two of those parameters ($A_{PEG/PEG}^{(0)}$ and $A_{DEX/DEX}^{(0)}$) were taken from Großmann et al. (1995). The remaining parameter ($A_{PEG/DEX}^{(0)}$) was taken from Tintinger (1999). Only binary parameters were considered for interactions between potassium ions and phosphate ions ($A_{K^+/H_2PO_4^-}^{(v)}$, $A_{K^+/HPO_4^{2-}}^{(v)}$, $A_{K^+/PO_4^{3-}}^{(v)}$ for $v = 0$ and 1). These parameters were fitted to new experimental results for the activity of water and the pH in aqueous solutions of the single salts K$_2$HPO$_4$ and KH$_2$PO$_4$ as well as of mixtures of those salts at 298.15 K. For example, the experimental results for the activity of water and for the pH in aqueous solutions of an equimolar mixture of K$_2$HPO$_4$ and KH$_2$PO$_4$ are compared to the correlation in Figures 16.1 and 16.2. The model represents the experimental data for the activity of water at 298.15 K within experimental uncertainty. The deviation between the experimental and the calculated results for the pH slightly exceeds the experimental uncertainty only at high salt concentrations (≥ 20 mass%).

Parameters for interactions between sodium and chloride ions as well as between potassium and chloride ions in aqueous solutions were taken from Pitzer (1991). Those between hydrogen and chloride ions from Kim and Frederick (1988). Model parameters for interactions between PEG 6000 and potassium phosphates ($A_{PEG/K^+}^{(0)}$, $A_{PEG/HPO_4^{2-}}^{(0)}$, $A_{PEG/PO_4^{3-}}^{(0)}$) were fitted to new experimental results for the liquid-liquid equilibrium in the aqueous two-

16.5 Experimental Results/Modelling

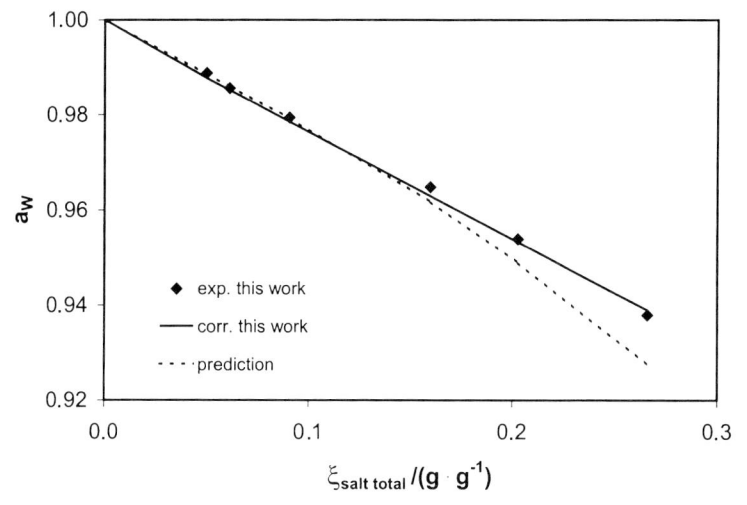

Figure 16.1: Activity of water in equimolar aqueous solutions of K$_2$HPO$_4$ and KH$_2$PO$_4$ at 298.15 K.

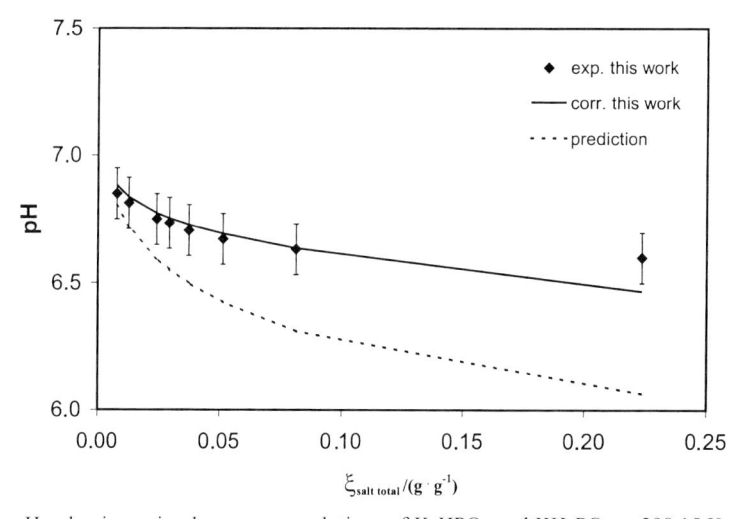

Figure 16.2: pH-value in equimolar aqueous solutions of K$_2$HPO$_4$ and KH$_2$PO$_4$ at 298.15 K.

phase system with PEG 6000 and K$_2$HPO$_4$ at 298.15 K (Brenneisen, 2001). Figure 16.3 shows a comparison between experimental results and calculations for the liquid-liquid equilibrium.

Deviations between calculated and measured concentrations in the coexisting phases are within experimental uncertainty. Figure 16.4 shows a similar comparison for the pH-numbers in the coexisting phases. The model predicts a small pH-difference between both phases.

Although that difference is nearly within the experimental uncertainty, it is confirmed by the experimental results. Model parameter $A^{(0)}_{\text{PEG/Cl}^-}$ for interactions between PEG and

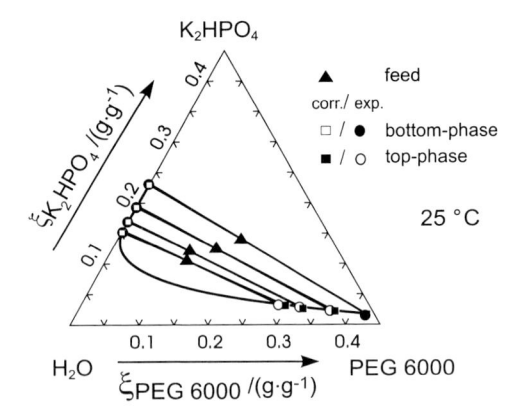

Figure 16.3: Liquid-liquid equilibrium in the ternary system water + PEG 6000 + K$_2$HPO$_4$ at 298.15 K.

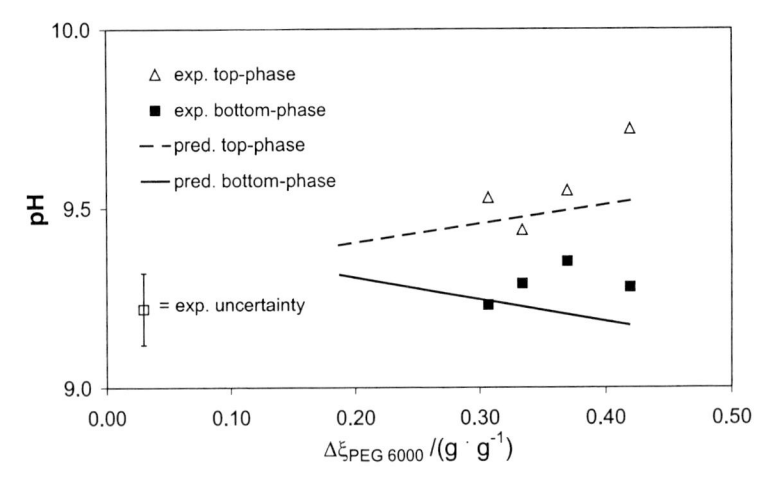

Figure 16.4: Comparison between experimental results and model predictions for the pH in coexisting phases of water + PEG 6000 + K$_2$HPO$_4$ at 298.15 K ($\Delta\xi_{PEG} = \xi_{PEG,top} - \xi_{PEG,bottom}$).

chloride was fitted to experimental results for the activity of water in aqueous solutions of PEG 6000 and potassium chloride whereas the parameter for interactions between DEX and chloride ions was fitted to new experimental results for the partitioning of small amounts of potassium chloride to coexisting liquid phases in the system PEG 6000 + DEX T500 + H$_2$O (Brenneisen, 2001). Potassium chloride reveals no preference for one of the coexisting phases at 298.15 K at least in the investigated concentration range (i.e. m_{KCl} = 0.2 and 0.5 molal). However, the salt increases the miscibility gap of the aqueous two-phase system. The partitioning of a pH-buffer might also have an influence on the liquid-liquid equilibrium. Mixtures of acetic acid and potassium acetate (molar ratio ≈ 1:1) and butylamine and hydrochloric acid (molar ratio ≈ 2:1) were used to adjust the pH to about 4.6 and 10.2, respectively. As the partitioning of the buffers might influence the liquid-liquid equilibrium

16.5 Experimental Results/Modelling

the model was extended to account for these components. Acetic acid is treated as a weak acid (pK_a = 4.756; Harned and Ehl, 1932). The binary parameter $A^{(0)}_{HAc/HAc}$ was taken from Kirsch et al. (1997). Yan and Han (1994) reported Pitzer parameters for potassium acetate in water. These parameters were adopted here. When potassium chloride was used to adjust the ionic strength and at the same time the acetate buffer was used to adjust the pH, the parameters for interactions between potassium and chloride ions by Pitzer (1991) were replaced by those reported by Yan and Han (1994) as these authors investigated aqueous solutions of potassium chloride and the acetate buffer. New experimental results for the partitioning of small amounts (0.02 molal) of the acetate buffer to aqueous two-phase systems of PEG 6000 and DEX T500 (in the presence of up to 0.5 molal KCl) were used to estimate the binary parameter $A^{(0)}_{PEG/HAc}$ (Brenneisen, 2001). In a similar way from the partitioning of butylamine/HCl buffer (concentrations: butylamine: 10 mmolal; HCl: 5.5 mmolal) to the same phase forming system (PEG 6000 + DEX T500 + KCl + H_2O) a binary parameter $A^{(0)}_{PEG/Butylamine}$ was estimated. All parameters for interactions between either PEG or DEX on one side and the buffers as well as potassium chloride on the other side are considered preliminary as the underlying experimental data base is very scarce. Only the interaction parameters mentioned in Table 16.4 were used, i.e. all other interaction parameters were set to zero. In particular no parameters for interactions with a protein were taken into account. The only information about the protein used for predicting the protein partitioning between the coexisting liquid phases is the protein's net charge, which is assumed to solely depend on the pH of the aqueous solution.

16.5.2 Protein Net Charge

Figure 16.5 shows a typical result of an acid-base titration of a protein. The titration results for bovine serum albumin (BSA) are adjusted to the isoelectric point (pI_{BSA} = 4.4). The isoelectric point was determined by isoelectric focussing. The results are given in Table 16.3 together with literature data. Detailed information is available elsewhere (Brenneisen, 2001). Usually a titration curve covered the range from pH ≈ 3 to pH ≈ 11. In the range of interest (4.5 < pH < 10) the results for the net charge number do not depend on the titration direction. Furthermore, in that range the new experimental results for the charge number agree within experimental uncertainty ($\Delta z = \pm 1$) with literature data (Tanford et al., 1955). Outside that pH-range the results from both investigations reveal larger differences, which might be due to an irreversible denaturation of BSA.

16.5.3 Protein Partitioning

The model for the excess Gibbs energy with parameters given in Table 16.4 was used together with the experimental results for the net charge of the proteins to predict the protein partition coefficient between the coexisting aqueous phases at 298.15 K. PEG 6000 was one of the phase-forming components in any two-phase system investigated in the present work. Therefore, in these figures the coexisting phases are characterized by the difference in the

16 Partitioning of Proteins in Aqueous Two-Phase Systems

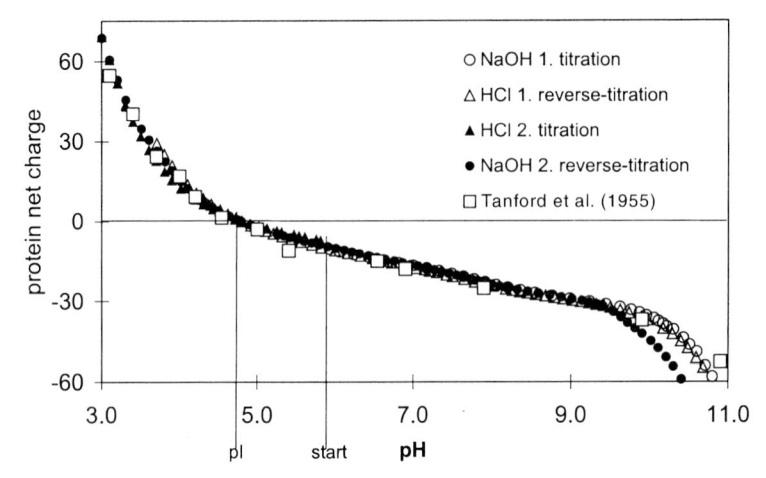

Figure 16.5: Influence of solution pH on the net charge number of BSA in aqueous solutions at 298.15 K.

mass fraction of PEG 6000 in both phases $\Delta\xi_{PEG} = \xi_{PEG,top} - \xi_{PEG,bottom}$. Figure 16.6 shows the comparison for the partition coefficients of lysozyme (at nearly infinite dilution) in water + PEG 6000 + K_2HPO_4.

The net charge number of lysozyme at the pH of that system (pH = 9) is $z_{Lysozyme}$ = (+1 to +2). For $z_{Lysozyme}$ = +2, predictions agree with the experimental results within experimental uncertainty. In the same two-phase system a similar agreement between prediction and experiment was found for the partitioning coefficients of cytochrome c and myoglobin. However, the predictions were only qualitatively correct for glucosidase and bovin serum albumin (BSA). The model predicted the correct preference of these proteins for one

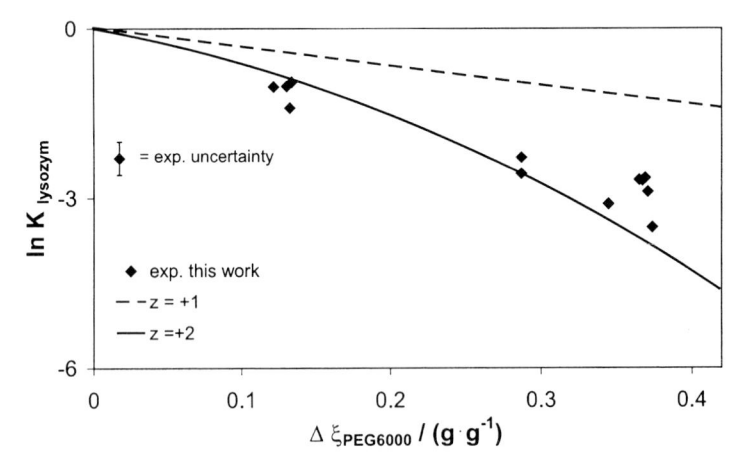

Figure 16.6: Comparison between experimental results and predictions for the partitioning of lysozyme to coexisting phases of water + PEG 6000 + K_2HPO_4 at 298.15 K.

16.5 Experimental Results/Modelling

of the coexisting phases, but there were large quantitative differences between prediction and experiment. There might be several reasons for that poor agreement. For example, the configuration of the proteins in the aqueous two-phase system can differ from that in the aqueous solution during titration and/or electrophoresis and therefore, the results of the titration/electrophoresis might not give the right influence of the pH on the charge number (particularly the surface charge) of the protein in that aqueous two-phase system. Such an influence might be due to the high ionic strength of the phosphate rich aqueous phase (protein titration was done at low ionic strength). However, also the assumption that the first term on the right hand side of Eq. (16.1) can be neglected might cause the large deviations between prediction and experiment for the partitioning of glucosidase and bovin serum albumin to the PEG/phosphate systems.

In systems of PEG 6000 and DEX T500, the distributions of BSA, chicken egg albumin, glucosidase, lysozyme, and myoglobin were examined at two ionic strengths (200 and 500 mmolal KCl) and three different pH values. pH was adjusted to 4.6 (by 10 mmolal potassium acetate and acetic acid each), to 6.8 (by 10 mmolal mono- and di-potassium hydrogenphosphate each), and to 10.2 (by 10 mmolal butylamine and 5.5 mmolal HCl). For BSA, glucosidase, and partly myoglobin the influence of the change of protein's net charge on the partition coefficient was predicted correctly. However, the charge numbers required to predict quantitatively the partition coefficient differed in some cases significantly from the experimental results (applying titration and isoelectric focusing). As typical examples the influences of solution pH and of ionic strength on the partition coefficient of bovin serum albumin in aqueous two-phase systems of PEG 6000 and DEX T500 are shown in Figures 16.7 (at $I = 500$ mmolal) and 16.8 (at pH = 6.8).

The protein's charge number determined experimentally is considerable smaller than that required for a good correlation. For example, at pH = 4.6 the experimental result is $z_{BSA} = -1$, whereas a good correlation of the partition coefficient at an ionic strength of

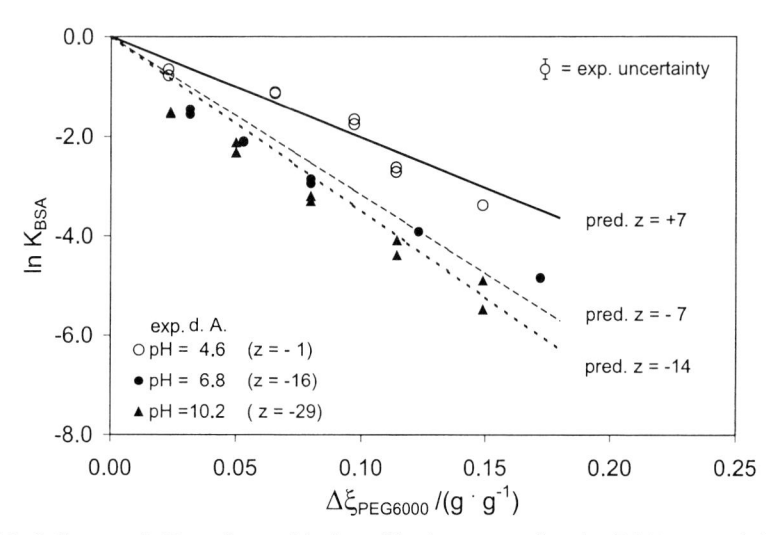

Figure 16.7: Influence of pH on the partitioning of bovine serum albumin (BSA) to coexisting phases of water + PEG 6000 + DEX T500 at 298.15 K and $I = 500$ mmolal.

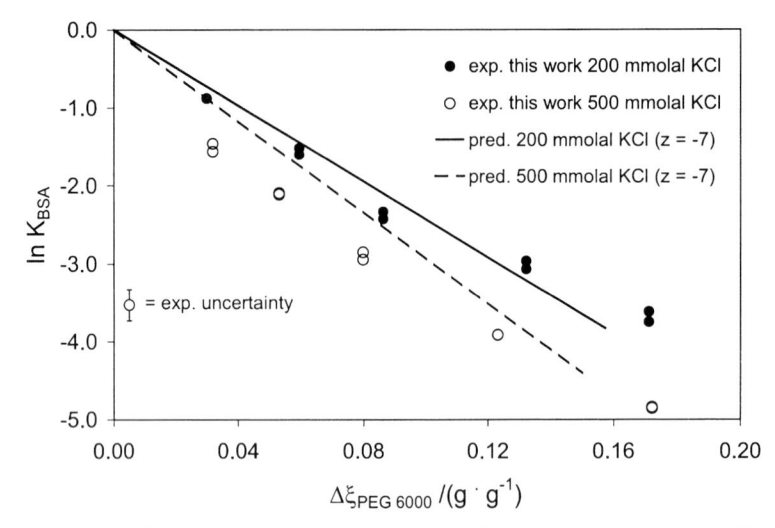

Figure 16.8: Influence of ionic strength on the partitioning of bovine serum albumin (BSA) to coexisting phases of water + PEG 6000 + DEX T500 at 298.15 K and pH = 6.8.

200 mmolal is achieved with $z_{BSA} = +7$. Similar differences were observed for the other pH-values. As is shown in Figure 16.8 with increasing ionic strength bovin serum albumin tends to prefer the dextran-rich over the PEG-rich phase. The model predicts that preference, but for a quantitative agreement the protein's charge number has to be adjusted to a number which does not agree with the independently performed experiments (titration and isoelectric focusing). The electrical potential difference in systems of water + PEG+ DEX is the result of the partitioning of the salt (here KCl) used for adjusting the ionic strength and the buffers (used to adjust the pH). The available experimental data do not allow a profound estimation of model parameters. Therefore, the discrepancies between the experimental results and the predictions for the partition coefficients of some proteins in such aqueous two-phase systems are at least partly due to the lack of sufficient information on the phase forming system.

16.6 Conclusions

A reliable, quantitative prediction of the partitioning of proteins to aqueous two-phase systems solely based on the currently available data for the phase-forming systems and some experimental results for the charge number of the proteins is possible only under favorable conditions. Additional experimental work on the influence of salts (required to adjust the ionic strength) and on buffers (required to adjust solution pH) on the phase behavior in aqueous two-phase systems as well as investigations on the behavior of proteins in aqueous solutions is needed to further test and improve methods for the prediction of partition coefficients of proteins in aqueous two-phase systems.

Acknowledgements

The authors gratefully acknowledge the advice and support by Prof. Kula and Dr. Rämsch in questions concerning enzyme technology. Furthermore they thank Dr. Großmann for assisting with his software source codes.

References

Albertsson, P.-Å. (1986): Partition of cell particles and macromolecules, 3rd edition. Wiley-Interscience, New York.
Brenneisen, J. (2001): Zur Verteilung von Proteinen auf wässrige Zwei-Phasen-Systeme. Dissertation, Universität Kaiserslautern.
Großmann, C. (1994): Untersuchungen zur Verteilung von Aminosäuren und Peptiden auf wäßrige Zwei-Phasen-Systeme. Dissertation, Universität Kaiserslautern.
Großmann, C.; Maurer, G. (1995): On the calculation of potential differences in aqueous two-phase systems containing ionic solutes. Fluid Phase Equilibria 106, p. 17–25.
Großmann, C.; Tintinger, R.; Zhu, J.; Maurer, G. (1995): Aqueous two-phase systems of poly(ethylene glycol) and dextran: experimental results and modeling of thermodynamic properties. Fluid Phase Equilibria 106, p. 111–138.
Harned, H. S.; Ehl, R. W. (1932): The dissociation constant of acetic acid from 0 to 35 °C. J. Am. Chem. Soc. 54, p. 1350–1357.
Kim, H.-T.; Frederick, W. J. (1988): Evaluation of Pitzer ion interaction parameters of aqueous electrolytes at 25 °C. 1. Single salt parameters. J. Phys. Chem. 33, p. 177–184.
Kirsch, T.; Ziegenfuß, H.; Maurer, G. (1997): Distribution of citric, acetic and oxalic acids between water and organic solutions of tri-n-octylamine. Fluid Phase Equilibria 129, p. 235 – 266.
Malamud, D.; Drysdale, J. W. (1978): Isoelectric points of proteins: A table. Anal. Biochemistry. 86, p. 620–647.
Manon, M. K.; Zydney, A. L. (1998): Measurement of protein charge and ion binding using capillary electrophoresis. Anal. Chem. 70, p. 1581–1584.
Marini, M. A.; Marti, G. E.; Berger, R. L.; Martin, C. J. (1980): Potentiometric titration curves of oxidized and reduced horse heart cytochrome c. Biopolymers 19(4), p. 885–898.
Pitzer, K. S. (1973): Thermodynamics of electrolytes. I. Theoretical basis and general equation. J. Phys. Chem. 77, p. 268–277.
Pitzer, K. S. (1991): Activity coefficients in electrolyte solutions, 2nd ed. CRC Press, Boca Raton, Florida.
Righetti, P. G.; Cravaggio, T. (1976): Isoelectric points and molecular weights of proteins. J. Chromat. 127, p. 1–28.
Rumpf, B.; Maurer, G. (1994): Solubility of ammonia in aqueous solutions of phosphoric acid: Model development and application. Journal of Solution Chemistry 23 (1), p. 37–51.
Tanford, C.; Wagner, M. L. (1954): Hydrogen ion equilibria of lysozyme. J. Am. Chem. Soc. 76, p. 3331–3336.
Tanford, C.; Swanson, S. A.; Shore, W. S. (1955): Hydrogen ion equilibria of bovine serum albumin. J. Am. Chem. Soc. 77, p. 6414–6421.
Theorell, H.; Åkesson, Å. (1941): Studies on cytochrome c. III. Titration curves. J. Am. Chem. Soc. 63, p. 1818–1827.
Tintinger, R. (1999): Thermodynamische Eigenschaften ausgewählter wäßriger Zwei-Phasen-Systeme. Dissertation, Universität Kaiserslautern.

Walter, D. E.; Brooks, Fisher, D. (1985): Partitioning in aqueous two-phase systems: Theory, methods, uses and applications to biotechnology. Academic Press, New York.

Yan, W.; Han, S. (1994): Estimation and comparison of the mixing parameters for mixed electrolyte solutions at 298.15 K: KCl + KAc + H$_2$O. Zhejiang Daxue Xuebao, Ziran Kexueban. 28, p. 376–384.

Nomenclature

a	activity
A_{ij}	binary interaction parameter
$A_{ij}^{(v)}$	binary interaction parameters ($v = 0$; 1)
A_φ	Debye-Hückel parameter for water
b	parameter ($= 1.2$ mol/kg)
B_{ijk}	ternary interaction parameter
BSA	bovine serum albumin
DEX	dextran
f	function
F	Faraday's constant
G	Gibbs energy
I	ionic strength
IEF	isoelectric focussing
pI	isoelectric point
K	partition coefficient
K_0	partition coefficient in the absence of an electrical potential difference
K_{chem}	chemical reaction equilibrium constant
m	molality
M	molecular mass
M_n	number averaged molecular mass
N	number of partitioning species
p	pressure
pK	$= -\log K_{chem}$
PEG	poly(ethylene glycol)
n	mole number
q	surface parameter
R	universal gas constant
T	thermodynamic temperature
z	charge number

Greek letters

α	constant
Δ	difference
Φ	electrical potential

Θ	relative surface fraction
γ	activity coefficient
μ	chemical potential
ν	number of groups
ξ	mass fraction

Subscripts

bottom	lower phase
BSA	bovine serum albumin
DEX	dextran
HAc	acetic acid
i	component, species
k	species
PEG	poly(ethylene glycol)
top	upper phase
w	water

Superscripts

I	phase I
II	phase II
E	excess
(m)	on molality scale
∞	infinite dilution
(0)	phase forming system

V Phase Equilibrium of Polymer Systems

17 Phase Behavior of Quaternary Polymer Solutions

Claudia Barth-Wiedmann, Matthias Wünsch, and Bernhard Anton Wolf[1]

17.1 Abstract

Phase diagrams and partial vapor pressures of the volatiles have been measured for different solvent + precipitant + polymer systems at various temperatures. Polysulfone (PSU) and polyethersulfone (PES) were the polymers and the mixed solvent consisted either of dimethylformamide + acetone (DMF + AC) or DMF + H_2O. With the ternary systems containing the very powerful precipitant water one observes a pronounced enrichment of this component in the gas phase, where the preferential evaporation increases exponentially with polymer concentration. All results, except this particular feature, can be modeled quantitatively on the basis of the Flory-Huggins theory if the binary interaction parameters are treated concentration dependent. However, the fact that a given set of parameters that describes vapor/liquid equilibria accurately cannot quantitatively forecast liquid/liquid demixing and *vice versa*, expresses a fundamental inadequacy of the Flory-Huggins theory, namely the neglect of inhomogeneous space filling by the macromolecules at low polymer concentrations. The necessity to introduce ternary interaction parameters for the description of ternary, water-containing systems constitutes a further indication of theoretical break-down. Notwithstanding these shortcomings the present modeling is of great value for obtaining otherwise inaccessible orienting information (i.e. the extension of unstable regions or other experimentally inaccessible data) on the phase behavior of multinary systems, as demonstrated for the quaternary system DMF + AC + H_2O + PSU.

17.2 Introduction

The phase state of polymer containing multicomponent systems plays a pivotal role for the synthesis of polymers as well as for their application. In the former case demixing may for instance lead to gelation of the system via cross-linking reactions. Another example of prac-

---
[1] Universität Mainz, Institut für Physikalische Chemie, Jakob-Welder-Weg 13, 55099 Mainz.

tical importance concerns the manufacturing of efficient membranes for which one requires a minimum of three components and usually employs at least four. In this case, information on the composition and temperature areas of unstable and metastable mixtures is of particular interest, in view of the fact that the morphology of the final product is determined by the particular mode of phase separation (nucleation and growth mechanism or spinodal decomposition).

Despite this obvious demand for reliable thermodynamic modeling of multinary mixtures, the facilities to reach that target are far from satisfying. The present work was undertaken to improve this situation by means of two comparatively new tools: A method for automated vapor pressure measurements, giving quick access to the activities of the volatile components over a broad concentration range, and a mathematical procedure for the calculation of equilibrium data that avoids the use of derivatives of the Gibbs energy of mixing. Like most literature work our theoretical consideration rest on the Flory-Huggins equation. To ensure practical relevance, we have chosen polyethersulfone (PES) and polysulfone (PSU), two highly temperature resistant and membrane forming polymers, as the high molecular weight components for the present work and common solvent-mixtures (DMF + H_2O; DMF + AC; DMF + AC + H_2O).

17.3 Experimental Part

17.3.1 Apparatus and Procedures

17.3.1.1 Static Light Scattering

Access to the Flory-Huggins interaction parameters in the limit of vanishing polymer concentration is easiest via light scattering or osmosis. We have used the apparatus Fica 50 (Sofica, Paris), modified by SLS (G. Baur, Freiburg). It was operated with polarized laser light of 632.8 nm and covers angles from 20 to 145°. The automatically registered data were evaluated in Zimm plots using refractive index increments determined by means of a Bodmann differential refractometer (Bodmann, 1957).

17.3.1.2 Headspace-Gaschromatography

Headspace-gaschromatography (HSGC) measurements were carried out with an apparatus consisting of a headspace-sampler and a normal gas chromatograph. The pneumatically driven thermostatted headspace-sampler (Dani HSS 3950, Milano, Italy) samples a constant amount of gas phase (in equilibrium with the liquid) and injects this mixture or pure vapor into the gas chromatograph (Shimadzu GC 14B, Kyoto, Japan). Helium was used as carrier gas. After separation of the volatile components in a capillary column they are detected individually by means of a thermal conductivity detector (TCD). The signals are sent to an inte-

17.3 Experimental Part

grator (Shimadzu Chromatopac C-R6A), which calculates the peak areas. Because these signals are proportional to the amount of gas in the sample volume (also proportional to the vapor pressures (Hachenberg and Beringer, 1996; Kolb, 1975; Kolb et al., 1977; Kolb and Ettre, 1997), the partial vapor pressures become accessible by means of a suitable calibration of the apparatus. A scheme of the experimental set-up is given in Figure 17.1. More information on the experimental procedure can be found in the literature (Petri, 1994; Petri et al., 1995, 1996; Petri and Wolf, 1994).

The separation of the volatile components in the gas chromatograph permits the determination of partial vapor pressures and consequently enables the calculation of the individual activities of the solvent and precipitant of the mixed solvents of present interest. Figure 17.2 shows how HSGC is used in these cases and gives an impression on the concentra-

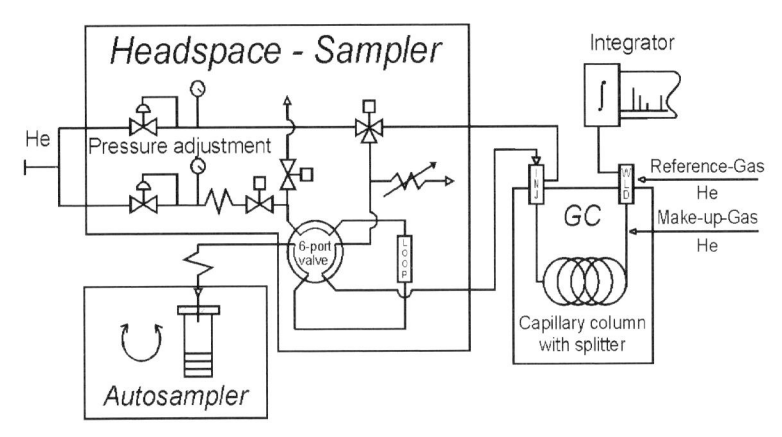

Figure 17.1: Scheme of the apparatus used for the determination of vapor pressures by means of HSGC.

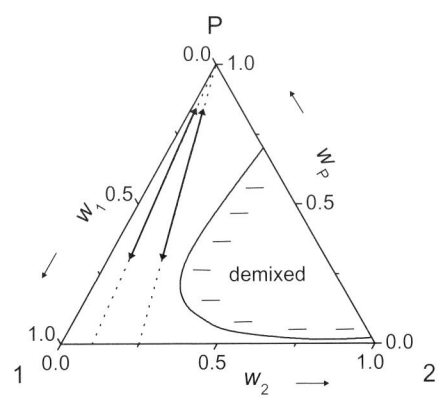

Figure 17.2: Gibbs phase triangle (1: solvent, 2: precipitant, P: polymer) demonstrating the experimental procedure for the determination of partial vapor pressures by means of HSGC. Measurements are performed for a given composition of the mixed solvent and variable polymer concentration (straight lines), where the solid parts indicate the composition range of reliable results.

tion range within which it can be applied reliably. For phase separated solutions or very low polymer concentrations, the differences between their vapor pressures and that of the polymer-free liquid become insufficient and for too high concentrations equilibria are hard to achieve as a result of the high viscosity of these systems.

17.3.1.3 Determination of Phase Diagrams

Cloud point curves were typically measured at 20, 30, 40, and 50 °C by adding drop-wise a proper mixture of solvent and precipitant to the stirred solutions of polymer in the mixed solvent until turbidity could be detected visually. Such titrations can often not be performed with the pure non-solvent particularly if they are so efficient that very few droplets suffice to cause turbidity. Furthermore the addition of pure precipitant causes the formation of local tough gel particles which need a long time to redissolve upon stirring.

In swelling experiments thin transparent films were prepared by dissolving the polymers, casting the solutions on glass, and evaporating the solvent slowly. These films were then stored in the precipitant until their weight (droplets of pure liquid on the surface of the films were cautiously dabbed away) did no longer change.

For the determination of tie lines the polymer solutions were titrated at constant temperature beyond the onset of turbidity in the manner described above. The starting composition was chosen such that one reaches the two-phase region approximately at the critical point of the ternary system. The highly turbid solutions were then homogenized by raising the temperature and subsequently cooled to the equilibrium temperature again at a constant cooling rate of typically 0.1 °C/min. The samples were kept at this temperature for 2 to 3 days to complete macroscopic phase separation. The coexisting phases (sol and gel) were then separated by means of a syringe and the volatile components removed by vacuum distillation. The remaining polymers were weighed and the collected liquids analyzed in terms of their refractive indices and by gas chromatography.

The information concerning the critical points was obtained from phase volume ratios. The procedure used to this end is identical with that described in the previous section except for the fact that the penetration into the miscibility gap is kept as small as possible for sub- and supercritical compositions. The critical point is obtained from these data by plotting the phase volume ratios as a function of polymer concentration (Krause and Wolf, 1997).

17.3.2 Materials

Polyethersulfone (PES) and polysulfone (PSU), two highly temperature resistant and membrane forming polymers, were chosen for the present study. Their characteristic data are collected in Table 17.1.

The polymer samples as obtained from the producer contain rests of water (PES 1.00 mass% and PSU 0.33 mass%). These residues do, however, not lead to detectable changes in the phase diagrams.

Table 17.1: Characteristic data of the polymers.

	PES	PSU
Brand name	Ultrason E 6020 P	Ultrason S 3010
Producer	BASF AG	BASF AG
Abbreviation	PES 49w	PSU 39w
M_w/kg mol^{-1}	49	39

The chemical formulae of the two polymers are given below

Polysulfone Ultrason S 3010

Polyethersulfone Ultrason E 6020 P

DMF (stored under nitrogen atmosphere) was selected as solvent for both polymers. The precipitants were water and acetone. The supplier of these liquids and their purity are given in Table 17.2.

Table 17.2: Solvents, their suppliers, and their purity.

Solvents/precipitants	Producer	Purity
DMF	Fluka	> 99%
acetone	Fluka	> 99.5%
water		distilled

17.4 Theoretical Background

17.4.1 Flory-Huggins Interaction Parameters

Despite well known deficiencies of the lattice theory of Flory and Huggins (Huggins, 1976, 1977; Kamide and Dobashi, 2000; Flory, 1944), this approach is still widely used for the description of polymer-solvent systems. The central equation for the composition dependence of $\overline{\overline{\Delta G}}$, the base molar Gibbs energy of mixing for m components, reads

$$\frac{\overline{\overline{\Delta G}}}{RT} = \sum_{i=1}^{m} \frac{\varphi_i}{N_i} \ln \varphi_i + \sum_{i=1}^{m-1} \sum_{j=i+1}^{m} g_{ij} \varphi_i \varphi_j + \sum_{i=1}^{m-2} \sum_{j=i+1}^{m-1} \sum_{k=j+1}^{m} g_{ijk} \varphi_i \varphi_j \varphi_k + \ldots \quad (17.1)$$

where φ = volume fractions, N = number of segments (ratio of the molar volume of the component, divided by the molar volume of a freely chosen segment), g_{ij} = (integral) binary interaction parameter between the components i and j, and g_{ijk} = the corresponding ternary interaction parameter.

In order to model experimental findings quantitatively, all shortcomings of the above approach are eliminated by treating the g parameters as concentration dependent, in contrast to the original concept. One of the consequences of that procedure consists in the necessity to distinguish between integral and differential interaction parameters. This feature results from the fact that the differentiation of the Gibbs energy with respect to one component to obtain its chemical potential also changes the interaction parameters due to their dependence on composition. For the most frequently determined chemical potential of the solvent, μ_1, in a binary polymer solution the corresponding equation reads

$$\mu_1 = \mu_1^o + RT \left[\ln \varphi_1 + \left(1 - \frac{1}{N_P}\right) \varphi_P + \chi \varphi_P^2 \right] \quad (17.2)$$

where μ_1^o is the chemical potential of the pure solvent and the index P stands for the polymer; χ represents the original (differential) Flory-Huggins parameter. Owing to the correlation between the reduced vapor pressure (in case of ideal gases, otherwise the ratio of the fugacities) of the solvent and its chemical potential we can also write

$$\ln \frac{p_1}{p_1^o} = \ln a_1 = \ln \varphi_1 + \left(1 - \frac{1}{N_P}\right) \varphi_P + \chi \varphi_P^2 \quad (17.3)$$

The following relations between χ and g hold true

$$\chi = g - (1 - \varphi_P) \frac{\partial g}{\partial \varphi_P} \quad (17.4)$$

and

$$g = \frac{1}{\varphi_1} \int_0^{\varphi_1} \chi d\varphi_1 \quad (17.5)$$

The necessity to distinguish between differential and integral parameters only vanishes for composition independent interaction parameters.

Several theoretically more or less justified mathematical expression are customary for the description of $g(\varphi)$. The following equations are used in the discussion:

$$g = \alpha + \frac{\beta}{1 - \gamma \varphi_P} \quad (17.6)$$

17.4 Theoretical Background

This expression, reported by Koningsveld and Kleintjens (1971), contains the three system specific parameters α, β, and γ, which can be interpreted theoretically and is particularly helpful for poorly interacting components, i.e. systems that tend to phase separation.

Another frequently used phenomenological equation is due to Redlich and Kister (1948)

$$g = b + c(2\varphi_1 - 1) + d(2\varphi_1 - 1)^2 + e(2\varphi_1 - 1)^3 + \ldots \qquad (17.7)$$

b to e represent adjustable parameters. Wilson (1964) reported the following expression

$$g = -\frac{1}{N_1 \varphi_P} \ln(1 - A_{21}\varphi_P) - \frac{1}{N_P \varphi_1} \ln(1 - A_{12}\varphi_1) \qquad (17.8)$$

where the system specific parameters are A_{12} and A_{21}.

Finally, we also checked the usefulness of approaches accounting for the non-random mixing. These were the so-called Non-Random Two Liquid (NRTL) theory of Renon and Prausnitz (Renon and Prausnitz, 1968),

$$g = \frac{\varepsilon_1}{RTN_P} \frac{\exp\left(\frac{-a\varepsilon_1}{RT}\right)}{\varphi_1 + \frac{\varphi_P}{N_P}\exp\left(\frac{-a\varepsilon_1}{RT}\right)} + \frac{\varepsilon_2}{RTN_P} \frac{\exp\left(\frac{-a\varepsilon_2}{RT}\right)}{\frac{\varphi_P}{N_P} + \varphi_1 \exp\left(\frac{-a\varepsilon_2}{RT}\right)} \qquad (17.9)$$

in which ε_1, ε_2, and a are adjustable parameters and the concept of quasi-chemical equilibria (Tompa, 1965; Fowler and Guggenheim, 1940), as expressed by

$$g_{qc} = g \frac{\sqrt{1 + 4\alpha_\omega} - 1}{2\alpha_\omega} \qquad (17.10)$$

where g, referring to random mixing, is transformed into g_{qc}, accounting for a further reduction in the Gibbs energy, and α_ω relates to the interaction energy ω as

$$\alpha_\omega = \varphi_1 \varphi_P \left[\exp\left(\frac{2\omega}{kT}\right) - 1\right] \qquad (17.11)$$

Temperature influences on g are usually modeled by an appropriate T-dependence of the different system specific parameters of the above relations, in some cases in addition to the already explicitly formulated T influences. In most cases it is by no means obvious which of them should be kept constant and which of them must vary with T.

17.4.2 Calculation of Phase Diagrams

The modeling of stable, metastable, and unstable temperature and composition ranges of mixtures based on the equations for $g(\varphi)$ presented in the last section in the usual manner, i.e. by means of chemical potentials, would be rather laborious and in some complex cases

17 Phase Behavior of Quaternary Polymer Solutions

even impossible. The method of direct minimization of the Gibbs energy (Horst, 1995, 1996; Horst and Wolf, 1996a,b) offers a way out of these difficulties (avoiding the algebraic derivatives). Its mode of functioning is briefly explained by means of Figure 17.3 describing the situation for the simplest possible case of binary mixtures. It should, however, be kept in mind that the feasibility to handle multinary systems is one of the strongest points of this approach.

The potential reduction in $\overline{\overline{G}}$ (or equivalently in $\Delta\overline{\overline{G}}$) is calculated by means of a series of "test tie lines". For this purpose a given over-all composition $\varphi^{o.a.}$ of the system within the two-phase regime is selected and tie lines are constructed such that their ends are floated along the curve representing homogeneous mixture until the maximum reduction of the Gibbs energy is achieved. The results of this procedure are identical with that of the conventional double tangent method, as can be visualized immediately by that graph. Repeating the calculation for different temperatures yields the binodal curve.

The direct minimization of the Gibbs energy also gives access to the spinodal lines. For such calculations one subdivides the composition axis into an arbitrarily large number n of tiny intervals and constructs within each of them very short "test tie lines" in the manner described above. Figure 17.3 demonstrates this proceeding for two compositions and $n = 5$ (Horst, 1995). For an overall "volume fraction of the polymer of 0.4" the resultant secant lies below the curve of ΔG corresponding to the hypothetic homogeneous state. This implies that the slightest concentration fluctuation leads to a reduction in Gibbs energy. Thence this composition lies inside the spinodal line where the mixtures are unstable. For an overall polymer volume fraction of 0.8, on the other hand, the secant is located above the line for the homoge-

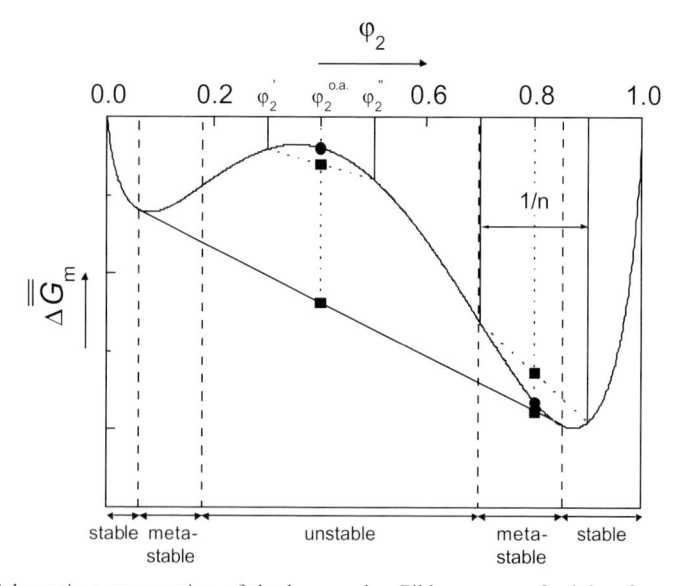

Figure 17.3: Schematic representation of the base molar Gibbs energy of mixing for a given constant temperature as a function of composition (volume fractions). The full curve represents the (sometimes hypothetical) homogeneous mixture and the straight line shows the double tangent to this dependence normally used to determine the composition of the coexisting equilibrium phases. The over-all composition of the mixture is indicated as o.a.

17.4 Theoretical Background

neous state, which means that work must be spent to produce a *local* demixing. For this reason the mixture is metastable. In actual calculations n is typically chosen >1000 in case of binary systems, >200 for ternary, and >100 for quaternary mixtures. Under these conditions the boarder line between instability and metastability, the spinodal curve, can be determined very accurately by performing the calculations for different temperatures. Calculation of critical points: looking for a test tie line tangential to the spinodal curve.

17.4.3 Calculation of Vapor Pressures

The method described in the last section is also able to deal with vapor-liquid equilibria enabling the calculation of the partial vapor pressures (Barth et al., 1998) of liquid mixtures. The main idea of that approach is visualized by means of the scheme in Figure 17.4 describing the experiment of thought underlying this approach.

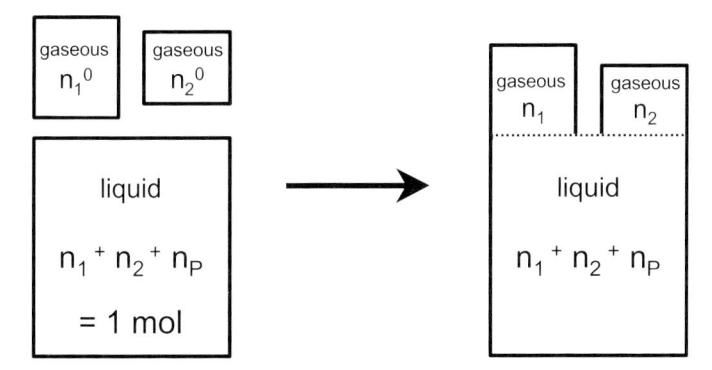

Figure 17.4: Diagram describing the experiment of thought underlying the calculation of vapor pressures of liquid mixtures for the example of polymer solutions in mixed solvents. These considerations are performed for one mole of the liquid mixture of given composition and – in separated reservoirs – minute amounts (e.g. 10^{-6} mol) of the pure vapor of the volatiles. Connecting the three containers enables the transport of the components that can cross the phase boundary. Under the present conditions (constant volume of the system) equilibrium is reached as the Helmholtz energy of the system assumes its minimum.

For the present calculation we have made some simplifications, which are justified for the systems of interest. First of all we neglect the differences between the Gibbs energy and the Helmholtz energy. This procedure enables the direct use of the equations formulated in Section 17.5.1 for the concentration dependence of the Flory-Huggins interaction parameters. Furthermore we treat the vapor phase as a mixture of ideal gases. One can therefore easily compute the partial vapor pressures for a given composition of the liquid mixture by means of the minimization routines. A comparing of the results for different expressions $g(\varphi)$ with the actual experimental data enables the judgment of the aptness of the individual approaches for the modeling of vapor pressure curves.

17.5 Results and Discussion

The modeling of multinary systems requires detailed information on all possible subsystems. For that reason the material is arranged according to the number of components the mixture contains, neglecting the molecular non-uniformity of the polymers. Most of the experimental data shown here refer to solutions of PES, since the corresponding results with PSU have already been shown (Barth and Wolf, 2000b).

17.5.1 Binary Systems

17.5.1.1 Mixed Solvents

An example for the partial vapor pressure data obtained by means of HSGC for a mixture of low molecular weight compounds is presented in Figure 17.5. This graph shows the results for the system DMF + water, together with the total vapor pressure calculated therefrom and corresponding literature data (Zielkiewicz and Oracz, 1990). The Flory-Huggins interaction parameters (cf. Table 17.3) calculated from the measured partial vapor pressures by means of an iteration process (to minimize the deviations) on the basis of the different concentration dependencies formulated in Section 17.5.1 are presented in Figure 17.6.

The differences in the interaction parameters and their variation with composition are marginal except for a large predominance of one component. This finding constitutes a common feature for the modeling of all systems and is probably due to the particularly large experimental uncertainties in this region as well as to the mathematical details of the theoretical expressions.

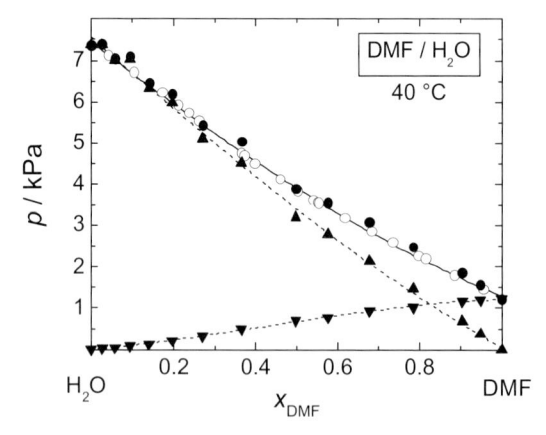

Figure 17.5: Partial vapor pressures of DMF and water determined at 40 °C by means of HSGC and the total vapor pressures calculated therefrom (full symbols). For comparison total vapor pressures reported in the literature (Zielkiewicz and Oracz, 1990) are also plotted (open symbols).

Table 17.3: Fit-parameter for the modeling of the integral binary interaction parameters of the system DMF + H$_2$O at 32 °C.

Equation	Parameter				
Koningsveld-Kleintjens	α 0.3223	β 0.0751	γ 0.9604		
Redlich-Kister	b 0.4658	c −0.1079	d 0.0348	e −0.2012	f 0.4462
Wilson	A_{21} 1.0054	A_{12} −11.2514			
NRTL	ε_1/RT 1.3846	ε_2/RT 1.9371	a 0.9772		

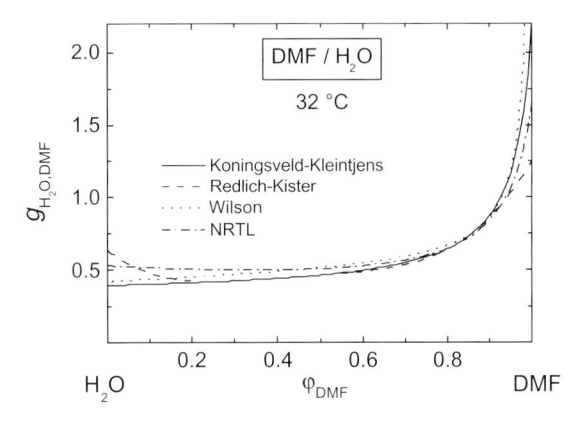

Figure 17.6: Dependence of the Flory-Huggins interaction parameter $g_{H2O,DMF}$ on the volume fraction of DMF at 32 °C as obtained from the iterations with Eqs. (17.6) – (17.9).

17.5.1.2 Polymer Solutions

In case of single solvents total and partial vapor pressures are identical. Figure 17.7 shows the results for the system DMF + PES at different temperatures. An example of the Flory-Huggins interaction parameters (Table 17.4) obtained therefrom by means of one of the relations of Section 17.5.1 is presented in Figure 17.8. The most obvious feature of that evaluation is an inversion in the sign of the temperature influence on solvent quality upon the variation of composition. Within the dilute regime g decreases as T is raised, indicative for endothermal mixing, whereas the opposite is the case (even if there is some uncertainty concerning the detailed shape) the mixing should proceed exothermally. For the system DMF + PSU the effects are similar and even more clear-cut. Such changes in the caloric effects with composition are not uncommon (Schotsch et al., 1984), but normally the systems are exothermal at low and endothermal at high polymer concentrations.

17 Phase Behavior of Quaternary Polymer Solutions

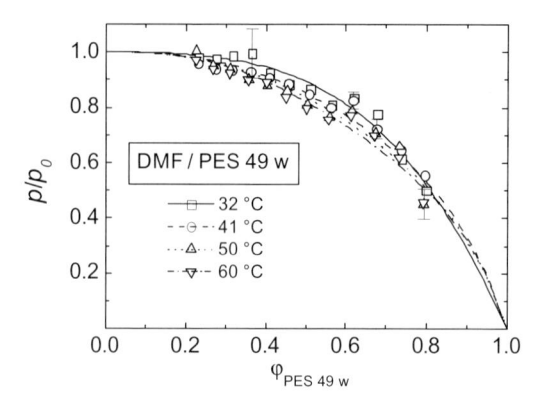

Figure 17.7: Reduced equilibrium vapor pressures of DMF above solutions of PES 49w as a function of the volume fraction of the polymer at different temperatures as indicated in the diagram. The curves are calculated according to Eq. (17.7).

Table 17.4: Fit-parameter for the modeling of the binary interaction parameters of the system DMF + PES and the indicated temperatures (Redlich and Kister, 1948).

$T/°C$	χ_0	b	c
32	0.3719	0.0793	−0.0236
41	0.3756	−0.1289	0.1000
50	0.3717	−0.1337	0.1144
60	0.3773	−0.1750	0.1473

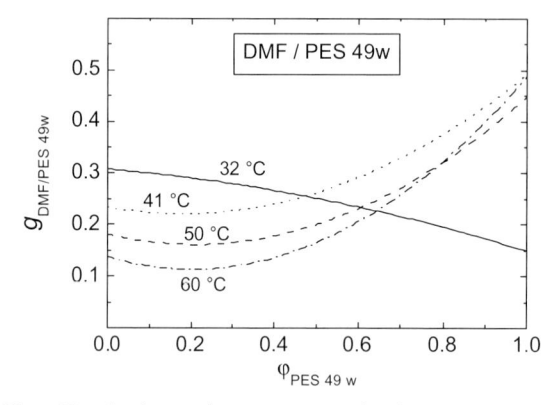

Figure 17.8: Integral Flory-Huggins interaction parameter g for the system DMF + PES 49w as a function of the volume fraction of the polymer according the expression of Redlich and Kister (Eq. (17.7)) for the different temperatures indicated in the graph.

17.5.2 Ternary Systems

One of the problems with polymer solutions in mixed solvents consists in the lack of data on the interaction between the precipitant and the polymer. Due to the unfavorable solvent quality the vapor pressures are too close to that of the pure liquid in case of demixed phases to enable the application of HSGC, as outlined in Section 17.4.1.2. Furthermore, only a small part of the concentration range of interest corresponds to homogeneous mixtures. For that reason one tries to obtain that figures from the modeling of measured phase diagrams. Owing to the impossibility of extracting the composition dependence of the interaction parameters from swelling experiments with the binary subsystems precipitant/polymer, one has to rely on the information contained in the ternary systems. Central points in that context are the knowledge of the critical conditions and of the position of tie lines (preferential solvation).

In the validation of the binary interaction parameters obtained by the evaluation of phase diagrams as outlined above one must keep in mind that possible ternary interaction parameters are automatically incorporated into the composition dependence of g_{2P}.

Another option to obtain access to the interaction parameters between the precipitants and polymer segments is supplied by the determination of the partial vapor pressures of solvent and precipitant for the ternary system. As with the previous approach the information is considerably less straight forward than the determination of interaction parameters between solvent and polymer, since it may depend on the details of the model used for the evaluation (Barth, 1999). In this case it is, however, in principle possible to investigate the necessity of ternary interaction parameters due to the fact that the partial vapor pressure of the precipitant yields additional information. In case $g_{2P}(\varphi_P)$ is already known from independent measurements with the corresponding binary subsystem, extra effects may be attributed to ternary interaction parameters (Figueruelo et al., 1985; Barth and Wolf, 2000a).

17.5.2.1 Phase Diagrams

An example for the determination of Flory-Huggins interaction parameters polymer/precipitant (Table 17.5) is in the following given for PES + AC and PES + H$_2$O, where the solvent was DMF in both cases. Figure 17.9 shows the measured ternary phase diagrams and Figure 17.10 displays the determination of the critical composition for one system.

The experimentally determined phase diagrams are modeled by means of a direct minimization of the Gibbs energy varying the expressions for the Flory-Huggins interaction parameter precipitant/polymer and its concentration dependence until the best fit is reached. For all systems studied here it was possible to find a set of parameters that reproduces the measured cloud point curves, the position of the critical point on it and the location of the tie lines within experimental error. Figure 17.11, referring to the system DMF + H$_2$O + PES, is typical for such calculations.

Table 17.5: Interaction parameters for the modeling of the ternary phase diagram DMF + H$_2$O + PES 49w (41 °C).

	Parameter				
$g_{H_2O/DMF}$	Koningsveld-Kleintjens α 0.3223	β 0.0751	γ 0.9604		
$g_{DMF/PES}$	Redlich-Kister b 0.4658	c −0.1079	d 0.0348	e −0.2012	f 0.4462
$g_{H2O/PES}$	$a_1 + a_2 \varphi^*_{PES}$ a_1 2.0	a_2 0.1			

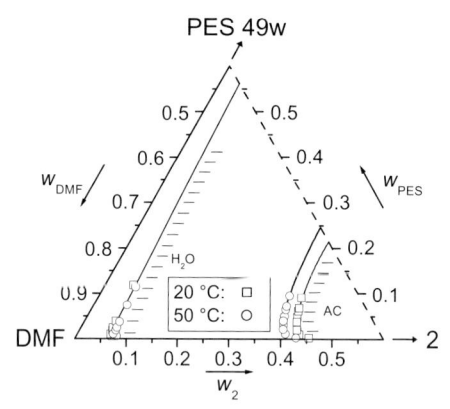

Figure 17.9: Measured two-phase regions of the systems DMF + H$_2$O + PES 49w and DMF + AC + PES 49w at 20 °C and 50 °C.

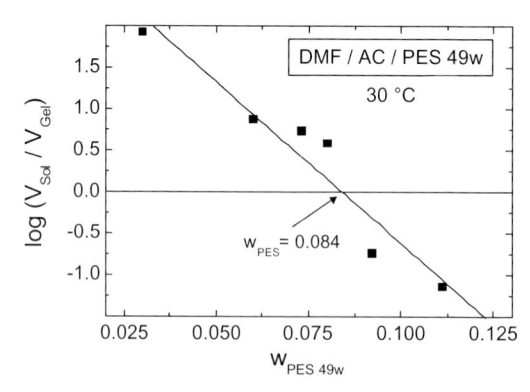

Figure 17.10: Phase volume ratio for incipient phase separation of the system DMF + AC + PES 49w at 30 °C as a function of polymer concentration. The critical composition is obtained from the condition that the phases volumes become identical.

17.5 Results and Discussion

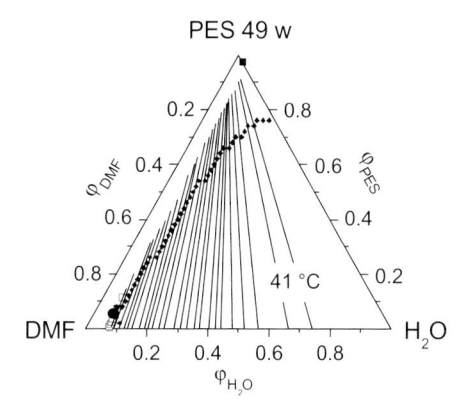

Figure 17.11: Tie lines and spinodal conditions (full diamonds) of the system DMF + H$_2$O + PES 49w at 41 °C calculated according to Eqs. (17.1), (17.6) ($g_{H_2O,DMF}$) and (17.7) ($g_{DMF/PES}$); $g_{H_2O,PES}$ adjustable. Also indicated are the swelling point of the polymer in the precipitant (full square), the measured cloud points (open squares), and the critical point of the system (full circle).

17.5.2.2 Vapor Pressures

The present section deals with the determination of the partial vapor pressures of the volatiles as a function of polymer concentration according to the experimental procedure outlined in Figure 17.2 and with their theoretical evaluation.

The results for the system DMF + AC + PES and constant weight ratio DMF : AC = 1.5 : 8.5 at 50 °C are given in Figure 17.12, together with some theoretical curves. The measured partial vapor pressures decline monotonically for both volatiles as the polymer content of the ternary mixture rises and can be described reasonably well by means of bin-

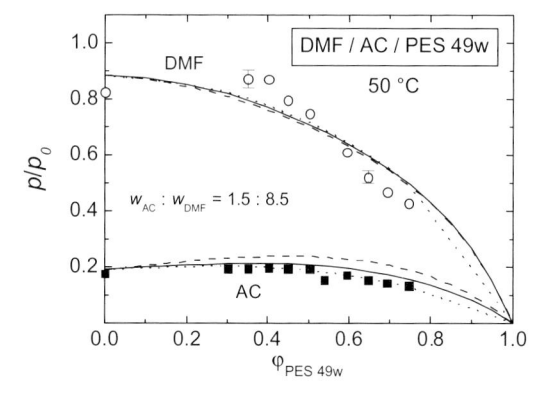

Figure 17.12: Reduced partial vapor pressures of DMF and AC for the ternary system DMF + AC + PES 49w at 50 °C and constant mass ratio of DMF : AC = 1.5 : 8.5. Also mapped are two theoretical curves calculated according to interaction parameters obtained by swelling experiments (full line), adjusted polynomial (dashed line), and Redlich and Kister (Eq. (17.7)).

395

ary interaction parameters. However, the reduction takes place considerably more rapidly for the solvent than for the precipitant. Furthermore, neither of the theoretical relations $g(\varphi)$ (Table 17.6) can describe the findings quantitatively.

Table 17.6: Binary interaction parameter $g_{AC/PES}$ for the modeling of the vapor pressures of the ternary system DMF + AC + PES 49w.

Full line	0.74, obtained from swelling experiments
Dashed line	$1.03 + 0.24\,\varphi^*_{PES} - 0.31\,\varphi^{*2}_{PES}$, parameter originates from modeling of the phase diagram
	$\left(\varphi^*_{PES} = \dfrac{\varphi_{PES}}{\varphi_{PES} + \varphi_{AC}}\right)$
Dotted line	0.492, adjusted to the vapor pressures

For the considerably more efficient precipitant water instead of acetone, the modeling becomes worse (cf. Fig. 17.13). In this case the partial pressure even rises with increasing polymer concentration to values that are almost three times that of the binary subsystem DMF/H$_2$O. In this case the discrepancies between the experimental observation and the behavior that can be modeled on the basis of binary interaction parameters (including that for $g_{2P}(\varphi_P)$, Table 17.7, obtained from the modeling of the phase diagram) becomes all the more pronounced. The results of further attempts to model the vapor pressures for this ternary system are shown in Figure 17.13. All observations indicate the necessity to include ternary interaction parameters for simultaneous contacts between all three components of a mixture. The situation is, however, rather complex, as discussed in more detail in the literature (Barth and Wolf, 2000a).

The striking enrichment of very powerful precipitants in the gas phase should be of industrial relevance in the context of separation processes, in particular for drying purposes. For a quantitative coverage of the phenomenon, we introduce a quantity π_2 as

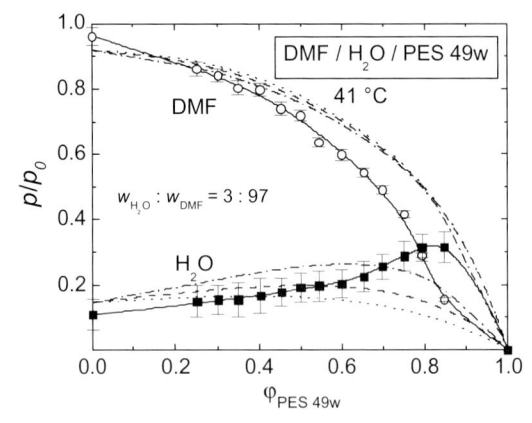

Figure 17.13: Reduced partial vapor pressures of DMF and water for the ternary system DMF + water + PES 49w at 41 °C and a constant mass ratio of DMF : water = 3 : 97. Full lines: guide for the eye; dashed lines: $g_{H_2O/PES}$ obtained from swelling experiments; dotted lines: $g_{H_2O/PES}$ from the modeling of phase diagram; dash-dotted lines: modeling of the vapor pressures.

17.5 Results and Discussion

Table 17.7: Binary interaction parameter $g_{H_2O/PES}$ for modeling the vapor pressure of the ternary system DMF + H$_2$O + PES 49w.

Full line	guide for the eye
Dashed line	2.7, obtained from swelling experiments
Dotted line	$2.15 + 0.13\,\varphi^*_{PES}$, parameter originates from modeling of the phase diagram $\left(\varphi^*_{PES} = \dfrac{\varphi_{PES}}{\varphi_{PES} + \varphi_{H_2O}}\right)$
Dash-dotted line	3.21, adjusted to the vapor pressures

$$\pi_2 = \frac{p_2/(p_1 + p_2)}{\varphi_2/(\varphi_1 + \varphi_2)} \tag{17.12}$$

It is defined as the quotient of the concentration of the precipitant in the gas phase and its concentration in the (polymer-free) mixed solvent. According to the present findings π increases markedly with rising polymer concentration. This dependence does, however, not start from unity, due to the different fugacities of the volatiles even in the absence of the polymer. Figure 17.14 demonstrates the extent of the preferential evaporation of the different precipitants and the influence of the composition of the mixed solvent.

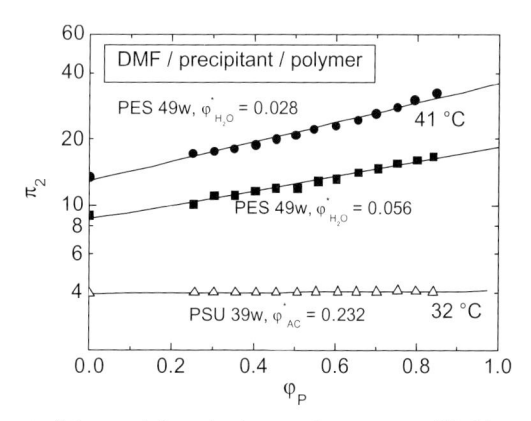

Figure 17.14: Enrichment of the precipitant in the gas phase as quantified by π_2 (cf. Eq. (17.12)) as a function of polymer concentration for the systems DMF + H$_2$O + PES 49w and DMS + AC + PSU 39w and the indicated temperatures.

According to the present results the augmentation of preferential evaporation by the addition of polymer is bound to a very unfavorable interaction of the precipitant with this solute. The effect is very pronounced for water, one of the most powerful non-solvents for PES, whereas it cannot be observed for AC, which is a less powerful precipitant for PSU. From a normalization of the π_2 data for the system DMF/H$_2$O/PES to their value in the absence of polymer reveals that the relative effects are independent of the composition of mixed solvent. In hindsight the qualitative explanation for this interesting effect is obvious: The Gibbs energy of the entire system is reduced if the unfavorably interacting precipitant is expelled from the liquid phase into the gas phase.

17.5.3 Quaternary Systems

Based on the total information on the phase behavior of all binary and ternary subsystems and the binary interaction determined therefrom and from vapor pressure measurements we can now model the spinodal surface of the quaternary system by means of the direct minimization of the Gibbs energy (Horst, 1996). Figure 17.15 shows the outcome for 32 °C for a given set of binary interaction parameters (Table 17.8).

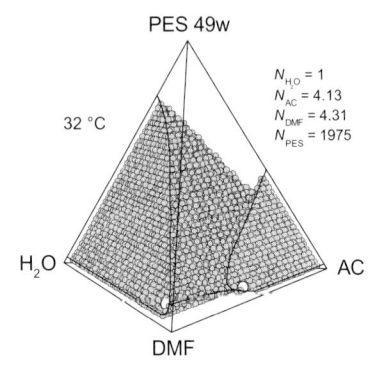

Figure 17.15: Spinodal planes for the quaternary system DMF + AC + H$_2$O + PES 49w calculated for 32 °C by means of the parameters given in Table 17.8.

Table 17.8: Binary interaction parameters for the modeling of the quaternary system DMF + H$_2$O + AC + PES 49w.

Interaction parameter	Equation
$g_{H_2O/AC}$	$0.43 + 0.51/(1 - 0.73\, \varphi_{AC}/(\varphi_{AC} + \varphi_{H_2O}))$
$g_{H_2O/DMF}$	$0.3223 + 0.0751/(1 - 0.9604\, \varphi_{DMF}/(\varphi_{DMF} + \varphi_{H_2O}))$
$g_{H_2O/PES}$	$\left(\dfrac{7.1446}{N_{DMF}}\right)\exp(-0.5955 \cdot 7.1446) \bigg/ \left(\varphi_{AC} + \dfrac{\varphi_{DMF}}{N_{DMF}}\exp(-0.5955 \cdot 7.1446)\right)$ $+ \left(\dfrac{0.033}{N_{DMF}}\right)\exp(-0.5955 \cdot 0.033) \bigg/ \left(\dfrac{\varphi_{DMF}}{N_{DMF}} + \varphi_{AC}\exp(-0.5955 \cdot 0.033)\right)$
$g_{AC/PES}$	$(0.92 + 0.1\, \varphi_{PES}/(\varphi_{AC} + \varphi_{PES}) + 0.05\, (\varphi_{PES}/(\varphi_{AC} + \varphi_{PES}))^2)/(N_{AC})$
$g_{DMF/PES}$	$(0.3719 - 3 \cdot 0.793 + 5 \cdot 0.0236 - 0.0793\, (2 \cdot \varphi_{PES}/(\varphi_{DMF} + \varphi_{PES})) - 1)$ $-0.0236\, (2 \cdot \varphi_{PES}/(\varphi_{AC} + \varphi_{PES}))^2/(N_{DMF})$

17.6 Conclusions

On the experimental side the most remarkable finding concerns the pronounced preferential evaporation of strong precipitants from polymer solutions in mixed solvents. The fact that the precipitant concentration in the vapor can be tripled by the presence of the polymer with respect to the vapor composition in its absence is theoretically well understood and is of technical relevance.

The modeling of phase diagrams for multinary systems by means of binary interaction parameters obtained from experiments with suitable (mostly binary) subsystems turned out to be possible with surprising precision. Such calculations are particularly helpful to examine whether one should expect general features for the multicomponent mixture under investigation, like cosolvency, co-nonsolvency, or three phase areas and so on. In the present case, for instance, the spinodal areas calculated for the quaternary system (Fig. 17.15) bear testimony to the fact that mixture of the precipitants water and acetone are considerably less efficient than either of the components alone.

A general conclusion that can be drawn from the present study pertains to the very fundamental question of the most practical mathematical expression one should use to describe the Gibbs energy of mixing for real systems. The Flory-Huggins equation used here has the advantage of being extremely simple in its original form, as compared with all more advanced theories (Flory-Orwoll-Vrij (Flory et al., 1964a,b), Sanchez-Lacombe (Lacombe and Sanchez, 1976; Sanchez and Lacombe, 1977, 1978), SAFT (Huang and Radosz, 1990, 1991; Banaszak et al., 1996)). However, this straightforwardness is often already lost due to the complicated concentration dependencies required. Furthermore the necessary parameters are sometimes physically meaningless. Last, but not least, even after accounting for all the deficiencies by incorporating them into composition dependent interaction parameters or by introducing extra interaction parameters for simultaneous contacts between more than two different species, it suffers from a general imperfection, namely the neglect of inhomogeneous space filling. Not the segments but the entire macromolecules are distributed over the total volume of the system at concentration below a certain characteristic value. It is probably this feature which leads to the observation that a set of parameters describing liquid/liquid phase equilibria quantitatively results in considerable deviations if used to predict liquid/gas equilibria and *vice versa*. A true solution to this problem requires a more adequate consideration of the "sea" of low molecular weight compounds that cannot interact with the dissolved macromolecules at low polymer concentrations.

Note added in proof:

Since the submission of this article a new approach has been established, which enables a quantitative description of liquid/liquid and liquid/gas phase equilibria by means of the same set of physically meaningful parameters.

Stryuk, S.; Wolf, B. A. (2003): Chain connectivity and conformational variability of polymers: Clues to an adequate thermodynamic description of their solutions III: Modeling of phase diagrams. Macromol. Chem. & Phys. 204, 1948–1955 and literature cited therein.

Acknowledgement

We would like to thank Dr. Roland Horst (Umass Amherst, USA) for fruitful discussions.

References

Banaszak, M.; Chen, C.K.; Radosz, M. (1996): Copolymer SAFT Equation of State. Thermodynamic Pertubation Theory Extended to Heterobonded Chains. Macromolecules 29, 6481–6486.
Barth, C. (1999): Untersuchungen zum thermodynamischen Phasenverhalten membranbildender Systeme. Dissertation, Johannes Gutenberg Universität, Mainz.
Barth, C.; Horst, R.; Wolf, B.A. (1998): (Vapour Plus Liquid) Equilibria of (Water Plus Dimethylformamide): Application of the Headspace Gas Chromatography for the Determination of Thermodynamic Interactions. J. Chem. Thermodyn. 30, 641–652.
Barth, C.; Wolf, B.A. (2000a): Evidence of Ternary Interaction Parameters for Polymer Solutions in Mixed Solvents from Headspace-Gaschromatography. Polymer 41, 8587–8596.
Barth, C.; Wolf, B.A. (2000b): Quick and Reliable Routes to Phase Diagrams for Polyethersulfone and Polysulfone Membrane Formation. Macromol. Chem. Phys. 201, 365–374.
Bodmann, O. (1957): Ein Differentialrefraktometer für Präzisionsmessungen. Chem.-Ing.-Tech. 7, 468–473.
Figueruelo, J.E.; Celda, B.; Campos, A. (1985): Predictability of Properties in Ternary Solvent (1)/Solvent (2)/Polymer (3) Systems from Interaction Parameters of the Binary Systems. 1. General Considerations and Evaluation of Preferential Solvation Coefficients. Macromolecules 18, 2504–2511.
Flory, P.J. (1944): Thermodynamics of Heterogeneous Polymers and their Solutions. J. Chem. Phys. 12, 425.
Flory, P.J.; Orwoll, R.A.; Vrij, A. (1964a): Statistical Thermodynamics of Chain Molecule Liquids. I. An Equation of State for Normal Paraffin Hydrocarbons. J. Am. Chem. Soc. 86, 3507–3514.
Flory, P.J.; Orwoll, R.A.; Vrij, A. (1964b): Statistical Thermodynamics of Chain Molecule Liquids. II. Liquid Mixtures of Normal Paraffin Hydrocarbons. J. Am. Chem. Soc. 86, 3515–3520.
Fowler, R.H.; Guggenheim, E.A. (1940): Statistical Thermodynamics of Super-lattices. Proc. Roy. Soc. A174, 189–206.
Hachenberg, H.; Beringer, K. (1996): Die Headspace-Gaschromatographie als Analysen- und Meßmethode. Vieweg, Braunschweig, Wiesbaden.
Horst, R. (1995): Calculation of Phase Diagrams Not Requiring the Derivatives of the Gibbs Energy Demonstrated for a Mixture of Two Homopolymers With the Corresponding Copolymer. Macromol. Theory Simul. 4, 449–458.
Horst, R. (1996): Calculation of Phase Diagrams Not Requiring the Derivatives of the Gibbs Energy for Multinary Mixtures. Macromol. Theory Simul. 5, 789–800.
Horst, R.; Wolf, B.A. (1996a): Calculation of Critical Points Not Requiring the Derivatives of the Gibbs Energy Demonstrated for a Mixture of Two Homopolymers With the Corresponding Copolymer. Macromol. Theory Simul. 5, 81–92.
Horst, R.; Wolf, B.A. (1996b): Calculation of Miscibility Behavior of Multinary Polymer Blends. Macromol. Symp. 112, 39–46.
Huang, S.H.; Radosz, M. (1990): Equation of State for Small, Large, Polydisperse, and Associating Molecules. Industrial and Engineering Chemical Research 29, 2284–2294.
Huang, S.H.; Radosz, M. (1991): Equation of State for Small, Large, Polydisperse, and Associating Molecules: Extension to Fluid Mixtures. Industrial and Engineering Chemical Research 30, 1994–2005.
Huggins, M.L. (1976): The Thermodynamic Properties of Liquids, Including Solutions. 13. Molecular and Intermolecular Properties from Excess Enthalpies. J. Chem. Phys. 80, 2732–2734.
Huggins, M.L. (1977): The Thermodynamic Properties of Liquids, Including Solutions. XIV. Solutions of Normal Alkanes: Models for Oligomer Solutions. Br. Polym. J. 9, 189–194.

Kamide, K.; Dobashi, T. (2000): Physical Chemistry of Polymer Solutions – Theoretical Background. Elsevier, Amsterdam.

Kolb, B. (1975): Application of Gas Chromatography Head-Space Analysis for the Characterization of Non-ideal Solutions by Scanning the Total Concentration Range. J. Chromatogr. 112, 287–295.

Kolb, B.; Popisil, P.; Auer, M. (1977): Die Bestimmung von Phasengleichgewichten in den Systemen Gas/Fest und Gas/Flüssig mittels der gas-chromatographischen Dampfraumanalyse. Ber. Bunsen-Ges. Phys. Chem. 81, 1067–1070.

Kolb, B.; Ettre, L.B. (1997): Static Headspace-Gas Chromatography – Theory and Practice. Wiley-VCH, New York.

Koningsveld, R.; Kleintjens, L.A. (1971): Liquid-Liquid Phase Separation in Multi-Component Polymer Systems. X. Concentration Dependence of the Pair-Interaction Parameter in the System Cyclohexane/Polystyrene. Macromolecules 4, 637.

Krause, C.; Wolf, B.A. (1997): Shear Effects on the Phase Diagrams of Solutions of Highly Incompatible Polymers in a Common Solvent 1: Equilibrium Behavior and Rheological Properties. Macromolecules 30, 885–889.

Lacombe, R.H.; Sanchez, I.C. (1976): Statistical Thermodynamics of Fluid Mixtures. J. Phys. Chem. 80, 2568–2580.

Petri, H.-M. (1994): Eine neue Methode zur Ermittlung von Wechselwirkungsparametern mit Hilfe von Dampfdruckmessungen. Dissertation, Johannes Gutenberg-Universität, Mainz.

Petri, H.M.; Horst, R.; Wolf, B.A. (1996): Determination of Interaction Parameters for Highly Incompatible Polymers. Polymer 37, 2709–2713.

Petri, H.-M.; Schuld, N.; Wolf, B.A. (1995): Hitherto Ignored Effects of Chain Length on the Flory-Huggins Interaction Parameters in Concentrated Polymer Solutions. Macromolecules 28, 4975–4980.

Petri, H.-M.; Wolf, B.A. (1994): Concentration-Dependent Thermodynamic Interaction Parameters for Polymer Solutions: Quick and Reliable Determination via Normal Gas Chromatography. Macromolecules 27, 2714–2718.

Redlich, O.; Kister, A.T. (1948): Algebraic Representation of Thermodynamic Properties and the Classification of Solutions. Industrial Engineering Chemistry 40, 345–348.

Renon, H.; Prausnitz, J.M. (1968): Local Compositions in Thermodynamic Excess Functions for Liquid Mixtures. AIChE J. 14, 135.

Sanchez, I.C.; Lacombe, R.H. (1977): An Elementary Equation of State for Polymer Liquids. Polymer Letters Edition 15, 71–75.

Sanchez, I.C.; Lacombe, R.H. (1978): Statistical Thermodynamics of Polymer Solutions. Macromolecules 11, 1145–1156.

Schotsch, K.; Wolf, B.A.; Jeberien, H.-E.; Klein, J. (1984): Concentration Dependence of the Flory-Huggins Parameter at Different Thermodynamic Conditions. Makromol. Chem. 185, 2169.

Tompa, H. (1965): Polymer Solutions. Butterworths Scientific Publications London.

Wilson, G.M. (1964): Vapor-Liquid Equilibrium. XI. A New Expression for the Excess Free Energy of Mixing. J. Am. Chem. Soc. 86, 127–130.

Zielkiewicz, J.; Oracz, P. (1990): Vapour-Liquid Equilibrium in the Ternary System N,N-Dimethylformamide+Methanol+Water at 313.15 K. Fluid Phase Equilib. 59, 279–290.

Nomenclature

a	parameter of Eq. (17.9)
A_{12}	parameters of Eq. (17.8)
A_{21}	parameters of Eq. (17.8)
a_i	activity of component i
b	constants of Eq. (17.7)
d	constants of Eq. (17.7)

e	constants of Eq. (17.7)
g	integral Flory-Huggins interaction parameter
G	Gibbs energy
M_i	molar mass of component i
N	number of segments
p	vapor pressure
R	universal gas constant
T	absolute temperature

Greek letters

α	constant of Eq. (17.6)
α_ω	constant related to the interaction energy ω
β	constant of Eq. (17.6)
γ	constant of Eq. (17.6)
ε	parameters of Eq. (17.9)
μ	chemical potential
π	parameter quantifying the enrichment of the non-solvent component in the gas phase (Eq. (17.12))
ρ	mass density of aqueous phase
φ	volume fraction
χ	differential (with respect to the solvent) Flory-Huggins interaction parameter
ω	interaction energy

Superscripts

=	base molar quantity
E	excess
o	pure solvent
o.a.	overall composition

Subscripts

1	solvent component
2	non-solvent component
i, j, k	component i, j, k
P	polymer
qc	referring to quasi-chemical equilibrium

Abbreviations

AC	acetone
DMF	dimethylformamid
HSGC	headspace-gaschromatography
PES	polyethersulfone
PSU	polysulfone

18 Calculation of the High Pressure Phase Equilibrium of Mixtures of Ethylene, Poly(ethylene-co-vinyl-acetate) Copolymers and Vinyl Acetate with a Cubic Equation of State

C. Browarzik, D. Browarzik[1], and H. Kehlen

18.1 Abstract

The cubic equation of state proposed by Sako, Wu, and Prausnitz (SWP-EoS) is applied to calculate the high pressure phase equilibrium for solutions of poly(ethylene-co-vinyl acetate) (EVA) in supercritical ethylene and vinyl acetate (VA) mixtures. The polydisperse character of copolymers, consisting of a large number of components differing in molecular mass and chemical composition, is taken into account by the use of a bivariate distribution function in the framework of continuous thermodynamics. Cloud-point curves, shadow curves and coexistence curves are calculated for several quasibinary ethylene + EVA systems with different averages of molecular mass and chemical compositions of the copolymer at five temperatures (393 to 473 K). The binary interaction parameter is found to be a function of the temperature and of the chemical composition of the copolymer but to be independent of molecular mass. Predictions of cloud-point curves, shadow curves, and coexistence curves for ethylene + EVA systems not included in parameter fit are successful. The binary interaction parameter of the ethylene + VA system is fitted to experimental vapor-liquid equilibrium data. For the EVA + VA system Henry constants were used for the parameter fit. Prediction of cloud-point pressures of the ternary system with the aid of SWP-EoS and the only use of binary parameters leads to a clear overestimation but the effect of lowering cloud-point pressure with increasing content of VA in the solvent mixture is described qualitatively correct.

18.2 Introduction

Phase equilibrium data of copolymer solutions are highly interesting both from an industrial and a scientific point of view. Poly(ethylene-co-vinyl acetate) copolymers (EVA) as investigated in this contribution are produced by radicalic high-pressure polymerization at temperatures of

---

[1] Martin-Luther-Universität Halle-Wittenberg, Institut für Physikalische Chemie, Geusaer Straße, 06217 Merseburg.

about 450 K. A detailed knowledge of the real phase boundaries (cloud-points) and the phase compositions (coexistence-points) especially their dependence on temperature and pressure is necessary for two reasons. At first, one has to be sure that the polymerization takes place in a homogenous phase; at second, only with this knowledge the downstream processes can be optimized. Appropriate high pressure experimental data are rare. Wagner (1983) studied the phase equilibria of the quasibinary system ethylene + EVA intensively, Nieszporek (1991) measured cloud-point pressures and coexistence data both for the binary system and for the ternary system with ethylene and VA as mixed solvent. Some coexistence data for the ternary system were also given by Finck et al. (1992). A new comprehensive collection of phase equilibrium data for copolymer systems was recently published by Wohlfahrt (2001).

Copolymers are multicomponent systems consisting of a large number of components differing in chain length and chemical composition. Therefore, bivariate distribution functions like the one proposed by Stockmayer (1945), which combines a Schulz-Flory-distribution function for the molecular mass polydispersity with a Gaussian-distribution function for the chemical polydispersity, are needed. Width and shape of both types of distribution functions influence the phase boundaries of the copolymer solution.

The influence of the molecular-mass distribution function on phase equilibrium has been investigated over the last thirty years by various authors. Koningsfeld and Staverman (1968 a, b) described qualitatively the differences in liquid-liquid phase separation between binary and multicomponent solutions and presented general calculation methods for liquid-liquid equilibrium in polymer solutions using a generalized Flory-Huggins model. Šolc (1970, 1975) extended these methods and studied the influence of the distribution functions on the cloud-point curve and shadow curve.

A very suitable procedure to calculate fluid phase equilibria of polydisperse systems like polymers is the continuous thermodynamics introduced by Kehlen and Rätzsch (1980) as well as applied and extended by numerous authors during the last two decades (Rätzsch and Kehlen, 1989; Rätzsch and Wohlfarth, 1990; Cotterman and Prausnitz, 1991; Bergmann et al., 1992; Hu et al., 1995; Browarzik and Kehlen, 2000). They consider the polymer as a continuous mixture replacing the summations for the components used in traditional thermodynamics by integrations for the total range of the distributed variable. Instead of the mass fractions of the polymer species there is a continuous distribution function.

The influence of the chemical heterogeneity on the phase equilibrium has been studied less intensively. Rätzsch et al. (1989, 1990) applied continuous thermodynamics to the liquid-liquid equilibrium of copolymer solutions using bivariate distribution functions in the framework of an improved Flory-Huggins model. An extensive review concerning calculation of phase equilibria in random copolymer systems has been given by Wohlfahrt (1993). Whereas Gibbs free-energy models are restricted to low pressure calculations, equation of state (EoS) approaches are suitable to estimate phase transitions (cloud points) over a wide range of pressure.

To meet the special requirements in polymer solutions Sako, Wu, and Prausnitz (1989) developed a cubic equation of state (further referred to as SWP-EoS) that considers a molecule as a chain of equally sized segments. Additionally to the two usual parameters there is a third one taking into account rotational and vibrational degrees of freedom due to the flexibility of the polymer chain. Sako et al. (1989) calculated the high-pressure phase equilibrium in polyethylene + ethylene solutions. More recently the SWP-EoS was applied to systems of ethylene with heavy alkanes and alkenes and to binary and ternary polyolefin systems by Tork et al. (1999a,b). Browarzik and Kowalewski (1999) calculated stability and phase equilibrium in the system polystyrene + cyclohexane + carbon dioxide using SWP-EoS in the framework of continuous thermodynamics.

The aim of this contribution is to extend that phase-equilibrium treatment to the calculation of phase equilibria in copolymer systems considering also chemical heterogeneity. Furthermore, parameter fit at five temperatures for eleven ethylene + EVA systems provides a set of parameters suitable to predict phase equilibrium of this system over a wide range of temperature, concentration, and copolymer composition. Vapor-liquid equilibrium of the monomer system ethylene + vinyl acetate and Henry constants of the system EVA + VA are the basis for the fit of the other binary parameters. On the basis of pure-component and binary parameters the phase equilibrium of the ternary system ethylene + EVA + VA should be predicted.

18.3 Theory

18.3.1 Segment-Molar Quantities and Polydispersity

Polymer solutions show enormous size differences between solvent and polymer molecules. Thus, instead of the very uncomfortable mole fractions mostly mass or volume fractions are used. A more general way to overcome this problem is the use of segment-mole fractions applied to g^E-models for a long time (Rätzsch and Kehlen, 1989) and recently introduced to EoS approaches (Browarzik and Kowalewski, 1999). In this, the molecules are thought to be divided into segments of the same size and, all quantities are related to them. The corresponding reference segment can be chosen arbitrarily, but often it has been proved to be very suitable assuming the smallest molecule in the mixture as the reference segment. Thus, in the system ethylene + EVA + vinyl acetate the ethylene molecule is chosen to be the reference segment with the segment number $r = 1$. The segment-mole fraction ψ_i of a component i is connected to the corresponding mole fraction x_i by:

$$\psi_i = \frac{x_i r_i}{\sum_j x_j r_j} \tag{18.1}$$

The denominator of Eq. (18.1) is the number-average $\bar{r}^M$ of the segment number of the system considered. To get the segment-molar quantity the corresponding molar quantity has to be divided through $\bar{r}^M$. On this basis, all the following equations are given with segment-molar quantities. The segment number of a polymer molecule r can be determined in different ways e.g. proportional to the volume, to the van der Waals volume, or to the molecular mass. For the sake of simplicity, we consider the segment number being proportional to the molecular mass as $r = M/M_{ethylene}$. In this case, the occurring segment-mole fractions of the mixture are mass fractions as mostly used to measure concentrations in polymer systems.

Besides the differences in size, the polydispersity of polymers is the second problem that requires a special treatment of polymer solutions. For homopolymers it is restricted to the heterogeneity in molecular mass and can be mathematically described by various analytical distribution functions (e.g. Schulz-Flory, Wesslau, or Hosemann-Schramek) depending on the asymmetry of the given molecular mass distribution. Copolymers show additionally chemical heterogeneity. For example, there are molecules with different contents of vinyl acetate in EVA. Hence, a bivariate distribution function is necessary to describe the composition of the copolymer. To the best of our knowledge, the only analytical expression for such a bivariate distribution was derived by Stockmayer (1945). The Stockmayer distribution function is a product of a Schulz-Flory-distribution with respect to the segment number and a Gaussian distribution with respect to the chemical composition. In segment molar notation it reads:

$$W(r, y) = \frac{k^k}{\Gamma(k)\bar{r}} \left(\frac{r}{\bar{r}}\right)^k \exp\left(-k\frac{r}{\bar{r}}\right) \cdot \sqrt{\frac{r}{2\pi\varepsilon}} \exp\left(-\frac{r(y-\bar{y})^2}{2\varepsilon}\right) \qquad (18.2)$$

Because of the supposed proportionality between the segment number r and the molecular mass M, $W(r, y)$ is a mass distribution function and $W(r, y)\,dy\,dr$ is the mass fraction of all species with r-values between r and $r + dr$ and y-values between y and $y + dy$. Thus, analogous to the discontinuous treatment, the integral for the total domain of definition of r and y must be one. The number average of the segment number of all copolymer species $\bar{r}$ and the average $\bar{y}$ of chemical composition of the copolymer y are related to the distribution function of the phase considered by

$$\frac{1}{\bar{r}} = \int_r \int_y \frac{1}{r} W(r, y)\,dy\,dr \qquad (18.3\text{a})$$

$$\bar{y} = \int_r \int_y y\, W(r, y)\,dy\,dr \qquad (18.3\text{b})$$

Here, and in the following treatment the integration is performed for the total domain of definition of r and y. That means concerning the Stockmayer-distribution for r: $0 \leq r < \infty$, and for y: $-\infty < y < \infty$.

The parameter k is connected with the non-uniformity U by the relation $k = 1/U$. It describes the width of the molecular mass distribution and is related to the mass average $\bar{M}_w$ and the number average $\bar{M}_n$ of the molecular mass:

$$\frac{1}{k} = \frac{\bar{M}_w}{\bar{M}_n} - 1 \qquad (18.4)$$

The parameter ε describes the width of the Gaussian-distribution and may be calculated according to

$$\varepsilon = \bar{y}(1-\bar{y})\left[1 - 4\bar{y}(1-\bar{y})(1 - q_\alpha q_\beta)\right]^{1/2} \qquad (18.5)$$

where q_α and q_β are the kinetic parameters of the copolymerization, with α denoting ethylene and β denoting vinyl acetate. For EVA-copolymers $q_\alpha = 1.07$ and $q_\beta = 1.08$ lead to small values of ε (cf. Table 18.1) resulting in a relatively small distribution function with respect to the chemical composition y. But the Stockmayer distribution only regards the one part of heterogeneity, the so called *"instantaneous heterogeneity"* (Bauer, 1985), resulting from the statistical nature of copolymerization. In this case all chains are assumed to be formed under identical conditions. The second type of heterogeneity particularly results from the temporal change of the monomer concentration caused by the depletion of the monomer from the reaction mixture in a rate different from the monomer concentration. Thus, the polymer species possess different compositions depending on the time of their formation. This *"conversion heterogeneity"* for the most copolymers is regarded to be the main part of heterogeneity. However, because for EVA copolymers the kinetic parameters q_α and q_β are nearly one also the *"conversion heterogeneity"* will be relatively small. Furthermore, there are no experimental values about this, so the Stockmayer distribution function with Eq. (18.5) for ε had to be used in the following calculations.

We studied the influence of the polydispersity on phase equilibrium in an earlier paper (Browarzik et al., 2001). In this we could show that the polydispersity with respect to the molecular mass influences the phase equilibrium very strongly. The influence of the chemical part of the polydispersity is smaller and decreases with increasing molecular mass of the copolymer molecules.

18.3.2 Calculation of the Cloud-Point Curves

To calculate cloud points for the pseudo ternary system ethylene (A) + EVA (B) + VA (C) we need the phase equilibrium conditions to a given phase *I* and an incipient phase *II*. In the framework of an EoS the isofugacity criterion has to be applied. In segment-molar notation this criterion reads:

$$\psi_i^I \bar{r}^{MI} \phi_i^I = \psi_i^{II} \bar{r}^{MII} \phi_i^{II}; \quad i = A, C \tag{18.6}$$

$$\psi_B^I \bar{r}^{MI} W^I(r, y) \phi_B^I(r, y) = \psi_B^{II} \bar{r}^{MII} W^{II}(r, y) \phi_B^{II}(r, y) \tag{18.7}$$

The first of these equations relates to the low-molecular components. In the following the segment-mole fraction of A in both phases is to be expressed by $\psi_A = 1 - \psi_B - \psi_C$. Equation (18.7) relates to the copolymer species. In Eqs. (18.6) and (18.7) ϕ_i and ϕ_B are the fugacity coefficients of the components A, C and of the copolymer species. The number average $\bar{r}^M$ of the segment number of the system may be written by:

$$\frac{1}{\bar{r}^M} = \frac{1 - \psi_B - \psi_C}{r_A} + \frac{\psi_C}{r_C} + \frac{\psi_B}{\bar{r}_B} \tag{18.8}$$

Here $\bar{r}_B$ is the number average of the segment number of the copolymer B defined by Eq. (18.3a). The segment number of the solvent r_A is set to one.

From an EoS the fugacity coefficient of a single component i is:

$$RT \ln \phi_i = \int_V^\infty \left(\frac{r_i \partial p}{\partial n_i} - \frac{RT}{V} \right) dV - RT \ln \left(\frac{pV\bar{r}^M}{nRT} \right) \tag{18.9}$$

where n and n_i are the segment-mole numbers of the mixture and of the component i obtained from the corresponding mole numbers by multiplication with $\bar{r}^M$ and r_i, respectively. V is the extensive volume. R is the universal gas constant, p is the pressure, and T is the temperature. For the copolymer species continuous thermodynamics leads to

$$RT \ln \phi_B(r, y) = \int_V^\infty \left(\frac{r \, \partial p}{\partial w(r, y)} - \frac{RT}{V} \right) dV - RT \ln \left(\frac{pV\bar{r}^M}{nRT} \right) \tag{18.10}$$

where $w(r, y)$ is the extensive distribution function obtained from W by multiplication with the segment-mole number of the copolymer. In the case of cubic equations of state assuming the segment-molar parameters to be independent of the chain-length (see Section 18.3.4) the fugacity coefficient of single component is:

$$\ln \phi_i = r_i A_i - \ln \bar{r}^M - \ln \left(\frac{pV_S}{RT} \right); \quad i = A, C \tag{18.11}$$

and, for the copolymer B:

$$\ln \phi_B(r, y) = B_1 r + B_2 ry - \ln \bar{r}^M - \ln \left(\frac{pV_S}{RT} \right) \tag{18.12}$$

The segment-molar volume V_S is obtained from the molar volume by division through $\bar{r}^M$. The equilibrium condition for the solvents Eq. (18.6) together with Eq. (18.11) can directly be applied to the cloud-point calculations. Because the equilibrium condition of the polymer Eq. (18.7) is a functional equation further mathematical treatment is necessary. At first, Eqs. (18.7) and (18.12) provide for the distribution function of the incipient phase II:

$$W^{II}(r, y) = \frac{\psi_B^I V_S^{II}}{\psi_B^{II} V_S^I} \exp(\Delta B_1 r + \Delta B_2 ry) W^I(r, y)$$

$$\Delta B_1 = B_1^I - B_1^{II} \qquad \Delta B_2 = B_2^I - B_2^{II} \tag{18.13}$$

For phase I we assume a Stockmayer-distribution Eq. (18.2). Inserting into Eq. (18.13) and integration leads to:

$$1 = \frac{\psi_B^I V_S^{II}}{\psi_B^{II} V_S^I} (\kappa^*)^{-(1+1/U)} \tag{18.14}$$

where the auxiliary quantity κ^* reads:

$$\kappa^* = 1 - \bar{r}_B^I U \left[\Delta B_1 + \bar{y}^I \Delta B_2 + (1/2) \varepsilon (\Delta B_2)^2 \right] \tag{18.15}$$

Dividing Eq. (18.13) through r, inserting Eq. (18.2) and integration gives:

$$\frac{1}{\bar{r}_B^{II}} = \frac{1}{\bar{r}_B^I} \frac{\psi_B^I}{\psi_B^{II}} \frac{V_S^{II}}{V_S^I} (\kappa^*)^{-1/U} \tag{18.16}$$

Combining Eqs. (18.14) and (18.16) in a way κ^* is eliminated leads to an equation for the average segment number of the phase II $\bar{r}_B^{II}$ where $\bar{y}^{II}$ no longer occurs.

$$\bar{r}_B^{II} = \bar{r}_B^I \left(\frac{\psi_B^I}{\psi_B^{II}} \frac{V_S^{II}}{V_S^I} \right)^{-U/(U+1)} \tag{18.17}$$

dividing Eq. (18.16) by Eq. (18.14) gives:

$$\frac{1}{\bar{r}_B^{II}} = \frac{\kappa^*}{\bar{r}_B^I} \tag{18.18}$$

Multiplying Eq. (18.13) by y and inserting the Stockmayer-distribution function integration leads to:

$$\bar{y}^{II} = \frac{\psi_B^I}{\psi_B^{II}} \frac{V_S^{II}}{V_S^I} (\bar{y}^I + \varepsilon \Delta B_2)(\kappa^*)^{-(1+1/U)} \tag{18.19}$$

The following division by Eq. (18.14) gives:

$$\bar{y}^{II} = \bar{y}^I + \varepsilon \Delta B_2 \tag{18.20}$$

With the aid of Eqs. (18.13) to (18.18) and Eq. (18.20) one can show that the distribution function of the phase II is also of the Stockmayer type with the parameters $\bar{r}_B^{II}, \bar{y}^{II}$, k, and ε where the parameters k and ε proved to have the same values in both phases. For a given temperature, cloud-point and shadow curve are obtained by solving the non-linear set of Eqs. (18.6) and (18.14) calculating p, ψ_B^{II}, and ψ_C^{II}. The pressure in dependence on the segment-mole fractions of the phase I gives the cloud-points. $p(\psi_B^{II}, \psi_C^{II})$ gives the so-called shadow points. In the case of polydisperse polymers the functions $p(\psi_B^I, \psi_C^I)$ and $p(\psi_B^{II}, \psi_C^{II})$ are quite different. In the monodisperse case they coincide (Rätzsch and Kehlen, 1989). The characteristic quantities of phase II, the number average of the segment number $\bar{r}_B^{II}$, and the averaged chemical composition $\bar{y}^{II}$ can be immediately calculated by Eqs. (18.17) and (18.20) and thus, $W^{II}(r, y)$ is known too. The segment-molar volumes V_S^I and V_S^{II} have to be computed by solving the EoS for both phases at all steps of iteration.

18.3.3 Calculation of the Coexistence Curves

For solutions of a polydisperse polymer in a solvent additionally to the cloud-point curve $p(\psi_B^I)$ and to the shadow curve $p(\psi_B^{II})$ there is an infinite number of coexistence curves (Rätzsch and Kehlen, 1989). They describe the phase split within the metastable or unstable region. In the monodisperse case the cloud-point curve, the shadow curve and all coexistence curves coincide forming the binodal curve.

We consider a feed phase F splitting into two phases *I* and *II*. Additionally to the equilibrium conditions between the phases *I* and *II* there are the following material balance equations:

$$\psi_B^F W^F(r, y) = (1 - \xi)\psi_B^I W^I(r, y) + \xi \psi_B^{II} W^{II}(r, y) \tag{18.21}$$

$$\psi_C^F = (1 - \xi)\psi_C^I + \xi \psi_C^{II} \tag{18.22}$$

Here ξ is the quotient of the amount of segments in the phase *II* and the amount of segments in the feed phase. For $\xi = 0$ and $\xi = 1$ Eqs. (18.21) and (18.22) will be trivial. Then the feed phase is identical with the phase *I* or *II* and, therefore, only the equilibrium conditions have to be considered as outlined in the previous section. However, the cases $\xi = 0$ and $\xi = 1$ are included and, therefore, the end points of all coexistence curves are located on the cloud-point curve or on the shadow curve.

In this contribution coexistence curves were calculated for the system ethylene + EVA only. Therefore, this case is discussed in more detail. Here, Eq. (18.22) is not needed. However, the application of Eq. (18.21) is complicated and does not enable an analytical solution. For this reason we generalize the so-called "method of moments" introduced by Cotterman and Prausnitz (1985). Here, the material balance equation is solved only approximately. The Stockmayer distribution is assumed to describe the composition of all phases. Then Eq. (18.21) is required only for some moments of the distribution function. For further simplification we assume the parameters k and ε of the Stockmayer distribution to possess the same values in the feed phase F and in the phases *I* and *II*. In the cases $\xi = 0$ and $\xi = 1$ this assumption is correct as mentioned in the previous section. In the general case this is an approximation only. Restricting to the three most important moments Eq. (18.21) may be replaced by the following three scalar equations:

$$\psi_B^F = (1 - \xi)\psi_B^I + \xi \psi_B^{II} \tag{18.23a}$$

$$\psi_B^F / \bar{r}_B^F = (1 - \xi)\psi_B^I / \bar{r}_B^I + \xi \psi_B^{II} / \bar{r}_B^{II} \tag{18.23b}$$

$$\psi_B^F \bar{y}^F = (1 - \xi)\psi_B^I \bar{y}^I + \xi \psi_B^{II} \bar{y}^{II} \tag{18.23c}$$

Equations (18.23) are obtained from Eq. (18.21) by multiplying with 1, 1/*r*, or *y* and double integration over *r* and *y* afterwards. With the aid of Eq. (18.23a) one finds

$$\xi = \frac{\psi_B^F - \psi_B^I}{\psi_B^{II} - \psi_B^I} \tag{18.24}$$

Now, we consider the quantities T, ψ_B^F, ψ_B^I, $\bar{r}_B^F$, $\bar{y}^F$ to be given. Then, there are the four equilibrium conditions (Eqs. (18.6), (18.14), (18.17), (18.20)) and two material balance equations (18.23b) and (18.23c) (eliminating ξ by Eq. (18.24)) to calculate the six unknowns p, ψ_B^{II}, $\bar{r}_B^{II}$, $\bar{y}^{II}$, $\bar{r}_B^I$, $\bar{y}^I$. The numerical procedure is performed step by step starting with a cloud point. Finally, the quantity ξ is known from Eq. (18.24).

18.3.4 Sako-Wu-Prausnitz Equation of State (SWP-EoS)

Sako et al. (1989) derived a cubic EoS suitable for polymers and applied it successfully to the high-pressure phase equilibrium of the system polyethylene + ethylene. The SWP-EoS is very similar to the Soave-Redlich-Kwong equation (SRK-EoS) but is taking into account the existence of flexible polymer chains via a third parameter c. Depending on the chain flexibility, c can take values between one and the segment number r of the polymer molecule. $3c$ is the total number of external degrees of freedom of a molecule. In this, vibration, rotation, and translation are considered to be equivalent external degrees of freedom. Using segment-molar quantities SWP-EoS is:

$$p = \frac{RT}{\bar{r}M V_s} + \frac{RT b_s c_s}{V_s(V_s - b_s)} - \frac{a_s}{V_s(V_s + b_s)} \tag{18.25}$$

The segment-molar parameters a_s, b_s, c_s are related to the corresponding molar quantities by:

$$a_s = \frac{a}{(\bar{r}M)^2}; \quad b_s = \frac{b}{\bar{r}M}; \quad c_s = \frac{c}{\bar{r}M} \tag{18.26}$$

Assuming these parameters to be independent of the chain length Eq. (18.12) is fulfilled. However, the segment-molar parameters depend on the chemical composition of the copolymer species. Because nothing is known about the type of this dependence, for the sake of simplicity, we set:

$$b_{s,B}(y) = b_r + y b_y \tag{18.27a}$$

$$c_{s,B}(y) = c_r + y c_y \tag{18.27b}$$

$$\sqrt{a_{s,B}}(y) = a_r + y a_y \tag{18.27c}$$

where polyethylene (PE) is characterized by a_r, b_r, c_r and poly(vinyl acetate) (PVAC) by $a_r + a_y$, $b_r + b_y$, $c_r + c_y$,.

Empirical mixing rules are required for mixture parameters a, b, and c. As usual, we assume arithmetic mixing rules for b and c and a geometric mixing rule for a with addition of a correctional term:

$$a_{ij} = \sqrt{a_{ii} a_{jj}}(1 - k_{ij}); \quad i,j = A, B, C; \quad k_{ii} = 0 \tag{18.28}$$

Taking into account the y-dependence of the pure-component parameters of the copolymer, the segment-molar parameters of the system ethylene + EVA copolymer are given by:

$$a_s = \left[(1 - \psi_B - \psi_C)\sqrt{a_{s,A}} + \psi_B(a_r + \bar{y} a_y) + \psi_C \sqrt{a_{s,C}}\right]^2$$
$$- 2\psi_B(1 - \psi_B - \psi_C)\sqrt{a_{s,A}}(a_r + \bar{y} a_y) k_{AB} - 2\psi_B \psi_C \sqrt{a_{s,C}}(a_r + \bar{y} a_y) k_{BC}$$
$$- 2(1 - \psi_B - \psi_C)\psi_C \sqrt{a_{s,A} a_{s,C}} k_{AC} \tag{18.29a}$$

$$b_s = (1 - \psi_B - \psi_C) b_{s,A} + \psi_B (b_r + \bar{y} b_y) + \psi_C b_{s,C} \tag{18.29 b}$$

$$c_s = (1 - \psi_B - \psi_C) c_{s,A} + \psi_B (c_r + \bar{y} c_y) + \psi_C c_{s,C} \tag{18.29 c}$$

The mixing parameters k_{ij} (i, j = A, B, C) have to be obtained by fitting the experimental data.

With Eqs. (18.25) to (18.29) the quantities A_i (i = A, C), B_1, B_2 become:

$$A_i = -c_{s,i} \ln\left(1 - \frac{b_s}{V_s}\right) + \frac{b_{s,i} c_s}{V_s - b_s} - \frac{b_{s,i} a_s}{RTb_s (V_s + b_s)} + \frac{\ln(1 + b_s/V_s)}{RTb_s}$$
$$\cdot \left\{ \frac{b_{s,i} a_s}{b_s} - 2\sqrt{a_{s,i}} \left[\psi_B (a_r + \bar{y} a_y)(1 - k_{iB}) + (1 - \psi_B - \psi_C)\sqrt{a_{s,A}} (1 - k_{iA}) + \right. \right.$$
$$\left. \left. \psi_C \sqrt{a_{s,C}} (1 - k_{iC}) \right] \right\} \tag{18.30}$$

$$B_1 = -c_r \ln\left(1 - \frac{b_s}{V_s}\right) + \frac{b_r c_s}{V_s - b_s} - \frac{b_r a_s}{RTb_s (V_s + b_s)} + \frac{\ln(1 + b_s/V_s)}{b_s RT}$$
$$\cdot \left\{ \frac{b_r a_s}{b_s} - 2 a_r \left[\psi_B (a_r + \bar{y} a_y) + (1 - \psi_C - \psi_B)\sqrt{a_{s,A}} (1 - k_{AB}) + \right. \right.$$
$$\left. \left. \psi_C \sqrt{a_{s,C}} (1 - k_{BC}) \right] \right\} \tag{18.31 a}$$

$$B_2 = -c_y \ln\left(1 - \frac{b_s}{V_s}\right) + \frac{b_y c_s}{V_s - b_s} - \frac{b_y a_s}{RTb_s (V_s + b_s)}$$
$$\cdot \left\{ \frac{b_y a_s}{b_s} - 2 a_y \left[\psi_B (a_r + \bar{y} a_y) + (1 - \psi_B - \psi_C)\sqrt{a_{s,A}} (1 - k_{AB}) + \right. \right.$$
$$\left. \left. \psi_C \sqrt{a_{s,C}} (1 - k_{BC}) \right] \right\} \tag{8.31 b}$$

18.3.5 Fit of Pure-Component Parameters

For volatile substances, the pure-component parameters are calculated from their critical data. For ethylene, because of $r = 1$ molar and segment-molar pure-component parameter are equal. The parameter $c_{s,A}$ is set to one. For $a_{s,A}$ and $b_{s,A}$ follows:

$$a_{s,A} = 0.467313 \text{ MPa} l^2 \text{ mol}^2 \, \alpha(T) \tag{18.32}$$

$$b_{s,A} = 0.040347 \text{ l/mol}$$

The temperature dependence of a is considered to be the same as proposed by Sako et al. (1989):

18.3 Theory

$$\alpha(T) = \frac{\alpha_0(1-T_R^2) + 2T_R^2}{1+T_R^2}; \quad T_R = T/T_{crit} \quad (18.33)$$

with $\alpha_0 = 1.55028$ and $T_{crit} = 282.34$ K for ethylene.
The pure-component parameters of vinyl acetate were found to be:

$$a_{s,C} = 0.209982 \text{ MPal}^2 \text{ mol}^2 \beta(T) \quad (18.34)$$

$$b_{s,C} = 0.026512 \text{ l/mol}$$

$$c_{s,C} = 0.4236$$

The parameter $c_{s,C}$ (together with the parameter k_{AC}) was fitted to the vapor-liquid equilibrium of the binary system ethylene + vinyl acetate (see Section 18.4.2). The temperature dependence of the parameter $a_{s,C}$ was chosen to be the same as in the Soave-Redlich-Kwong EoS resulting in:

$$\beta(T) = [1 + 0.992487(1-T_R)]^2 \quad (18.35)$$

with $T_{crit} = 524.0$ K.
Firstly, the pure component parameters of the copolymer or, respectively, of the homopolymers polyethylene (PE) and poly(vinyl acetate) (PVAC) were determined using the relations proposed by Sako et al. (1989). In this, a bases on the molar polarization, the van der Waals volumes and the first ionization potentials of the corresponding saturated monomers ethane and ethyl acetate. b is correlated to the van der Waals volumes of higher n-alkanes and c to the density data of some polymers. However, calculations for the system polyethylene + ethylene showed, that using these parameters a_r, b_r, c_r the experimental cloud-point curves (Rätzsch et al., 1980) could not properly be described. Especially, the location of the maximum and the width of the cloud-point curve were incorrect. Fitting a_r, b_r, c_r to the experimental data of polyethylene + ethylene it is obtained:

$$a_r = 0.21590 \exp(-0.52266 \cdot 10^{-3} T/K)\sqrt{\text{MPa}} \text{ l/mol} \quad (18.36)$$

$$b_r = 0.011268 \text{ l/mol} \quad c_r = 0.137$$

The temperature dependence of a_y was assumed to be the same as proposed by Sako et al. (1989). The parameter a_y, b_y, c_y according to Sako et al. (1989) proved totally unsuitable (Sako et al. (1989) also had problems describing the properties of PVAC) to describe the experimental cloud-point curves measured by Wagner (1983). A set of parameters derived from the parameters of Eq. (18.36) and from data of the high-pressure equilibrium of the system ethylene + EVA (see Section 18.4.1) is successfully used:

$$a_y = a_r \cdot 2; \quad b_y = b_r; \quad c_y = c_r \quad (18.37)$$

18.4 Results and Discussion

18.4.1 The Binary System Ethylene (A) + EVA (B)

After the determination of the pure component parameters, the mixing parameter k_{ij} of the binary system considered had to be adjusted to experimental phase equilibrium data. The most comprehensive high-pressure study of the system ethylene + EVA dates back to Wagner (1983) who measured cloud-point curves at five temperatures between 393.15 K and 473.15 K for five ethylene + EVA copolymer (CP1 to CP5) systems with different molecular masses and chemical compositions of the copolymer. To eliminate the influence of the chemical composition, systems with copolymers (F1 to F6) resulting from molecular mass fractionation of the copolymer CP4 were studied. Table 18.1 gives the characterization data of these polymers used for parameter fit and, additionally, the data for the polymers examined by Nieszporek (1991) used here to test the quality of our cloud-point calculation basing on the SWP-EoS.

Table 18.1: Characterization of EVA-copolymers used for parameter fit and calculations.

EVA	$\overline{M}_n$/g mol^{-1}	$\overline{M}_w$/g mol^{-1}	U	$\bar{y}$	ε
CP1 [a,c]	20,400	145,000	6.1	0.075	0.071
CP2 [a,c]	19,700	120,000	5.1	0.127	0.115
CP3 [a,c]	13,700	60,800	3.4	0.273	0.211
CP4 [a,c]	16,200	45,800	1.8	0.318	0.231
CP5 [a,c]	19,800	44,000	1.3	0.427	0.263
F1 [a,c]	5430	8220	0.51	0.330	0.236
F2 [a,c]	9590	16,900	0.76	0.334	0.238
F3 [a,c]	15,000	28,300	0.89	0.332	0.237
F4 [a,c]	15,500	40,800	1.63	0.323	0.233
F5 [a,c]	21,800	46,400	1.13	0.321	0.233
F6 [a,c]	27,500	60,300	1.19	0.314	0.230
EVA2 [b,c]	33,400	137,700	3.12	0.175	0.151
EVA3–1 [b,c]	13,000	22,100	0.7	0.325	0.234
EVA3–7 [b,c]	83,600	231,000	1.76	0.26	0.204

a) Wagner, 1983; b) Nieszporek, 1991; c) Wohlfahrt, 2001

The designation of the copolymers is the same as proposed by Wagner and Nieszporek. Besides the number average of the molecular mass $\overline{M}_n$ and the averaged mass fraction of VA within the copolymer $\bar{y}$, the quantities characterizing the double polydispersity of the copolymer (the non-uniformity U and the parameter ε) are listed. U describes the width of the molecular mass distribution that has decisive influence on shape and position of the phase boundary curve. So, the binodal curve characteristic for a monodisperse polymer solution is separated into cloud-point curve and shadow curve for a polydisperse one. The cloud-point pressure for a given polymer concentration increases significantly with increasing U and the critical points are shifted to lower mass fractions of the polymer. Chemical

18.4 Results and Discussion

heterogeneity has a similar effect on the phase equilibrium of a polymer solution, but, because of the small chemical distribution for EVA-copolymers (corresponding to the small ε values) the effect is not very strong. Both kinds of polydispersity were more detailed discussed in our previous paper (Browarzik et al., 2001).

At first, calculations were performed for 393.15, 413.15, 433.15, 453.15, and 473.15 K. The binary parameter k_{AB} was adjusted to the experimental cloud-point data at each temperature separately. Typical results for some ethylene + EVA systems at 433.15 K are shown and compared with the experimental data (Wagner, 1983) in Figure 18.1. The CP-type systems reveal an increase of the equilibrium pressure with increasing non-uniformity and decreasing VA content of the copolymer, whereas the F-type systems (not shown here) reveal a similar increase of the equilibrium pressure with increasing molecular mass. Altogether, the description for the most systems is remarkably good although the slopes of the calculated curves are a little too steep at their right-hand branches. Only the system ethylene + CP1 shows larger differences between calculated and experimental cloud-point data due to the high non-uniformity of this copolymer sample. The characteristic shoulder experimentally found around the critical point for ethylene + EVA systems and well known from polyethylene + ethylene systems (Folie and Radosz, 1995) too, cannot be adequately described by the model. Perhaps the Stockmayer distribution is too simple and a more sophisticated molecular-mass part of the distribution function might solve this problem. Because the EVA

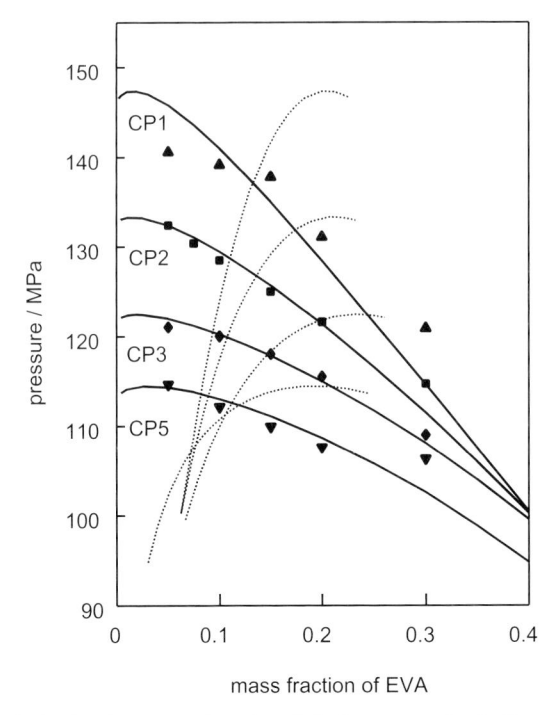

Figure 18.1 Calculated cloud-point curves (———) and shadow curves (···) of some ethylene + EVA systems at 433.15 K compared with experimental data (Wagner, 1983); ethylene + CP1 (▲), ethylene + CP2 (■), ethylene + CP3 (◆), ethylene + CP5 (▼).

copolymers studied by Wagner (1983) cover a wide range of the molecular mass and of the VA content, k_{AB} may be analyzed in its dependence on $\overline{M}_n$ and $\bar{y}$. We found k_{AB} to be independent of the molecular mass. Similar results were obtained by Tork et al. (1999 b) for LDPE + ethylene systems using SWP-EOS and a pseudocomponent approach. On the other hand, k_{AB} decreases nearly linear with increasing $\bar{y}$. A simultaneous fit (including data for all 11 systems) at 433.15 K resulted in:

$$k_{AB} = 0.11413 - 0.36691 \cdot \bar{y} \qquad (18.38)$$

In this way, reliable predictions of the cloud-point curve for other ethylene + EVA systems (with different $\overline{M}_n$ and $\bar{y}$) should be possible. To test this, the cloud-point curve of some ethylene + EVA systems studied experimentally by Nieszporek (1991) was predicted and compared with the experimental data. Figure 18.2 shows a remarkable good agreement.

The phase equilibrium at the other temperatures could be described similarly good by adjusting k_{AB} to the experimental data from Wagner (1983) for each temperature. In all cases k_{AB} proved to be nearly independent of the molecular mass and linearly dependent on $\bar{y}$. Considering this, we found the following expression for the temperature function of k_{AB}:

$$k_{AB} = \left[-33.91327 + 0.342644\, T/K + (244.75932 - 1.413512\, T/K)\bar{y}\right]/1000 \qquad (18.39)$$

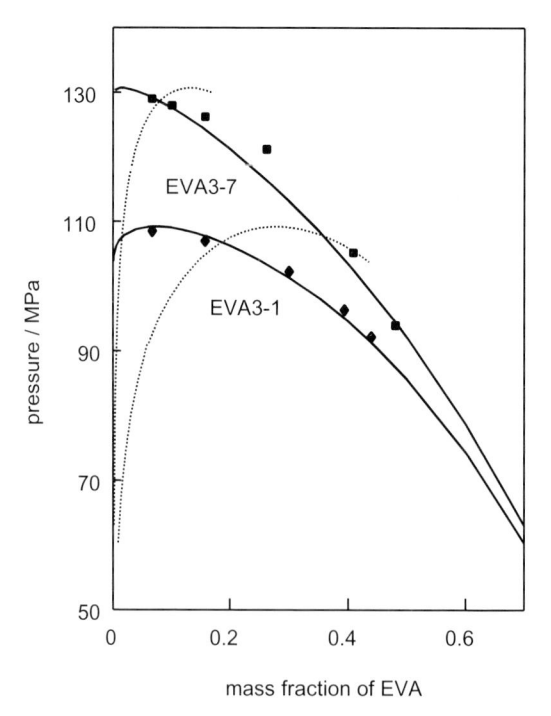

Figure 18.2: Calculated cloud-point curves (———) and shadow curves (· · ·) of the systems ethylene + EVA3–1 and ethylene + EVA3–7 at 433.15 K compared with experimental data (Nieszporek, 1991); ethylene + EVA3–7 (■), ethylene + EVA3–1 (◆).

18.4 Results and Discussion

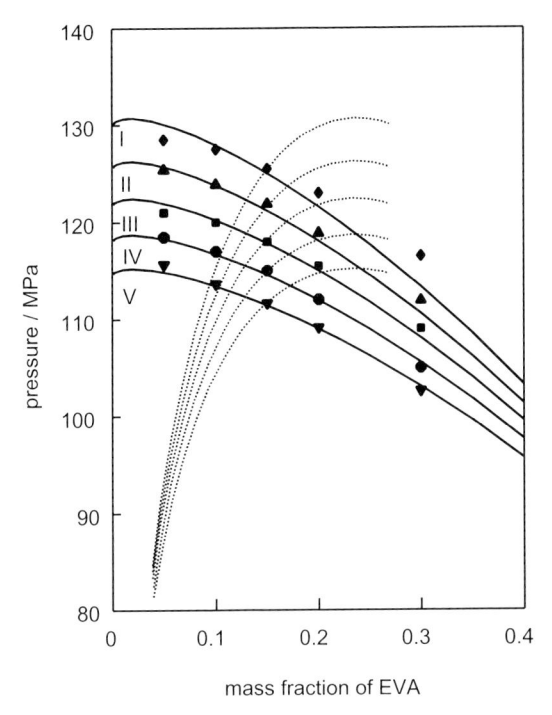

Figure 18.3: Calculated cloud-point curves (———) and shadow curves (· · ·) of the system ethylene + CP3 at different temperatures compared with experimental data (Wagner, 1983); I: (◆) 393.15 K, II: (▲) 413.15 K, III: (■) 433.15 K, IV: (●) 453.15 K, V: (▼) 473.15 K.

Using this equation for k_{AB} the phase equilibrium for all systems of ethylene with co-polymers from Table 18.1 was calculated at 5 temperatures. As an example, Figure 18.3 shows the calculated cloud-point curves and shadow curves of the system ethylene + CP3 compared with the experimental data.

The SWP-EoS describes the phase equilibrium of the ethylene + CP3 system properly at all temperatures. The model predictions were again compared with the systems studied experimentally by Nieszporek (1991). Figure 18.4 shows the comparison between calculated and experimental cloud-point curves for the system ethylene + EVA2. The predictions are a little poorer than those with k_{AB} parameters being specific to the temperature considered. Nevertheless, the error of the cloud-point pressure is lower than 5%.

A characteristic phenomenon for ethylene + EVA systems is a cross-over of cloud-point curves in the range of higher copolymer mass fractions. For the systems studied here this occurs between the EVA mass fractions 0.6 and 0.7. On the left-hand side of the cross-over the cloud-point pressure decreases with increasing temperature whereas on the right-hand side the cloud-point pressure increases with increasing temperature. This phenomenon was also predicted from the COR-EoS and Flory's free volume model by Finck et al. (1992) and from the SAFT EoS by Folie et al. (1996). In contrast to Folie et al. (1996) who calculated the cross-over pressure considerably too low, our predictions are in good agreement with the experimental data (cf. Fig. 18.4).

18 Calculation of the High Pressure Phase Equilibrium

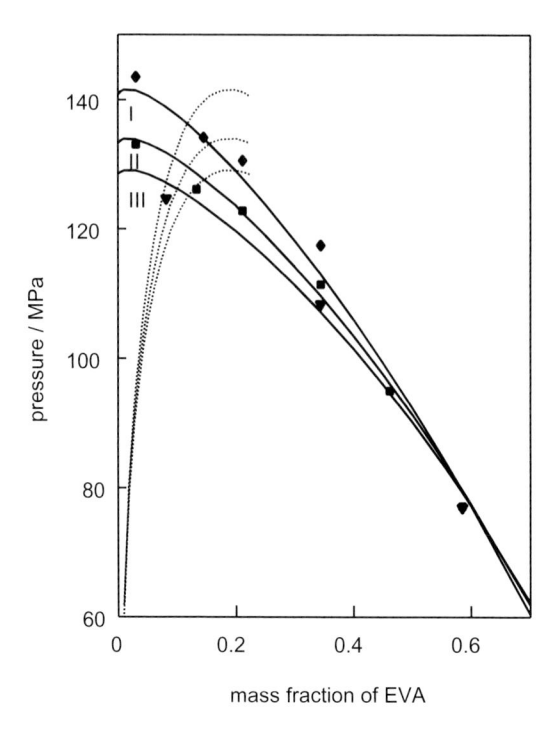

Figure 18.4: Calculated cloud-point curves (———) and shadow curves (· · ·) of the system ethylene + EVA2 at different temperatures compared with experimental data (Nieszporek, 1991); I: (♦) 403.15 K, II: (■) 433.15 K, III: (▼) 453.15 K.

The influence of polydispersity discussed above causes, that in polymer solutions the critical point no longer may expected to be an extreme value of the binodal but to be the intersecttion point of the cloud-point curve and the shadow curve. Thus, it is the only point on the cloud-point curve corresponding to an incipient phase with the same distribution function of the molecular mass and of the chemical composition as the feed phase. The position of the critical point is mainly influenced by the molecular mass of the polymer, by the temperature, and by the polydispersity that has already been discussed. Increasing $\overline{M}_n$ leads to an increase of the critical pressure p_{crit} and shifts the critical mass fraction ψ_{crit} to lower values. An increase of the temperature results in a decrease of the critical pressure p_{crit} for all systems, but, the critical mass fraction ψ_{crit} practically does not change with temperature.

In contrast to the monodisperse case, where the concentrations of the coexisting phases in the two-phase region exclusively lie on the binodal curve, the situation in the polydisperse case is much more complicated. Cloud-point curve and shadow curve correspond to the binodal curve of the monodisperse case. They describe the formation of an infinite small amount of a new phase. But from a practical point of view the phase split within the metastable or unstable region yielding significant amounts of both phases is of special interest. To describe this region correctly, an infinitely number of coexistence curves is necessary. For the calculation of the coexistence curves besides the equations for the phase equilibrium, the material balances (Eqs. (18.21)) or at least Eqs. (18.23) have to be considered. Only the

18.4 Results and Discussion

coexistence curve with the feed concentration being the critical concentration has a closed shape. All the other curves consist of two branches. The branch that describes the concentration of the polymer-rich phase starts at the cloud-point curve and the branch that describes the corresponding concentration of the polymer-lean phase starts at the shadow curve. Figure 18.5 shows two pairs of coexistence curves for the ethylene + EVA2 system at 433.15 K together with the appropriate cloud-point and shadow curves. For the sake of clearness only the coexistence curve starting at the feed mass fraction $\psi_B^F = 0.133$, near the critical point is compared with experimental data.

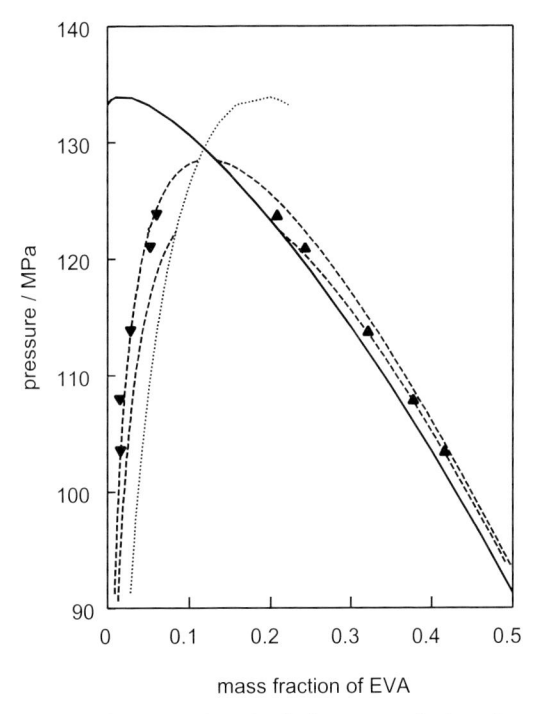

Figure 18.5: Calculated cloud-point curves (——), shadow curves ($\cdots$), and coexistence curves (----) of the system ethylene + EVA2 at 433.15 K. The coexistence curve starting at the feed mass fraction of EVA $\psi_B^F = 0.133$ is compared with experimental data (Nieszporek, 1991); (▲) polymer-rich phase, (▼) polymer-lean phase.

All coexistence curves approach the cloud-point curves or shadow curves respectively. This effect becomes stronger the higher the feed mass fraction of the polymer is and so, also all experimental cloud-point data lie very close together. Figure 18.6 shows coexistence curves for the same system at three temperatures all starting at the feed weight fraction $\psi_B^F = 0.133$. The right hand branches of the coexistence curves show a cross-over in the same region as the cloud-point curves of the same system do.

Figures 18.5 and 18.6 prove, that the SWP-EoS is able to describe the experimental coexistence data for the ethylene + EVA systems as accurately as the cloud-point curves.

18 Calculation of the High Pressure Phase Equilibrium

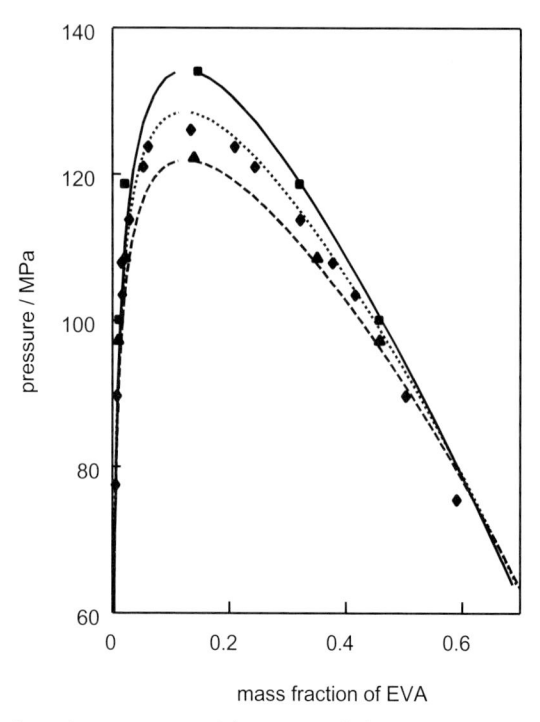

Figure 18.6: Calculated coexistence curves of the system ethylene + EVA2 compared with experimental data (Nieszporek, 1991); (——, ■) 403.15 K, (···, ◆) 433.15 K, (----, ▲) 463.15 K.

In the following a binary system under isothermal conditions is discussed. Starting in a homogeneous region and decreasing the pressure results in a phase separation. During this process fractionation occurs, i.e. along the coexistence curves with decreasing pressure the molecular mass distributions and the chemical distributions too are continuously changing. Taking into account the double polydispersity of the copolymer, coexistence curve calculations give information about the change of the number averaged molecular mass and the content of vinyl acetate in the copolymer during the phase separation process. As an example, for the ethylene + EVA2 system $\left(\psi_B^F = 0.133, \overline{M_n^F} = 33.400 \text{ g/mol}, \overline{y^F} = 0.175\right)$ at 433.15 K the cloud-point pressure is 128.5 MPa. Decreasing the pressure by 50 MPa leads to an increase of the number averaged molecular mass in the polymer-rich phase to $\overline{M_n^I} = 93.700$ g/mol and a decrease of $\overline{M_n}$ in the polymer-lean phase to $\overline{M_n^{II}} = 1500$ g/mol. The chemical fractionation is not that strong (Rätzsch and Wohlfahrt 1990), the content of vinyl acetate in the polymer-rich phase hardly differs from that of the feed phase but $\bar{y}$ of polymer-lean phase slightly decreases to $\overline{y^{II}} = 0.158$. For other ethylene + EVA systems and feed concentrations the fractionation happens in a similar manner with enrichment of long chained polymers in the polymer-rich phase and of short chained and somewhat vinyl acetate poorer polymers in the polymer-lean phase.

18.4.2 The Binary System Ethylene (A) + VA (C)

The vapor-liquid equilibrium of this monomer system was calculated between 363 and 423 K and compared with experimental data measured by Nieszporek (1991). The binary parameter k_{AC} was adjusted to three temperatures separately and found to be linear dependent on temperature:

$$k_{AC} = (-216.2758 + 0.61667 \ T/K)/1000 \tag{18.40}$$

Simultaneously, the parameter c_s was fitted. Equation (18.40) represents the best fit obtainned for $c_s = 0.4236$ (see Eq. (18.34)). Figure 18.7 presents the results for all temperatures.

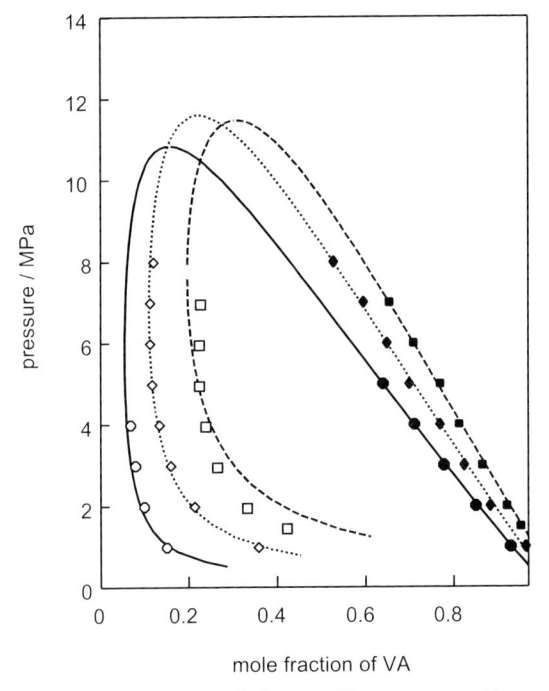

Figure 18.7: Calculated VLE of the system ethylene + VA compared with experimental data (Nieszporek, 1991). The open symbols are experimental dew points and the filled symbols experimental bubble points; (——, ●) 363.15 K, (· · ·, ◆) 393.15 K, (----, ■) 423.15 K.

The calculated liquidus curves are in good agreement with the experimental data and also the vapor pressure curves for 363 and 393 K are correct. Only the slope of the vapor pressure curves for 423 K is somewhat too steep

18.4.3 The Binary System EVA (B) + VA (C)

Experimental data for this binary system are rare. Such data are required to fit the parameter k_{BC}. Only Henry's constants are available in the literature (Maloney and Prausnitz, 1976; Liu and Prausnitz, 1976; Wohlfarth, 1994). Considering polymer B (EVA) being the nonvolatile solvent and component C (VA) the volatile solute, Henry's constant is defined by:

$$H_C = \lim_{\psi_B^L \to 1} \frac{f_C^L}{\psi_C} \qquad (18.41)$$

Here, f_C^L is the fugacity of the volatile component in the liquid phase L. At low pressures, the fugacitiy of C in the vapor phase, f_C^V, is the pressure p. Because of the equality of the fugacities in both phases, f_C^L is replaced by the pressure too:

$$H_C = \lim_{\psi_B^L \to 1} \frac{p}{\psi_C} \qquad (18.41\,a)$$

From Eq. (18.6) (for i = C) yields with notations I = L; II = V and $p \to 0$; $\psi_B^L \to 1$; $\psi_C^V \to 1$; $\bar{r}^{ML} = \bar{r}_B$; $\bar{r}^{MV} = r_C$:

$$\phi_C^L = \frac{r_C}{\bar{r}_B} \frac{1}{\psi_C} \qquad (18.42)$$

Besides, Eq. (18.11) for C yields for these limits:

$$\phi_C^L = \frac{RT}{pV_s^L} \frac{1}{\bar{r}_B} \exp\left(r_C A_{0,C}^L\right) \qquad (18.43)$$

Combining Eqs. (18.41), (18.42), and (18.43) gives:

$$H_C = \frac{RT}{r_C V_{s,B0}} \exp\left[r_C \lim_{\psi_B^L \to 1} A_{0,C}^L\right] \qquad (18.44)$$

where, $V_{s,B0}$ is the segment-molar volume of the polymer B with $p \to 0$. With Eqs. (18.29) and (18.30) for C one obtains:

$$\lim_{\psi_C^L \to 0} A_{0,C}^L = -c_{s,C} \ln\left(1 - \frac{b_r + \bar{y}b_y}{V_{s,B0}}\right) + \frac{b_{s,C}(c_r + \bar{y}c_y)}{V_{s,B0} - (b_r + \bar{y}b_y)} - \frac{b_{s,C}(a_r + \bar{y}a_y)^2}{RT(b_r + \bar{y}b_y)[V_{s,B0} + (b_r + \bar{y}b_y)]}$$

$$+ \frac{\ln\left[1 + \frac{(b_r + \bar{y}b_y)}{V_{s,B0}}\right]}{RT(b_r + \bar{y}b_y)} \left\{ \frac{b_{s,C}(a_r + \bar{y}a_y)^2}{b_r + \bar{y}b_y} - 2\sqrt{a_{s,C}}\left[(a_r + \bar{y}a_y)(1 - k_{BC})\right]\right\} \qquad (18.45)$$

With Eqs. (18.44) and (18.45) the binary parameter k_{BC} can be calculated from Henry's constant. However, to do so, one has to know $V_{s,B0}$. This volume is obtained from Eq.

(18.25) considering the limits $p \to 0$; $\psi_B^L \to 1$. In this way the following quadratic equation is found:

$$(V_{s,B0})^2 + p^* V_{s,B0} + q^* = 0 \tag{18.46}$$

$$p^* = \bar{r}_B \left\{ (b_r + \bar{y}b_y)(c_r + \bar{y}c_y) - \frac{(a_r + \bar{y}a_y)^2}{RT} \right\} \tag{18.47}$$

$$q^* = \bar{r}_B (b_r + \bar{y}b_y) \left\{ (b_r + \bar{y}b_y)(c_r + \bar{y}c_y) + \frac{(a_r + \bar{y}a_y)^2}{RT} \right\} - (b_r + \bar{y}b_y)^2 \tag{18.48}$$

Because here the solution with the lower volume is of interest $V_{s,B0}$ is given by:

$$V_{s,B0} = -\frac{p^*}{2} - \sqrt{\frac{(p^*)^2}{4} - q^*} \tag{18.49}$$

Henry's constants for the polethylene + VA system and for three EVA+VA systems at five temperatures between 398 and 498 K were used to fit the binary parameters. The EVA copolymers have different VA contents $\bar{y}$. The influence of T and $\bar{y}$ on k_{BC} is described by:

$$k_{BC} = [230.23592 - 0.50365 \cdot T/K - (164.47921 - 0.26105 \cdot T/K)\bar{y}]/1000 \tag{18.50}$$

18.4.4 The Ternary System

After fitting the k_{ij} parameter for the three binary systems, the phase behavior of the ternary system was predicted. Figure 18.8 represents the calculated cloud-point pressures in dependence on the temperature for the ethylene + EVA3–7 + VA system compared with the experimental data.

All curves relate to $\psi_B \approx \psi_C$. The VA mass fraction $\eta_C = \psi_C/(1 - \psi_B)$ on a polymer-free basis varies between 0.1 and 0.5. The description of the p-T isopleths is not satisfactory. The cloud-point pressures were clearly overestimated. This difference increases with increasing content of vinyl acetate in the solvent. Despite the inaccuracy in the cloud-point calculations, the general tendencies were described qualitatively correct. The cloud-point pressure decreases with increasing amount of VA in the solvent, as VA acts as a cosolvent (Folie and Radosz, 1995). The weak temperature dependence of the cloud-point pressures (p is slightly falling with increasing T for VA-lean solvent mixtures and is slightly increasing with increasing T for the VA-richest solvent mixture with $\eta_C \to 0.5$) is reasonably described. However, the VA mass fraction corresponding to the inversion of the temperature dependency is not predicted exactly. The fractionation proved to be hardly temperature dependent. The chemical composition of the incipient phase stays quite unchanged and the number averaged molecular mass slightly decreases with increasing temperature. Figure 18.9 shows the cloud-point and shadow curves for the same system at 403.15 K for some values of the VA content η_C (on a polymer-free basis) in the solvent mixture.

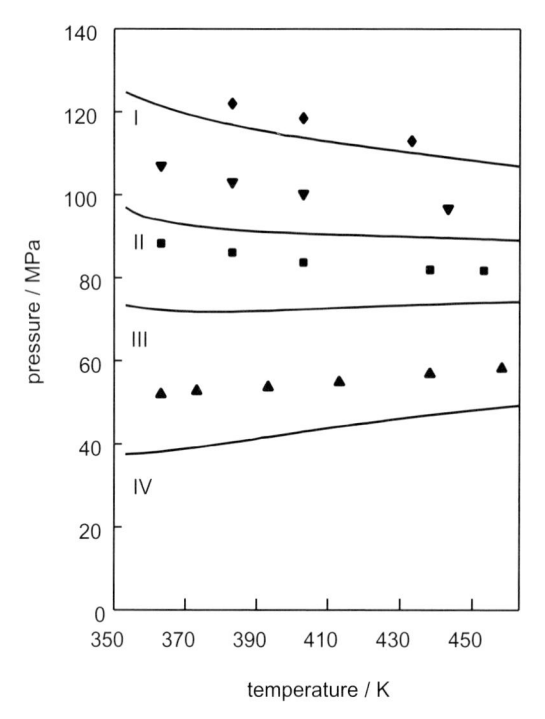

Figure 18.8: Calculated cloud-point pressures (p-T isopleths with $\psi_B \approx \psi_C$) of the pseudo-ternary system ethylene + EVA3–7 + VA compared with experimental data (Nieszporek, 1991); VA mass fraction on a polymer-free basis I: (◆) $\eta_C = 0.11$; II: (◆) $\eta_C = 0.22$; III: (■) $\eta_C = 0.32$; IV:(▲) $\eta_C = 0.50$.

This diagram is similar to Figures 18.1 to 18.6 but, Figure 18.9 gives the results for a solvent mixture instead of a single solvent. The curve I represents the pseudo-binary system ethylene + EVA. The effect of decreasing cloud-point pressure with increasing VA content in the solvent mixture is obvious again but it weakens with increasing mass fraction of EVA. Furthermore, an increasing VA content η_C shifts the critical points to lower polymer mass fractions and decreases the number averaged molecular mass of the incipient phase whereas its chemical composition stays nearly unchanged.

Unfortunately, the available experimental cloud-point data are not sufficient to test the shape and the position of the extreme of the cloud-point curves. Overall, the calculations in the ternary system are not quite satisfactory. Probably, the reason is some uncertainties in the determination of the pure-component parameters of EVA. Furthermore, it is not clear if the Henry's constants for EVA + VA are a suitable source for the k_{BC} parameter of this system.

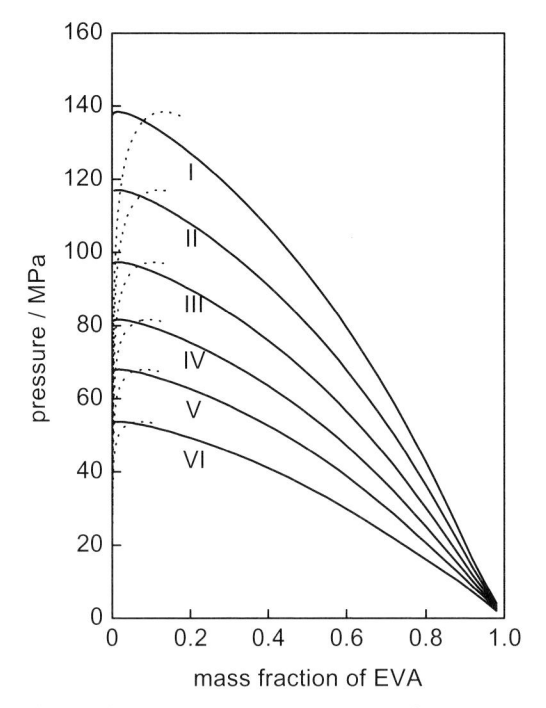

Figure 18.9: Calculated cloud-point curves and shadow curves for the ternary system ethylene + EVA3–7 + VA at 403.15 K in dependence on the VA mass fraction η_C on a polymer-free basis; I: $\eta_C = 0$, II: $\eta_C = 0.11$, III: $\eta_C = 0.22$, IV: $\eta_C = 0.32$, V: $\eta_C = 0.41$, VI: $\eta_C = 0.50$.

18.5 Conclusions

The treatment of high-pressure phase equilibrium of polymer systems based on continuous thermodynamics and using the equation of state of Sako, Wu, and Prausnitz (SWP-EOS) was extended to copolymer systems. In this, the double polydispersity of copolymers with respect to molecular mass and to chemical composition were considered using the bivariate Stockmayer distribution function. Even for solutions of a polymer in a solvent, polydispersity causes a separation of the binodal curve into a cloud-point curve, a shadow curve, and an infinite number of coexistence curves. The molecular-mass polydispersity influences the phase equilibrium considerably stronger than the polydispersity with respect to chemical composition.

The equations derived were applied to ethylene + EVA + VA systems over a wide range of temperature. The pure-component parameters of ethylene and VA were fitted to critical data. Only the (pure component) parameter c_s of VA was fitted to binary vapor-liquid equilibrium data. As for the copolymer EVA no critical data are available, the pure-component parameters of EVA were fitted together with the binary interaction parameter k_{AB} to experimental cloud-point pressures of the pseudo-binary system ethylene + EVA. The binary interaction parameter k_{AB} was found to be independent of the molecular mass but to depend linearly both on the VA content of the copolymer and on the temperature. The temperature dependent binary interaction parameter k_{AC} (for the binary system ethylene + VA) was adjusted to binary VLE data. The final model gives a good description of the experimental VLE data. The binary parameter k_{BC} was determined from experimental results for Henry's constants of VA in EVA. k_{BC} depends on temperature and composition.

The high-pressure phase equilibrium (cloud-point curves, shadow curves, and coexistence curves) of the pseudo-binary system ethylene + EVA as calculated is in good agreement with experimental data. This good agreement is also observed for systems which were not included in the parameter determination. Thus, the proposed method basing on the SWP-EoS enables the reliable prediction of cloud-point and shadow curves for ethylene + EVA systems in a wide range of the molecular mass, the chemical composition, and temperature.

Furthermore, cloud-point pressures of the ternary system were calculated and compared with experimental data. Because ternary data were not included in the parameter fit these calculations are predictions. The predicted *p-T* isopleths are described qualitatively correct. However, the quantitative agreement with the experimental data is not quite satisfactory. Probably, the deviations from the experimental data originate from uncertainties in the determination of the pure-component parameters of EVA.

References

Bauer B. J. (1985): Equilibrium phase compositions of heterogeneous copolymers. Polym. Eng. Sci. 25, 1081–1087.

Bergmann J.; Teichert H.; Kehlen H.; Rätzsch M.T. (1992): Application of continuous thermodynamics to the stability of polymer system. IV. J. Macrom. Sci. A 29, 371–379.

Browarzik C.; Browarzik D.; Kehlen H. (2001): Phase-equilibrium calculations for solutions of poly (ethylene-co-vinyl acetate) copolymers in supercritical ethylene using a cubic equation of state. J. of Supercritical Fluids 20, 73–88.

Browarzik D.; Kehlen H. (2000): Polydisperse fluids. In: J.V. Sengers, R.F. Kayser, C.J. Peters, and H.J. White, Jr. (Eds.), Equations of state for fluids and fluid mixtures. Amsterdam, Lausanne, New York, Oxford, Shannon, Singapore, Tokyo.

Browarzik D.; Kowalewski M. (1999): Calculation of the stability and of the phase equilibrium in the system polystyrene+cyclohexane+carbon dioxide based on equations of state. Fluid Phase Equilibria 163, 43–60.

Cotterman R.L.; Prausnitz J.M. (1985): Flash calculations for continuous mixtures using an equation of state. Ind. Eng. Chem. Proc. Des. Dev. 24, 434–443.

Cotterman R.L.; Prausnitz J.M. (1991): Continuous thermodynamics for phase-equilibrium calculations in chemical process design. In: G. Astarita; S.I. Sandler (Eds.), Proceedings of an ACS Symposium on Kinetic and Thermodynamic Lumping of Multicomponent Mixtures, p. 229–275.

Finck U.; Wohlfahrt C.; Heuer T. (1992): Calculation of high pressure phase equilibria of mixtures of ethylene vinyl acetate and an (ethylene-vinyl acetate) copolymer. Ber. Bunsen-Ges. Chem. 96, 179–188.

Folie B.; Radosz M. (1995): Phase equilibria in high pressure polyethylene technology. Ind. Eng. Chem. Res. 34, 1501–1516.

Folie B.; Gregg C.; Luft G.; Radosz M. (1996): Phase equilibria of poly(ethylene-co-vinyl acetate) copolymers in subcritical and supercritical ethylene and ethylene-vinyl acetate mixtures. Fluid Phase Equilibria 120, 11–37.

Hu Y.; Ying X.; Wu D.T.; Prausnitz J. M. (1995): Continuous thermodynamics for polydispers polymer solutions. Fluid Phase Equilibria 104, 229–252.

Kehlen H.; Rätzsch M.T. (1980): Continuous thermodynamics of multicomponent mixtures. Proceedings of the 6th International Conference on Thermodynamics, p. 41–51.

Koningsfeld R.; Staverman A.J. (1968 a): Liquid-liquid phase separation in multicomponent polymer solutions. I. Statement of the problem and description of methods of calculation. J. Polym. Sci. A-2 6, 305–323.

Koningsfeld R.; Staverman A.J. (1968 b): Liquid-liquid phase separation in multicomponent polymer solutions. III. Cloud-point curves. J. Polym. Sci. A-2 6, 349–366.

Liu D.D.; Prausnitz J.M. (1976): Solubilities of gases and volatile liquids in polyethylene and ethylene-vinyl acetate copolymers in the region 125–225°C. Ind. Eng. Chem., Fundam. 15, 330–335.

Maloney D.P.; Prausnitz J.M. (1976): Solubilities of ethylene and other organic solutes in liquid, low-density polethylene in the region 124 to 300°C. AIChE Journal 22, 74–82.

Nieszporek B. (1991): Untersuchungen zum Phasenverhalten quasibinärer und quasiternärer Mischungen aus Polyethylen, Ethylen-Vinylacetat-Copolymeren, Ethylen und Vinylacetat unter Druck. Dissertation, TH Darmstadt.

Rätzsch M.; Findeisen R.; Sernov V.S. (1980): Untersuchungen zum Phasenverhalten von Monomer-Polymer-Systemen unter hohem Druck. Z. Phys. Chemie (Leipzig) 261, 995–1000.

Rätzsch M.T.; Kehlen H. (1989): Continuous thermodynamics of polymer solutions. Prog. Polym. Sci. 14, 1–46.

Rätzsch M.T.; Browarzik D.; Kehlen H. (1989): Refined continuous thermodynamic treatment for the liquid-liquid equilibrium of copolymer solutions. J. Macromol. Sci.-Chem. A 26, 903–920.

Rätzsch M.T.; Browarzik D.; Kehlen H. (1990): Liquid-liquid equilibrium of copolymer solutions with broad and asymmetric chemical distributions. J. Macromol. Sci.-Chem. A 27, 809–830.

Rätzsch M.T.; Wohlfahrt C. (1990): Continuous thermodynamics of copolymer systems. Adv. Polym. Sci. 98, 49–114.

Sako T.; Wu A.H.; Prausnitz J.M. (1989): A cubic equation of state for high-pressure phase equilibria of mixtures containing polymers and volatile fluids. J. Appl. Polym. Sci. 38, 1839–1858.

Šolc K. (1970): Cloud-point curves of polymer solutions. Macromolecules 3, 665–673.

Šolc K. (1975): Cloud-poit curves of polymers with logarithmic-normal distribution of molecular weight. Macromolecules 8, 819–827.

Stockmayer, W.H. (1945): Distribution of chain lengths and compositions in copolymers. J. Chem. Phys. 13, 199–207.

Tork T.; Sadowski G.; Arlt W.; de Haan A.; Krooshof G. (1999 a): Modeling of high pressure phase equilibria using the Sako-Wu-Prausnitz equation of state I. Pure-components and heavy n-alkane solutions. Fluid Phase Equilibria 163, 61–77.

Tork T.; Sadowski G.; Arlt W.; de Haan A.; Krooshof G. (1999 b): Modeling of high pressure phase equilibria using the Sako-Wu-Prausnitz equation of state II. Vapour-liquid equilibria and liquid-liquid equilibria in polyolefine systems. Fluid Phase Equilibria 163, 79–98.

Wagner P. (1983): Zum Phasengleichgewicht im System Ethylen+Ethylen-Vinylacetat-Copolymer unter hohem Druck. Dissertation, TH Merseburg.

Wohlfahrt C. (1993): Calculation of phase equilibria in random copolymer systems. Makromol. Chem., Theory Simul. 2, 605–635.

Wohlfahrt C. (1994): Vapour-liquid-equilibrium data of binary polymer solutions. physical science data 44. Amsterdam, Lausanne, New-York, Oxford, Shannon, Tokyo.

Wohlfahrt C. (2001): CRC handbook of thermodynamic data of copolymer solutions. Boca Raton, London, New York, Washington, DC.

Nomenclature

A_i	quantity (cf. Eq. (18.11))
a	attraction parameter in SWP-EoS
b	repulsion parameter in SWP-EoS
B_1, B_2	quantities (cf. Eq. (18.12))
c	chain flexibility parameter in SWP-EoS
k	parameter of the Schulz-Flory distribution
k_{ij}	binary parameter for attraction parameter a
M	molecular mass
$\overline{M}_n$	number average molecular mass
$\overline{M}_w$	mass average molecular mass
n	segment-mole number
p	pressure
q_α, q_β	kinetic parameter of the copolymerization
R	universal gas constant
r	segment number
$\bar{r}_B$	number-averaged segment number of polymer
$\bar{r}^M$	number-averaged segment number of mixture
T	temperature
U	polydispersity parameter $U = \overline{M}_w / \overline{M}_n - 1$
V	volume
VA	vinyl acetate
$W(r, y)$	bivariate segment-molar distribution function
$w(r, y)$	extensive bivariate segment-molar distribution function
x	mole fraction
y	mass fraction of vinyl acetate in a copolymer species
$\bar{y}$	averaged mass fraction of vinyl acetate in copolymer

Abbreviations

EoS	equation of state
EVA	poly(ethylene-co-vinyl acetate)
PE	polyethylene
PVAC	poly(vinyl acetate)
SWP-EOS	equation of state introduced by Sako, Wu, and Prausnitz (1989)
VA	vinyl acetate
VLE	vapor-liquid equilibrium

Greek letters

ε	parameter of the Gausssian-distribution
η_C	mass fraction of VA on a polymer-free basis ($\eta_C = \psi_C/(1 - \psi_B)$)
κ^*	quantity (cf. Eq. (18.14))
ξ	quotient of the amount of segments in phase II and in the feed F

Nomenclature

ϕ fugacity coefficient
ψ mass fraction

Subscripts

i, j	component i, j
A	ethylene
B	poly(ethylene-co-vinyl acetate) copolymer (EVA)
C	vinyl acetate (VA)
r	pure-component parameters of the copolymer species for $y = 0$
s	segment-molar quantity
y	change of the pure-component parameters of the copolymer species with y

Superscripts

F, I, II	phases (feed phase F splits into the two phases I and II)
M	mixture

19 Modeling of Copolymer Phase Equilibria Using the Perturbed-Chain SAFT Equation of State

Feelly Tumakaka and Gabriele Sadowski[1]

19.1 Abstract

The Perturbed-Chain SAFT equation of state is applied to mixtures of copolymers and solvents. The phase behavior of copolymer systems is calculated using the information of the two respective homopolymers that build the copolymer chains. One additional parameter describes the interactions between the different segment types of different copolymer chains. Using only one temperature-*in*dependent binary parameter for each binary system, the model reveals strong predictive capabilities and allows safe correlation and interpolation over a wide range of pressure, temperature, concentration, molecular mass, and copolymer composition. This is demonstrated for ethylene-based copolymers like poly(ethylene-*co*-propylene), poly(ethylene-*co*-1-butene), as well as for copolymers with non-polar and polar segments such as poly(ethylene-*co*-vinyl acetate) and poly(ethylene-*co*-methyl acrylate).

19.2 Introduction

The phase equilibria of polymer systems are of vital importance for design and optimization of polymer production, separation, and purification processes. Modeling the thermodynamic properties of polymer mixtures is demanding in several ways: polymer systems often exhibit a pronounced density dependence, where Gibbs excess energy models fail. An equation of state needs to be used instead. Furthermore, experimental data is often scarce. Considerable experimental effort is generally required for determining the phase equilibria of polymer systems. As experimental data are in many cases not available for a certain condition of interest, it is important from a practical point of view, that an equation of state is robust for extrapolations beyond the region, where parameters were identified. The modeling of copolymers, especially copolymers with polar and non-polar comonomers, presents a new challenge to model due to their

---

[1] Universität Dortmund, Lehrstuhl für Thermodynamik, Emil-Figge-Str. 70, 44227 Dortmund.

complex structure and phase behavior. The physical properties of those copolymers vary not only with molecular mass and degree of branching, but also with the copolymer composition.

Physically based equations of state derived by applying principles of statistical mechanics have continuously been developed and improved over the past three decades. In 1988 Chapman et al. developed a very successful molecular theory (Statistical Associating Fluid Theory (SAFT) equation of state) which assumed molecules to be chains of connected spherical segments by applying Wertheim's Thermodynamic Perturbation Theory of first order (TPT1) (Wertheim, 1984a,b, 1986a,b) and extending it to mixtures. Huang and Radosz (1990, 1991) proposed a modification of the SAFT equation of state and determined the pure-component parameters for numerous low molecular as well as polymeric substances. Various modifications of the SAFT model were subsequently suggested (for a review see Müller and Gubbins, 2001).

19.3 Theory

19.3.1 Copolymer Concept

Amos and Jackson (1991) first applied the SAFT concept to triatomic hard-sphere molecules. Banaszak et al. (1996) and Shukla and Chapman (1997) then presented the extended SAFT version to heteronuclear hard chains, referred to as Copolymer-SAFT, in which each segment in a chain can have different properties. In addition to the three pure-component parameters, each copolymer is characterized in terms of segment fractions that describe the amount of segments α and β in the copolymer chain, and the bond fractions which define to a certain extend the arrangement of the segments in chains.

Figure 19.1: Molecular model of a copolymer chain with two different segment types α and β.

The segment fractions and the bond fractions can be estimated on the basis of the molecular structure. For block copolymers, the values of the respective bond fractions are clearly defined. A block copolymer has only one bond of type $\alpha\beta$, i.e. $B_{\alpha,\beta} = \dfrac{1}{(m_i - 1)}$, where m_i is the number of segments in a chain of component i. The bond fraction of type $\alpha\alpha$ is simply given by $B_{\alpha\alpha} = \dfrac{z_{i,\alpha}(m_i - 2)}{(m_i - 1)}$ (the same for $B_{\beta\beta}$), where $z_{i\alpha}$ is the fraction of segments α on a chain of component i (Shukla and Chapman, 1997). In case of an alternating copolymer, there is only bond type, namely $B_{\alpha\beta}$ in the chains.

The bond fractions in a random copolymer can only be determined experimentally, e.g. using NMR-spectroscopy. If only the segment fractions are available, one can estimate the bond fractions as proposed here: if segment type α is in majority, it is assumed that there is no bond between $\beta\beta$-segments. Hence, the bond fractions for this copolymer are:

$$B_{\beta\beta} = 0$$

$$B_{\alpha\beta} = \frac{2 \cdot z_{i,\alpha} \cdot m_i}{(m_i - 1)}$$

$$B_{\alpha\alpha} = 1 - B_{\alpha\beta}$$

If segment type α and β are equally in total, the copolymer is assumed to be an alternating one.

This copolymer concept was compared to molecular dynamics simulation data for pure hard chain copolymer systems (Shukla and Chapman, 1997), and was applied to real copolymer system of poly(ethylene-co-1-butene) in propane (Banaszak et al., 1996).

19.3.2 Perturbed-Chain SAFT and its Extension to Copolymer Systems

The Perturbed-Chain SAFT model is a theoretical-based equation of state which is suitable for small molecules like gases and solvents as well as for macromolecules such as polymers and copolymers. It was shown to accurately describe vapor pressures, densities, and calorimetric properties of pure-components, and was successfully applied to binary and ternary mixtures of polymers, solvents, and gases (Gross and Sadowski, 2001 a, b).

The model development is in detail described by Gross and Sadowski (2000). The summary of equations for calculating pressure, density, fugacity coefficients, and caloric properties using the Perturbed-Chain SAFT model can be found in the literature (Gross and Sadowski, 2001 a). Only the basic ideas and the model extension to copolymers, referred to as Copolymer Perturbed-Chain SAFT equation of state, are presented here.

In the framework of Perturbed-Chain SAFT, molecules are assumed to be chains of freely jointed spherical segments which exhibit attractive forces among each other. The model development is based on a perturbation theory, where the interactions of molecules are divided into a repulsive part and a contribution due to the attractive part of the potential. The repulsive interactions of the Perturbed-Chain SAFT equation of state are described with a hard-chain expression derived by Chapman et al. (1988). The attractive interactions are again separated into dispersive interactions and contributions due to association and polarity. The compressibility factor Z is finally calculated as the sum of the ideal gas contribution ($Z^{id} = 1$), the hard-chain term (Z^{hc}), the dispersive part ($Z^{disp.\ chain}$), and contributions due to association (Z^{assoc}) and polarity (Z^{pol}), as

$$Z = 1 + Z^{hc} + Z^{disp,chain} + Z^{assoc} + Z^{pol} \qquad (19.1)$$

with $Z = pv/(RT)$, where p is the pressure, v is the molar volume, R denotes the gas constant, and T is the temperature.

19.3 Theory

In contrary to the original SAFT equation of state, the Perturbed-Chain SAFT model accounts for the chain-like structure of the molecules also in the dispersion term $Z^{disp,\,chain}$.

The effects of association and multipole interactions (like dipole-dipole forces and quadrupole-quadrupole forces) are not accounted for here. These will be considered in subsequent investigations.

Based on Wertheim's thermodynamic perturbation theory of first order and following Amos and Jackson (1992), the final expression of the compressibility factor for the fluid mixtures of heteronuclear hard chain molecules consisting of two different segment type is in terms of the bond fractions $B_{\alpha\beta}$ (Banaszak et al., 1996; Shukla and Chapman, 1997):

$$Z^{hc} = \bar{m} Z^{hs} - \sum_i x_i (m_i - 1) \sum_{\alpha=1}^{2} \sum_{\beta=1}^{2} B_{\alpha\beta} \cdot \rho \frac{\partial \ln g^{hs}_{i\alpha, i\beta}(d_{i\alpha, i\beta})}{\partial \rho} \tag{19.2}$$

$$\bar{m} = \sum_i x_i m_i \sum_\alpha z_{i\alpha} = \sum_i x_i m_i \tag{19.3}$$

where $\bar{m}$ is the mean segment number, x_i is the mole fraction of chains of component i, $g^{hs}_{i\alpha,i\beta}$ is the radial pair distribution function for segment type α and β of component i in the hard sphere system. Superscript hs indicates quantities of the hard-sphere system, ρ is the total number density of molecules as

$$\rho = \frac{6}{\pi} \eta \cdot \sum_i x_i m_i \sum_\alpha^2 z_{i\alpha} d_{i\alpha}^3 \tag{19.4}$$

In Eq. (19.4), η is the packing fraction of the hard-sphere mixture, $d_{i\alpha}$ is the temperature-dependent segment diameter of segment α on a chain of component i as the function of the temperature-independent segment diameter $\sigma_{i\alpha}$ and the depth of the potential $\varepsilon_{i\alpha}$ according to

$$d_{i\alpha} = \sigma_{i\alpha}\left(1 - 0.12 \cdot \exp\left(-3\frac{\varepsilon_{i\alpha}}{kT}\right)\right) \tag{19.5}$$

Expressions of Mansoori et al. (1971) are used for mixtures of the hard-sphere mixture, given by

$$Z^{hs} = \frac{\zeta_3}{(1-\zeta_3)} + \frac{3\zeta_1\zeta_2}{\zeta_0(1-\zeta_3)^2} + \frac{3\zeta_2^3 - \zeta_3\zeta_2^3}{\zeta_0(1-\zeta_3)^3} \tag{19.6}$$

where ζ_n is

$$\zeta_n = \frac{\pi}{6}\rho \sum_i x_i m_i \cdot \sum_\alpha z_{i\alpha} d_{i\alpha}^n \quad n \in \{0, 1, 2, 3\} \tag{19.7}$$

Note that ζ_3 is the packing fraction η of the hard sphere mixture. The hard sphere mixture distribution function at contact used in Eq. (19.2) is according to Mansoori et al. (1971):

$$g_{i\alpha,j\beta}^{hs}(d_{i\alpha,j\beta}) = \frac{1}{(1-\zeta_3)} + \left(\frac{d_{i\alpha} d_{j\beta}}{d_{i\alpha}+d_{j\beta}}\right)\frac{3\zeta_2}{(1-\zeta_3)^2} + \left(\frac{d_{i\alpha} d_{j\beta}}{d_{i\alpha}+d_{j\beta}}\right)^2 \frac{2\zeta_2^2}{(1-\zeta_3)^3} \quad (19.8)$$

The perturbation theory of Barker and Henderson (1967) is applied to calculate the attractive part of the chain interactions by using van der Waals one-fluid mixing rules as

$$Z^{disp,chain} = -2\pi\rho \cdot \frac{\partial(\eta \cdot I_1(\bar{m},\eta))}{\partial\eta} \cdot \sum_i\sum_j x_i x_j m_i m_j \sum_\alpha^2 \sum_\beta^2 z_{i\alpha} z_{j\beta} \left(\frac{\varepsilon_{i\alpha,j\beta}}{kT}\right) \sigma_{i\alpha,j\beta}^3$$

$$- \pi\rho \cdot \bar{m} \cdot \left[C_1 \frac{\partial(\eta \cdot I_2(\bar{m},\eta))}{\partial\eta} + C_2 \cdot \eta \cdot I_2(\bar{m},\eta)\right] \cdot \quad (19.9)$$

$$\sum_i\sum_j x_i x_j m_i m_j \sum_\alpha^2 \sum_\beta^2 z_{i\alpha} z_{i\beta} \left(\frac{\varepsilon_{i\alpha,j\beta}}{kT}\right)^2 \sigma_{i\alpha,j\beta}^3$$

where $I_1(\bar{m},\eta)$ and $I_2(\bar{m},\eta)$ are integrals of the perturbation theory, and C_1 and C_2 are abbreviations as defined by Gross and Sadowski (2001 a).

Conventional combining rules are employed to determine the parameters for a pair of unlike segments

$$d_{i\alpha,i\beta} = \frac{(d_{i\alpha}+d_{i\beta})}{2} \quad (19.10)$$

$$\varepsilon_{i\alpha,i\beta} = (1-k_{i\alpha,i\beta}) \cdot \sqrt{\varepsilon_{i\alpha} \cdot \varepsilon_{j\beta}} \quad (19.11)$$

In the concept of the Copolymer Perturbed-Chain SAFT model, the copolymer chains are built by segments of the considered homopolymers. Consequentially, the modeling of copolymer-solvent mixture requires the pure-component parameters of the respective homopolymers (the segment diameter σ, the segment number m, and the dispersion energy ε). Only one additional parameter accounts for the interactions between the different segment types of different chains. For example, the solubility of poly(ethylene-co-vinyl acetate) in cyclopentane is calculated from the pure-component parameters of polyethylene and poly(vinyl acetate). The binary parameters for this system are k_{ij} of polyethylene + cyclopentane, k_{ij} of poly(vinyl acetate) + cyclopentane, and the third parameter k_{ij} for the interaction between ethylene and vinyl acetate segments (denoted as $k_{i\alpha,i\beta}$, see Eq. (19.11)).

19.4 Pure-Component Polymer Parameters

A well established approach to determine pure-component parameters for compounds of low molecular mass is to fit the parameters to vapor pressure and liquid density data simultaneously. This is not possible for macromolecules: neither vapor pressure data nor heats of vaporization are accessible for polymers. Moreover, calorimetric data (liquid heat capacity

19.4 Pure-Component Polymer Parameters

data) can not be used for parameter regression, because the ideal gas contribution to the heat capacity can only be estimated with considerable uncertainty.

It has been reported, that pure-component parameters adjusted solely to liquid densities often lead to unsatisfying results, when subsequently used in phase equilibrium calculations of mixtures (Behme, 2000; Koak et al., 1999; Lee et al., 1996; Sadowski et al., 1997). The reason for this is the low sensitivity of the equation of state's energy parameter towards liquid densities.

Two approaches have been introduced to meet the difficulties of determining the pure-component parameters for polymers. The first one obtains pure-component parameters by extrapolating the pure-component parameters of a series of low molecular-mass components. Parameters for linear polyethylene can for example be determined from the n-alkane series. This route was proposed by Sako et al. (1989) and was briefly explored for the Perturbed-Chain SAFT model (Gross, 2000). However, molecular effects such as entanglement, self-interactions, and shielding effects are not properly accounted for when pure-component parameters are determined from such extrapolations. Moreover, only those polymers are accessible, for which a homologous series of low molecular-mass component exists.

Alternatively, pure-component parameters can be determined by simultaneously fitting liquid-density data and binary phase-equilibrium data. Thus, a total of four parameters needs to be optimized, when a set of pure-component parameters of a non-associating polymer is fitted simultaneously to pure liquid density data and to binary data. Besides the three pure-component parameters, it is the k_{ij} parameter of the binary system that also has to be adjusted. This approach is clearly of pragmatic nature, but has successfully been applied in modeling polymer systems (Behme, 2000; Koak et al., 1999; Lee et al., 1996; Sadowski et al., 1997).

Figure 19.2 shows cloud points of polyethylene (LDPE) in mixtures with ethane, propane, butane, ethylene, propylene, and 1-butene. The measurements were performed by Hasch (1994) and Dietzsch (1999) for LDPE + ethylene mixtures. The molecular mass of the polymer samples used are M_w = 106 kg/mol with a polydispersity index of M_w/M_n = 5.1 (Hasch) and M_w = 198 kg/mol with a polydispersity index of M_w/M_n = 2.9 (Dietzsch). The calculations were performed assuming monodisperse polymers of molecular mass $M = M_w$. The liquid-density data of LDPE and the solubility data of LDPE in ethane were considered here to estimate the pure-component parameters of LDPE. The results are shown in Table 19.1. The same set of parameters was then used to calculate properties of all the other LDPE + solvent mixtures. Using one temperature-independent k_{ij} parameter for each binary system, the Perturbed-Chain SAFT equation of state matches the experimental data well for all temperatures. It becomes apparent, that pure-component parameters determined this way are suitable for different mixtures and can thus be regarded characteristic for a polymer.

Table 19.1 also gives the pure-component parameters for all the homopolymers considered in this work, as well as the specifications of the binary systems which were used for parameter characterization.

19 *Modeling of Copolymer Phase Equilibria*

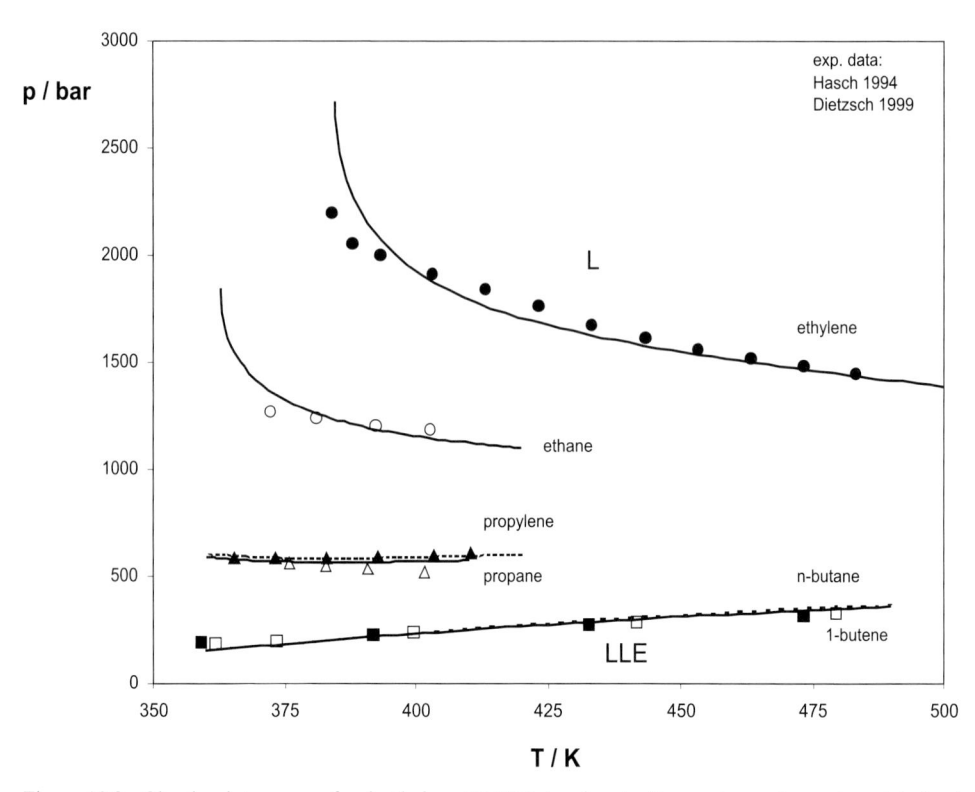

Figure 19.2: Cloud-point curves of polyethylene (LDPE) + solvents. Comparison of experimental cloud points (Hasch, 1994; Dietzsch, 1999) with Copolymer Perturbed-Chain SAFT calculations (LDPE + ethane: $k_{ij} = 0.0325$, LDPE + propane: $k_{ij} = 0.02$, LDPE + n-butane: $k_{ij} = 0.011$, LDPE + ethylene: $k_{ij} = 0.04$, LDPE + propylene: $k_{ij} = 0.0257$, LDPE + 1-butene: $k_{ij} = 0.013$).

Table 19.1: Pure-component parameters of the Perturbed-Chain SAFT equation of state for polymers. *) Gross and Sadowski (2001 b), +) Spuhl (2000), #) Witteman (2001).

Polymer	m/M [mol/g]	σ [Å]	ε [K]	Binary system Solvent	Reference
polyethylene (HDPE)*	0.0263	4.0217	252.0	ethylene	de Loos et al., 1983
polyethylene (LDPE)	0.0263	4.0217	249.5	ethane	Hasch, 1994
polypropylene[+]	0.02305	4.1	217.0	n-pentane	Martin et al., 1999
poly(1-butene)[+]	0.014	4.2	230.0	1-butene	Koak et al., 1999
poly(vinyl acetate)	0.03211	3.3972	204.65	cyclopentane	Beyer, 1999
poly(methyl acrylate)[#]	0.03088	3.5	275.0	2-octanone	Witteman, 2001

19.5 Results for Copolymer Systems

19.5.1 Poly(ethylene-*co*-propylene) (PEP) Systems

By copolymerizing the ethylene monomer with an alkyl-branched comonomer such as propylene and 1-butene, one gets amorphous linear macromolecules which have short hydrocarbon branches and properties like branched polyethylene.

According to the copolymer concept described above, the calculation of properties of PEP-solvent mixtures requires the pure-component parameters of polyethylene (HDPE) and polypropylene (PP), the binary parameters k_{ij} for HDPE + solvent, PP + solvent, and for the interaction between ethylene and propylene segments. HDPE was chosen here due to the linear structure of the polyethylene segments in the PEP chains. The pure-component parameters of the two homopolymers were identified independently from different systems (see Table 19.1). Experimental cloud points for PP + *n*-pentane mixtures at several temperatures and varying molecular mass of polypropylene are given by Martin et al. (1999). These data were used for the parameter identification of polypropylene. Modeling results obtained from the Perturbed-Chain SAFT equation of state are compared in Figure 19.3 with experimental

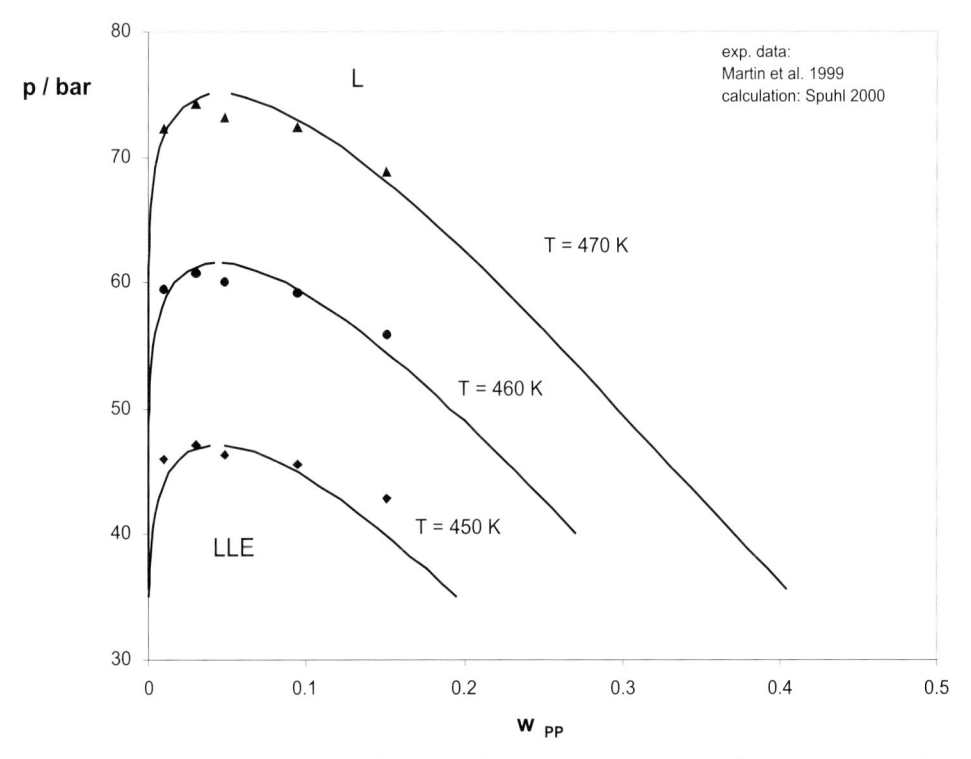

Figure 19.3: Liquid-liquid equilibria of polypropylene (PP) + *n*-pentane at three temperatures. (PP: M_w = 50.4 kg/mol, M_w/M_n = 2.2). Lines: calculations with the Perturbed-Chain SAFT (Spuhl, 2000, k_{ij} = 0.0137). The polymer was assumed to be monodisperse.

liquid-liquid equilibria data of that system at three temperatures. Since the molecular mass distribution of polypropylene is of moderate width ($M_w/M_n = 2.2$), the calculations were performed using the experimental molecular mass $M = M_w = 50.4$ kg/mol of the polymer. The correlation results are in good agreement with the experimental phase behavior.

The k_{ij} value for HDPE + 1-butene (Table 19.2) and for the interaction between ethylene and propylene segments (Table 19.3) were also independently determined from the HDPE + 1-butene and PEP + n-pentane systems, respectively (Spuhl, 2000). Table 19.2 lists all k_{ij} values for the binary homopolymer + solvent system determined for this study. Since no PP + 1-butene binary data were available, the binary parameter for PP + 1-butene had to be obtained from the copolymer + solvent, i.e. PEP + 1-butene mixtures (Spuhl, 2000).

Table 19.2: Binary parameters of the homopolymer + solvent systems.

Polymer–solvent	k_{ij}
LDPE–ethane	0.0325
LDPE–propane	0.02
LDPE–butane	0.011
LDPE–ethylene	0.04
LDPE–propylene	0.0257
LDPE–1-butene	0.013
LDPE–cyclopentane	–0.016
HDPE–propane	0.021
HDPE–1-butene	0.01
PP–n-pentane	0.0137
PP–1-butene	0.0335 [a]
PB–propane	0.025 [b]
PVAc–cyclopentane	0.0233
PMA–ethylene	0.08 [c]
PMA–propylene	0.078 [c]
PMA–1-butene	0.082 [c]

a) fitted to PEP-1-butene
b) fitted to EB 94-propane
c) fitted to the EMA-solvent data,
i.e. EMA-ethylene, EMA-propylene, and EMA-1-butene, respectively

Table 19.3: Segment-segment interaction parameters.

Segment–segment	Interaction parameter
ethylene–propylene	–0.009
ethylene–1-butene	0.008
ethylene–vinyl acetate	$0.1431 - 0.1548 \cdot w_{VA}$ [a]
ethylene–methyl acrylate	$0.6933 \cdot x_{MA}^2 + 0.0297 \cdot x_{MA} + 0.0972$ [b]

a) vinyl acetate content in EVA is in mass%,
b) amount of methyl acrylate segments in EMA in mol%.

19.5 Results for Copolymer Systems

Figure 19.4 shows experimental solubility data of alternating PEP with 50 mol% propylene content in the copolymer in the solvent 1-butene (Chen and Radosz, 1992) for four different molecular masses as well as the modeling results with the copolymer version of the Perturbed-Chain SAFT. Using pure-component parameters of HDPE and PP obtained from other binary systems (HDPE + ethylene and PP + *n*-pentane, respectively), and only one additional parameter which considers the interactions between ethylene and propylene segments, the model is able to describe the influence of the temperature and molecular mass very well.

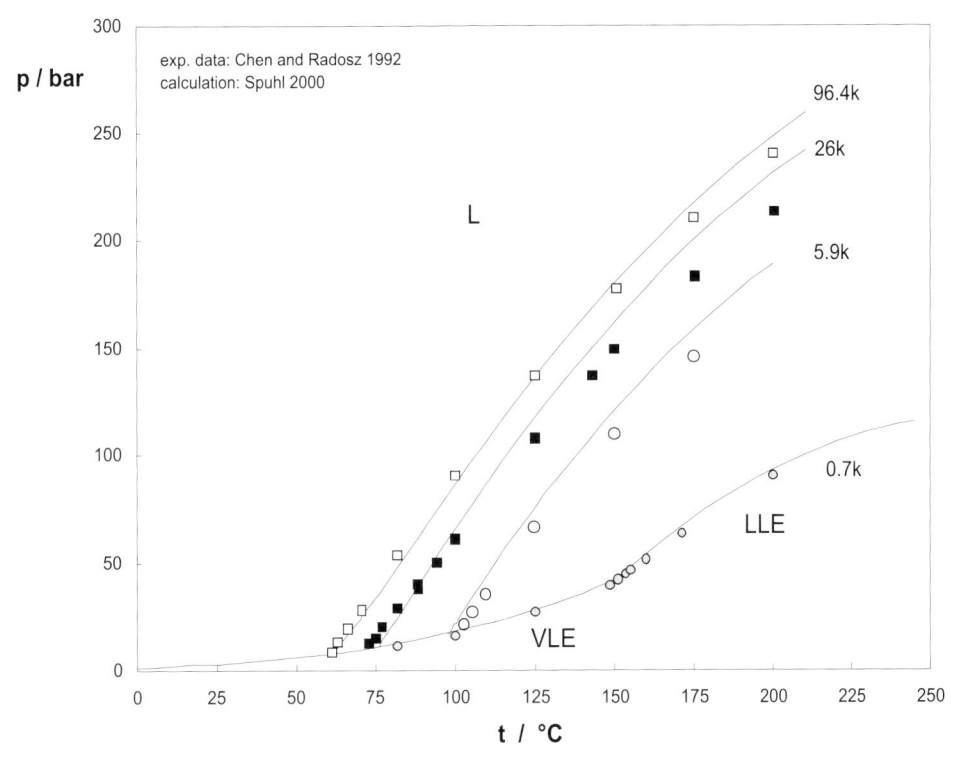

Figure 19.4: Phase behavior of alternate PEP + 1-butene ($w_{PEP} = 0.15$) for different molecular masses. Experimental cloud point data (Chen and Radosz, 1992) compared to calculations with the Copolymer Perturbed-Chain SAFT model.

19.5.2 Poly(ethylene-*co*-1-butene) (EB) Systems

Chen et al. (1995) investigated the influence of the degree of branching on the polymer solubility. For this purpose, they measured the phase behavior of 5 mass% poly(ethylene-*co*-1-butene) (EB) in propane. The degree of branching is varied by changing the contents of 1-butene in EB. All the copolymers have a very low polydispersity ($M_w/M_n = 1.1$). These systems are outstandingly suitable to estimate the ethylene + 1-butene segment interactions since the content of 1-butene in EB varies from zero to 94 mol%.

The pure-component parameters required to model the properties of these copolymer systems are the parameters for HDPE and poly(1-butene) (PB). The latter ones were determined from liquid densities of PB and PB + 1-butene mixtures (Spuhl, 2000).

The k_{ij} value for PB + propane had to be fitted on the binary data of EB 94 + propane due to lack of experimental data of PB + propane. Again, the copolymer model describes the influence of temperature, molecular mass, and copolymer compositions correctly (Fig. 19.5). The polymers were assumed to be monodisperse with $M = M_w$ in the calculations.

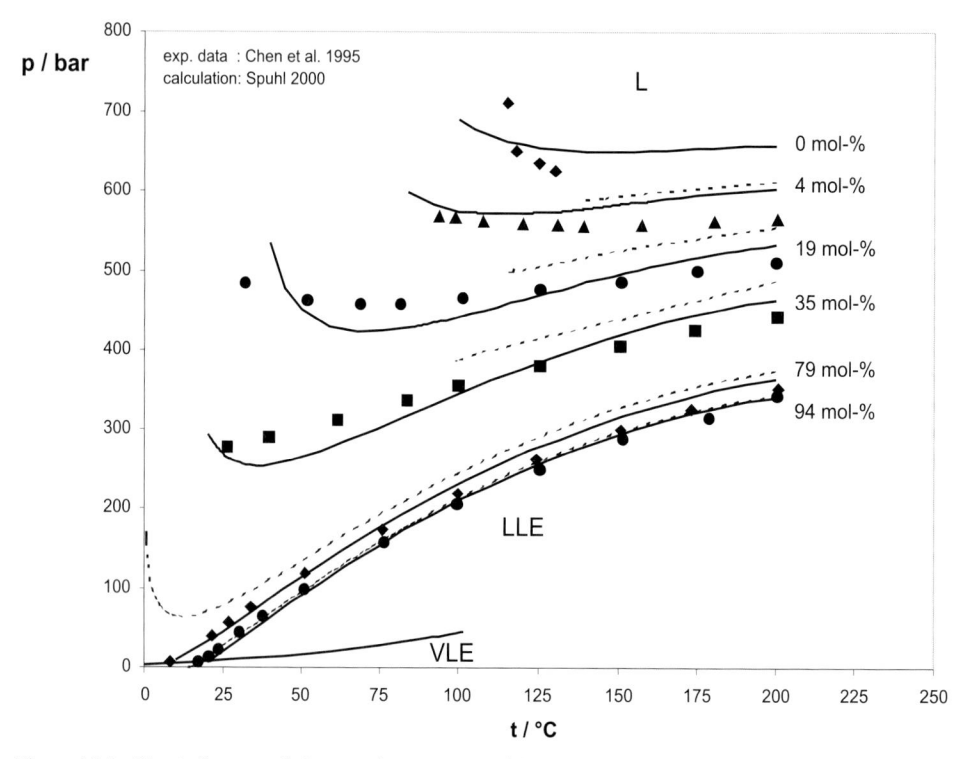

Figure 19.5: The influence of the copolymer composition on solubility of poly(ethylene-co-1-butene) (EB) + propane ($w_{EB} = 0.05$). HDPE: $M_w = 120$ kg/mol, EB 04: $M_w = 62$ kg/mol, EB 19: $M_w = 96$ kg/mol, EB 35: $M_w = 85$ kg/mol, EB 79: $M_w = 91$ kg/mol, EB 94: $M_w = 90$ kg/mol (Chen et al., 1995). HDPE + propane: $k_{ij} = 0.021$, PB + propane: $k_{ij} = 0.025$, ethylene and 1-butene segment: $k_{ij} = 0.008$. Solid lines: modeling results using the Copolymer Perturbed-Chain SAFT. Dashed lines: predictions, i.e. the k_{ij} for the interaction between ethylene and 1-butene segment was set to zero. The calculations were performed using only both, the HDPE and PB pure-component parameters and the k_{ij} of HDPE + propane = 0.021 and the k_{ij} of PB + propane = 0.025.

19.5.3 Poly(ethylene-co-vinyl acetate) (EVA) Systems

In the last years, a new class of ethylene-based copolymers has been developed which incorporate polar or hydrogen-bonding groups into the backbone of the polymer chain. Vinyl acetate is a typical example. These copolymers present a new challenge to produce and model due to their complex structure and phase behavior. They contain both, non-polar and polar segments. Poly(ethylene-co-vinyl acetate) (EVA) was investigated in this study because of the availability of experimental data for various comonomer compositions. Beyer et al. (1999) presented experimental cloud points for the homopolymers polyethylene (LDPE) and poly(vinyl acetate) (PVAc) with cyclopeantane. Furthermore, the phase behavior of EVA at three different compositions (11, 25, and 43 mol% of vinyl acetate monomers in EVA's) with different molecular masses in cyclopentane was also investigated. An interesting phase behavior is observed here: an increase of the vinyl acetate content up to 11 mol% of vinyl acetate in the copolymer first shifts the demixing curve to lower pressures, whereas with further addition of vinyl acetate groups in the copolymer the solubility again decreases. Figure 19.6 shows, that the Copolymer Perturbed-Chain SAFT model is able to capture even this ob-

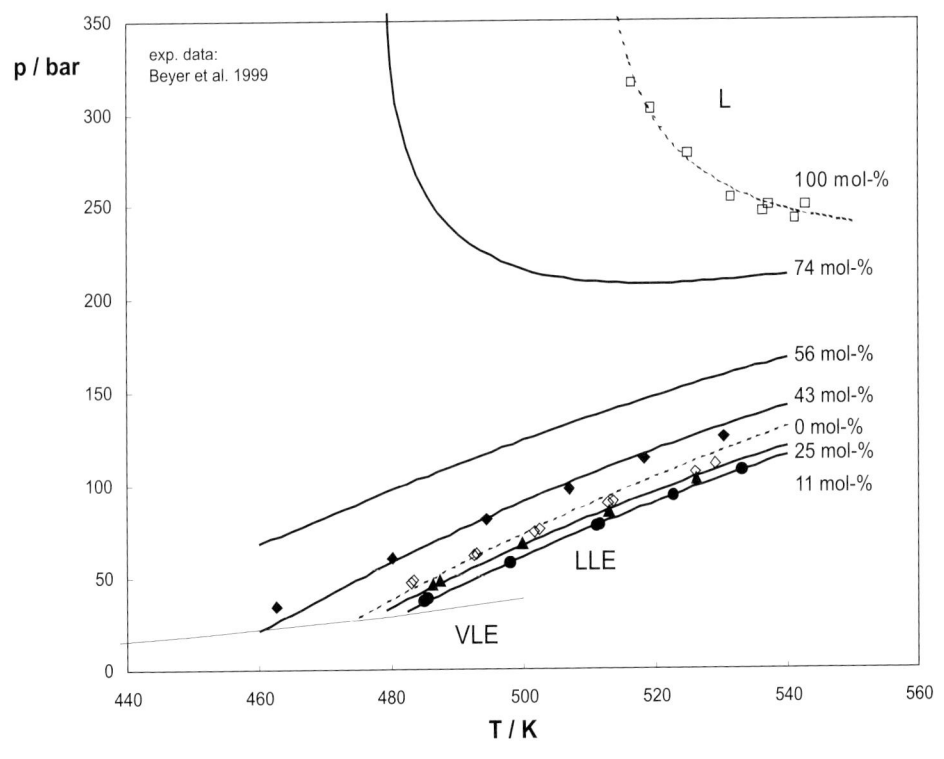

Figure 19.6: Solubility data of cyclopentane in LDPE (M_w = 106 kg/mol), PVAc (M_w = 125 kg/mol) and EVA's (EVA 11: M_w = 256 kg/mol, EVA 25: M_w = 254 kg/mol, EVA 43: M_w = 285 kg/mol). Symbols: experimental data; dashed and solid lines: modeling results with the Copolymer Perturbed-Chain SAFT. Mol% indicates the vinyl acetate content in poly(ethylene-co-vinyl acetate).

served behavior and the correct dependence of temperature, molecular mass, and comonomer composition. The pure-component parameters for PVAc and the binary parameter of PVAc + cyclopentane ($k_{ij} = 0.0233$) were determined by fitting the binary data of PVAc + cyclopentane and PVAc pressure-volume-temperature data simultaneously. The k_{ij} value of -0.016 for LDPE + cyclopentane was obtained from the experimental cloud points (0 mol% of vinyl acetate in copolymer, Fig. 19.6). Only the binary parameter for the interaction between ethylene and vinyl acetate segments was fitted to the copolymer data. It is a linear function of the vinyl acetate mass fraction in EVA ($k_{i,\text{ethylene},i,\text{vinyl acetate}} = 0.1431 - 0.1548 \cdot w_{VA}$).

Using this simple linear function, the Copolymer Perturbed-Chain SAFT model gives good representation of the observed non-linear dependence of comonomer content. The advantage of this copolymer concept is that the calculation or prediction for any comonomer composition and any molecular mass is in general possible. As an example, the prediction of the cloud-point curves for two copolymer compositions of 56 and 74 mol% vinyl acetate content with $M_w = 250$ kg/mol is also shown in Figure 19.6.

Once all the pure-component and binary parameters required to calculate a copolymer-solvent system are obtained, a study of the influence of molecular mass and copolymer

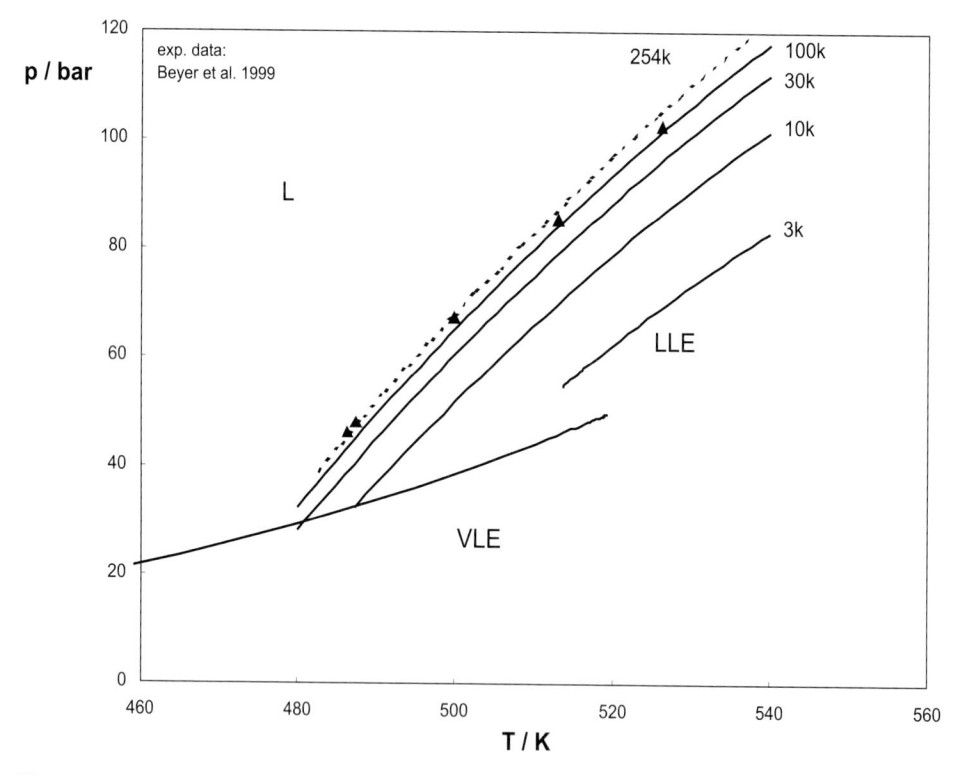

Figure 19.7: Influence of molecular mass on the phase behavior of EVA (25 mol% vinyl acetate) + cyclopentane system. Symbols: experimental cloud points of EVA (25 mol% vinyl acetate $M_w = 254$ kg/mol) + cyclopentane. Dashed line: calculation for $M_w = 254$ kg/mol. Solid lines: predictions for different molecular masses.

composition is conceivable. Some calculations were performed for different molecular masses at EVA with 25 mol% of vinyl acetate. Figure 19.7 gives the expected influence of molecular mass on the solubility of a polymer in a solvent: With increasing of molecular mass, the solubility decreases. For polymers with high molecular mass (>100 kg/mol), the dependence becomes smaller.

19.5.4 Poly(ethylene-*co*-methyl acrylate) (EMA) Systems

The phase behavior of poly(ethylene-*co*-methyl acrylate) EMA in different solvents and co-solvents has been extensively studied experimentally (see for example Hasch, 1994; Dietzsch, 1999). Like poly(ethylene-*co*-vinyl acetate), EMA consists of polar and non-polar segments in its chains.

In Figure 19.8, the cloud-point curves of EMA + ethylene system are displayed. As poly(methyl acrylate) (PMA) does not dissolve in ethylene up to 3000 bar and 500 K, the

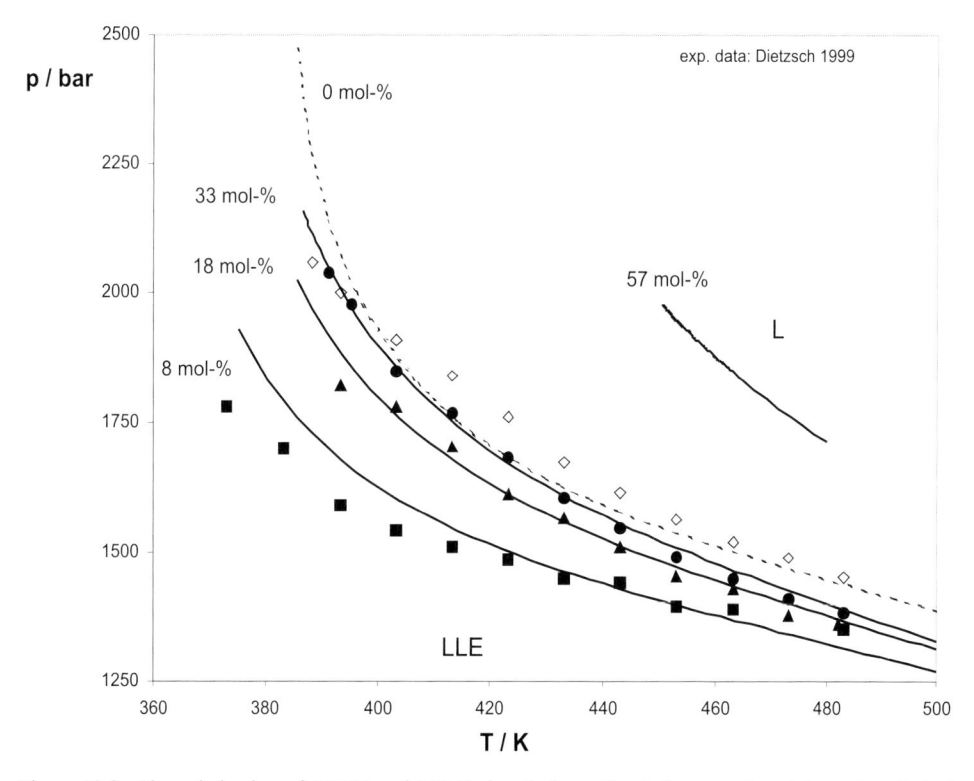

Figure 19.8: Phase behavior of LDPE and EMA's in ethylene. Symbols: experimental results; dashed and solid lines: modeling results with Copolymer Perturbed-Chain SAFT. Mol% expresses the methyl acrylate content in poly(ethylene-*co*-methyl acrylate). LDPE: M_w = 198 kg/mol, EMA 08: M_w = 165 kg/mol, EMA 18: M_w = 150 kg/mol, EMA 33: M_w = 158 kg/mol.

19 Modeling of Copolymer Phase Equilibria

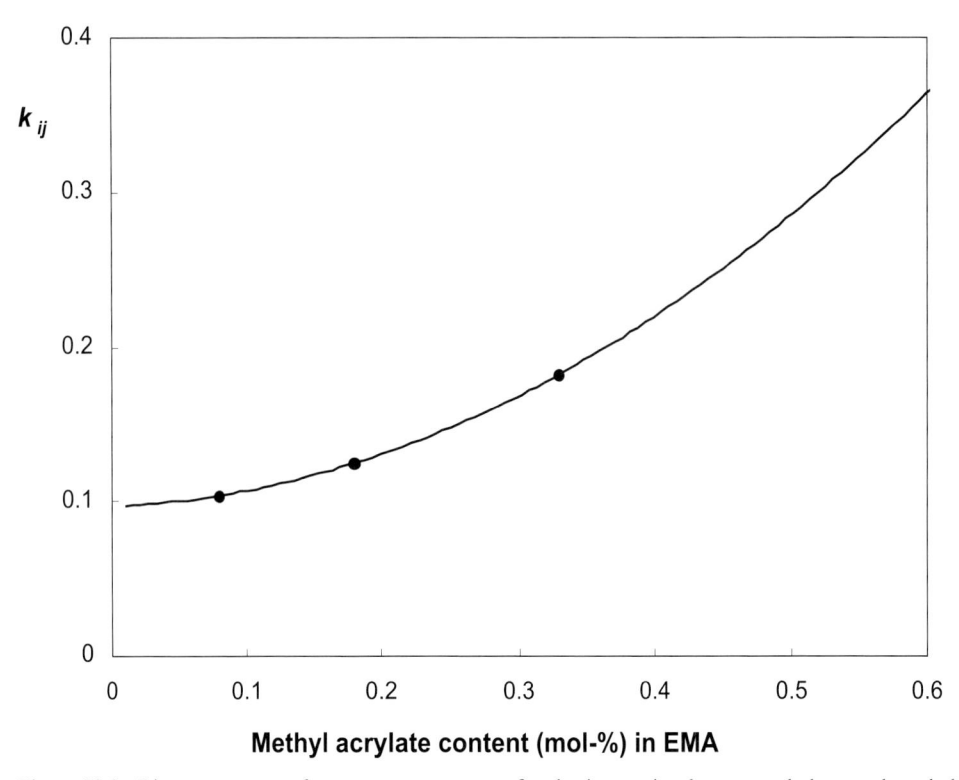

Figure 19.9: Binary parameter $k_{i,\text{ethylene}, i,\text{methyl acrylate}}$ for the interaction between ethylene and methyl acrylate segments as a function of methyl acrylate content (mol%) in EMA

pure-component parameters of PMA were determined from PMA + 2-octanone binary data and PMA liquid-density data (Witteman, 2001). The data are given in Table 19.1. The binary parameters for PMA + ethylene and for ethylene + methyl acrylate ($k_{ij} = 0.6933 \cdot x_{MA}^2 + 0.0297 \cdot x_{MA} + 0.0972$, see Fig. 19.9) were fitted to the copolymer data, i.e. EMA + ethylene binary data. Like in the case for EVA system, the later one is a simple function of the polar segment, i.e. methyl acrylate content in mol% in the EMA copolymer. The amount of methyl acrylate in the copolymer is expressed in mole per cent in the function (Fig. 19.9), because the best result of the regression was obtained by the use of this concentrations scale (instead of mass%). Considering that the polar contribution was neglected in all calculations, the results are considered satisfying. Subsequent studies will investigate the effects on the calculation if the polar contribution is included in the equation of state. Moreover, the prediction for the solubility of 57 mol% methyl acrylate in EMA (M_w = 150 kg/mol) in ethylene is also shown in Figure 19.8.

The pure-component parameters for LDPE and PMA, the binary parameter for LDPE + propylene and LDPE + 1-butene, as well as that for the interactions between ethylene and methyl acrylate segments were then used to calculate the solubility of EMA in propylene (Fig. 19.10) and 1-butene (Fig. 19.11). Experimental data were presented by Hasch (1994) for different methyl acrylate content in EMA (10 mol% with M_w = 75 kg/mol, 31 mol% with

19.5 Results for Copolymer Systems

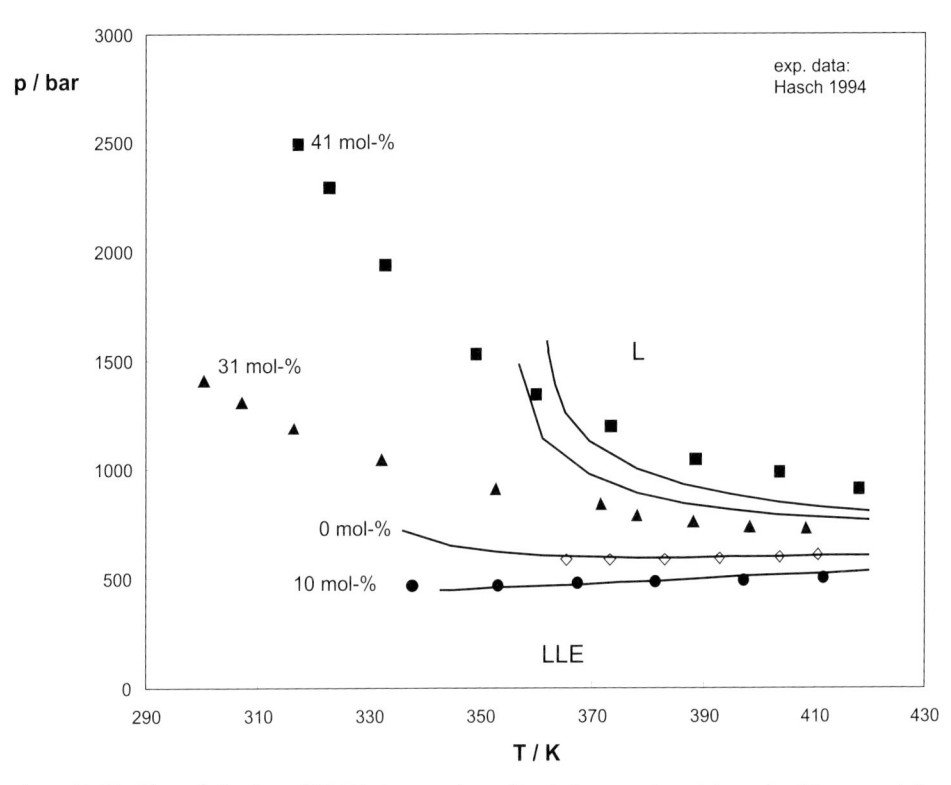

Figure 19.10: Phase behavior of EMA's in propylene. Symbols: experimental results. Lines: modeling results with $k_{ij} = 0.078$ for PMA + propylene. Mol% denotes the methyl acrylate content in poly(ethylene-*co*-methyl acrylate).

$M_w = 109$ kg/mol, and 41 mol% with $M_w = 110$ kg/mol). However, the EMA copolymers used by Hasch (1994) were different from the ones used by Dietzsch (1999). There was no information about the bonding fractions in the copolymers available. For the calculations, all the copolymers were therefore assumed to be random copolymers.

Due to the fact that no binary data were existing, the binary parameters for PMA + propylene and for PMA + 1-butene had to be fitted to the copolymer data, i.e. EMA + propylene and EMA + 1-butene, respectively. The results (Figs. 19.10 and 19.11) show good qualitatively results taking into consideration that only one temperature-independent binary parameter was adjusted to the copolymer data. Furthermore, both authors investigated different EMA copolymers, whereas in the modeling all copolymers were regarded to be of the same random type.

19 *Modeling of Copolymer Phase Equilibria*

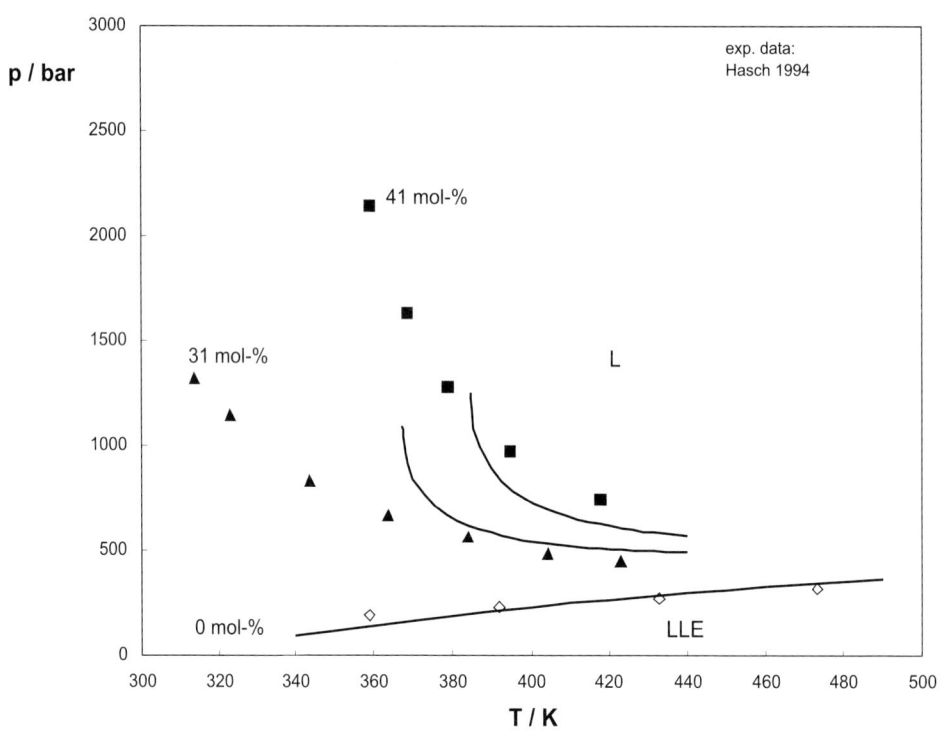

Figure 19.11: Solubility data of EMA's in 1-butene. Symbols: experimental results. Lines: calculation with the Copolymer Perturbed-Chain SAFT. $k_{ij} = 0.082$ for PMA + propylene was fitted to the EMA + 1-butene data. Mol% indicates the methyl acrylate content in poly(ethylene-*co*-methyl acrylate).

19.6 Conclusions

The Copolymer Perturbed-Chain SAFT equation of state was applied for modeling the phase equilibria of ethylene-based copolymer systems like poly(ethylene-*co*-propylene) (PEP) and poly(ethylene-*co*-1-butene) (EB), as well as for copolymers of polar and non-polar segments such as poly(ethylene-*co*-vinyl acetate) (EVA) and poly(ethylene-*co*-methyl acrylate) (EMA). The pure-component parameters of the homopolymers were determined by simultaneously fitting to pure-component liquid-density data and binary phase-equilibrium data. The parameters obtained were suitable for different mixtures and can thus be regarded characteristic for a polymer. Only one temperature-independent k_{ij} parameter for each binary system was used for all considered systems. In the modeling concept of the Copolymer Perturbed-Chain SAFT, the calculation of phase behavior of copolymer systems requires the pure-component parameters of the respective homopolymers and only one additional parameter that describes the interactions between different segment types in chains. Using only these data, the modeling results show that the equation of state is able to capture the temperature and the molecular-mass dependence for various copolymer compositions. For systems with PEP and EB, the parameter which accounts for the interaction between the different segments in copolymer chains (ethylene and propylene segments, ethylene and butene segments, respectively) was a constant, i.e. it is temperature-, molecular-mass-, and copolymer-composition-independent. Good agreement with experimental phase behavior was achieved for PEP + 1-butene and EB + propane binary systems.

In the case of copolymers with polar segments like EVA and EMA, the segment-segment interaction parameter is a simple function of the polar segment content. Nevertheless, the Copolymer Perturbed-Chain SAFT model was found to give good results in describing the observed non-monotonous and non-linear dependence of the solubility on the polar segment content in the copolymer. This was shown for EVA + cyclopentane, EMA + ethylene, EMA + propylene, and EMA + 1-butene systems. The advantage of this copolymer concept is that it allows predictions for any molecular mass and copolymer composition once all the parameters are determined.

Acknowledgements

This study was accomplished when both authors were at the Department of Thermodynamics at the Technical University of Berlin. Both authors gratefully thank Prof. Wolfgang Arlt for supporting this work.

References

Amos, M.D.; Jackson, G. (1991): BHS theory and computer simulations of linear heteronuclear triatomic hard-sphere molecules. Mol. Phys., 74, 191–210.

Amos, M.D.; Jackson G. (1992): Bonded hard-sphere (BHS) theory for the equation of state of fused hard-sphere polyatomic molecules and their mixtures. J. Chem. Phys., 96, 4604–4618.

Banaszak, M.; Chen, C.K.; Radosz, M. (1996): Copolymer SAFT equation of state. Thermodynamic perturbation theory extended to heterobonded chains. Macromolecules, 29, 6481–6486.

Barker, J.A.; Henderson, D. (1967): Perturbation theory and equation of state for fluids: The square-well potential. J. Chem. Phys., 47, 2856–2861.

Behme, S. (2000): Thermodynamik von Polymersystemen bei hohen Drücken. Dissertation, Technische Universität Berlin, Berlin, Germany.

Beyer, C.; Oellrich, L.R.; McHugh, M.A. (1999): Effect of copolymer composition and solvent polarity on the phase behavior of mixtures of poly(ethylene-*co*-vinyl acetate) with cyclopentane and cyclopentene. Chem. Ing. Tech., 71, 1306–1310.

Chapman, W.G.; Jackson, G.; Gubbins, K.E. (1988): Phase equilibria of associating fluids. Chain molecules with multiple bonding sites. Mol. Phys., 65, 1057–1079.

Chen, S.-J.; Radosz, M. (1992): Density-tuned polyolefin phase equilibria. 1. Binary solution of alternating poly(ethylene-propylene) in subcritical and supercritical propylene, 1-butene, and 1-hexene. Experiment and Flory-Patterson model. Macromolecules, 25, 3089–3096.

Chen, S.-J.; Banaszak, M.; Radosz, M. (1995): Phase behavior of poly(ethylene-1-butene) in subcritical and supercritical propane: Ethyl branches reduce segment energy and enhance miscibility. Macromolecules, 28, 1812–1817.

de Loos, Th.W.; Poot, W.; Diepen, G.A.M. (1983): Fluid phase equilibria in the system polyethylene + ethylene. 1. Systems of linear polyethylene + ethylene at high pressure. Macromolecules, 16, 111–117.

Dietzsch, H. (1999): Hochdruck-Copolymerisation von Ethen und (Meth)Acrylsäureestern: Entmischungsverhalten der Systeme Ethen/Cosolvens/poly(Ethen-co-Acrylsäureester) – Kinetik der Ethen-Methylmethacrylat-Copolymerisation. Dissertation, Georg August-Universität zu Göttingen, Göttingen, Germany.

Gross, J.; Sadowski, G. (2000): Application of perturbation theory to a hard-chain reference fluid: an equation of state for square-well chains. Fluid Phase Equilib., 168, 183–199.

Gross, J.; Sadowski, G. (2001a): Perturbed-Chain SAFT: An equation of state based on a perturbation theory for chain molecules. Ind. Eng. Chem. Res., 40, 1244–1260.

Gross, J.; Sadowski, G. (2001b): Modeling polymer systems using the Perturbed-Chain SAFT Equation of State. Ind. Eng. Chem. Res., 41, 1084–1093.

Gross, J. (2000): Entwicklung einer Zustandsgleichung für einfache, assoziierende und makromolekulare Stoffe. Dissertation, Technische Universität Berlin, Berlin, Germany.

Hasch, B.M. (1994): Hydrogen bonding and polarity in ethylene copolymer-solvent mixtures: Experiment and modeling. Dissertation, Johns Hopkins University, Baltimore, Maryland, USA.

Huang, S.H.; Radosz, M. (1990): Equation of state for small, large, polydisperse and associating molecules. Ind. Eng. Chem. Res., 29, 2284–2294.

Huang, S.H.; Radosz, M. (1991): Equation of state for small, large, polydisperse and associating molecules: Extensions to fluid mixtures. Ind. Eng. Chem. Res., 30, 1994–2004.

Koak, N.; Visser, R.M.; de Loos, Th.W. (1999): High-pressure phase behavior of the systems polyethylene + ethylene and polybutene + 1-butene. Fluid Phase Equilib., 158–160, 835–846.

Lee, S.-H.; Hasch, B.M.; McHugh, M.A. (1996): Calculating copolymer solution behavior with Statistical Assoziating Fluid Theory. Fluid Phase Equilib., 117, 61–68.

Mansoori, G.A.; Carnahan, N.F.; Starling, K.E.; Leland, T.W. (1971): Equilibrium thermodynamic properties of the mixture of hard spheres. J. Chem. Phys., 54, 1523–1525.

Martin, T.M.; Lateef, A.A.; Roberts, C.B. (1999): Measurements and modeling of cloud point behavior for polypropylene/n-pentane and polypropylene/n-pentane/carbon dioxide mixtures at high pressures. Fluid Phase Equilib., 154, 241–259.

Müller, E.A.; Gubbins, K.E. (2001): Molecular-based equations of state for associating fluids: A review of SAFT and related approaches. Ind. Eng. Chem. Res., 40, 2193–2211.

Sadowski, G.; Mokrushina, L.V.; Arlt, W. (1997): Finite and infinite dilution activity coefficients in polycarbonate systems. Fluid Phase Equilib., 139, 391–403.

Sako, T.; Wu, A.H.; Prausnitz, J.M. (1989): A cubic equation of state for high-pressure phase equilibria of mixtures containing polymers and volatile fluids. J. Appl. Polym. Sci., 38, 1839–1858.

Shukla, K.P.; Chapman, W.G. (1997): SAFT equation of state for fluid mixtures of hard chain copolymers. Mol. Phys., 91, 1075–1081.

Spuhl, O. (2000): Berechnung von Copolymer-Lösungsmittel-Phasengleichgewichten mit einer Zustandsgleichung, Diplomarbeit, Technische Universität Berlin, Berlin, Germany.

Wertheim, M.S. (1984a): Fluids with highly directional attractive forces: I. Statistical thermodynamics. J. Stat. Phys., 35, 19–34.

Wertheim, M.S. (1984b): Fluids with highly directional attractive forces: II. Thermodynamic perturbation theory and integral equations. J. Stat. Phys., 35, 35–47.

Wertheim, M.S. (1986a): Fluids with highly directional attractive forces: III. Multiple attraction sites. J. Stat. Phys., 42, 459–476.

Wertheim, M.S. (1986b): Fluids with highly directional attractive forces: IV. Equilibrium polymerization. J. Stat. Phys., 42, 477–492.

Witteman, R. (2001): Experimental phase behavior and thermodynamic modeling of polymer and co-polymer solutions using the Perturbed-Chain SAFT equation of state. Diploma thesis, Delft University of Technologie, Delft, The Netherlands.

Nomenclature

$B_{\alpha\alpha}$, $B_{\alpha\beta}$, $B_{\beta\beta}$	bond fraction of type $\alpha\alpha$, $\alpha\beta$, and $\beta\beta$
C_1, C_2	abbreviations in Eq. (19.9)
$d_{i\alpha}$	temperature-dependent segment diameter of segment α on a chain of component i
$g^{hs}_{i\alpha,i\beta}$	radial pair distribution function for segment type α and β of component i in the hard-sphere system
$g^{hs}_{i\alpha,i\beta}(d_{i\alpha,i\beta})$	radial pair distribution function for segment type α and β of component i in the hard-sphere system at contact
$I_1(\bar{m},\eta)$, $I_2(\bar{m},\eta)$	integrals of the perturbation theory in Eq. (19.9)
k	Boltzmann's constant
k_{ij}	binary parameter for the interaction between component i and j
$k_{i\alpha,i\beta}$	binary parameter for the interaction between different segment types α and β of copolymer molecules i
$\bar{m}$	mean segment number
M	molecular mass
m_i	number of segments in a chain of component i
M_n	number-average molecular mass of polymer
M_w	mass-average molecular mass of polymer
p	total pressure
R	gas constant
T	temperature in Kelvin
t	temperature in °C
v	molar volume
w_{VA}	mass fraction of vinyl acetate monomer in poly(ethylene-*co*-vinyl acetate)
x_i	mole fraction of component i
x_{MA}	mole fraction of methyl acrylate monomer in poly(ethylene-*co*-methyl acrylate)
Z	compressibility factor
Z^{assoc}	association contribution to the compressibility factor
$Z^{disp,\,chain}$	dispersion contribution to the compressibility factor with hard-chain reference fluid
Z^{hc}	hard-chain contribution to the compressibility factor
Z^{hs}	hard-sphere contribution to the compressibility factor
$z_{i,\alpha}$	fraction of segments α on a chain of component i
Z^{id}	ideal-gas contribution to the compressibility factor
Z^{pol}	polar contribution to the compressibility factor

Greek letters

$\varepsilon_{i\alpha}$	dispersion energy of segment type α on a chain of component i
η	segment packing fraction ($\eta = \zeta_3$)
ρ	total number density of molecules
$\sigma_{i\alpha}$	temperature-independent segment diameter of segment α on a chain of component i
$\zeta_0, \zeta_1, \zeta_2$	functions of the molar density in Eq. (19.7)
ζ	segment packing fraction ($\zeta_3 = \eta$)

Superscripts

assoc	association
disp, chain	dispersion, reference fluid is hard-chain
hc	hard chain
hs	hard sphere
id	ideal gas
pol	polar

Subscripts

i, j	component i, j
α, β	specific segments within copolymer chain molecules

Abbreviations

EB	poly(ethylene-*co*-1-butene)
EMA	poly(ethylene-*co*-methyl acrylate)
EVA	poly(ethylene-*co*-vinyl acetate)
HDPE	high density polyethylene
LDPE	low density polyethylene
MA	methyl acrylate
PB	poly(1-butene)
PEP	poly(ethylene-*co*-propylene)
PMA	poly(methyl acrylate)
PP	polypropylene
PVAc	poly(vinyl acetate)
VA	vinyl acetate

20 Cloud Point Pressures of Ethene + Acrylate + Poly(ethene-*co*-acrylate) Systems

Michael Buback[1] and Markus Busch

20.1 Abstract

The cloud point behavior of ethene-acrylic acid ester and ethene-methacrylic acid copolymers in ethene and partly in ethene-comonomer mixtures has been measured in an extended pressure and temperature range up to 300 MPa and 260 °C, respectively. Copolymer synthesis is carried out in a continuously operated stirred tank reactor (CSTR) which ensures production of chemically homogeneous copolymer material. Part of the phase behavior studies are performed under flow-through conditions in a second CSTR which is penetrated by the mixture from the polymerization reactor. Alternatively, the copolymer is isolated after synthesis and, after preparation of a polymer-monomer mixture, the phase behavior is mapped out in a discontinuously operated high-pressure cell. The influence of copolymer composition on the cloud point behavior in mixtures with ethene (E) has been systematically studied for E + methyl acrylate (MA), E + butyl acrylate (BA), E + ethylhexyl acrylate (EHA), E + acrylic acid (AA), and E + methacrylic acid (MAA) copolymers. For E + MA and E + BA copolymers also the variation of cloud point pressure (CPP) with the comonomer content of the monomer mixture has been investigated and numbers for the specific change in CPP have been deduced. The CPPs of the E + BA + poly(E-*co*-BA) system are primarily determined by entropic effects. At lower temperature, interactions of polar segments with the solvent come into play. They are even more important with the E + MA + poly(E-*co*-MA) system. At high copolymer MA contents, the phase behavior of E + MA + poly(E-*co*-MA) is additionally affected by intra-segmental interactions of the methyl ester groups. With the E + poly(E-*co*-MAA) systems, strong intra-segmental interactions via hydrogen bonds are operative which largely reduce copolymer solubility in ethene. The variation of CPP with both copolymer composition and with the composition of the monomer mixture has been determined within extended regions. The data allow to estimate homogeneous reaction conditions for ethene copolymerizations in wide pressure, temperature, and composition regions. The data obtained at systematic variation of the type and amount of polar comonomer within both monomer mixture and copolymer should be perfectly suited for modeling purposes.

---

[1] Universität Göttingen, Institut für Physikalische Chemie, Tammannstr. 6, 37077 Göttingen.

20.2 Introduction

Precisely knowing the phase behavior of fluid mixtures is extremely important in order to identify reaction conditions where technical fluid-phase processes may be carried out in homogeneous phase. Moreover, information on the phase behavior is required for the optimal design of separation processes. Experimental devices for the study of cloud point pressures and for the analysis of coexisting fluid phases have been reviewed by Fornari et al. (1990) and by Dohrn and Brunner (1995) with particular emphasis on systems that do not contain polymeric material. The complexity of phase diagrams of supercritical fluid solutions has been outlined by Schneider (1972, 1994). An attractive aspect of the experimental work by the Schneider group consists of using near-infrared spectroscopy for quantitative in-line determination of concentrations within each of the coexisting phases under high-pressure and high-temperature conditions. An important advantage of this method relates to the fact that no sampling is required. As a consequence, cloud point pressures and equilibrium data may be mapped out over an extended pressure and temperature range with a reactor of relatively small size.

Investigations into fluid phase systems which contain high molecular weight polymer are associated with additional problems, especially in cases where the size distribution is broad, but also in studies into copolymers that are chemically inhomogeneous. In addition to the dependence on temperature, pressure, and concentration, the impact on phase behavior of differences in the (average) size and in the distribution of molecular weight and of a (potential) distribution of copolymer composition need to be considered.

A significant scientific interest centers around the understanding of the thermodynamics of fluid phase polymer solutions including supercritical fluid phases. Fluid-phase polymerizations are of enormous technical relevance due to the tunability of polymer properties by polymerization in a wide range of fluid phase conditions. Measuring cloud point pressures and equilibrium compositions for the multitude of potential reaction conditions and for polymers which differ in size, size distribution, degree of short-chain and long-chain branching mostly is beyond reach because of the associated enormous amount of experimental work. On the other hand, careful analysis of a limited number of systems is required in order to obtain a general understanding of polymer solubility in fluid solutions. One such important type of systems are copolymers with a nonpolar and a polar repeat unit, such as ethene + acrylic acid ester (E + acrylate) copolymers in supercritical ethene or in a fluid mixture of the two monomers. It appears mandatory to study the dominant effects on cloud point behavior of well-defined such copolymers. The cloud point pressure is defined as the point at which the mixture becomes opaque to such an extent that it is no longer possible to see the stirrer inside the fluid mixture or to detect reflections of the incident light on the polished internal surfaces of the high-pressure cell. Lee et al. (1996) compared results obtained according to this definition with data from laser light turbidity measurements where the cloud point is defined by the situation of a 90% decrease in transmitted light intensity. Data from the two methods were found to be identical within experimental accuracy.

Although several groups have addressed the phase behavior of fluid phase copolymer systems, the thermodynamics of these complex mixtures is far from being fully understood. Hasch et al. (1993), Lo Stracco et al. (1994), Lee et al. (1994), and Meilchen et al. (1992) measured cloud point pressures of 5 mass% poly(ethene-*co*-methyl acrylate) at MA copolymer contents between 10 and 35 mol% with propane, acetone, ethanol, and chlorodifluoromethane being the low molecular weight solvent material. Poly(ethene-*co*-acrylic acid) with AA copolymer contents between 2.4 and 9.2 mol% acid has been investigated in butane and butene solutions by Hasch and McHugh (1995). Folie and Radosz (1995) studied *p-T* isopleths at fixed copolymer concentrations of 5 mass% for poly(ethene-*co*-methyl acrylate), poly(ethene-*co*-butyl acrylate), poly(ethene-*co*-vinyl acetate), poly(ethene-*co*-acrylic acid), and poly(ethene-*co*-methacrylic acid) in ethene. A systematic investigation into cloud point *p-T* isopleths for poly(ethene-*co*-methyl acrylate) and poly(ethene-*co*-butyl acrylate) in ethene was presented by Byun et al. (1996). They report the first series of cloud point curves at systematic variation of the acrylate copolymer content between the limiting situations of the homopolymers, polyethylene and polyacrylate. With the poly(ethene-*co*-methyl acrylate) system particularly strong effects of copolymer composition on the location of the cloud point curve are seen. The data show that the interaction of the polar segments strongly influences cloud point behavior. The ethene + poly(ethene-*co*-butyl acrylate) system is indicative of some shielding action of the alkyl moiety of the ester group. Müller and Oellrich (1996) demonstrated that proper stabilization of the potentially reactive ethene monomer needs to be carefully considered. With the concentration of the stabilizing agent being too low, further polymerization may occur during the thermodynamic experiment which is highly undesirable. Higher concentrations of stabilizing agent, on the other hand, may affect cloud point pressures. Polymerization during the cloud point measurement needs to be specially monitored and, as far as possible, suppressed in studies with the highly reactive acrylate monomers. Luft and Wind (1992) succeeded to measure cloud point pressures for ethene + poly(ethene-*co*-acrylic acid) systems where they took advantage of some inhibitor activity of the acrylic acid monomer on free-radical polymerization. Such an inhibition effect has also been observed by Wittkowski (1998) in kinetic studies into ethene + acrylic acid and ethene + methacrylic acid copolymerizations. The Luft and Wind (1992) investigations were carried out at 15 mass% copolymer. The acrylic acid (AA) copolymer content was varied between 0 and 9 mass%. With increasing AA content of the copolymer, the cloud point pressure is enormously enhanced. So far, in only two studies cloud point pressures of copolymers have been investigated in the associated monomer-comonomer mixture. Luft and Subramanian (1987) measured data for ethene + methyl acrylate + poly(ethene-*co*-methyl acrylate) and Folie et al. (1996) reported data for ethene + vinyl acetate + poly(ethene-*co*-vinyl acetate). In both studies significant amounts of inhibitor had to be added in order to prevent polymerization. Banaszak et al. (1996) showed that branching of α-olefin copolymers may significantly affect the phase behavior. Using such nonpolar copolymers, these authors demonstrated the influence of branching density and of the average length of branches on cloud point pressure.

The literature data strongly suggest that cloud point studies should be primarily carried out on well-defined copolymer materials which are chemically homogeneous, that is the individual copolymer molecules should be very similar in overall composition and in distribution of the two (or more) comonomer-derived units along the polymer backbone. The availability of suitable copolymers which are best prepared by polymerization in a continuously stirred tank reactor (CSTR) may be a major problem. With respect to solubility studies on chemically homogeneous copolymers, it is particularly advantageous to use a combined experimental set-up which consists of two CSTRs that are arranged in series (Buback et al., 1996). The first continuously operated stirred tank reactor, CSTR1, serves for polymer preparation and the second one, CSTR2, is used for the thermodynamic experiment (see Fig. 20.1, below). CSTR1 is maintained at polymerization pressure and temperature for extended periods of time, typically for up to 12 h, whereas CSTR2 is adjusted to several temperatures at each of which the pressure is varied and the cloud point pressure is identified by permanent monitoring of the internal volume in CSTR2 via a boroscope. Measurements of phase behavior under continuous flow conditions have already been carried out by Simnick et al. (1977). By this strategy, the authors succeeded to circumvent the problem of tetralin decomposition in studying hydrogen + tetralin systems.

The set-up with two CSTRs, for continuous polymer production and for cloud-point measurement, respectively, offers several advantages: (i) Under flow-through conditions, the copolymer is at high-temperature and high-pressure conditions only for a few minutes which eliminates polymer decomposition or polymer modification reactions. (ii) With the temperature of CSTR2 being lower than the one of CSTR1, no or only negligible polymerization takes place at the short residence times. As a consequence, cloud point curves may be measured without the addition of stabilizers. (iii) The system under thermodynamic investigation closely resembles the one met in technical polymerizations. It should be particularly noted that both types of monomers are present. (iv) A further attractive feature relates to the fact that reaction conditions in CSTR1 may be varied such as to change polymer properties in a well-defined way and the resulting effects on the location of the cloud point curve may be mapped out with the other properties of the monomer + polymer system remaining unchanged. (v) Additional monomer or solvent material may be introduced into the flowing reaction mixture in between CSTR1 and CSTR2, thus enabling measurement of the resulting effects on cloud point pressure, again with the other parameters of the thermodynamic experiment being kept constant.

It goes without saying that running a device with two CSTRs requires significant expertise. Moreover, this experiment is tedious and time-consuming. It is for this reason that, in addition to several series of fundamental studies into the phase behavior of ethylene + acrylate copolymers, an extended number of investigations has been carried out by using two separate experimental set-ups. Copolymer is produced in CSTR1 and is isolated. In a discontinuously operated autoclave equipped with a sapphire window, cloud point pressures are measured after preparing a polymer + monomer mixture. The latter experiment (see Fig. 20.3, below) requires a stabilizing agent.

The intention of experiments within the present study was to systematically vary the copolymer mole fraction, F, and the mole fraction of the monomer mixture, f. The cloud point investigations were performed on several ethene + acrylate + poly(ethene-*co*-acrylate) systems. The comonomers to ethene were methyl acrylate (MA), butyl acrylate (BA), and ethylhexyl acrylate (EHA). They differ considerably in the size of the alkyl part of the ester moiety. The influence of strong hydrogen-bonded interactions between comonomer segments will be illustrated by a series of cloud point studies into ethene + poly(ethene-*co*-methacrylic acid) systems.

20.3 Experimental Methods

The scheme of the continuously operated experimental set-up for copolymer production and subsequent cloud point measurements at pressures up to 300 MPa and temperatures up to 300 °C is depicted in Figure 20.1.

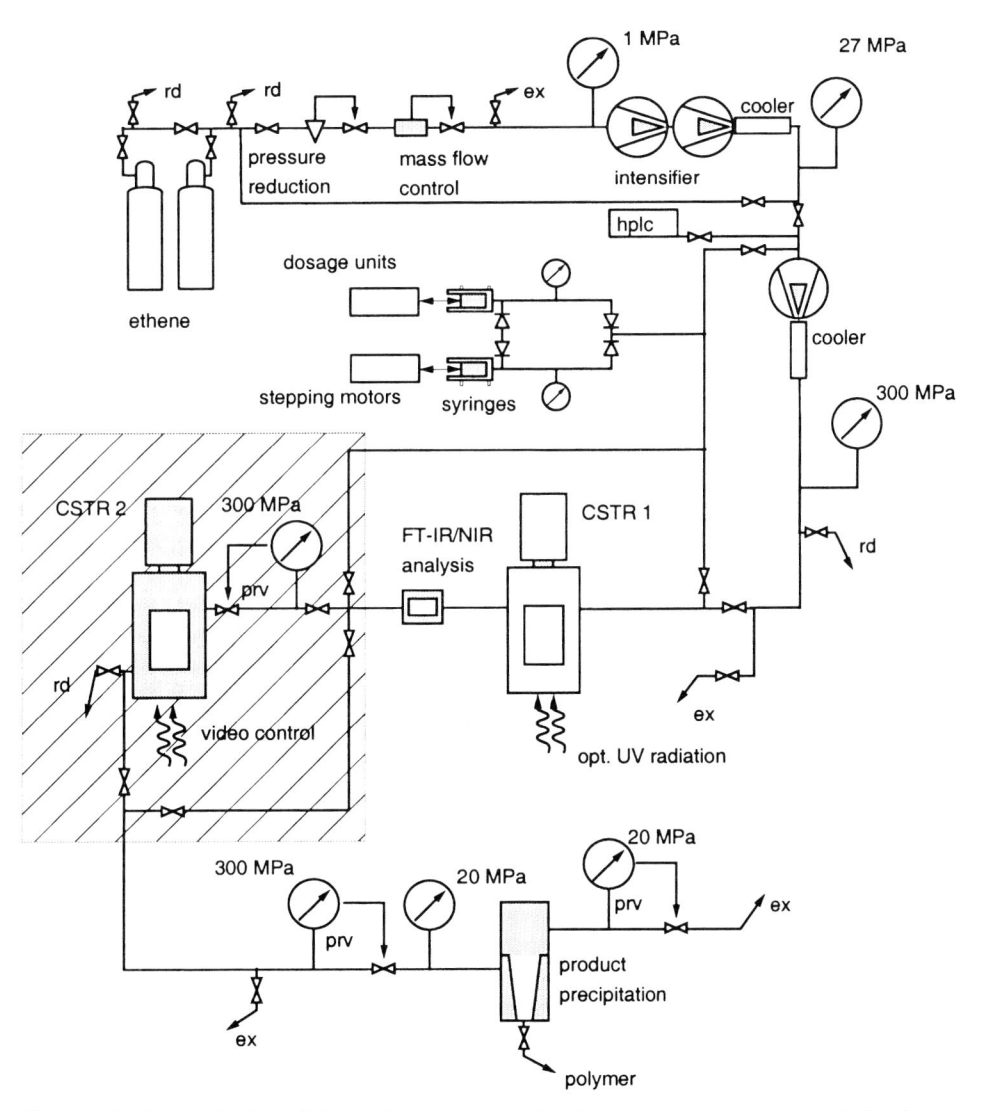

Figure 20.1: Schematic view of the continuously operated high-pressure high-temperature device for copolymer synthesis and subsequent measurement of cloud point pressure (CPP) curves. The part of the device required for CPP determination is indicated by the shaded area. rd: rupture disc; hplc: high-pressure liquid chromatography pump; prv: pressure release valve; ex: exhaust; CSTR: continuously operated stirred tank reactor.

The shaded area encompasses that part of the set-up which serves for cloud point investigations. The remaining part of the mini-plant device has already been extensively used for high-pressure high-temperature syntheses of ethene homo- and copolymers. More detailed descriptions of measuring mass flows, of pressure generation via the three-stage unit, of in-line Fourier transform analysis of the reaction mixture after passing CSTR1, and of the pressure release and polymer precipitation units are presented elsewhere (Buback et al., 1994, 1995, 1996a,b). Via the dosage units, the acrylate comonomer and, if required, chain-transfer agents and initiator may be introduced into the ethene fluid before or after passing through the third compressor stage. Comonomer (or any other material of interest) may also be added to the reaction mixture after passing through CSTR1. The latter procedure, e.g., allows for mapping out the dependence of cloud point pressure on comonomer concentration.

The design of the high-temperature high-pressure vessel, CSTR1 and CSTR2, is illustrated in Figure 20.2. The cell body is made from a nickel-based alloy of high tensile strength. A sapphire window of 22.9 mm height and 38.1 mm diameter (at an aperture of 21.2 mm) enables continuous visual inspection of the internal volume under high-pressure high-temperature conditions. The mixture inlet is located close to the sapphire window. The outlet boring sits on the opposite side of the internal volume. The stirrer is driven from outside via permanent CoSm-magnets. The associated equipment is not shown in Figure 20.2. The configuration of the stirrer was optimized until close-to-ideal mixing conditions were reached. The quality of mixing was checked by CO_2 tracer pulse experiments under high-pressure high-temperature conditions (Buback et al., 1992; Lovis, 1995). The internal diameter of the autoclave is 42 mm, the outer diameter is 150 mm.

Mass flows were typically of the order of 1 kg/h. The mini-plant was operated for up to 16 h with the operation time being primarily limited by the ethene supply. Running

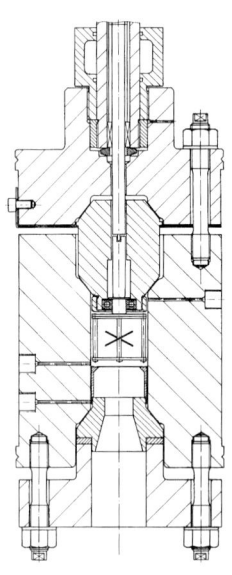

Figure 20.2: Stirred tank reactor for operation up to 300 MPa and 300 °C. The stirrer is driven by a magnetic coupling system (not shown in the figure). Visual or video observation of the reaction mixture under pressure and temperature is easily carried out through the sapphire window of 21.2 mm aperture.

20.3 Experimental Methods

CSTR1 under stationary conditions ensures production of chemically homogeneous statistical copolymer dissolved in a mixture of both monomers. The stationarity of the product mixture is controlled by in-line infrared/near infrared (IR/NIR) spectroscopy. The homogeneity of the mixture in CSTR1 is visually checked through the sapphire window.

Measurement of cloud point pressures in CSTR2 requires significant variation of pressure at different levels of the CSTR2 temperature without affecting the pressure in CSTR1. The procedure is carried out manually, essentially by fine-tuning the pressure release valve which sits behind the autoclave used for product separation. The valve has been specially designed to allow for very small apertures which enable expansion of the high-pressure fluid to ambient pressure. Slight variations of the pressure in CSTR1 can not always be avoided. They do not disturb the thermodynamic experiment as cloud point pressures are not determined unless stationary conditions are reached (again) in both CSTR1 and CSTR2 with p and T of CSTR1 being maintained at the particular level which has been selected for the entire experimental series. The polymer is collected during the entire course of the experiment. Depending on the degree of monomer conversion that has been chosen for a particular study, copolymer is obtained in quantities up to a few hundred grams. Thus sufficient material is available for various types of subsequent analytical studies. Moreover, the copolymer may be re-dissolved and cloud point pressures may be determined via the classical "synthetic" method using the discontinuously operated high-pressure high-temperature variable-volume cell equipped with a sapphire window (see Fig. 20.3).

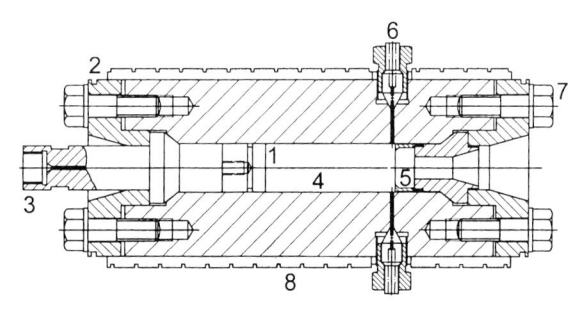

Figure 20.3: High-pressure high-temperature cell for discontinuous cloud point experiments via the so-called synthetic method. 1: movable piston; 2: glant; 3: inlet for pressurizing fluid; 4: sample volume; 5: steel ram with high-pressure sapphire window (of 7 mm aperture); 6: thermocouple; 7: bolt; 8: heating.

The system under investigation (4) is separated from the pressure-generating part by a movable piston which is sealed by an O-ring made from Viton®. In studies at temperatures above 220 °C or in the presence of acrylic acid, Kalrez® is used as O-ring material. The internal volume is illuminated through the sapphire window (5) and cloud point pressures are detected through the same window by means of a boroscope. Polishing the flat surface of the piston enormously improves the quality by which the internal volume may be monitored and cloud point pressures be detected. Temperature is recorded via sheathed thermocouples (6) which are introduced into the system under investigation. Copolymer, stabilizer and, if desired, liquid comonomer or a cosolvent are fed into the sample volume and the system is pressurized with a precisely measured (by mass) amount of ethene. Cloud point pressures are determined by the procedure described elsewhere (Byun et al., 1996; Hasch et al., 1991, 1993, 1995; Dietzsch, 1999).

In the discontinuous cloud point experiments carried out in the apparatus shown in Figure 20.3, typically 0.2 mass% of 2,6-di-*tert*-butyl-4-methyl-phenol are applied as stabilizer. The amount of copolymer was mostly of the order of 5 mass%. The cloud point experiments were performed either by using the combined flow-through set-up (Fig. 20.1) or by using the non-shaded part of this set-up for copolymer production and, after isolation of the polymeric product, measuring phase behavior with the variable-volume cell. The second method allows to select copolymer concentration without being forced to adjust polymerization conditions in CSTR1 such as to yield a particular degree of monomer conversion. Using both techniques in parallel enables detailed thermodynamic studies.

20.4 Results and Discussion

The most extensive set of cloud point pressure (CPP) data for ethylene + acrylate copolymers in fluid ethene exists for poly(ethene-*co*-butyl acrylate). Figure 20.4 displays this data for a wide range of copolymer compositions, F, and includes the two homopolymer systems. As with all subsequent data that refer to polymer solubility in ethene (without comonomer), the measurements have been carried out by polymer production in CSTR1 and CPP meas-

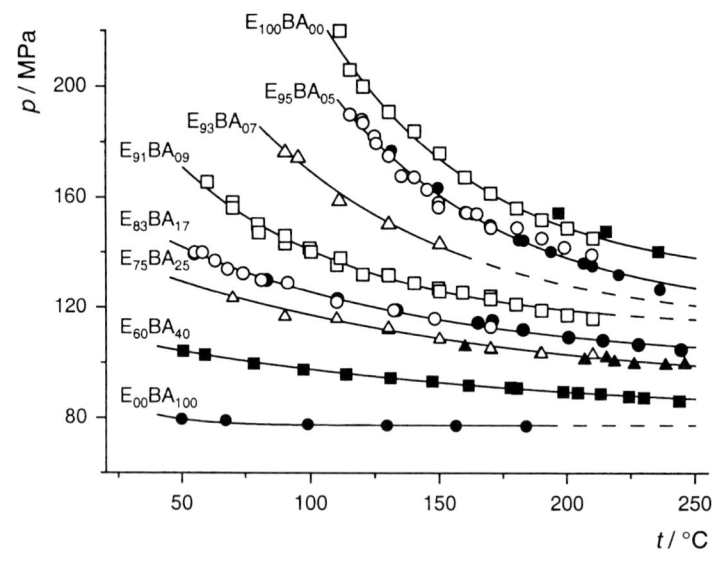

Figure 20.4: Cloud point pressure curves of E + BA copolymers in ethene. The filled symbols are literature data reported by Byun et al. (1996). The subscripts characterize copolymer composition in mol%. The copolymers were synthesized in CSTR1. The CPPs were measured in the discontinuously operated device shown in Fig. 20.3 at polymer concentrations of 5 mass%.

20.4 Results and Discussion

urement in the device shown in Figure 20.3. The subscripts indicate the mole fractions of monomer-derived units within the polymer. The homogeneous single-phase region extends from the *p-T* isopleths toward higher pressure. The filled symbols refer to data reported by Byun et al. (1996). At identical copolymer composition, an excellent agreement of the literature data with the results from the present study (open symbols) is seen. Copolymers of the two studies with identical composition differ in molecular weight (distribution) and in copolymer microstructure with these properties being tabulated elsewhere (Byun et al., 1996; Dietzsch, 1999). Thus the number average of the $E_{95}BA_{05}$, for example, has been 40,700 g/mol in the Byun et al. study and was 44,300 g/mol within the present study. As Figure 20.4 demonstrates, this modest difference in polymer size is not associated with a variation in CPP within the experimental accuracy of ± 3 MPa.

Cloud point pressures are lowered toward increasing temperature and increasing copolymer BA content. Both effects are more pronounced at lower temperatures. The continuous and significant lowering of CPP with BA copolymer content at lower temperature is assigned to the favorable polar interactions of the BA segments of the polymer with the quadrupole moment of the ethene solvent. The importance of these interactions on phase separation decays toward higher temperature.

Data for the E + poly(E-*co*-MA) system are plotted in Figure 20.5. The E + MA copolymers were produced under conditions as chosen for E + BA synthesis. As with the E + poly(E-*co*-BA) system, the CPPs are lowered toward higher temperature and the CPPs at low copolymer content, e.g., at 8 mol% MA units, $E_{92}MA_{08}$, are significantly below the polyethylene data. In clear contrast to what has been seen with the E + BA system, further increase of MA copolymer content reverses this trend. The variation of CPP with MA content is indicated by the arrows in Figure 20.5.

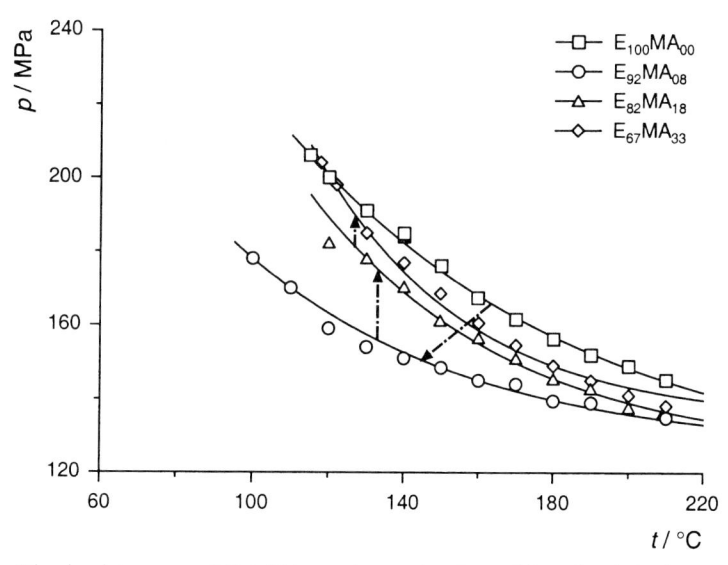

Figure 20.5: Cloud point curves of E + MA copolymers in ethene. The subscripts characterize copolymer composition in mol%. The copolymers were synthesized in CSTR1. The CPPs were measured in the discontinuously operated device shown in Fig. 20.3 at polymer concentrations of 5 mass%. The change in copolymer composition is indicated by the arrows.

The cloud point behavior of the E + poly(E-co-MA) system at higher MA content is understood (Byun et al., 1996) as being due to dominant contributions of polar interactions between MA segments within a copolymer molecule. They disfavor solubility of the copolymer in ethene. It should be noted that homogeneity of 5 mass% poly(MA) in fluid ethene can not be reached even at pressures and temperatures as high as 300 MPa and 250 °C, respectively.

As has been discussed by Byun et al. (1996), increasing the size of the alkyl group within the ester moiety results in a shielding of the intra-segmental interactions of polar segments which enhances the importance of the interactions between polar segments and the monomer. As a consequence, the increase of BA copolymer content lowers CPP with this effect continuing to operate up to the maximum BA content, in poly(BA). Along these lines it comes as no surprise that the CPP behavior of the E + poly(E-co-ethylhexyl acrylate) where such shielding should also operate, is very close to the one of the E + poly(E-co-BA) system.

CPP data for E + poly(E-co-EHA) are presented in Figure 20.6. The CPPs are lowered with EHA copolymer content during the entire composition range. The data for the E + poly(EHA) system are from Lora et al. (1999).

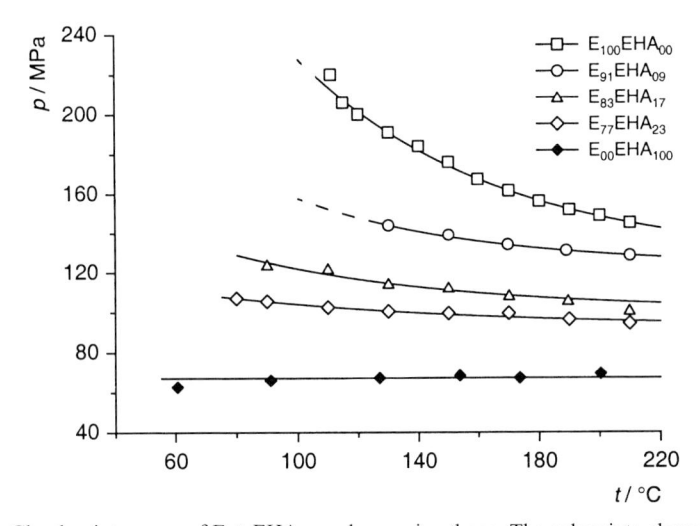

Figure 20.6: Cloud point curves of E + EHA copolymers in ethene. The subscripts characterize copolymer composition in mol%. The copolymers were synthesized in CSTR1. The CPPs were measured in the discontinuously operated device shown in Fig. 20.3 at polymer concentrations of 5 mass%. The EHA homopolymer CPP data are from Lora et al. (1999).

The CPPs of the E + poly(E-co-BA) and E + poly(E-co-EHA) systems are almost the same as can be seen (Fig. 20.7) from the common plot of CPP vs. comonomer content, F, for the three E + poly(E-co-acrylate) systems at 200 °C. The homogeneous region appears to be slightly more extended with the EHA system. This effect is, however, difficult to be safely established with the limits of experimental accuracy. Moreover, slight differences between the E + BA and E + EHA copolymers in molecular weight and in microstructure may contribute to the observed minor effect. E + poly(E-co-MA) behaves distinctly different.

20.4 Results and Discussion

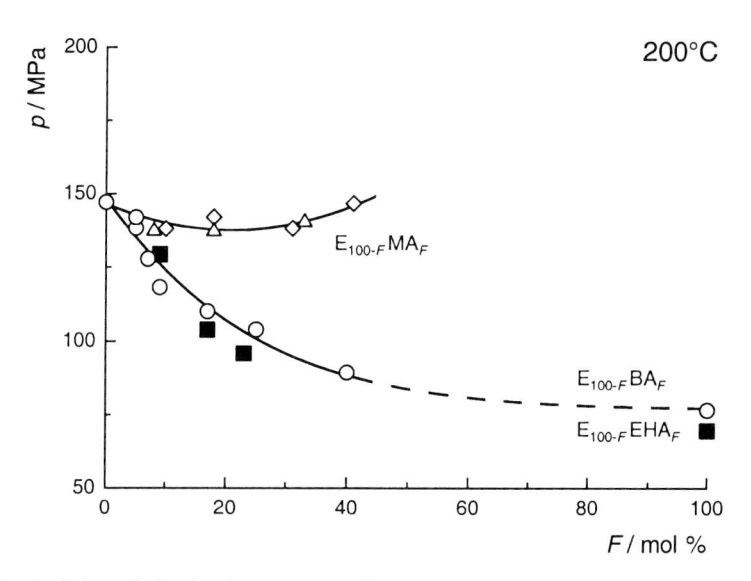

Figure 20.7: Variation of cloud point pressure with copolymer composition, F, for the three systems: △,◇ E + MA + poly(E-MA); ■ E + BA + poly(E-BA); ○ E + EHA + poly(E–EHA). All data refer to 200 °C.

A minimum in the CPP vs. F correlation is clearly seen at 200 °C, although it is less pronounced as compared to the situation at temperatures below 150 °C (see Fig. 20.5).

Support for the assumption that intra-segmental interactions of polar groups on the (same) copolymer molecule are responsible for the reduced solubility at higher F_{MA} is provided by investigations into the cloud point behavior of E + poly(E-co-methacrylic acid) (Latz, 1999). Shown in Figures 20.8 and 20.9 are p-T isopleths for E + poly(ethene-co-acrylic acid) and E + poly(ethene-co-methacrylic acid) systems, respectively. Even minor copolymer methacrylic acid contents result in a pronounced enhancement of CPP. A steep increase in CPP with methacrylic acid copolymer content occurs from $F_{AA} = 0$ on. This effect is larger for the acrylic acid (AA) than for the methacrylic acid (MAA) system. At comonomer contents as low as 10 mol% methacrylic acid, CPP exceeds 300 MPa which prevents CPP studies in the available apparatus even at temperatures as high as 240 °C. The intra-segmental hydrogen-bonded interactions of –COOH groups on the copolymer molecule obviously are strong enough to effectively prevent enhanced solubility originating from attractive interactions between polymer segments and the ethene solvent. As a consequence, even the initial decrease of CPP with comonomer content F, that is seen with the E + poly(E-co-MA) system at low F, does not occur with the E + poly(E-co-(meth)AA) systems.

That the solubility reduction with E + poly(E-co-AA) slightly exceeds the effect seen with E + poly(E-co-MAA) is most likely due to some shielding effect or steric contribution of the α-methyl group by which the quality of hydrogen bonding between the –COOH groups is reduced.

Differences in CPP of E + poly(E-co-methacrylate) and of E + poly(E-co-methacrylic acid) systems may be determined for copolymers of identical polymer microstructure and molecular weight distribution by applying polymer modification reactions which, e.g., trans-

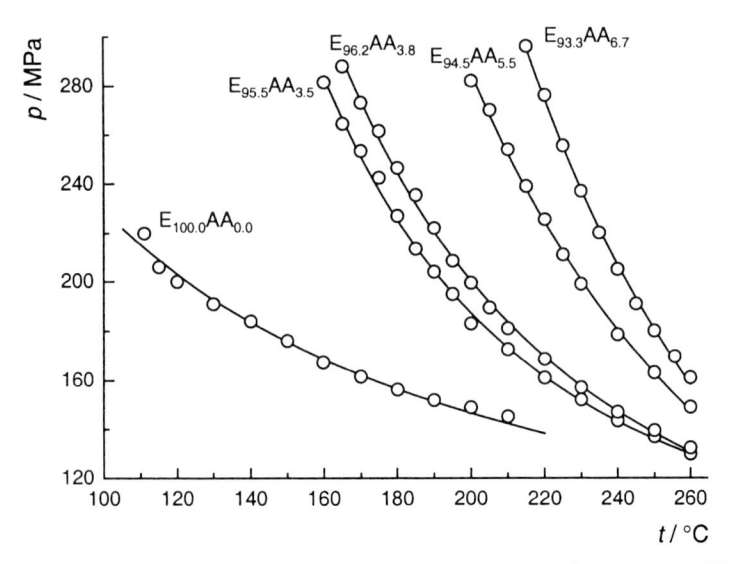

Figure 20.8: Cloud point curves of E + acrylic acid (AA) copolymers in ethene (Latz, 1999). The polyethylene data refer to material with molecular weight being close to the ones of the acrylic acid copolymers. The subscripts characterize copolymer composition in mol%. The copolymers were synthesized in CSTR1. The CPPs were measured in the discontinuously operated device shown in Fig. 20.3 at polymer concentrations of 5 mass%.

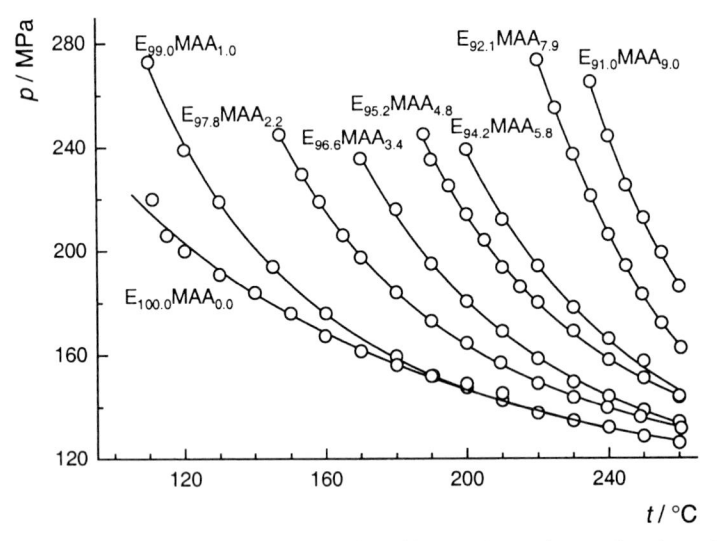

Figure 20.9: Cloud point curves of E + methacrylic acid (MAA) copolymers in ethene (Latz, 1999). The polyethylene data refer to material with molecular weight being close to the ones of the methacrylic acid copolymers. The subscripts characterize copolymer composition in mol%. The copolymers were synthesized in CSTR1. The CPPs were measured in the discontinuously operated device shown in Fig. 20.3 at polymer concentrations of 5 mass%.

20.4 Results and Discussion

form –COOH groups into –COOCH$_3$ groups. With E + AA copolymers, this polymer modification reaction may be associated with minor variation of polymer microstructure whereas no such change seems to occur with E + MAA copolymers (Lacik et al., 2001). Figure 20.10 shows CPP curves for an E + poly(E$_{93.3}$AA$_{6.7}$) system and for an E + poly(E$_{93}$-MAA$_{07}$) system. The two copolymers have been obtained by synthesis in CSTR1. According to the literature procedure (Möller, 2000), both samples have also been esterified to yield an E$_{93.3}$MA$_{6.7}$ and an E$_{93}$MMA$_7$ copolymer, respectively. The CPP curves for the associated poly(E$_{93.3}$MA$_{6.7}$) and E + poly(E$_{93}$MMA$_{07}$) systems have also been measured. As can be seen from Figure 20.10 and as is to be expected from data as in Figure 20.4, the E + MA copolymer (from polymer modification reaction) has a significantly lower CPP pressure than the associated E + AA copolymer. Comparison of the CPPs for the E + MA and the E + methyl methacrylate (MMA) systems indicates that the α-methyl group enhances solubility also for ester moieties, although to a weaker extent than with acid comonomers.

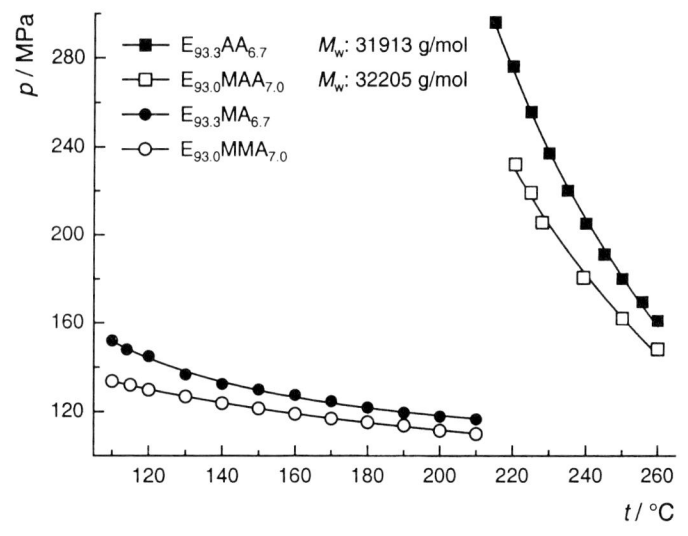

Figure 20.10: Cloud point pressure curves in ethene of an E$_{93.3}$AA$_{6.7}$ copolymer (full squares) and an E$_{93}$MAA$_{07}$ copolymer (open squares). Shown as circles are the cloud point pressures measured after methyl esterification of the ethene + methacrylic acid copolymers which yields an E$_{93.3}$MA$_{6.7}$ copolymer (full circles) and an E$_{93}$MMA$_{07}$ copolymer (open circles), respectively; data from Latz (1999).

The data presented so far refer to effects on CPP (in mixtures with ethene) resulting from variations of the type and of the content of comonomer units in the ethene copolymer. As the solution behavior is controlled by the interplay of intra-segmental interactions of polar groups within the copolymer and of interactions of the polar segments with the solvent, also variation of solvent polarity should result in major effects. An interesting and also technically relevant measure of the impact of solvent polarity on cloud point pressure is provided by investigations into the variation of CPP with comonomer content, f (in mol%). The specific cloud point pressure change, SCPPC, represents the change in CCP induced by an enhancement of comonomer content of the (monomer-comonomer) solvent mixture by one

463

mole percent. With the ethene + acrylate copolymers studied so far, the CPPs always decreased upon increasing f. The acrylate monomers thus are acting as cosolvent materials.

Shown in Figure 20.11 are CPP curves of an ethene + butyl acrylate copolymer, $E_{78}BA_{22}$, that is contained at a concentration of 5 mass% in E + BA monomer mixtures containing up to f_{BA} = 3.6 mol% BA. Rather modest BA monomer concentrations largely affect CPP. The SCPPC is slightly more pronounced at lower temperature. As detailed elsewhere (Dietzsch, 1999), the CCP drop with BA content, f_{BA}, closely follows a linear correlation in the concentration range under investigation, up to f_{BA} = 4 mol%. Thus a single number for SCPPC is sufficient to characterize CPP as a function of f_{BA}. For a few copolymer materials, CPP vs. f_{BA} correlations have been studied up to 10 mol% and linear dependences have been found even for this extended range. It goes without saying that this linearity should only be used for interpolation within the f range of the underlying experiments.

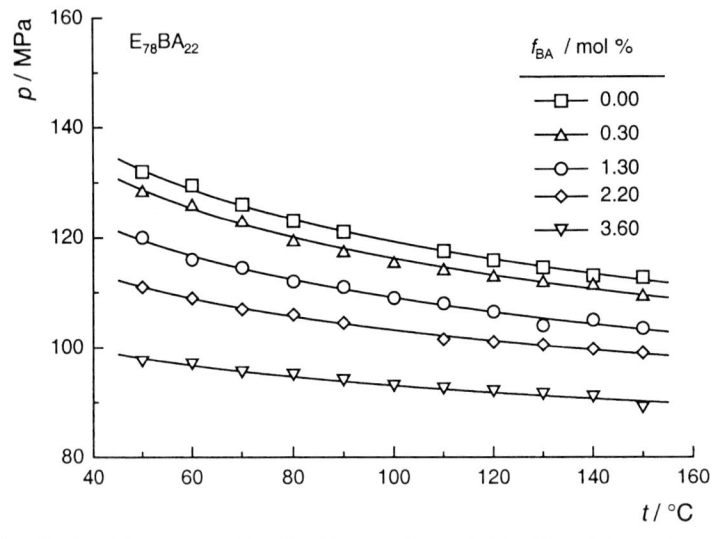

Figure 20.11: Cloud point pressures of an $E_{78}BA_{22}$ copolymer in E + BA mixtures of varying BA content, f_{BA}. Data are presented for temperatures between 50 and 150 °C.

An extended series of SCPPC data for ethene + butyl acrylate copolymers, with BA copolymer contents up to F_{BA} = 25 mol%, is presented in Figure 20.12. The SCPPC is expressed in terms of the variation of CPP with comonomer content, $\partial p/\partial f_{BA}$. The full symbols indicate experimental data deduced from the continuously operated device that includes two CSTRs (Fig. 20.1) whereas the open symbols are from experiments where copolymer is produced in CSTR1 and the CPPs are mapped out in the discontinuously operated high-pressure cell shown in Figure 20.3. The data sets from the two types of experiments are in excellent agreement. Within the experiments underlying the data of Figure 20.12, the BA content of the monomer mixture was extended up to about f_{BA} = 4 mol%. This range fully covers the technically relevant area. It should be noted that the reactivity ratios of E + BA free-radical copolymerization are such that, e.g., reaction of an E + BA monomer mixture with 4 mol% BA at 200 °C yields a copolymer which contains close to F_{BA} = 40 mol% BA units.

20.4 Results and Discussion

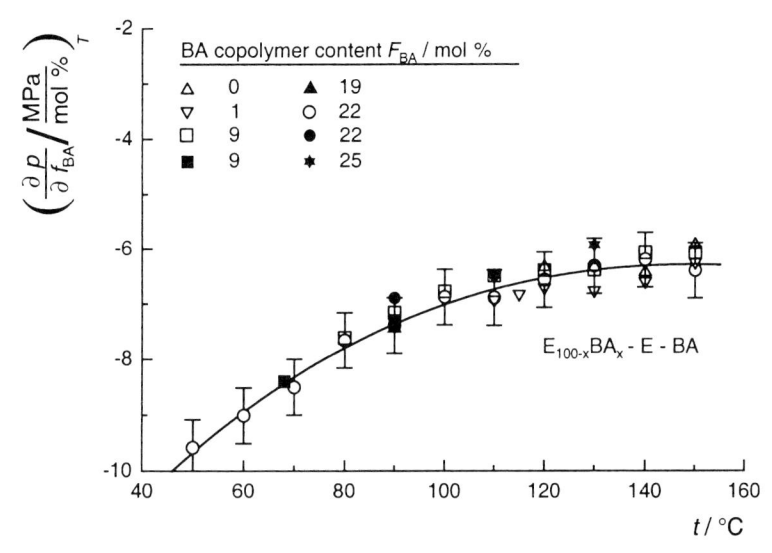

Figure 20.12: Temperature dependence of $\partial p/\partial f_{BA}$, the specific cloud point pressure change (SCPPC), for the E + BA + poly(E–BA) system. The maximum BA copolymer content is 25 mol%. Open symbols refer to CPP data obtained by the synthetic method (with the device shown in Fig. 20.3) whereas the filled symbols are from experiments using the continuously operated device depicted in Fig. 20.1.

The specific CPP drop increases toward lower temperature and exceeds 9 MPa per mol% BA at 50 °C and high comonomer content, $F_{BA} = 22$ mol%. Above 120 °C, the specific CPP drop approaches a value of 6 MPa per mol% BA. It is particularly noteworthy that the SCPPC at the higher temperatures is insensitive toward BA copolymer content. The SCPPC data for BA copolymer contents between $F_{BA} = 0$ and 25 mol% more or less fall on top of each other. The weak temperature dependence and, more importantly, the insensitivity of SCPPC toward F_{BA} suggest that polar interactions are not dominant at the higher temperatures.

On the other hand, the clear variation of SCPPC with temperature, seen below 100 °C, indicates that polar segment-solvent interactions operate at lower temperature. SCPPC data for polyethylene and for copolymer that contains small amounts of BA are not yet available for temperatures below 100 °C. Such data are difficult to obtain because of the associated very high CPPs.

Both increasing copolymer BA content, F_{BA}, at constant f_{BA}, as has been demonstrated for $f_{BA} = 0$ (Fig. 20.4), and increasing f_{BA} at constant F_{BA} (Figs. 20.11 and 20.12) induce a drop in CPP. Within the narrow region of f_{BA} variation, up to 4 mol% BA, the correlation of CPP with f_{BA} is linear and thus may be represented by a single number of SCPPC. The variation of CPP with F_{BA} has been studied over the entire range from polyethylene to poly(BA). As can be seen from Figure 20.7, no linear CPP vs. F_{BA} correlation holds within this wide composition range. On the other hand, fitting the CPP vs. F_{BA} data for the narrow range between $F_{BA} = 0$ and 10 mol% to a straight line poses no problem. Comparison of the specific CPP drop per mol% of F_{BA}, obtained from the E + BA + poly(E-co-BA) data at low F_{BA} (Fig. 20.7), with the SCPPC data in Figure 20.12 indicates that the lowering of CPP associated with increasing f_{BA} exceeds the corresponding effect brought upon by increasing F_{BA} by about a factor of two. Within this comparison it is assumed that the SCPPC at 200 °C is close to the value measured for 150 °C.

The specific lowering of CPP for the E + MA + poly(E-co-MA) system measured at two copolymer contents, F_{MA} = 8 and 20 mol%, is illustrated in Figure 20.13. The data from experiments with CPP detection under flow conditions (full symbols) are in satisfactory agreement with those from phase behavior measurements under discontinuous conditions (Fig. 20.3). The results for E + MA + poly(E-co-MA) significantly differ from the E + BA + poly(E-co-BA) data (represented by the dashed line) in that the specific cloud point pressure drop is very small at the higher temperatures, but largely increases toward lower temperature. The SCPPCs of both systems are almost identical at 90 °C.

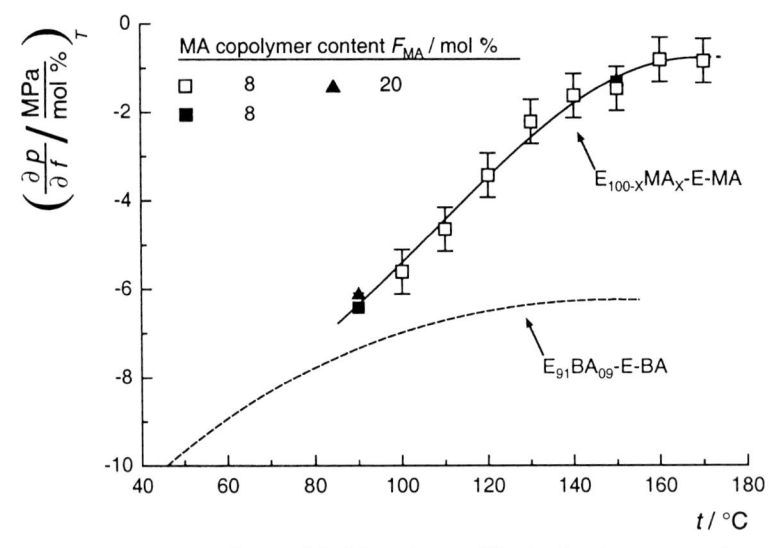

Figure 20.13: Temperature dependence of $\partial p/\partial f_{MA}$, the specific cloud point pressure change (SCPPC) for E + MA + poly(E–MA) systems with copolymer contents of 8 and 20 mol% MA units. Open symbols refer to CPP data obtained by the synthetic method (with the device shown in Fig. 20.3) whereas the filled symbols are from experiments using the continuously operated device depicted in Fig. 20.1. The E + BA + poly(E–BA) data from Fig. 20.12 are given for comparison (dashed line).

Only tentative explanations of the behavior shown in Figure 20.13 can be given at present. The low values of SCPPC seen with the E + MA + poly(E-co-MA) system at the higher temperatures are probably due to the small size of the MA molecule. The number of segments which may be used to characterize monomer size is much larger with BA than with MA. The entropic contributions provided by MA solvent molecules thus are significantly below the ones due to BA. As a consequence, under conditions where enthalpic effects are less important, BA has a much stronger solubilizing effect than MA.

That entropic effects originating from the size of solvent molecules strongly affect polymer solubility has been demonstrated by Chen et al. (1992) by investigations into poly(-ethene-co-propene) solubility. Cloud point pressure experiments at 130 °C on an $E_{92}MA_{08}$

20.4 Results and Discussion

copolymer (Dietzsch, 1999) support this argument. Summarized in Table 20.1 are the SCPPCs for this copolymer induced by MA, acetone, BA, and *n*-heptane. With the two smaller solvent molecules, SCPPC is around –2 MPa/mol% whereas for BA and *n*-heptane this number is close to –5 MPa/mol%.

Table 20.1: Specific cloud point pressure changes (SCPPCs) measured for an ethene + methyl acrylate copolymer ($E_{92}MA_{08}$) at 130 °C in several ethene + cosolvent mixtures.

Cosolvent	$\partial p/\partial f$ [MPa/mol%]
MA	– 2.25
acetone	– 2.09
BA	– 4.92
heptane	– 5.10

At lower temperature, polar segment-solvent interactions come into play. As is to be expected, this effect is much more pronounced with the E + MA + poly(E-*co*-MA) than with the E + BA + poly(E-*co*-BA) system. This is clearly seen from the E + MA + poly(E-*co*-MA) data at high F_{BA} and temperatures below 120 °C.

As was suggested by the data in Figure 20.7, the phase behavior of the E + poly(E-*co*-MA) system reflects intra-segmental interactions of MA units at high F_{MA} with this effect being particularly pronounced at lower temperature. Figure 20.14 demonstrates that the

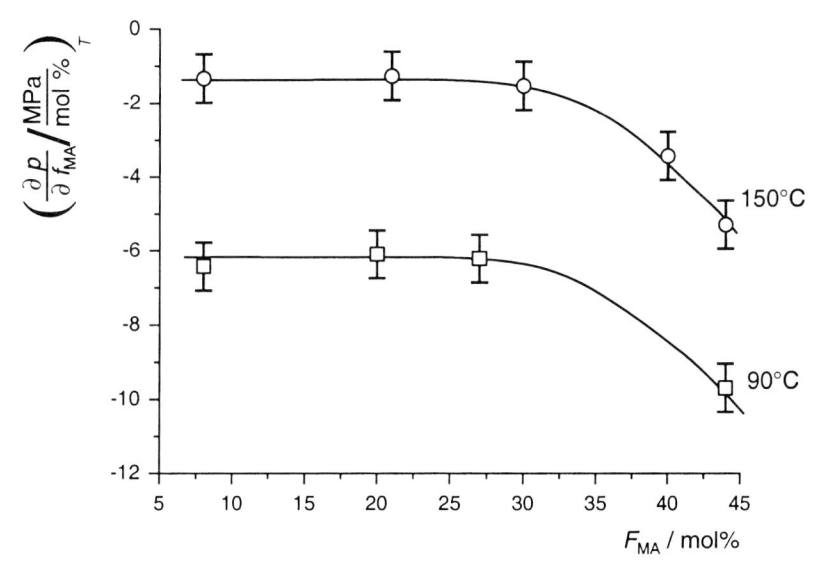

Figure 20.14: Variation with copolymer composition, F_{MA}, of $\partial p/\partial f_{MA}$, the specific cloud point pressure change (SCPPC), in E + MA + poly(E–MA) systems at 90 and 150 °C.

SCPPC data are also indicative of such an effect (Dröge, 1997). At F_{MA} contents above 30 mol% a strong increase in specific CPP drop is seen. At about the same copolymer composition, CPPs (Fig. 20.7) are enhanced upon further increase in F_{MA}. The observation in Figure 20.14 for higher F_{MA} is understood as being due to polar intra-segmental interactions of MA units being replaced by segment-solvent interactions which is associated with a significant enhancement of copolymer solubility.

20.5 Conclusions

The investigations into the CPP behavior of E + acrylate + poly(E-*co*-acrylate) systems reveal that, with BA being the acrylate component, entropic effects play a dominant role at higher temperature. Thus increasing the monomer mole fraction, f_{BA}, and increasing copolymer content, F_{BA}, both lower CPP. Enthalpic contributions come into play at lower temperatures (and higher BA comonomer contents). With the E + MA + poly(E-*co*-MA) system, in addition, intra-segmental interactions of the (less-shielded) methyl ester units on the polymer chain need to be taken into account. They are associated with a pronounced dependence of specific CPP drop on copolymer composition, F_{BA}. The effects of polar segment-solvent interactions observed at lower temperature are more pronounced with MA than with BA.

Combination of the reported CPP and SCPPC data provides access to a large amount of information on the E + acrylate + poly(E-*co*-acrylate) phase behavior. The copolymer material from reaction in a CSTR that is used within the present study is of excellent chemical homogeneity, which makes the measured CPP and SCPPC data perfectly suited for modeling purposes, e. g., by applying the statistical associating fluid theory (SAFT). The experimental studies should be extended to SCPPC determinations on ethene + methacrylic acid copolymers, but also further systematic studies into the effect of ester groups with different sizes of the alkyl moiety seem to be rewarding.

References

Banaszak, M.; Chen, C. K.; Radosz, M. (1996): Copolymer SAFT equation of state. Thermodynamic perturbation theory extended to heterobonded chains. *Macromolecules* **29**, 6481

Buback, M.; Busch, M.; Lovis, K.; Mähling, F.-O. (1994): Entwicklung eines kontinuierlich betriebenen Hochdruck-Hochtemperatur-Rührkessels mit Lichteinkopplung. *Chem.-Ing.-Tech.* **66**, 510

Buback, M.; Busch, M.; Lovis, K.; Mähling, F.-O. (1995): Mini-Technikumsanlage für Hochdruck-Polymerisationen bei kontinuierlicher Reaktionsführung. *Chem.-Ing.-Tech.* **67**, 1652

Buback, M.; Busch, M.; Dietzsch, H.; Dröge, T.; Lovis, K. (1996): Cloud point curves in ethylene-acrylate-poly(ethylene-*co*-acrylate) systems. *Process Technology Proceedings* **12**, „High Pressure Chemical Engineering", R. v. Rohr, Ch. Trepp (Hrsg.), Elsevier

Buback, M.; Busch, M.; Lovis, K.; Mähling, F.-O. (1996a): High pressure free radical copolymerization of ethene and butyl acrylate. *Macromol. Chem. Phys.* **197**, 303

Buback, M.; Busch, M.; Panten, K.; Vögele, H.-P. (1992): Direkte spektroskopische Bestimmung von Verweilzeitverteilungen bei Fluidphasenreaktionen bis zu hohem Druck. *Chem.-Ing.-Tech.* **64**, 352

Buback, M.; Dröge, T.; van Herk, A.; Mähling, F.-O. (1996b): High-pressure free-radical copolymerization of ethene and 2-ethylhexyl acrylate. *Macromol. Chem. Phys.* **197**, 4119

Byun, H.-S.; Hasch, B. M.; McHugh, M. A.; Mähling, F.-O.; Busch, M.; Buback, M. (1996): Poly(ethylene-*co*-butyl acrylate). 1. Phase behaviour in ethylene compared to the poly(ethylene-*co*-methyl acrylate)-ethylene system. 2. Aspects of copolymerization kinetics at high pressures. *Macromolecules* **29**, 1625

Dietzsch, H. (1999): Hochdruck-Copolymerisationen von Ethen und (Meth)Acrylsäureestern: Entmischungsverhalten der Systeme Ethen/Cosolvens/poly(Ethen-*co*-Acrylsäureester; Kinetik der Ethen-Methylmethacrylat-Copolymerisation. *Ph.D. Thesis*, Göttingen

Dohrn, R.; Brunner, G. (1995): High pressure fluid phase equilibria: Experimental methods and systems investigated (1988–1993). *Fluid Phase Equilibria* **106**, 213

Dröge, T. (1997): Radikalische Hochdruck-Copolymerisation von Ethen mit (Meth)Acrylsäureestern: Kinetik und Phasenverhalten. *Ph.D. Thesis*, Göttingen

Folie, B.; Gregg, C.; Luft, G.; Radosz, M. (1996): Phase equilibria of poly(ethylene-*co*-vinyl acetate) copolymers in subcritical and supercritical ethylene and ethylene-vinyl acetate mixtures. *Fluid Phase Equilibria* **120**, 11

Folie, B.; Radosz, M. (1995): Phase equilibria in high-pressure polyethylene technology. *Ind. Eng. Chem. Res.* **34**, 1501

Fornari, R. E.; Alessi, P.; Kikic, I. (1990): High pressure fluid phase equilibria: experimental methods and systems investigated. *Fluid Phase Equilibria* **57**, 1

Hasch, B. M.; McHugh, M. A. (1991): Phase behavior of the ethylene-methyl acrylate system. *Fluid Phase Equilibria* **64**, 251

Hasch, B. M.; McHugh, M. A. (1995): Calculating poly(ethylene-*co*-acrylic acid)-solvent phase behavior with the SAFT equation of state. *J. Polym. Sci.: Part B: Polym. Phys.* **33**, 715

Hasch, B. M.; Meilchen, M. A.; Lee S.-H.; McHugh, M. A. (1993): Cosolvency effects on copolymer solutions at high pressure. *J. Polym. Sci.: Part B: Polym. Phys.* **31**, 429

Lacik, I.; Beuermann, S.; Buback, M. (2001): Aqueous phase size-exclusion-chromatography used for PLP-SEC studies into free-radical propagation rate of acrylic acid in aqueous solution. *Macromolecules* **34**, 6224

Latz, H. (1999): Entmischungsverhalten der Systeme Ethen/poly(Ethen-*co*-Acrylsäure) und Ethen/poly(Ethen-*co*-Methacrylsäure) bis zu hohen Drücken und Temperaturen. *Diploma Thesis*, Göttingen

Lee, S.-H.; LoStracco, M. A.; McHugh, M. A. (1994): High-pressure, molecular weight-dependent behavior of (co)polymer-solvent mixtures: Experiments and modeling. *Macromolecules* **27**, 4652

Lee, S.-H.; LoStracco, M. A.; McHugh, M. A. (1996): Cosolvent effect on the phase behavior of poly(ethylene-*co*-acrylic acid) butane mixtures. *Macromolecules* **29**, 1349

Luft, G.; Subramanian, N. S. (1987): Phase behavior of mixtures of ethylene, methyl acrylate, and copolymers under high pressures. *Ind. Eng. Chem. Res.* **26**, 750

Luft, G.; Wind, R. W. (1992): Phase behavior of ethylene/ethylene-acrylic acid copolymer mixtures under high pressure. *Chem.-Ing.-Tech.* **64**, 1114

Lora, M.; Rindfleich, F.; McHugh, M. A. (1999): Influence of the alkyl tail on the solubility of poly (alkyl acrylates) in ethylene and CO_2 at high pressures: Experiments and modeling. *J. Appl. Polym. Sci.* **73**, 1979

LoStracco, M. A.; Lee, S.-H.; McHugh, M. A. (1994): Comparison of the effect of density and hydrogen bonding on the cloud-point behavior of poly(ethylene-*co*-methyl acrylate)-propane-cosolvent mixtures. *Polymer* **35**, 3272

Lovis, K. (1995): Photoinduzierte radikalische Hochdruckpolymerisation von Ethen bei kontinuierlicher Reaktionsführung. *Ph.D. Thesis*, Göttingen

Meilchen, M. A.; Hasch, B, M.; Lee, S.-H.; McHugh M. A. (1992): Poly(ethylene-*co*-methyl acrylate)-solvent-cosolvent phase behavior at high pressures. *Polymer* **33**, 1922

Möller, O. H. (2000): Umsetzungen von Ethen-Acrylat-Copolymeren in nahekritisch fluidem Wasser bei Variation von Druck und Temperatur. *Ph.D. Thesis*, Göttingen

Müller, C.; Oellrich, L. R. (1996): The influence of different inhibitors and ethylenes of different origin on the location of cloud points due to thermal polymerization of ethylene. *Acta Polym.* **47**, 404

Schneider, G. M. (1972): Phase behavior and critical phenomena in fluid mixtures under pressure. *Ber. Bunsen-Ges. Phys. Chem.* **76**, 325

Schneider, G. M. (1994): Physico-chemical properties and phase equilibria of pure fluids and fluid mixtures at high pressures. *J. Supercrit. Fluids*, 91

Simnick, J. J.; Lawson, C. C.; Lin, H. M.; Chao, K. C. (1977): Vapor-liquid equilibrium of hydrogen-tetralin system at elevated temperatures and pressures. *AIChE Journal* **23**, 469

Wittkowski, L. (1998): Experimentelles Studium und Modellierung der radikalischen Hochdruck-Polymerisation von Ethen und (Meth)Acrylsäure. *Ph.D. Thesis*, Göttingen

Nomenclature

AA	acrylic acid
BA	butyl acrylate
CPP	cloud point pressure
CSTR	continuously stirred tank reactor
E	ethene
EHA	2-ethyl hexyl acrylate
f_X	mole fraction of monomer X
F_X	mole fraction of comonomer units from X in the copolymer
IR	infrared
MA	methyl acrylate
MAA	methacrylic acid
MMA	methyl methacrylate
NIR	near infrared
poly(X-*co*-Y)	copolymer containing units from monomers X and Y
poly(X_{xx}-*co*-Y_{yy})	copolymer with xx mol% of monomer X units and yy mol% of monomer Y units
p	pressure
SAFT	statistical associating fluid theory
SCPPC	specific cloud point pressure change
t	temperature in °C
T	temperature in K
$X_{xx}Y_{yy}$	copolymer with xx mol% of monomer X units and yy mol% of monomer Y units
X–Y–poly(X-*co*-Y)	mixture of monomers X and Y together with a copolymer composed of units from monomers X and Y

Greek letters

$\partial p/\partial f_X$	specific cloud point pressure change induced by the cosolvent X

Subscripts

AA	acrylic acid
BA	butyl acrylate
E	ethene
EHA	2-ethyl hexyl acrylate
MA	methyl acrylate
MAA	methacrylic acid
MMA	methyl methacrylate
poly(X_{xx}-co-Y_{yy})	copolymer with xx mol% of monomer X units and yy mol% of monomer Y units
$X_{xx}Y_{yy}$	copolymer with xx mol% of monomer X units and yy mol% of monomer Y units

21 Gas Expanded Polymer Solutions

Bernd Bungert and Wolfgang Arlt [1]

21.1 Abstract

By using thermodynamic phenomena in systems with dense gases, product properties such as morphology, size of particles, and fibers can be tailored to industrial needs. Separations of complex mixtures can be performed. The thermodynamical basic properties were assessed in this project for the sample system polystyrene + cyclohexane + carbon dioxide with monodisperse polymer ($M_w/M_n = 1.09$) at temperatures up to 250 °C and pressures up to 16 MPa. Small amounts of carbon dioxide shift the lower critical solution temperature (LCST) to dramatically lower temperatures. When looking at the pressure-concentration representtation, there is a striking similarity between the effect of temperature and the effect of gas content on the shape of the two-phase region. Pressure has a pronounced effect on the chain partitioning of polystyrenes of different molecular weights in the two liquid phases.

21.2 Introduction

Compressed gases can cause the following phenomena in solutions of polymers and solvent (Bungert, 1998):

- adjustment of the solvent power with the gas as an antisolvent
- control of supersaturation and thus kinetics of phase transitions by the parameters pressure and/or antisolvent concentration
- change of properties by dissolution of compressed gases: swelling of polymers, decrease of glass transition points, and melting points

---

[1] Technische Universität Berlin, Institut für Verfahrenstechnik, Fachgebiet Thermodynamik und Thermische Verfahrenstechnik, Straße des 17. Juni 135, 10623 Berlin.

21.2 Introduction

- reduction of viscosities and increase of mass transfer and diffusion coefficients in the presence of compressed gases

Many technical applications are demonstrated in Table 21.1.

Table 21.1: Processes using compressed or supercritical gases.

Abbreviation	Process	Related conventional process	Phenomena[a]	Application
GAS (GAS batch)	gas antisolvent precipitation	drowning-out crystallization	I, II	microparticles
PCA (GAS contin.)	precipitation with a compressed fluid antisolvent	spray drying, drowning-out crystallization, SCF extraction	I, II	microparticles
SEDS	solution-enhanced dispersion of solids	spray drying, drowning-out crystallization, SCF extraction	I, II	microparticles
RESS	rapid expansion from supercritical solution	spray drying	I, II	microparticles
PGSS	particle generation from gas-saturated solution		II, III, IV	microparticles
SAS	supercritical antisolvent fractionation	extraction	I, II, III, IV	separation
nonwovens			I, II, IV	microfibers
TIPS	thermal-induced phase separation		I, II, III, IV	microfoams
UNICARB	Union Carbide spray painting process	spray drying	IV	spray coating
SFI	supercritical fluid infiltration		II, IV	dyeing of polymer fibers
	polymerization in SCF		I, IV	reaction influenced by phase equilibria

[a] I, adjusting of solvent power; II, kinetics of phase transition; III, swelling of polymers, decreasing of melting points and glass-transition points; IV, reduced viscosity, increased diffusion and mass transfer.

The scope of this project was the investigation of the influence of the compressed gas CO_2 on solutions of the polymer polystyrene in the solvent cyclohexane. All systems containing small and large molecules show an interesting feature: Besides a liquid-liquid demixing at low temperatures (UCST) and a complete miscibility at medium temperatures, another liquid demixing appears at higher temperatures (LCST). This principle behavior is shown in Figure 21.1.

It was the aim of this project to show the influence of compressed gases to the shape of this diagram. Further, the distribution of molar mass of the polymer in the different phases should be determined.

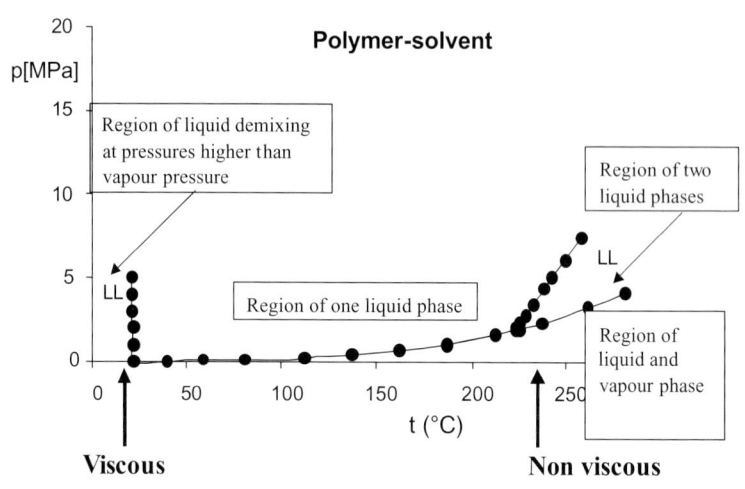

Figure 21.1: General phase behavior for a polymer solution in pressure-temperature-composition space.

21.3 Fundamentals

21.3.1 Separation Processes

21.3.1.1 Antisolvent Crystallization

Antisolvent crystallization with compressed gases is performed quite analogously to the conventional process. The addition of an antisolvent to the solution of a crystallizable compound reduces the solubility. When the saturation conditions are reached, crystallization occurs. When using a compressed gas as the antisolvent, an increased pressure is needed to keep the gas in solution as follows from the solvent-gas binary vapor-liquid equilibrium. With increasing pressure and thus gas concentration in the solution, the solubility of the solid decreases and approaches the solubility level in the pure gas as a limiting value. At the same time a volume increase occurs that is only moderately dependent on the density of the solvent-gas mixture but rather depends simply on the amount of dissolved gas. Nevertheless, the term "expansion" has been widely used for this phenomenon. A good solvent-antisolvent combination thus follows the criteria of high solubility of solid and gas in the solvent and little solubility of the solid in the gas. A detailed study on the solvent-antisolvent selection has been given by Kordikowski et al. (1994).

The particle size achieved in supercritical antisolvent crystallization depends strongly on the degree of supersaturation. The latter is the key parameter to influence critical particle size and number of critical nuclei. A direct measure for the degree of supersaturation is the gas pressure or the antisolvent concentration. By the possibility of high pressure-buildup rates, extremely high degrees of supersaturation can be achieved. Therefore, the gas pressure can be used to control the particle size as well as the particle size distribution.

21.3.1.2 Liquid-Liquid-Phase Separation of Polymer Solutions

Almost all polymer solutions show a region of two liquid phases at low temperatures (upper critical solution temperature, UCST) due to enthalpic effects. When increasing the temperature, a homogeneous solution forms. Upon further heating of the system the volume of the solvent increases since it is approaching its critical point while the polymer, being far below its hypothetical critical point, essentially does not expand. This difference in "free volume" eventually leads to a liquid-liquid-phase separation called the lower critical solution temperature (LCST). The presence of a dissolved gas as a nonsolvent is comparable to an increase in free volume caused by a temperature increase. This way the "solvent power" can be continuously adjusted. As a result, the LCST is shifted to lower temperatures with increasing amounts of dissolved gas while the UCST is only slightly influenced. Finally, both miscibility gaps merge to one region of two liquid phases over the whole temperature range. Figure 21.1 shows this behavior qualitatively in the pressure-temperature composition space.

The LCST depends on the molecular mass of the polymer: it is shifted to lower temperatures and higher pressures with increasing molecular mass. A solution of a polymer in a solvent is only a quasibinary system since the polymer underlies a chain-length distribution. Therefore, when liquid-liquid demixing of polymer solutions occurs, this is always accompanied by a partitioning of short and long chains onto the polymer-lean and the polymer-rich phases, respectively. This partitioning is strongly influenced by the mass balance. Since the miscibility gap for a polymer solution usually is very asymmetric, the latter can be translated into the "depth of penetration" into the miscibility gap.

In the two-phase region the binodal limits the homogeneous and inhomogeneous regions whereas the spinodal marks the transition from the metastable to the unstable region. In the critical point both lines coincide. While in the unstable region phase separation occurs immediately, in the metastable region phase separation takes place by nucleation and growth when there is sufficient supersaturation. At low polymer concentrations nuclei of a second polymer-rich phase form in a continuous solvent-rich phase. For high polymer concentrations nuclei of a second polymer-lean phase form. When the unstable region is reached – typically close to the critical concentration – a continuous phase separation occurs.

21.3.1.3 Liquid-Liquid-Vapor to Vitrified Liquid-Vapor Equilibrium

While dissolved crystalline solids can be precipitated in one step by increasing the gas pressure, for most polymers there first occurs a liquid-liquid (LL) phase separation before at sufficiently high gas concentrations the liquid polymer-rich phase vitrifies. With the help of Figure 21.2 the formation of solid polymer particles upon addition of a gaseous antisolvent to a polymer solution can be explained. The cross-hatched area denotes those conditions where the gas and the solvent concentrations in the polymer-rich phase are no longer sufficient to plasticize the polymer (i.e., where the polymer is below its glass-transition conditions). The path of gas addition is marked by an arrow. Starting from the binary polymer solution, adding more and more gas to the system increases the pressure and at the same time changes the phase behavior dramatically. First, only a homogeneous liquid and a vapor or gas phase are present (a). At this state the actual tie line does not touch the

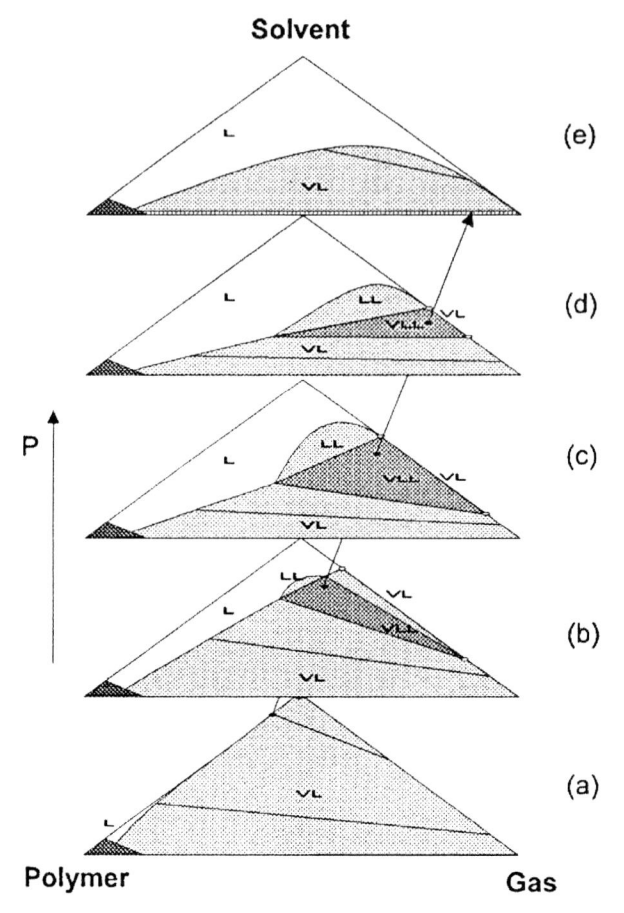

Figure 21.2: Pressure dependence of the phase behavior of a ternary polymer + solvent + gas system at constant temperature. (The arrow shows the path followed when adding a compressed gas to a binary polymer solution. The hatched area at high polymer concentrations denotes the conditions where the polymer-rich phase vitrifies.)

vitrified area. When the pressure is high enough to dissolve sufficient amounts of gas – as dictated by the binary vapor-liquid equilibrium – LL-phase separation is induced and a VLL three-phase area evolves (b, c). Upon further addition of the gas, the solvent + gas binary system approaches its critical point (d) until above the binary critical point the polymer-lean phase vanishes (e). The solvent of the polymer-rich phase is now being extracted until the gas and solvent concentrations are sufficiently low to cause vitrification of that phase.

21.3.1.4 Change of Properties by the Dissolution of Compressed Gases

A number of properties of solids and solutions can be changed by dissolution of compressed gases. These effects are being exploited in a beneficial way by the processes discussed here. Dissolved gases have a plasticizing effect on amorphous polymers. While hydrostatic pressure usually increases the glass transition temperature (T_g), there can be a minimum in T_g when the pressurizing medium is a gas that dissolves in the polymer. Detailed studies have been given by Chow (1980), Wang et al. (1982), Wissinger and Paulaitis (1987), Condo and Johnston (1992), and Condo et al. (1992). Since dissolved gas enhances the chain mobility, it can change the degree of crystallinity (Lambert and Paulaitis, 1991). Also, it decreases the crystallization temperature of melts of polymers as well as of nonpolymeric materials (Weidner et al., 1996; Freund and Steiner, 1997). As more gas dissolves in the polymer, considerable swelling occurs. While in the unswollen state the diffusion coefficient for large molecules such as additives or dyes is negligible, it increases by orders of magnitude in the swollen state. Then, it is almost independent of the size of the penetrating molecule (Berens et al., 1992). It is obvious that gases dissolved in melts or solutions reduce the viscosity of these systems. A decrease by an order of magnitude or even more is possible (Nielsen et al., 1991; Lee et al., 1990). A systematic study of the viscosity of homogeneous mixtures of poly(dimethylsiloxanes) with CO_2 has been performed by Mertsch and Wolf (1994).

21.3.2 Thermodynamic Modeling of Phase Equilibrium

The prior discussed phase diagrams at high pressure can be modeled using the thermodynamic phase equilibrium conditions. For this purpose an equation of state is needed which is suitable to describe the systems of interest.

The starting point is the equality of chemical potentials

$$T^\alpha = \ldots = T^{Ph}$$
$$p^\alpha = \ldots = p^{Ph} \qquad (21.1\text{--}21.3)$$
$$\mu_i^\alpha = \ldots = \mu_i^{Ph}$$

with $\alpha = 1 \ldots$ Ph phases und $i = 1 \ldots$ K components.

The equality of chemical potentials is equal to the equality of fugacities:

$$\mu_i(T, p, x_i) = \mu_i^+(T, p^+) + RT \ln\left(\frac{f_i}{f_i^+}\right) \qquad (21.4)$$

Fugacities are not subject of thermodynamical models (equation of states) but fugacitiy coefficients as defined by Eq. (21.5)

$$\varphi_i = \frac{f_i}{x_i P} \qquad (21.5)$$

Since many of the applications comprise high-molecular-weight molecules or polymers, the equation of state has explicitly to account for the deviation of the molecules from spherical shape. This can be done by a segmentation of the long-chain molecules in the lattice theory (e.g., Flory et al., 1964). To account for density changes with pressure, lattice-hole theories were developed (Sanchez and Lacombe, 1976; Kleintjens and Koningsveld, 1980; Yoo and Lee, 1996; Yoo et al., 1996). More recently, starting from the pioneering work of Prausnitz' group, a number of perturbation theories were developed, which consider the chain formation of a molecule as perturbation of hard-sphere behavior, e.g., PHCT (Beret and Prausnitz, 1975), SAFT (Huang and Radosz, 1990, 1991), PHSC (Song et al., 1994). These theories were successfully applied to model systems containing long-chain molecules and supercritical substances. Examples are the systems polystyrene + (propane or ethane) (Pradhan et al., 1994), polyethylene + (butane or pentane) (Xiong and Kiran, 1995), poly(ethylene-*co*-acrylic acid) + alkanes (Hasch and McHugh, 1995), polyethylene + hexene + ethylene (Chen et al., 1993, 1994), and polystyrene + cyclohexane + carbon dioxide (Behme et al., 1997). Moreover, well-known cubic equations of state such as the van der Waals EoS or the Peng-Robinson EoS were used for the description of polymer systems (e.g., Kontogeorgis et al., 1994; Orbey and Sandler, 1994). A cubic equation of state which was especially developed for application to polymer systems is the Sako-Wu-Prausnitz EoS (Sako et al., 1989). It was recently proved to be very suitable for the description of high-pressure phase equilibria with long-chain *n*-alkanes and supercritical polymer solutions as well (Tork et al., 1999 a,b).

A special problem occurs for polymers, because these chemical substances have a molecular mass distribution. This distribution can be modeled by a distribution function or by moments of the distribution function. In general, only three related moments are sufficient to characterize a polymer. Equation (21.6) shows the general formulation of the k. moment.

$$M^k = \int w(M) M^k \, dM \qquad (21.6)$$

It follows

$$M_n = M^1$$
$$M_w = \frac{M^2}{M^1} \qquad (21.7\text{a–c})$$
$$M_z = \frac{M^3}{M^2}$$

21.4 Experimental

21.4.1 Apparatus and Chemicals

The experiments were performed in a variable-volume autoclave of about 1000 ml designed for vapor-liquid-liquid (VLL) and vapor-liquid-solid (VLS) experiments at pressures up to 20 MPa and temperatures up to 250 °C. Details are given by Bungert (1997 b). Experiments were performed using cyclohexane (Fluka, purity >99.5%) and carbon dioxide (Linde, technical grade, purity >99.5%). The solvent was dried over molecular sieves. The water content was checked by Karl-Fischer-analysis. The polystyrenes were ionically synthesized samples that were obtained from BASF (PS40k: M_w = 42,700 g/mol, M_n = 37,800 g/mol; PS100k: M_w = 101,400 g/mol, M_n = 93,000 g/mol; PS160k: M_w = 159,600 g/mol, M_n = 147,900 g/mol). Samples from the autoclave were placed into stainless steel containers while the pressure in the autoclave was kept constant by adding more hydraulic fluid into the metal bellows. The samples were weighed to better than 0.0005 g. After cooling the bombs to 0 °C, the gas content was determined by p, V, T measurement and additional weighing of the sample bombs. Polymer concentration was determined by stripping the solvent in a rotary evaporator under vacuum and weighing the flask. The molecular mass distribution was established by gel-permeation-chromatography.

21.5 Results

Figure 21.3 shows the phase behavior of the polystyrene + cyclohexane + carbon dioxide system in the pressure-temperature representation as determined in the present work (cf. Bungert et al., 1997a,b) as well as reported in the literature by de Loos (1994) and Saeki et

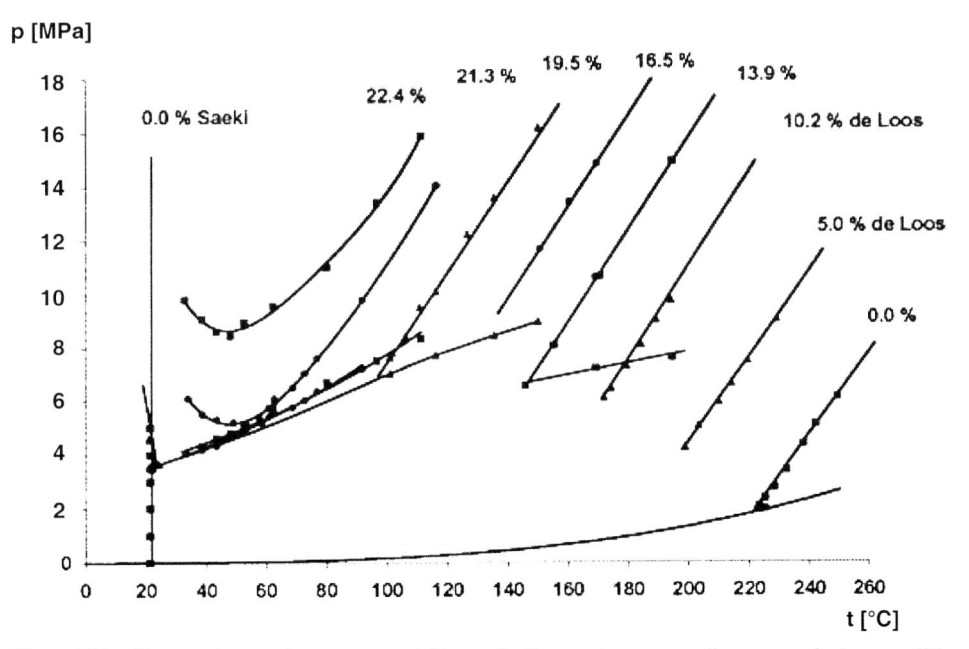

Figure 21.3: Pressure-temperature representation of the system monodisperse polystyrene (M_w 100,000 g/mol) + cyclohexane + carbon dioxide (Bungert et al., 1997a,b; de Loos, 1994; Saeki et al., 1975). % denotes mass% of CO_2.

21 Gas Expanded Polymer Solutions

al. (1975). Dissolution of increasing amounts of carbon dioxide shifts the LCST to low temperatures. The influence on the UCST is much less pronounced. The UCST is first shifted to slightly lower temperatures upon an increase of the carbon dioxide concentration. Only shortly before the merging of the two miscibility gaps at around 21.3 mass% CO_2 the UCST temperature rises to merge with the LCST. Similar results were found by Koak et al. (1998) for the UCST of the polystyrene + methylcyclohexane + carbon dioxide system. It is noteworthy that with increasing amounts of gas the lower critical end point (LCEP) pressure at first increases and then decreases to a relatively low value.

The ternary phase behavior of the system polystyrene 100k + cyclohexane + CO_2 at 170 °C and different pressures becomes evident when looking at Figure 21.4. In pure polystyrene only about 4 mass% CO_2 dissolves. When increasing amounts of cyclohexane are present, more gas dissolves in the system until finally the vapor-liquid-liquid three-phase area appears. This is dictated by the cyclohexane + CO_2 binary VLE and the LLE of the ternary system. The tie-lines in the LL-region are almost parallel with slightly less CO_2 in

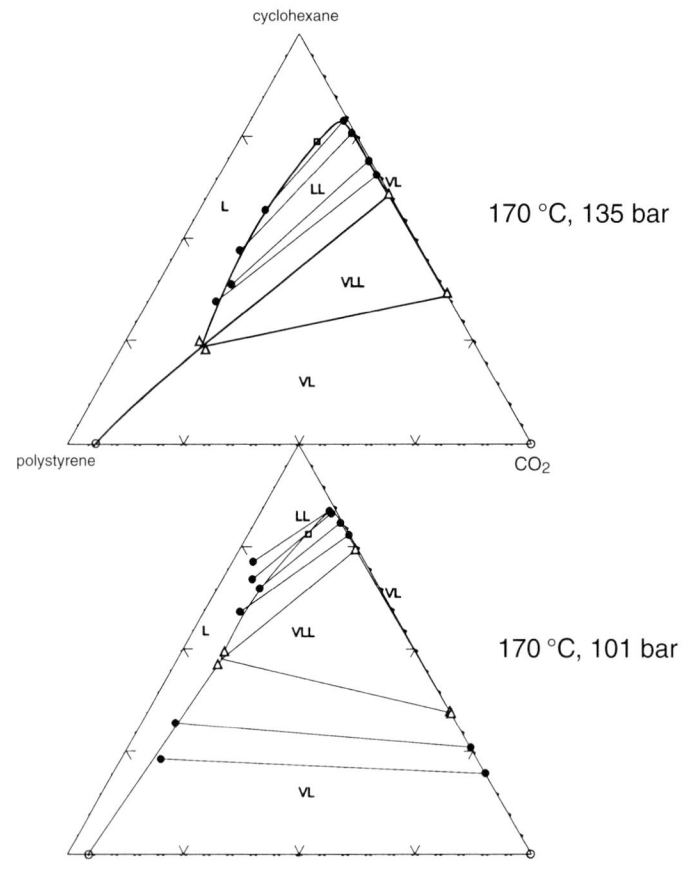

Figure 21.4: Ternary system polystyrene 100k + cyclohexane + CO_2 for different pressures (black circles and white triangles: samples; white squares: interpolated cloud pressures; white circles: calculated gas solubilities in the pure polymer melt).

21.5 Results

the polystyrene-rich phase. According to Figure 21.4 at 170 °C a minimum pressure of 6 MPa and a minimum gas content of 10 mass% CO_2 is required to cause LL-demixing and hence a three-phase area. Increasing pressures cause a decreasing two-phase region because the binary subsystem cyclohexane + CO_2 approaches its mixture critical point.

The samples of the liquid phases sampled in the experiments shown in Figure 21.4 were analyzed by gel-permeation chromatography for a 50:50 mixture of PS160k and PS40k. Figure 21.5 shows the molecular mass distributions. Starting of the bidisperse feed solution, the polystyrene-rich phase contains most of the long chains whereas the polystyrene-lean phase contains most of the short chains after phase separation at higher pressures. When lowering the pressure further, a light phase is obtained that is essentially lean of long chains. Since the fraction of this phase is going to low values, most of the polymer will be found in the heavy phase. Thus, the molecular mass distribution of that phase is almost identical to that of the feed. In the heavy phase, the data show a slight deviation from the initial distribution. This is most likely due to chain breaking in the course of the experiments.

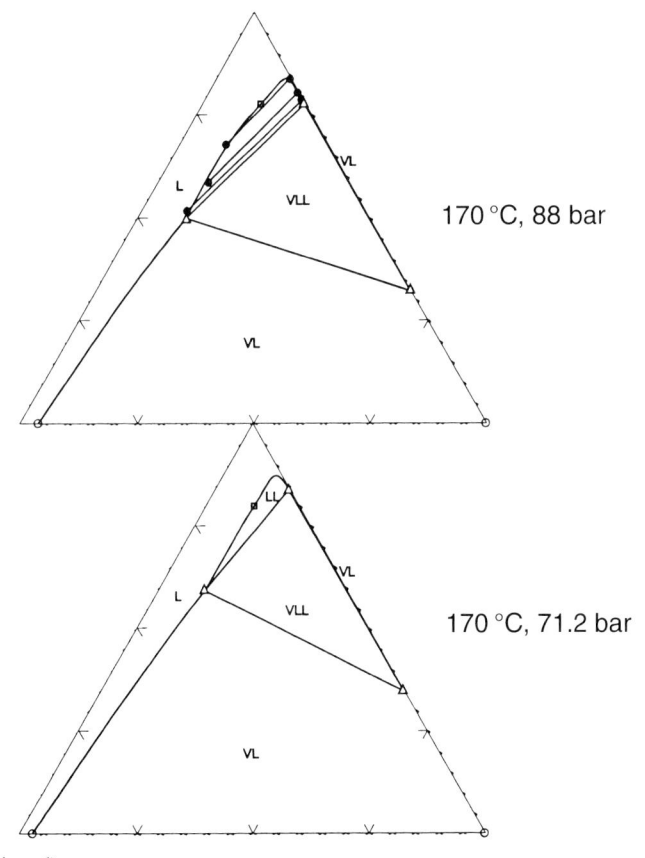

Figure 21.4 (continued)

21 Gas Expanded Polymer Solutions

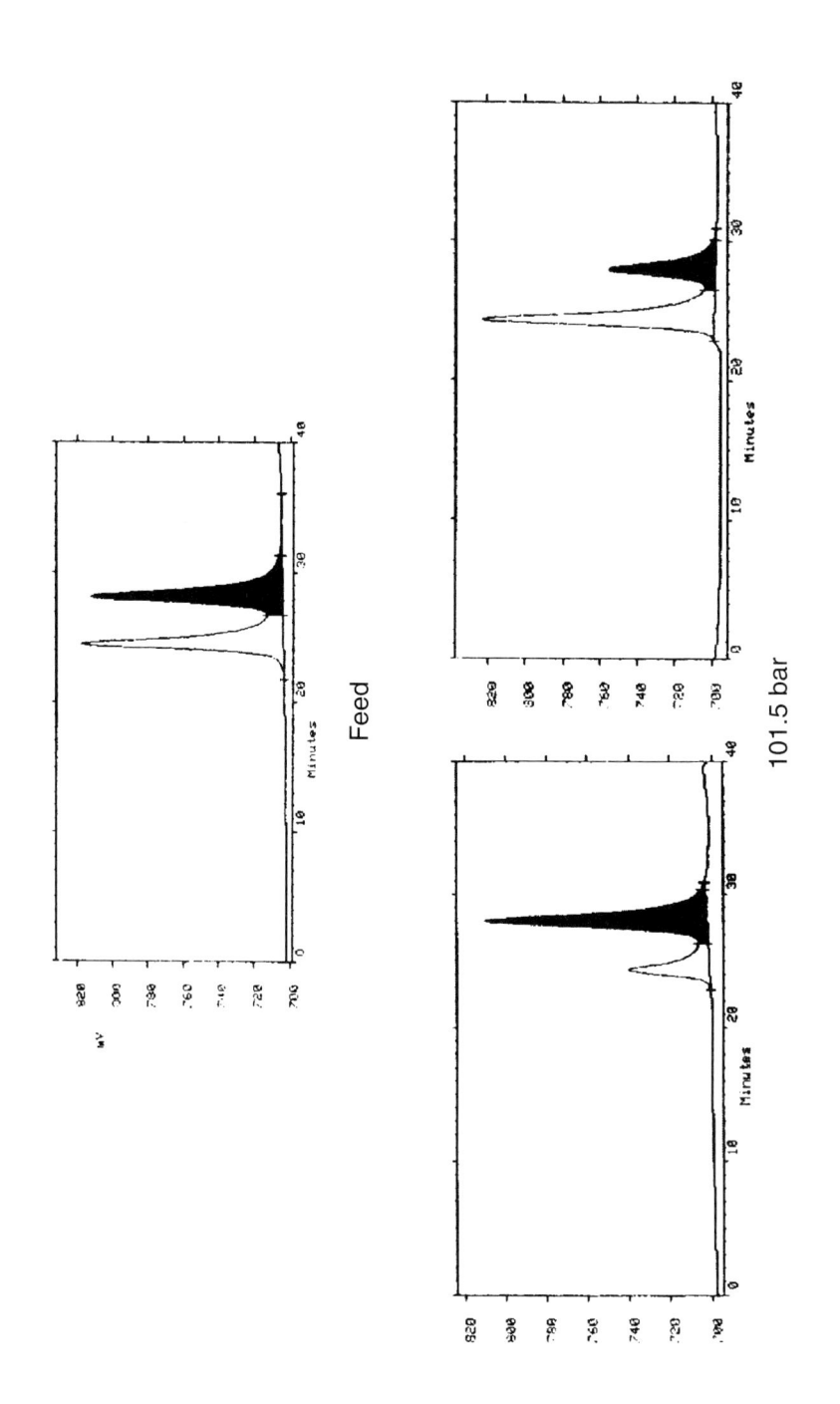

Feed

101.5 bar

21.5 Results

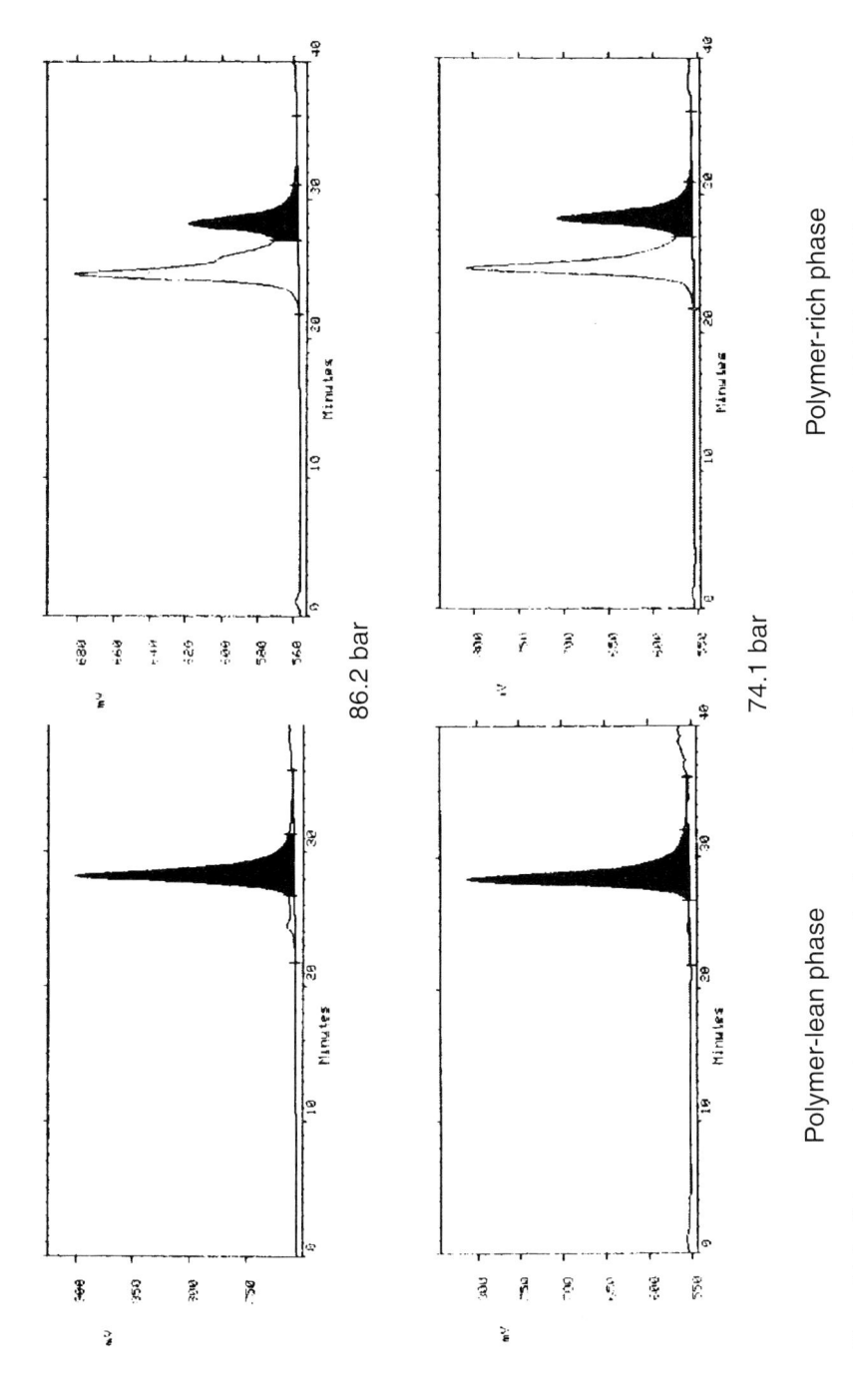

Polymer-lean phase   86.2 bar   Polymer-rich phase

Polymer-lean phase   74.1 bar   Polymer-rich phase

Figure 21.5: Gel-permeation chromatograms showing detector signals versus elution time for the system PS160k + PS40k + cyclohexane + CO_2.

21.6 Conclusion

Liquid-liquid phase separation in ternary polymer + solvent + gas systems results in an almost polymer-free solvent-rich phase. At the time, the polymer-rich phase is obtained that has an increased polymer content on a gas-free basis, reducing the energy requirement of industrial drying steps by solvent evaporation. Furthermore, this separation can take place at lower temperatures reducing the possibility of polymer degradation. Depending on pressure and initial polymer concentration a separation of the polymers by chain length can be accomplished. Especially, it is possible to quantitatively "cut off" the low-molecular-mass and to give a short-free molecular mass distribution that is important to industrial application. Chain partitioning is most pronounced within a window of 2 MPa the maximum of the cloud point curve. This is in contrast to the known separation by temperature change without a gas where the chain partitioning takes place within an interval of up to only a few Kelvin. Thus, pressure can be used as an additional process parameter in the molecular weight fractionation of polymers.

Acknowledgement

The cooperation of Prof. Gabriele Sadowski, University of Dortmund, former scientific coworker in the Technical University of Berlin, is gratefully acknowledged.

References

Behme, S.; Sadowski, G.; Arlt, W. (1997): Calculation of Polymer Phase Equilibria Using the SAFT Equation of State. AIChE Spring National Meeting, March 9–13, 1997, Houston, TX.

Berens, A. R.; Huvard, G. S.; Korsmeyer, R. W.; Kunig, F. W. (1992): Application of Compressed Carbon Dioxide in the Incorporation of Additives into Polymers. Appl. Polym. Sci., 46, 231.

Beret, S.; Prausnitz, J. M. (1975): Perturbed Hard-Chain Theory: An Equation of State for Fluids Containing Small or Large Molecules. AIChE J., 21, 1123.

Bungert, B.; Sadowski, G.; Arlt, W. (1997a): New Processes with Compressed Gases. Chem.-Ing.-Tech., 69, 298.

Bungert, B.; Sadowski, G.; Arlt, W. (1997b): Supercritical Antisolvent Fractionation: Measurements in the Systems Monodisperse and Bidisperse Polystyrene-Cyclohexane-Carbon Dioxide. Fluid Phase Equilib., 139, 349.

Bungert, B; Sadowski, G.; Arlt, W. (1998): Separations and Material Processing in Solutions with Dense Gases, Ind. Eng. Chem. Res., 37, 3208.

Chen, C.-K.; Duran, M. A.; Radosz, M. (1993): Phase Equilibria in Polymer Solutions. Block Algebra, Simultaneous Flash Algorithm Coupled with SAFT Equation of State, Applied to Single-Stage Supercritical Antisolvent Fractionation of Polyethylene. Ind. Eng. Chem. Res., 32 (12), 3123.

References

Chen, C.-K.; Duran, M. A.; Radosz, M. (1994): Supercritical Antisolvent Fractionation of Polyethylene Simulated with Multistage Algorithm and SAFT Equation of State: Staging Leads to High Selectivity Enhancements for Light Fractions. Ind. Eng. Chem. Res., 33, 306.

Chow, T. S. (1980): Molecular Interpretation of the Glass-Transition Temperature of Polymer-Diluent Systems. Macromolecules, 13, 362.

Condo, P. D.; Johnston, K. P. (1992): Retrograde Vitrification of Polymers with Compressed Fluid Diluents: Experimental Confirmation. Macromolecules, 25 (24), 6730.

Condo, P. D.; Sanchez, I. C.; Panayiotou, C. G.; Johnston, K. P. (1992): Glass-Transition Behavior Including Retrograde Vitrification of Polymers with Compressed Fluid Diluents. Macromolecules, 25, 6119.

de Loos, T. W. (1994): Personal communication.

Flory, P. J.; Orwoll, R. A.; Vrij, A. (1964): A Statistical Thermodynamics of Chain Molecule Liquids. I. An Equation of State for Normal Paraffin Hydrocarbons. J. Am. Chem. Soc., 86, 3507.

Freund, H.; Steiner, R. (1997): Determination of Solidus Lines of Binary Melts Under Gas Pressure. Chem.-Ing.-Tech., 69, 1409.

Hasch, B. M.; McHugh, M. A. (1995): Calculating Poly(ethylene-*co*-acetic acid)- Solvent Phase Behavior with the SAFT Equation of State. J. Polym. Sci., Part B: Polym. Phys., 3, 715.

Huang, S. H.; Radosz, M. (1990): Equations of State for Small, Large, Polydisperse and Associating Molecules. Ind. Eng. Chem. Res., 29 (11), 2284.

Huang, S. H.; Radosz, M. (1991): Equation of State for Small, Large, Polydisperse, and Associating Molecules: Extension to Fluid Mixtures. Ind. Eng. Chem. Res., 30, 1994.

Kleintjens, L. A.; Koningsveld, R. (1980): Liquid-Liquid-Phase Separation in Multicomponent Polymer Systems. Colloid Polym. Sci., 258, 711.

Koak, N.; de Loos, T. W.; Heidemann, R. A. (1998): UCST Behavior of the System Polystyrene + Methylcyclohexane. Influence of CO_2 on the LLE. Fluid Phase Equilib., 145, 311.

Kontogeorgis, G. M.; Harismiadis, V. I.; Fredenslund, A.; Tassios, D. P. (1994): Application of the van der Waals Equation of State to Polymers. I. Correlation. Fluid Phase Equilib., 96, 65.

Kordikowski, A.; Peters, C. J.; de Swaan Arons (1994): J. Thermodynamic Analysis of the Gas-Antisolvent Crystallization Process. Proceedings of the 3rd International Symposium on Supercritical Fluids, Strasbourg, France, 1994; ISASF: Nancy, France, 1994; Vol. 3, 217.

Lambert, S. M.; Paulaitis, M. E. (1991): Crystallization of Poly(ethylene terephthalate) induced by Carbon Dioxide Sorption at Elevated Pressures. J. Supercrit. Fluids, 4, 15.

Lee, C.; Hoy, K. L.; Donohue, M. D. (1990): Supercritical Fluids as Diluents in Liquid Spray Applications of Coatings. U.S. Patent 4923720.

Mertsch, R.; Wolf, B. A. (1994): Solutions of Poly(dimethylsiloxane) in Supercritical CO_2: Viscosimetric and Volumetric Behavior. Macromolecules, 27, 3289.

Nielsen, K. A.; Glancy, C. W.; Hoy, K. L.; Perry, K. M. (1991): A New Atomization Mechanism for Airless Spraying: The Supercritical Fluid Spray Process. 5th International Symposium on Liquid Atomization and Spray Systems, Gaithersburg, MD, 367.

Orbey, N.; Sandler, S. I. (1994): Vapor-Liquid Equilibrium Using a Cubic Equation of State. AIChE J., 40, 1203.

Pradhan, D.; Chen, C.-K.; Radosz, M. (1994): Fractionation of Polystyrene with Supercritical Propane and Ethane: Characterization, Semibatch Solubility Experiments, and SAFT Simulations. Ind. Eng. Chem. Res., 33, 1984.

Saeki, S.; Kuwahara, N.; Nakata, M.; Kaneko, M. (1975): Pressure Dependence of Upper Critical Solution Temperatures in the Polystyrene-Cyclohexane System, Polymer, 16 (6), 445.

Sako, T.; Wu, A. H.; Prausnitz, J. M. (1989): A Cubic Equation of State for High-Pressure Phase Equilibria of Mixtures Containing Polymers and Volatile Fluids. J. Appl. Polym. Sci., 38, 1839.

Sanchez, I. C.; Lacombe, R. H. (1976): An Elementary Molecular Theory of Classical Fluids. Pure Fluids. J. Phys. Chem., 80, 2352.

Song, Y.; Lambert, S. M.; Prausnitz, J. M. (1994): Equation of State for Mixtures of Hard Sphere Chains Including Copolymers. Macromolecules, 27, 441.

Tork, T.; Sadowski, G.; Arlt, W.; de Haan, A.; Krooshof, G. (1999a): Modelling of High-Pressure Phase Equilibria Using the Sako-Wu-Prausnitz Equation of State. I. Pure-Component Parameters and Heavy n-Alkane Solutions. Fluid Phase Equil., 163, 61.

Tork, T.; Sadowski, G.; Arlt, W.; de Haan, A.; Krooshof, G. (1999b): Modelling of High-Pressure Phase Equilibria Using the Sako-Wu-Prausnitz Equation of State. II. Ethylene/Polyethylene (LDPE) Systems. Fluid Phase Equil., 163, 79.

Wang, W.-C. V.; Kramer, E. J.; Sachse, W. H. (1982): Effects of High-Pressure CO_2 on the Glass-Transition Temperature and Mechanical Properties of Polystyrene. J. Polym. Sci, 20, 1371.

Weidner, E.; Steiner, R.; Knez, Z. (1996): Powder Generation from Poly(ethylene glycols) with Compressible Fluids. High Pressure Chemical Engineering Process, Technology Proceedings; Elsevier: Amsterdam, The Netherlands, 223.

Wissinger, R. G.; Paulaitis, M. E. (1987): Swelling and Sorption in Polymer-CO_2 Mixtures at Elevated Pressures. J. Polym. Sci., Part B: Polym. Phys., 25, 2497.

Xiong, Y.; Kiran, E. (1995): Comparison of Sanchez-Lacombe and SAFT Model in Predicting Solubility of Polyethylene in High-Pressure Fluids. J. Appl. Polym. Sci., 55,1805.

Yoo, K.-P.; Lee, C. S. (1996): A New Lattice-Fluid Equation of State and its Group-Contribution Applications for Predicting Phase Equilibria of Mixtures. Fluid Phase Equilib., 117, 48.

Yoo, S. J.; Shin, H. Y.; Yoo, K.-P.; Lee, C. S.; Arlt, W. (1996): Predictive Quasilattice Equation of State for Unified High-Pressure Phase Equilibria of Pure Fluids and Mixtures. High Pressure Chemical Engineering Process Technology Proceedings; Elsevier: Amsterdam, The Netherlands, 385.

Nomenclature

Latin letters

f	[Pa]	fugacity
M	[kg/kmol]	molecular weight
P	[Pa]	pressure
T	[K]	temperature
w	[–]	weight fraction
x	[–]	liquid phase mol fraction

Greek letters

γ	[–]	activity coefficient
μ	[J/mol]	chemical potential
φ	[–]	fugacity coefficient
ϕ	[–]	volume fraction

Subscripts

0	pure substance
i	component i
n	number average
m	mass average

Superscripts

+	reference state
L	liquid
V	vapor
S	solid

22 Calculation of the Stability and of the Phase Equilibrium of the System Methylcyclohexane + Polystyrene Based on an Equation of State

Dieter Browarzik[1] and Mario Kowalewski

22.1 Abstract

Cloud-point curves, spinodal curves, and critical points of the system methylcyclohexane (MCH) + polystyrene (PS) are calculated with the Sako-Wu-Prausnitz equation of state (SWP-EoS) taking into account the molecular-mass polydispersity of the polymer by continuous thermodynamics. The pressure-concentration diagram of the system MCH + PS shows both a curve with a lower critical point (LCP) and a curve with an upper critical point (UCP) in a limited temperature region. At low temperatures there is an hour-glass shaped two-phase region. Firstly, we proved the capability of the SWP-EoS to describe the experimental data for a solution of nearly monodisperse polystyrene. Secondly, we performed computer simulations showing the transition of the LCP/UCP behavior into the hour-glass behavior for the polydisperse case never studied before. This transition is demonstrated to be quite different from that of solutions of monodisperse PS in MCH. Thirdly, we calculated spinodal and cloud-point curves for a real system MCH + PS with a widely distributed PS sample.

22.2 Introduction

Experimental data for the cloud-point and spinodal curves of polymer + solvent systems under high pressures reveal in some cases two curves in a pressure-concentration diagram (at constant temperature). If the polymer is monodisperse, there is an upper curve with a lower critical point (LCP) as minimum and a lower curve with an upper critical point (UCP) as maximum. Changing the temperature both curves can become closer and merge, finally forming an hour-glass two-phase region. The system methylcyclohexane (MCH) + polystyrene (PS) is one of the best studied systems of this type (Saeki et al.,1973; Wells et al.,

---

[1] Martin-Luther-Universität Halle Wittenberg, Institut für Physikalische Chemie, Geusaer Straße, 06217 Merseburg.

1993; Hosokawa et al., 1993; Vanhee et al., 1994; Enders and de Loos, 1997; Koak et al., 1998). Particularly, Vanhee et al. (1994) studied the transition of the LCP + UCP behavior into the hour-glass diagram very thoroughly using a nearly monodisperse PS sample. They measured both spinodal and cloud-point pressures applying a pressure-pulse-induced critical scattering (PPICS) technique (Wells et al., 1993). So, they could prove that the LCP curve and the UCP curve coincide at their critical points which, in the monodisperse case, are the extreme values of these curves. After forming the hour-glass shaped curves critical points no longer exist. However, the knowledge about the monodisperse case does not permit conclusions about how the transformation is going on in the polydisperse case. One knows that the polydispersity can considerably influence phase equilibrium and stability. So, it is well known that in the polydisperse case the critical point is located neither in an extreme value of the cloud-point curve nor in an extreme value of the spinodal curve. Fortunately, Enders and de Loos (1997) investigated the same system but using a polymer sample with a broad molecular-mass distribution. They also measured many spinodal and cloud-point data applying the PPICS technique. However, all these data belong to the UCP curves. So, there is no explanation what happens in detail if the LCP curve and the UCP curve coincide. Enders and de Loos (1997) tried to understand the formation of the hour-glass two-phase region by performing calculations. The calculations showed that when the temperature decreases the spinodals (UCP + LCP) become closer and closer, finally forming a cross and, after that, the hour-glass curves are formed. The spinodals merge at their extreme values. However, it is not clear what happens to the critical points, the cloud-point curves, and the shadow curves. Furthermore, Enders and de Loos applied a Gibbs excess energy model instead of an equation of state (EoS) that should be preferred in modeling high-pressure phase equilibrium.

The goal of the present work was to calculate both spinodals and cloud-point curves from an EoS. The cubic EoS introduced by Sako, Wu, and Prausnitz (1989) (SWP-EoS) was selected as it had proven to be suitable for polymer systems. The calculations were aimed to explain the transition of the UCP + LCP behavior into the hour-glass behavior, particularly, in the polydisperse case by discussing spinodal curves, cloud-point curves, shadow curves, and critical points.

To take into account the molecular-mass polydispersity, continuous thermodynamics is a powerful tool. In continuous thermodynamics the composition of the polymer is described by a continuous distribution function with respect to the molecular mass. So, instead of hundreds of mass fractions of the polymer species only a distribution function is needed. First introduced by Kehlen and Rätzsch (1980) already a few years later continuous thermodynamics was applied to calculate cloud-point curves (Rätzsch and Kehlen, 1985) and stability (Browarzik et al., 1990) of polymer systems. However, the stability treatment was firstly restricted to Gibbs excess energy models. More recently, the stability theory of continuous thermodynamics was extended to equations of state (Browarzik and Kehlen, 1996; Browarzik et al., 1998) including also polymer systems (Browarzik and Kowalewski, 1999). In this contribution, the influence of the molecular-mass polydispersity of the polymer on the phase equilibrium and on the stability of the system MCH + PS is treated with continuous thermodynamics.

22.3 Theory

22.3.1 Calculation of the Cloud-Point and the Shadow Curves

In calculations of phase equilibrium by means of equations of state, the fugacity coefficients are the most important quantities. Continuous thermodynamics results in the following expressions for the fugacity coefficients of the solvent A and the species of the polymer B identified by their segment number r (Browarzik and Kowalewski; 1999):

$$\ln \phi_A = r_A B_A - \ln \bar{r}^M - \ln \left(\frac{pV_s}{RT} \right) \tag{22.1a}$$

$$\ln \phi_B(r) = r B_B - \ln \bar{r}^M - \ln \left(\frac{pV_s}{RT} \right) \tag{22.1b}$$

R is the universal gas constant, T is the temperature, and p is the pressure. The quantity r_A denotes the segment number of the solvent molecules. V_s is the segment-molar volume which is obtained from the molar volume by division through the number average $\bar{r}^M$ of the segment number of the system. It is assumed that the segment numbers are proportional to the molecular mass. Equations (22.1a) and (22.1b) are general expressions valid for any EoS. Only, the quantities B_i ($i = A, B$) depend on the nature of the EoS. Equations (22.1a) and (22.1b) are not very different from the corresponding notation in traditional thermodynamics. The only difference is that Eq. (22.1b) is not related to a real discrete component but rather to a polymer species identified by the segment number r that can take any value between zero and infinity. If an EoS is chosen the quantities B_i ($i = A, B$) can be derived in the usual way applying either traditional thermodynamics or continuous thermodynamics. The latter is thoroughly outlined Browarzik and Kehlen (2000). In Section 22.3.4 of the present contribution expressions for the SWP-EoS are presented. Here, at first the general treatment based on continuous thermodynamics is discussed.

Once the fugacity coefficients are known the equilibrium conditions for the phases I and II in segment-molar notation (Browarzik et al., 1998) read

$$(1 - \psi^I) \phi_A^I \bar{r}^{MI} = (1 - \psi^{II}) \phi_A^{II} \bar{r}^{MII} \tag{22.2a}$$

$$\psi^I W_s^I(r) \phi_B^I(r) \bar{r}^{MI} = \psi^{II} W_s^{II}(r) \phi_B^{II}(r) \bar{r}^{MII} \tag{22.2b}$$

ψ is the total mass fraction of the polymer. $W_s(r)$ is the segment-molar distribution function of the polymer. If there is no solvent ($\psi = 1$), integration of this distribution function gives the mass fraction of all polymer species possessing segment numbers between the limits of the integral. Corresponding to that, integration for the total domain of definition of the segment number r yields one (normalization condition). The number average of the segment number $\bar{r}^M$ of the mixture is governed by:

$$\frac{1}{\bar{r}^M} = \frac{1-\psi}{r_A} + \frac{\psi}{\bar{r}_B}; \quad \frac{1}{\bar{r}_B} = \int \frac{W_s(r)}{r} \, dr \tag{22.3}$$

$\bar{r}_B$ is the number average of the segment number of the polymer molecules. The integral in Eq. (22.3) and all following ones of this type are related to the whole domain of definition of the segment number r.

With the aid of Eqs. (22.1a) and (22.2a) one finds:

$$\frac{1}{r_A} \ln\left(\frac{(1-\psi^I) V_S^{II}}{(1-\psi^{II}) V_S^I}\right) + B_A^I - B_A^{II} = 0 \tag{22.4}$$

For further treatment the distribution function of the phase I has to be known. We presume a generalized Schulz-Flory distribution

$$W^I(r) = \frac{k^k}{\Gamma(k)\bar{r}_B^I} \left(\frac{r}{\bar{r}_B^I}\right)^k \exp\left[-k\frac{r}{\bar{r}_B^I}\right] \tag{22.5}$$

Γ is the gamma function and k describes the width of the molecular mass distribution. This parameter is connected to the so-called non-uniformity U by $k = 1/U$. In this, $U = \bar{M}_w/\bar{M}_n - 1$ holds and $\bar{M}_w$ and $\bar{M}_n$ are the mass average and the number average of the molecular mass distribution. Originally, the Schulz-Flory distribution relates to the molecular mass but, assuming the segment number to be proportional to the molecular mass $W_s(r)$ is the same type of function. With the aid of Eq. (22.5) integration of Eq. (22.2b) from zero to infinity yields:

$$\ln\left(\frac{\psi^I V_S^{II}}{\psi^{II} V_S^I}\right) - \left(1 + \frac{1}{U}\right) \ln\left\{1 - \bar{r}_B^I U (B_B^I - B_B^{II})\right\} = 0 \tag{22.6}$$

Multiplying Eq. (22.2b) by r, using Eqs. (22.1b) and (22.5), and integrating this expression from zero to infinity gives an equation for the number average of the segment number of the polymer in the incipient phase II.

$$\bar{r}_B^{II} = \bar{r}_B^I \left(\frac{\psi^{II} V_S^I}{\psi^I V_S^{II}}\right)^{\frac{U}{U+1}} \tag{22.7}$$

One can show that the distribution function for the phase II is also one of the Schulz-Flory type with the parameters $\bar{r}_B^{II}$ und $k = 1/U$ where, indeed, the non-uniformity U has in both phases the same value. The segment-molar volumes V_S^I and V_S^{II} have to be computed by solving the EoS for both phases. For given temperature the equilibrium pressure p and the mass fraction ψ^{II} of the polymer in the phase II have to be calculated solving the non-linear set of Eqs. (22.4) and (22.6). The function $p(\psi^I)$ is the cloud-point curve. $p(\psi^{II})$ is the so-called shadow curve that coincides with the cloud-point curve only in the monodisperse case ($U = 0$).

22.3.2 Calculation of the Spinodal Curve

The procedure for calculating the spinodal curve of a system consisting of a polydisperse polymer, a solvent, and a gas from an equation of state by continuous thermodynamics was described recently (Browarzik and Kowalewski, 1999). The polydispersity of the polymer is taken into account by applying the principles of continuous thermodynamics tool (Kehlen and Rätzsch, 1980; Rätzsch and Kehlen, 1985, 1989). Here, that research is extended to solutions of a polydisperse polymer B in a solvent A. The non-linear part of the segment-molar Helmholtz energy is given by:

$$\Delta A_s = RT \left\{ \int \frac{\psi}{r} W_s(r) \ln \left[\psi W_s(r) \right] dr + \frac{1-\psi}{r_A} \ln(1-\psi) \right\} - \frac{RT}{\bar{r}M} \ln V_s + J_s \quad (22.8)$$

A detailed derivation of Eq. (22.8) and why the segment-molar notation should be preferred for polymer systems can be found in a previous publication (Browarzik and Kowalewski, 1999). The integral in Eq. (22.8) is again related to the whole domain of definition of the segment number r. The quantity J_s includes the information about the equation of state:

$$J_s = \int_{V_s}^{\infty} \left(p - \frac{RT}{\bar{r}M V_s} \right) dV_s \quad (22.9)$$

Equation (22.8) involves only non-linear terms with respect to the mass fractions and to the distribution function. For the spinodal curve the linear terms do not matter. The terms in braces of Eq. (22.8) present the ideal athermic contribution of the segment-molar Helmholtz energy. In the framework of Gibbs excess energy models this term is well known as the Flory-Huggins contribution. It describes the entropic effect originating from the size differences of the molecules due to a statistical background. However, using an equation of state the calculation of the segment-molar Helmholtz energy from the corresponding molar quantity yields this contribution to Eq. (22.8) with no need for any statistical treatment.

To calculate the spinodal curve, firstly, one has to form $\delta^2 A_s$, i.e. the second differential of the segment-molar Helmholtz energy with respect to the variations $\delta\psi$ and $\delta[\psi W_s(r)]$. A simple technique to form this derivative was given by Browarzik and Kehlen (2000). Assuming J_s depends only on V_s and ψ yields:

$$\delta^2 A_s = RT \left\{ \frac{(\delta\psi)^2}{r_A(1-\psi)} + \int \frac{\{\delta[\psi W_s(r)]\}^2}{r \psi W_s(r)} dr \right\} - \frac{2RT}{V_s} \delta V_s \int \frac{1}{r} \delta[\psi W_s(r)] dr +$$

$$+ \left[J_{VV} + \frac{RT}{(V_s)^2 \bar{r}M} \right] (\delta V_s)^2 + 2 \left(J_{V\psi} + \frac{RT}{r_A V_s} \right) \delta V_s \delta\psi + J_{\psi\psi} (\delta\psi)^2 \quad (22.10)$$

$J_{VV}, J_{V\psi}, J_{\psi\psi}$ are the second derivatives of J_s with respect to V_s and ψ.

The next step is to find that variation function $\delta[\psi W_s(r)]^*$ which minimizes $\delta^2 A_s$. In this step, the following side condition which originates from the normalization condition of the distribution function, has to be taken into account.

$$\int \delta [\psi W_s(r)] \, dr = \delta \psi \tag{22.11}$$

Therefore, the minimization procedure Lagrange's method of undetermined multipliers is applied. In this way the following variation function is found:

$$\delta [\psi W_s(r)]^* = \frac{r}{\bar{r}_B^{(1)}} W_s(r) \, d\psi + \frac{\psi W_s(r)}{V_s} \left(1 - \frac{r}{\bar{r}_B^{(1)}}\right) dV_s \tag{22.12}$$

Here, $\bar{r}_B^{(1)} = \bar{r}_B(1 + U)$ is the first moment of the distribution function $W_s(r)$. Generally, the i-th moment is defined as

$$\bar{r}_B^{(i)} = \int r^i W_s(r) \, dr \tag{22.13}$$

According to Eq. (22.13) the quantity $1/\bar{r}_B$ is the -1-th moment of the distribution function $W_s(r)$. Once the variation function $\delta [\psi W_s(r)]^*$ is known this expression has to be introduced into the second differential $\delta^2 A_s$ giving $\delta^2 A_s^* = \delta^2 A_s \{\delta [\psi W_s(r)]^*\}$ resulting in

$$\delta^2 A_s^* = \alpha_{VV} (\delta V_s)^2 + 2 \alpha_{V\psi} \delta V_s \delta\psi + \alpha_{\psi\psi} (\delta\psi)^2 \tag{22.14}$$

where α_{VV}, $\alpha_{V\psi}$, and $\alpha_{\psi\psi}$ are defined as

$$\alpha_{VV} = J_{VV} + \frac{RT}{(V_s)^2} \left\{\frac{1 - \psi}{r_A} + \frac{\psi}{\bar{r}_B(1 + U)}\right\} \tag{22.15a}$$

$$\alpha_{V\psi} = J_{V\psi} + \frac{RT}{V_s} \left\{\frac{1}{r_A} - \frac{1}{\bar{r}_B(1 + U)}\right\} \tag{22.15b}$$

$$\alpha_{\psi\psi} = J_{\psi\psi} + RT \left\{\frac{1}{r_A(1 - \psi)} + \frac{1}{\bar{r}_B(1 + U)\psi}\right\} \tag{22.15c}$$

The following steps are the same as in the monodisperse case. The variation $\delta\psi^*$ (as a function of δV_s) which minimizes $\delta^2 A^*$ has to be found. Differentiating Eq. (22.14) with respect to $\delta\psi$ and setting the result equal to zero gives $\delta\psi^* = -(\alpha_{V\psi}/\alpha_{\psi\psi}) \delta V_s$. Introducing this variation into Eq. (22.14) and setting the expression of the second differential obtained in this way equal to zero gives the spinodal condition

$$K_0 = \alpha_{VV} \alpha_{\psi\psi} - (\alpha_{V\psi})^2 = \begin{vmatrix} \alpha_{VV} & \alpha_{V\psi} \\ \alpha_{V\psi} & \alpha_{\psi\psi} \end{vmatrix} = 0 \tag{22.16}$$

This condition is similar to that for the monodisperse case. The only difference is the occurrence of $\bar{r}_B(1 + U)$ (corresponding to the mass average $\overline{M}_w$) instead of $\bar{r}_B$ (corresponding to the number average $\overline{M}_n$).

22.3.3 Calculation of Critical Points

A further important question is how to find an additional necessary condition for the critical point. For this purpose, δK_0 has to be formed starting from Eq. (22.16).

$$\delta K_0 = L_V \delta V_s + L_\psi \delta\psi + L_r \delta\left[\psi \bar{r}_B^{(1)}\right] \tag{22.17}$$

with

$$L_V = \alpha_{\psi\psi}\left\{J_{VVV} - \frac{2RT}{(V_s)^3}\left[\frac{1-\psi}{r_A} + \frac{\psi}{\bar{r}_B^{(1)}}\right]\right\} +$$

$$+ \alpha_{VV}J_{V\psi\psi} - 2\alpha_{V\psi}\left\{J_{VV\psi} - \frac{RT}{(V_s)^2}\left[\frac{1}{r_A} - \frac{1}{\bar{r}_B^{(1)}}\right]\right\} \tag{22.18a}$$

$$L_\psi = \alpha_{\psi\psi}\left\{J_{VV\psi} + \frac{RT}{(V_s)^2}\left[-\frac{1}{r_A} + \frac{2}{\bar{r}_B^{(1)}}\right]\right\} +$$

$$+ \alpha_{VV}\left\{J_{\psi\psi\psi} + \frac{RT}{r_A(1-\psi)^2}\right\} - 2\alpha_{V\psi}\left\{J_{V\psi\psi} - \frac{RT}{V_s}\frac{1}{\psi\bar{r}_B^{(1)}}\right\} \tag{22.18b}$$

$$L_r = -\frac{RT}{\left[\bar{r}_B^{(1)}\right]^2}\left\{\frac{\alpha_{\psi\psi}}{(V_s)^2} + \frac{\alpha_{VV}}{\psi^2} + \frac{2\alpha_{V\psi}}{\psi V_s}\right\} \tag{22.18c}$$

Here, again $\bar{r}_B^{(1)} = \bar{r}_B(1+U)$ holds. Furthermore, $\delta\left[\psi \bar{r}_B^{(1)}\right]$ is defined as

$$\delta\left[\psi \bar{r}_B^{(1)}\right] = \int r\delta\left[\psi W_s(r)\right]dr \tag{22.19}$$

Introducing Eq. (22.12) for $\delta\left[\psi W_s(r)\right]$ into Eq. (22.19) gives:

$$\delta\left[\psi \bar{r}_B^{(1)}\right]^* = \frac{\bar{r}_B^{(2)}}{\bar{r}_B^{(1)}}\delta\psi + \frac{\psi}{V_s}\left[\bar{r}_B^{(1)} - \frac{\bar{r}_B^{(2)}}{\bar{r}_B^{(1)}}\right]\delta V_s \tag{22.20}$$

Additionally, introducing $\delta\psi^* = -(\alpha_{V\psi}/\alpha_{\psi\psi})\delta V_s$ instead of $\delta\psi$ into Eqs. (22.17) and (22.20) results in an expression for δK_0 corresponding to the minimizing variations $\delta[\psi W_s(r)]^*$ and $\delta\psi^*$. This expression has the form $\delta K_0 = K_1 \delta V_s$. With $K_1 = 0$ the second necessary critical condition (the first condition is from Eq. (22.16)) is obtained as follows:

$$L_V \alpha_{\psi\psi} - L_\psi \alpha_{V\psi} + L_r \bar{r}_B^{(1)}\left\{\frac{\psi}{V_s}\left[1 - \frac{\bar{r}_B^{(2)}}{\left(\bar{r}_B^{(1)}\right)^2}\right]\alpha_{\psi\psi} - \frac{\bar{r}_B^{(2)}}{\left(\bar{r}_B^{(1)}\right)^2}\alpha_{V\psi}\right\} = 0 \tag{22.21}$$

Whereas the spinodal curve depends only of the number average $\bar{r}_B$ and on the first moment $\bar{r}_B^{(1)} = \bar{r}_B(1 + U)$ the critical point is additionally influenced by the second moment $\bar{r}_B^{(2)}$. All analytical expressions for the distribution function depend on parameters $\bar{r}_B$ and U. The spinodal curve does not depend on the expression chosen for the distribution function, but, the critical point does. Assuming the generalized Schulz-Flory distribution function (Eq. (22.5)) results in:

$$\bar{r}_B^{(2)} = (\bar{r}_B)^2 (1 + U)(1 + 2U) \tag{22.22}$$

22.3.4 Sako-Wu-Prausnitz Equation of State (SWP-EoS)

For calculating an equation of state (EoS) has to be chosen. As polydispersity of polymers is a complicated problem to handle, the EoS should be relatively simple but appropriate to polymers. These requirements are fulfilled by the cubic EoS introduced in 1989 by Sako, Wu, and Prausnitz (1989) to describe the high-pressure equilibrium of the system ethylene + polyethylene. In a preceeding paper the SWP-EoS was applied to calculate the stability criterion and the phase equilibrium of the system cyclohexane + carbon dioxide + polystyrene at high pressures (Browarzik and Kowalewski, 1999). The SWP-EoS of the mixture reads in segment-molar notation:

$$p = \frac{RT}{\bar{r}^M V_s} + \frac{RT\, b_s\, c_s}{V_s (V_s - b_s)} - \frac{a_s}{V_s (V_s + b_s)} \tag{22.23}$$

The segment-molar parameters a_s, b_s, c_s are obtained from molar parameters a, b, and c by division through $(\bar{r}^M)^2$ (for a) or through $\bar{r}^M$ (for b and c). If all molecules can be presented by a single segment Eq. (22.23) reduces to the Soave-Redlich-Kwong EoS. $3c$ is the total number of external degrees of freedom per molecule. In this, the external rotational and vibrational degrees of freedom are considered as equivalent translational degrees of freedom. Depending on the chain flexibility the parameter c of a pure species can take values between one and the number r of segments of the polymer molecule. According to that the segment-molar parameter c_s ranges between zero and one.

Mixing rules are needed for calculating the parameters a_s, b_s, c_s of the mixture from those of the pure components. As usually for cubic equations of state arithmetic mixing rules are applied for b and c, whereas parameter a is expressed using a geometric averaging and introducing a binary parameter:

$$a_{ij} = \sqrt{a_{ii}\, a_{jj}}\, (1 - k_{ij}) \tag{22.24}$$

Thus, the segment-molar parameters of the system MCH (A) + PS (B) are given by

$$a_s = \left\{ (1 - \psi)\sqrt{a_{s,A}} + \psi\sqrt{a_{s,B}} \right\}^2 - 2\psi(1 - \psi)\sqrt{a_{s,A}\, a_{s,B}}\, k_{AB} \tag{22.25a}$$

$$b_s = (1 - \psi)\, b_{s,A} + \psi\, b_{s,B} \tag{22.25b}$$

$$c_s = (1-\psi)c_{s,A} + \psi\, c_{s,B} \tag{22.25c}$$

A standard segment has to be chosen for characterizing the segment numbers. Here that segment is that of a methylcyclohexane molecule. Furthermore, it is assumed that the segment number is proportional to the molecular mass:

$$r = 10.1852\, \frac{M}{1000\ \mathrm{gmol^{-1}}} \tag{22.26}$$

The determination of the other pure-component parameters for methylcyclohexane is relatively simple. The segment-molar quantities do not differ from the molar ones because of $r_A = 1$. Furthermore, for the sake of simplicity we set $c_A = c_{s,A} = 1$. Then the procedure is the same as in the case of the Soave-Redlich-Kwong EoS. To determine the parameters a_s, b_s for a volatile component as methylcyclohexane the critical temperature (572.19 K) and the critical pressure (3.4714 MPa) have to be available. These critical data were taken from Daubert and Danner (1992), resulting in

$$b_{s,A} = 0.118737\ \mathrm{3\ l\ mol^{-1}} \tag{22.27a}$$

and at the critical temperature $a_{s,A} = 2.787113\ \mathrm{MPa\ l^2\ mol^{-2}}$. The temperature dependence for the parameter $a_{s,A}$ is calculated following a proposal by Sako et al. (1989). The van der Waals volume needed in this equation was taken from Bondi's tables (Bondi, 1968). In this way we obtained

$$a_{s,A} = 2.787113\, \frac{1.686389 + 0.313611\,(T/572.19K)^2}{1 + (T/572.19K)^2}\ \mathrm{MPa\ l^2\ mol^{-2}} \tag{22.27b}$$

The determination of the pure-component parameters of the polymer is very difficult as no critical data are available. Therefore, these parameters are adjusted to the experimental spinodal data of the pseudobinary system MCH + PS assuming for $a_{s,B}$ the same temperature dependence as for $a_{s,A}$.

In calculating the spinodal curve the quantity J_s plays an important role. Introducing the SWP-EoS into Eq. (22.9) gives:

$$J_s = RT c_s \ln\left(\frac{V_s}{V_s - b_s}\right) + \frac{a_s}{b_s} \ln\left(\frac{V_s}{V_s + b_s}\right) \tag{22.28}$$

Based on this relation the derivatives J_{VV}, $J_{V\psi}$, $J_{\psi\psi}$ of Eqs. (22.15a–c) are calculated also taking into account that the dependence of the quantity ψ is hidden in the segment-molar parameters a_s, b_s, c_s.

The fugacities are calculated with the help of the quantities B_i ($i = A, B$) (cf. Eqs. (22.1a) and (22.1b)):

$$B_i = -c_{s,i} \ln\left(1 - \frac{b_s}{V_s}\right) + \frac{c_s b_{s,i}}{V_s - b_s} - \frac{a_s b_{s,i}}{RT b_s (V_s + b_s)} + \frac{\ln(1 + b_s/V_s)}{RT b_s} \cdot$$

$$\cdot \left\{ \frac{a_s b_{s,i}}{b_s} - 2\sqrt{a_{s,i}} \left[\sum_j \psi_j \sqrt{a_{s,j}} (1 - k_{ij}) \right] \right\} \tag{22.29}$$

In Eq. (22.29) the subscripts i, j relate to A, B where: $\psi_A = 1 - \psi$; $\psi_B = \psi$.

22.4 Results

22.4.1 Parameter Fit

There are four parameters which have to be fitted to experimental data. These are the three pure component parameters of PS and the mixing parameter k_{AB}. The parameter $a_{s,B}$ is needed only at the critical temperature because, for simplicity, we assume the temperature dependence of a_s to be the same as for methylcyclohexane. Vanhee et al. (1994) measured spinodal and cloud-point pressures of the system MCH + PS immediately describing the formation of the hour-glass behavior. The polystyrene sample used was a nearly monodisperse one ($U = 0.06$) with a number averaged molecular mass of 16,500 g/mol. Firstly, the parameters were fitted to the experimental spinodal pressures at 20.8 °C, particularly, to the experimental UCP points shown in Figure 22.1:

$$a_{s,B}(T) = 0.95316\, a_{s,A}(T); \quad b_{s,B} = 0.1260\,\text{lmol}^{-1}; \quad c_{s,B} = 0.028; \quad k_{AB} = 0.00296 \tag{22.30}$$

As shown in Figure 22.1 the LCP curve is not perfectly described. Particularly, the calculated LCP curves are too small. The reason is that three parameters are needed to describe the UCP curve with respect to maximum and width. One additional parameter is necessary to generate a LCP curve, the minimum of which possesses the correct pressure difference from the maximum of the UCP curve. For the correct description of the shape of the LCP curves there is no parameter left. Unfortunately, the parameters a_s, b_s, c_s for PS do not take the same values as estimated in a previous paper (Browarzik and Kowalewski, 1999) for the system cyclohexane + carbon dioxide + polystyrene. The parameter c_s for PS is remarkably small which could originate from the low flexibility of the polystyrene chains. Higher numbers for c_s would result in smaller cloud-point and spinodal curves in the pressure-concentration diagram.

Furthermore, to get the correct influence of temperature on the parameter k_{AB} the experimental spinodal points at 20.7 and 20.6 °C and the cloud-point pressures at 20.6, 21, and 21.2 °C were included in the parameter fit.

$$k_{AB} = 0.00296 - 0.00017(T/K - 293.95) \tag{22.31}$$

22 *Calculation of the Stability and of the Phase Equilibrium*

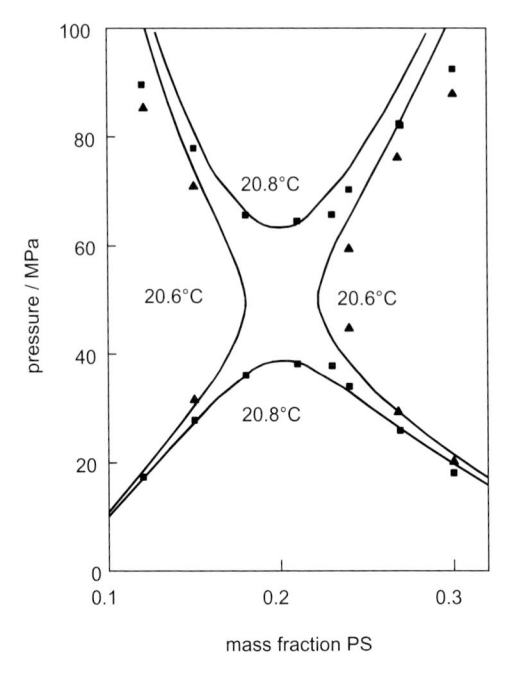

Figure 22.1: Calculated spinodal curves of the system MCH + PS at 20.6 and 20.8 °C compared with experimental data: (▲) 20.6 °C; (■) 20.8 °C.

Figure 22.1 shows the results for the spinodal curve at 20.8 and 20.6 °C. At 20.8 °C there is still an UCP + LCP behavior. When the temperature is reduced by only 0.2 K the hour-glass shaped curves are formed. Corresponding to that, very small changes of the k_{AB}-parameter result in a considerable pressure change. Therefore, it is not surprising that at 20.6 °C the calculated pressures deviate somewhat from the experimental results. In Figure 22.2 the cloud-point curves are compared with the experimental data at three temperatures. Considering that the parameters were mainly fitted to the spinodal data the agreement with experimental results is satisfactory.

22.4.2 Formation of the Hour-Glass Curves in the Polydisperse Case

Once the parameters are determined computer simulations for the polydisperse case are possible. So, the same system with the same number average of the molecular weight (16,500 g/mol) for polystyrene is considered but, with a non-uniformity $U = 1$ (corresponding to a mass averaged molecular mass of 33,000 g/mol). The pure component parameters of PS and the mixing parameter as function of the temperature were taken as given by Eqs. (22.30) and (22.31).

Figures 22.3 to 22.9 show the transition of the LCP + UCP behavior into the hour-glass shaped curves due to decreasing temperatures for the spinodal curves, the cloud-point

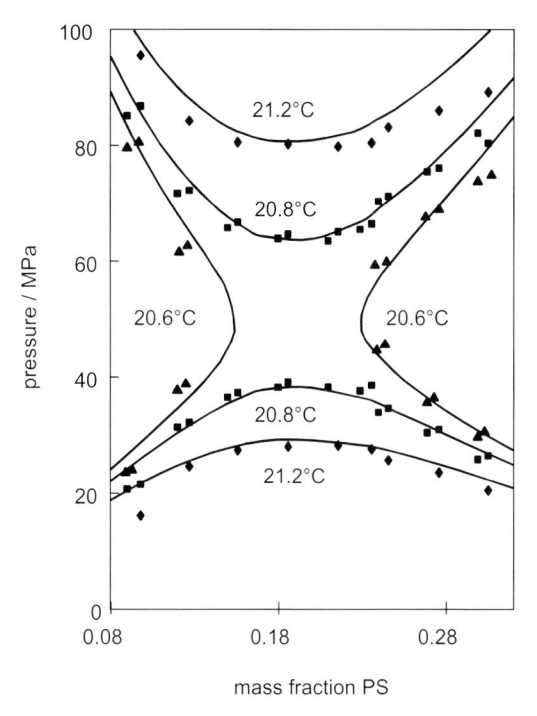

Figure 22.2: Calculated cloud-point curves of the system MCH + PS at 20.6, 20.8, and 21.2 °C compared with experimental data: (▲) 20.6 °C; (■) 20.8 °C; (◆) 21.2 °C.

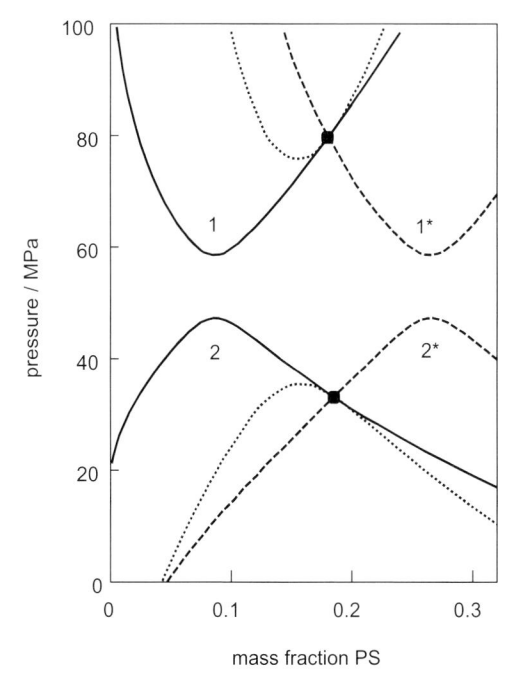

Figure 22.3: LCP + UCP behavior at 23.5 °C in the system MCH + polydisperse PS ($\overline{M}_n$ = 16500 g/mol, U = 1): (—) cloud-point curves, (- - -) shadow curves, (···) spinodal curves, (■) critical points.

22 Calculation of the Stability and of the Phase Equilibrium

curves, and for the shadow curves. Even the temperature range is very small (from 23.5 to 22.35 °C) essential changes in the phase diagram occur.

Figure 22.3 shows the results for 23.5 °C. At this temperature there is a LCP curve as well as an UCP curve in each case. Because of the polydispersity of the polymer ($U = 1$) the extremes of the spinodal curves neither coincide with the extremes of the cloud-point curves nor with the critical points. There is a large pressure difference between the extremes of both spinodal curves, but both cloud-point curves are relatively close to each other. Additionally, there are shadow curves originating from the polydispersity of the polymer. If the weight fraction ψ^I of the polymer in the phase I is given, the incipient phase formed at the same temperature and pressure has a polymer weight fraction ψ^{II}. In the monodisperse case the curves $p(\psi^I)$ and $p(\psi^{II})$ coincide forming the binodal curve. In the polydisperse case there is a cloud-point curve $p(\psi^I)$ and a shadow curve $p(\psi^{II})$. The critical point is the intersection of the cloud-point curve and the shadow curve.

When the temperature decreases, the cloud-point curves approach each other and, finally, merge at their extremes. This happens at 23.473 °C (cf. Fig. 22.4). Then they form a cross of straight lines. The shadow curves show the same behavior.

Further lowering of the temperature results in hour-glass shaped cloud-point curves as presented in Figure 22.5 for 23.35 °C. However, the spinodal curves still show an UCP + LCP behavior. Both critical points are located on the cloud-point curve in the polymer-rich region. It is interesting to have a look at the corresponding shadow curves. There is a cloud-

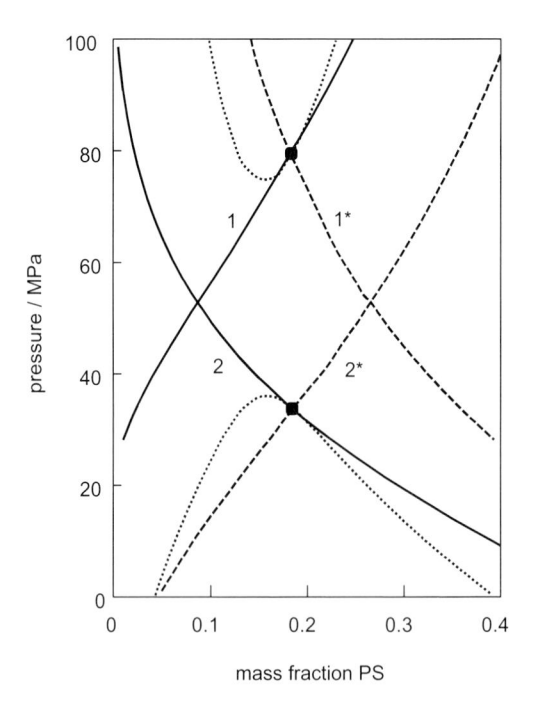

Figure 22.4: Coinicidence of the cloud-point curves at 23.473 °C in the system MCH + polydisperse PS ($\bar{M}_n$ = 16500 g/mol, $U = 1$): (——) cloud-point curves, (- - -) shadow curves, (···) spinodal curves, (■) critical points.

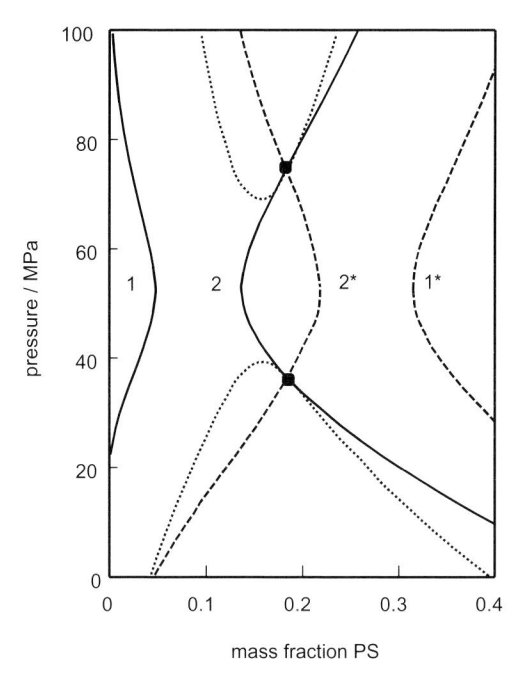

Figure 22.5: Hour-glass behavior for the cloud-point curves at 23.35 °C in the system MCH + polydisperse PS ($\overline{M}_n$ = 16500 g/mol, U = 1): (—) cloud-point curves, (- - -) shadow curves, (···) spinodal curves, (■) critical points.

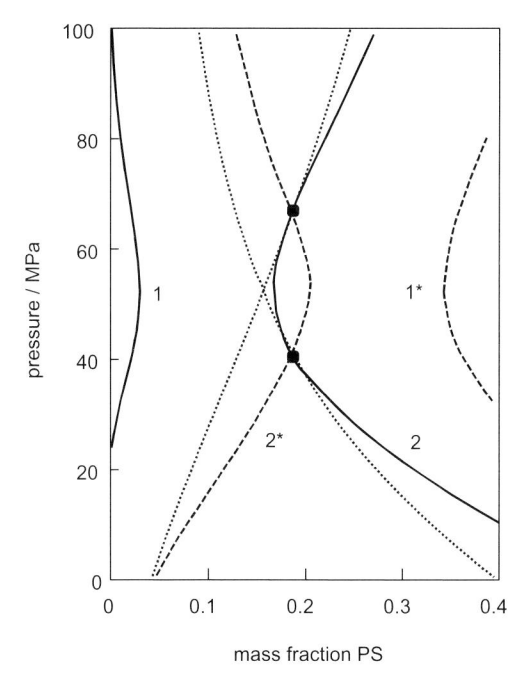

Figure 22.6: Coinicidence of the spinodal curves at 23.168 °C in the system MCH + polydisperse PS ($\overline{M}_n$ = 16500 g/mol, U = 1): (—) cloud-point curves, (- - -) shadow curves, (···) spinodal curves, (■) critical points.

501

22 Calculation of the Stability and of the Phase Equilibrium

point curve 1 and shadow curve 1* which do not cross and there is a cloud-point curve 2 and a shadow curve 2* with two intersection points being the critical points.

When the temperature is reduced to 23.168 °C (cf. Fig. 22.6) the spinodal curves merge forming a cross of straight lines similar to the merging of the cloud-point curves at 23.473 °C.

At 23.15 °C the cloud-point curves as well as the spinodal curves show the hour-glass shape (cf. Fig. 22.7). The critical points still exist. The cloud-point curve and the spinodal curve of the polymer-rich region are very close to each other within a wide pressure range. This behavior is forced by the critical points which have to be located both on the spinodal curve and on the cloud-point curve.

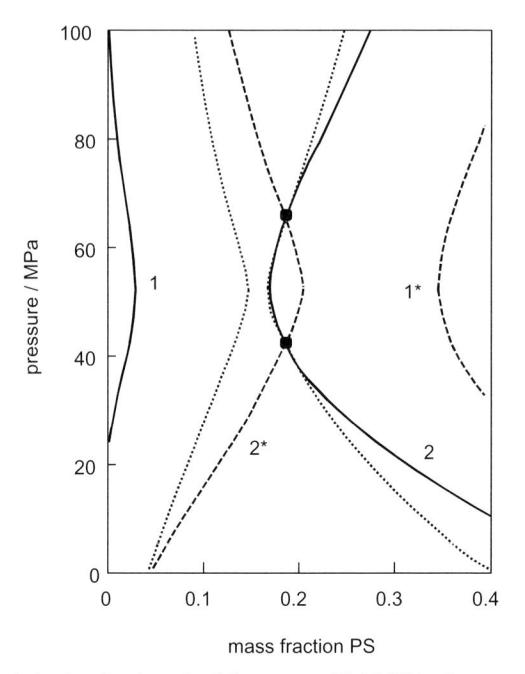

Figure 22.7: Hour-glass behavior for the spinodal curves at 23.15 °C in the system MCH + polydisperse PS ($\overline{M}_n$ = 16500 g/mol, U = 1): (—) cloud-point curves, (- - -) shadow curves, (···) spinodal curves, (■) critical points.

Figure 22.8 shows the situation at 23.058 °C. Here, the critical points coincide forming a double critical point. The cloud-point curve, the shadow curve, and the spinodal curve touch each other in the double critical point.

Finally, at 22.35 °C (cf. Fig. 22.9), there are no longer critical points. According to that, the spinodal and the cloud-point curve of the polymer-rich region are separated. The cloud-point curve 2 and the shadow curve 2* no longer cross.

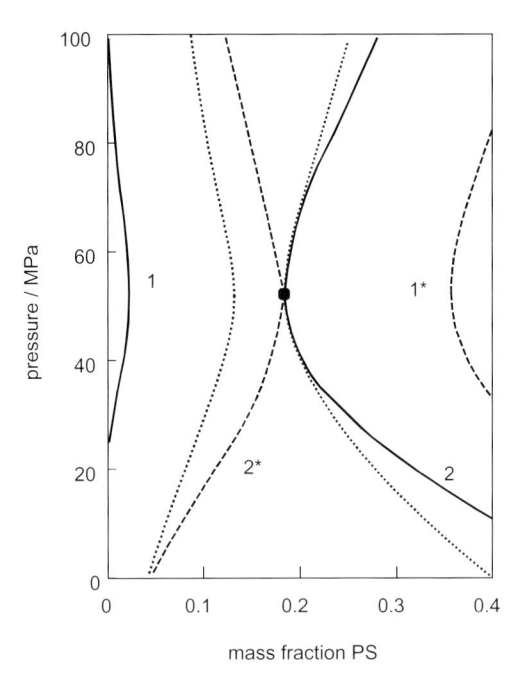

Figure 22.8: Formation of a double critical point at 23.058 °C in the system MCH + polydisperse PS ($\overline{M}_n$ = 16500 g/mol, U = 1): (—) cloud-point curves, (- - -) shadow curves, (···) spinodal curves, (■) critical points.

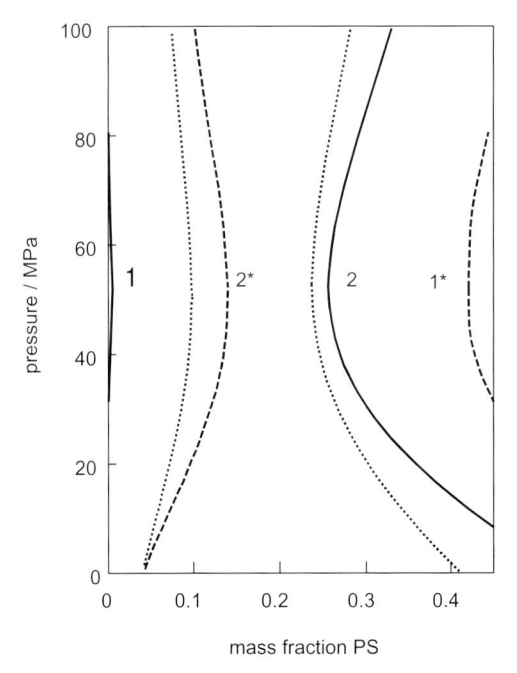

Figure 22.9: Hour-glass behavior without critical points at 22.35 °C in the system MCH + polydisperse PS ($\overline{M}_n$ = 16500 g/mol, U = 1): (—) cloud-point curves, (- - -) shadow curves, (···) spinodal curves.

22.4.3 Application of the EoS to a Real Polydisperse System MCH + PS

The calculations of the last section are to illustrate the complexity of the formation of hourglass diagrams in polydisperse systems. However, there are no experimental data for comparison available. So, it remains an open question if the predictions by the EoS for a real polydisperse system agree with reality. Fortunately, Enders and de Loos (1997) measured numerous spinodal and cloud-point pressures for a polydisperse system MCH + PS. The PS sample had a number averaged molecular mass of 143,000 g/mol which is essentially higher than that of the sample considered before. Corresponding to the mass averaged molecular mass (405,000 g/mol) the non-uniformity is $U = 1.83$. The pure component parameters of MCH and PS are known from the preceding calculations. The mixing parameter k_{AB} was refitted because k_{AB} might depend on the molecular mass. Furthermore, k_{AB} is needed in a wide temperature range because experimental data are only available in two temperature regions widely separated from each other. The low temperature region is about 316 to 332 K and the high temperature region is about 500 to 520 K. There are experimental results for the cloud-point in both temperature ranges measured in a Cailletet apparatus and spinodal data in the low temperature range measured with a PPICS equipment. The low temperature range corresponds to curves with an upper critical solution temperature (UCST) and the high temperature range corresponds to a lower critical solution temperature (LCST). We tried to find a temperature function for the binary parameter k_{AB} suitable for both temperature regions. At first, k_{AB} was fitted separately to the UCST spinodal data and to the LCST cloud-point pressures. In this way two linear temperature functions valid either in the small temperature ranges from 316 to 322 K or between 500 and 522 K were found. Finally, we searched for a common function $k_{AB}(T)$ for both temperature regions. The calculations revealed that such a function needs at least four parameters. The following function covering both temperature regions was found:

$$k_{AB} = -1.5772578 + 1.2493527 \cdot 10^{-2} (T/K) - 3.1444661 \cdot 10^{-5} (T/K)^2 +$$
$$+ 2.444442 \cdot 10^{-8} (T/K)^3 \tag{22.32}$$

Figure 22.10 shows a comparison of the calculated spinodal curves with experimental data for two polymer mass fractions (0.018 and 0.0682 mass%). The agreement between the calculated and the experimental values it good. The experimental data represent UCP curves. Experimental information on the corresponding LCP curves is not available. In a plot of temperature versus pressure there would be two curves with minima. These minima indicate the temperature and the pressure at which the spinodal curves begin to be hour-glass shaped. Enders and de Loos (1997) reported also experimental results for the cloud-point pressures (LCST) at 500 and 520 K for two polymer mass fractions ($\psi = 0.065$ and $\psi = 0.187$). The data at $\psi = 0.187$ were used for fitting $k_{AB}(T)$ (cf. Eq. (22.32)). Figure 22.11 shows a comparison between experimental data and the calculations for the cloud-point pressure. The calculations represent the experimental cloud-point pressures very well at $\psi = 0.187$, whereas the calculated cloud-point pressures are overestimated by about 2 MPa for $\psi = 0.065$. The reason for the unsatisfactory result is that the calculation locate the critical point at a very low polymer concentration and the parameter k_{AB} is not able to force the critical point to be shifted to higher polymer concentrations. Such a shift can only be achieved when the pure component parameters of PS are changed. However, pure component para-

22.4 Results

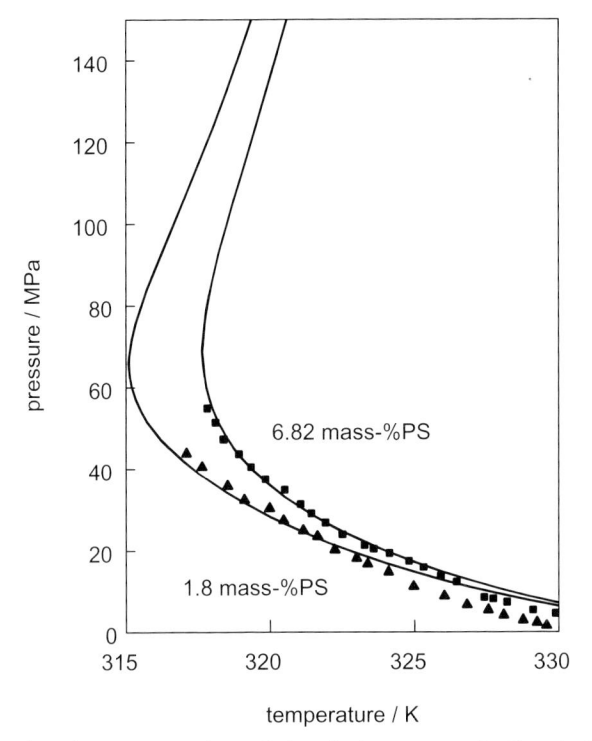

Figure 22.10: Comparison between experimental data (polymer mass fraction ψ: (▲) $\psi = 0.018$; (■) $\psi = 0.0682$) and calculations for the influence of temperature on the spinodal pressure (UCST) of the system MCH + PS.

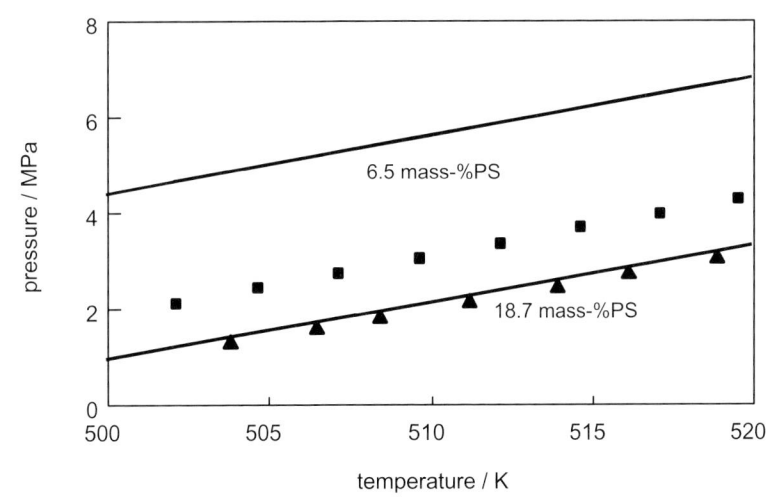

Figure 22.11: Comparison between experimental data (polymer mass fraction ψ: (■) $\psi = 0.065$; (▲) $\psi = 0.187$) and calculations for the influence of temperature on the cloud-point pressure (LCST) of the system MCH + PS.

meters b_s, c_s should not be dependent on temperature. Therefore only the influence of temperature on parameter a_s (assumed to be the same as for MCH) might be changed. However, that path was not followed as it would increase the number of adjustable parameters.

22.5 Conclusions

Spinodal and cloud-point curves of the system methylcyclohexane (MCH) + polystyrene (PS) can be described by the Sako-Wu-Prausnitz equation of state (SWP-EoS). The pressure-concentration diagrams of the system MCH + PS shows – in a limited temperature region – a curve with an upper critical point (UCP) as well as a curve with a lower critical point (LCP). A small temperature decrease induces a qualitative change of the diagram producing hour-glass shaped curves. The SWP-EoS is able to describe that transition. If the polystyrene is polydisperse the hour-glass shape for the spinodal and for the cloud-point curves are formed at different temperatures. This phenomenon was studied by combining the SWP-EoS with the concept of continuous thermodynamics. With decreasing temperature, the cloud-point curves merge at their extreme values forming a cross of straight lines. After that the cloud-point curves are hour-glass shaped. For the spinodal curves the formation of the hour-glass shape happens at a lower temperature. Before, the spinodal curves merge at their extremes too forming also straight lines near the crossing. In contrast to monodisperse polystyrene the critical points still exist after the formation of the hour-glass curves. When the temperature decreases further the critical points become closer, form a double critical point, and finally, disappear.

The authors expect that this treatment may be successfully applied to a real polydisperse system. However, there might be difficulties if the calculations are to be extended to a wide temperature range.

References

Bondi, A. (1968): Physical properties of molecular crystals, liquids and glasses, Wiley, New York.
Browarzik, D.; Kehlen, H.; Rätzsch, M.T. (1990): Application of continuous thermodynamics to the stability of polymer systems. III. J. Macromol. Sci.-Chem. A 27, 549–561.
Browarzik, D.; Kehlen, H. (1996): Stability of polydisperse fluid mixtures. Fluid Phase Equilib. 123, 17–28.
Browarzik, D.; Kowalewski, M.; Kehlen, H. (1998): Stability calculations of semicontinuous mixtures based on equations of state. Fluid Phase Equilib. 142, 149–162.
Browarzik, D.; Kowalewski, M. (1999): Calculation of the stability and of the phase equilibrium in the system polystyrene + cyclohexane + carbon dioxide based on equations of state. Fluid Phase Equilib. 163, 43–60.
Browarzik, D.; Kehlen, H. (2000): Polydisperse fluids. In: Sengers, J.V.; Kayser, R.F.; Peters, C.J.,

White, H.J. (Jr.) (Eds.), Equations of state for fluids and fluid mixtures (IUPAC), Elsevier, Amsterdam, 479–521.

Cotterman, R.L.; Prausnitz, J.M.(1991): Continuous thermodynamics for phase equilibrium calculations in chemical process design. In: Astarita, G. and Sandler, S.I. (Eds.), Kinetic and thermodynamic lumping of multicomponent mixtures, Elsevier, Amsterdam, 229–275.

Daubert, T.E.; Danner, R.P. (1992): Physical and thermodynamical properties of pure chemicals, Taylor and Francis, Washington.

Enders, S.; de Loos, Th.W. (1997): Pressure dependence of the phase behaviour of polystyrene in methylcyclohexane. Fluid Phase Equilib. 139, 335–347.

Hosokawa, H.; Nakata, M.; Dobashi, J. (1993): Coexistence curve of polystyrene in methylcyclohexane. VII. Coexistence surface and critical double point of binary system in $T-P-\phi$ space. Fluid Phase Equilib. 98, 10,078–10,084.

Kehlen, H.; Rätzsch, M.T. (1980): Continuous thermodynamics of multicomponent mixtures. Proc. 6th Int. Conf. Thermodyn., Merseburg, 41–51.

Koak, N.; de Loos, Th.W.; Heidemann, R.A. (1998): Upper-critical-solution-temperature behaviour of the system polystyrene + methylcyclohexane. Influence of CO_2 on the liquid-liquid equilibrium. Fluid Phase Equilib. 145, 311–323.

Rätzsch, M.T.; Kehlen, H. (1985): Continuous thermodynamics of polymer solutions: The effect of polydispersity on the liquid-liquid equilibrium. J. Macromol. Sci. Chem. A 22, 323–334.

Rätzsch, M.T.; Kehlen, H. (1989): Continuous thermodynamics of polymer systems. Prog. Polym. Sci. 14, 1–46.

Saeki, S.; Kuwahara, N.; Konno, S.; Kaneko, M. (1973): Upper and lower critical solution temperatures in polystyrene solutions. Macromolecules 6, 246–250.

Sako, T.; Wu, A.H.; Prausnitz, J.M. (1989): A cubic equation of state for high-pressure phase equilibria of mixtures containing polymers and volatile fluids. J. Appl. Polym. Sci. 38, 1839–1858.

Vanhee, S.; Kiepen, F.; Brinkmann, D.; Borchard, W.; Koningsveld, R.; Berghmans, H. (1994): The system methylcyclohexane/polystyrene. Experimental critical curves, cloud-point and spinodal isopleths, and their description with a semi-phenimenological treatment. Macromol. Chem. Phys. 195, 759–780.

Wells, P.A.; de Loos, Th.W.; Kleintjens, L.A. (1993): Pressure pulse induced critical scattering spinodal and binodal curves for the system polystyrene + methylcyclohexane. Fluid Phase Equilib. 83, 383–390.

Nomenclature

A	Helmholtz energy
a	parameter of the SWP-EoS
B_i	quantity introduced by Eqs. (22.1a) and (22.1b)
b, c	parameters of the SWP-EoS
J_s	quantity to calculate the spinodal cf. Eq. (22.9)
J_{xy}	second derivative of J_s with respect to x, y (V_s, ψ)
J_{xyz}	third derivative of J_s with respect to x, y, z (V_s, ψ)
K_0	quantity introduced by Eq. (22.16)
k	parameter of the Schulz-Flory distribution, $k = 1/U$
k_{AB}	binary parameter for interactions between components i and j
L_V, L_ψ, L_r	auxiliary quantities introduced by Eqs. (22.17) and (22.18)

22 Calculation of the Stability and of the Phase Equilibrium

M	molecular mass
$\overline{M}_n$	number averaged molecular mass
$\overline{M}_w$	mass averaged molecular mass
p	pressure
R	universal gas constant
r	segment number
$\bar{r}^M$	number-averaged segment number of the mixture
$\bar{r}_B$	number-averaged segment number of polymer B
$\bar{r}_B^{(i)}$	i-th moment of the distribution function $W_s(r)$
T	temperature
U	non-uniformity of the polymer, $U = \overline{M}_w/\overline{M}_n - 1$
V	volume
$W_s(r)$	continuous segment-molar (mass) distribution function depending on segment number r

Abbreviations

EoS	equation of state
LCP	lower critical point
LCST	lower critical solution temperature
MCH	methylcyclohexane
PPICS	pressure-pulse-induced-critical-scattering
PS	polystyrene
SWP-EoS	equation of state introduced by Sako, Wu, Prausnitz (1989)
UCP	upper critical point
UCST	upper critical solution temperature

Greek letters

α_{ij}	quantity of the spinodal determinant
Γ	gamma function
ϕ_A	fugacity coefficient of the solvent A
$\phi_B(r)$	fugacity coefficient of the species of the polymer B with segment number r
ψ	total polymer mass fraction

Superscripts

I, II	phases (I is considered to be the feed phase)
M	mixture

Subscripts

A	solvent, methylcyclohexane (MCH)
B	polymer, polystyrene (PS)
i	component i
s	segment-molar quantity
s, i	segment-molar quantity of pure component i

VI High-Pressure Phase Equilibria

23 Phase Equilibrium (Solid-Liquid-Gas) in Binary Systems of Poly(ethylene glycols), Poly(ethylene glycol) dimethyl ether with Carbon Dioxide, Propane, and Nitrogen

Eckhard Weidner[1] and Veronika Wiesmet

23.1 Abstract

Phase behavior of binary polymer + gas systems containing two types of polymers (either poly(ethylene glycols) (PEG) with molecular masses from 200 to 8000 g/mol, respectively poly(ethylene glycol) dimethyl ether (PEGDME) with molecular masses of 200 and 2000 g/mol) and three gases (carbon dioxide, nitrogen, and propane) was investigated. Experiments were performed at temperatures from 298 to 393 K and in a pressure range from 0.1 to 30 MPa. Melting points and phase behavior were determined in a high-pressure optical cell. Vapor-liquid and liquid-liquid equilibria were measured in an autoclave using a static analytic method. Phase behavior with liquid polymers depends on the gas and its state of aggregation (liquid/supercritical). Open and closed miscibility gaps (liquid – gas, liquid – liquid) as well as three phase areas (solid – liquid – gas, liquid 1 – liquid 2 – gas) were observed. The experimental data of the system PEG + carbon dioxide are described successfully by the SAFT-equation. Above a certain molecular mass (approx. >600 g/mol) the solubility of carbon dioxide in PEG depends mainly on the chain length of the polymer. The influence of the functional endgroups (OH or methyl ether) of the polymer chain on the solubility of CO_2 and C_3H_8 is dominant for short-chained polymers. For polymers with molecular mass above 1500 g/mol the influence of the functional end group is less pronounced. The melting points of the polymers increase with pressure in the presence of nitrogen, while with carbon dioxide and propane melting points either increase or decrease. This behavior can be explained from the solubility of gas in the molten polymer, which is illustrated by p, T- and p, x-plots.

---

1 Ruhr-Universität Bochum, Institut für Thermo- und Fluiddynamik, Lehrstuhl für Verfahrenstechnische Prozesse, IB6/126, Universitätsstr. 150, 44780 Bochum.

23.2 Introduction

Poly(ethylene glycols) (PEG) are polymers, which, due to their chemical and thermophysical properties, are used in large quantities in the cosmetic, pharmaceutical, chemical, and food industries. Due to their specific fluiddynamic properties, which can be adjusted by the molar mass obtained during polymerisation, liquid PEGs are e.g. applied as lubricants in compressors. Poly(ethylene glycol) dimethyl ether (PEGDME) is used as absorbent in gas cleaning. In both applications high pressure and sub-/supercritical gases are involved. Several authors (Ali, 1983; Kiran et al., 1989; Kumar et al., 1986; Mc Hugh and Krukonis, 1989; Tuminello et al., 1995) studied the solubility of polymers in supercritical fluids (SCF). For the solubility of SCF in polymers and solid-liquid phase transitions in polymer + gas systems only a limited number of experimental data is available (Daneshvar et al., 1989; Daneshvar et al., 1990; Bonner, 1977; Davis et al., 1993).

23.3 Experimental Equipment, Methods, and Substances

23.3.1 Polymers

The general formula of PEG is $H(OCH_2CH_2)_nOH$, where n is the number of ethylene oxide groups. By etherification of the hydroxyl-group at the beginning and the end of the PEG-chain poly(ethylene glycol dimethyl ethers) $CH_3(OCH_2CH_2)_nOCH_3$ are formed. Both polymers are synthesized with various mean molecular masses, which results e.g. in different melting points. Due to the narrow size distribution melting typically occurs in a temperature range of one Kelvin. In the present work PEGs with molar masses of 200, 1500, 4000, and 8000 g/mol were used. PEG is liquid up to a molar mass of 600 g/mol, otherwise PEG is a solid at ambient temperature. PEGDMEs with a molar mass of 200 (liquid) and 2000 g/mol (solid) were used. PEG and PEGDME were purchased from Hoechst AG and used without further purification. The gas purity was >99.9%. Gases were supplied by AGA.

23.3.2 Determination of Melting Point

The high-pressure optical cell shown in Figure 23.1 was used to investigate the phase behavior and for measuring the melting point at elevated pressure. To ensure, that the whole cell content is visible, the windows have the same diameter as the bore ot the steel body (see Fig. 23.1).

The cell is designed for pressures up to 30 MPa and a maximum temperature of 423 K. The volume of the cell is 15.7 cm^3. To lower the time to reach equilibrium a magnetic stirrer is fixed below the cell. The solid polymers are introduced in a vertical mounted

23.3 Experimental Equipment, Methods, and Substances

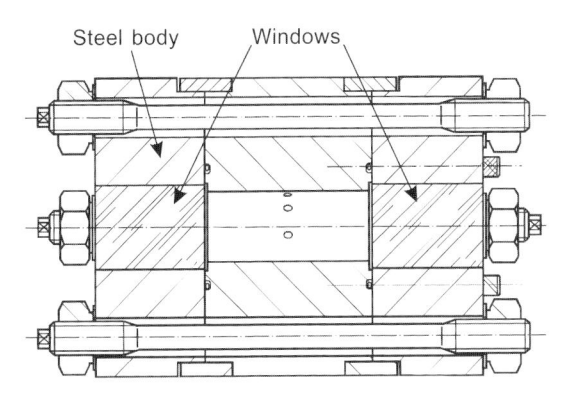

Figure 23.1: High-pressure optical cell with insight into the complete pressurized volume.

glass tube into the optical cell (modified Thiele method). A certain pressure is established by discharging gas. Then the temperature is slowly increased (which causes a change in pressure) in order to observe melting of the polymer in the glass tube. Subsequently the temperature is reduced to the temperature and the pressure at which solidification occurs. The measurements were repeated at least two times and the averages are listed as melting conditions. Typically the data vary by ±0.5 K and ±0.03 MPa. To verify these results some measurements were repeated by a different operator in another optical cell. Typical deviations were within the before described range.

23.3.3 Determination of Solubility

The phase behavior of mixtures of liquid polymer and gas was studied in the optical cell additionally. The construction allows to observe the complete content (see above), which is necessary to detect the formation/existence of phases of even small volumes. From those experiments pressure and temperature regions where phase transitions occur were determined qualitatively. Based on this information quantitative measurements are planned. The solubility of gases in liquid polymers was measured by means of an autoclave (volume: about one liter) designed for pressures up to 50 MPa and temperatures between room temperature and 423 K. The polymer was filled into the autoclave, which was then flushed with the gas under consideration to remove air. The thermostated autoclave was filled with the gas up to a certain pressure. Mixing of the content was established by shaking the autoclave by means of an oscillating device. The time until equilibrium is reached was determined by taking and analyzing samples after different time periods of mixing. In a typical experiment about 30 min were required to reach phase equilibrium. To be on the safe side, the minimum mixing time was increased to approx. 90 min. Pressure was measured with an accuracy of ±0.02 MPa and temperature with an accuracy of ±0.1 K. Another 30 to 60 min were allowed for phase separation (where the autoclave was kept in a vertical position) before samples were taken from the upper and lower phase by expanding into glass vials. The volume of the gas set free during that expansion and the mass of the precipitated polymers in the glass vial were determined. For more details of the experimental procedure see Wiesmet (1998).

23.3.4 Correlation of the Experimental Results with SAFT

Some isotherms of the system PEG + carbon dioxide, PEG + nitrogen, and PEG + propane were correlated with the Statistical Associating Fluid Theory (SAFT) by means of a computer code developed by the group of Arlt and Sadowski at the Institute of Verfahrenstechnik, TU Berlin. The SAFT-model assumes that small molecules or parts of molecules (e.g. functional groups) are spherical. The spheres are characterized as segments. For example CO_2 consists of a single segment. A polymer is characterized by a row of segments. The number of segments increases linear with the length of the chain. For computing the phase equilibrium, the number of segments, the volume (which is not depending on the temperature), and parameters describing the interaction between segments are required (Arlt, 1996; Mier, 1995; Huang and Radosz, 1990).

23.4 Results and Discussion

The measurements were performed at pressures ranging from 0.1 to 30 MPa, at temperatures ranging from 298 to 393 K. A survey on the investigated conditions is given in Table 23.1. Some selected but typical results will be presented here. The full results are available elsewhere (Wiesmet, 1998).

Table 23.1: Phase behavior of PEG and PEGDME with carbon dioxide, nitrogen, and propane: Investigated temperatures.

Gas			Carbon dioxide			
			T/K			
PEG 200	303	323	–	353	373	
PEG 600	–	323	–	353	373	
PEG 1500	316	323	333	353	–	
PEG 4000	–	328	333	353	373	
PEG 8000	–	–	–	353	–	
PEGDME 200	298	311	–	351	373	393
PEGDME 2000	–	–	331	351	373	393
			Nitrogen			
PEG 1500	323	333	351	373		
PEG 4000	–	338	353	373		
PEG 8000	–	–	353	373		

Table 23.1 (continued)

Gas			Propane			
PEG 200	303	323	–	–	373	393
PEG 600	–	–	–	–	373	393
PEG 1500	–	323	–	353	373	393
PEG 4000	–	–	333	353	373	393
PEG 8000	–	–	–	353	373	393
PEGDME 200	298	311	–	351	373	393
PEGDME 2000	–	–	331	351	373	393

23.4.1 PEG + Carbon Dioxide

23.4.1.1 Liquid + Gas Systems

Figure 23.2 (and Table 23.2) show experimental results for the solubility of carbon dioxide in PEG 4000.

For all investigated temperatures CO_2 is supercritical. With increasing temperature the solubility of CO_2 in the polymer-rich liquid phase decreases, thus the miscibility gap between liquid and gas increases. The solubility of PEG in supercritical CO_2 is smaller than the limit of detection of the applied analytical method (approx. 0.1 mass% polymer in carbon dioxide). Therefore, the composition of the vapor phases are not shown in the quantitative diagrams. Open mixing gaps are observed for all investigated PEG + CO_2 systems. The

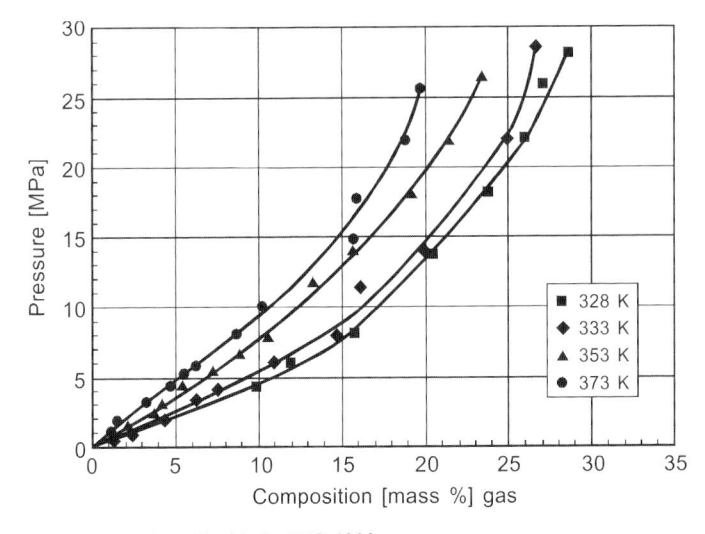

Figure 23.2: Solubility of carbon dioxide in PEG 4000.

Table 23.2: Solubility of carbon dioxide in PEG 4000.

T = 328 K		T = 333 K		T = 353 K		T = 373 K	
p/MPa	mass% CO_2	p/MPa	mass% CO_2	p/MPa	mass% CO_2	p/MPa	mass% CO_2
4.3	9.8	0.5	1.3	0.6	1.1	1.1	1.1
6.0	11.9	0.9	2.2	1.6	2.2	2.0	1.5
8.1	15.8	2.0	4.4	2.4	3.7	3.2	3.2
13.8	20.5	3.4	6.2	3.1	4.2	4.0	4.2
18.2	23.8	4.0	7.3	4.4	5.4	4.4	4.7
22.1	26.0	6.1	10.9	5.4	7.3	5.3	5.4
26.0	27.1	8.0	14.7	6.6	8.9	5.8	6.2
28.2	28.6	11.5	16.2	7.9	10.6	8.1	8.7
		13.9	19.8	11.7	13.3	10.1	10.2
		22.0	24.9	14.0	15.7	14.9	15.7
		28.6	26.6	18.1	19.3	17.8	15.9
				21.9	21.5	22.0	18.8
				26.5	23.4	25.7	19.7

solubility of CO_2 in PEG (1500 or 4000 or 8000) was correlated with the SAFT-model (Wiesmet et al., 2000). A comparison of the experimental data with the correlation is shown in Figure 23.3 for 353 K, additionally the numerical comparison is given in Table 23.3.

The SAFT model gives a slightly higher gas solubility than determined experimentally. The calculations confirm the experimental findings, that the gas solubility depends only weakly on the molecular mass (and thus on the chain length) of the polymer. From comparison with literature data and own measurements with PEG of lower molecular weight, it was found, that this behavior changes considerably in a range of molecular mass between

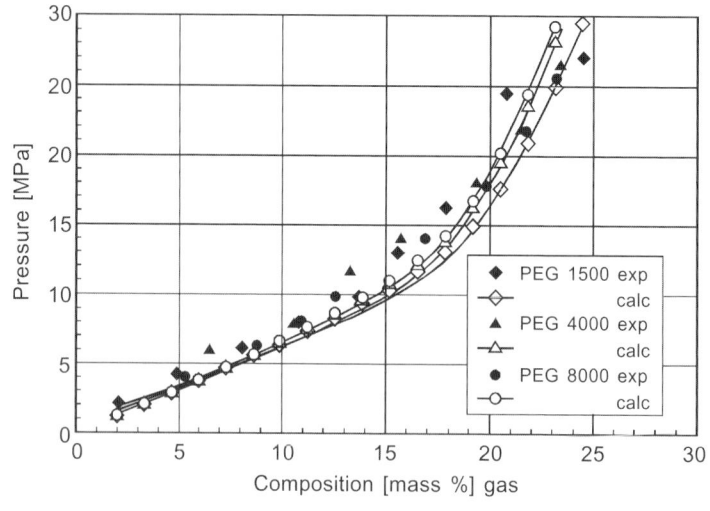

Figure 23.3: Solubility of carbon dioxide in PEG 4000 at 353 K: Comparison of experimental results and calculations using SAFT.

23.4 Results and Discussion

Table 23.3: Solubility of carbon dioxide at 353 K in different PEGs as calculated with SAFT.

\multicolumn{2}{c}{PEG 1500}	\multicolumn{2}{c}{PEG 4000}	\multicolumn{2}{c}{PEG 8000}
p/MPa	mass% CO_2	p/MPa	mass% CO_2	p/MPa	mass% CO_2
1.31	2.00	1.27	2.00	1.28	2.00
2.05	3.32	2.11	3.32	2.13	3.32
2.88	4.64	2.97	4.64	2.99	4.64
3.72	5.96	3.83	5.96	3.87	5.96
4.57	7.28	4.72	7.28	4.76	7.28
5.45	8.60	5.62	8.60	5.68	8.60
6.34	9.92	6.56	9.92	6.62	9.92
7.27	11.2	7.52	11.2	7.60	11.2
8.23	12.6	8.54	12.6	8.64	12.5
9.25	13.9	9.63	13.9	9.75	13.9
10.4	15.2	10.8	15.2	10.9	15.2
11.6	16.5	12.2	16.5	12.4	16.5
13.0	17.8	13.9	17.8	14.2	17.8
14.9	19.2	16.3	19.2	16.7	19.2
17.5	20.5	19.5	20.5	20.1	20.5
20.9	21.8	23.5	21.8	24.4	21.8
24.9	23.1	28.2	23.1	29.3	23.1
29.5	24.4				

600 and 1500 g/mol. The solubility of carbon dioxide in the liquid polymer is considerably lower in PEG 200 and 600 than in PEG 1500, whereas the solubility is very similar in PEGs 1500, 4000, and 8000. This behavior was also found when carbon dioxide was replaced by propane. The solubility of a gas in a long-chained polymer liquid seems to be correlated with the solvation of the functional chain group by the dissolved gas. Each chain segment is solvated by a certain number of gas molecules, resulting in a specific "local" gas concentration. By increasing the number of functional groups in the polymer chain, the ratio between each segment and "locally" surrounding gas molecules seems to be constant and thus the measured overall gas solubility is practically unchanged.

23.4.1.2 Solid-Liquid Transition

If mechanical pressure is applied, it is expected, that the melting point of the pure polymer increases. This behavior changes if the pressure is applied by means of a gas, that is soluble in the polymer (Weidner, 1997). Figure 23.4 and Table 23.4 show the influence of pressure on the melting point of PEG 4000.

At low pressures (up to approx. 1 MPa) a certain increase in the melting temperature is observed, followed by a pronounced reduction (up to approx. 10 MPa). At even higher pressures the melting temperature is almost constant and then rises with increasing pressure. These phenomena can be understood from the competing effects of melting point depression by a solute (here: carbon dioxide) and the increase by the "mechanic" effect of pressure. At low pressures the "mechanic" effect seems to prevail, while the solute effect becomes dominant in the pressure range up to 10 MPa. As can be seen from Figure 23.3 the solubility of

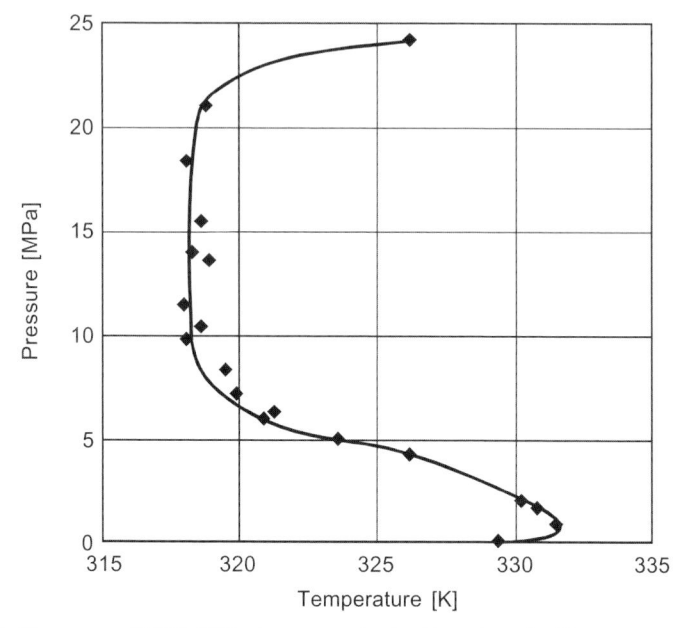

23 *Phase Equilibrium (Solid-Liquid-Gas) in Binary Systems*

Figure 23.4: Melting curve of PEG 4000 when pressurized with carbon dioxide.

Table 23.4: Melting temperature of PEG 4000 pressurized by carbon dioxide.

p/MPa	T/K	p/MPa	T/K	p/MPa	T/K
24.3	326.2	10.4	318.6	5.02	323.6
21.1	318.8	9.8	318.1	4.27	326.2
18.4	318.1	8.4	319.5	2.05	330.2
15.5	318.6	7.2	319.9	1.7	330.8
14.0	318.3	6.3	321.3	0.95	331.5
13.6	318.9	6.0	320.9	0.11	329.4
11.5	318.0				

carbon dioxide in the polymer-rich phase rises strongly at lower pressures, causing a reduction in the melting temperature. At higher pressures, the pressure dependency of the gas solubility in the polymer is less pronounced. In that range, the "mechanic" effect becomes more important, while at still higher pressures an upper critical end point might by reached in the system PEG 4000 + carbon dioxide. Nevertheless, it has to be admitted, that the sharp increase at pressures below approx. 1 MPa (which was measured in independent apparatuses by different authors) is not yet completely understood.

23.4 Results and Discussion

23.4.2 PEG + Nitrogen

23.4.2.1 Liquid + Gas Systems

The solubility of nitrogen in PEG 4000 (cf. Fig. 23.5 and Table 23.5) is much lower than that of carbon dioxide.

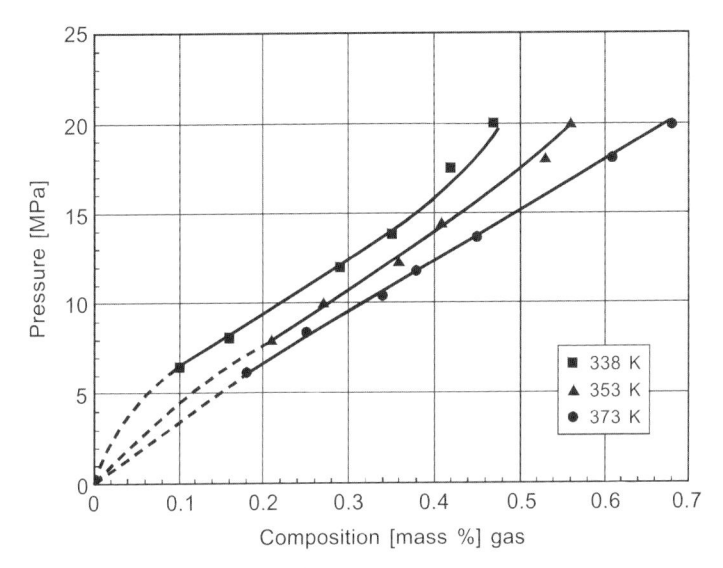

Figure 23.5: Solubility of nitrogen in PEG 4000.

Table 23.5: Solubility of nitrogen in PEG 4000.

T = 338 K		T = 353 K		T = 373 K	
p/MPa	mass% N_2	p/MPa	mass% N_2	p/MPa	mass% N_2
6.5	0.1	8.0	0.21	6.2	0.18
8.1	0.16	10.0	0.27	8.4	0.25
12.0	0.29	12.3	0.36	10.4	0.34
12.0	0.29	14.5	0.41	11.8	0.38
13.8	0.35	18.0	0.53	13.7	0.45
13.9	0.35	20.0	0.56	18.1	0.61
17.5	0.42			19.9	0.68
20.0	0.47				

The maximum solubility of nitrogen determined in the present work is 0.7 mass%, whereas it is around 25 mass% for CO_2. Additionally the PEG + nitrogen system reveals a different behavior when the temperature is varied. At constant pressure the solubility of nitrogen in PEG rises with increasing temperature for PEG samples of molar masses from 1500–8000 g/mol.

For carbon dioxide as well as for nitrogen the gas solubility is almost independent of the polymer chain length, which is a further indication for the above given explanation for gas solubility in polymers with different chain lengths.

23.4.2.2 Solid-Liquid Transition

As it could be assumed from the low solubilities, pressurization of PEG with nitrogen influences the melting point in a different way than pressurization with carbon dioxide (cf. Fig. 23.6).

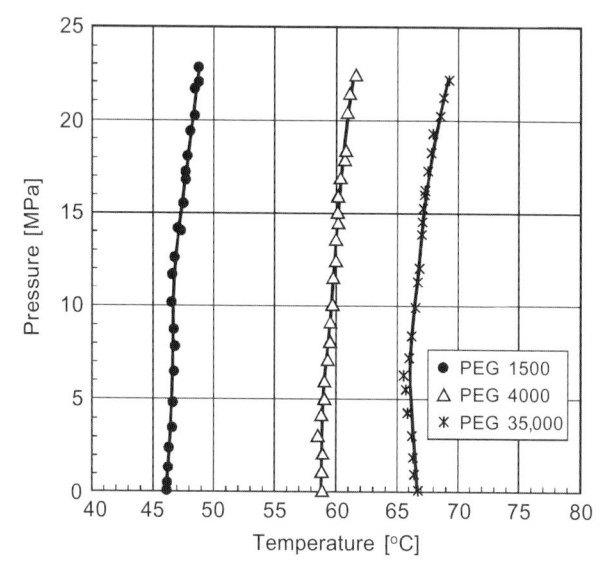

Figure 23.6: Melting temperature of PEG when pressurized with nitrogen (Weidner et al., 1997a).

For PEGs with medium molar mass the melting point increases slightly with pressure, while for high molecular PEG (e.g. PEG 35,000) a small decrease followed by a slight increase is observed. This behavior indicates, that – due to the low solubility – the solute effect is very small and the "mechanic" effect dominates. The slight decrease for the high molecular PEGs might be attributed to morphological changes of the polymers. This effect is subject to further investigations.

23.4.3 PEG + Propane

23.4.3.1 Liquid + Gas Systems

Figure 23.7 and Table 23.6 show the experimental data for the solubility of propane in PEG 4000. The maximal solubility of C_3H_8 in PEG in the investigated range is about 9 mass% and thus is between those measured for carbon dioxide and nitrogen.

23.4 Results and Discussion

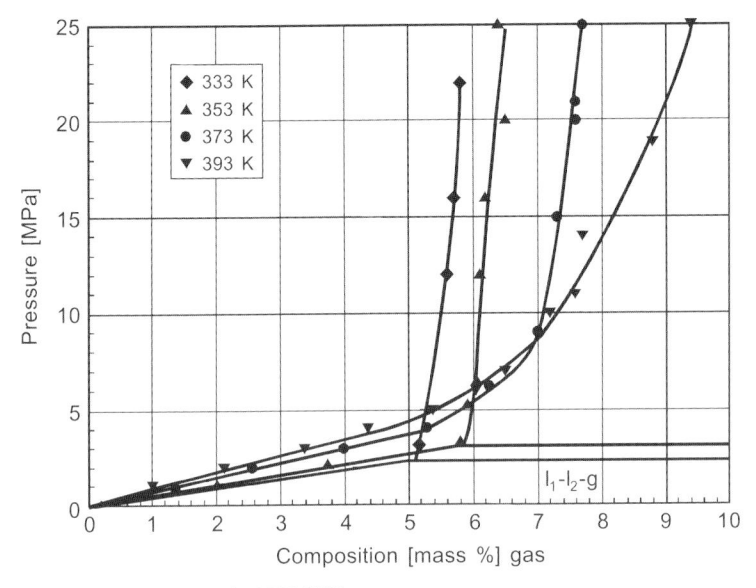

Figure 23.7: Solubility of propane in PEG 4000.

Table 23.6: Solubility of propane in PEG 4000.

T = 333 K		T = 353 K		T = 373 K		T = 393 K	
p/MPa	mass% C_3H_8	p/MPa	mass% C_3H_8	p/MPa	mass% C_3H_8	p/MPa	mass% C_3H_8
2.0	4.2	1.1	2.0	0.9	1.4	1.1	1.0
2.3	5.0	2.1	3.8	2.0	2.6	2.0	2.1
3.1	5.2	5.2	5.9	3.0	4.0	3.0	3.4
5.0	5.3	6.4	6.1	4.0	5.3	4.0	4.4
12.0	5.6	8.0	5.8	6.2	6.2	5.0	5.4
16.0	5.7	11.0	6.4	9.0	7.0	6.0	6.0
22.0	5.8	12.0	6.1	15.0	7.3	7.0	6.5
		16.0	6.2	20.0	7.6	10.0	7.2
		20.0	6.5	21.0	7.6	11.0	7.6
		25.0	6.4	25.0	7.7	14.0	7.7
						19.0	8.8
						25.0	9.4

When the temperature is below the critical temperature of propane, the solubility of propane in the polymer-rich liquid phase increases strongly with increasing pressure. In the vicinity of the vapor pressure of propane a sharp bend in the isotherm is observed, indicating the coexistence of three phases: a liquid polymer-rich phase (l_1), a liquid propane-phase (l_2), and gaseous propane (g). At higher pressures a liquid-liquid area exists. The influence of pressure on the solubility of propane in PEG is less pronounced than in liquid-gas domains. Below the three phase line, the density of gaseous propane increases strongly with

23 Phase Equilibrium (Solid-Liquid-Gas) in Binary Systems

pressure, resulting in a significant rise of solubility in the liquid polymer. The influence of pressure on the density of the propane-rich liquid phase is much lower. Therefore also the influence of pressure on the solubility of liquid propane in liquid PEG is lower than that of gaseous propane.

The three-phase behavior disappears at temperatures above the critical temperature of propane. The 373 K- and 393 K-isotherms no longer exhibit a sharp bend. Over the whole pressure range a liquid polymer-rich phase (l) coexists with a gaseous/supercritical phase of almost pure propane. The solubility of PEG in propane (either liquid, gaseous, or supercritical) is – similar to the solubility in carbon dioxide – in the range of the detection limit of the applied method. Therefore the gas phase lines are not plotted. As is shown in Figure 23.7 the isotherms intersect. At low pressures the solubility of propane in liquid PEG decreases with rising temperature (similar to the solubility of CO_2). At higher pressures, the solubility increases with increasing temperature (similar to the solubility of N_2). The influence of the molecular mass of the polymer on the solubility of propane is depicted in Figure 23.8 and Table 23.7.

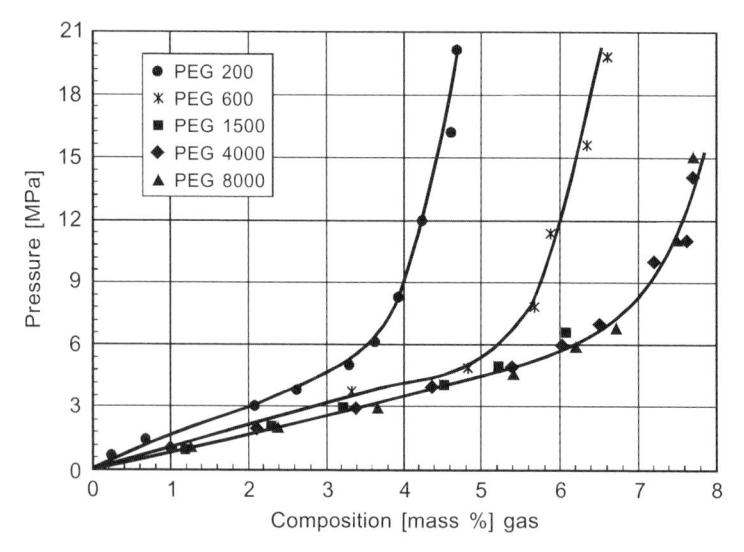

Figure 23.8: Solubility of supercritical propane at 393 K in poly(ethylene glycols) of different molecular mass.

The solubility of propane (at constant temperature and pressure) decreases with increasing molecular mass of PEG. At high molecular masses, the influence of the chain length is diminished. The higher polarity of the smaller polymers caused by the ratio between chain length and OH-end gives cause to the low solubility of the almost nonpolar propane (dipole moment: 0.084 Debye). For longer chains, the influence of the OH-end groups on the solubility of propane is reduced.

23.4 Results and Discussion

Table 23.7: Solubility of supercritical propane in liquid PEG at 393 K.

PEG 200		PEG 600		PEG 1500		PEG 4000		PEG 8000	
p/MPa	mass% C_3H_8	p/MPa	mass% C_3H_8	p/MPa	mass% C_3H_8	p/MPa	mass% C_3H_8	p/MPa	mass% C_3H_8
0.7	0.23	3.8	3.3	1.0	1.2	1.1	1.0	1.1	1.3
2.0	1.13	4.9	4.8	2.1	2.3	2.0	2.1	2.0	2.4
3.1	2.09	7.8	5.7	3.0	3.2	3.0	3.4	3.0	3.7
3.9	2.62	11.4	5.9	4.1	4.5	4.0	4.4	4.6	5.4
5.1	3.30	15.6	6.3	5.0	5.2	5.0	5.4	5.9	6.2
6.2	3.64	19.8	6.6	6.6	6.1	6.0	6.0	6.8	6.7
8.3	3.94					7.0	6.5	11.0	7.5
12.0	4.24					10.3	7.2	15.0	7.7
16.2	4.60					11.0	7.6		
20.1	4.68					14.0	7.7		

23.4.3.2 Solid-Liquid Transition

Figure 23.9 (and Table 23.8) show the experimental results for the melting temperature of the PEG 4000 + propane system.

The melting point depends on the state of aggregation of propane. At low pressure propane is gaseous and the melting point of PEG 4000 decreases with increasing pressure. This is due to the large solute effect (the sharp increase of the solubility in the l-g area cf. Fig. 23.7). At the transition from gaseous propane to liquid propane there is a sharp bend in

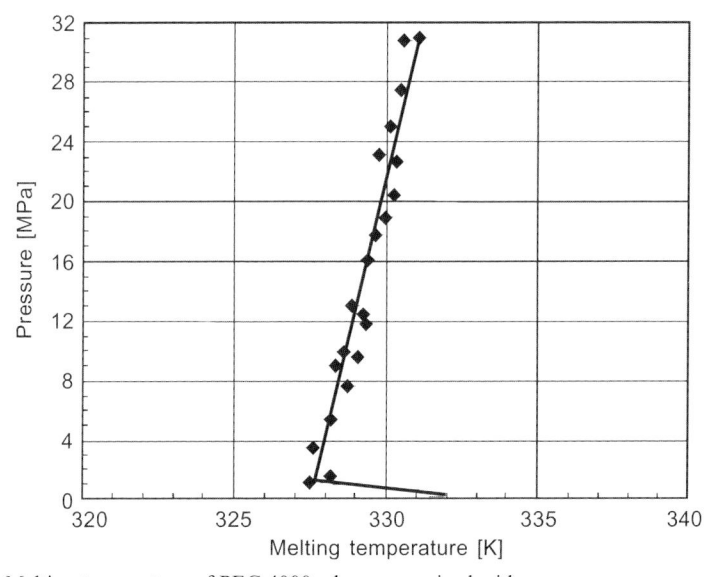

Figure 23.9: Melting temperature of PEG 4000 when pressurized with propane.

Table 23.8: Melting points of PEG 4000 in pressurized propane.

p/MPa	T/K	p/MPa	T/K	p/MPa	T/K
27.48	330.5	16.05	329.4	3.48	327.6
25.0	330.2	12.90	328.9	1.20	328.0
23.13	329.8	12.40	329.3	1.15	327.5
22.73	330.4	11.70	329.4	1.13	328.1
20.43	330.3	9.50	329.1	0.85	328.5
18.90	330.0	7.55	328.8	0.5	330.9
17.75	329.7	5.30	328.2	0.1	331.7

the melting curve above which the melting temperature increases with increasing pressure. As mentioned above, when propane is liquid, the solubility of propane in PEG rises only slightly with pressure, thus the "mechanic" effect dominates.

23.4.4 Comparison of the Phase Behavior of Liquid and Solid PEGs With Different Gases

As shown in Figure 23.10, the phase behavior of a mixture of (liquid) PEG and a gas depends on the nature of the gas and its state of aggregation.

Systems with supercritical carbon dioxide possess a liquid-gas behavior with an open mixing gap over the whole investigated pressure range. The same is observed for nitrogen as the gaseous component. In contrary to carbon dioxide, the solubility of nitrogen in PEG increases with rising temperature. Propane is both sub- and supercritical in the temperature range under consideration. Its phase behavior with PEG is characterized by two-phase (vapor-liquid and liquid-liquid) and three-phase (liquid-liquid-vapor) equilibria. The concentra-

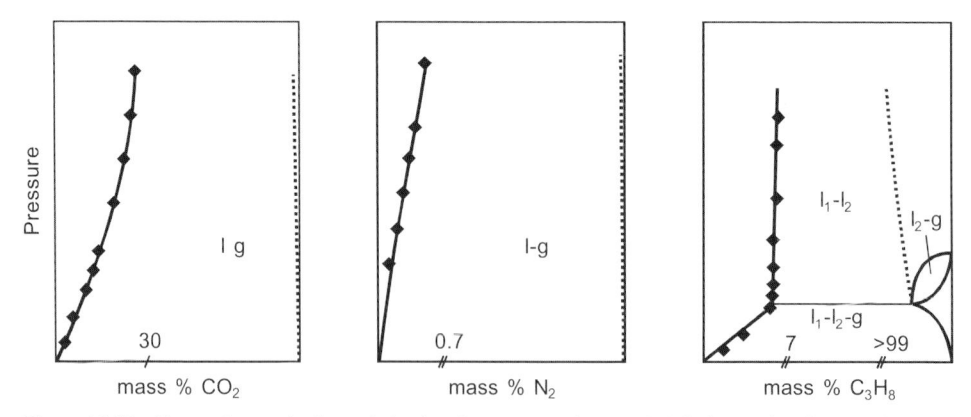

Figure 23.10: Comparison of phase behavior between molten poly(ethylene glycols) (molar mass >600 g/mol) and carbon dioxde, nitrogen, and propane (pressure-composition plots).

tion of PEG in the vapor phase as well as in the propane-rich liquid phase is so small, that it could not be measured in the present investigation. Therefore these phase boundaries are only drawn qualitatively in Figure 23.10. The influence of the polymer chain length on the solubility for all three gases is rather small as long as the molecular mass of PEG is between 1500 and 8000 g/mol. The solubility of carbon dioxide and propane in liquid PEG of lower molecular mass is considerably smaller then the solubility of those gases in PEG of higher molecular mass.

The melting behavior for the three systems is compared qualitatively in Figure 23.11.

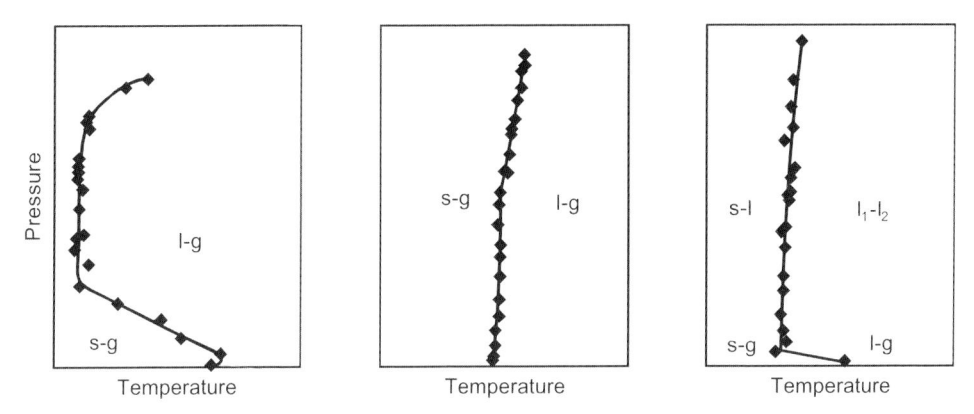

Figure 23.11: Comparison of melting behavior of solid poly(ethylene glycols) pressurized by carbon dioxde, nitrogen, and propane (pressure-temperature plots).

The maximum reduction of the melting temperature of PEG + CO_2 mixtures is about 10 K. The minimum melting temperature is reached at pressures between 10 and 17 MPa. In PEG + nitrogen mixtures the melting temperature increases with increasing pressure. For example pressurization to 30 MPa increases the melting temperature by 2 to 3 K. When PEG is pressurized by gaseous propane, the melting temperature at first decreases with increasing pressure. The minimum temperature is reached in the vicinity of the vapor pressure of pure propane, where it is about 4 to 6 K lower than the melting temperature of pure PEG.

23.4.5 Influence of Functional Groups

The results shown above indicate that the functional groups have a significant effect on the phase behavior of PEG + gas systems. To verify this observation, experimental investigations on molecules containing the same functional groups in the chain, but different end groups (methyl ether- or hydroxy-group) were performed. The results are presented first for short-chained and then for long-chained polymers.

23 *Phase Equilibrium (Solid-Liquid-Gas) in Binary Systems*

23.4.5.1 Short Polymer Chains

Figure 23.12 (and Table 23.9) show the solubility of carbon dioxide in PEG 200.

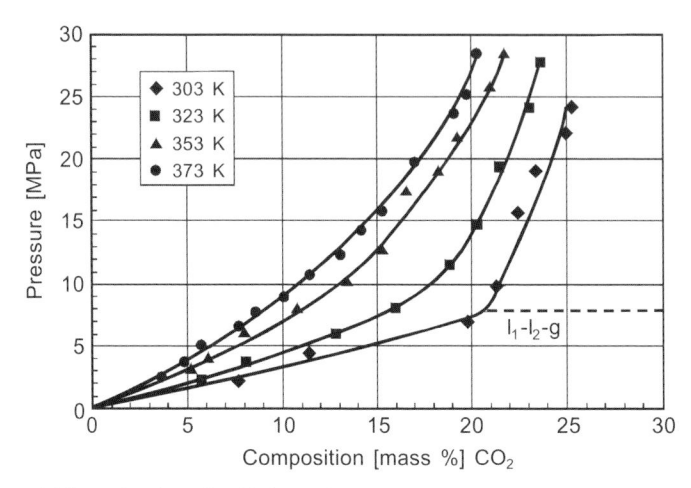

Figure 23.12: Solubility of carbon dioxide in PEG 200.

Table 23.9: Solubility of carbon dioxide in PEG 200.

T = 303 K		T = 323 K		T = 353 K		T = 373 K	
p/MPa	mass% CO_2	p/MPa	mass% CO_2	p/MPa	mass% CO_2	p/MPa	mass% CO_2
2.2	7.7	2.3	5.7	3.2	5.2	2.5	3.6
4.4	11.4	3.7	8.1	4.0	6.1	3.7	4.8
6.9	19.8	6.0	12.8	6.1	7.9	5.1	5.8
7.8	20.8	8.1	16.0	8.0	10.7	6.6	7.7
9.8	21.4	11.6	18.8	10.2	13.4	7.8	8.5
15.7	22.4	14.8	20.3	12.7	15.2	9.0	10.0
19.0	23.3	19.4	21.5	17.4	16.5	10.7	11.4
22.1	25.0	24.1	23.0	19.0	18.2	12.3	13.0
24.2	25.3	27.8	23.6	21.8	19.2	14.3	14.2
				25.8	21.0	15.8	15.2
				28.5	21.7	19.8	17.0
						23.7	19.0
						25.2	19.7
						28.5	20.3

The system has a large miscibility gap. The solubility of CO_2 in PEG 200 increases with increasing pressure to about 25 mass% at 30 MPa. The phase behavior changes when the OH-end group in PEG is substituted by the methyl ether group. This is shown in Figure 23.13.

23.4 Results and Discussion

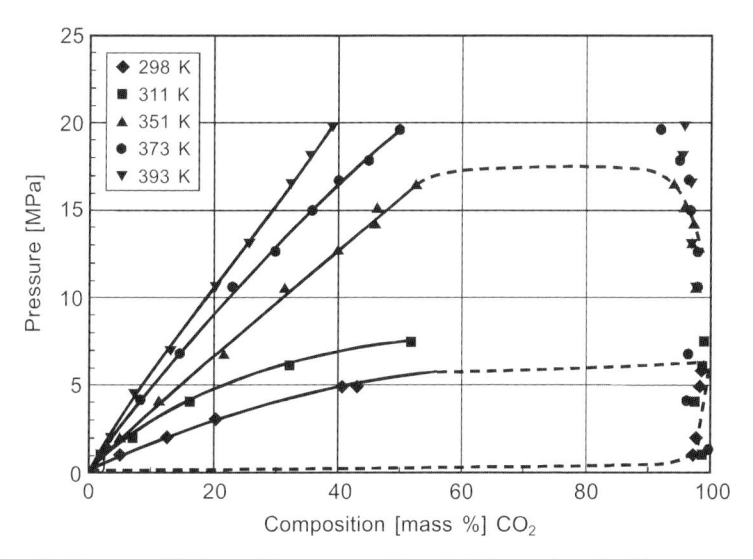

Figure 23.13: Liquid-gas equilibrium of the system PEGDME 200 + carbon dioxide.

Table 23.10: Liquid-gas equilibrium of the system PEGDME 200 + carbon dioxide.

$T = 298$ K		$T = 311$ K		$T = 351$ K		$T = 373$ K		$T = 393$ K	
p/MPa	mass% CO_2	p/MPa	mass% CO_2	p/MPa	mass% CO_2	p/MPa	mass% CO_2	p/MPa	mass% CO_2
				Liquid phase					
1.0	5.17	1.0	1.85	2.0	5.07	1.4	2.88	2.0	3.64
2.0	12.55	2.0	6.89	4.1	11.27	4.2	8.10	4.5	7.33
3.0	20.34	4.0	16.26	6.7	21.64	6.8	14.51	7.0	13.04
4.1	30.64	6.1	32.20	10.5	31.52	10.6	23.00	10.6	20.14
4.9	40.73	7.5	51.75	12.7	39.91	12.6	29.88	13.1	25.56
5.7	56.04			14.2	45.93	15.0	35.73	16.5	32.40
				15.1	46.45	16.7	40.08	18.1	35.45
				16.5	52.6	17.8	44.90	19.8	39.09
						19.6	49.77		
				Gas phase					
1.0	97.31	1.0	98.64	10.5	97.57	1.3	99.55	10.6	97.68
2.0	97.87	4.0	97.54	12.7	97.98	4.1	96.14	13.1	96.96
3.0	96.32	6.1	98.70	14.2	97.35	6.8	96.28	16.5	96.75
4.1	98.12	7.5	98.85	15.1	95.67	10.6	97.75	18.1	95.33
4.9	98.43			16.5	93.97	12.6	97.89	19.8	95.57
5.7	98.63					15.0	96.67		
						16.7	96.38		
						17.8	94.90		
						19.6	91.78		

527

23 *Phase Equilibrium (Solid-Liquid-Gas) in Binary Systems*

The mutual solubility of carbon dioxide in PEGDME 200 is much larger than in PEG 200. The miscibility gap closes at rather low temperatures and pressures. Considerable amounts of ethers are found in the gas phase. The influence of temperature on the phase behavior is similar for both polymers, e.g., the miscibilty gap increases with increasing temperature.

23.4.5.2 Longer Polymer Chains

The phase behavior of two mixtures (carbon dioxide + PEG 1500 and carbon dioxide + PEGDME 2000) are compared in Figure 23.14.

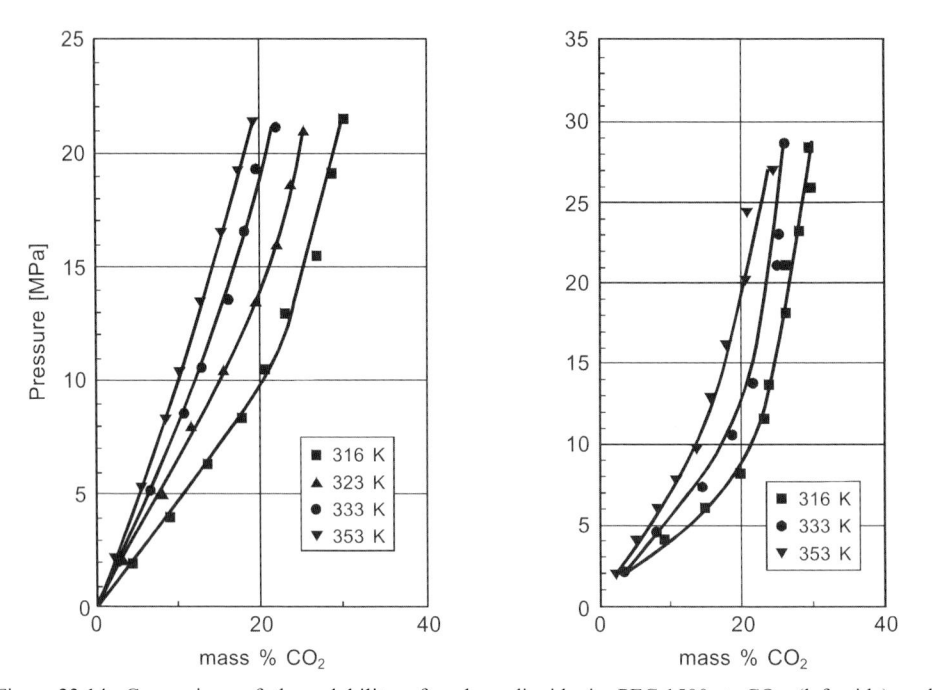

Figure 23.14: Comparison of the solubility of carbon dioxide in PEG 1500 + CO_2 (left side) and PEGDME 2000 (right side).

In opposite to the system with short-chained molecules (see Section 23.4.5.1) the phase behavior of the polymers with the two different end groups does not change qualitatively. Both systems with the longer-chained polymers have an open mixing gap with carbon dioxide. For example, the solubility of CO_2 in PEG 1500 is about 24 mass% at 331 K and about 29 mass% at 333 K, both at 30 MPa. The influence of the end group on the gas solubility is much less pronounced then for short-chained polymers.

Table 23.11: Solubility of carbon dioxide in PEGDME 2000.

T = 316 K		T = 333 K		T = 353 K	
p/MPa	mass% CO_2	p/MPa	mass% CO_2	p/MPa	mass% CO_2
2.0	4.1	2.1	3.4	2.0	2.1
4.1	9.0	4.5	8.0	4.1	5.
6.1	14.8	7.4	14.3	6.1	8.1
8.2	19.8	10.6	18.7	7.9	10.8
11.6	23.2	13.8	21.5	9.8	13.7
13.7	23.8	21.1	24.9	13.0	15.6
18.1	26.2	23.0	25.3	16.2	17.9
21.1	26.2	28.6	26	20.2	20.6
23.2	28.1			24.4	20.8
				27.0	24.5

Table 23.12: Solubility of carbon dioxide in PEG 1500.

T = 316 K		T = 323 K		T = 333 K		T = 353 K	
p/MPa	mass% CO_2	p/MPa	mass% CO_2	p/MPa	mass% CO_2	p/MPa	mass% CO_2
2.0	4.1	2.1	3.4	2.1	3.4	2.0	2.1
4.1	9.0	4.1	9.0	4.5	8.0	4.1	5.0
6.1	14.8	5.4	11.7	7.4	14.3	6.1	8.1
8.2	19.8	7.8	16.9	10.6	18.7	7.9	10.8
11.6	23.2	11.8	20.0	13.8	21.5	9.8	13.7
13.7	23.8	15.7	21.5	21.1	24.9	13.0	15.6
18.1	26.2	20.3	26.3	23.0	25.3	16.2	17.9
21.1	26.2	24.0	29.8	28.6	26.0	20.2	20.6
23.2	28.1	27.1	30.0			24.4	20.8
25.9	29.7					27.0	24.5
28.4	29.5						

23.4.6 Classification of Phase Behavior for (Solid and Liquid) PEGs and Pressurized Gases

Scott and van Konynenborg proposed a classification system for high pressure vapor liquid equilibria (Scott and van Konynenborg, 1970; Scott, 1972; van Konynenborg and Scott, 1980). The investigated systems containing liquid and gaseous/supercritical phases are assigned to that classification. As an example, the p-T projection of the phase behavior of the system PEG 200 + carbon dioxide is presented in Figure 23.15.

The projection show the l-g lines of the pure component 1 (carbon dioxide) and pure component 2 (PEG 200) which start at the triple point (TP) and end in the critical points (KP_1 and KP_2). The liquid-liquid demixing is represented by the three phase line l_1-l_2-g, which – due to the boiling point depression caused by PEG – is slightly below the vapor

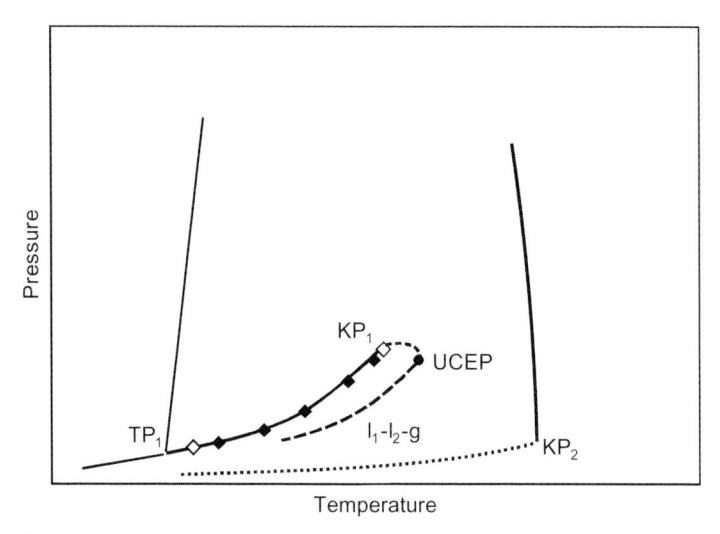

23 Phase Equilibrium (Solid-Liquid-Gas) in Binary Systems

Figure 23.15: Phase boundaries in the system PEG 200 + CO_2, class III according to Scott/Van Konyenborg.

pressure curve of pure carbon dioxide. At pressures and temperatures between the two curves two liquid phases coexist. The l_1-l_2-g curve ends in an upper critical end point (UCEP). Above the UCEP a large area exits, where a polymer-rich liquid phase coexists with an almost pure CO_2-gas phase. Scott and van Konynenborg classified this type of phase behavior as number III. The same phase behavior for example is found in the systems CO_2 + H_2O and ethane + H_2O.

The classification was extended to systems with solid phases by Mc Hugh, who suggests two further classes (McHugh et al., 1984; Mc Hugh and Krukonis, 1986). According to this classification, the system PEG 4000 + CO_2 shows a phase behavior of class A (cf. Fig. 23.16).

Beside the two l-g lines of the pure components a s-l-g three-phase line exists. It starts at the triple point of PEG, reveals a temperature-maximum (near the triple point of PEG) and a temperature minimum at elevated pressures (cf. also Fig. 23.4). On the right hand side of the s-l-g line a liquid polymer phase coexists with almost pure carbon dioxide. On the left hand side of the s-l-g line solid PEG coexists with either gaseous or supercritical carbon dioxide. This type of phase behavior was also found for the *n*-butanol + H_2O system (McHugh, 1989). Further details on the classification were given by (Wiesmet, 1998).

23.5 Application

The knowledge on phase behavior forms the basis for designing and understanding of physical, chemical, and technical processes. Two recently developed examples, which illustrate some technical conclusions drawn from the above described phase behavior, are briefly described below.

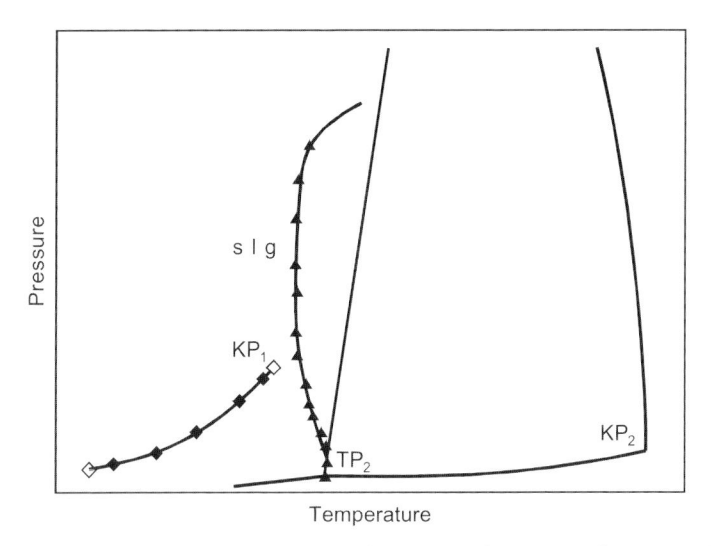

Figure 23.16: Phase boundaries in the system PEG 4000 + CO_2, class A according to Mc Hugh.

23.5.1 Powder Generation from Poly(ethylene glycols)

The melting point depression allows to liquify PEG without ever heating it over its melting temperature at normal pressure. This phenomenon could be named as "gas induced liquefaction". The effect is of interest for liquifying thermolabile substances, which decompose before melting. The pressurized, gas-saturated solutions have interesting properties, that are subject of running investigations (Koukova, 2001). Melt viscosity and interfacial tension are for instance significantly reduced. This allows e.g. to carry out filtrations, that without the presence of compressed gases are difficult or impossible. The low viscosity and the pressure dependency of gas solubility also form favorable conditions for processes for particle generation. Due to the specific properties, the gas-saturated melts can be sprayed through nozzles. By the volume increase of the expanding gas after the nozzle the liquid substance to be powdered bursts into very fine droplets. Depending on the amount and state of aggregation of the dissolved gas, the temperature during expansion may fall significantly due to Joule-Thomson phenomenon or evaporation/sublimation of the gas. The temperature may reach values considerably below the solidification temperature of the melt, which then solidifies and is obtained as a fine-dispersed powder. An example, where micronized particles are generated from gas-saturated solutions of PEG and CO_2 is presented in Weidner et al. (1996).

23.5.2 Polyglycols as Lubricants in Climatisation Systems

Knowledge on phase behavior is also of high relevance for designing lubrication systems and cooling circuits. As mentioned in the introduction, PEG and PEG derivatives like copolymers of poly(ethylene glycols) and poly(propylene glycols) or ethers of such compounds are used as lubricants. In the last years carbon dioxide as environmentally benign refrigerant is subject of intensive international research. Due to the high number of revolution of refrigerant compressors in automobiles, lubrication is the basis for assuring sufficient life-time of such "mobile" climatisation systems. Polyglycols are considered as a promising candidate for such applications (e.g. Fahl, 1998). Due to the mutual solubility between carbon dioxide and PEG, the lubricants cannot be kept in the compressor, but are transported with the refrigerant through the entire cooling circuit. During the circulation, the conditions change from supercritical on the high-temperature/high-pressure side of the compressor to subcritical after the expansion into the "cold" side. All of the above described phase behavior may occur in such a circuit. The knowledge of phase equilibria is essential for the proper design and operation of such climatisation circuits. It must e.g. be assured, that at no place in the circuit solid lubricant forms.

Depending on the temperature- and pressure dependency of the mutual solubility, the heat exchangers in the cooling circuit are either operated in the one-phase area or at conditions where liquid-gas demixing or liquid-liquid demixing or even three phases may occur. For designing and optimizing those important components of climatisation systems, the number of existing phases is essential to know (Hauk, 2001; Hauk and Weidner, 2000).

The energetic design and optimization of cooling systems is typically performed under the assumption, that pure refrigerant circulates. The boiling curve and the evaporation enthalpy of the pure fluid determine the minimum temperature to be reached, the pressure level in the high- and low-pressure part, and the specific cooling power per mass unit of refrigerant. In typical automotive applications the mass of lubricant in the cooling system is around 20 wt% of the refrigerant. From the high solubility of CO_2 in polyglycols can be derived, that the results of the energetic optimization obtained under the "pure refrigerant assumption" are subject to further improvement. The principles of phase behavior in polyglycol systems described above may help in developping energetically optimized climatisation systems with environmentally benign refrigerants like carbon dioxide.

23.6 Conclusion

The solid-liquid-gas behavior of binary systems containing a gas (carbon dioxide, nitrogen, or propane) and a polymer (poly(ethylene glycol) or poly(ethylene glycol) dimethyl ether) with various molar masses was studied and classified both in p,T- and p-x diagrams. The phase behavior depends on the kind of gas and its state of aggregation. The solubility of gas in the liquid polymer phase increases as follows: nitrogen ⇒ propane ⇒ carbon dioxide. The influence of functional groups (either hydroxy or methyl ether) at the beginning and the end of the polymer chain is strong for short chained polymers like PEG 200, PEG 600, and PEGDME 200. For polymers with molecular masses higher than 1500 g/mol the influence of the functional group on gas solubility in the polymer is small. The melting temperature of the polymers under pressure correlate with the solubility of the gas in the molten polymer. The sometimes complex melting behavior under pressure can be understood from two competing effects: "Mechanically" applied pressure increases the melting point of the pure substance, while pressure established via a soluble gas causes a decrease of the melting point.

Acknowledgements

The authors wish to thank Deutsche Gesellschaft für Luft-und Raumfahrt (travel grants), Hoechst/Clariant and AGA (suppliers of polymers and gases), Prof. Arlt and his group in Berlin, Prof. Sadowski (Dortmund), and Prof. Knez and his group from the University in Maribor.

References

Ali S. (1983): Thermodynamic properties of polymer solutions in compressed gases. Zeitschrift für Physikalische Chemie Neue Folge 137, 13.
Arlt W. (1996): Textbook on lectures in thermodynamics. TU Berlin, Germany.
Bonner D.C. (1977): Löslichkeit überkritischer Gase in Polymeren – eine Übersicht. Polym. Eng. Sci. 17(2), 65.
Daneshvar M., Gulari E. (1989): Supercrit. Fluid Sci. Technol. ACS Symp. Ser. 406, 72.
Daneshvar M., Kim S., Gulari E.J. (1990): High-pressure phase equilibria of poly(ethylene glycol)-carbon dioxide systems. Phys. Chem. 94, 2124.
Davis R.A., Menéndez R.E., Sandall O.C.J. (1993): Physical, thermodynamic, and transport properties for carbon dioxide and nitrous oxide in solutions of diethanolamine or di-2-propanolamine in poly(ethylene glycol). Chem. Eng. Data 38, 119.
Fahl J. (1998): Schmierstoffe für den Einsatz mit Kohlendioxid als Kältemittel. KI Luft- und Kältetechnik 8, 375.
Garg A., Gulari E., Manke C.W. (1994): Thermodynamics of polymer melts swollen with supercritical gases. Macromolecules 27, 5643.
Hauk A., Weidner E. (2000): Thermodynamic and fluid-dynamic properties of carbon dioxide with different lubricants in cooling circuits for automobile application. Ind. Eng. Chem. Res. Vol. 39, No. 12, 4646.
Hauk A. (2001): Thermo- und Fluiddynamik von synthetischen Schmierstoffen mit Kohlendioxid als Kältemittel in PKW-Klimaanlagen. Dissertation, Bochum.
Huang S., Radosz M. (1990): Equation of state for small, large, polydisperse and associating molecules. Ind. Chem. Res. 29, 2284.
Kiran E., Saraf V.P., Sen Y.L. (1989): Solubility of polymers in supercritical fluids. Int. J. Thermophys. 10, 437.
Konynenborg van P.H., Scott R.L. (1980): Critical lines and phase equilibria in binary van der Waals mixtures. Phil. Trans. Royal Soc. of London 298, 495.
Koukova E., Petermann M., Weidner E. (2001): Measurements of phase equilibrium and fluid properties of the binary system squalane – CO_2 by various methods. Proc. 2nd Int. Meeting on High Press. Chem. Eng., Hamburg, published on CD.

Kumar S.K., Suter U.W., Reid R.C. (1986): Fractionation of polymers with supercritical fluids. Fluid Phase Equilibria 29, 373.
McHugh M.A., Seckner A.J., Yogan T.J. (1984): High pressure phase behavior of binary mixtures of octacosane and carbon dioxide. Ind. Eng. Chem. Fund. 23, 493.
McHugh M.A., Krukonis V.J. (1986): Supercritical fluid extraction: Principles and practice. Butterworths, Boston, 23–49.
McHugh M.A., Krukonis V.J. (1989): „Supercritical fluids", Encyclopedia of Polymer Science and Engineering. Vol. 16, 368, 2nd ed., Mark-Bikales-Overberger-Menges (eds.), John Wiley & Sons, New York.
Mier M. (1995): Modellierung von Hochdruckphasengleichgewichten mit der SAFT-Zustandsgleichung. Diplomarbeit, TU Berlin, 9–29.
Scott R.L., van Konynenbouo? o?rg P.H. (1970): Van der Waals and related models for hydrocarbon mixtures. Discus. Faraday Soc. 40, 87.
Scott R.L. (1972): The thermodynamics of critical phenomena in fluid mixtures. Ber. Bunsen-Ges. Phys. Chem. 76, 296.
Tuminello W.H., Dee G.T., McHugh M.A. (1995): Dissolving perfluoropolymers in supercritical carbon dioxide. Macromolecules 28, 1506.
Weidner E., Knez Z., Steiner R. (1996): Powder generation from poly(ethylene glycols) with compressible fluids. Preprints 3rd Int. Symp. on High Pressure Chem. Eng., 223–228.
Weidner E., Wiesmet V., Knez Z., Skerget M. (1997): Phase equilibrium (solid-liquid-gas) in poly(ethylene glycol) – carbon dioxide systems. J. Supercrit. Fluids 10, 139.
Weidner E., Knez Z., Wiesmet V., Kokot K. (1997a): Phase equilibrium (solid-liquid-gas) in the system poly(ethylene glycols) – nitrogen. Preprints 4th Italian Conf. on Supercrit. Fluids and their Applications, 95–103.
Wiesmet V. (1998): Phasenverhalten binärer Systeme aus Polyethylenglykolen, Polyethylenglykoldimethylethern und verdichteten Gasen. Dissertation, Erlangen.
Wiesmet V., Weidner E., Behme S., Sadowski G., Arlt W. (2000): Measurement and modelling of high-pressure phase equilibria in the systems PEG-propane, PEG-nitrogen and PEG-carbon dioxide. J. Supercrit. Fluids 17, 1–12.

Nomenclature

g	gas
l	liquid
p	total pressure
s	solid
T	absolute temperature
PEG	poly(ethylene glycol)
PEGDME	poly(ethylene glycol) dimethyl ether

Subscripts

1	phase 1
2	phase 2

24 Measurements and Modeling of High-Pressure Fluid Phase Equilibrium of Systems Containing Benzene Derivatives and CO$_2$

Gerd Brunner[1], Oliver Pfohl, and Stanimir Petkov

24.1 Abstract

Phase equilibria in binary and ternary systems containing *o*-cresol, *p*-cresol, carbon dioxide, water, and ethanol have been investigated experimentally at temperatures between 323.15 and 473.15 K and pressures ranging from 10 to 35 MPa. The experimental results provide a systematic basis of phase equilibrium data, yielding the effect of temperature on the influence of the position of the methyl groups of cresols that are in phase equilibrium with carbon dioxide. Based on the different solubilities of the cresol isomers in carbon dioxide, the separation of *o*-cresol and *p*-cresol was investigated. The dependence of the separation factor between both cresol isomers on concentration, temperature, and pressure is obtained from experiments in the ternary system *o*-cresol + *p*-cresol + carbon dioxide. The influence of ethanol added to each of the binary systems cresol isomer + carbon dioxide in order to enhance the solubility of the cresols in the carbon dioxide-rich phase is also shown.

The experimental data have been correlated using seven different equations of state (EoS): Peng-Robinson, two modifications of the Peng-Robinson EoS, one by Yu and Lu, the other with a volume translation, and four EoS explicitly accounting for intermolecular association (Statistical Association Fluid Theory (SAFT) by Chapman, Gubbins, Huang, and Radosz, the SAFT modification by Pfohl and Brunner for near critical fluids, a modified „cubic plus association" (CPA) EoS according to the ideas by Tassios and coworkers, and one of the EoS by Anderko). The mixing rule proposed by Mathias, Klotz, and Prausnitz with two binary interaction parameters per binary system influencing intermolecular attractive forces is used for all EoS as a basis for an objective comparison of the EoS. The phase behavior of ternary systems determined experimentally, can be predicted well, using two binary interaction parameters per binary system, optimized to reproduce the phase equilibrium of the binary subsystems.

---

1 Technische Universität Hamburg-Harburg, Thermische Verfahrenstechnik, Eissendorfer Str. 38, 21073 Hamburg.

24.2 Introduction

Carbon dioxide is preferred as a supercritical solvent for Supercritical Fluid Extraction (SFE) processes due to its low critical temperature, low cost, chemical inert behavior, and availability in purified form. Food and health industries greatly appreciate carbon dioxide because it is nontoxic and has a natural character. Extractions of aromas, caffeine, coloring, drugs, flavors, oils, and perfumes from plant material are only some of the numerous applications. Furthermore supercritical fluids are used for processing heavy hydrocarbons. Used lubricants can be recovered and purified, heavy petroleum fractions can be processed and tar sands can be extracted. In chemical industry supercritical fluids are used for separations and purifications like the extraction of organic material from water, the separation of aromatic and paraffinic hydrocarbons and aromatic isomer separation (Brunner, 1994).

This work provides a systematic data base of experimentally determined phase equilibria covering binary and ternary systems including ten benzene derivatives with hydroxy and methyl groups, carbon dioxide, ethanol, and water at temperatures and pressures relevant to supercritical fluid extraction (SFE). The results clearly show the influence of number and position of hydroxy groups of benzene derivatives that are in high-pressure phase equilibrium with carbon dioxide and water. This knowledge allows initial estimates on the applicability of an extraction of naturally occurring aromatic compounds with hydroxy and methyl groups using supercritical carbon dioxide. Finally, the data obtained here allow the investigation of equations of state and mixing rules incorporating group contribution methods because the aromatic compounds investigated here only differ by number and position of hydroxy and methyl groups. Furthermore equations of state incorporating association, which proved to be very useful at low pressure, can be evaluated at high pressure.

24.3 Experimental/Methods

24.3.1 Apparatus

The high-pressure apparatus used in this work is shown in Figure 24.1 and is described in detail elsewhere (Pfohl et al., 1997; Pfohl and Brunner, 1998). The apparatus consists of two high-pressure stainless steel bombs (300 and 1100 cm^3) immersed in a temperature-controlled oil bath. The larger one, the equilibrium autoclave is equipped with a motor-driven stirrer and allows sampling of up to three phases. Both autoclaves can be filled with molten high boiling phenols after evacuation. A pressure of up to 35 MPa can be attained by liquefying carbon dioxide and pumping it into the autoclaves. Fluids like toluene and water which can not be sucked into the hot autoclaves because of their high vapor pressures, are pumped with a syringe pump. For liquid-liquid equilibrium without carbon dioxide this pump is also used to attain the desired system pressure. Liquids and gases can be withdrawn from the autoclaves via valves. The heavier liquid phases of both autoclaves can be connected. The

24.3 Experimental/Methods

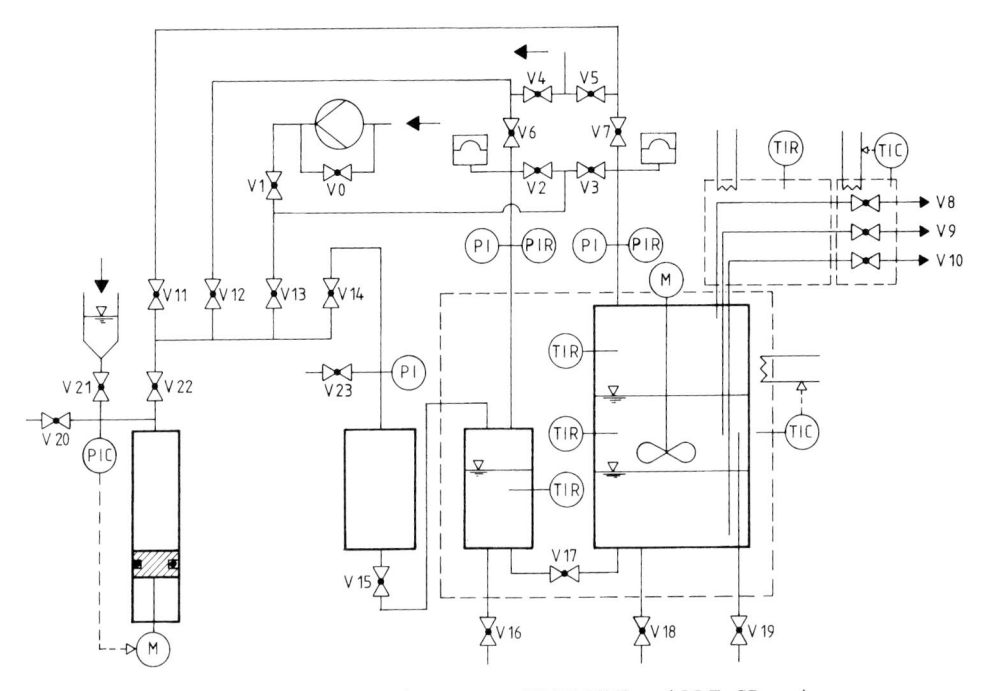

Figure 24.1: High-pressure apparatus used to measure VLLE, VLE, and LLE. SP: syringe pump = pressure controlled syringe pump with 260 cm^3 internal volume. B1: buffer 1 and B2: buffer 2 = 300 cm^3 autoclaves. EQ: equilibrium cell = 1100 cm^3 autoclave with stirrer and three sampling capillaries. PM101 = pump for liquid carbon dioxide. Buffer and equilibrium cell are inside a temperature controlled oil bath (dashed line).

gas phase in the smaller autoclave, the buffer autoclave, can be connected to the bottom of a third autoclave (300 cm^3) not inside the oil bath, which can be kept at constant pressure. This third autoclave only serves as additional buffer minimizing the amount of aromatic compounds diffusing from the hot autoclaves into the pump when valves are open.

The samples are expanded through a minimum of two cold traps immersed in water/ice and acetone/dry ice mixtures. In this way each sample is split into a condensable part caught in the cold traps and gaseous carbon dioxide. The difference in weight of the cold traps before and after sampling caused by the condensate gives the mass of the components other than carbon dioxide. In case of binary condensates the relative amounts of the components are determined either by Karl Fischer titration or gas chromatography.

24.3.2 Procedure

- *Equilibration*: First, the equilibrium cell and buffer are evacuated below 1 hPa. Then, appropriate amounts of the molten cresols are sucked into the equilibrium cell before appropriate amounts of ethanol are added to the autoclave with the syringe pump. Finally, the temperature of the bath is adjusted and carbon dioxide is added to the cell in order to achieve the desired pressure. The necessary amounts of cresols and (if applicable) ethanol are estimated so that the ends of the different sampling capillaries come to lie in different coexisting phases. The buffer is filled in a way that the pressure is the same as in the equilibrium cell and that the lower phase should have the same composition as the lower phase in the equilibrium cell. After starting the stirrer, further carbon dioxide is continuously added to the cells in order to make up for pressure drops caused by solvation effects. After the pressure in the equilibrium cell is constant without adding further carbon dioxide, the stirrer is operated for ca. six additional hours. Afterwards the phases are allowed to settle ca. eight hours before samples are taken.

- *Sampling*: Sampling generally begins with the lower phase. In order to avoid pressure drops when sampling, the valves between the pressure controlled pump, the third autoclave, the buffer, and the middle of the equilibrium cell are opened. In this way the pressure controlled syringe pump holds constant the pressure inside the equilibrium cell by pumping carbon dioxide into the buffer and pressing the equilibrated lower liquid phase of the buffer into the middle of the equilibrium cell. Before taking the first sample, the capillary is rinsed with an appropriate amount of the equilibrium cell content to minimize errors owing to the internal volumes of the capillary and sampling valve. Then three samples from the lower phase are taken. Because only relatively small samples from the lower phase have to be taken in order to receive enough sample condensate in the cold traps and because the syringe pump pumps equilibrated liquid from the buffer into the equilibrium cell, the equilibrium inside the equilibrium cell is not disturbed by this procedure. Afterwards three samples from the upper phase are taken in the same way as described for the lower phase but with the buffer connected to the bottom of the equilibrium cell. When sampling the carbon dioxide-rich phase at pressures where the solubility of cresols in carbon dioxide is very small (10 MPa), huge samples have to be taken in order to receive enough condensate in the cold traps. Then, the liquid content of the buffer often only lasts for the first sample. But experiments showed that in the cases where carbon dioxide passing the buffer enters the equilibrium cell, the second and third sample from the gas phase give the same results as the first sample. The carbon dioxide entering the equilibrium cell at the bottom becomes saturated when passing the liquid phase on the way up to the gas phase like in the apparatuses constructed for dynamic measuring methods.

24.3.3 Analysis

Weighing the cold traps before and after experiment gives the total mass of condensate. When the condensate in the cold traps consists of two different components, an analysis is carried out to determine the relative amounts of the different components in the condensate. The exact procedure is described in detail in Pfohl et al. (1997) and Pfohl and Brunner (1998).

24.3.4 Materials

Carbon dioxide was obtained with a purity of better than 99.99%, o-Cresol and m-cresol with a guaranteed purity of 99%, 2,4-xylenol 98%, toluene 99.5%, p-cresol 99%, re-distilled phenol >99%, 2,6-xylenol 99%, o-xylene 98%, ethanol 99.8%. Water was deionized at the university with ion exchangers and reverse osmosis to a resistance of 18 mΩ cm and degassed under vacuum. No efforts were made to further purify any of the other components.

24.4 Results and Discussion

24.4.1 Binary Systems: Benzene Derivative + Carbon Dioxide

The compositions of the coexisting phases of the systems CO_2 + o-cresol, CO_2 + m-cresol, CO_2 + p-cresol, CO_2 + phenol, CO_2 + 2,4-xylenol, and CO_2 + 2,6-xylenol are shown in Figures 24.2 to 24.4. Tabulated values are presented in Tables 24.1 to 24.6 (Pfohl et al., 1997; Pfohl and Brunner, 1998; Pfohl, 1998). The mean relative reproducibility $\frac{1}{n_{exp}} \sum_{n_{exp}} \sqrt{\frac{1}{n} \sum_{i=1}^{n} \left(\frac{x_i - \bar{x}}{\bar{x}}\right)^2}$ of all n_{exp} mole fractions of n samples of one phase is 2.0% for the liquid phase and 2.9% for the gaseous phase.

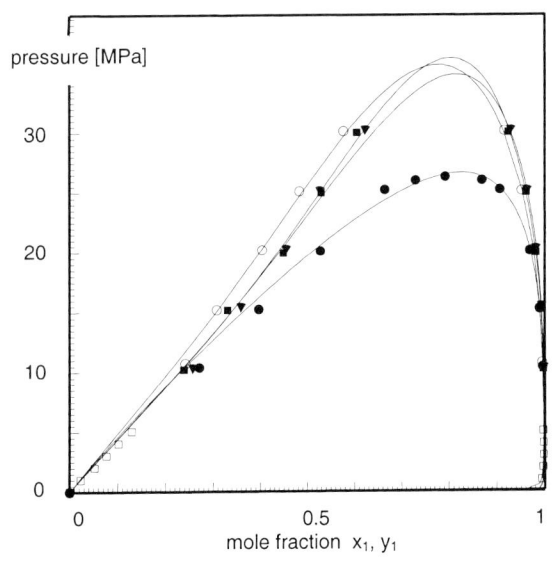

Figure 24.2: p,x Phase diagram for four binary carbon dioxide (1) + phenol derivative (2) systems at 373.15 K. Experimental data: ● = o-cresol, ■ = m-cresol, ▼ = p-cresol, ○ = phenol; □ = phenol: data (Yau and Tsai, 1992). Curves = calculations with the Peng-Robinson EoS and parameters from Table 24.15.

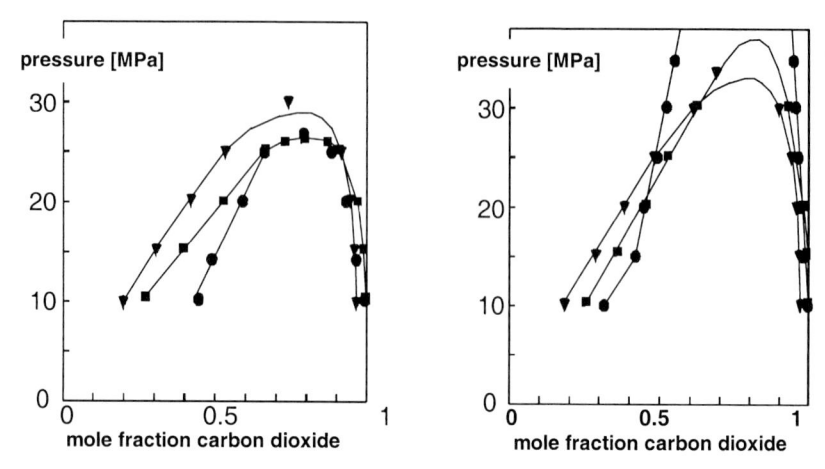

Figure 24.3: p,x,y Diagrams for o-cresol + carbon dioxide (a) and p-cresol + carbon dioxide (b) at ● = 323.15 K, ■ = 373.15 K, and ▼ = 473.15 K.

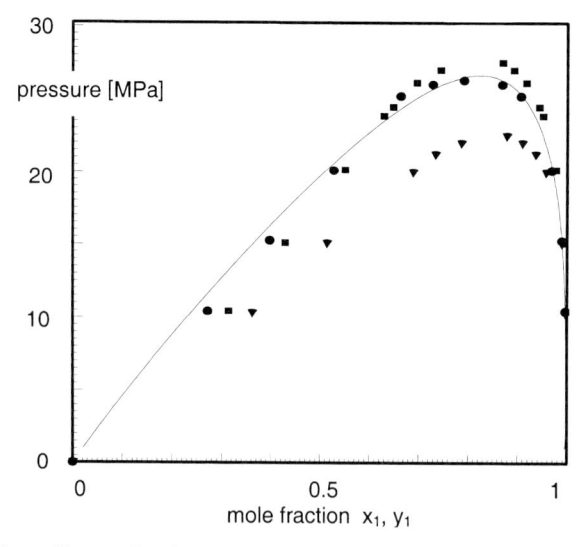

Figure 24.4: p,x Phase diagram for three binary carbon dioxide (1) + o-cresol derivative (2) systems at 373.15 K. Experimental data: ● = o-cresol, ■ = 2,4-xylenol, ▼ = 2,6-xylenol. Curve: calculation with the Peng-Robinson EoS and parameters from Table 24.15 for carbon dioxide + o-cresol.

24.4 Results and Discussion

Table 24.1: Phase compositions for the system phenol + carbon dioxide at 373.15 K.

p [MPa]	x Phenol [mol/mol]	y Phenol [mol/mol]
10.7	0.758	0.0056
15.2	0.692	0.0109
20.2	0.596	0.0251
25.1	0.517	0.0482
30.1	0.423	0.0843

Table 24.2: Phase compositions for the system o-cresol + carbon dioxide.

T [K]	p [MPa]	x o-Cresol [mol/mol]	y o-Cresol [mol/mol]
323.15	10.2	0.554	0.0072
	14.2	0.511	0.0355
	20.1	0.410	0.0703
	25.0	0.339	0.1184
	26.8	one phase:	0.2100
373.15	10.4	0.729	0.0045
	15.3	0.603	0.0112
	20.1	0.472	0.0318
	25.2	0.336	0.0939
	26.0	0.271	0.1312
	26.3	one phase:	0.2082
473.15	9.9	0.803	0.0364
	15.2	0.693	0.0418
	20.2	0.579	0.0561
	25.0	0.467	0.0859
	30.0	one phase:	0.2610

Table 24.3: Phase compositions for the system m-cresol + carbon dioxide at 373.15 K.

p [MPa]	x m-Cresol [mol/mol]	y m-Cresol [mol/mol]
10.2	0.761	0.0031
15.2	0.669	0.0075
19.9	0.551	0.0192
25.0	0.471	0.0396
30.0	0.395	0.0765

Table 24.4: Phase compositions for the system p-cresol + carbon dioxide.

T [K]	p [MPa]	x p-Cresol [mol/mol]	y p-Cresol [mol/mol]
323.15	10.0	0.684	0.0031
	15.0	0.580	0.0178
	20.0	0.553	0.0271
	25.0	0.509	0.0355
	30.1	0.479	0.0443
	34.8	0.452	0.0520
373.15	10.3	0.743	0.0032
	15.4	0.641	0.0076
	20.2	0.546	0.0199
	25.1	0.474	0.0386
	30.2	0.378	0.0718
473.15	10.1	0.817	0.0283
	15.1	0.712	0.0326
	20.0	0.618	0.0427
	25.0	0.520	0.0608
	29.9	0.391	0.1028
	33.6	one phase:	0.3150

Table 24.5: Phase compositions for the system 2,4-xylenol + carbon dioxide at 373.15 K.

p [MPa]	x 2,4-Xylenol [mol/mol]	y 2,4-Xylenol [mol/mol]
10.4	0.686	0.0028
15.1	0.571	0.0075
20.1	0.449	0.0236
23.8	0.370	0.0494
24.5	0.351	0.0568
26.1	0.303	0.0828
27.0	0.255	0.1077
27.5	one phase:	0.1300

Table 24.6: Phase compositions for the system 2,6-xylenol + carbon dioxide at 373.15 K.

p [MPa]	x 2,6-Xylenol [mol/mol]	y 2,6-Xylenol [mol/mol]
10.3	0.639	0.0048
15.0	0.486	0.0118
19.9	0.311	0.0437
21.2	0.266	0.0648
21.9	0.214	0.0915
22.5	one phase:	0.1226

24.4 Results and Discussion

24.4.2 Ternary Systems

- *Ternary system o-cresol + p-cresol + carbon dioxide:* The phase equilibrium compositions for this system are listed in Table 24.7 (Pfohl et al., 1997; Pfohl and Brunner, 1998; Pfohl, 1998). The mean relative reproducibility for the cresol mole fractions is 1.4% for the liquid phase and 1.5% for the gas phase. Figure 24.5 shows one Gibbs triangle with most of these data. Diagrams with the separation factors and gas phase loads are shown in Figure 24.6.

Table 24.7: Phase compositions for the system o-cresol + p-cresol + carbon dioxide.

T [K]	p [MPa]	x o-Cresol [mol/mol]	x p-Cresol [mol/mol]	y o-Cresol [mol/mol]	y p-Cresol [mol/mol]	$\hat{\rho}^L$ [g/cm^3]	$\hat{\rho}^V$ [g/cm^3]
323.15	15.1	0.2832	0.2789	0.0164	0.0117	1.06	0.78
	19.9	0.2514	0.2524	0.0242	0.0184	1.02	0.85
	24.9	0.2256	0.2290	0.0319	0.0258	1.02	0.92
356.15	18.3	0.4719	0.0106	0.0311	0.000541	1.02	0.63
	18.3	0.4279	0.0769	0.02583	0.00357	0.94	0.63
	18.3	0.0841	0.4650	0.00372	0.0149	1.00	0.59
	18.3	0.0109	0.5438	0.000464	0.0166	1.04	0.58
373.15	18.9	0.4916	0.0109	0.0255	0.000445	0.89	0.51
	19.0	0.4295	0.0778	0.0217	0.00304	0.94	0.51
	19.0	0.2778	0.2745	0.0119	0.00885	0.86	0.50
	19.0	0.0843	0.4697	0.00337	0.0139	1.07	0.49
	19.0	0.0118	0.5575	0.00046	0.0156	0.89	0.48
	21.9	0.4185	0.0093	0.0467	0.000866	0.82	0.63
	22.1	0.3731	0.0680	0.0391	0.00582	0.86	0.62
	22.0	0.2434	0.2432	0.0201	0.0158	0.83	0.61
	22.0	0.0781	0.4346	0.00538	0.0230	0.87	0.59
	22.0	0.0105	0.5077	0.000677	0.0249	0.87	0.59
	24.1	0.2263	0.2210	0.0265	0.0217	n.a.	0.59
	25.8	0.2027	0.1959	0.0338	0.0290	n.a.	0.63
	28.2	0.1683	0.1645	0.0474	0.0442	n.a.	0.66

The densities calculated here are based on unpublished material.

- *Ternary systems benzene derivative + water + carbon dioxide*: The phase compositions of the coexisting phases in the systems phenol + water + CO_2, m-cresol + water + CO_2, and toluene + water + CO_2 are listed in Tables 24.8 to 24.12 (Pfohl et al., 1997; Pfohl and Brunner, 1998; Pfohl, 1998). Experimental results are also shown in Figures 24.7 to 24.9. The mean relative reproducibility for the benzene derivative-rich phase is 1.5%, whereas the water fractions reported for the aromatic-rich phase of the system toluene + water + carbon dioxide can only be reproduced to within a mean absolute deviation of 0.2 mol%.

- *Ternary systems cresol isomer + ethanol + carbon dioxide:* The experimental results for the compositions of coexisting phases in these systems are listed in Tables 24.13 and 24.14 (Pfohl et al., 1997; Pfohl and Brunner, 1998; Pfohl, 1998) and shown in Figure 24.10.

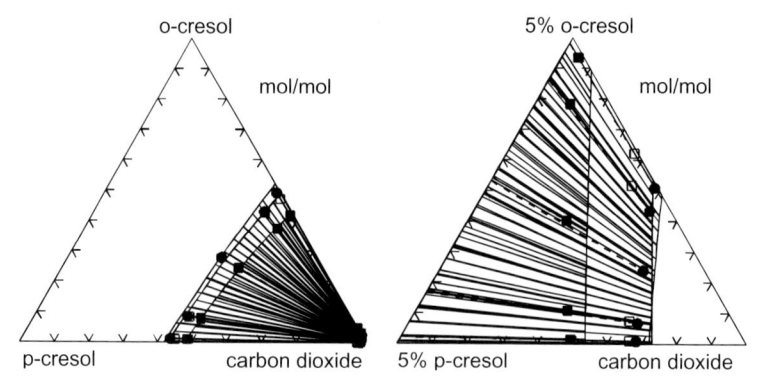

24   Measurements and Modeling of High-Pressure Fluid Phase Equilibrium

Figure 24.5: *Left side*: Gibbs triangle for *o*-cresol + *p*-cresol + carbon dioxide with experimental data at ☐ = 356.15 K, 18.3 MPa; ● = 373.15 K, 19 MPa; ■ = 373.15 K, 22 MPa. Solid lines are predictive calculations with the Peng-Robinson EoS based on binary equilibria, only. *Right side:* The corner with the gas phase is enlarged.

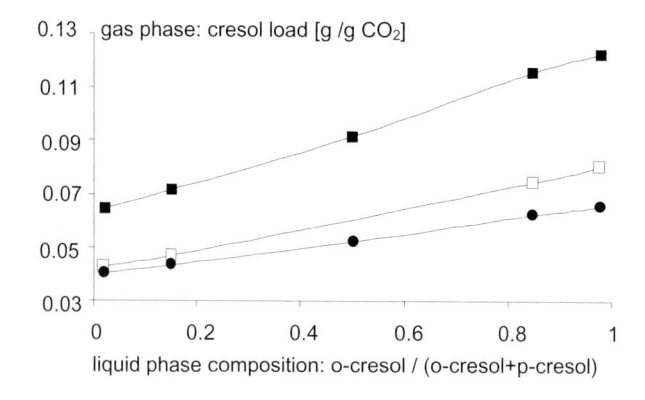

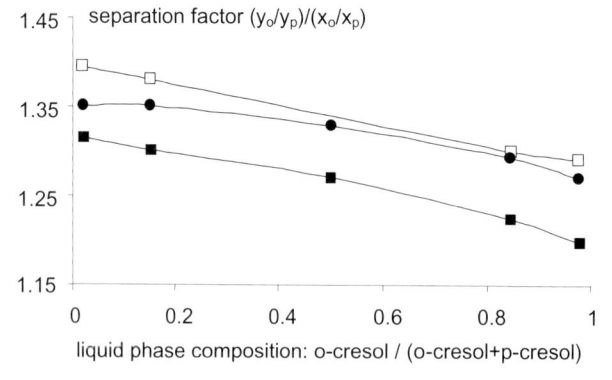

Figure 24.6: Cresol load of carbon dioxide and separation factor between the cresol isomers determined experimentally in the ternary system *o*-cresol + *p*-cresol + carbon dioxide at ☐ = 356.15 K, 18.3 MPa; ● = 373.15 K, 19 MPa; and ■ = 373.15 K, 22 MPa.

24.4 Results and Discussion

Table 24.8: Phase compositions for the system phenol + water + carbon dioxide at 373.15 K.

p [MPa]	x CO_2 [mol/mol]	x Phenol [mol/mol]	y Water [mol/mol]	y Phenol [mol/mol]
10.0	0.024	0.068	0.0168	0.0029
10.0	0.102	0.313	0.0172	0.0025
10.1	0.178	0.563	0.0102	0.0035
20.1	0.038	0.064	0.0202	0.0100
20.0	0.187	0.289	0.0216	0.0108
20.1	0.324	0.469	0.0174	0.0173
30.0	0.044	0.064	0.0267	0.0221
30.0	0.241	0.266	0.0298	0.0268
30.1	0.431	0.384	0.0250	0.0484

Table 24.9: Phase compositions for the system o-cresol + water + carbon dioxide at 373.15 K.

p [MPa]	x^{L1} CO_2 [mol/mol]	x^{L1} o-Cresol [mol/mol]	x^{L2} CO_2 [mol/mol]	x^{L2} o-Cresol [mol/mol]	y Water [mol/mol]	y o-Cresol [mol/mol]
10.0	0	0.0095	0	0.426	–	–
10.2	0.0112	0.0088	0.139	0.349	0.0179	0.0026
10.0	–	–	0.176	0.492	0.0163	0.0033
20.0	0	0.0092	0	0.423	–	–
20.0	0.0125	0.0086	0.141	0.358	–	–
20.1	0.0168	0.0076	0.278	0.313	0.0214	0.0142
20.0	–	–	0.363	0.392	0.0197	0.0197
29.9	0	0.0092	0	0.427	–	–
30.0	0.0085	0.0086	0.150	0.359	–	–
30.0	0.0163	0.0070	0.410	0.279	0.0417	0.0526

Table 24.10: Phase compositions for the system m-cresol + water + carbon dioxide at 373.15 K.

p [MPa]	x^{L1} CO_2 [mol/mol]	x^{L1} m-Cresol [mol/mol]	x^{L2} CO_2 [mol/mol]	x^{L2} m-Cresol [mol/mol]	y Water [mol/mol]	y m-Cresol [mol/mol]
10.0	0	0.0097	0	0.363	–	–
9.9	0.0130	0.0092	0.113	0.318	0.0167	0.0016
10.1	–	–	0.195	0.531	0.0129	0.0023
20.0	0	0.0096	0	0.369	–	–
20.0	0.0186	0.0080	0.240	0.299	0.0157	0.0094
20.0	–	–	0.350	0.427	0.0139	0.0137
19.9	0.0053	0.0098	0.037	0.340	–	–
30.0	0	0.0097	0	0.369	–	–
30.1	0.0155	0.0085	0.153	0.299	–	–
30.1	0.0194	0.0072	0.315	0.267	0.0308	0.0271
30.0	–	–	0.494	0.334	0.0246	0.0454

Table 24.11: Phase compositions for the system p-cresol + water + carbon dioxide at 373.15 K.

p [MPa]	x^{L1} CO$_2$ [mol/mol]	x^{L1} p-Cresol [mol/mol]	x^{L2} CO$_2$ [mol/mol]	x^{L2} p-Cresol [mol/mol]	y Water [mol/mol]	y p-Cresol [mol/mol]
10.0	0	0.0092	0	0.333	–	–
10.0	0.0096	0.0096	0.104	0.280	0.0199	0.0011
20.0	0	0.0091	0	0.337	–	–
20.0	0.0197	0.0078	0.210	0.256	0.0236	0.0086
30.0	0	0.0086	0	0.341	–	–
30.0	0.0218	0.0070	0.296	0.244	0.0351	0.0250

Table 24.12: Phase compositions for the system toluene + water + carbon dioxide at 373.15 K.

p [MPa]	x^{L1} CO$_2$ [mol/mol]	x^{L1} Toluene [mol/mol]	x^{L2} Water [mol/mol]	x^{L2} Toluene [mol/mol]
10.0	0	164	0.0261	0.9739
10.0	0.0022	90	0.0297	0.8527
10.0	0.0064	85	0.0462 (*)	0.5646
10.0	0.0091	65	0.0265	0.4699
VLLE, y =			0.0194	0.0340
10.0	–	–	0	0.488
VLE, y =			0	0.0313
20.0	0	148	0.0225	0.9775
20.0	0.0071	113	0.0362	0.6868
20.0	0.0099	56	0.0339	0.4636
19.9	0.0112	65	0.0298	0.2045
30.0	0	149	0.0236	0.9764
30.0	0.0090	83	0.0298	0.6706
30.0	0.0150	76	0.0333	0.3833
30.0	0.0119	48	0.0286	0.1685

(*) Outlier, not referred to in calculation

Table 24.13: CO$_2$ + ethanol + o-cresol at 373.15 K.

p [MPa]	x Ethanol [mol/mol]	x o-Cresol [mol/mol]	y Ethanol [mol/mol]	y o-Cresol [mol/mol]
10.1	0.0450	0.6848	0.0029	0.0039
10.1	0.1810	0.5478	0.0087	0.0031
10.0	0.3863	0.3482	0.0253	0.0021
10.0	0.5424	0.1717	0.0449	0.0012
20.0	0.0362	0.4430	0.0046	0.0323
20.0	0.1056	0.3560	0.0170	0.0312
20.0	0.1807	0.2460	0.0436	0.0333

24.4 Results and Discussion

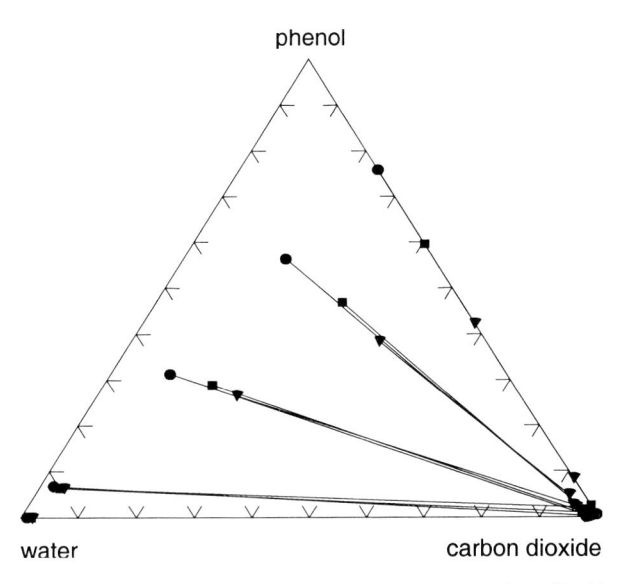

Figure 24.7a: Phase diagram for the ternary system phenol + water + carbon dioxide at 373.15 K. Pressure: ● = 10 MPa, ■ = 20 MPa, ▼ = 30 MPa.

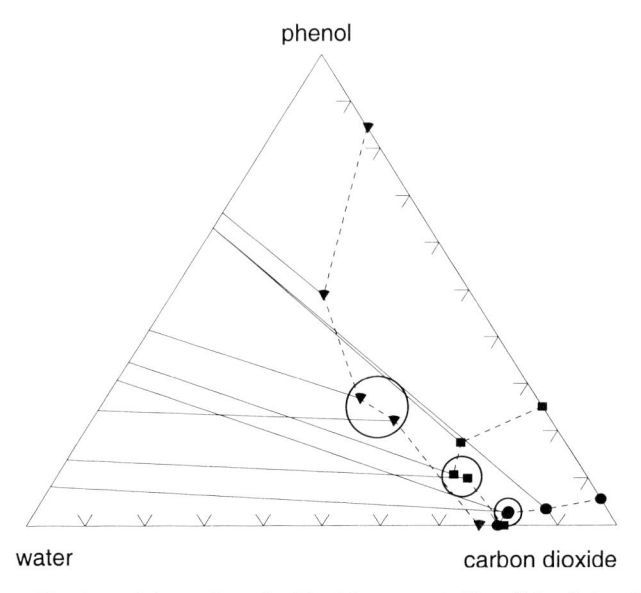

Figure 24.7b: Magnification of the carbon dioxide-rich corner (>90 mol%) of the phase diagram for the ternary system phenol + water + carbon dioxide at 373.15 K (Fig. 24.7a). Pressure: ● = 10 MPa, ■ = 20 MPa, ▼ = 30 MPa. For explanation of the circles, see text.

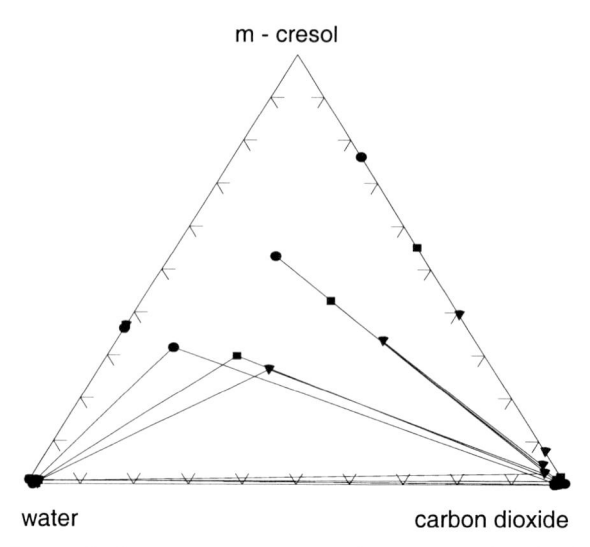

24 *Measurements and Modeling of High-Pressure Fluid Phase Equilibrium*

Figure 24.8: Phase diagram for the ternary system *m*-cresol + water + carbon dioxide at 373.15 K. Pressure: ● = 10 MPa, ■ = 20 MPa, ▼ = 30 MPa.

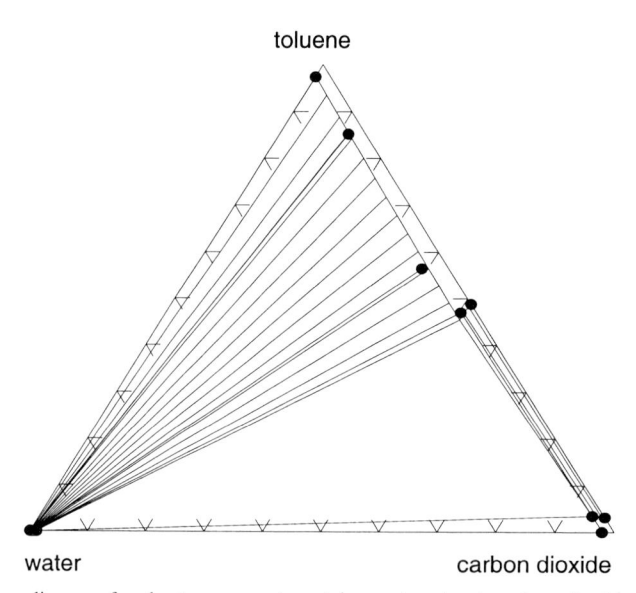

Figure 24.9: Phase diagram for the ternary system toluene + water + carbon dioxide at 373.15 K and 10 MPa. Points = experimental results, lines = predictive calculations with the Peng-Robinson EoS and parameters from Table 24.15.

24.4 Results and Discussion

Table 24.14: CO_2 + ethanol + p-cresol at 373.15 K.

p [MPa]	x Ethanol [mol/mol]	x p-Cresol [mol/mol]	y Ethanol [mol/mol]	y p-Cresol [mol/mol]
10.0	0.1679	0.5882	0.0081	0.0020
10.2	0.3865	0.3624	0.0237	0.0013
10.0	0.5431	0.1863	0.0420	0.00086
20.0	0.1205	0.4075	0.0137	0.0181
20.0	0.2373	0.2361	0.0468	0.0213
30.0		one phase:	0.0650	0.2250

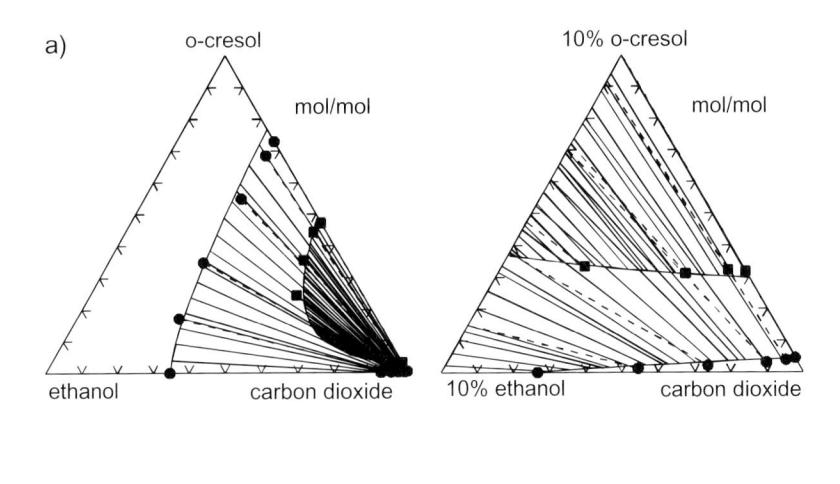

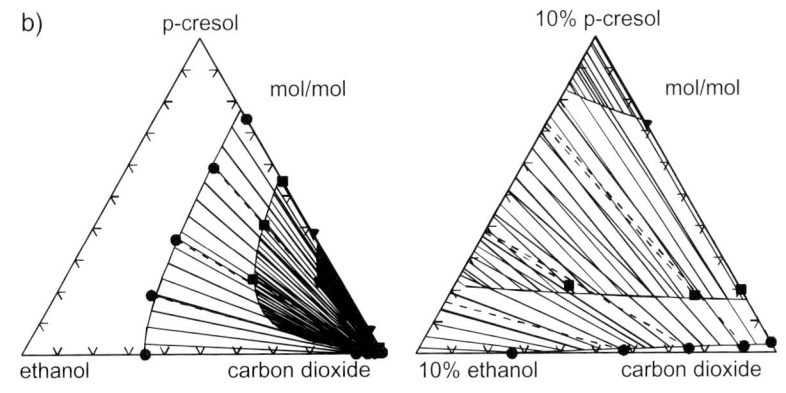

Figure 24.10: *Left side*: Gibbs triangles for ternary systems o-cresol + ethanol + carbon dioxide (a) and p-cresol + ethanol + carbon dioxide (b) at 373.15 K with experimental data at ● = 10 MPa, ■ = 20 MPa, and ▼ = 30 MPa. Solid lines are predictive calculations with the Peng-Robinson EoS based on binary equilibria, only (see text). *Right side*: The corners with the gas phases are enlarged.

24.4.3 Discussion of Experimental Results

24.4.3.1 Binary Systems Benzene Derivative + Carbon Dioxide

The available literature data have been evaluated to determine the influence of number and position of methyl groups of benzene derivatives that are in equilibrium with carbon dioxide. The size of the two-phase region slightly increases with the number of methyl groups bonded to the benzene ring (Pfohl et al., 1997). This behavior can be explained looking at the pure component properties of the benzene derivatives and comparing them with the properties of carbon dioxide: The substitution of hydrogen atoms with methyl groups causes an increase of molecular weight, critical temperature, and acentric factor, and a decrease of critical pressure and saturation pressure at 373.15 K. The differences between *o*-cresol and *p*-cresol become more pronounced at low temperature. The reason is that – contrary to entropic effects – enthalpic effects like hydrogen bonding are more pronounced at low temperatures. Estimated critical points for the systems *m*-cresol + carbon dioxide and *p*-cresol + carbon dioxide based on the data by Wäterling (1992), Pfohl et al. (1997) and the data from Pfohl (1998) agree well with each other.

- *Phenol*: Benzene becomes completely miscible with carbon dioxide at 13–14 MPa, whereas phenol is not completely miscible with carbon dioxide at pressures as high as 30 MPa. The decrease in miscibility is expected because carbon dioxide is non-polar and has only a very limited solvating power for polar compounds.

- *Cresols*: A methyl group bonded to phenol causes a decrease of the size of the two-phase region. That decrease is minor when the methyl group is in meta or para position to the hydroxy group. At 373.15 K, neither phenol, nor *m*-cresol nor *p*-cresol are completely miscible with carbon dioxide below 30 MPa. If the methyl group is in ortho position to the hydroxy group the miscibility with carbon dioxide is enhanced significantly: *o*-cresol becomes completely miscible with carbon dioxide at about 26 MPa. This behavior can be explained with the methyl group shielding the hydroxy group, i.e. steric hindrance. Iwai et al. (1996) applied Monte Carlo simulation for binary mixtures of aromatic compounds and carbon dioxide using a group contribution site model. The results of that study are in agreement with our results.

- *Xylenols*: The influence of the position of the second methyl group bonded to phenol follows the same rule as that of the first methyl group. A second methyl group bonded to *o*-methylphenol (*o*-cresol) which is in ortho position to the hydroxy group, increases the miscibility of the aromatic compound with carbon dioxide more than a second methyl group not in ortho position: 2,6-dimethylphenol (2,6-xylenol) is better miscible with carbon dioxide than o-methylphenol and 2,4-dimethylphenol (cf. Fig. 24.4).

24.4.3.2 Ternary System *o*-Cresol + *p*-Cresol + Carbon Dioxide

The binary system *o*-cresol + carbon dioxide is completely miscible at pressures higher than 26 MPa whereas *p*-cresol + carbon dioxide are not completely miscible below 30 MPa (at 373.15 K). Inspired by the different sizes of the miscibility gaps in the systems *o*-cresol +

carbon dioxide and p-cresol + carbon dioxide, the separation of o-cresol and p-cresol by means of supercritical carbon dioxide was investigated. Most of the experiments have been carried out at I) 373.15 K/19 MPa, II) 373.15 K/22 MPa, and III) 356.15 K/18.3 MPa. A comparison of I and II yields the effect of pressure at constant temperature (373.15 K), a comparison of I and III yields the effect of temperature at similar loading of the gas phase (40 g p-cresol/kg CO_2), and, finally, a comparison of II and III yields the effect of temperature at constant CO_2 density (530 kg/m^3).

The separation factor $\alpha_{o-/p\text{-cresol}} = (y_{o\text{-cresol}}/y_{p\text{-cresol}})/(x_{o\text{-cresol}}/x_{p\text{-cresol}})$ represents the difficulty of separating o-cresol from p-cresol and can serve as a direct measure for the required number of theoretical stages of a counter current column for a required separation. The gas phase loading is a measure of the required solvent to feed ratio and thus operating costs and column diameter. Figure 24.6 shows the separation factor and the gas phase loading for binary cresol mixtures with 2 to 98% of each cresol isomer. Comparing I and III shows that an operation at the higher temperature would not only require a taller column but also more solvent. The comparison between II and III shows that the better separation factors for III are achieved on the expense of a higher amount of solvent necessary. The measurements at 323.15 K at high pressure show that the differences in the phase densities become small, not allowing to benefit from the high gas phase loadings and good separation factors in countercurrent columns because of bad hydrodynamics. The separation is discussed in detail in Pfohl (1998).

24.4.3.3 Ternary Systems Benzene Derivative + Water + Carbon Dioxide

Phenol, cresols, and toluene exhibit a different phase behavior when mixed with water and carbon dioxide, based on the different phase behavior of the binary systems benzene derivative + carbon dioxide and benzene derivative + water.

- *Cresols*: In the ternary systems cresol + water + carbon dioxide none of the binary subsystems is completely miscible at the conditions investigated experimentally, with the exception of the system o-cresol + carbon dioxide at 30 MPa. This leads to a three-phase region in every system at any pressure with three (o-cresol at 30 MPa: two) adjacent two-phase regions of the binary subsystems. Since the solubility of carbon dioxide in the cresols increases with pressure, the composition of the cresol-rich upper liquid phase of the three-phase equilibrium also approaches the carbon dioxide corner with an increase in pressure.

- *Phenol*: Carbon dioxide is neither completely miscible with water nor with phenol at the conditions investigated experimentally. In contrast to the cresols, phenol is completely miscible with water at 373.15 K, leading to tie lines ending in the carbon dioxide-rich corner of the phase diagram (Figs. 24.7a and 24.7b). The miscibility of the binary subsystem phenol + carbon dioxide changes significantly with pressure, affecting the size of the region of immiscibility in the ternary system. A close look at the measured mole fractions in the gas phase of this system (cf. Fig. 24.7b) reveals that for the two tie lines with the least amount of phenol the gas phase composition is nearly identical although their liquid phase compositions are significantly different (cf. Fig. 24.7a). This indicates the existence of a three-phase region similar to the three-phase regions found in the systems cresol + water + carbon dioxide. But additional experiments carried out in subsequent work (Pagel, 1997) could not prove this assumption. Calculations with equations of state described below, reproduce tie lines with nearly identical gas phase compositions in this region at low pressure also and show a three-phase region at 30 MPa/373.15 K (cf. Fig. 24.14).

- *Toluene*: The phase behavior of the ternary system toluene + water + carbon dioxide (cf. Fig. 24.9) is similar to that of the ternary system *o*-cresol + water + carbon dioxide – but with toluene already becoming completely miscible with carbon dioxide at about 14 MPa, turning the three phase region which exists at 10 MPa to a two-phase region at high pressure.

24.4.3.4 Ternary Systems Cresol Isomer + Ethanol + Carbon Dioxide

The phase equilibrium measurements in the systems cresol isomer + ethanol + carbon dioxide give evidence of the suitability of ethanol as a modifier in order to increase the solubility of the cresols in carbon dioxide. A higher solubility would lower the necessary solvent to feed ratio (see above) and might lead to lower separation costs. Brunner (1994) discusses the use of modifiers in supercritical fluid extraction in detail.

Figure 24.10 clearly shows, that adding up to 20% ethanol at 373.15 K and up to 20 MPa does not increase the cresol load of the gas phase at all. On the contrary, the addition of ethanol at 10 MPa leads to a reduced uptake of cresols in the vapor: the interactions between ethanol and cresols in the gas phase are not even strong enough to compensate at low ethanol concentrations that the binary system ethanol + carbon dioxide is subcritical.

24.5 Theory/Methods

24.5.1 Equations of State

The phase equilibria determined experimentally in this study have been correlated using seven different equations of state (EoS): three simple cubic EoS and four elaborated EoS explictly accounting for intermolecular association.

24.5.1.1 Cubic EoS

The Peng-Robinson (PR) EoS and the 3P1T EoS as proposed by Peng and Robinson (1976) and by Yu and Lu (1987), respectively, have been used.

$$p = \frac{RT}{v-b} - \frac{a(T)}{v(v+b)+b(v-b)} \tag{24.1}$$

$$p = \frac{RT}{v-b} - \frac{a(T)}{v(v+c)+b(3v+c)} \qquad (24.2)$$

In order to obtain the parameters a, b, and c for each phase (itself a mixture of up to three components with mole fractions x_i) the pure-component parameters $a_i(T)$ have been averaged according to the one-fluid mixing rule proposed by Mathias et al. (1991):

$$a(T) = \sum_i \sum_j \left(x_i x_j \sqrt{a_i(T) * a_j(T)} (1-k_{ij}) \right) + \sum_i x_i \left(\sum_j x_j \left(\sqrt{a_i(T) * a_j(T)} \right)^{1/3} \lambda_{ji}^{1/3} \right)^3 \qquad (24.3)$$

This mixing rule has two adjustable binary interaction parameters for each binary subsystem $i-j$: $k_{ij} = k_{ji}$ and $\lambda_{ij} = -\lambda_{ji}$. It reduces to the original van der Waals mixing rule with one empirical correction parameter per binary system if all λ_{ij} are set to zero. This mixing rule has very recently been proved to be the best simple mixing rule under investigation for the prediction of phase equilibria in the ethanol + chloroform + hexane + acetone system and its ternary subsystems based on the binary subsystems (Solórzano et al., 1996) which does not suffer the so-called Michelsen-Kistenmacher syndrome (Michelsen and Kistenmacher, 1990). Pfohl et al. (1996) showed that this mixing rule in combination with the Soave-Redlich-Kwong equation of state (Soave, 1972) also leads to good predictions of phase equilibria in the system acetone + water + carbon dioxide, which exhibits a phase behavior similar to the benzene derivative + water + carbon dioxide systems investigated here. For the mixture, parameters b and c have been obtained by averaging the pure-component covolume parameters b_i and c_i arithmetically:

$$b = \sum_i \sum_j x_i x_j \frac{b_i + b_j}{2} \qquad (24.4)$$

$$c = \sum_i \sum_j x_i x_j \frac{c_i + c_j}{2} \qquad (24.5)$$

The pure-component parameters a_i, b_i, and c_i have been determined according to the formulas given by Peng and Robinson (1976) and Yu and Lu (1987) based on the pure-component critical pressure p_c, critical temperature T_c, and acentric factor ω. This way, the calculated point of inflection of the critical isotherm appears at the experimentally observed critical temperature and critical pressure and the slope of the vapor pressure curve is reproduced sufficiently for all pure compounds. Incorporation of additional fit parameters for the pure compounds by introducing other temperature dependencies of parameter $a(T)$ in order to fine-tune the description of the pure component properties was omitted. This way, both equations of state which had been derived for systems with no specific interactions like hydrogen bonding were kept as simple as possible.

The pure-component properties for all components were taken from the data bank by Reid et al. (1987), with the exception of the acentric factor for carbon dioxide, ω_{CO_2} (the value given by Reid et al. (0.239) is not in agreement with the high accuracy data from the IUPAC monography (Angus et al., 1976) which leads to $\omega_{CO_2} = 0.225$). The values are listed in

Table 24.15. It is worthwhile to mention that – because results for liquid-liquid equilibrium calculations are very sensitive to these values – slightly different pure-component properties already lead to quite different calculation results. In order to overcome the limitations of the corresponding states principle with three parameters when modeling polar and associating fluids, a modification of the PR EoS with a volume-translation according to Peneloux (1982) and a modified alpha function according to Mathias (1983) – cf. Eqs. (24.6) to (24.10) – were also included in the comparison: The Mathias alpha function (Eq. (24.10)) offers two parameters, κ and p_1, for fine tuning the shape of the vapor pressure curve and additionally leads to a physically sound behavior of the attractive forces at high reduced temperatures (Boston and Mathias, 1980). For $\tilde{c} = p_1 = 0$, this modification, called PR-DVT hereafter, reduces to the simple PR EoS at subcritical temperatures.

$$p = \frac{RT}{\tilde{v} - b} - \frac{a(T)}{\tilde{v}^2 + 2b\tilde{v} - b^2} \quad \text{with} \quad \tilde{v} = v + \tilde{c} \tag{24.6}$$

$$a(T) = a(T'_c)\alpha(T'_r, \omega') \quad \text{with} \quad a(T'_c) = 0.45724 \frac{R^2 T'^2_c}{p'_c} \quad \text{and} \quad T'_r = T/T'_c \tag{24.7}$$

$$b = 0.0778 \frac{RT'_c}{p'_c} \tag{24.8}$$

$$\kappa = 0.37464 + 1.54226\,\omega' - 0.26992\,\omega'^2 \tag{24.9}$$

$$\text{if } T'_r < 1: \alpha(T'_r) = \left(1 + \kappa(1 - \sqrt{T'_r}) - p_1(1 - T'_r)(0.7 - T'_r)\right)^2 \tag{24.10a}$$

$$\text{if } T'_r \geq 1: \alpha(T'_r) = \left(\exp\left(c(1 - T'^d_r)\right)\right)^2 \quad \text{with} \quad c = 1 + \frac{\kappa}{2} + 0.3 p_1 \quad \text{and} \quad d = \frac{c-1}{c} \tag{24.10b}$$

The primes (′) indicate that the pure-component EoS parameters T'_c, p'_c, and ω' do not necessarily *have to* be set equal to T_c, p_c, and ω from experiment. T'_c and p'_c are the calculated critical values for the PR-DVT EoS (but not for the PR-CPA EoS if association is included, see below).

24.5.1.2 Association EoS

Four EoS explicitly accounting for association have been included in the investigation. Statistical Associating Fluid Theory (SAFT) by Chapman et al. (1989), and Huang and Radosz (1990, 1991, 1993) has been used as proposed by Huang and Radosz (1991, 1993). In SAFT, all molecules are modeled as chains of covalent-bonded spheres with the possibility to associate according to the physical theory. Because this sphere chain model was shown to give poor results near critical points of pure compounds (Pfohl et al., 1998) and some experimental results cover the near critical state of carbon dioxide, the SAFT modification by Pfohl et al. (1999) is also included in the comparison. In this modification the supercritical fluid is not modeled as a sphere chain but as a convex body (CB) in order to allow an excellent representation of its near critical region. In addition to SAFT and SAFT-CB, two EoS have been investigated where an association part is added to a simple cubic EoS. These EoS reduce to the original cubic EoS for non-associating compounds. Anderko (1989a,b) proposed to add an association part according to the chemical theory to the cubic 3P1T EoS by Yu and Lu. Anderko's "AeoS" that incorporates the linear association scheme according to Kempter and Mecke (1940) for alcohols and phenols was used here. Finally, a self made "Cubic-plus-Association" (CPA) EoS using the SAFT association part as proposed by Tassios' group (Kontogeorgis et al., 1996; Voutsas et al., 1997; Yakoumis et al., 1997) was investigated here (Eq. (24.11) to (24.16)). The four differences of this EoS compared to Tassios' CPA EoS are summarized (Pfohl et al., 1999):

$$p = \frac{RT}{v-b} - \frac{a(T)}{v^2 + 2bv - b^2} + \frac{RT}{v}\left(\frac{\partial \tilde{a}^{asso}}{\partial \rho}\right)\rho \quad (24.11)$$

with $\tilde{a}^{asso} = \frac{a^{asso}}{RT} = \sum_{i=1}^{N} x_i \sum_{A_i=1}^{sites(i)} \left[\ln X^{A_i} + \frac{1-X^{A_i}}{2}\right]$ (24.12)

with $X^{A_i} = \dfrac{1}{1 + N_A \rho \sum_{j=1}^{N} \sum_{B_j=1}^{sites(j)} x_j X^{B_j} \Delta^{A_i B_j}}$ (24.13)

with $\Delta^{A_i B_j} = g_{ij}\left[\exp\left(\frac{\varepsilon^{A_i B_j}}{kT}\right) - 1\right]\frac{2\pi D_{ij}^3}{3}\beta^{A_i B_j}$ (24.14)

with $g_{ij} = \dfrac{1}{1-\xi_3} + \dfrac{3\xi_2}{(1-\xi_3)^2}\dfrac{D_i D_j}{D_i + D_j} + \dfrac{2\xi_2^2}{(1-\xi_3)^3}\left(\dfrac{D_i D_j}{D_i + D_j}\right)^2, \quad D_{ij} = \dfrac{D_i + D_j}{2}$ (24.15)

$D_i = \sqrt[3]{\dfrac{3b_i}{2\pi N_A}}, \quad \xi_k = \dfrac{\pi}{6}N_A \rho \sum_{i=1}^{N} x_i D_i^k, \quad \varepsilon^{A_i B_j} = \sqrt{\varepsilon^{A_i}, \varepsilon^{B_j}}, \quad \beta^{A_i B_j} = \sqrt{\beta^{A_i}\beta^{B_j}}$ (24.16)

with the pure-component parameters a and b adapted from the PR EoS, i.e. from Eqs. (24.6) to (24.10) after setting $\tilde{c} = p_1 = 0$. For zero values of ε or κ, the PR-CPA EoS reduces to the PR-DVT modification. Comparing the above equations with the equations by Huang and Radosz (1991, 1993) and Gupta und Johnston (1994) and taking into account $(\kappa^{A_i}/m)_{SAFT} \approx 2.1\,(\beta^{A_i})_{CPA}$, the association parameters of the PR-CPA EoS can be related to enthalpy and entropy of hydrogen bonding:

$$\Delta h^{HB} = -N_A \varepsilon \quad \text{for} \quad \Delta v^{HB} = 0 \quad (24.17)$$

$$\Delta s^{HB} \approx R\ln(2.1\,m\beta), \text{ where } m \text{ is the chain length in SAFT} \quad (24.18)$$

24.5.1.3 Mixing Rules

In order to obtain the mixture parameters for each phase, the pure-component parameters $a_i(T)$ (and u_i/k for SAFT) have been averaged according to the one-fluid mixing rule proposed by Mathias, Klotz, and Prausnitz (MKP) (Mathias et al., 1991):

$$a(T) = \sum_i \sum_j \left(x_i x_j \sqrt{a_i(T) a_j(T)} (1 - k_{ij}) \right) + \sum_i x_i \left(\sum_j x_j \left(\lambda_{ji} \sqrt{a_i(T) a_j(T)} \right)^{1/3} \right)^3 \quad (24.19)$$

This mixing rule has two adjustable interaction parameters for each binary subsystem because $k_{ij} = k_{ji}$ and $\lambda_{ij} = -\lambda_{ji}$. It reduces to the original van der Waals mixing rule with one empirical correction parameter per binary system if all λ_{ij} are set to zero. In combination with the SRK EoS and PR EoS this mixing rule led to good predictions of phase equilibria in previous work in ternary systems containing carbon dioxide and associating compounds at high pressure (Pfohl et al., 1996, 1997; Pfohl, 1998). In addition, this mixing rule has recently been proved to be the best simple mixing rule under investigation for the prediction of phase equilibria in the ethanol + chloroform + hexane + acetone system and its ternary subsystems based on the binary subsystems which does not suffer the so-called Michelsen-Kistenmacher syndrome. Michelsen and Kistenmacher (1990) pointed out that previous mixing rules – like the well known one by Panagiotopoulos and Reid, 1985 ('PaRe') – yield incorrect results because subdividing one fraction of molecules x_i in arbitrary sub-fractions x_i' and x_i'' with $x_i = x_i' + x_i''$ leads to different calculation results. This finding is important for the separation of nearly identical compounds as encountered in this study. This disadvantage of the PaRe mixing rule is directly related to its failure in describing the phase behavior of the ternary system o-cresol + p-cresol + carbon dioxide. Subdividing the o-cresol fraction in the system o-cresol + carbon dioxide leads to bent binodals in contrast to theory which enforces straight lines. The same wrong curvature is calculated for the system o-cresol + p-cresol + carbon dioxide. The MKP mixing rule gives straight lines for both systems (Pfohl et al., 1999). The parameters b and c for the mixture have been obtained by averaging the pure-component co-volume parameters b_i and c_i arithmetically.

24.5.2 Parameter Estimation

24.5.2.1 Pure Component Parameters

The pure-component EoS parameters have been adjusted in such a way that the EoS represents selected pure-component properties from literature best. Details on the procedure are available elsewhere (Pfohl et al., 1997; Pfohl, 1998). The EoS parameters T_c' and p_c' have been set equal to the experimental values. Literature data for the acentric factors were also used as input for the PR EoS and 3P1T EoS. The parameters ω', $\tilde{c}$, and p_1 for the PR-DVT EoS have been optimized by minimizing the deviation between calculated and experimental values of vapor pressure and liquid volume:

$$dev = \frac{1}{2} \sqrt{ \frac{1}{n_v} \sum_{i=1}^{n_v} \left(\frac{p_{Exp.}^{SAT}(T_i) - p_{EOS}^{SAT}(T_i)}{p_{Exp.}^{SAT}(T_i)} \right)^2 } + \frac{1}{2} \sqrt{ \frac{1}{n_v} \sum_{i=1}^{n_v} \left(\frac{v_{L,Exp.}^{SAT}(T_i) - v_{L,EOS}^{SAT}(T_i)}{v_{L,Exp.}^{SAT}(T_i)} \right)^2 }$$

(24.20)

Table 24.15: Pure-component EoS parameters.[1]

	CO_2	Phenol	o-Cresol	m-Cresol	p-Cresol	Toluene	Ethanol	Water
			Literature source for data regression [2]					
	[37]	[38]	[38]	[38]	[38]	[38]	[38]	[39]
			Cubic EoS (PR, PR-DVT, 3P1T)					
T_c	304.13	694.25	697.55	705.85	704.65	591.79	516.25	647.3
p_c	7.377	6.13	5.006	4.56	5.15	4.109	6.384	22.12
ω	0.225	0.426	0.434	0.449	0.513	0.264	0.637	0.344
			PR-DVT EoS					
ω'	0.22	0.42784	0.4314	0.44634	0.51882	0.26185	0.65708	0.3437
c/b	–[2]	–0.10277	–0.08279	0.068943	–0.12185	0.026637	0.069322	0.2252
p_1	–[2]	–0.09214	–0.09343	–0.23011	–0.0594	–0.04738	0.13667	0.06882
			PR-CPA EOS					
Sites	0	2	2	2	2	0	2	3 [2]
T'_c	304.13	600.05	687.89	710.53	696.66	591.79	404.62	545.82
p'_c	7.377	5.0406	4.6052	4.7915	4.6547	4.109	5.5241	22.1248
ω'	0.22	0.304 [2]	0.339 [2]	0.339 [2]	0.339 [2]	0.264	0.15 [2]	0 [2]
ε/k	–	2152.1	1953.4	2510.9	2333.8	–	2812.3	2258.6
β	–	0.044842	0.004622	0.001213	0.003653	–	0.009757	0.018517
$-\Delta h^{HB}$ [3]	–	17900	16200	20900	19400	–	23400	18800
$-\Delta s^{HB}$ [3]	–	9.5	26.9	38.1	28.9	–	38.5	27.0
			AEOS					
T'_c	304.13	652.61	634.50	692.09	675.26	591.79	443.84	517.43
p'_c	7.377	5.6158	4.3819	4.7189	4.6083	4.109	6.5476	25.706
ω'	0.225	0.286 [2]	0.315 [2]	0.315 [2]	0.315 [2]	0.264	0.15 [2]	0 [2]
$-\Delta h^{HB}$	–	20958	20397	22458	23121	–	24560	24994
$-\Delta s^{HB}$	–	89.004	82.278	95.282	92.268	–	96.263	94.195
			SAFT					
Sites	0 [4]	2	2	2	2	0	2	3
u^0/k	216.08	290 [2]	290 [2]	290 [2]	290 [2]	251.72	213.87	574.96
v^{00}	13.578	14 [2]	14 [2]	14 [2]	14 [2]	12.608	12.449	12.304
m	1.417	3.3788	4.0032	4.0461	4.0480	4.1684	2.4384	1 [2]
E/k	40 [2]	10	10	10	10	10	10	1 [2]
ε/k	–	2329.3	1861.8	2263.5	2329.3	–	2791.2	1940.4
κ	–	0.07172	0.07486	0.05019	0.04252	–	0.025819	0.011717
$-\Delta h^{HB}$ [3]	–	19400	15500	18800	19400	–	23200	16100
$-\Delta s^{HB}$ [3]	–	21.9	21.6	24.9	26.3	–	30.4	37.0

[1] Units: T_c, T'_c, e/k, ε/k, u^0/k [K], p_c, p'_c [MPa], ω, ω', m, κ, c/b, p_1, β [–], v^{00} [cm^3 mol^{-1}], Δh [J mol^{-1}], Δs [J mol^{-1} K^{-1}].
[2] Details: Pfohl (1998).
[3] Calculated from ε/k and β (κ) using Eqs. 24.17 and 24.18, respectively.
[4] SAFT sphere chain parameters for CO_2 by Huang and Radosz (1990). The SAFT-CB (convex body) parameters for CO_2 have been optimized by Pfohl and Brunner (1998).

Table 24.16: Binary interaction parameters.

	PR		PR-DVT		3P1T		SAFT		SAFT-CB		PR-CPA		AEOS	
	k_{ij}	λ_{ij}	k_{ij}	λ_{ij}	k_{ij}	λ_{ij}	k_{ij}	λ_{ij}	k_{ij}	λ_{ij}	k_{ij}	λ_{ij}	k_{ij}	λ_{ij}

Parameters regressed in order to receive optimum representation at 373.15 K and 10–35 MPa

CO_2 + o-cresol	0.1211	0.1325	0.1219	0.1320	0.1223	0.1451	0.0658	0.0713	0.0723	0.0776	0.1235	0.1689	0.1350	0.2177
CO_2 + m-cresol	0.1378	0.1057	0.1366	0.1187	0.1388	0.1212	0.0717	0.0844	0.0788	0.0938	0.1137	0.1493	0.1183	0.1785
CO_2 + p-cresol	0.0998	0.1000	0.1008	0.1013	0.0980	0.1130	0.0703	0.0800	0.0779	0.0884	0.1144	0.1628	0.1202	0.1852
CO_2 + phenol	0.0906	0.0822	0.0910	0.0848	0.0892	0.0941	0.0674	0.0880	0.0743	0.0965	0.1247	0.1992	0.0981	0.1685
CO_2 + toluene	0.0798	–	0.0866	–	0.0710	–	0.1097	–	0.1310	–	0.0871	–	same as 3P1T	
CO_2 + water	0.0774	−0.2151	0.0909	−0.2233	0.0704	−0.2125	−0.0158	−0.0220	0.0056	−0.0116	−0.0258	0.0035	−0.0477	−0.0206
CO_2 + ethanol	0.1111	0.0759	0.1160	0.0790	0.1046	0.0830	0.0666	0.1152	0.0786	0.1490	0.0859	0.2131	0.0630	0.1847
H_2O + o-cresol	−0.0520	0.0569	−0.0469	0.0557	−0.0487	0.0577	0.0016	−0.0035	same as SAFT		−0.0606	0.0736	−0.0310	0.2000
H_2O + m-cresol	−0.0721	0.0812	−0.0626	0.0777	−0.0672	0.0829	−0.0069	−0.0147	same as SAFT		−0.0755	0.0213	−0.0414	0.1253
H_2O + p-cresol	−0.0903	−0.0435	−0.0856	−0.0449	−0.0828	−0.0377	−0.0128	−0.0256	same as SAFT		−0.0957	0.0008	−0.0614	0.1051
H_2O + toluene	0.1008	0.3700	0.1059	0.3706	0.0986	0.3699	0.0972	0.1332	same as SAFT		−0.0054	−0.1963	−0.0336	0.1724

Parameters regressed in order to receive optimum representation at 323.15 K and 10–35 MPa

CO_2 + o-cresol	0.1160	0.0870	0.1136	0.0909	0.1153	0.0919	0.0603	0.0493	0.0726	0.0636	0.1225	0.1408	0.1376	0.1553
CO_2 + p-cresol	0.1020	0.0530	0.0992	0.0570	0.0978	0.0581	0.0708	0.0636	0.0823	0.0666	0.1227	0.1397	0.1312	0.1468

Parameters regressed in order to receive optimum representation at 473.15 K and 10–35 MPa

CO_2 + o-cresol	0.0673	0.1031	0.0922	0.0931	0.0569	0.1186	0.0672	0.1358	0.0644	0.1575	0.0901	0.0933	0.0604	0.1943
CO_2 + p-cresol	0.0585	0.1169	0.0821	0.1095	0.0489	0.1408	0.0692	0.1241	0.0677	0.1446	0.0825	0.1200	0.0570	0.1809

CO_2 + cresol isomer parameters in order to receive an *exact* reproduction of (extrapolated) mole fractions for binaries at T [K] and P [MPa]

o-, 356.15/18.3	0.1275	0.1432	0.1269	0.1454	0.1341	0.1725	0.0683	0.0705	0.0739	0.0905	0.1324	0.1937		
o-, 373.15/19.0	0.1214	0.1439	0.1224	0.1436	0.1280	0.1782	0.0705	0.0832	0.0731	0.1082	0.1221	0.1745		
o-, 373.15/22.0	0.1269	0.1459	0.1266	0.1432	0.1346	0.1752	0.0679	0.0707	0.0740	0.0855	0.1304	0.1835		
p-, 356.15/18.3	0.1065	0.1200	0.1060	0.1241	0.1095	0.1556	0.0695	0.0885	0.0724	0.1123	0.1253	0.2119		
p-, 373.15/19.0	0.1047	0.1297	0.1059	0.1318	0.1074	0.1700	0.0699	0.1071	0.0705	0.1298	0.1185	0.1986		
p-, 373.15/22.0	0.1047	0.1171	0.1052	0.1171	0.1086	0.1511	0.0702	0.0868	0.0746	0.1035	0.1231	0.1934		

24.5.2.2 Binary Systems: Determining Binary Interaction Parameters

The binary interaction parameters k_{ij} and λ_{ij} have been adjusted in order to optimize the reproduction of the compositions of the coexisting phases in the binary systems at 323.15, 373.15, and 473.15 K by minimizing the deviation:

$$deviation(T) = \sqrt{\frac{1}{n_f} \sum_{i=1}^{n_f} \left(\frac{x_{EOS}(T, p_i) - x_{Exp.}(T, p_i)}{x_{Exp.}(T, p_i)} \right)^2}, \text{ where } x = y, x^{L1}, x^{L2}$$

(24.21)

The binary interaction parameters and averaged deviations are listed in the first three sections of Tables 24.16 and 24.17, respectively.

Further, the interaction parameters for o-cresol + carbon dioxide and p-cresol + carbon dioxide determined at 373.15 K have been used to predict the phase behavior of these systems at 323.15 and 473.15 K in order to compare the predictions with experiments. Some of the comparisons are shown in Figure 24.11. The average deviations (calculated from Eq. (24.21)) are given in the last section of Table 24.17. Because all predictions for 323.15 K (except SAFT and SAFT-CB) yield miscibility gaps that are *far* too small, a rating is given instead of numbers.

Table 24.17: Average deviations of phase composition reproductions and predictions (Eq. (24.21)) in binary systems using different EoS [%].

	PR	Cubic EoS PR-DVT	3P1T	SAFT	EoS with association term SAFT-CB	PR-CPA	AEoS
		Reproductions of systems with carbon dioxide at 373.15 K					
$(\Delta x/x)_{liquid}$	3.0	3.3	4.5	5.4	6.1	7.3	9.7
$(\Delta x/x)_{gas}$	4.9	5.3	6.6	18.3	18.9	10.0	14.6
		Reproductions of LLE in systems water + benzene derivative (both phases)					
373.15 K	1.4	1.4	1.4	2.6		1.5	2.5
		Reproductions of systems cresol isomer + carbon dioxide (both phases)					
323.15 K	11.2	12.6	14.6	14.6	19.3	18.5	29.4
473.15 K	3.1	2.7	3.0	3.9	2.4	3.6	3.7
		Predictions of systems cresol isomer + carbon dioxide based on parameter from 373.15 K (both: o-cresol and p-cresol. both phases)					
323.15 K	poor	poor	poor	good	good	poor	poor
473.15 K	12.7	7.0	15.3	10.7	11.2	7.7	15.6

24.5.2.3 Ternary Systems

Using the binary interaction parameters from above, the phase equilibrium in multicomponent systems can be predicted at similar temperatures. The phase equilibria of the ternary systems o-cresol + p-cresol + carbon dioxide and cresol isomer + ethanol + carbon dioxide (investigated in the experimental part of this work) of systems hydroxymethylbenzene deri-

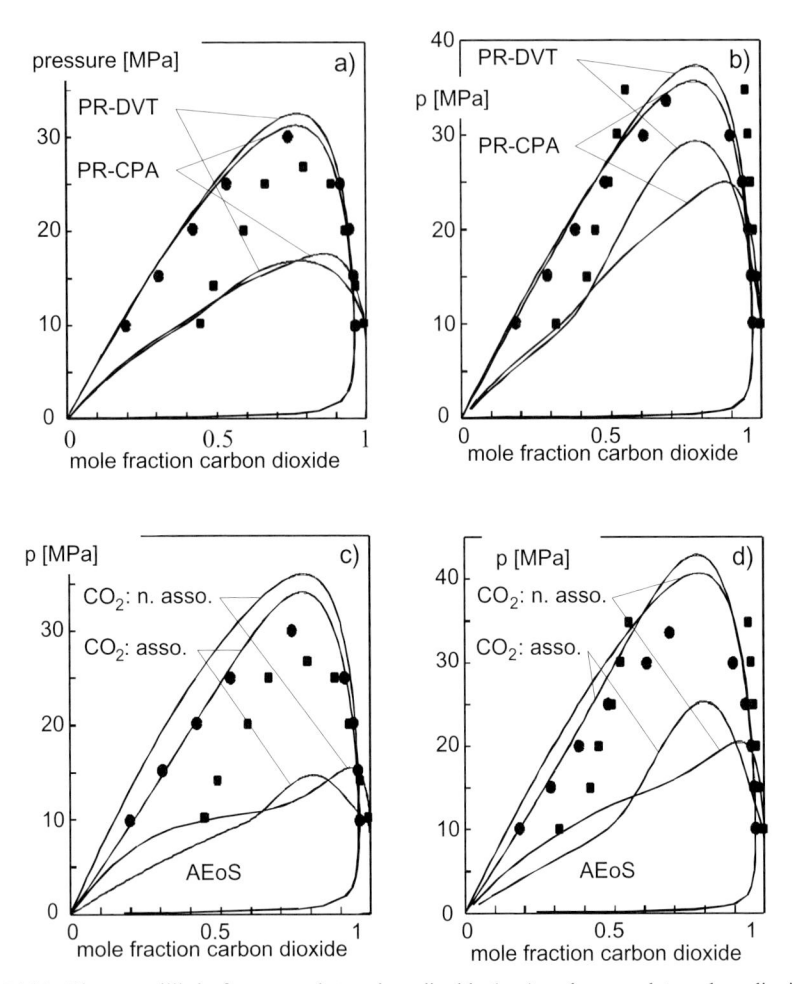

Figure 24.11: Phase equilibria for *o*-cresol + carbon dioxide (a, c) and *p*-cresol + carbon dioxide (b, d) at ■ = 323.15 K and ● = 473.15 K compared to predictions with I) PR-DVT EoS, II) PR-CPA EoS, III) AEoS, and IV) AEoS where CO_2 is modeled as associating compound [4]. The binary parameters had been adjusted to the equilibria measured at 373.15 K, before (see text).

vative + water + carbon dioxide were predicted this way using the seven different EoS described above. The binary interaction parameters for the binary systems *o*-cresol + *p*-cresol and cresol isomer + ethanol have been set to zero, because these systems are completely miscible at the conditions investigated experimentally and an ideal mixture *in the sense of the EoS investigated* could give information if accounting for hydrogen bonds yields advantages, here.

The phase behavior of all ternary systems benzene derivative + water + carbon dioxide was predicted using only the binary interaction parameters mentioned. The deviations between predictions and experimental results for the compositions of the coexisting phases are listed in Table 24.18. In general, the predictions are significantly less accurate than the re-

24.5 Theory/Methods

productions of the binary systems but they are still very good. The *relative* deviations surmount 10% only for small mole fractions of at least one component in the phase. The *absolute* deviations are always smaller than 0.3 mol% for the water-rich phase and 0.8 mol% for the carbon dioxide-rich phase. Both equations lead to very similar predictions. Graphical comparisons for some representative ternary systems are shown in Figures 24.12 and 24.13. The predictions are very satisfactory except for the ternary system phenol + water + carbon dioxide at 30 MPa where both EoS predict a three-phase region in contrary to experiments (carried out later), cf. also Figure 24.14.

Table 24.18: Average relative deviations in predicted mole fractions (Eq. (24.21)) and separation factors in ternary systems (based on binary systems) [%].

		Cubic EoS			EoS with association term		
	PR	PR-DVT	3P1T	SAFT	SAFT-CB	PR-CPA	AEoS

Predicted mole fractions in systems benzene derivative + H_2O + CO_2 at 373.15 K (all phases)

$\Delta x/x$	12.7	13.2	13.2	24.7	25.1	15.6	22.2

Predicted separation factors $\alpha_{ethanol-cresol}$ in systems with CO_2 at 373.15 K

$\Delta\alpha/\alpha$			> 50%				

Predicted separation factors $\alpha_{o\text{-cresol-}p\text{-cresol}}$ in the ternary system with CO_2
1. line: $k, \lambda \neq f(T,p)$ 2. line: $k, \lambda = f(T,p)$

$\Delta\alpha/\alpha$	9.4	9.4	13.5	11.2	13.8	15.1	1.6 *)
$\Delta\alpha/\alpha$	4.5	4	4.1	2.3	1.6	5.4	

*) Results from a cancellation of errors: $\Delta y_{cresol}/y_{cresol} \approx 50\%$.

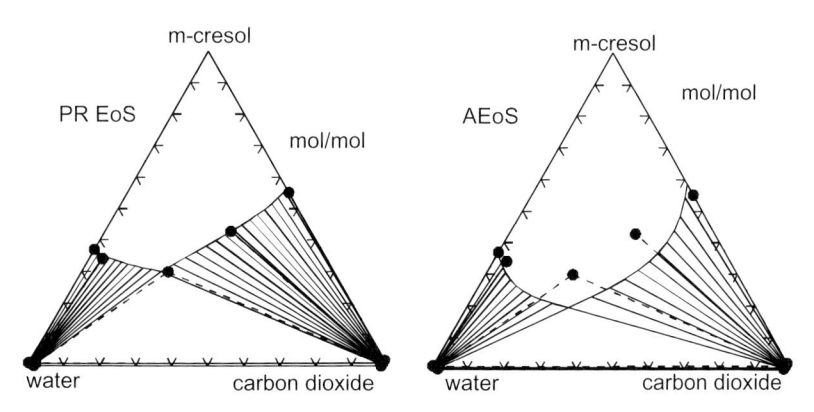

Figure 24.12: Gibbs triangles for the ternary system *m*-cresol + water + carbon dioxide at 373.15 K and 20 MPa with experimental data by Pfohl et al. [1] (1997). Solid lines are predictive calculations with the Peng-Robinson EoS and the Anderko EoS using binary parameters regressed from the binary subsystems (see text).

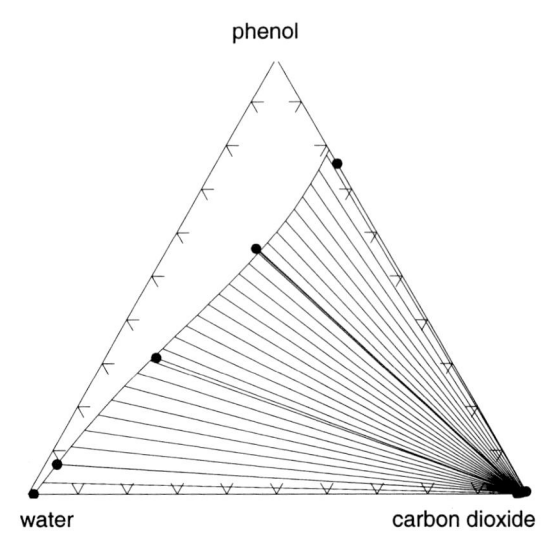

24   *Measurements and Modeling of High-Pressure Fluid Phase Equilibrium*

Figure 24.13: Phase diagram for the ternary system phenol + water + carbon dioxide at 373.15 K and 10 MPa. Points = experimental data, lines = semi-predictive calculations with the Peng-Robinson EoS and parameters from Table 24.15.

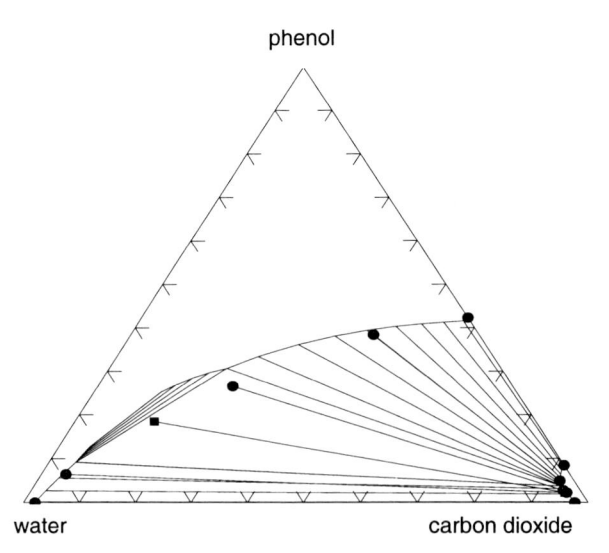

Figure 24.14: Phase diagram for the ternary system phenol + water + carbon dioxide at 373.15 K and 30 MPa. ● = experimental data from Table 24.8. Lines = semi-predictive calculations with the Peng-Robinson EoS and parameters from Table 24.15. ■ = experiment carried out after prediction of three-phase region with EoS in order to verify prediction (30.1 MPa, $x_{CO_2} = 0.142$, $x_{Phenol} = 0.185$, $y_{Water} = 0.0297$, $y_{Phenol} = 0.0222$ mol/mol).

24.6 Results and Discussion

Pure Component Behavior: With the exception of the phenol parameters for the PR-CPA EoS ($-\Delta s^{HB} \ll 20$ J mol^{-1} K^{-1}), the resulting association parameters are very reasonable as discussed in detail by Pfohl (1998). The association energy for *o*-cresol is lower than the association energy for the other cresol isomers for all EoS. This finding is in agreement with the methyl group's shielding the hydroxy group in ortho position as mentioned above. Furthermore, the values for association energy and entropy for SAFT and the PR-CPA EoS are in good agreement with data published by Książczak and Moorthi (1985). The larger values of the entropy for the AEoS determined here are in good agreement with entropies for the AEoS determined by Anderko (1989a,b).

The vapor pressures and the liquid densities of a pure associating component are much better correlated when association is explicitly accounted for in the three cubic EoS (cf. Pfohl et al., 1999) – similar to what is often shown in literature. But, this is mainly due to *fitting* a high number of parameters of the association model to these properties rather than *predicting* the properties based on the corresponding states principle and cannot be used as a criterion for the quality of an EoS. For the sake of an "adequate" comparison, all pure-component parameters of the simple cubic EoS have been treated as adjustable parameters. The results indicate (Pfohl et al., 1999) that the simple cubic EoS can reproduce the pure-component properties as well as the EoS which take association into account as long as the number of adjustable parameters is similar: The PR-DVT EoS with four ($\omega' \equiv \omega$) or five adjustable parameters works as well or better than any of the association EoS discussed here. It is remarkable that none of the association EoS manages to predict the critical points of the pure associating components well. In particular SAFT fails to reach this aim – similar as observed for non-associating compounds by Pfohl et al. (1998). These overpredictions are a measure for the model's inability to sufficiently describe all p,v,T properties of pure components. Pfohl and Brunner (1998) showed that this leads to calculated gas phase densities being remarkably off the experimental data. The simple cubic EoS which do not account for association manage to predict the critical point of pure water as well or better than the three EoS accounting for association (Pfohl et al., 1999). As was shown before (Pfohl et al., 1998) fixing the critical point under such circumstances either leads to high dviations (in Eq. (24.21)) or physically unreasonable parameters, the association EoS are not necessarily to be seen much superior in the description of the pure-component behavior, here.

Representation of Binary Equilibria: With the exception of some $\lambda_{H_2O\text{-toluene}}$ values, the optimum interaction parameters listed in Table 24.16 are of the order of 0.1 and indicate that none of the EoS is an absolutely inadequate model for the investigated systems. For a given system, the binary parameters for the three cubic EoS are nearly identical, indicating that these EoS are very similar. The interaction parameters for the cubic equations with additional association part (PR-CPA and AEoS) are not smaller than the interaction parameters for the simple cubic EoS, indicating that although association is incorporated, the same order of corrections is necessary. The magnitudes of the interaction parameters for SAFT and SAFT-CB are only about 50% of those for the other EoS, but this is a result of the SAFT interaction parameters being defined on segment basis, thus influencing the behavior of a molecule with a chain length of m ≠ 1 multiple times. The optimum interaction parameters for

SAFT-CB are insignificantly larger than the values for SAFT, in agreement with prior findings and probably due to the larger differences in the intersegmental potentials (Pfohl and Brunner, 1998). Table 24.17 shows that none of the association EoS manages to reproduce the behavior of the binary systems with carbon dioxide as good as the simple cubic EoS: the relative error when describing the liquid phase is about the two-fold, the relative error when describing the carbon dioxide-rich phase is about the three-fold. This finding is in agreement with the findings by Jennings et al. (1993) who found that with SAFT relative errors were 2–3 times larger than those with a cubic EoS for carbon dioxide + 1-alkanol systems at 315 to 337 K and 5 to 12 MPa. SAFT and SAFT-CB give very similar results, indicating that SAFT does not fail because of a poor description of the properties of pure carbon dioxide (compare Pfohl and Brunner, 1998; Pfohl et al., 1998). The PR-CPA EoS and the AEoS give least accurate results than the same EoS without accounting for association (PR/PR-DVT and 3P1T), not allowing to attribute any positive results to the incorporation of association, here. Using the AEoS and modeling carbon dioxide as weakly associating according to Anderko (1989b) by using his parameters for carbon dioxide and/or using the second interaction parameter in order to fine-tune the cross association does not improve the performance of the AEoS significantly (Pfohl, 1998), minimizing hopes that a more complicated approach yields much better results.

Table 24.17 shows that the liquid-liquid equilibrium of aqueous systems and the equilibrium at 473.15 K are described within experimental accuracy by all EoS. The former is probably due to two interaction parameters for the representation of equilibria with mole fractions that are nearly independent of pressure in the range investigated, while the latter is probably due to reduced specific interactions at high temperatures in experiment and calculations.

Prediction of Binary Equilibria: Table 24.17 and Figure 24.11 show that the predictions for the cresol + carbon dioxide system at 323.15 K (using only interaction parameters derived at 373.15 K) are poor except when using SAFT/SAFT-CB. The failure of the AEoS and PR-CPA EoS indicates that the benefits of SAFT should not be attributed to its association term which is identical to that of the PR-CPA EoS but should be attributed to the different reference system, i.e. hard spheres instead of van der Waals theory.

Prediction of Ternary Equilibria: Results of PR EoS predictions of the phase equilibrium of ternary systems are shown in Figures 24.9, 24.10, 24.12, and 24.14 for some systems with hydroxymethylbenzenes, carbon dioxide, and water. The predicted binodals of the three-phase regions are generally in good agreement with the experimental results. The predictions from the other two cubic EoS do not differ much from these predictions. In most cases, the differences between calculated and experimental results cannot be seen when simply looking at the binodals alone. Instead, the slopes of the tie lines, i.e. separation factors, have also to be considered (enlargements in Figs. 24.5, 24.7b, and 24.10). The predictions for the separation factors α_{ij} (i,j ≠ CO_2) are poor – independent of the EoS used: more than 50% error for cresol + ethanol and about 10% for o-cresol + p-cresol + carbon dioxide. Because the values of $\alpha_{o\text{-cresol}/p\text{-cresol}}$ are very close to unity an error of 10% would lead to an error in the height of a multi stage countercurrent column of about 30% and is therefore unacceptable. In modeling the ternary systems with water using the AEoS, predictions with bent binodals are observed (cf. Fig. 24.12). Although it is possible that these bent lines result from modeling water with the linear association model on the one hand, it is unlikely on the other hand, because SAFT does not yield these bent binodals if water is modeled with two instead of three association sites (Pfohl, 1998).

Because the binary systems, i.e. the borders of the ternary systems, are poorly described with the EoS incorporating association, predictions for the ternary systems are also less accurate than predictions with the cubic EoS. The first line in Table 24.17 shows that the relative errors in the predicted mole fractions of the ternary systems using these association EoS is twice of that using the simple cubic EoS – in agreement with the different quality of the reproductions of the binary subsystems.

Further Predictions of Ternary Equilibria: In order to eliminate the poor representation, a modified procedure was applied to predict the separation factors in the *o*-cresol + *p*-cresol + carbon dioxide system. For these predictions, the binary interaction parameters for the cresol isomer + carbon dioxide binaries were adjusted in such a way that the (extrapolated) mole fractions of both phases in each binary are *exactly* reproduced at the temperature and pressure of the ternary measurements. The determination of such parameters for the AEoS was not possible due to the (predicted) occurance of three-phase splitting. The resulting sets of binary parameters for the different conditions investigated experimentally are listed in the last section of Table 24.16. The relative errors in the predicted separation factors of the ternaries this way, are shown in the last line of Table 24.18. The predictions using SAFT/SAFT-CB are best, but again the bad performance of the PR-CPA EoS (which also uses the SAFT association term) implies that this good performance of SAFT is at least to a great extent caused by the different *non*-associating reference system.

24.7 Conclusion

High-pressure phase equilibria in binary and ternary systems with *o*-cresol, *p*-cresol, ethanol, water, and carbon dioxide were investigated experimentally at 323.15 to 473.15 K and pressures between 10 and 35 MPa. The separation of *o*-cresol and *p*-cresol using supercritical carbon dioxide was investigated, as was the effect of adding ethanol in order to enhance the solubility of the cresols in supercritical carbon dioxide. The dependencies of the gas phase solubilities and the separation factor between the two cresols on temperature, pressure, and overall composition were discussed. The separation factor is small and clearly indicates that the even more difficult separation of *m*-cresol and *p*-cresol – which still is a challenging task in industry – will not be economic with supercritical carbon dioxide at the conditions investigated. It was shown that ethanol cannot be used as an entrainer, as adding ethanol does not increase the uptake of cresols in carbon dioxide.

The data base provided in this study was used to evaluate the capabilities of seven different EoS to describe and predict such phase equilibria. The three investigated simple EoS *not* accounting for association generally allowed better reproductions of the experimental results than the four investigated EoS which explicitly account for association. These association-EoS could *not* prove to be superior in modeling or predicting any of the equilibria measured here. Two slight benefits of SAFT/SAFT-CB in predicting the measured equilibria cannot be attributed to the incorporation of association, as can be seen form the failure of the PR-CPA EoS which accounts for association in the same way as SAFT.

References

Anderko, A. (1989a): A simple equation of state incorporating association. Fluid Phase Equilibria, 45, 39.

Anderko, A. (1989b): Extension of the AEOS model to systems containing any number of associating and inert compounds. Fluid Phase Equilibria, 50, 21.

Angus, S.; Armstrong, B.; de Reuck, K.M. (1976): International thermodynamic tables of the fluid state – 3. Carbon dioxide. Pergamon Press, Oxford.

Boston, J.F.; Mathias, P.M. (1980): 2nd Int. Conf. on Phase Equilibria and Fluid Properties in the Chemical Process Industries. Berlin, March 17–21, p. 823.

Brunner G. (1994): Gas Extraction – An introduction to fundamentals of supercritical fluids and the application to separation processes. Springer, New York.

Chapman, W.G.; Gubbins, K.E.; Jackson, G.; Radosz, M. (1989): SAFT: Equation-of-state solution model for associating fluids. Fluid Phase Equilibria, 52, 31.

Daubert, T.E.; Danner, R.P. (1989): Physical and thermodynamic properties of pure chemicals: Data compilation. Hemisphere Publications, New York.

Gupta, R.B.; Johnston, K.P. (1994): Lattice fluid hydrogen bonding model with a local segment density. Fluid Phase Equilibria, 99, 135.

Huang, S.H.; Radosz, M. (1990): Equation of state for small, large, polydisperse, and associating molecules. Ind. Eng. Chem. Res., 29, 2284.

Huang, S.H.; Radosz, M. (1991): Equation of state for small, large, polydisperse, and associating molecules: Extension to fluid mixtures. Ind. Eng. Chem. Res., 30, 1994.

Huang, S.H.; Radosz, M. (1993): Additions and corrections. Ind. Eng. Chem. Res., 32, 762.

Iwai, Y.; Uchida, H.; Koga, Y.; Arai, Y.; Mori Y. (1996): Monte Carlo simulation of solubilities of aromatic compounds in supercritical carbon dioxide by a group contribution site model. Ind. Eng. Chem. Res., 35, 3782.

Jennings, D.W.; Gude, M.T.; Teja, A.S.: High-pressure vapor-liquid equilibria in carbon dioxide and 1-alkanol mixtures, in E. Kiran and J.F. Brennecke (Eds.), Supercritical Fluid Engineering Science – Fundamentals and Applications, ACS Symposium Series 514, American Chemical Society, Washington, DC, 1993, p. 10.

Kempter, H.; Mecke, R. (1940): Spektroskopische Bestimmung von Assoziationsgleichgewichten. Z. Phys. Chem. B., 46, 229.

Kontogeorgis, G.M.; Voutsas, E.C.; Yakoumis, I.V.; Tassios, D.P. (1996): An equation of state for associating fluids. Ind. Eng. Chem. Res., 35, 4310.

Książczak, A.; Moorthi, K. (1985): Prediction of thermodynamic properties of alkylphenol + hydrocarbon binary systems on the basis of pure liquid properties. Fluid Phase Equilibria, 23, 153.

Mathias, P.M. (1983): A versatile phase equilibrium equation of state. Ind. Eng. Chem. Process Des. Dev., 22, 385.

Mathias, P.M.; Klotz, H.C.; Prausnitz, J.M. (1991): Equation-of-state mixing rules for multicomponent mixtures: the problem of invariance. Fluid Phase Equilibria, 67, 31.

Michelsen, M.L.; Kistenmacher, H. (1990): On composition-dependent interaction coefficients. Fluid Phase Equilibria, 58, 229.

Pagel, A. (1997): Diplom Thesis, Technische Universität Hamburg-Harburg, Arbeitsbereich Thermische Verfahrenstechnik.

Panagiotopoulos, A.Z.; Reid, R.C. (1985): High-pressure phase equilibria in ternary fluid mixtures with a supercritical component. ACS Division of Fuel Chemistry, Preprints 30/3, 46.

Peneloux, A.; Rauzy, E.; Freze, R. (1982): A consistent correction for Redlich-Kwong-Soave volumes. Fluid Phase Equilibria, 8, 7.

Peng, D.-Y.; Robinson, D.B. (1976): A new two-constant equation of state. Ind. Eng. Chem. Fundam., 15, 59.

Pfohl, O. (1998): Messung und Berechnung von Phasengleichgewichten mit nahe- und überkritischem Kohlendioxid sowie assoziierenden Komponenten im Hochdruckbereich. VDI Fortschritt-Bericht, Reihe 3, Nr. 572, ISBN 3-18-357203-6. VDI Verlag, Düsseldorf.

Pfohl, O.; Avramova, P.; Brunner G. (1997): Two- and three-phase equilibria in systems containing benzene derivatives, carbon dioxide, and water at 373.15 K and 10–30 MPa. Fluid Phase Equilibria, 141, 179.

Pfohl, O.; Brunner, G. (1998): II. Using BACK to modify SAFT in order to enable density and phase equilibrium calculations connected to gas-extraction processes. Ind. Eng. Chem. Res., 37, 2966.

Pfohl, O.; Giese, T.; Dohrn, R.; Brunner, G. (1998): I. Comparison of 12 equations of state with respect to gas-extraction processes: Reproduction of pure-component properties when enforcing the correct critical temperature and pressure. Ind. Eng. Chem. Res., 37, 2957.

Pfohl, O.; Pagel, A.; Brunner, G. (1999): Phase equilibria in systems containing o-cresol, p-cresol, carbon dioxide, and ethanol at 323.15–473.15 K and 10–35 MPa. Fluid Phase Equilibria, 157, 53.

Pfohl, O.; Timm, J.; Dohrn, R.; Brunner, G. (1996): Measurement and correlation of vapor-liquid-liquid equilibria in the glucose + acetone + water + carbon dioxide system. Fluid Phase Equilibria, 124, 221.

Reid, R.C.; Prausnitz, J.M.; Poling, B.E. (1987): The properties of gases and liquids. McGraw Hill, New York.

Schmidt, E. (1969): Properties of water and steam in SI-units. Springer, Berlin.

Soave G. (1972): Equilibrium constants from a modified Redlich-Kwong equation of state. Chem. Eng. Science, 27, 1197.

Solórzano-Zavala, M.; Barragán-Aroche, F.; Bazúa, E.R. (1996): Comparative study of mixing rules for cubic equations of state in the prediction of multicomponent vapor-liquid equilibria. Fluid Phase Equilibria, 122, 99.

Span, R.; Wagner, W. (1996): A new equation of state for carbon dioxide covering the fluid region from the triple-point temperature to 1100 K at pressures up to 800 MPa. J. Phys. Chem. Ref. Data, 25/6, 1509.

Voutsas, E.C.; Kontogeorgis, G.M.; Yakoumis, I.V.; Tassios, D.P. (1997): Correlation of liquid-liquid equilibria for alcohol/hydrocarbon mixtures using the CPA equation of state. Fluid Phase Equilibria, 132, 61.

Wäterling, U. (1992): Ein Beitrag zur experimentellen Untersuchung und Berechnung von Phasengleichgewichten weitsiedender Gemische bei erhöhten Temperaturen und Drücken. VDI Fortschritt-Bericht, Reihe 3, Nr. 275. VDI Verlag, Düsseldorf.

Yakoumis, I.V.; Kontogeorgis, G.M.; Voutsas, E.C.; Tassios, D.P. (1997): Vapor-liquid equilibria for alcohol/hydrocarbon systems using the CPA equation of state. Fluid Phase Equilibria, 130, 31.

Yau, J.-S.; Tsai, F.-N. (1992): Solubility of carbon dioxide in phenol and in catechol. J. Chem. Eng. Data, 37, 141.

Yu, J.-M.; Lu, B.C.-Y. (1987): A three-parameter cubic equation of state for asymmetric mixture density calculations. Fluid Phase Equilibria, 34, 1.

Nomenclature

a, b	EoS parameters in some cubic equations
a^{asso}	association part of Helmholtz energy
$\tilde{a}^{asso}$	association part of Helmholtz energy reduced by RT
A_i, B_j	characterizes association site A (B) on molecule i (j)
C	EoS parameter in Yu and Lu EoS
c, d	pure-component EoS parameters for Mathias alpha function
$\tilde{c}$	EoS parameter for volume translation according to Peneloux
D_i, D_{ij}	sphere diameters
e	pure-component EoS parameter: temperature dependence of u^0 in SAFT
g_{ij}	radial distribution function at contact
Δh^{HB}	enthalpy change when hydrogen bonding
k	Boltzmann constant

k_{ij}	binary interaction parameter
N	number of compounds
N_A	Avogadro number
n	number of samples taken from one phase (typically: 3)
n_{exp}	number of determined mole fractions
n_f	number of (averaged) mole fractions
n_v	number of points on vapor pressure curve
m	pure-component EoS parameter: chain length in SAFT
p	pressure
p_c	critical pressure
p'_c	pure-component EoS parameter, that can be set equal to p_c
p_1	pure-component EoS parameter for Mathias alpha function
R	universal gas constant
Δs^{HB}	entropy change when hydrogen bonding
T	temperature
T_c	critical temperature
T'_c	pure-component EoS parameter that can be set equal to T_c
T_r	reduced temperature
T'_r	reduced temperature (scaling temperature: T_c')
u^0	pure-component EoS parameter: intermolecular potential depth in SAFT
v	molar volume
v^{00}	pure-component EOS parameter: sphere volume in SAFT
Δv^{HB}	volume change when hydrogen bonding
$\tilde{v}$	translated molar volume
x_i	mole fraction of component i (x^{L1}, x^{L2}: in liquid phases L1 and L2)
$\bar{x}$	mean mole fraction
X^{A_i}	*fraction of non associating sites of type A on molecule i*
y_i	mole fraction of component i in gas phase

Greek letters

α_{ij}	relative volatility, separation factor: $(y_i/y_j)/(x_i/x_j)$
β^{A_i}	pure-component association parameter for site A on molecule i
$\beta^{A_i B_j}$	association parameter between site A on molecule i and B on molecule j
$\Delta^{A_i B_j}$	association strength between site A on molecule i and B on molecule j
ε^{A_i}	pure-component association parameter for site A on molecule i
$\varepsilon^{A_i B_j}$	association potential depth between site A on molecule i and B on molecule j
κ	EoS parameter influencing the slope of the vapor pressure curve
κ^{A_i}	pure-component association parameter for site A on molecule i in SAFT
$\kappa^{A_i B_j}$	association volume between site A on molecule i and B on molecule j
λ_{ij}	*binary interaction parameter*
ρ	molar density
$\hat{\rho}$	density estimate determined experimentally under some assumptions [g/cm^3]
ω	acentric factor
ω'	EoS parameter that can be set equal to ω
ξ_i	abbreviations defined in Eq. (24.16)

VII Phase Equilibrium in Microemulsion Systems

25 The Potential of Surfactants Modified Supercritical Fluids for Dissolving Different Classes of Substances

Uta Lewin-Kretzschmar and Peter Harting[1]

25.1 Abstract

The solubility of some nitroaromatic compounds, polar dyes, and adamantane in the supercritical fluids carbon dioxide and ethane ($scCO_2$ and scC_2H_6), unmodified and modified with various surfactants and solvents was determined at temperatures from 323 to 373 K and a pressure range from 10 to 45 MPa.

The experimental results for the solubility of the nitroaromatics in the unmodified system are compared to predictions from a group contribution equation of state (GC-EOS). The solubility data for the dyes were correlated with some empirical equations. Furthermore, dyeing experiments with wool and cotton dyes were conducted in order to characterize the impact of modifiers on the dyeing procedure and the coloring properties and to prove the practicability of modified fluid systems for commercial procedures.

25.2 Introduction

The role of separation techniques using compressed gases has increased rapidly in the fields of chemical analysis and industrial processes in the last few years. One of the reasons for this development is the use of modified fluids in extraction techniques. Classical solvents such as methanol, ethanol, and acetone (Camel et al., 1995) were used in most cases as modifiers. However, the class of surfactants is also of interest for that purpose because of its amphiphilic character and the formation of micelles (Consani and Smith, 1990; Smith et al., 1989; Johnston et al., 1989). Therefore, perfluorinated surfactants with very high solubilities in supercritical carbon dioxide which are able to form carbon dioxide-water emulsions were synthesized and used, e.g., for the extraction of proteins or heavy metals and for dispersion (Johnston et

---

1 Institut für Nichtklassische Chemie e.V. an der Universität Leipzig, Permoserstr. 15, 04318 Leipzig.

al., 1996; Yazdi et al., 1996; McClain et al., 1996; Triolo et al., 2000). The disadvantages of these surfactants are their very high price and a low biodegradation rate.

The solubility of commercial nonfluorinated surfactants, which are considerably cheaper and involve lower degradation risks, was qualitatively investigated by several groups (Consani and Smith, 1990; Bartle et al., 1991) and first systematically quantified by Blasberg et al. (1998). In the present work the solubility of some substances in such supercritical solutions was investigated. Furthermore, the potential of such surfactant-modified supercritical fluids for new innovative procedures and technologies was studied.

The class of nitroaromatic compounds was selected because of its high environmental relevance, the high toxic potential (Lewin, 1994; Schneider et al., 1994) and the widespread occurrence at sites contaminated by such substances from e.g. an intensive use for military purposes in the past (Lewin-Kretzschmar et al., 2000). More than 2000 such sites in present-day Germany are suspected to be contaminated by explosives, their by-products, or metabolites (Preuß and Wiegandt, 1992). These sites are often located in the vicinity of large water supply areas, as the production of explosives is a water-intensive process. Therefore, the situation needs to be remedied by suitable procedures based on analytically supervision. Supercritical fluid extraction (SFE) with the possibility of varying the selectivity or the solubility of a solute by several orders of magnitude by adjusting the system pressure or temperature or the adding modifiers seems to be a suitable technique for this purpose. However, the confines of this technique became apparent in an analytical application (Pörschmann et al., 1998) at increasing polarity and with the greater strength of the matrix binding of the analyte. This is attributed either to strong matrix interactions by the contaminant or to insufficient solubility in the supercritical fluid. These aspects are investigated in this paper. Adamantane (cf. Fig. 25.1) is a favorable model substance for examining the impact of surfactant modification on the solubility of nonpolar compounds. It is the first member in a homologous series of structurally diamond-like compounds (Wingert, 1992). These compounds are precipitated during the production of natural gas from associated gas and petroleum wells and lead to corrosion and plugging of pipelines (King, 1988).

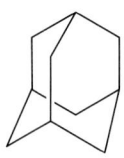

Figure 25.1: Structure of adamantane.

A new approach to isolate adamantane from petroleum is the extraction by means of supercritical fluids. However, former experiments with pure supercritical fluids methane, ethane, and carbon dioxide (Smith and Teja, 1996) showed that the solubility of adamantane needs to be increased by using a modifier (such as a surfactant) to achieve a reasonable process from an economical point of view.

Another interesting application is the use of surfactant-modified fluids in dyeing processes. The solubility of polar dyes in supercritical fluids might be enhanced by a surfactant. From an ecological point of view, dyeing in $scCO_2$ is a very favorable process, as no pollution occurs and the amount of energy required is much lower than in a conventional dyeing process using aqueous solutions (Poulakis et al., 1991; Knittel and Schollmeyer, 1992).

In addition, excess dyes can be easily recovered. Moreover, the procedure does not need neither any wetting nor any dispersing agents (Gießmann et al., 1999). The inert and physiologically harmless carbon dioxide is especially suitable because of its favorable critical data, its availability at reasonable prices and its good large-scale handling. However, a basic condition for the dyeing process is a sufficiently high solubility of the dye in the supercritical fluid. This is usually the case for hydrophobic disperse dyes used in polyester dyeing (Poulakis et al., 1991; Ebner and Schelz, 1989). However, the solubility in supercritical carbon dioxide is small for many highly polar hydrophilic acidic, reactive and direct dyes which are used in the conventional dyeing of wool, cotton and cellulose fibers (Giehl et al., 1998; Gießmann, 1998). Therefore, the work deals with using surfactant-modified supercritical carbon dioxide as solvents for such dyes. The practicability of the best suitable modifying systems was verified by means of dyeing tests.

25.3 Experimental/Methods

25.3.1 Equipment

The solubility experiments were conducted applying the analytical-static method using the autoclave system (HPM-3; Premex Reactor AG, Lengnau, Switzerland) shown in Figure 25.2. This equipment had already proven successfully for determining the solubility of surfactants in supercritical fluids (Blasberg et al., 1998). The autoclave is supplied with a stirrer. It has a volume of 300 ml and can be used at pressures up to 50 MPa at temperatures up to 473 K.

The concentrations of nitroaromatics, dyes, and surfactants were determined by HPLC analysis. The HPLC equipment was a commercial device (Knauer, Berlin, Germany) consisting of a pump (Maxi-Star K-1000), a degasser, a dynamic mixing chamber, a thermostat (Jetstream 2), a six-port valve (A0634), an UV detector (K-2500), and a light scattering detector (Sedex 55, ERC, Alteglofsheim, Germany; for details cf. Dreux et al., 1996).

For the analysis of adamantane a gas chromatograph CP 9000 (Chrompack/Varian, Middelburg, Netherlands) with split/splitless injector and flame ionisation detector was used.

25.3.2 Procedure

First the evacuated autoclave was filled with the pure solute or a solute mixture. The amount of substance varied between 3 and 45 g, dependent on the solubility in a supercritical solvent. That solubility was estimated from preliminary experimental data. The nitroaromatics and the dyes were investigated as mixtures (at each case three components) to reduce the

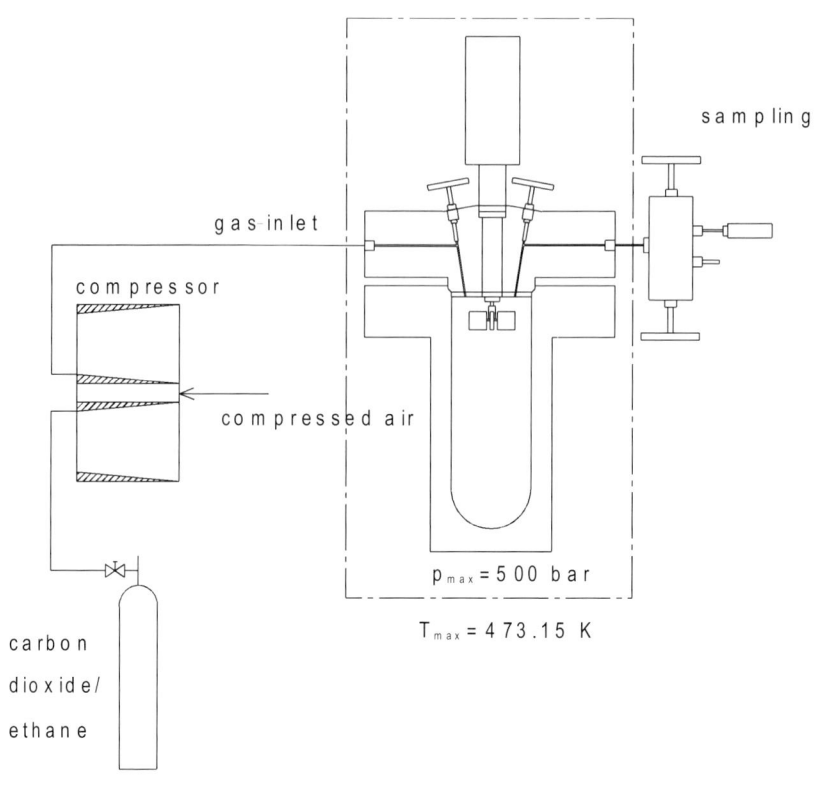

Figure 25.2: Scheme of autoclave system.

number of experiments. For this purpose it was clarified that the mixture does not significantly influence the solubility values found for the single compounds and that all components can well be detected by chromatography also in the presence of the different modifiers. The supercritical solvent (carbon dioxide or ethane) was then added, and temperature and pressure were set as desired and kept constant. After establishing the phase equilibrium (about 4 h, 2.5 h with stirring, then 1.5 h without stirring, established on the basis of preliminary experiments), an aliquot of the fluid was taken manually via a three-way valve and flushed into an evacuated vessel (5 ml) made of stainless steel. The volume of the aliquot taken is regarded to be small enough not to disturb the phase equilibrium and large enough to be within the detection limits of the successive analytical determinations. This sampling is considered to be adiabatic. The volume of the vessel was expanded into 20 ml methanol, water, or n-hexane (depending on the solubility properties of the investigated compounds) at room temperature via a capillary for a few hours. The resulting solution was filled up to a defined volume with the corresponding solvent and was analyzed according to the procedures described below.

The solubility of the solutes in supercritical fluids was determined at 323, 348, and 373 K at pressures from 10 to 45 MPa. Some experiments (in most cases at temperatures/pressures where the solute concentration reached a maximum) were repeated three times to

25.3.3 Analysis

Nitroaromatics were analyzed by isocratic HPLC (50% methanol/50% water, 1 ml/min, room temperature) at an Eurospher RP-18 column (250 x 4 mm, 5 µm; Knauer) and UV detection at 254 nm.

The concentration of adamantane was determined by capillary gas chromatography using a nonpolar column CP-SIL 5CB (l = 10 m, ID = 0.32 mm; Chrompack/Varian GmbH; nitrogen carrier gas (25 kPa), 2 min isotherm at 328 K, then with 8 K/min to 533 K and 10 min isotherm postrun; injection of 1 µl with split ratio 1:50) and a flame ionization detector.

The first dye mixture (cf. Table 25.1) was analyzed by HPLC (column: Eurospher RP-18; mobile phase: 60% methanol/40% aqueous 0.01 M phosphate buffer (pH 3); flow rate: 1 ml/min; temperature: 303 K; UV-detection at 340 nm.) The second dye mixture was analyzed by HPLC at the Eurospher column with a methanol/aqueous 0.01 M acetic buffer (pH 4.5) mobile phase (8 min isocratic 45%/55%, then within 5 min to 100% methanol and 15 min isocratic postrun with 100% methanol; flow: 1 ml/min, temperature: 303 K) and UV-detection at 580 nm.

The determination of the surfactant Texin DOS 75 in the experiments for the optimization of the dye procedure was carried out using a Kromasil C4-column (250 x 4 mm, 5 µm; Knauer) with 80% methanol/20% water as mobile phase (flow: 1 ml/min, temperature: 303 K) and light scattering detection (evaporator 313 K, 0.2 MPa nitrogen pressure to ensure the aerosol formation).

25.3.4 Dyeing Experiments

In order to determine the influence of the modifiers on the dyeing process, dyeing experiments were carried out with two wool dyes (Lissamine Yellow AE and Doractive Blue WRL), two cotton dyes (primuline and Doramine Light Blue BR 200%), and the polyester-wool-dye Domalane Blue BL 150%. The focus of these experiments was on questions like the competition between fiber and modifier for the dye, the increase or decrease in the affinity of the dye to the fiber, and the fastness of dyeing to walking and rubbing in the case of surfactant addition.

These experiments were carried out in a 500 ml autoclave (HPM-P 500; Premex Reactor AG, Lengnau, Switzerland) at 373 K and 45 MPa in aqueous Texin DOS 75/ethanol modified $scCO_2$ (the conditions, where the highest solubilities were detected) as well as in the pure fluid and in water, ethanol and water/ethanol modified fluids for comparison (the amount of these modifiers was optimized in preliminary experiments. The wool or cotton cloth (approx. 8 x 10 cm) was tightly wound around a stainless steel tube, which was fixed on the top of the autoclave. After the dyeing experiment, the cloth was divided in three parts. One part was dried in air, the others were rinsed in hot water (323 K, 1 h) and in acetone (room temperature, 1 h), respectively. Then the samples were visually evaluated as to depth of color, levelness of dyeing, fastness to washing and rubbing.

25.3.5 Materials

All nitroaromatics (2-methylnitroaniline (2-M-3-NA), 1,3-dinitrobenzene (1,3-DNB), and 2-nitrotoluene (2-NT)) and adamantane were from Merck Darmstadt, Germany (cf. Table 25.1). The amount of mass of a nitroaromatic in the autoclave was about 2 g (2-M-3-NA), 3 g (1,3-DNB), 5 g (2-NT). For adamantane it was 8 g (in pure $scCO_2$) and 17 g (in pure scC_2H_6 and modified $scCO_2$) respectively.

Table 25.1: Nitroaromatics investigated.

Nitroaromatics	$M/g \cdot mol^{-1}$	Structural formula
2-methyl-3-nitroaniline (2-M-3-NA)	152.2	O_2N–(ring with CH_3)–NH_2
1,3-dinitrobenzene (1,3-DNB)	168.1	O_2N–(ring)–NO_2
2-nitrotoluene (2-NT)	137.1	(ring with CH_3)–NO_2

The dyes primuline and parafuchsine hydrochloride were from Aldrich GmbH (Taufkirchen, Germany), all other dyes (cf. Table 25.2) were gifts by M. Dohmen GmbH Korschenbroich, Germany. The amount of mass of a dye in the autoclave was 3 g.

The surfactant modifiers were selected in a preliminary investigation (Blasberg et al., 1998): Dehyquart C (Henkel KGaA Düsseldorf, Germany), Texin DOS 75 and Präwozell FCE 1214 (both from Buna GmbH Schkopau, Germany, cf. Table 25.3) in quantities of 6, 18, and 15 g. These surfactants were also used before in supercritical fluid extraction experiments with nitroaromatics (Pörschmann et al., 1998).

Besides these surfactants, Fluowet OTN, a perfluorinated surfactant (Clariant GmbH) was used as modifier for adamantane in amounts of 17 g. The solvents ethanol (18.3 g), water (7 g), methanol (12.5 g), pentanol (20.5 g) were purchased from Merck Darmstadt.

(When optimizing the dyeing experiments, the amount of substances varied from those given here).

25.3 Experimental/Methods

Table 25.2: Dyes investigated.

Trade name (application)	$M/\text{g} \cdot \text{mol}^{-1}$	Structural formula
First group:		
primuline (for cotton)	475.6	
Lissamine Yellow AE (for wool)	452.4	
parafuchsine hydrochloride (for microscopy)	323.9	
Second group:		
Doractive Blue WRL (for wool)	643.5	
Doramine Light Blue BR 200% (for cotton)	927.8	
Domalane Blue BL 150% (for poly-ester/wool blend)		dye mixture (detailed composition not available)

577

Table 25.3: Surfactant modifier investigated.

Systematic name (trade name)	$M/\text{g·mol}^{-1}$	Structural formula
sodium diisooctyl-sulfosuccinate (Texin DOS 75, AOT)	444.5	$H_{17}C_8OOC$–CH–SO$_3$Na, CH$_2$, $H_{17}C_8OOC$
N-dodecyl-pyridiniumchloride (Dehyquart C)	283.9	pyridinium ring with N$^+$–C$_{12}H_{25}$, Cl$^-$
dodecyl-oxethpropylate (Präwozell FCE 1214)	638.9	$H_{25}C_{12}$–O–[CH$_2$–CH$_2$–O]$_n$–[CH(CH$_3$)–CH$_2$–O]$_m$–H

25.3.6 Statistical Methods

The statistical testing methods "F-test" and "t-test" (Doerffel, 1987; Doerffel et al., 1994) were used for the objective evaluation of the influence of the various modifiers on the solubility experiments. This leads to statements "systematical <" or "systematical >" in the case of a significant change of the average concentration values caused by the surfactant in a permissible t-test. The differences are described as stochastic if, according to the t-test, the difference between the average concentration values is in the range of random spreading. In some cases the series of measurements with and without surfactants could not be compared because of serious stochastic errors detected in a F-test.

25.4 Correlation and Prediction of the Solubility of Dyes in Supercritical Fluids

Due to the high experimental expenditure in determining solubilities in supercritical fluids, it was aimed to use empirical correlation methods according to Mitra and Wislon (1991) and to Del Valle and Aguilera (1988) to predict the solubility of dyes under non-examined conditions of state on the basis of the experimentally determined values. For the other compounds examined, the number of experimentally determined sets of data was too small to form a base for the correlation methods. However, a greater number of experiments could not be realized within the scope of this work, because of the considerable time that would have been needed.

Therefore the attempt was continued to calculate the solubility of the compounds of interest directly by means of a group contribution equation of state (GC-EOS) proposed by Skjold-Jørgensen et al. (1988).

25.4.1 Method of Mitra/Wilson

The method of Mitra and Wilson (1991) allows for the prediction of the influence of density ρ and temperature T on the solubility c in a supercritical fluid:

$$\ln c = A\rho + BT + C \tag{25.1}$$

A, B, C are constants depending on the supercritical fluid and the solute. They are obtained by a multivariable linear regression of experimental data.

Thus the general experience is taken into account that the solubility will increase with the growing density of the fluid due to the stronger interactions between the solvent and the solute, as well as with rising temperature due to the higher vapor pressure of the solute.

25.4.2 Method of del Valle/Aguilera

Del Valle and Aguilera proposed a modification of the solvation model of Chrastil (1982), according to which a solute molecule is solvated by k solvent molecules. They presented a four-parameter equation for describing the influence of density and temperature on the solubility:

$$\ln c = l + m/T + n/T^2 + k \ln \rho \tag{25.2}$$

where l, m, n, k are constants depending on both the supercritical solvent and the solute.

Due to the chemical and physical background of the model, further properties can be derived. For example, the constant k corresponds to the number of solvent molecules in the solvation shell and the reaction enthalpy Q_R is the sum of contributions from solvation and evaporation of the solute:

$$Q_R/R = m + 2\,n/T \tag{25.3}$$

where R is the universal gas constant.

25.4.3 Group Contribution Equation of State

The group contribution equation of state (GC-EOS) of Skjold-Jørgensen et al. (Skjold-Jørgensen, 1988; Schmelzer et al., 1989, 1990) was used to predict the solubility of 2-nitrotoluene and 1,3-dinitrobenzene in pure $scCO_2$. For all other solutes (2-methyl-3-nitroaniline, adamantane, the dyes, and the surfactant modifiers) that method could not be applied as the necessary group interaction parameters are not available. The calculations were performed using the software code GCPXY (Hytof et al., 1994) for the binary systems carbon dioxide + 2-nitrotoluene as well as carbon dioxide + 1,3-dinitrobenzene at 323, 348, and 373 K. Interaction parameters were taken from PAREST (Skjold-Jørgensen, 1984) – cf. also Lewin-Kretzschmar et al. (2002).

25.5 Results and Discussion

25.5.1 Solubility of Nitroaromatics in Pure and Surfactant-Modified $scCO_2$

The solubilities of the pure nitroaromatics in the $scCO_2$ are listed in Table 25.4 (average value of three experiments and variation coefficients). The maximum pressure in those experiments was about 15 MPa, as at 30 or 45 MPa the solute samples were completely dissolved in the supercritical phase. This has consequences for the assessment of the results of supercritical fluid extraction experiments by Pörschmann et al. (1998), which were carried out at pressures above 20 MPa.

Table 25.4: Solubility of nitroaromatics in $scCO_2$ at about 15 MPa.

Nitroaromatics	c (323 K)/mol · dm^{-3} [v_c/%]	c (348 K)/mol · dm^{-3} [v_c/%]	c (373 K)/mol · dm^{-3} [v_c/%]
2-M-3-NA	0.026 [6.7]	0.005 [7.8]	0.002 [4.4]
1,3-DNB	0.042 [4.7]	0.009 [5.0]	0.004 [4.1]
2-NT	0.119 [2.3]	0.072 [0.7]	0.044 [4.3]

v_c variation coefficient

As expected, the concentrations determined reflect the influence of the polarity of the analytes: An increase in the solubility with decreasing polarity in the order 2-M-3-NA, 1,3-DNB, and 2-NT was found at all temperatures. Furthermore, the density is evidently the most important factor for the solubility of those nitroaromatics. The solubility increases with increasing density and decreasing temperature.

25.5 Results and Discussion

The results imply that the solubility of nitroaromatics in supercritical carbon dioxide was not the limiting factor for the extraction, but that the interactions with the matrix were the limiting factors. Therefore, the solubility of nitroaromatics cannot be improved by surfactants. However, the surfactants also show no displacing effect regarding the matrix, instead they cause the formation of a supplementary desorption barrier and prevent desorption from the matrix. These results indicate that the experiments with the surfactants must be restricted to low pressures, because any positive effect by the surfactant is only possible within this range of least solubility.

The experimental data for the solubility at 15 MPa (cf. Table 25.4) were compared to calculated data. However, it should be noted that the calculations were always performed for a binary system, while the experiments were carried out simultaneously with the three nitroaromatics in line with SFE practice.

The calculated molecular fractions in the gas phase are shown as an example for 2-NT graphically in Figure 25.3, while Table 25.5 comparing the calculated and experimental values of 2-NT and 1,3-DNB under the relevant conditions of state.

It is apparent that in five out of six cases (cf. Table 25.5), calculation tallied closely with the experimental values. Only in the case of the highest solubility of 2-nitrotoluene at 15 MPa and 323 K did calculations deliver a significantly higher value; the reasons for this are currently unknown. Moreover, the calculations confirm the high capacity of supercritical

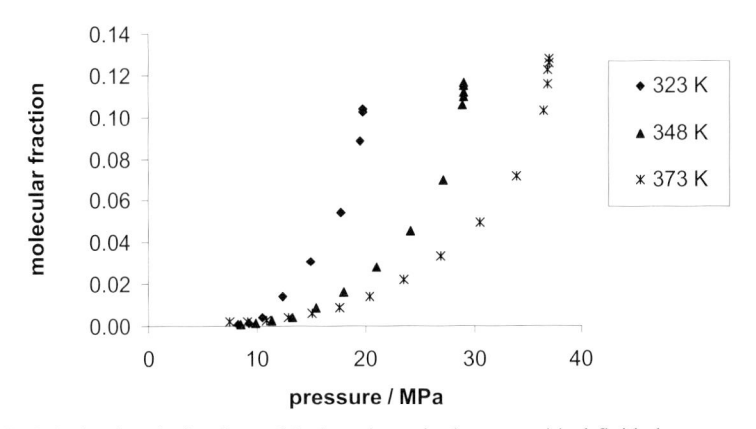

Figure 25.3: Calculated mole fractions of 2-nitrotoluene in the supercritical fluid phase.

Table 25.5: Comparison between calculated (GCEOS) and experimental data for the solubility of 2-NT and 1,3-DNB in supercritical carbon dioxide at 15 MPa.

T/K	2-NT y_{calc}	2-NT y_{exp}	1,3-DNB y_{calc}	1,3-DNB y_{exp}
323	0.030	0.006	0.002	0.002
348	0.006	0.005	0.001	0.001
373	0.004	0.004	0.0006	0.0003

25 The Potential of Surfactants Modified Supercritical Fluids

carbon dioxide for these compounds at higher pressures. This conforms with the experimentally observed complete dissolution of the assigned amount of nitroaromatics. For 2-nitrotoluene, even unlimited miscibility was calculated at approx. 20 MPa (323 K), 30 MPa (348 K), and 40 MPa (373 K).

The experimental results reveal that the nitroaromatics of interest are highly soluble in the density range considered for extraction by supercritical carbon dioxide. Thus the decrease in solubilities which was observed in experiments where nitroaromatics were extracted from a matrix, is caused by strong interactions with that matrix. Therefore, the influence of a surfactant on the solubility of a nitroaromatic compound in scCO$_2$ was studied only at the lowest pressure considered (15 MPa). This is primarily an epistemological issue and is described below in more detail. The experimental results are given in Tables 25.6 to 25.8. The influence of substance polarity, solvent density, and temperature on the solubility of a nitroaromatic compound is the same as without the cationic surfactant Dehyquart C.

Table 25.6: Solubility of nitroaromatics in supercritical carbon dioxide modified with Dehyquart C at 15 bar.

		$T = 323$ K	$T = 348$ K	$T = 373$ K
2-M-3-NA	c/mol · dm^{-3}	0.005	0.002	0.001
	v_c/%	12.7	6.2	15.2
	statistic	systematical<	systematical <	systematical <
1,3-DNB	c/mol · dm^{-3}	0.035	0.012	0.004
	v_c/%	6.0	2.5	8.6
	statistic	systematical <	systematical >	stochastical
2-NT	c/mol · dm^{-3}	0.105	0.077	0.047
	v_c/%	0.1	4.5	5.5
	statistic	not comparable	not comparable	stochastical

For 2-methyl-3-nitroaniline, a statistically significant decrease in the solubility was found in the presence of the surfactant at all temperatures. This is attributed to strong specific interactions between the polar part of the surfactant and the nitroaromatics (Kordikowski, 1992; Scheidgen, 1997) which might result in the formation of larger and thus less soluble surfactant/nitroaromatic clusters. However, the solubility of 1,3-DNB as well as that of 2-NT was not changed significantly by Dehyquart C, i.e. the surfactant does not have a significant effect on the solubility of the less polar nitroaromatics. Texin DOS 75 decreases the solubility of 2-M-3-NA and 2-NT (cf. Table 25.7). However, 1,3-dinitrobenzene, which is less soluble at 323 K and much better soluble at 348 and 373 K, does not exhibit an uniform solubility behavior. But, the experimental results reveal only a very small increase of the solubility which is of no practical significance. The experimental finding might also be due to a misinterpretation as the pressure could only be adjusted within about 0.5 MPa and the difference in the pressure (and not the influence of the surfactant) might be the reason for this finding.

Among the commercial surfactants tested by Blasberg et al. (1998), the non-ionic surfactant Präwozell FCE 1214 was the most soluble in scCO$_2$. That surfactant decreases solubility (cf. Table 25.8). This behavior confirms the postulated hypothesis of clustering effects

25.5 Results and Discussion

Table 25.7: Solubility of nitroaromatics in supercritical carbon dioxide modified with Texin DOS 75 at 15 MPa.

		$T = 323$ K	$T = 348$ K	$T = 373$ K
2-M-3-NA	$c/\text{mol} \cdot \text{dm}^{-3}$	0.012	0.004	0.002
	$v_c/\%$	7.5	4.9	13.9
	statistic	systematical <	systematical <	stochastical
1,3-DNB	$c/\text{mol} \cdot \text{dm}^{-3}$	0.035	0.011	0.005
	$v_c/\%$	5.9	3.5	13.0
	statistic	systematical <	systematical >	systematical >
2-NT	$c/\text{mol} \cdot \text{dm}^{-3}$	0.089	0.058	0.037
	$v_c/\%$	1.7	1.8	9.3
	statistic	systematical <	systematical <	systematical <

Table 25.8: Solubility of nitroaromatics in supercritical carbon dioxide modified with Präwozell FCE 1214 at 15 MPa.

		$T = 323$ K	$T = 348$ K	$T = 373$ K
2-M-3-NA	$c/\text{mol} \cdot \text{dm}^{-3}$	0.008	0.003	0.001
	$v_c/\%$	10.2	12.6	5.0
	statistic	systematical <	systematical <	systematical <
1,3-DNB	$c/\text{mol} \cdot \text{dm}^{-3}$	0.024	0.008	0.003
	$v_c/\%$	7.4	5.1	5.4
	statistic	systematical <	stochastical	systematical <
2-NT	$c/\text{mol} \cdot \text{dm}^{-3}$	0.098	0.051	0.023
	$v_c/\%$	2.1	0.4	4.5
	statistic	systematical <	systematical <	systematical <

in supercritical fluid, because the highest number of molecules for the formation of clusters is available by the higher solubility of Präwozell FCE 1214.

The results of these experimental investigations clearly reveal that the selected surfactants do not increase the solubility of nitroaromatics in scCO$_2$. The slight increase observed for the solubility of 1,3-dinitrobenzene by Texin DOS 75 is insignificant. The complexity of the systems considered here and the variety of interactions do not yet allow a complete interpretation of these findings. Moreover it must be noted that the experiments were limited to rather low pressures because of the relatively high solubility of nitroaromatics in pure supercritical carbon dioxide, even though surfactants generally exhibit an extremely low solubility at these pressures.

The limited results from calculations using the GC-EOS correspond well with the experimental data and encourage to test that method in future work.

The interactions between the solutes and their matrix (e.g. soils) seem to be far more important for the success of a supercritical extraction process than the (comparatively high) solubility of such solutes in scCO$_2$. Therefore, using surfactants as modifiers for enhancing the solubility of nitroaromatics did not result in any improvement of the supercritical extraction of such substances from a matrix. Thus the analysis and remediation of soils by supercritical fluid extraction are impractical unless procedures to overcome the strong interaction of nitroaromatics with the solid matrix are found.

25.5.2 Solubility of Adamantane in Pure and Surfactant-Modified Supercritical Fluids

The experimental results for the solubility of adamantane in supercritical carbon dioxide and ethane are compiled in Tables 25.9 and 25.10, respectively. Due to the low solubility of adamantane in $scCO_2$ and its high solubility in scC_2H_6 (Smith and Teja, 1996), the experiments were carried out at high pressures (20 and 30 MPa) with $scCO_2$ and lower pressures (7.5 and 10 MPa) with scC_2H_6.

Table 25.9: Solubility of adamantane in $scCO_2$ at 373 K.

		$p = 20$ MPa	$p = 30$ MPa
Without surfactant	c/mol · dm^{-3}	0.096	0.165
	v_c/%	4.06	16.1
Dehyquart C	c/mol · dm^{-3}	0.150	0.215
	v_c/%	1.4	0.7
	statistic	systematical >	not comparable
Texin DOS 75	c/mol · dm^{-3}	0.131	0.201
	v_c/%	2.2	1.09
	statistic	systematical >	not comparable
Präwozell FCE 1214	c/mol · dm^{-3}	0.135	0.211
	v_c/%	13.8	6.3
	statistic	not comparable	stochastical
Fluowet OTN	c/mol · dm^{-3}	0.114	0.311
	v_c/%	2.5	1.56
	statistic	systematical >	not comparable

The experimental results reveal that the surfactants increase the solubility of adamantane in $scCO_2$. The increase was particularly high for Fluowet OTN (approx. double at 30 MPa). However, the (already high) solubilities in ethane were not increased by the surfactants. On the contrary, a slight decrease was observed in some experiments. It is concluded that surfactants can exert a significant modifier effect even without the formation of reverse micelles. This is primarily significant for dissolving and extraction processes which also involve nonpolar compounds. Further investigations, especially into the practical use of these

Table 25.10: Solubility of adamantane in scC$_2$H$_6$ at 373 K.

		p = 7.5 MPa	p = 10 MPa
Without surfactant	c/mol · dm^{-3}	0.039	0.106
	v_c/%	1.4	3.7
Dehyquart C	c/mol · dm^{-3}	0.040	0.100
	v_c/%	1.4	2.7
	statistic	stochastical >	stochastical
Texin DOS 75	c/mol · dm^{-3}	0.041	0.114
	v_c/%	0.2	1.1
	statistic	not comparable	systematical >
Präwozell FCE 1214	c/mol · dm^{-3}	0.035	0.098
	v_c/%	1.2	4.6
	statistic	systematical <	systematical <

surfactant-modified systems for the removal of adamantane and other diamondoits and for the prevention of their deposition during natural gas and petroleum production could not be carried out during this work.

25.5.3 Solubility of Polar Dyes in Pure and Surfactant-Modified scCO$_2$ and scC$_2$H$_6$

A survey of the solution behavior of the dyes of the first group (cf. Table 25.2) in modified scCO$_2$ at 373 K is shown in Table 25.11. Table 25.11 also shows a comparison of the results for pure scCO$_2$ and modified scCO$_2$ by means of the results of the F- and t-test. The solubility determined in unmodified scCO$_2$ and in scCO$_2$ modified with aqueous Texin DOS 75 and ethanol is presented in Tables 25.12 and 25.13. Furthermore, the reaction enthalpy Q_R (as the sum of the solvation and evaporation enthalpy) and the association number k (from the model of del Valle and Aguilera) are reported. In addition, the correlation constants A, B, and C of Eq. (25.1) are listed. The correlation coefficients for Eq. (25.1) are between 0.90 and 0.99. In both cases, this method proved adequate for interpolating the experimental results for the solubility in scCO$_2$.

The solubility of a dye in the supercritical fluid increases with rising temperature and increasing pressure or density, respectively. But the various dyes behave differently. For the most part, parafuchsine hydrochloride, i.e. the most polar dye, is the least soluble one. The results are in accordance with the studies by Blasberg et al. (1998) for the solubility behavior of the surfactants used. The solubility in unmodified scCO$_2$ is too small to realize that phenomenon for a supercritical dyeing process. This again is in agreement with published data (Gießmann, 1998; Giessmann et al., 1999). A slight increase of the solubility (by a factor of 2.5 at maximum), but in a few cases even a reduction of it, was found when either

Table 25.11: Solubility of dyes (1. group cf. Table 25.2) in scCO$_2$ at 373 K and 45 MPa.

		Lissamine Yellow AE	Primuline	Parafuchsin hydrochloride
Without modifier	c/mmol · dm^{-3}	0.096	0.038	0.017
	v_c/%	4.5	5.9	4.2
Texin DOS 75/water	c/mmol · dm^{-3}	0.106	0.027	0.017
	v_c/%	3.2	4.7	4.9
	statistic	stochastical >	systematical <	stochastical <
Texin DOS 75/ethanol/ water	c/mmol · dm^{-3}	1.418	1.379	0.276
	v_c/%	4.8	3.7	5.9
	statistic	not comparable	not comparable	not comparable
Texin DOS 75/pentanol/ water	c/mmol · dm^{-3}	0.115	0.060	0.050
	v_c/%	4.5	3.7	5.3
	statistic	systematical >	systematical >	systematical >
Water	c/mmol · dm^{-3}	0.179	0.078	0.040
	v_c/%	6.6	5.3	9.1
	statistic	systematical >	systematical >	not comparable
Ethanol	c/mmol · dm^{-3}	0.168	0.043	0.028
	v_c/%	5.3	4.5	4.7
	statistic	systematical >	systematical >	systematical >
Methanol[a]	c/mmol · dm^{-3}	0.162	0.053	0.053
Dehyquart C (dry)[a]	c/mmol · dm^{-3}	0.134	0.134	0.021
Dehyquart C/water	c/mmol · dm^{-3}	0.263	0.182	0.231
	v_c/%	3.5	7.7	8.0
	statistic	systematical >	not comparable	not comparable
Dehyquart C/ethanol[a]	c/mmol · dm^{-3}	0.134	0.166	0.191
Präwozell FCE 1214 (dry)[a]	c/mmol · dm^{-3}	0.124	0.033	0.017
Präwozell FCE 1214/water[a]	c/mmol · dm^{-3}	0.197	0.087	0.035
Präwozell FCE 1214/ethanol[a]	c/mmol · dm^{-3}	0.189	0.051	0.023
Präwozell FCE 1214/ethanol/ water[a]	c/mmol · dm^{-3}	0.193	0.084	0.034
Texin DOS 75/Präwozell FCE 1214/ethanol/water[a]	c/mmol · dm^{-3}	0.824	0.769	0.139
Dehyquart C/Präwozell FCE 1214/ethanol/water[a]	c/mmol · dm^{-3}	0.228	0.164	0.199
Dehyquart C/Präwozell FCE 1214/water[a]	c/mmol · dm^{-3}	0.246	0.178	0.212

a) twofold determination

25.5 Results and Discussion

Table 25.12: Solubility of dyes (1. group) in unmodified scCO$_2$.

$T = 323$ K		$T = 348$ K		$T = 373$ K	
p/MPa	c/mmol · dm^{-3}	p/MPa	c/mmol · dm^{-3}	p/MPa	c/mmol · dm^{-3}
Lissamine Yellow AE					
15.2	0.024	15.0	0.020	15.5	0.025
30.2	0.032	29.6	0.045	30.0	0.075
44.7	0.071	44.8	0.103	44.4	0.136
44.8	0.070	44.9	0.090	45.0	0.142
				45.1	0.130
$Q_R = -18.6$ kJ · mol^{-1}		$Q_R = -27.0$ kJ · mol^{-1} $k = 2$		$Q_R = -34.3$ kJ · mol^{-1}	
$A = 3.938$		$B = 0.028$		$C = -22.616$	
Primuline					
15.2	0.005	15.0	0.010	15.5	0.015
30.2	0.006	29.6	0.015	30.0	0.026
44.7	0.006	44.8	0.022	44.4	0.038
44.8	0.006	44.9	0.024	45.0	0.035
				45.1	0.039
$Q_R = -57.8$ kJ · mol^{-1}		$Q_R = -37.7$ kJ · mol^{-1} $k = 1$		$Q_R = -20.2$ kJ · mol^{-1}	
$A = 1.771$		$B = 0.039$		$C = -26.058$	
Parafuchsine hydrochloride					
15.2	0.001	15.0	0.002	15.5	0.007
30.2	0.004	29.6	0.004	30.0	0.012
44.7	0.009	44.8	0.008	44.4	0.017
44.8	0.008	44.9	0.007	45.0	0.016
				45.1	0.017
$Q_R = 3.8$ kJ · mol^{-1}		$Q_R = -34.9$ kJ · mol^{-1} $k = 2$		$Q_R = -68.5$ kJ · mol^{-1}	
$A = 3.279$		$B = 0.035$		$C = -26.750$	

one of the single surfactants (anionic surfactant Texin DOS 75, cationic surfactant Dehyquart C, non-ionic surfactant Präwozell FC 1214) was added or when a surfactant mixture (with Präwozell FC 1214 as one component) was used.

In all, there is no significant influence of the surfactant modifiers on the solubilization process. This seems to be the more the case as under the examined conditions the different surfactants partially show considerably different solubilities (Blasberg et al., 1998) and thus they are available in very different concentrations for the solubilization process.

However, a distinguished increase in the solubility of a dye was achieved by adding a mixture of Texin DOS 75 and ethanol. In particular at higher temperatures and pressures the solubility of a dye in the supercritical fluid was increased by a factor between about 10 (Lissamine Yellow AE and parafuchsine hydrochloride) and about 50 (primuline). This finding is in agreement with the important influence of the cosolvent ethanol on the solubility of Texin DOS 75 in supercritical fluids (cf. Table 25.14).

This is caused by the formations of reverse-micellar structures under the influence of an alcohol (cf. Hutton et al., 1999) for the increase of the solubility of methylorange and ri-

Table 25.13: Solubility of dyes (1. group) in scCO$_2$ modified with aqueous Texin DOS 75 and ethanol.

T = 323 K		T = 348 K		T = 373 K	
p/MPa	c/mmol · dm^{-3}	p/MPa	c/mmol · dm^{-3}	p/MPa	c/mmol · dm^{-3}
Lissamine Yellow AE in scCO$_2$					
15.2	0.024	15.0	0.020	15.5	0.025
30.2	0.032	29.6	0.045	30.0	0.075
44.7	0.071	44.8	0.103	44.4	0.136
44.8	0.070	44.9	0.090	45.0	0.142
				45.1	0.130
Q_R = −75.1 kJ · mol^{-1}		Q_R = −47.9 kJ · mol^{-1} k = 5		Q_R = −32.0 kJ · mol^{-1}	
A = 8.278		B = 0.050		C = −22.616	
Primuline					
15.2	0.005	15.0	0.010	15.5	0.015
30.2	0.006	29.6	0.015	30.0	0.026
44.7	0.006	44.8	0.022	44.4	0.038
44.8	0.006	44.9	0.024	45.0	0.035
				45.1	0.039
Q_R = −100.8 kJ · mol^{-1}		Q_R = −54.6 kJ · mol^{-1} k = 6		Q_R = −14.5 kJ · mol^{-1}	
A = 9.801		B = 0.057		C = −35.856	
Parafuchsine hydrochloride					
15.2	0.001	15.0	0.002	15.5	0.007
30.2	0.004	29.6	0.004	30.0	0.012
44.7	0.009	44.8	0.008	44.4	0.017
44.8	0.008	44.9	0.007	45.0	0.016
				45.1	0.017
Q_R = −59.1 kJ · mol^{-1}		Q_R = −39.1 kJ · mol^{-1} k = 3		Q_R = −21.8 kJ · mol^{-1}	
A = 3.279		B = 0.035		C = −26.750	

Table 25.14: Solubility of Texin DOS 75 (c in mmol dm^{-3}) in scCO$_2$.

Conditions	Without modifier[a]	Water	Ethanol/water
373 K/45 MPa	4.81	5.29	40.05
373 K/30 MPa	0.92	0.69	20.08
348 K/45 MPa	3.66	3.86	36.46

a) Blasberg 1998

boflavine). Ethanol acts as a cosolvent for Texin DOS 75, increasing the concentration so that reverse-micellar structures are possible. When ethanol is acting as a cosolvent, the solubility of a dye is increased to such levels, that it can be utilized for a dyeing process with scCO$_2$. The concentration of the dye in the supercritical phase is higher than e.g. than the solubility of dispersive reactive dyes which were developed for the dyeing of wool in scCO$_2$ (Gießmann et al., 2000), and nearly as high as in conventional bath-dyeing (Römpp, 1995).

25.5 Results and Discussion

However, there was no significant change (in comparison to unmodified scCO$_2$) when ethanol was replaced by pentanol as cosolvent for Texin DOS 75 or when the non-ionic surfactant Präwozell FCE 1214 was additionally added to the Texin DOS 75 + ethanol + water system. Likewise, the addition of the various solvents to the surfactants Dehyquart C and Präwozell FCE 1214 or their combination resulted in only slight solubility increases.

With solvents ethanol, methanol, and water only a minor increase of the solubility of all three dyes was observed in pure scCO$_2$. The effect was largest with water as modifier and with parafuchsine hydrochloride as the dye.

This solubility increase, however, was considerably smaller than the one observed in the aqueous Texin DOS 75 + ethanol-modified fluid, which is another proof that the drastic solubility increases are not due to the influence of the pure solvents.

As the formation of reverse micelles in supercritical ethane + Texin DOS 75 has already been described (Blasberg, 1998; Cason and Roberts, 2000; Roberts and Thompson, 1998), the solubility of the three dyes was additionally investigated in this system. Table 25.15 shows a typical example for 373 K and 45 MPa. The solubility in unmodified scC$_2$H$_6$ and in scC$_2$H$_6$ modified with aqueous Texin DOS 75 are given in Tables 25.16 and 25.17.

Table 25.15: Solubility of dyes (1. group) in scC$_2$H$_6$ at 373 K and 45 MPa.

		Lissamine Yellow AE	Primuline	Parafuchsin hydrochloride
Without modifier	c/mmol · dm^{-3}	0.616	0.115	0.095
	v_c/%	6.3	5.2	6.5
Texin DOS 75/water	c/mmol · dm^{-3}	4.232	1.085	0.105
	v_c/%	4.3	7.0	4.5
	statistic	not comparable	not comparable	systematical >
Texin DOS 75/ethanol/ water	c/mmol · dm^{-3}	3.441	1.071	0.163
	v_c/%	9.7	8.9	9.0
	statistic	not comparable	not comparable	not comparable
Ethanol	c/mmol · dm^{-3}	0.441	0.112	0.104
	v_c/%	7.4	6.8	8.4
	statistic	systematical <	stochastical <	stochastical >
Water (0.4 mol) [a]	c/mmol · dm^{-3}	0.505	0.103	0.093
Water (1.0 mol) [a]	c/mmol · dm^{-3}	0.390	0.090	0.102

a) twofold determination

The solubility of the dyes is larger in supercritical ethane than in supercritical carbon dioxide. Using ethanol or water as a cosolvent had no significant influence on the solubility of the dyes in supercritical ethane. As expected, the solubility of Lissamine Yellow AE and primuline increased drastically with Texin DOS 75 (cf. Table 25.18). This behavior is again attributed to the formation of reverse micelles. The only small increase of the solubility of parafuchsine hydrochloride results from the polar and anionic character of this dye and also

Table 25.16: Solubility of dyes (1. group) in unmodified scC$_2$H$_6$.

T = 323 K		T = 348 K		T = 373 K	
p/MPa	c/mmol · dm^{-3}	p/MPa	c/mmol · dm^{-3}	p/MPa	c/mmol · dm^{-3}

Lissamine Yellow AE

15.3	0.035	15.1	0.036	15.2	0.069
30.0	0.076	30.2	0.167	29.9	0.296
45.1	0.164	44.3	0.333	44.8	0.623
				44.8	0.575
				44.9	0.651

$Q_R = -33.2$ kJ · mol^{-1} $Q_R = -46.5$ kJ · mol^{-1} $k = 6$ $Q_R = -58.1$ kJ · mol^{-1}
$A = 12.935$ $B = 0.037$ $C = -28.126$

Primuline

15.3	0.013	15.1	0.013	15.2	0.025
30.0	0.019	30.2	0.019	29.9	0.054
45.1	0.033	44.3	0.033	44.8	0.120
				44.8	0.108
				44.9	0.116

$Q_R = -21.9$ kJ · mol^{-1} $Q_R = -36.8$ kJ · mol^{-1} $k = 4$ $Q_R = -49.7$ kJ · mol^{-1}
$A = 10.972$ $B = 0.037$ $C = -27.179$

Parafuchsine hydrochloride

15.3	0.008	15.1	0.010	15.2	0.019
30.0	0.019	30.2	0.024	29.9	0.035
45.1	0.034	44.3	0.048	44.8	0.099
				44.8	0.097
				44.9	0.088

$Q_R = -15.6$ kJ · mol^{-1} $Q_R = -36.6$ kJ · mol^{-1} $k = 4$ $Q_R = -54.7$ kJ · mol^{-1}
$A = 9.006$ $B = 0.036$ $C = -28.126$

from the anionic character of Texin DOS 75. In contrary to carbon dioxide, adding a mixture of ethanol and aqueous Texin DOS 75 to supercritical ethane gave only a small increase for the solubility of parafuchsine hydrochloride and a decrease of the other two dyes. This is in agreement with the finding that the solubility of Texin DOS 75 in scC$_2$H$_6$ is not significantly increased by ethanol.

The maximum concentrations of Lissamine Yellow AE in modified scC$_2$H$_6$ and in modified scCO$_2$, differ by a factor of about three, while the solubilities for parafuchsine hydrochloride and primuline were smaller. Therefore, ethane is no alternative for carbon dioxide as supercritical solvent for dyeing.

Table 25.19 gives a survey of the solution behavior for the second group of dyes (cf. Table 25.2) in scCO$_2$.

The results for Doractive Blue WRL and Doramine Light Blue BR 200% are very similar to those described before. For example, the highest solubility is observed at 373 K and 45 MPa, but the solubility is always rather small in unmodified scCO$_2$. Adding a pure surfactant results only in very small changes. But adding aqueous Texin DOS 75 + ethanol mixtures causes a drastic increase of the solubility. But, Doractive Blue WRL and Doramine Light

25.5 Results and Discussion

Table 25.17: Solubility of dyes (1. group) in scC$_2$H$_6$ modified with aqueous Texin DOS 75.

| T = 323 K | | T = 348 K | | T = 373 K | |
p/MPa	c/mmol · dm^{-3}	p/MPa	c/mmol · dm^{-3}	p/MPa	c/mmol · dm^{-3}
Lissamine Yellow AE					
15.7	0.055	15.1	0.048	15.3	0.360
30.3	0.791	29.9	0.924	30.2	0.987
45.1	1.930	45.0	3.172	44.8	4.220
				44.9	4.058
				44.9	4.419
Q_R = –23.2 kJ · mol^{-1}		Q_R = –55.7 kJ · mol^{-1} k = 9		Q_R = –83.9 kJ · mol^{-1}	
A = 27.242		B = 0.056		C = –37.229	
Pimuline					
15.7	0.012	15.1	0.028	15.3	0.053
30.3	0.107	29.9	0.094	30.2	0.158
45.1	0.653	45.0	1.101	44.8	1.123
				44.9	0.998
				44.9	1.133
Q_R = –44.3 kJ · mol^{-1}		Q_R = –54.3 kJ · mol^{-1} k = 9		Q_R = –62.9 kJ · mol^{-1}	
A = 27.875		B = 0.056		C = –38.654	
Parafuchsine hydrochloride					
15.7	0.014	15.1	0.012	15.3	0.020
30.3	0.037	29.9	0.057	30.2	0.031
45.1	0.088	45.0	0.126	44.8	0.101
				44.9	0.105
				44.9	0.110
Q_R = –31.7 kJ · mol^{-1}		Q_R = –23.5 kJ · mol^{-1} k = 5		Q_R = –16.4 kJ · mol^{-1}	
A = 15.680		B = 0.025		C = –24.538	

Table 25.18: Solubility of Texin DOS 75 (c in mmol dm^{-3}) in scC$_2$H$_6$.

Conditions	Without modifier[a]	Water	Ethanol/water
373 K/45 MPa	48.48	50.58	49.54
373 K/30 MPa	16.26	19.42	20.79
348 K/45 MPa	20.71	23.01	23.42

a) Blasberg 1998

Blue BR 200% are less soluble in the supercritical phases as the other dyes. Domalane Blue BL 150% is an exception from that rule. Domalane Blue BL 150% is a mixture of polar and non polar dyes used for the dyeing of polyester/wool fabrics. However, the exact composition and thus a molecule mass were not available. Therefore, the experimental results are presented in unit mg g^{-1}. In general, the solubility of Domalane Blue BL 150% was much higher than of the other dyes, but showing the same tendencies like the other dyes: The highest solu-

Table 25.19: Solubility of dyes (2. group) in scCO$_2$ at 373 K and 45 MPa.

		Doractive Blue WRL	Doramine Light Blue BR 200%	Domalane Blue BL 150%
Without modifier	c/mmol · dm^{-3}	0.0163	0.00904	1.656[b]
	v_c/%	6.5	7.2	3.2
Texin DOS 75/water	c/mmol · dm^{-3}	0.0161	0.00902	2.348[b]
	v_c/%	7.6	7.3	6.2
	statistic	stochastical <	stochastical <	systematical >
Texin DOS 75/ethanol/water	c/mmol · dm^{-3}	0.102	0.069	4.578[b]
	v_c/%	6.5	8.6	4.8
	statistic	not comparable	not comparable	systematical >
Water	c/mmol · dm^{-3}	0.037	0.020	2.818[b]
	v_c/%	6.9	9.7	4.9
	statistic	systematical >	systematical >	systematical >
Ethanol	c/mmol · dm^{-3}	0.028	0.015	3.510[b]
	v_c/%	8.6	10.5	6.1
	statistic	systematical >	systematical >	systematical >
Ethanol/water[a]	c/mmol · dm^{-3} [a]	0.032	0.016	2.398[b]

a) twofold determination
b) scale unit: mg · g^{-1} (because no exact molecule mass was available)

bility was observed at 373 K and 45 MPa and the solubility is slightly increased by a pure surfactant and strongly increased by aqueous Texin DOS 75 + ethanol mixtures.

25.5.4 Dyeing Tests

Altogether, the highest solubility of a dye was achieved with aqueous Texin DOS 75 + ethanol-modified scCO$_2$ at 373 K and 45 MPa for all dyes. Therefore, for the dyeing tests of this system and, additionally, for reference purposes the pure fluid and the water-, ethanol-, and water + ethanol-modified fluids were also used under the same conditions. Initially, the amount of surfactant and solvents was optimized to achieve a high solubility of the dye at a low amount of surfactant (cf. Table 25.20).

Than the dyeing tests were carried out in a 500 ml autoclave containing (besides the supercritical fluid) 13 g Texin DOS 75, 28 g ethanol, 15 g water, and 3 g of the single dye.

As expected, dyeing results were unsatisfactory when no modifiers were used. The best dyeing (and good rubbing fastness) was always achieved with aqueous Texin DOS 75 +

Table 25.20: Solubility of dyes (1. group) and Texin DOS 75 in scCO$_2$ at 373 K and 45 MPa.

Conditions			Lissamine yellow AE	Primuline	Parafuchsine hydrochloride	Texin DOS 75
$m_{Ethanol}$ /g	m_{Water} /g	$m_{TexinDOS\,75}$ /g	c/mmol·dm^{-3}	c/mmol·dm^{-3}	c/mmol·dm^{-3}	c/mmol·dm^{-3}
4.0	4.0	9.0	0.143	0.074	0.044	8.988
10.0	4.0	9.0	0.155	0.078	0.051	10.743
18.3	4.0	9.0	1.253	1.213	0.226	26.087
18.3	10.0	4.0	0.929	0.994	0.157	24.542
18.3	10.0	9.0	1.475	1.452	0.280	42.172
18.3	7.0	18.0	1.418	1.379	0.276	40.050
18.3	10.0	18.0	1.465	1.432	0.273	36.338
18.3	20.0	9.0	1.444	1.387	0.290	39.860
30.0	10.0	9.0	1.479	1.457	0.282	44.597
45.0	10.0	9.0	1.511	1.611	0.326	38.656

ethanol mixtures. However, the dyeing results were not accompanied by good washing fastness and, as a rule, fastness to water was lower than fastness to acetone. This means that the dye only attached to the surface of the fabric, but it did not penetrate the fibers. This may be explained by diffusion barriers caused by the adsorption of the surfactant on the fiber surface. But it is more probable (and also revealed by the lower degree of fastness to water) that the dye is deposited in a reverse-micellar structure on the surface of the fabric and washing with water destroys that structure and dissolves the dye in the aqueous phase.

The dyeing with ethanol + water mixtures as modifier was not as deep as with the surfactant-modified systems, but the so dyed fabrics revealed a better washing fastness and thus showed the best dyeing results. As far as dyeing depth and color is concerned, it was also found out that wool, in principle, is better suitable for dyeing with supercritical solvents than cotton, which is in full agreement with literature findings (Schäfer et al., 1999). Altogether the levelness of dyes, especially in cases of deeper dyeing, was not yet satisfactory; but this is mainly due to wrapping or winding techniques used for fixing the material in the autoclave. A more even dyeing seems to be possible by the use of so-called perforated dyeing trees as they are used in conventional dyeing from aqueous liquors and in which the dyeing medium is pressed through the device. However, the method cannot be used in the autoclave system applied here.

25.6 Conclusion

The experiments show the potentials but also the confines of commercial surfactants as modifiers for increasing the solubility of compounds of three very different classes of substances in supercritical fluids.

For nitroaromatics (which have a high solubility already in an unmodified supercritical fluid) the application of surfactants does not result in higher concentrations in the supercritical fluid when the nitroaromatics are extracted from a matrix (e.g. from soils). Thus the analysis and remediation of soils is not favored by supercritical fluid extraction unless the strong interactions of the nitroaromatics with the matrix are eliminated by other procedures (e.g. the use of microwaves). However, in the case of sparingly soluble analytes like adamantane, a drastic solubility enhancement can be achieved by a suitable surfactant. It can be concluded that surfactants can develop a significant modifying effect even without the formation of reverse micelles. This might be important for dissolving and extracting non-polar compounds. For the group of polar dyes, a drastic solubility increase (by a factor >50) was achieved by modifying $scCO_2$ with aqueous Texin DOS 75 + ethanol or scC_2H_6 with aqueous Texin DOS 75, respectively, which can be explained by the formation of reverse-micellar structures. The thus obtained solubility of dyes are sufficient for dyeing with supercritical fluids as solvents, which also resulted in acceptable dyeing depths in the dyeing experiments. However, the levelness and fastness were not yet satisfactory which will necessitate additional investigations, e.g. with regard to the design of a material-holding device or to preliminary fabric treatment.

In principle, the formation of micellar structures in a supercritical fluid offers good chances for dissolving highly polar or ionic compounds. While in the scope of this project the orientation was on the use of the very reasonable commercial surfactants, further investigations should also deal with other surfactants (e.g. perfluorinated compounds) with a high solubility in $scCO_2$ and a high potential for micelle formation.

Acknowledgement

We would like to thank Buna GmbH (Schkopau) and Henkel KgaA (Düsseldorf) for providing the surfactants and M. Dohmen GmbH (Korschenbroich) for providing dyes. Our thanks also go to Dr. A. Al-Hallak for his assistance in the calculations using the codes PARAEST and GCPXY.

References

Bartle, K. D.; Clifford, A. A.; Jafar, S. A.; Shilstone, G. F. (1991): Solubilities of Solids and Liquids of Volatility in Supercritical Carbon Dioxide. J. Phys. Chem. Ref. Data 20, 713–756
Blasberg, L. (1998): Modifizierung überkritischen Kohlendioxids und Ethans mit handelsüblichen Tensiden – Löslichkeiten von Tensiden in überkritischen Fluiden und Anwendung zur Extraktion von Nitroaromaten aus Böden. Thesis, Leipzig
Blasberg, L.; Harting, P.; Quitzsch, K. (1998): Löslichkeit ausgewählter handelsüblicher Tenside in

überkritischem Kohlendioxid und Ethan-Bestimmung, Korrelation und Vergleich. Tenside Surf. Det. 35, 439–446

Camel, V.; Tambute, A; Caude, M. (1995): Influence of Ageing on the Supercritical Fluid Extraction of Pollutants in Soil. J. Chromatogr. A 693, 101–111

Cason, J. P.; Roberts, C. B. (2000): Metallic Copper Nanoparticles Synthesis in AOT Reverse Micelles in Compressed Propane and Supercritical Ethane Solution. J. Phys. B 104, 1217–1221

Chrastil, J. (1982): Solubility of Solids and Liquids in Supercritical Gases. J. Phys. Chem. 86, 3016–3021

Consani, K. A.; Smith, R. D. (1990): Observations on the Solubility of Surfactants and Related Molecules in Carbon Dioxide at 50 °C. Journal of Supercritical Fluids 3, 51–65

del Valle, J. M.; Aguilera, J. M. (1988): An Improved Equation for Predicting the Solubility of Vegetable Oils in Supercritical CO_2. Ind. Eng. Chem. Res. 27, 1551–1553

Doerffel, K. (1987): Statistik in der analytischen Chemie. 4. Auflage, Dt. Verlag für Grundstoffindustrie, Leipzig

Doerffel, K.; Geyer, R.; Müller, H. (1994): Analytikum. 9. Auflage, Dt. Verlag für Grundstoffindustrie, Leipzig

Dreux, M.; Lafosse, M.; Morin-Allory L. (1996): Der Verdampfung-Lichtstreudetektor. LC-GC Internat. 9, 148–156

Ebner, G.; Schelz, D. (1989): Textilfärberei und Farbstoffe. 1. Auflage. Springer-Verlag, Heidelberg

Giehl, A.; Gießmann, M.; Heine, E.; Schäfer, K.; Höcker, H. (1998): Färben von Wolle und Polyester/Wollmischungen aus überkritischem CO_2. Tagungsband des 2. Anwendertreffens SFE-SFC-XSE, 8. – 9.10.1997, Siegen, 51–52

Gießmann, M. (1998): Unkonventionelle Verfahren zum Färben von Polyester-Wolle. Thesis, RWTH Aachen

Gießmann, M.; Schäfer, K.; Höcker, H. (1999): Färben aus überkritischem CO_2 von Wolle und Polyester/Wolle-Mischungen. Textilveredlung 34, 12–19

Gießmann, M.; Schäfer, K.; Höcker, H. (2000): Färben von Polyester/Wolle aus überkritischem CO_2 mit Dispersions- und dispersiven Reaktivfarbstoffen. DWI Reports 123, 465–469

Hutton, B. H.; Perera, J. M.; Gieser, F.; Stevens, G. W. (1999): Investigation of AOT reverse microemulsions in supercritical carbon dioxide. Colloids and Surfaces A: Physicochem. Eng. Aspects 146, 227–241

Hytoft, G.; Christensen, L.; Skjold-Jørgensen, S. (1994): GCFLASH, GCENVEL, GCPXY – Description of Some User Programs for the GC-EOS, Lyngby

Johnston, K. P.; McFann, G. J.; Lemert, R. M. (1989): Pressure Tuning of Reverse Micelle for Adjustable Solvation of Hydrophiles in Supercritical Fluids. In: Supercritical Fluid Science and Technology, Amer. Chem. Soc., Washington, 140–164

Johnston, K. P.; Harrison, K. L.; Clarke, M. J.; Howdle, S. M.; Heitz, M. P.; Bright, F. V.; Carlier, C.; Randolph, T. W. (1996): An Environment for Hydrophiles Including Proteins. Science 271, 624–626

Lewin, U. (1994): Analytische Bestimmung von Sprengstoffen aus dem Gebiet der Rüstungsaltlast Elsnig sowie ökochemische und toxikologische Charakterisierung der Verbindungen. Abschlußarbeit zum Postgradualstudium Toxikologie, Leipzig

Lewin-Kretzschmar, U.; Efer, J.; Engewald, W. (2000): Analysis of Explosives by Liquid Chromatography. In: Cooke, M.; Poole, C.F. (eds.). Encyclopaedia of Separation Science, Academic Press, London, 2767–2782

Lewin-Kretzschmar, U.; Blasberg, L.; Harting, P. (2002): Solubility of Nitroaromatics and Adamantane in Pure and Surfactant-Modified Compressed Gases. Tenside Surf. Det. 39, 29–35

King, W. J. (1988): Operating Problems in the Hanlan Swan Hills Gas Field. Proceedings Gas Technology Symposium. Dallas, SPE: Brookfield, CT, 469–474

Knittel, D.; Schollmeyer, E. (1992): Farbstoffe in überkritischem CO_2 und supramolekulare Chemie. GIT 12, 993–996

Kordikowski, A. (1992): Fluidphasengleichgewichte ternaerer und quaternaerer Mischungen schwerflüchtiger organischer Substanzen mit überkritischem Kohlendioxid bei Temperaturen zwischen 298 K und 393 K und Drücken zwischen 10 MPa und 100 MPa. Thesis, Bochum

McClain, J. B.; Betts, D. E.; Canelas, D. A.; Samulski, E. T.; DeSimone; J. M., Londono, J. D.; Cochran, H. D.; Wignall, G. D.; Chillura-Martino, D.; Triolo, R. (1996): Design of Nonionic Surfactants for Supercritical Carbon Dioxide. Science 274, 2049–2052

Mitra, S.; Wilson, N. K. (1991): An Empirical Method to Predict Solubility in Supercritical Fluids. J. Chromatog. Sci. 29, 305–309

Poulakis, K.; Spee, M.; Knittel, D.; Buschmann, H.-J.; Schneider, G. M.; Schollmeyer, E. (1991): Färbung von Polyester in überkritischen CO_2. Chemiefasern/Textilindustrie 41/93, 142–146

Pörschmann, J.; Blasberg, L.; Mackenzie, K.; Harting, P. (1998): Application of Surfactants to the Supercritical Fluid Extraction of Nitroaromatics Compounds from Sediments. J. Chromatogr. A 816, 221–232

Preuß, J.; Wiegandt, C.-C. (1992): Rüstungsaltlasten des Deutschen Reiches auf dem Gebiet der Bundesrepublik Deutschland. Geogr. Rundschau 44, 175–178

Roberts, C. B.; Thompson, J. B. (1998): Investigation of Solvent Effects on the Solvation of AOT Reverse Micelles in Supercritical Ethane Solution. J. Phys. B 102, 9074–9080

Römpp (1995): Chemie Lexikon, CD – Version 1.0. Georg Thieme Verlag, Stuttgart/New York

Schäfer, K.; Arndt, M.; Dechesne, M.; Höcker, H. (1999): Ultraschall-unterstütztes Färben von Cellulosefaser/Wollmischungen. DWI Reports 122, 555–560

Scheidgen, A. (1997): Fluidphasengleichgewichte binärer und ternärer Kohlendioxidmischungen mit schwerflüchtigen organischen Substanzen bis 100 MPa. PhD-Thesis, Ruhr-Universität Bochum

Schmelzer, J.; Stimming, R.; Kalek, C. (1989): Berechnungen von Phasengleichgewichten mit der Gruppenbeitragszustandgleichung (GC-EOS). Chem. Tech. 41, 196–199

Schmelzer, J.; Stimming, R.; Kalek, C. (1990): Vorausberechnungen von Hochdruck-phasengleichgewichten mit der Gruppenbeitragszustandgleichung (GC-EOS) im Vergleich zu herkömmlichen Methoden. Chem. Tech. 42, 520–522

Schneider, K.; Hassauer, M.; Kalberlah, F. (1994): Toxikologische Bewertung von Rüstungsaltlasten. Z. Umweltchem. Ökotox. 6, 271–276

Skjold-Jørgensen, S. (1988): Group Contribution Equation of State (GC-EOS): A Predictive Method for Phase Equilibrium Computations over Wide Ranges of Temperature and Pressures up to 30 MPa. Ind. Eng. Chem. Res. 27, 110–118

Skjold-Jørgensen, S. (1984): Manual for PARAEST: Parameter Estimation in the Group Contribution Equation of State. Institut for Kemiteknik, Danmarks Tekniske Universitet, Lyngby

Smith, R. D.; Blitz, J. P.; Fulton, J. L. (1989): Structure of Reverse Micelles and Microemulsion Phases in Near-Critical and Supercritical Fluids as Determined from Dynamic Light Scattering Studies. In: Supercritical Fluid Science and Technology, Am. Chem. Soc., Washington, 165–183

Smith, V. S.; Teja, A. S. (1996): Solubilities of Diamondoids in Supercritical Solvents. J. Chem. Eng. Data 41, 923–925

Triolo, F.; Triolo, A.; Triolo, R. (2000): Critical Micelle Density for the Self-Assembly of Block Copolymer Surfactants in Supercritical Carbon Dioxide. Langmuir 16, 416–421

Wingert, W. S. (1992): GC-MS Analysis of Diamondoid Hydrocarbons in Smackover Petroleums. Fuel 71, 37–43

Yazdi, A. V.; Lepilleur, C.; Singley, E. J.; Liu, W.; Adamsky, F. A.; Enick, R. M.; Beckman, E. J. (1996): Highly Carbon Dioxide Soluble Surfactants, Dispersants, Chelating Agents. Fluid Phase Equilibria 117, 297–303

Nomenclature

A, B, C	constants (cf. Eq. (25.1))
calc	calculated
c	concentration
$scCO_2$	supercritical carbon dioxide
scC_2H_6	supercritical ethane
exp	experimental
l, m, n, k	constants (cf. Eqs. (25.2) and (25.3))
M	molecular mass
m	mass
p	total pressure
Q_R	enthalpy of reaction
R	universal gas constant
T	absolute temperature
v_c	variation coefficient
V	volume
Y	molecular fraction

Greek letters

ρ	density

26 Liquid-Liquid Phase Equilibria in Microemulsion Forming Systems Based on Carbohydrate Surfactants

Sabine Enders, Dirk Häntzschel, Heike Kahl, and Konrad Quitzsch[1]

26.1 Abstract

Liquid-liquid and liquid-liquid-liquid phase equilibria of quaternary mixtures composed of a carbohydrate surfactant, water, an oil component, and a polar organic solvent (i.e., a cosurfactant), as well as related ternary and binary subsystems were investigated experimentally and by different modeling concepts. For this purpose experimental investigations were performed using various instrumental tools (turbidimetric titration, HPLC analytics, vibrating tube densimetry, differential-scanning calorimetry, maximum bubble pressure and spinning drop tensiometry, dynamic light scattering, low-stress rheometry, freeze-fracture transmission electron microscopy, polarizing microscopy), and exchanging either the surfactant or the cosurfactant respectively the oil in the systems. The main activities were directed to the microemulsion state spontaneously forming by self-assembly of surfactant molecules to colloidal domains in one equilibrium phase of the mixture.

Furthermore the utility of some computational procedures, applying micelle formation theory, simple activity coefficient model, and the Ginzburg-Landau functional were checked extensively in order to correlate the experimental data of phase equilibria in two-, three-, and four-component systems and to understand the aggregation and interfacial properties that have been observed.

26.2 Introduction

The wide-spread application of surfactants has its origin in the amphiphilic molecular characteristics of these substances, namely, they consist of a polar part (head group or head groups) that is hydrophilic and a non-polar part (tail) that is hydrophobic but is compatible

---

[1] Universität Leipzig, Wilhelm-Ostwald-Institut für Physikalische und Theoretische Chemie, Linnéstr. 2, 04103 Leipzig.

26.2 Introduction

with oil. Resulting from an extensive experimental study of the solubility limits in many systems containing surfactant, water, and an oily component, Winsor (1948, 1954) distinguishes among three general types of phase coexistence which may be all realized successively in one and the same of these systems by variation of an intensity parameter (temperature, pressure, composition). A sequence set up as follows or the version vice versa will then be observed: surfactant containing aqueous phase in equilibrium with excess oil nearly free of the other components (Winsor I); middle phase containing surfactant and all the other components, in equilibrium with oil and water both in excess (Winsor III), and surfactant containing oil phase in equilibrium with excess water (Winsor II). Solutions containing at least surfactant, water, and an oily component can show typical features of the microemulsion state. Microemulsions appear as thermodynamically stable, isotropic, and optically translucent phases, although they depend on the individual self-organisation of surfactant molecules into colloidal domains (swollen micelles, vesicles). Nevertheless they exhibit a Newtonian flow behavior and attain a very low interfacial tension against coexisting oil-rich and water-rich phases. At well defined compositions microemulsions can form bicontinuous structures in which the oil and water regions interpenetrate. With respect to the solubilizing power of the surfactant aggregates which depends on the realized hydrophile-lipophile balance in the surfactant molecules further components should be added to the main compounds surfactant, water, oil, in order to prepare stable microemulsions (Schulman et al., 1959). These are either organic solvents, such as alcohols, which act as cosurfactants, and/or electrolytes. Therefore experimental investigations of microemulsion systems happen usually under isothermal and isobaric conditions in a tetrahedral composition space which may be regarded as three-dimensional simplex.

Figure 26.1 exhibits an example of such a tetrahedron (Kahlweit et al., 1995). A tie line (dotted lines) triangle illustrates the three-phase behavior in the interior of the tetrahedron. Any point in this triangle represents a mixture that will split into three phases whose compositions are represented by the vertice of the triangle. A large number of experiments is necessary to describe the shape of the whole three-phase region and the extent and nature of the surrounding two- and single-phase states. Such a program would ordinarily not be carried out except to delineate the phase behavior of systems that have considerable importance. Most frequently the studies of the phase behavior are prompted by a need to screen or compare various amphiphilic compounds that are candidates for a particular application. The investigations may very well settle for phase information within a subspace of the full phase diagram. Generally the subspace demands fixing a ratio of oil to water which is maintained constant in the overall system. This gain in simplicity is offset by the disadvantage that although the phase boundaries can be correctly delineated, the compositions of the conjugate phases can not be shown since these compositions do not necessarily lie in the plane of investigations. The position of a pseudoternary can be fixed e.g. at a water to oil mass ratio of 1:1. Starting from a point near the water-oil axis on the bottom line of the pseudoternary triangle an increase of alcohol content in the mixture will effectuate a significant change from an oil-in-water microemulsion coexisting with oil excess ($\underline{2}$) over a three-phase region (3) with a middle-phase microemulsion, rich in surfactant, in equilibrium with both oil and water excess to a water-in-oil microemulsion coexisting with water excess ($\bar{2}$). Experimentally established feed compositions are located on boundary lines covering the different states of phase coexistence which form in total a fish-like figure ("Kahlweit fish"; Kahlweit et al., 1990). The body of the fish is given by a closing envelope due to all feed compo-

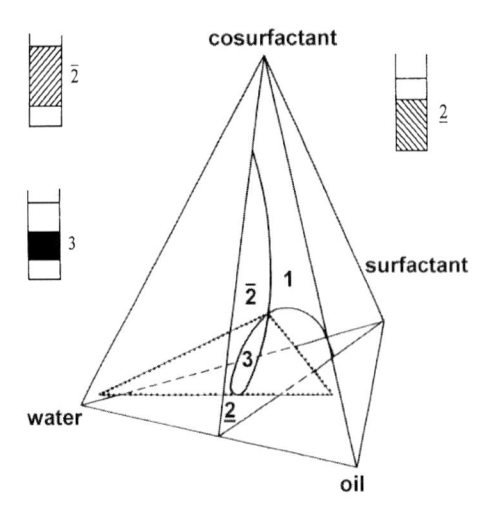

26 Liquid-Liquid Phase Equilibria in Microemulsion Forming Systems

Figure 26.1: Tetrahedral composition space for schematic visualisation the phase behavior of a quaternary system at constant temperature and at constant pressure. Types of liquid-liquid equilibrium: 2 = Winsor I; 2̄ = Winsor II; 3 = Winsor III.

sitions of the three-phase region. In the vicinity of the three-phase region, the microemulsion indicates the highest mutual solubility of water and oil. Only a small change of an intensive variable will then be necessary in order to enter an one-phase microemulsion region in the tetrahedron.

Surfactants, derived from carbohydrates, such as n-alkyl-β-D-glucopyranosides, n-alkyl-β-D-maltopyranosides, or 1-alkanoyl-N-methylglucamides belong to the class of nonionic amphiphiles. However, they produce an unexpected solvation behavior in aqueous solution (Balzer, 1991) combining typical properties of nonionic surfactants (e.g., insensitiveness to the admixing of electrolytes, resistance to alkaline media, demixing in the region of low surfactant concentration) with those exhibited by anionic surfactants (e.g., high solubility in water, low solubility in oil, extremely small capability to solubilize hydrophobic substance into aqueous solution, lyotropic liquid-crystalline structures until relatively high temperatures). With view on ecology it is important to know that they easily may be obtained by synthesis from native educts. Besides they prove completely biodegradable and they act in solution without toxicity. Recently opened fields of application make use of sugar amphiphiles as wetting or foam generating agents, drug delivery systems, emulsifying or creaming additives and as essential compounds of detergents (Hill et al., 1997; Kiwada et al., 1985). 1-Alkanoyl-N-methylglucamides (Hildreth, 1982) have received attention due to their effectiveness in solubilizing and crystallizing membrane proteins without denaturation (Hanatani et al., 1984; Jones, 1992; Sulthana et al., 2000; Nanna and Xia, 2001), and because they enable biochemists to perform analytical operations with high selectivity and precision (Pfüller, 1986; Lawrence and Rees, 2000).

Recipes for preparing microemulsions, included the observation of occurring phase equilibrium sequences, in the quaternary systems n-dodecyl-β-D-glucopyranoside + water + n-dodecane + pentan-1-ol (Kahlweit et al., 1992) or n-octyl-β-D-glucopyranoside + water + n-octane + butan-1-ol (Ryan et al., 1997, Stubenrauch et al., 1997) have been developed but the quantitative determination of the compositions and the volumes of equilibrium phase has been omitted. The reported experimental studies dealing with the phase behavior of such systems have been mostly done using commercial blends of alkylpolyglucosides (Balzer, 1991; von Rybinski and Hill, 1998; Kutschmann et al., 1995; Hill et al., 1997; Fukuda et al., 1993; Kahlweit et al., 1995, 1997) with only a few data of systematic studies using pure n-alkyl-β-D-glucopyranosides (Kahlweit et al., 1991, 1995, 1996; Ryan and Kaler, 1997; Aveyard et al., 1998; Ryan et al., 1997). Therefore, we focused our research efforts on a comprehensive inspection on some quaternary systems of carbohydrate surfactant + water + oily compound + cosurfactant and on the related binary and ternary subsystems which are of a predominant importance in this context but not yet explored. One main goal was to test the ability of these mixtures to form microemulsions and to analyze the phase behavior according to Winsor's ideas. As well, carbohydrate surfactant with different alkyl chain length and head groups of different structure and size as different cosurfactants or different oil compounds served to vary molecular interaction in the liquid mixtures.

This contribution is dealing with a selection of facts and problems obtained from an extensive experimental and theoretical work (Kahl, 1996; Schulte, 1998; Häntzschel, 1999; Lippold, 1999; Enders, 2000; Ngiruwonsanga, 2000), the results of which have been compiled over a longer period and partly published in preceding papers (Kahl et al., 1996, 1997, 2001; Enders and Häntzschel, 1998; Schulte et al., 1999; Häntzschel et al., 1999a,b; Lippold et al., 1998; Lippold and Quitzsch, 2000; Kahl and Enders, 2002; Ngiruwonsanga et al., 2001).

26.3 Experimental

26.3.1 Apparatus

26.3.1.1 Phase Equilibrium

The isothermic and isobaric phase behavior was studied at 298 K and atmospheric pressure applying the synthetic method. Samples of defined feed composition were weighed into glass tubes with Teflon coated screw lids, shaken, equilibrated thoroughly and then the appearance of the phases and the extent of their volumes were registered. Constant surfactant mass fractions (mostly 0.04 mass fraction) and equal amounts of water and oily component were chosen permanently. The cosurfactant concentration was changed step by step within a wide limit. After approaching the phase equilibrium, the coexisting phases were separated carefully from each other and analyzed using high performance liquid chromatography (HPLC). The experiments were carried out with Bischoff equipment (Leonberg, Germany). A RP-18 column was used for separating the components. The signal was detected refractometrically or by UV spectrometry. The measurements were performed either with pure methanol eluent to determine the content of n-octane and isopropyl tetradecanoate or a methanol + water 75/25 V/V mixture to determine all other components. Figure 26.2 shows a typical example (for the system n-decyl-β-D-maltopyranoside + water + n-octane + butan-

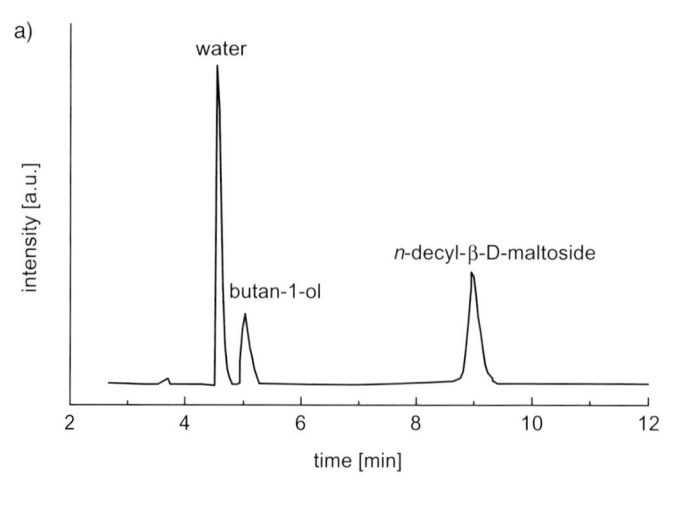

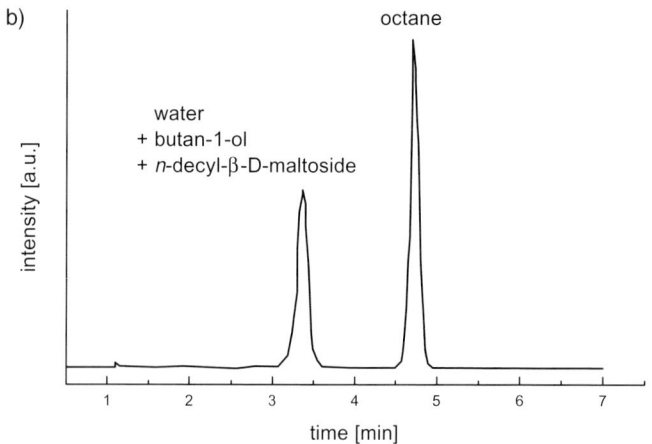

Figure 26.2: Example for analyzing the coexisting phases of the system n-decyl-β-D-maltopyranoside + water + n-octane + butan-1-ol at 298 K by HPLC: a) separation with mixed eluent (methanol + water 75/25 v/v), b) separation with pure eluent (methanol).

1-ol). Mass fractions were determined with a mean accuracy of 5×10^{-3}. More detailed information about the experimental procedure can be found elsewhere (Kahl, 1996; Häntzschel, 1999b).

26.3.1.2 Surface Tension

The critical micelle concentration (CMC) was determined at 298 K from surface tension measurements applying the maximum bubble pressure method. The surface tension was measured by a dynamic maximum bubble pressure tensiometer BP-2 from Krüss GmbH (Hamburg, Germany) and monitored continuously as a function of time. The capillary diameter was 0.250 mm. Confirmation of surface tension of pure water at 298 K was used to calibrate the tensiometer and to check the cleanliness of the glassware. The recordings were stopped when the reading did no longer change with time.

The CMC was obtained by plotting the surface tension against the logarithm of the surfactant mass fraction. The criterion of maximum adsorbed amount was used to determine the position of the discontinuity corresponding to the CMC, the maximum slope being calculated using a least-squares fit. The effective cross-sectional area of the polar head group can be obtained from the slope of the curve at concentrations below the CMC in combination with the Gibbs adsorption equation.

26.3.1.3   Interfacial Tension

The interfacial tension between the pre-equilibrated phases was determined with a spinning drop tensiometer SITE-40 from Krüss GmbH (Hamburg, Germany). The temperature of the samples in the tensiometers was controlled within ± 0.1 K. Reliable numbers were obtained down to 8×10^{-4} mN m^{-1}.

26.3.1.4   Dynamic Light Scattering

Dynamic laser light scattering experiments allow for the characterization of the aggregation behavior of the surfactant component. All samples were examined at the Institut für Experimentelle Physik I (Universität Leipzig) with a laser light spectrometer/goniometer DLS 5000 from ALV (Langen, Germany). The apparatus was equipped with a diode pumped ND:YAG-laser (type DPY 315 II; wavelength: 532 nm). The measurement angle was normally 90°, except where otherwise stated. Samples were continuously filtered through a 450 nm Teflon membrane filter and equilibrated with an accuracy of ± 0.1 K. The experimentally available time autocorrelation function $G_2(t)$ was analyzed using the CONTIN procedure developed by Provencher (1976).

26.3.2   Materials

All surfactants were commercially available products (abbreviations in brackets): *n*-octyl-β-D-glucopyranosid (C_8G_1), *n*-nonyl-β-D-glucopyranoside (C_9G_1), and 1-decanoyl-*N*-methylglucamide (MEGA-10) all from Bachem (Heidelberg, Germany), *n*-decyl-β-D-glucopyranoside ($C_{10}G_1$) and *n*-dodecyl-β-D-glucopyranoside ($C_{12}G_1$) both from Sigma Chemical (St. Louis, MO USA), *n*-decyl-β-D-maltopyranoside ($C_{10}G_2$) from Anatrace (Maumie, OH USA) and Calbiochem-Novabiochem Corporation (La Jolla, CA USA). All surfactants were free of isomers and had a purity >98 mass% as was checked according to Hill et al. (1997). *n*-Octane (used as oil component) and the cosurfactant compounds butan-1-ol, octan-1-ol, phenol were purchased from Acros Organics (Nidderau, Germany), isopropyl tetradecanoate (used as an alternative oil component) from Merck-Schuchardt (Hohenbrunn, Germany). Water was distilled twice over potassium permanganate. All other substances were used without further treatment.

26.4 Theory

26.4.1 Micelle Formation Model

Among the most noteworthy characteristics of surfactants is their behavior in dilute aqueous solutions in which they self-assemble to form aggregates so as to achieve segregation of their hydrophobic moieties from water. Depending upon the type of surfactant and the solution conditions, the aggregates may have a closed structure with spherical, globular, or rod-like shape or have the structure of spherical bilayers. The closed aggregates with hydrophobic interiors are called micelles while the spherical bilayers containing an encapsulated aqueous phase are called vesicles. A molecular thermodynamic model of surfactant aggregation in aqueous solutions was developed by Nagarajan and Ruckenstein (2000), as well by Puvvada and Blankschtein (1990, 1992a, 1992b), to predict the critical micelle concentration, the average size and shape of aggregates as well as the size distribution of aggregates. By these approaches, the physicochemical factors controlling self-assembly were identified by examining all the changes experienced by a single dispersed surfactant molecule when it becomes part of an aggregate. Relatively simple, explicit equations were then formulated to calculate the contribution to the Gibbs energy of aggregation associated with each of these factors. Since the molecular structure of the surfactants and the solution conditions are favorable for an estimation of the molecular constants appearing in the Gibbs energy expressions, the equations could be used to make predictions. The most general description is provided by the multicomponent solution model which visualizes the surfactant solution as a multicomponent system consisting of the solvent, surfactant monomers, and aggregates of various sizes and shapes. Each of these aggregates is treated as a distinct chemical component described by a characteristic chemical potential.

The total Gibbs energy of such a complex mixture G, is modeled as the sum of three contributions: the free energy of formation G_f, the ideal free energy of mixing G_{ideal}, and the free energy of interactions between the various components G_{res}:

$$\frac{G}{RT} = \frac{G_f}{RT} + \frac{G_{ideal}}{RT} + \frac{G_{res}}{RT} \tag{26.1}$$

The ideal Gibbs energy of mixing is given by:

$$\frac{G_{ideal}}{RT} = N_W \ln X_W + \sum_{g=1}^{\infty} N_g \ln X_g \tag{26.2}$$

where X_W is the molar fraction of water and X_g is the molar fraction of aggregates with the aggregation number g. The interactions between micellar aggregates, surfactant monomers and water molecules are approximated by the following mean-field expression:

$$\frac{G_{res}}{RT} = -\frac{1}{2}\left(C_1 + \frac{C_2}{T}\right) N_S \frac{\gamma X_S}{1 + (\gamma - 1) X_S} \tag{26.3}$$

where X_S is the surfactant molar fraction, given by the surfactant mass balance equation:

$$X_S = \sum_{g=1}^{\infty} g X_g \qquad (26.4)$$

C_1, C_2, and γ in Eq. (26.3) are adjustable parameters. A particularly important consequence of Eq. (26.3) is that intermicellar interactions do not affect the micellar size distribution, since G_{res} depends only on the total amount of amphiphile N_S (or X_S) and not on the distribution of the amphiphiles. The Gibbs energy of formation summarizing the molecular interactions responsible for self-association in a dilute reference solution that lacks intermicellar interactions takes the following form:

$$G_f = N_W \mu_W^0 + \sum_{g=1}^{\infty} N_g \mu_g^0 \qquad (26.5)$$

where μ_W^0 is the molar Gibbs energy of water in its standard state (pure component) and μ_g^0 reflects the molar Gibbs energy of a single aggregate g in a fictive standard state, characterized by aggregation number g and shape described above. The difference of molar standard Gibbs energies $\mu_g^0 - g\mu_1^0$ between a surfactant molecule in an aggregate of number g and one in the single dispersed state can be split into several contributions:

$$\Delta \mu_g^0 = \mu_g^0 - g\mu_1^0 = \left(\mu_g^0\right)_{tr} + \left(\mu_g^0\right)_{def} + \left(\mu_g^0\right)_{ster} + \left(\mu_g^0\right)_{int} \qquad (26.6)$$

$(\mu_g^0)_{tr}$ describes the contribution when the hydrophobic tail is removed from contact with water and transferred to the aggregate core which behaves as a hydrocarbon liquid. This contribution is estimated from independent experimental data on the solubility of hydrocarbons in water. $(\mu_g^0)_{def}$ results from "deformations", as the surfactant tail inside the aggregate core is subjected to packing constraints as the polar head group should remain at the aggregate-water interface. The third contribution $(\mu_g^0)_{ster}$ takes into account that the surfactant head groups are brought to the aggregate surface, giving rise to steric repulsion between them. If the head groups are compact in nature pointing out a definable hard-core area, then the steric interactions can be estimated as hard-particle interactions by using the van der Waals approach. The fourth contribution $(\mu_g^0)_{int}$ includes the effects from the creation of an interface between the aggregates hydrophobic domain and water. $(\mu_g^0)_{int}$ is calculated as the product of the surface area in contact with water and the macroscopic interfacial tension of the aggregate core-water interface (for more details cf. Nagarajan and Ruckenstein, 2000, and Enders and Häntzschel, 1998).

From Eq. (26.1) the molar Gibbs energies of aggregates and water are calculated by differentiating of G with respect to the number of moles (N_g and N_w):

$$\frac{\mu_g}{RT} = \frac{\mu_g^0}{RT} + 1 + \ln X_g - g\left(X_W + \sum_{g=1}^{\infty} X_g\right) - \frac{C(T)\gamma g X_S(2X_W + \gamma X_S)}{2(1 + (\gamma - 1)X_s)^2} \qquad (26.7)$$

$$\frac{\mu_W}{RT} = \frac{\mu_W^0}{RT} + 1 + \ln X_W - \left(X_W + \sum_{g=1}^{\infty} X_g\right) - \frac{C(T)\gamma g X_S^2}{2(1 + (\gamma - 1)X_s)^2} \qquad (26.8)$$

Applying the principle of multiple chemical equilibrium (Mukerjee, 1972) between aggregates of different sizes and monomers, that is $\mu_g = g\,\mu_1$, the following expression for the micellar size distribution is obtained:

$$X_g = \exp\left(g\left(1 + \ln X_1 - \frac{\Delta \mu_g^0}{RT}\right) - 1\right) \tag{26.9}$$

where X_1 is the molar fraction of monomer which can be evaluated from the simultaneous solution of the surfactant mass balance (Eq. (26.4)) and Eq. (26.9). The equation for the size distribution of aggregates (Eq. (26.9)), in conjunction with the geometrical characteristics of the aggregates and the expressions for the different contributions to the Gibbs energy of micellization, allows to calculate the size distribution of aggregates, the CMC, and various average properties of the surfactant. Two molecular constants (the number of carbon atoms in the hydrophobic tail n_t and the effective cross-sectional area of the polar head group a_p) enter the predictive calculations. The a_p-value can be obtained experimentally using the surface tension at different surfactant concentrations below the CMC. Table 26.1 gives the molecular specific data used here. The minimization of the Gibbs energy of formation yields information on the geometric properties of the aggregate shape. Following Nagarajan and Ruckenstein (2000) we distinguish between four types of surfactant aggregate (spherical micelles, prolate ellipsoids, rodlike micelles, and spherical bilayer vesicles) which all were treated previously (Nagarajan and Ruckenstein, 2000; Enders and Häntzschel, 1998).

Table 26.1: Molecular parameters of surfactants and cosurfactants.

Substance	n_t	a_p/nm^2	Reference
C_8G_1	8	0.49	Matsumura et al. (1991)
$C_{10}G_1$	10	0.49	Matsumura et al. (1991)
$C_{12}G_1$	12	0.49	Matsumura et al. (1991)
$C_{10}G_2$	10	0.652	Kahl et al. (2001)
MEGA-10	10	0.34	Naragajan and Ruckenstein (2000)
alkan-1-ol	C_3-C_7	0.07	Rao and Ruckenstein (1987)

Many applications, ranging from laundry detergents to microemulsions for enhanced oil recovery, make use of combinations of surfactants and cosurfactants. Among the very large number of additives revealed by a literature survey (Zana, 1995), alcohols have occupied a special place, being by far the most frequently used. Having in prospect e.g. a representative of n-alkyl-β-D-glucopyranosides as surfactant, then the application of cosurfactants is cogent necessary to generate microemulsions (Kahlweit et al., 1995, 1996; Ryan and Kaler, 1997; von Rybinski et al., 1998). An approach to describing the addition of a third component to a binary water + surfactant system has to consider its influence on micellization. The influence of alcohol on micellization can be treated by applying a thermodynamic model developed for surfactant mixtures (Puvvada and Blankschtein, 1992a,b; Rao and

Ruckenstein, 1986; Nagarajan and Ruckenstein, 2000) presuming that alcohols are nonionic amphiphiles with a polar head group. But a single alcohol alone does not form micelles in aqueous solutions as do nonionic surfactants. Hence, the starting point for the development of the surfactant mixture theory is the generalization of theories for binary systems. Therefore principally the various contributions to the total Gibbs energy of the system referring to Eq. (26.1) are the same as in the preceding treatment but they differ in formula expressions (Nagarajan and Ruckenstein, 2000). The mixture consists of N_W moles of water, N_S moles of surfactant, and N_C moles of cosurfactant at temperature T and pressure P. If the concentration of such a surfactant mixture exceeds its CMC, the surfactant molecules will self-assemble to form a distribution of mixed micelles, each of them characterized by a total aggregation number g (which is the sum of g_s molecules of the surfactant and g_c molecules of the cosurfactant) and the chemical composition α. In a mixed micelle there are $g\alpha$ surfactant molecules and $g(1-\alpha)$ cosurfactant molecules. In the spirit of the multiple-chemical equilibrium description, mixed micelles of different sizes and compositions are treated as distinct species in chemical equilibrium with each other as well with the free monomers in the solutions. The Gibbs energy of formation is expressed as

$$G_f = N_W \mu_W^0 + \sum_{g=1}^{\infty} g \int_0^1 N_{g,\alpha} \mu^0(g,\alpha) \, d\alpha \qquad (26.10)$$

where $N_{g,\alpha}$ is the molar number of aggregates formed by $g = g_S + g_C$ molecules and with the composition α. Therefore in addition to the polydispersity with respect to aggregation size, mixed micelles show chemical polydispersity too. Thus, the micelles considered may not be adequately identified by a single discrete variable g. In addition a divariant distribution function has to be applied for describing the mixed micelles. In contrary to the aggregation number g which is always an integer number, the chemical property α of the mixed micelles is continuously distributed, because α can take all values from 0 to 1. For this reasons there is a sum with respect to the aggregation number g and an integral with respect to the chemical composition α in Eq. (26.10). The function $\mu^0(g,\alpha)$ gives a relation between the standard chemical potential of an aggregate of size g with the portions g_S/g surfactant and g_C/g cosurfactant, and those of surfactant and cosurfactant in their singly dispersed states in water. A definite value of $\mu^0(g,\alpha)$ summarizes the many complex physicochemical factors responsible for mixed micelle formation such as hydrophobic effects, conformational changes associated with hydrophobic-tail packing in the micellar core, steric interactions between the hydrophilic head groups, and the entropy of mixing of the two different tails in the mixed micelle. It is calculated according to the micelle formation model (Nagarajan and Ruckenstein, 2000) for different geometrical shapes. In contrary to Nagarajan and Ruckenstein (2000) we account only for the entropy of mixing of the surfactant tails and the cosurfactant tails in the hydrophobic core of the micelle, and neglect the enthalpy of mixing. The extension to systems *n*-alkyl-β-D-glucopyranoside + water + alcohol requires only two additional parameters. The first one is the number of carbon atoms in the alcohol molecules, and the second parameter is the effective cross-sectional area of the head group, a_p (cf. Table 26.1).

The Gibbs energy of mixing of the systems composed of mixed micelles, free monomers, and water is described by

26 Liquid-Liquid Phase Equilibria in Microemulsion Forming Systems

$$G^M = RT\left(N_W \ln X_W + \sum_{g=1}^{\infty} \int_0^1 N_{g,\alpha} \ln X_{g,\alpha}\, d\alpha\right) \qquad (26.11)$$

where $X_{g,\alpha}$ is the mole fraction of aggregates consisting of $g\alpha$ surfactant molecules and $g(1-\alpha)$ cosurfactant molecules. The chemical potentials of water and the aggregates are calculated from the expressions for the Gibbs energy. For water the following expression is obtained:

$$\mu_W = \mu_W^0 + RT\left(1 + \ln X_W - X_W - \sum_{g=1}^{\infty} \int_0^1 X_{g,\alpha}\, d\alpha\right) + \mu_W^E \qquad (26.12)$$

where μ_W^E results from G_{res}.

For the chemical potential $\mu_{g,\alpha}$ of a mixed micelle becomes:

$$\mu_{g,\alpha} = g\alpha\mu_S^0 + g(1-\alpha)\mu_C^0 + g\mu_{g,\alpha}^0 +$$

$$+ RT\left[1 + \ln X_{g,\alpha} - g\left(X_W + \sum_{g=1}^{\infty} \int_0^1 X_{g,\alpha}\, d\alpha\right)\right] + g\alpha\mu_S^E + g(1-\alpha)\mu_C^E \qquad (26.13)$$

The excess quantities μ_S^E and μ_C^E are essential parts of the interaction term G_{res}. Replacing the integral in Eqs. (26.12) and (26.13) by the sum over the chemical composition leads to identical expressions given by Puvvada and Blankschtein (Puvvada and Blankschtein, 1992 a,b). The conditions of multiple chemical equilibrium reads for the formation of mixed micelles:

$$\mu_{g,\alpha} = g\alpha\mu_S + g(1-\alpha)\mu_C \qquad (26.14)$$

with

$$\mu_S = \mu_S^0 + RT\left(1 + \ln X_S^1 - X_W - \sum_{g=1}^{\infty} \int_0^1 X_{g,\alpha}\, d\alpha + \mu_S^E\right) \qquad (26.15)$$

and

$$\mu_C = \mu_C^0 + RT\left(1 + \ln X_C^1 - X_W - \sum_{g=1}^{\infty} \int_0^1 X_{g,\alpha}\, d\alpha\right) + \mu_C^E \qquad (26.16)$$

where X_S^1 is the surfactant monomer mole fraction and X_C^1 represents the cosurfactant monomer mole fraction. Inserting Eqs. (26.15) and (26.16) in Eq. (26.14) leads to the following expression for the equilibrium micellar size and composition distribution:

$$X_{g,\alpha} = \exp\left(g\left[\alpha \ln X_S^1 + (1-\alpha)\ln X_C^1 + 1 - \frac{\mu_{g,\alpha}^0}{RT}\right] - 1\right) \tag{26.17}$$

The knowledge of the distribution function $X_{g,\alpha}$ allows for the calculation of different moments $M_{i,j}$ of the distribution function:

$$M_{i,j} = X_S^1 + X_C^1 + \sum_{g=2}^{\infty} n^i \int_0^1 \alpha^j X_{g,\alpha}\, d\alpha \tag{26.18}$$

Average values of properties characterizing the mixed micelle, like number-average aggregation number g_N, mass-average aggregation number g_W, and number-average chemical composition α_N, are then given by:

$$g_N = \frac{M_{1,0}}{M_{0,0}} \qquad g_W = \frac{M_{2,0}}{M_{1,0}} \qquad \alpha_N = \frac{M_{0,1}}{M_{0,0}} \tag{26.19}$$

A "theoretical" definition of the critical micelle concentration was proposed (Nagarajan and Ruckenstein, 2000) as that concentration of the single dispersed surfactant at which the shape of the micelle size distribution function exhibits a transition from a monotonically decreasing one to a function possessing a maximum and a minimum. From the mathematical point of view this gives different conditions. The following conditions were applied in connection with this treatment:

$$\left(\frac{\partial^2 g_N(X_S^1)}{\partial(X_S^1)^2}\right) = 0 \quad \text{and} \quad \left(\frac{\partial^2 g_W(X_S^1)}{\partial(X_S^1)^2}\right) = 0 \tag{26.20}$$

All investigations of the present work were performed respecting the conditions in related binary and ternary subsystems at 298 K.

26.4.2 G^E-Model

An alternative way of correlating and predicting the liquid-liquid equilibrium in microemulsion forming multiphase systems applies a rather simple excess Gibbs energy model suggested by Koningsveld and Kleintjens (1971). In order to take the difference in the molecular size of different components into account segment-molar quantities were introduced. All molecules of the system are imagined to be divided into segments of equal size. If M^o represents the molar mass of an arbitrary chosen standard segment, then the segment number r of a species of kind i with molar mass M_i is:

$$r_i = \frac{M_i}{M^o} \tag{26.21}$$

A water molecule usually is taken as a standard segment. The segmental related compositions in the mixture can be quantified by segment molar fractions y_i as follows:

$$y_i = \frac{r_i X_i}{\sum_{i=1}^{n} r_i X_i} \tag{26.22}$$

The Flory-Huggins theory (Flory, 1953) gives the Gibbs energy of mixing G^M in systems containing molecules of different sizes:

$$\frac{G^M}{RT} = \sum_{i=1}^{n} \frac{y_i}{r_i} \ln y_i + \frac{G^E}{RT} \tag{26.23}$$

where the first term exhibits a combinatorial contribution of random mixture and the last term results from contributions by intermolecular interactions. Koningsveld and Kleintjens (1971) suggested a simple but very flexible G^E model which reads for a binary system:

$$\frac{G^E}{RT} = \frac{y_i y_j \beta_{ij}(T)}{1 - \gamma_{ij} y_j} \quad \text{with} \quad \beta(T) = \beta_1 + \frac{\beta_2}{T} + \frac{\beta_3}{T^2} \tag{26.24}$$

This model is capable of representing phase diagrams of systems exhibiting upper and lower critical solution temperature, as individually as both combined with open gaps, and even closed loop or hourglass behavior. The Koningsveld-Kleintjens model extended to a quaternary system reads:

$$\frac{G^E}{RT} = \frac{y_A y_B \beta_{AB}}{1 - \gamma_{AB} y_B} + \frac{y_A y_C \beta_{AC}}{1 - \gamma_{AC} y_C} + \frac{y_A y_D \beta_{AD}}{1 - \gamma_{AD} y_D} + \frac{y_B y_C \beta_{BC}}{1 - \gamma_{BC} y_C} + \frac{y_B y_D \beta_{BD}}{1 - \gamma_{BD} y_D} + \frac{y_C y_D \beta_{CD}}{1 - \gamma_{CD} y_D} \tag{26.25}$$

This equation includes contributions from each binary system represented by two binary parameters β and γ. Every pair of parameters can be fitted to the equilibrium data of a related subsystem, if a miscibility gap at room temperature exists. The parameters for all other binary subsystems have to be fitted to one of the tie line in a corresponding ternary system. The experimental data used for this purpose were taken from our previous results (Kahl, 1996; Kahl et al., 1997; Häntzschel, 1999; Häntzschel et al., 1999b) in combination with data for some binary systems from the literature (Sørensen and Arlt, 1979).

For a quantitative description of equilibrium properties in a microemulsion forming system, it seems to be indispensable to use additionally a molecular thermodynamic theory for a partially-ordered fluid, as a microemulsion is not even approximately a random mixture, as assumed by Eq. (26.23). Widom (1984, 1986, 1996) employed a simple cubic lattice whose sites carry oil (AA), water (BB), and amphiphile (AB) molecules. Widom reported an empirical expression for the Gibbs energy of microemulsions that is based on results by Talmon and Prager (1977), as later modified by de Gennes and Taupin (1982). In this model, a three-component microemulsion (oil + water + surfactant) is imagined to consist of microdomains of oil and water surrounded by a surfactant film, and the Gibbs energy of the solution is replaced by the entropy of mixing of those domains, and terms which account for the properties of the surfactant film and the interaction at the interface between unlike domains. Hu and Prausnitz (1988) used Guggenheim's quasichemical theory to evaluate the combinatorial contribution that corresponds to Widom's picture. The application of this approach to the quaternary systems results in:

$$\frac{G^M}{RT} = y_1 \ln \frac{y_1}{y_1 + \frac{y_2}{2} + \frac{y_3}{2}} + y_4 \ln \frac{y_4}{y_4 + \frac{y_2}{2} + \frac{y_3}{2}} +$$

$$+ y_2 \ln \frac{y_2/2}{\sqrt{y_1 + \frac{y_2}{2} + \frac{y_3}{2}}\sqrt{y_4 + \frac{y_2}{2} + \frac{y_3}{2}}} + y_3 \ln \frac{y_3/2}{\sqrt{y_1 + \frac{y_2}{2} + \frac{y_3}{2}}\sqrt{y_4 + \frac{y_2}{2} + \frac{y_3}{2}}}$$

(26.26)

where subscripts 1, 2, 3, and 4 stand for water, surfactant, cosurfactant, and oil respectively. The calculation of the phase diagram in the microemulsion forming system can be carried out by using the G^E expression given by Eq. (26.25) and the combinatorial contribution given by Eq. (26.26). The spinodals and critical points were calculated using stability determinants. A method introduced by Solc (Solc, 1969) was applied to avoid trivial solutions for compositions near the critical point.

26.4.3 Landau-Theory

Kahlweit et al. (1990) emphasized the close analogy between the three-phase region of microemulsion phase equilibria and the phase equilibria near the tricritical points of mixtures. The three-phase split near a tricritical point in a ternary system should follow the Landau-type model (Rudolph et al., 1997; Griffiths, 1974; Kaufman and Griffiths, 1982; Kleinert 1986a,b). Rudolph et al. (1997) used a Landau-type model which expands the Helmholtz free energy in a power series of an order parameter, aiming at the modeling of phase equilibrium data of the system 2-butoxyethanol + water + n-alkane. The order parameter quantifies the distance from the critical point in terms of one of its variables, e.g. the temperature, as in the classical theory of Landau. Fitting the parameters for a pseudo binary cross section at a constant oil-to-water ratio leads to a very good representation of the experimental results. With the same parameters the model is able to predict the phase equilibrium pattern of the system under conditions of another oil-to-water ratio. However, the experimental data in the systems studied here are far away from the tricritical point (Kahl, 1996; Kahl et al., 1997). The tricritical point is located at a much higher surfactant content, than that at which the experiments were carried out.

Nevertheless the Landau-Ginzburg functional can be used for a study of some physical properties, like concentration fluctuations and interfacial tensions, near the tricritical point, at least in principle. Kleinert (1986a,b) proposed to describe ternary systems, built from water, oil, and surfactant, by a Landau free energy functional:

$$f_{Pot} = -\frac{3}{8}\tau x + \frac{9}{16}x^2 + \frac{1}{2}\tau x^3 - \frac{3}{2}x^4 + x^6 \tag{26.27}$$

where x is the order parameter related to the oil in water concentration, while x^2 gives the surfactant concentration. τ is related to a reduced temperature or to the relative salinity. Using Eq. (26.27) we determined the shape of the three-phase regime of a ternary mixture of amphiphile, water, and oil in a space of concentration and temperature. Following Kleinert (1986a) a study of fluctuations around the three phase equilibria was performed adding a gradient energy and postulating the local energy density:

$$f = \frac{1}{2}(\partial x(r))^2 + f_{Pot}(x(r)) \tag{26.28}$$

where the order parameter x is now a function of the space variable r and the length scale has been normalized to unity. Then the long-wavelength correlation functions in momentum space are given by the inverse curvatures at the three minima of f_{pot}:

$$\langle x(k)x(k)\rangle|_{k=0} = \left(\frac{1}{2}f_{Pot}^{II}(x_{1,2,3})\right)^{-1} \tag{26.29}$$

The inverse curvatures in the minima i = 1, 2, 3 are observable via the intensities I of scattered light (or the relaxation time) (Kleinert, 1986a). For the middle phase one gets:

$$I_3 \propto (x_3 - x_1)^2(x_3 - x_2)^2 \tag{26.30}$$

In the oil and water rich phases, on the other hand, one finds

$$I_1 \propto (x_1 - x_2)^2(x_1 - x_3)^2 \tag{26.31}$$

and

$$I_2 \propto (x_2 - x_1)^2(x_2 - x_3)^2 \tag{26.32}$$

A further set of experimental data delivers the interfacial tension between any two of the three coexisting liquid phases with concern to the free energy of mixing. The latter will be gained per area of a "domain wall solution" $x_d(z)$ of the field equation:

$$\frac{d^2 x_d(z)}{d z^2} = f_{pot}^I(x_d(z)) \tag{26.33}$$

which connects the minimum at i with that at j. The interfacial tension between states at the minima i and j results from:

$$\sigma_{i,j} = \int_{z_i}^{z_j} dz \left[\frac{1}{2}\left(\frac{d x_d(z)}{d z}\right)^2 + f_{pot}(x_d(z))\right] \tag{26.34}$$

In conclusion we see that the expressions for the Gibbs energy of the micellar mixture developed numerically by a gradient term is capable of explaining the temperature behavior of the shape of the three-phase coexistence regime, the size of the fluctuations in each phase, and the interfacial tensions.

26.5 Results and Discussion

26.5.1 Strategy and Concepts of the Investigations

The main intention of the present study was to gather comprehensive knowledge how to prepare microemulsions on the basis of the self-assembly of carbohydrate surfactant molecules. The influences of the hydrophilic and lipophilic moieties in the molecular structure of this kind of amphiphiles, of the polarity and structure of the solvent, acting as cosurfactant, and of the peculiarity of the oil compound altogether taken into account, the following quaternary systems were chosen as objects of research

C_8G_1 + water + *n*-octane + butan-1-ol

C_9G_1 + water + *n*-octane + butan-1-ol

$C_{10}G_1$ + water + *n*-octane + butan-1-ol

$C_{10}G_1$ + water + *n*-octane + octane-1-ol

$C_{10}G_1$ + water + *n*-octane + phenol

$C_{10}G_1$ + water + isopropyl tetradecanoate + butan-1-ol

$C_{12}G_1$ + water + *n*-octane + butan-1-ol

$C_{10}G_2$ + water + *n*-octane + butan-1-ol

MEGA-10 + water + *n*-octane + butan-1-ol

MEGA-10 + water + *n*-heptane + phenol

The phase equilibrium behavior of these systems was characterized in a quasi triangular subspace of the composition tetrahedron, as described in Section 26.2.2. The feed compositions were established keeping the water-to-oil mass ratio at 1:1, 1.5:1 etc. and the surfactant mass fraction constant (w_s = 0.02, 0.04, 0.06). If the cosurfactant content will then be varied, one may search all over the different states of phases and phase coexistence and their extensions both in a representative spacial part of the tetrahedron. As an essential result of that procedure the position of the "Kahlweit fish" (cf. Fig. 26.1) could be located, and the phases coexisting in a liquid-liquid or liquid-liquid-liquid equilibrium following a Winsor sequence became available for further investigations. The phases were characterized with regard to many properties (cf. Kahl, 1996; Schulte, 1998; Häntzschel, 1999; Lippold, 1999; Enders, 2000; Ngiruwonsanga, 2000), including the quantitative analyses of phase compositions by HPLC, the determinations of volume, density, dynamic viscosity, surface tension of single phases as well as interfacial tensions between phases in equilibrium. Additionally particular configurations and structures as well as mobilities of the colloidal domains in micellar solutions respectively microemulsions were identified by dynamic light scattering experiments and by transmission electron microscopy.

The development of methods for calculating the complex phase behavior of the surfactant + cosurfactant systems was started with a discussion of the phase behavior of the binary and ternary subsystems.

26.5.2 Binary Subsystems

Information on the self-assembly of the carbohydrate surfactants in aqueous solution was of prior interest. In diluted aqueous solution all surfactants form micelles. C_XG_1 + water binaries, for $X \geq 10$ show a miscibility gap that exists throughout the experimental temperature range at low surfactant concentration. When heated the micellar solutions splits into two coexisting phases where the lower critical point is located below 273 K. At high surfactant concentration the observed lyotropic behavior follows the classical pattern including hexagonal, cubic, and lamellar phases. At temperatures from 288 to 328 K the n-octyl-β-D-glucopyranoside and n-decyl-β-D–maltopyranoside as well as 1-decanoyl-N-methylglucamide are well soluble in water up to very high concentrations, where they form lyotropic mesophases (Kahl, 1996; Häntzschel, 1999; Lippold, 1999).

The applied thermodynamic micelle formation model predicts that the carbohydrate surfactants in water predominantly assemble to spherical bilayer vesicles (Enders and Häntzschel, 1998; Kahl and Enders, 2002). The vesicles were found to swell linearly upon increasing the total surfactant concentration. This result was confirmed by an experimental study of a similar system by Regev and Guillemet (1999). They could observe a structural evolution of bilayers in a single-tail nonionic surfactant (n-alkyl-cocodiethanolamide) + water system using cryo-transmission electron microscopy, small-angle X-ray scattering, light scattering, and rheological measurements. With increasing surfactant concentration, unilamellar vesicles turn to multilamellar vesicles while at very high concentrations a fully expanded lamellar phase appears. Putlitz et al. (2000) found by small angle neutron scattering and freeze-fracture electron microscopy that sulfonium surfactant with a single surfactant tail form also vesicles of narrow size distribution. Additional support for the predictions based on the micelle formation model can be gained by molecular dynamics simulation (Bogusz et al., 2000). The critical micellar concentration predicted on this base as a function of temperature runs through a minimum in agreement with experimental results (Enders and Häntzschel, 1998; Enders, 2000; Kahl and Enders, 2002).

Combining the micellar formation model with phase separation thermodynamics permits the calculation of the critical point, the spinodal curve, and the cloud point curve. The model is able to describe the experimental cloud point curves for $C_{10}G_1$ + water and $C_{12}G_1$ + water systems when three parameters (C_1, C_2, and γ of Eq. (26.3)) are adjusted (Enders and Häntzschel, 1998). Additionally, the micelle formation model allows for the calculation of the aggregation size distribution functions in both phases. The results are illustrated in Figure 26.3. These results show clearly that aggregates exist in both phases but the total and the average numbers are different in both phases. These results were also confirmed by transmission electron microscopy (Häntzschel, 1999b).

26.5 Results and Discussion

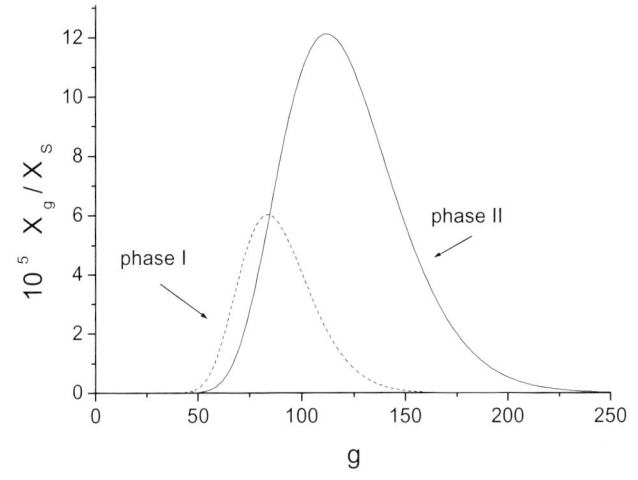

Figure 26.3: Distributions of the aggregation number in the coexisting phases of the binary system n-decyl-β-D-glucopyranoside + water at 330 K. The calculations were carried out using micelle formation model with the following parameters $C_1 = 0.7661$, $C_2 = 200$; and $\gamma = 310$.

Furthermore the application of the Koningsveld-Kleintjens G^E-model permits the fitting of the liquid-liquid equilibrium (Enders, 2000; Kahl and Enders, 2002). However, this approach does not allow for the calculation of the size distribution functions in the coexisting phases. When all binary parameters are known (from fitting the experimental data of the relevant binary subsystems) the phase behavior of the ternary systems can be predicted.

The binary parameters of Eq. (26.25) were fitted to the experimental results for the binodal curve for all those binary systems which reveal liquid-liquid immiscibility. Experimental LLE data were taken from the compilation of Sørensen and Arlt (1979). Kahl and Enders (2002) reported the results of the parameter evaluation for 298.15 K.

26.5.3 Ternary Subsystems

The ternary phase diagram of the system $C_{10}G_1$ + water + n-octane mixtures shows a very narrow one-phase region emerging from the surfactant + water binary (Kahl et al., 1997). Replacing $C_{10}G_1$ by C_8G_1, C_9G_1, or $C_{12}G_1$ or MEGA-10 did not substantially increase the one-phase region. Only 2 mass% of n-octane can be dissolved in an aqueous solution of n-decyl-β-D-glucopyranoside at 298 K and at relatively high surfactant concentration (Kahl et al., 1997). In an aqueous solution of 1-decanoyl-N-methylglucamide the solubility of n-octane rises to a mass fraction $w_0 = 0.04$ (Häntzschel, 1999). Therefore, the only way to form a microemulsion, especially a middle phase microemulsion, with these two surfactants is to add a cosurfactant, i.e. to establish a quaternary system. Consequently, the investigations must be extended to ternary systems with and without carbohydrate surfactant compounds. Figure 26.4 gives an example for the influence of four ternary subsystems on the one-, two-, and three-phase behavior of the quaternary mixture. The central triangle shows the poor solubili-

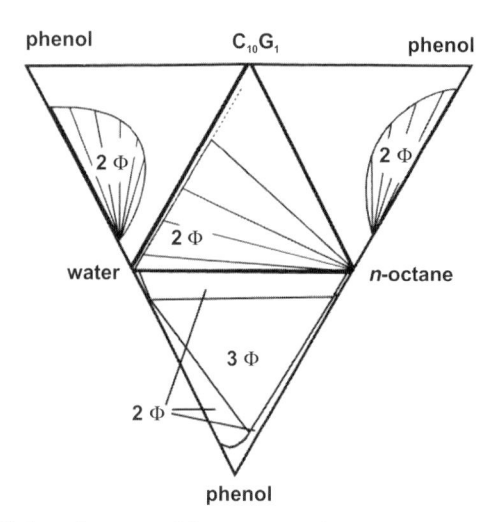

26 Liquid-Liquid Phase Equilibria in Microemulsion Forming Systems

Figure 26.4: Phase equilibrium diagrams of four ternary subsystems at 298 K as unfolded tetrahedron areas which belong to the space of quaternary system n-decyl-β-D-glucopyranoside + water + n-octane + phenol.

zation of n-octane by $C_{10}G_1$ + water combination. For simplicity, the solid and liquid-crystalline phases appearing at high surfactant concentration were not illustrated in Figure 26.4.

The efficiency of the additive in modifying the water structure would be reflected by a change in the hydration of the hydrocarbon tail, and thus, ultimately, in the concentration at which the micellization process starts. The CMCs of some combinations of C_8G_1 + water + alkan-1-ol are shown in Figure 26.5 as a function of the molar fraction of the cosurfactant. Figure 26.5 was calculated utilizing the mixed micelle modeling concept (cf. Eq. (26.20)). The curves effectuated by pentan-1-ol or long-chain alcohols have the same shape: The addition of alcohol produces a sigmoidal descent of the CMC and the increase of the alcohol carbon number produces a downward shift to lower values of the molar fraction of the cosurfactant. This behavior of course reflects the experience that the change of CMC is predominantly caused by the penetration of the alcohol in the micelles. Adding a short-chain alcohol, however creates in detail an uncommon CMC behavior of the ternary system. The predictions for the CMC in systems containing butan-1-ol and propan-1-ol reveal that the CMC increases at very low molar fraction of the cosurfactant. At higher cosurfactant concentration the CMC then decreases again. These (predicted) phenomena may be explained by a modification of the properties of water caused by the short-chain alcohol participation in the micellization process. The short-chain alcohol + water mixtures therefore may be considered as better solvents for carbohydrate surfactants than pure water resulting in micelle formation only at higher surfactant concentration. Such a behavior was observed indeed experimentally for different surfactant + alcohol + water mixtures (Zana, 1995). The predictions are in good agreement with the data given in this study which are shown in Figure 26.5 as solid symbols. The largest deviations between calculated and experimental data are observed for the system C_8G_1 + water + octan-1-ol. That might be due to two phenomena. At first, the effective cross-sectional area of the head group, a_p, perhaps depends on the carbon number of the alcohol. At second, the rather low solubility of octanol in water renders more difficult to

26.5 Results and Discussion

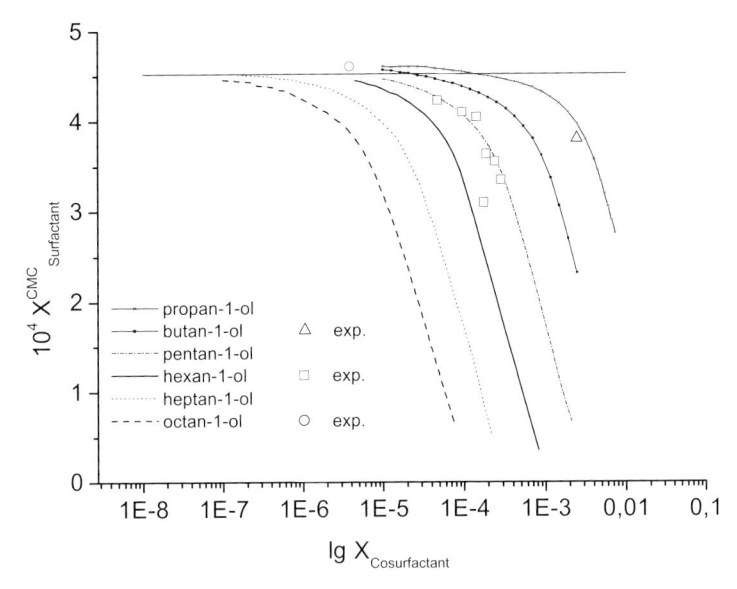

Figure 26.5: Comparison between predicted and experimental critical micelle concentration in the system n-decyl-β-D-glucopyranosid + water + alkan-1-ol at 298 K as function of alkan-1-ol molar fraction.

measure the surface tension in this mixing system with high precision. Though a short-chain alcohol promotes the self-assembly in ternary systems it inhibits the lyotropic formation of liquid crystals in high concentrated binary surfactant + water systems (Kahl, 1996; Enders, 2000).

In Figure 26.6 the experimental phase diagram for the system $C_{10}G_1$ + water + butan-1-ol at 298 K is compared with the theoretical calculations using the Koningsveld-Kleintjens model (Eq. (26.25)). As no phase splitting occurs in the binary systems surfactant + alcohol at low surfactant concentration the parameters for binary interactions had to be fitted to an experimental tie line of the ternary system surfactant + water + alcohol. The G^E-model allows the calculation of the phase equilibrium compositions which are close to the experimental data. Especially the slopes and the end points of the tie lines were excellently reproduced (Enders, 2000; Kahl and Enders, 2002). With respect to two miscibility gaps in two of the binary subsystems (water + butan-1-ol and $C_{10}G_1$ + water) a three-phase equilibrium was to be expected and also predicted (cf. Fig. 26.6). Some experimental observations (obtained after the calculations were performed) revealed such a three-phase equilibrium, where two very small phases coexisting with a large phase could be observed visually.

However, in those experiments the small phases remained segregated in the large phase, the compositions of the coexisting phases could not be determined. The phase diagram for the system $C_{10}G_1$ + water + phenol shows a similar phase equilibrium behavior (Häntzschel, 1999; Häntzschel et al., 1999b). The binary subsystem $C_{10}G_1$ + octane reveals no liquid-liquid phase split but a liquid-solid coexistence of octane and the surfactant both as nearly pure components. For this reason the interaction parameters were set to zero. In order to predict the phase diagram for the system $C_{10}G_1$ + water + octane the parameters ob-

26 *Liquid-Liquid Phase Equilibria in Microemulsion Forming Systems*

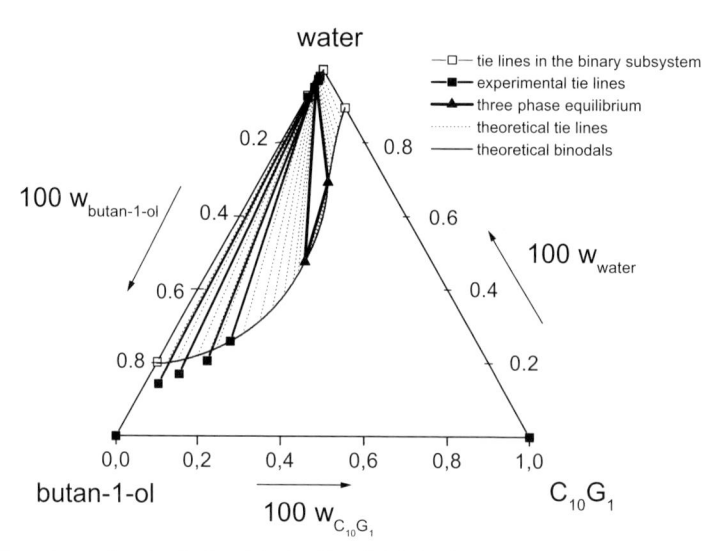

Figure 26.6: Experimental and calculated (using Eq. (26.24)) phase equilibrium data (mass fractions) of the system *n*-decyl-β-D-glucopyranosid + water + butan-1-ol at 298 K.

tained from describing the behavior of the binary systems were used (Enders, 2000; Kahl and Enders, 2002). Two phases coexist over a wide-spread region within the phase triangle. Similar to the system discussed above, a three-phase split is also observed here. In contrast to the system $C_{10}G_1$ + water + butan-1-ol, the three-phase state is located very close to the water-surfactant side of the phase diagram. Subsequently, this phase behavior could be observed experimentally (Kahl and Enders, 2000). However analyzing the coexisting phases was impossible, as the volume of one phase was too small. The ternary subsystem water + octane + butan-1-ol is characterized by a large two-phase region (Sørensen and Arlt, 1979), reaching from the butan-1-ol + water side to the water + octane side. Butan-1-ol and octane are completely miscible. The binary parameters for that system parameters were fitted to the experimental data for a tie line in the ternary system. The behavior of the ternary subsystem phenol + water + octane (Häntzschel, 1999) behaves very different from that described before. All binary subsystems show a miscibility gap. The phase behavior of that ternary system could be predicted nicely using parameters which were fitted to the binary systems only (Enders, 2000; Kahl and Enders, 2002).

The phase equilibrium properties of the system water + octane + octan-1-ol can also be calculated using the segmental G^E-model (Kahl and Enders, 2002). Similar to the system water + octane + butan-1-ol the predictions for the tie lines are in good agreement with the experimental findings. However the lengths of the tie lines are overestimated.

In summary one has to state that the concepts for modelling the liquid-liquid phase equilibrium data described above, could be employed for ternary systems with partial success. The best results were obtained, if any detail of self-assembly in a colloidal dimension was neglected (Enders, 2000; Kahl and Enders, 2002).

26.5.4 Quaternary Systems

Although the investigations to this study are not yet completed, relevant results can already be presented. Special attention must be called to the distinct behavior of the systems developing a Winsor-like sequence, to the volume extension and composition of the equilibrium phases, to the microstructures of surfactant containing phases, and to interfacial properties.

As is shown schematically in Figure 26.7, the three quaternary systems which all consist of $C_{10}G_1$, water, and n-octane, but differ in the cosurfactant component, show a very individual phase equilibrium sequence when at constant oil-to-water mass ratio (1:1) as well as a constant surfactant mass fraction ($w_s = 0.04$) either butan-1-ol or octan-1-ol or phenol are added. For the butan-1-ol containing system, one finds at room temperature a generalizing Winsor pattern. Starting at low butan-1-ol portions ($w_c = 0.02$) in the feed composition a two-phase system will be established with an oil-rich upper phase (45–50 vol%) in equilibrium with an aqueous surfactant solution (45–50 vol%). The increase of the butan-1-ol concentration in the feed leads to a three-phase state with a surfactant-rich middle phase (6–20 vol%) in equilibrium with an oil-rich upper phase (50–60 vol%) and a water-rich lower phase (25–35 vol%). Three phases coexist within a small range of cosurfactant mass fraction ($w_c = 0.03$–0.07) in the feed. At a cosurfactant mass fraction of $w_c = 0.055$ the maximum solubility power for incorporating n-octane into an aqueous surfactant solution is achieved. The two-phase system of type Winsor II with surfactant in the upper oil-rich phase (65–70 vol%) and a lower aqueous phase (30–35 vol%) appears at cosurfactant mass fractions $w_c > 0.07$. In addition the compositions of the coexisting phases are shown for the $C_{10}G_1$ + water + n-octane + butan-1-ol system in Figure 26.8.

When butan-1-ol is replaced by octan-1-ol a quite different phase equilibrium behavior is found (Häntzschel, 1999). Only two-phase states were realized by variation of the cosurfactant mass fraction in the feed from $w_c = 0.02$ to 0.15. As octan-1-ol is more hydrophobic than butan-1-ol a relatively extended upper phase (65–70 vol%) rich in hydrocarbon and oc-

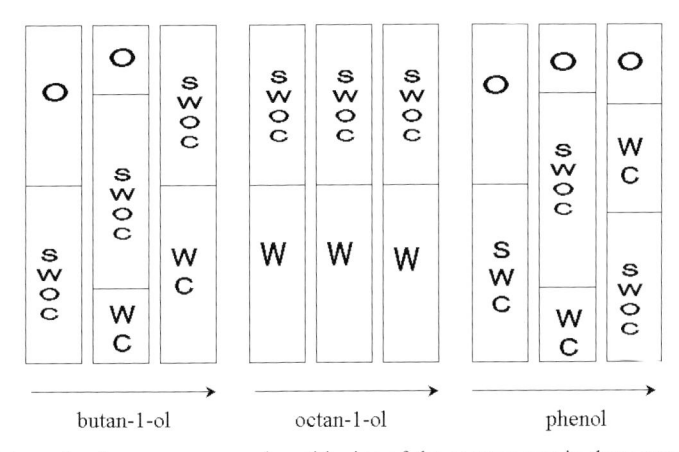

Figure 26.7: Schematic phase sequence and partitioning of the components in three quaternary systems n-decyl-β-D-glucopyranosid + water + n-octane + cosurfactant at 298 K; S = surfactant, W = water, O = oil, C = cosurfactant.

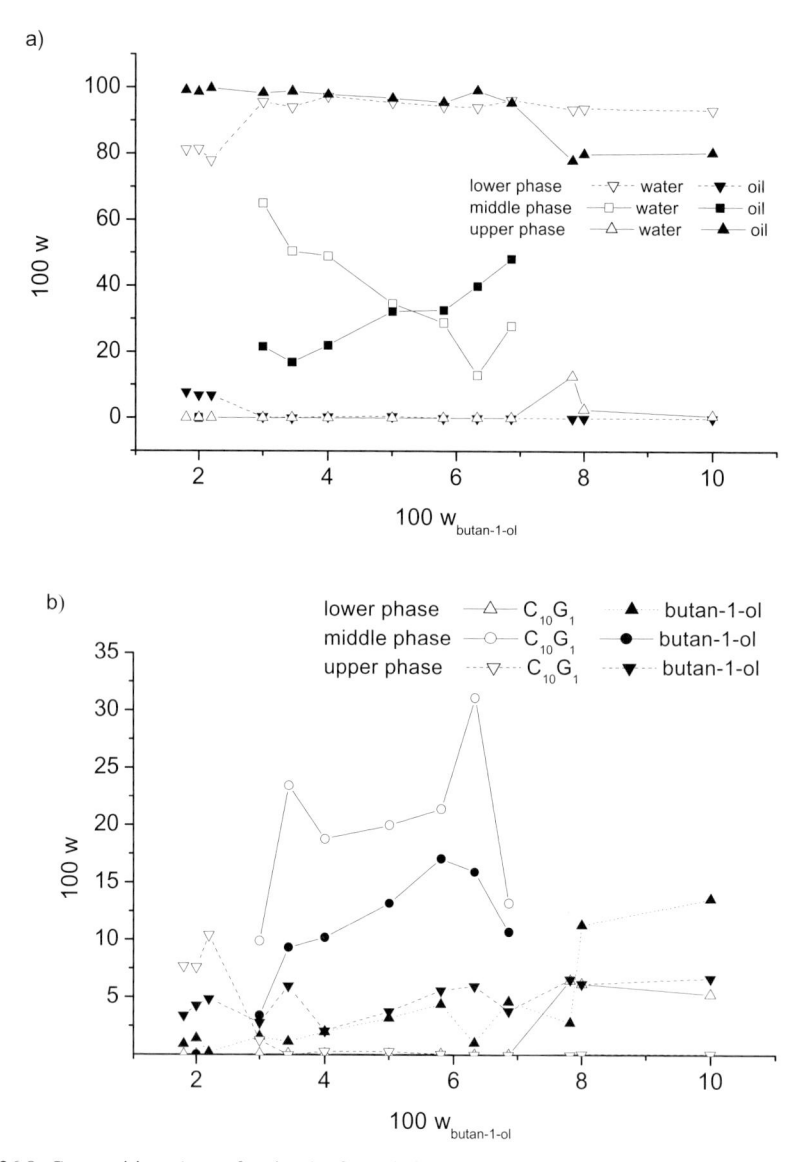

26 *Liquid-Liquid Phase Equilibria in Microemulsion Forming Systems*

Figure 26.8: Compositions (mass fractions) of coexisting phases following a Winsor sequence in the system *n*-decyl-β-D-glucopyranosid + water + *n*-octane + butan-1-ol (from HPLC experiments) at 298 K; a) water and oil; b) surfactant and cosurfactant.

tan-1-ol and containing nearly all surfactant ($w_s = 0.04$) coexists with almost pure water (30–35 vol%) in the lower phase. The $C_{10}G_1$ surfactant probably is aggregated into inverse micelles (Häntzschel et al., 1999b). Analysis by HPLC showed that only small amounts of water are solubilized in the oil-rich phase. With view on a Winsor pattern Stubenrauch et al. (1996) examined a system of similar composition phase sequence: $C_{10}G_1$ + water + *n*-de-

620

cane + decan-1-ol. At a surfactant mass fraction below 0.02 a three-phase region was found. The middle-phase microemulsion was of type Winsor III. In our case presumably three phases may coexist too at a very low surfactant concentration in the feed.

Figure 26.7 presents also the phase equilibrium sequence of the system $C_{10}G_1$ + water + n-octane + phenol. A two-phase region was observed up to cosurfactant mass fraction $w_c = 0.02$ in the feed. The upper phase (60 vol%) is nearly pure n-octane, the lower one (40 vol%) a water-rich solution of $C_{10}G_1$ and phenol. Adding more phenol (up to 20 mass%) results in three coexisting phases which exhibit moreover a very interesting inversion phenomenon. An upper phase (35–50 vol%) of nearly pure octane coexists with a microemulsion (middle phase: 15–40 vol%) which contains nearly even amounts of surfactant water, oil, and phenol and an aqueous solution of phenol (the lower phase: 15–30 vol%). When the phenol concentration in the feed is increased the volume of the middle phase grows significantly and finally becomes the lower phase (20–30 vol%). This density inversion between the former middle and lower phases appears when the mass fraction of phenol in the feed passes 0.145.

The four-component system was modified by replacing the surfactant $C_{10}G_1$ by C_8G_1, C_9G_1, and $C_{12}G_1$ respectively. The special interest thereby was directed to the influence of the lipophilic alkyl chain of the surfactant on the conditions favoring the development of a phase sequence. From the results obtained one may learn that with increasing chain length the three-phase body is shifted to lower cosurfactant concentration in the tetrahedron. Therefore in comparison to the systems with $C_{10}G_1$ or $C_{12}G_1$, in systems with C_8G_1 system a considerably higher amount of cosurfactant must be added in order to produce a middle-phase microemulsion. The extension of the closed-loop body area of the "Kahlweit fish", however, was found to be independent on the lipophilic moiety of the surfactant: Furthermore, the phenomenology of the Winsor phase sequence in all systems involving a homologous C_nG_1 series proved to be very uniform. Replacing $C_{10}G_1$ by MEGA-10 or $C_{10}G_2$ results in a change of the properties of the head group (chemical structure and volume) whereas the length of the alkyl chain remained unchanged. The question was, if these variations will cause a change in the phase behavior sequence. However, the macroscopic phase equilibrium properties were not very different from those of the $C_{10}G_1$ system. The middle phase microemulsions of these systems, however, revealed differences in the ordered colloidal domains in their microspheres. This was perceived by means of transmission electron microscopy (Talmon, 1996; Häntzschel, 1999; Häntzschel et al., 1999b). While in the microemulsions based on $C_{10}G_1$ bicontinuous structures are predominant, the MEGA-10 containing middle phases appear in a lamellar-like order, characterized by manifoldly broken layers.

In order to consider at last the activity of the oily component on the formation of the phase sequence in the investigated types of four-component systems, n-octane was replaced by isopropyl tetradecanoate. Kahlweit et al. (1997) had already employed isopropyl tetradecanoate as oily component in two systems (Epikuron® + water + oil + pentane-1,2-diol and alkyl polyglucoside ($C_{12-14}G_{1.3}$) + water + n-octane + octan-1-ol). Although isopropyl tetradecanoate involves polar ester groups into the solubilizing process, the finally resulting phase volume ratios of the system $C_{10}G_1$ + water + n-octane + butan-1-ol and the system $C_{10}G_1$ + water + isopropyl tetradecanoate + butan-1-ol do not differ in their phase equilibrium characteristics (Häntzschel et al., 1999b). With the alternative oil the middle-phase microemulsion started at the same cosurfactant concentration in the feed. For the nonionic surfactant class of alkylpolyglycol ethers Kahlweit et al. (1983) found that the three-phase temperature interval shrinks for a given surfactant with decreasing hydrophobicity of the oil, and for a given oil with increasing hydrophobicity of the surfactant. With carbohydrate surfactants (instead of alkylpolyglycol ethers) this phenomenon is less emphasized (Häntzschel, 1999).

In order to model the equilibrium compositions in a multiphase quaternary system with the $C_{10}G_1$ + water + octane + butan-1-ol system in a first approach the Koningsveld-Kleintjens formalism with the parameter given by Kahl and Enders (2002) was checked neglecting any properties of the self-organized system. A typical result for the partitioning is given in Table 26.2 in comparison with experimental data. A variation of the parameters of Eq. (26.25) did not lead to a better approximation. Introducing additional terms into the Gibbs energy relation – screening the enthalpic repulsion (Kilpatrick et al., 1986) or improving the entropic contribution (Eq. (26.26); Hu and Prausnitz, 1988) – did not result in a better agreement with experimental data. That finding supports the statement that one cannot avoid the numerical expense of the molecular thermodynamic approach, involving the micellization process (Nagarajan and Ruckenstein, 2000).

Table 26.2: Comparison between experimentally realized and calculated compositions (mass fractions) of coexisting phases at 298.15 K.

Substance	w_{feed}	w^{upper}exp.	w^{upper}cal.	w^{middle}exp.	w^{middle}cal.	w^{lower}exp.	w^{lower}cal.
$C_{10}G_1$	0.04	0.001	6.37×10^{-6}	0.200	0.128	0.003	0.000775
water	0.461	0.000	0.01235	0.346	0.804	0.954	0.9746
octane	0.461	0.967	0.9598	0.322	0.00338	0.005	0.001028
butan-1-ol	0.038	0.032	0.027837	0.132	0.0634	0.038	0.023591

The study of the phase equilibrium in the quaternary systems was accompanied by dynamic light scattering experiments (Kahl, 1996; Häntzschel, 1999; Lippold, 1999). Figure 26.9 shows some results for the system $C_{10}G_1$ + water + *n*-octane + butan-1-ol. The average scattering intensities of surfactant containing phases are presented in dependence of the butan-1-ol feed mass fraction. The intensity increases rapidly with transition from the Winsor I two-phase to the three-phase region. Thus the turbidities that were to be observed by light scattering must be interpreted in connection with a lower consolute critical point which is characterized by large concentration fluctuations (Chu, 1991). The increase of scattering intensity would then simply be a manifestation of the critical opalescence observed also for simple fluids near their critical points. In order to decide if the rapid increasing is connected to the critical phenomena, model calculations were performed on the basis of the Landau functional (Eqs. (26.30) to (26.32)). The results are shown in Figure 26.9b. Comparing both diagrams leads to the conclusion that the rapid increase in light scattering intensity really corresponds with the transition from Winsor I to Winsor III caused by critical fluctuations (Burstyn, 1983). A tricritical point presumably exists not far from the covered region of state. All over the Winsor III region the average scattering intensities decrease rapidly in order to attain minimal values, if the system enters Winsor II region. Similar observations as in the system $C_{10}G_1$ + water + octane + phenol were made in the system MEGA-10 + water + phenol + *n*-octane before the middle-phase inversion occurs (Häntzschel et al., 1999b; Lippold, 1999).

The phenomenon of ultralow interfacial tension is well known and definitely correlated to the Winsor phase equilibrium pattern. Certainly, as two phases approach a critical

26.5 Results and Discussion

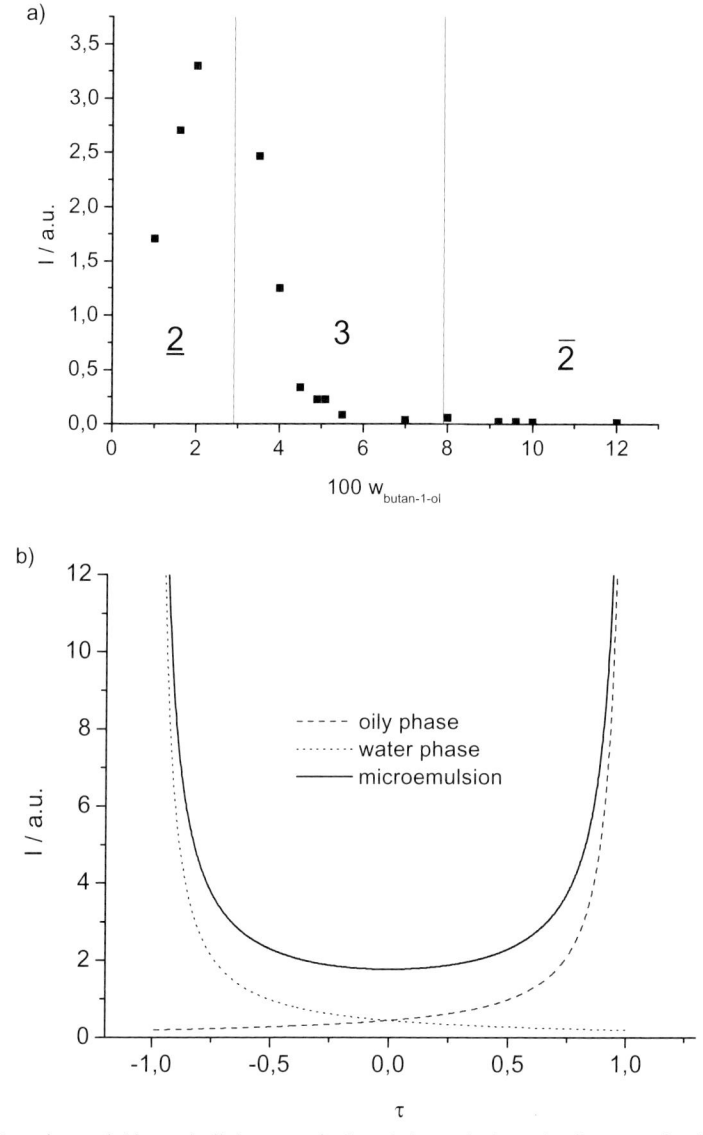

Figure 26.9: Experimental (dynamic light scattering) and theoretical results for scattering intensities in the system n-decyl-β-D-glucopyranosid + water + n-octane + butan-1-ol at 298 K; a) experimental; b) calculated using the Landau functional (Eqs. (26.30) to (26.32)).

state, the interfacial tension between them vanishes. Therefore, micellar solutions which solubilize the largest quantities of oil and water are expected to exhibit the lowest interfacial tensions. Interfacial tensions between the respective phases, experimentally determined at 298 K, are plotted against the butan-1-ol feed concentration in Figure 26.10a. When the butan-1-ol concentration in the feed is increased, the interfacial tension between the micellar

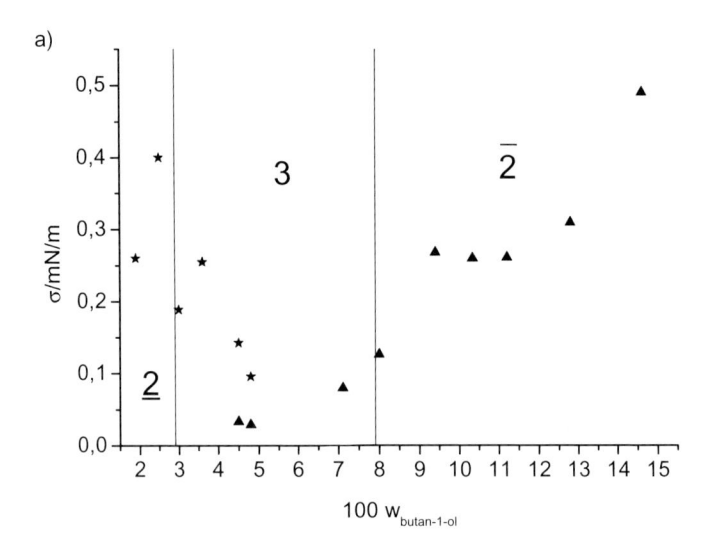

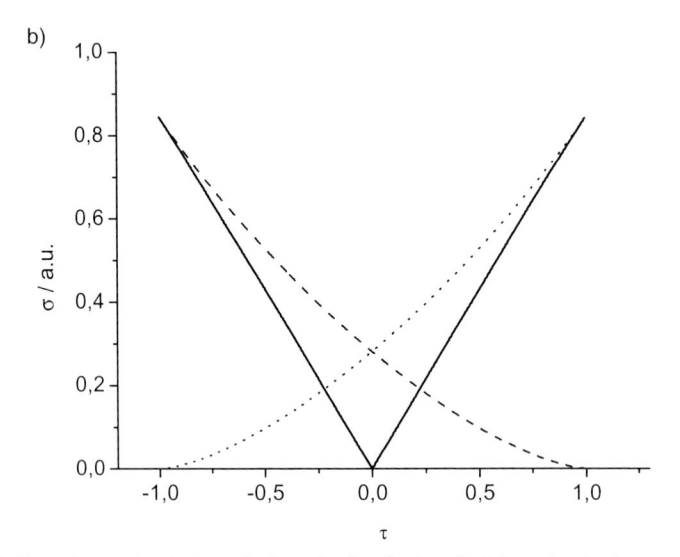

Figure 26.10: Experimental and theoretical results for the interfacial tension in the two- and three-phase states of the system n-decyl-β-D-glucopyranosid + water + n-octane + butan-1-ol at 298 K; a) experimental; b) calculated using the Landau functional (Eq. (26.34)); dotted line: interfacial tension between the microemulsion and excess water phase; dashed line: interfacial tension between the microemulsion and excess oil phase; solid line: interfacial tension between both excess phases.

solution (or the microemulsion) and the excess hydrocarbon phase decreases, whereas the interfacial tension between the micellar solution and the excess aqueous phase increases. The two curves appear to intersect inside the three-phase region. Predictions by the Landau-Ginzburg theory reveal only very small numbers for the interfacial tension as long as the chemical potentials of the components do not differ markedly from their value at the critical point (cf. Fig. 26.10b). Provided that such a condition is held and a middle-phase microemulsion is in equilibrium simultaneously with an oil-rich upper phase and a water-rich lower phase (Borzi et al., 1986), then the interfacial tension σ_{ow} of the oil-water interface is below the sum of the tensions σ_{om} and σ_{mw} of the oil-microemulsion and microemulsion-water interfaces. This implies that the microemulsion does not perfectly wet (spread at) the oil-water interface; rather the three phases meet with non-zero contact angles. From Cahn's theory of wetting follows (see Borzi et al., 1986) that close to any critical endpoint the microemulsion should spread at the interface between the other two phases. In the non-wet regime the interface presumably consists of a microscopically thin, concentrated layer of oriented surfactant (perhaps even a monolayer), while in the wetting regime it consists of a thick layer of the bulk microemulsion phase.

26.6 Conclusion

This study confirmed by experimental results and modeling the assumption that carbohydrate surfactants self-organize in aqueous solution predominantly to vesicular aggregates. This explains decisively, why such surfactants are not able to solubilize hydrocarbons (i.e. an oil) to an extent of practical interest. Nevertheless the investigations have given evidence too that n-decyl-β-D-glucopyranoside or 1-decanoyl-N-methylglucamide e.g. prove to be well suitable for the preparation of a microemulsion incorporating large amounts of water and oil if a polar organic solvent – an alkan-1-ol (carbon number 3 to 7) or phenol – is added as cosurfactant. Varying the cosurfactant content in the feed permits to produce various types of phase coexistence, in all of which a globular or bicontinuous microemulsion will appear in one of the phases. It has been found that the phase equilibrium pattern and the interfacial behavior of this kind of microemulsion forming mixture is not influenced crucially by moderate variations in the chain length of the lipophilic tail ($C_8 - C_{12}$) or the structure and the volume of (gluco- or maltopyranoside group) surfactant molecules. The specific amphiphilicity of a cosurfactant, however, was ruling the rise of phase equilibrium in an individual sequence and the corresponding partitioning of components between the phases. The thermodynamic approach of Nagarajan and Ruckenstein allows for the prediction of the specific aggregation behavior of a homologous series of carbohydrate surfactants in aqueous and aqueous-alcoholic solution, the formation of vesicles of definite size, the size distribution function of the vesicles, and the composition of mixed vesicles. The predictions for the influence of the concentration and the chain length of an alcohol on the critical micelle concentration of the carbohydrate surfactants agree satisfactory with experimental findings. A correct description of the liquid-liquid phase equilibrium data using this approach was possible in binary surfactant + water systems. In comparison with that the corresponding properties of ternary or quaternary systems containing carbohydrate surfactants could be modeled only with the semiempirical relations proposed by Koningsveld and Kleintjens.

Acknowledgement

The authors thank Priv.-Doz. Dr. M. Helmstedt, Institut für Experimentelle Physik I, Universität Leipzig, for gratefully supporting the light scattering experiments.

References

Aveyard, R.; Binks, B.P.; Chen, J.; Esquena, J.; Fletcher, P.D.I.; Buscall, R.; Davies, S. (1998): Surface and Colloid Chemistry of Systems Containing Pure Sugar Surfactant. Langmuir 14, 4699–4709.
Balzer, D. (1991): Alkylpolyglucosides their Physico-Chemical Properties and their Uses. Tenside Surf. Det. 28, 419–427.
Bogusz, S.; Venable, R.M.; Pastor, R.W. (2000): Molecular Dynamics Simulations of Octyl Glucoside Micelles: Structural Properties. J. Phys. Chem. B. 104, 5462–5470.
Borzi, C.; Lipowsky, R.; Widom, B. (1986): Interfacial Phase Transitions of Microemulsions. J. Chem. Soc., Faraday Trans. 2, 82, 1739–1752.
Burstyn, H.C.; Sengers, J.V.; Bhattacharjee, J.K.; Ferrell, R.A. (1983): Dynamic Scaling Function for Critical Fluctuations in Classical Fluids. Phys. Rev. A 28, 1567–1578.
Chu, B. (1991): Laser Light Scattering, 2nd ed., Academic Press, Boston.
Enders, S. (2000): Phasen- und Grenzflächenverhalten von komplexen fluiden Systemen. Habilitationsschrift, Universität Leipzig.
Enders, S.; Häntzschel, D. (1998): Thermodynamics of Aqueous Carbohydrate Surfactant Solutions. Fluid Phase Equilibria 153, 1–21.
Flory, P.J. (1953): Principles of Polymer Chemistry, Cornell Univ. Press, Ithaca.
Fukuda, K.; Söderman, O.; Lindman, B.; Shinoda, K. (1993): Microemulsions Formed by Alkyl Polyglucosides and an Alkyl Glycerol Ether. Langmuir 9, 2921–2925.
de Gennes, P.G.; Taupin, C. (1982): Microemulsions and the Flexibility of Oil/Water Interfaces. J. Phys. Chem. 86, 2294–2304.
Griffiths, R.B. (1974): Thermodynamic Model for Tricritical Points in Ternary and Quaternary Mixtures. J. Chem. Phys. 60, 195–206.
Hanatani, M.; Nishifuji, K.; Futani, M.; Tsuchiya, T. (1984): Solubilization and Reconstitution of Membrane Proteins of *Escherichia coli* Using Alkanoyl-N-methylgluconamid. J. Biochem. 95, 1349–1353.
Häntzschel, D. (1999): Phasenverhalten mikroemulsionsbildender Mischsysteme auf der Basis von Kohlenhydrattensiden. Dissertation, Universität Leipzig.
Häntzschel, D.; Schulte, J.; Enders, S.; Quitzsch, K. (1999a): Thermotropic and Lyotropic Properties of n-Alkyl-β-D-glucopyranoside Surfactants. Phys. Chem. Chem. Phys. 1, 895–904.
Häntzschel, D.; Enders, S.; Kahl, H.; Quitzsch, K. (1999b): Phase Behavior of Quaternary Systems Containing Carbohydrate Surfactants-Water-Oil-Cosurfactant. Phys. Chem. Chem. Phys. 1, 5703–5710.
Hildreth, J.E.K. (1982): N-D-Gluco-N-methylalkanamide Compounds, a New Class of Nonionic Detergents for Membrane Biochemistry. Biochem. J. 207, 363–366.
Hill, K.; v. Rybinski, W.; Stoll, G.; (1997): Alkyl Polyglycoside, Technology, Properties and Application, VCH, Weinheim, New York, Basel, Cambridge, Tokyo.
Hu, Y.; Prausnitz, J.M. (1988): Molecular Thermodynamics of Partielly-Ordered Fluids: Microemulsions. AIChE J. 34, 814–824.

Jones, M.N. (1992): Surfactant Interactions with Biomembrans and Proteins. Chem. Soc. Rev. Band 127–136.

Kahl, H. (1996): Mischphasenthermodynamische und grenzflächenchemische Untersuchungen an dem mikroemulsionsbildenden System n-Decyl-β-D-glucopyranosid/Wasser/n-Octan/Butan-1-ol. Dissertation, Universität Leipzig.

Kahl, H.; Quitzsch, K.; Stenby, E. (1996): Über das Phasengleichgewichtsverhalten des mikroemulsionsbildenen Systems n-Decyl-β-D-glucopyranosid/Wasser/n-Octan/Butan-1-ol. Festschrift zum 70. Geburtstag von Professor Konrad Bier, Karlsruhe. 1–15

Kahl, H.; Quitzsch, K.; Stenby, E.H. (1997): Phase Equilibria of Microemulsion Forming System n-Decyl-β-D-glucopyranoside/Water/n-Octane/1-Butanol. Fluid Phase Equilibria 139, 295–309.

Kahl, H.; Enders, S. (2002): Thermodynamics of Carbohydrate Surfactant Containing Systems. Fluid Phase Equilibria 194–197, 739–753.

Kahl, H.; Enders, S.; Quitzsch, K. (2001): Experimental and Theoretical Studies of the System n-Decyl-β-D-maltopyranoside + Water. Colloids and Surfaces A: Physicochem. and Eng. Aspects 183–185, 665–683.

Kahlweit, M.; Lessner, E.; Strey, R. (1983): Influence of the Properties of the Oil and the Surfactant on the Phase Behavior of Systems of the Type H_2O-Oil-Nonionic Surfactant. J. Phys. Chem. 87, 5032–5040.

Kahlweit, M.; Strey, R.; Busse, G. (1990): Microemulsions: A Qualitative Thermodynamic Approach. J. Phys. Chem. 94, 3881–3894.

Kahlweit, M.; Strey, R.; Aratono, M.; Busse, G.; Jen, J.; Schubert, K.V. (1991): Tricritical Points in H_2O-Oil-Amphiphile Mixtures. J. Chem. Phys. 95, 2842–2853.

Kahlweit, M.; Jen, J.; Busse, G. (1992): On the Stability of Microemulsions, I. Langmuir 7, 2928–2936.

Kahlweit, M.; Busse, G.; Faulhaber, B. (1995): Preparing Microemulsions with Alkyl Monoglucosides and the Role of n-Alkanols. Langmuir 11, 3382–3387.

Kahlweit, M.; Busse, G.; Faulhaber, B. (1996): Preparing Nontoxic Microemulsions With Alkyl Monoglucosides and the Role of Alkanediols as Cosolvents. Langmuir 12, 861–862.

Kahlweit, M.; Busse, G.; Faulhaber, B. (1997): Preparing Nontoxic Microemulsions. 2. Langmuir 13, 5249–5251.

Kaufman, M.; Griffiths, R.B. (1982): Thermodynamic Model for Tricritical Mixtures with Application to Ammonium Sulfate + Water + Ethanol + Benzene. J. Chem. Phys. 76, 1508–1524.

Kilpatrick, P.K.; Gorman, C.A.; Davis, H.T.; Scriven, L.E.; Miller, W.C. (1986): Patterns of Phase Behavior in Ternary Ethoxylated Alcohol-n-Alkane-Water Mixtures. J. Phys. Chem. 90, 5292–5299.

Kiwada, H.; Niimura, H.; Fujisaki, Y.; Yamada, S.; Kato, Y. (1985): Application of Synthetic Alkyl Glycoside Vesicles as Drug Carriers. I. Preparation and Physical Properties. Chem. Pharm. Bull. 33, 753–759.

Kleinert, H. (1986a): The Neighborhood of the Three-Phase Regime in Microemulsions. J. Chem. Phys. 85, 4148–4152.

Kleinert, H. (1986b): Microemulsions in the Three-Phase Regime. J. Chem. Phys. 84, 964–968.

Koningsveld, R.; Kleintjens, L. (1971): Liquid-Liquid Phase Separation in Multicomponent Polymer Systems. X. Concentration Dependence of the Pair-Interaction Parameter in the System Cyclohexane-Polystyrene. Macromolecules 4, 637–641.

Kutschmann, E.M.; Findenegg, G.H.; Nickel, D.; von Rybinski, W. (1995): Interfacial-Tension of Alkylglycosides in Different APG/Oil/Water Systems. Colloid & Polym. Sci. 273, 565–571.

Lawrence, M.J.; Rees, G.D. (2000): Microemulsion-Based Media as Novel Drug Delivery Systems. Adv. Drug Deliv. Rev. 45, 89–121.

Lippold, H.; Findeisen, M.; Quitzsch, K.; Helmstedt, M. (1998): Micellar Incorporation Without Solubilizing Effect. A Study on the System Water-Phenol-Decanoyl-N-Methylglucamide. Colloids and Surfaces A: Physicochem. and Eng. Aspects 135, 235–244.

Lippold, H. (1999): Untersuchungen zur mizellaren und hemimizellaren Inkorporation am Beispiel des Systems Wasser/Phenol/1-Desoxy-1-N-methyldecanamido-D-sorbit/Silicagel. Dissertation, Universität Leipzig.

Lippold, H.; Quitzsch, K. (2000): Simultaneous Surfactant Aggregation in Aqueous Solution and at the Solid-Liquid Interface. Coadsorption of Decanoyl-N-methylglucamide and Phenol on Silica Gel. Colloids and Surfaces A: Physicochem. and Eng. Aspects 172, 1–6.

Matsumura, S.; Imai, K.; Yoshikawa, S.; Kawada, K.; Uchibori, T. (1991): Surface Activities, Foam Suppression, Biodegradability and Antimicrobial Properties of s?-Alkyl Glucopyranosides. J. Japan. Oil Chem. Soc. 40, 709–714.

Mukerjee, P. (1972): The Size Distribution of Small and Large Micelles: A Multiple Equilibrium Analysis. J. Phys. Chem. 76, 565–570.

Nagarajan, R.; Ruckenstein, E. (2000): Self-Assembled systems, in Sengers, J.V.; Kayser, R.F.; Peters, C.J.; White, H.J. (Eds.) Equation of State for Fluids and Fluid Mixtures, Elsevier, 589–749.

Nanna, J.A.; Xia, J. (Eds.) (2001): Protein-Based Surfactants. Synthesis, Physicalchemical Properties, and Applications. Surfactant Science Series No. 101. Marcal Dekker, Inc., New York and Basel.

Ngiruwonsanga, Th. (2000): Phasenzustände und Phasenkoexistenz im Zusammenhang mit Strukturbildung und -wandlung in Mehrkomponentensystemen aus Soja Phosphatidylcholin, Wasser, Alkan-1-ol und einem Ölbestandteil. Dissertation, Universität Leipzig.

Ngiruwonsanga, T.; Quitzsch, K.; Findeisen, M.; Winkler, U.; Kärger, J. (2001): Self-Aggregation of a Native Soybean Phosphatidylcholin in n-Butanol. Tenside Surf. Det. 38, 111–115:

Pfüller, U. (1986): Mizellen, Vesikel, Mikroemulsionen. Verlag Volk und Gesundheit, Berlin.

Provencher, S.W. (1976): An Eigenfunction Expansion Method for the Analysis of Exponential Decay Curves. J. Chem. Phys. 64, 2772–2777.

zu Putlitz, B.; Landfester, K.; Förster, S.; Antonietti, M. (2000): Vesicle-Forming Single-Tail Hydrocarbon Surfactants with Sulfonium Headgroup. Langmuir 16, 3003–3005.

Puvvada, S.; Blankschtein, D. (1990): Molecular-Thermodynamic Approach to Predict Micellization, Phase Behavior and Phase Separation of Micellar Solutions. I: Application to Nonionic Surfactants. J. Chem. Phys. 92, 3710–3724.

Puvvada, S.; Blankschtein, D. (1992a): Thermodynamics Description of Micellization, Phase Behavior, and Phase Separation of Aqueous Solutions of Surfactant Mixtures. J. Phys. Chem. 96, 5567–5579.

Puvvada, S.; Blankschtein, D. (1992b): Theoretical and Experimental Investigations of Micellar Properties of Aqueous Solutions Containing Binary Mixtures of Nonionic Surfactant. J. Phys. Chem. 96, 5579–5592.

Rao, I.V.; Ruckenstein, E. (1986): Micellization Behavior in the Presence of Alcohols. J. Colloid Interface Sci. 113, 375–387.

Regev, O.; Guillemet, F. (1999): Various Bilayer Organizations in a Single-Tail Nonionic Surfactant: Unilamellar Vesicles, Multilamellar Vesicles, and Flat-Stacked Lamellae. Langmuir 15, 4357–4364.

Rudolph, E.S.J.; Cacao Pedroso, M.A.; de Loos, T.W.; de Swaan Arons, J. (1997): Phase Behavior of Oil + Water + Nonionic Surfactant Systems for Various Oil-to Water Ratios and the Representation by a Landau-Type Model. J. Phys. Chem. B. 101, 3914–3918.

Ryan, L.D.; Kaler, E.W. (1997): Role of Oxygenated Oils in n-Alkyl-β-D-monoglucoside Microemulsion Phase Behavior. Langmuir 13, 5222–5228.

Ryan, L.D.; Schubert, K.V.; Kaler, E.W. (1997): Phase Behavior of Microemulsions Made with n-Alkyl Monoglucosides and n-Alkyl Polyglycol Ethers. Langmuir 13, 1510–1518.

von Rybinski, W.; Hill, K. (1998): Alkyl Polyglycosides – Properties and Applications of a New Class of Surfactants. Angew. Chem. Int. Ed. 37, 1328–1345.

von Rybinski, W.; Guckenbiehl, B.; Tesmann, H. (1998): Influence of Co-Surfactants on Microemulsions with Alkyl Polyglycosides. Colloids and Surfaces A: Physicochem. and Eng. Aspects 142, 333–342.

Schulman, J.H.; Stoeckenius, W.; Prince, L.M. (1959): Mechanism of Formation and Structure of Microemulsions by Electron Microscopy. J. Phys. Chem. 63, 1677–1680.

Schulte, J. (1998): Rheologische Charakterisierung von Kohlenhydrattensiden in Mehrkomponentensystemen. Dissertation, Universität Leipzig.

Schulte, J.; Enders, S.; Quitzsch, K. (1999): Rheological Studies of Aqueous Alkylpolyglucoside Surfactant Solutions. Colloid & Polym. Sci. 277, 827–836.

Sørensen, J.M.; Arlt, W. (1979): Liquid-Liquid Equilibrium Data Collection Bd. 1, Binary Systems. DECHEMA, Behrens, D.; Eckermann, R. (Eds.); Vol. V, Part I, Frankfurt/Main.

Solc, K. (1969): Limiting Polydispersity of a Fraction Separated from a Polymer Solution. Collec. Czech. Chem. Commun. 34, 992–1001.

Stubenrauch, C.; Kutschmann, E.M.; Paeplow, B.; Findenegg, G.H. (1996): Phase Behavior of the Quaternary System Water-Decane-Decyl Monoglucoside-Decanol. Tenside Surf. Det. 33, 237–241.

Stubenrauch, C.; Paeplow, P.; Findenegg, G.H. (1997): Microemulsions Supported by Octyl Monoglucoside and Geraniol. 1. The Role of the Alcohol in the Interfacial Layer. Langmuir 13, 3652–3658.

Sulthana, S.B.; Rao, P.V.C.; Bhat, S.G.T.; Nakano, T.Y.; Sugihara, G.; Rakshit, A.K. (2000): Solution Properties of Nonionic Surfactants and Their Mixtures: Polyoxyethylene (10) Alkyl Ether [C_nE_{10}] and MEGA-10. Langmuir 16, 980–987.

Talmon, Y. (1996): Transmission Electron Microscopy of Complex Fluids: The State of the Art. Ber. Bunsen-Ges. Phys. Chem. 100, 364–372.

Talmon, Y.; Prager S. (1977): Statistical Mechanics of Microemulsions. Nature 267, 333–335.

Widom, B. (1984): Lattice-Gas Model of Amphiphiles and of Their Orientation at Interface. J. Phys. Chem. 88, 6508–6514.

Widom, B. (1986): A Model Microemulsion. J. Chem. Phys. 81, 1030–1046.

Widom, B. (1996): II. Theoretical Modeling: An Introduction. Ber. Bunsen-Ges. Phys. Chem. 100, 242–251.

Winsor, P.A. (1948): I. Hydrotropy, Solubilization, and Related Emulsification Processes. IV. Solubilization with Miscellaneous Amphiphiles. Trans. Faraday Soc. 44, 376–398, 451–471.

Winsor, P.A. (1954): Solvent Properties of Amphiphilic Compounds. Butterworths, London.

Zana, R. (1995): Aqueous Surfactant-Alcohol Systems: A Review. Advances in Colloid and Interfaces Science 57, 1–64.

Nomenclature

1	one-phase system
$\underline{2}$	Winsor I system (two phases, surfactant in the lower phase)
$\overline{2}$	Winsor II system (two phases, surfactant in the upper phase)
3	Winsor III system (three phases, surfactant normally in the middle phase)
A, B	types of a site in a lattice model
a	area of a surfactant head group
C	cosurfactant or interaction parameter (cf. Eq. (26.3))
C_8G_1	n-octyl-β-D-glucopyranoside
C_9G_1	n-nonyl-β-D-glucopyranoside
$C_{10}G_1$	n-decyl-β-D-glucopyranoside
$C_{12}G_1$	n-dodecyl-β-D-glucopyranoside
$C_{10}G_2$	n-decyl-β-D-maltopyranoside
f	Landau free energy functional
G	Gibbs energy (free enthalpy)
g	number of molecules in a micelle or vesicle
I	light scattering intensity
l	length of a hydrocarbon chain
M	molar mass or moment of the distribution function
MEGA-10	1-decanoyl-N-methylglucamide
N	number of moles or molecules

26 Liquid-Liquid Phase Equilibria in Microemulsion Forming Systems

n	number of carbon atoms in a hydrocarbon chain
O	oil
R	universal gas constant
r	number of segments in a molecule
S	surfactant
T	temperature
V	volume
W	water
w	mass fraction
X	molar fraction
x	Landau order parameter
y	segment molar fraction
z	field variable (Landau functional)

Greek letters

α	molecular fraction of surfactant in mixed micellar aggregates
β	parameter (cf. Eqs. (26.24), (26.25))
Δ	difference
γ	interaction parameter (cf. Eq. (26.3))
ϕ	phase symbol
μ	chemical potential resp. molar Gibbs energy
ρ	density
σ	interfacial tension
τ	Landau temperature (or other intensity parameter)

Subscripts

agg	aggregate core water interface
C or c	cosurfactant
d	domain
def	deformation
f	formation
g	aggregate size number
i, j, k	component indices
1	monomer
ideal	indicating a property of an ideal liquid mixture
int	interface
N	number-average property
o	oil
p	polar head group
pol	polar
pot	potential (Landau free energy functional)
res	residual property concerning interactions
S or s	surfactant
ster	steric
t	tail

tr	transfer
W	water
w	mass-average property
z	z-average property

Superscripts

E	excess property
i, j	integer of observables (Landau theory)
M	mixing property
o	standard state property

27 Interactions of Polyelectrolytes and Zwitterionic Surfactants in Aqueous Solution

Heinz Hoffmann[1], Holger Lauer, and Klaus Redlich

27.1 Abstract

The properties of aqueous solutions of (modified) polyacrylic acids (PAA) and the zwitterionic surfactant tetradecyldimethylaminoxide ($C_{14}DMAO$) were investigated with various experimental methods including rheology, surface tension measurements, and electric birefringence (EBF). In aqueous mixtures of PAA and $C_{14}DMAO$ the viscosity strongly depends on the pH. At pH < 6.9 mixtures of 1 mass% PAA and 100 mM $C_{14}DMAO$ phase separate with a solid, white precipitate. At pH values slightly above 6.9 highly viscoelastic solutions occur. The solutions show viscosities up to 100 Pa s and the storage moduli are of the order of about 50 Pa. By increasing the pH drastic changes in the behavior of the solutions are observed. The viscosity decreases about four orders of magnitude down to 0.01 Pa s and the solutions loose their viscoelasticity. At pH ≥ 8 no interactions between the polyelectrolyte and the surfactant can be observed. The viscosity is the same as in the surfactant-free polyelectrolyte solutions. In addition the surface tension of $C_{14}DMAO$ in water and in an 1 mass% aqueous PAA solution is the same, both below and above the CMC (critical micelle concentration). The surfactant exists as zwitterionic species and therefore no interactions with the negatively charged polyelectrolyte occur.

The behavior of the mixtures can be explained by the protonation of the surfactant at pH < 8. Mixed micelles of mainly zwitterionic and weakly protonated species are formed. The protonated molecules in the micelles interact with the carboxylate groups of the polyelectrolyte and a three dimensional network with highly viscoelastic properties is formed in which the micelles are interconnected by the polyelectrolyte. At pH < 6.9 the degree of protonation of the surfactant and therefore the interactions become even stronger and a water insoluble polyelectrolyte-surfactant complex precipitates.

The second part of this study deals with modified PAAs which were used in unprotonated form. Investigations on modified PAA + $C_{14}DMAO$ mixtures showed a change in viscosity only at the transition of spherical to rod-like micelles. Therefore no interactions between the polyelectrolyte in its totally titrated form and $C_{14}DMAO$ are present in solution. The modified PAAs behave similarly as the non-modified compounds. The surface tension

---
[1] Universität Bayreuth, Physikalische Chemie I, Universitätsstr. 30, 95447 Bayreuth.

curves of the mixtures never cross the curve of the pure surfactant. It can be concluded that interactions between the polyelectrolytes and the surfactant monomers are absent.

Also relative to the addition of HCl the modified PAAs show the same behavior as the non-modified compounds.

27.2 Introduction

Within this project the interactions of (modified) polyelectrolytes of different molecular mass and neutralization degrees (e.g. polyacrylic acid, polymethacrylic acid, copolymers of acrylic and maleic acid and of acrylic acid and olefins) and zwitterionic surfactants in aqueous solution have been studied with various physical-chemical methods. The binding of surfactant molecules and of micelles to polyelectrolytes and the structure of the polyelectrolyte-surfactant complexes have been examined. The influence of the nature of the components and of parameters like concentration, pH value, or ionic strength was studied. In addition the influence of these complexes on the macroscopic behavior of the systems was examined. From the experimental results a model for complex formation between zwitterionic surfactants and polyelectrolytes has been developed.

27.3 Experimental/Methods

Within the scope of the project examinations by means of surface and interfacial tension measurements, pH-titration, conductivity, static light scattering, electric birefringence, and rheology were carried out on the following systems:

- I: Polyacrylic acids (PAA) of molecular mass ranging from 1000–250,000 and different degrees of neutralization (fixed by addition of NaOH) in aqueous solution without as well as with the addition of the zwitterionic surfactant tetradecyldimethylaminoxide (C_{14}DMAO). The abbreviations of these samples include the molecular mass of the polymer, which was determined by light scattering experiments (e.g. PAA 230 characterizes a polyacrylic acid with a molecular mass of 230,000).

- II: Modified polyacrylic acids (Sokalanes) and copolymers of maleic acid and olefin components in aqueous solution and with addition of C_{14}DMAO. The composition of these polyelectrolytes, which were made available by BASF/Germany, are given in Table 27.1. We investigated the aqueous solutions of the sodium salts of the completely neutralized acids. The aqueous solutions show pH values between 9 and 10; changes in the pH value were realized by addition of HCl.

Table 27.1: Classifying of the Sokalanes.

Product name	Composition	Concentration mass%	M [a]	M [b]
Sokalan CP 9	maleic acid diisobutylene	25.7	12,000	100,000
Sokalan PM 10 I	hydrophobically modified PAA	47.4	–	4300
Sokalan CP 2	maleic acid/methyl vinyl ether	37.9	70,000	92,000
Sokalan CP 5	maleic acid/acrylic acid	45.5	70,000	55,000

a) manufacturer specification
b) experimental determination by light scattering

27.4 Results and Discussion

27.4.1 Aqueous Solutions of Polyacrylic Acids

Examinations on PAAs showed that the titration curves of the aqueous solutions have a higher slope than that of the monomeric acrylic acid due to charge interactions. The titration curve of the multibasic acid PAA should be considered as a superposition of titration curves of compounds with different pK_a values (cf. Fig. 27.1). Moreover the polyelectrolytes exist in a coiled conformation and the carboxylic groups are therefore better accessible for NaOH. Addition of salt (NaCl) shifts the titration curves to slightly lower pH values because the activity of the H^+-ions is increased. Moreover the NaCl screens the charges of the carboxyl groups and therefore polyacrylic acid can be neutralized more easily by NaOH.

By measuring the zero shear viscosity as function of the polyelectrolyte concentration two typical overlap concentrations c_1 and c_2 can be detected (cf. Fig. 27.2). At c_1 the ion clouds of separate polyelectrolyte molecules start to interact. With further increasing concentration the coiled polymer molecules finally start to overlap sterically at c_2 which is associated with a strong increase of viscosity. By increasing the degree of neutralization c_1 shifts to lower and c_2 to higher concentrations which can be explained by an increase in inter- and intramolecular charge interactions.

With increasing pH the zero shear viscosity passes through a maximum (plateau) which reflects the conformational changes of the PAA with increasing degree of neutralization (cf. Fig. 27.3).

27.4 Results and Discussion

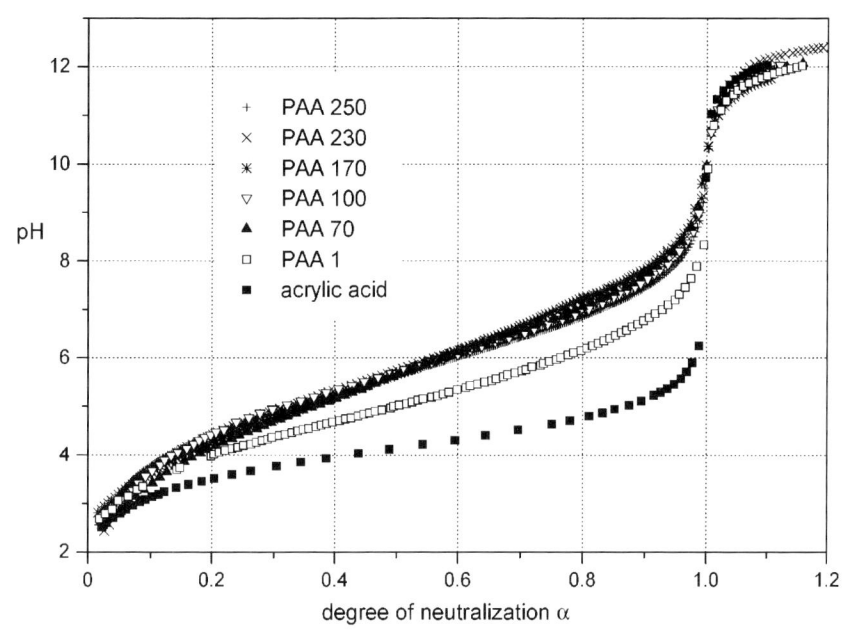

Figure 27.1: Titration curves of aqueous solutions of different polyacrylic acids (each 1 mass%, 139 mM) with 1 M NaOH.

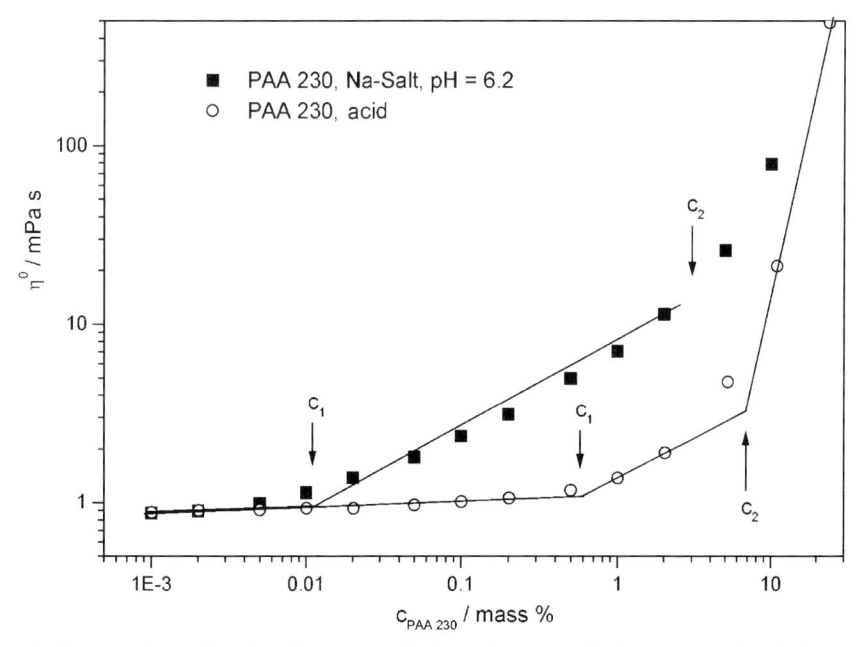

Figure 27.2: Zero shear viscosity of aqueous solutions of PAA 230 in the protonated and deprotonated form as function of the polymer concentration.

635

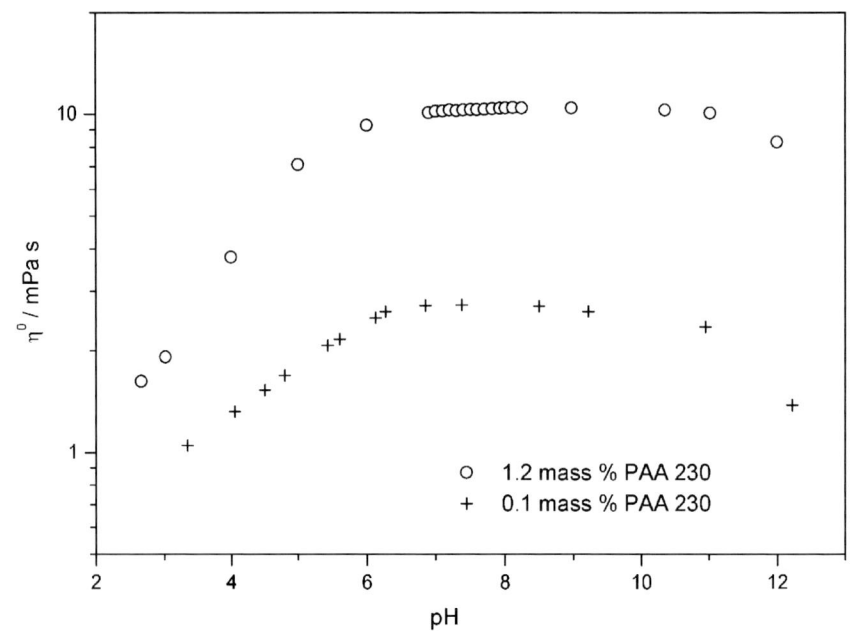

Figure 27.3: Zero shear viscosity of aqueous solutions of PAA 230 as function of the degree of protonation.

27.4.2 Aqueous Solutions of C_{14}DMAO

The CMC of C_{14}DMAO is determined to be 0.14 mM, investigated by surface tension measurements of aqueous solutions (cf. Fig. 27.4). All other samples discussed in this section have C_{14}DMAO concentrations of two orders of magnitude above the CMC.

A solution of 1 mass% C_{14}DMAO has a pH of 8.39 which indicates the basic character of the negatively charged oxygen. Addition of an equimolar amount of NaCl shifts the titration curve to higher pH, resulting in a slightly increased pK_a value (cf. Fig. 27.5). The positive charges of the protonated head groups are screened by the salt, which facilitates further protonation (Maeda et al., 1997).

The effect of added salt is much more significant in case of the zero shear viscosity. For the salt-free samples it remains constant in the range of water viscosity, independent of the degree of protonation. By addition of an equimolar amount of NaCl the zero shear viscosity passes through a maximum when it is plotted as a function of the degree of protonation. At the maximum the viscosity is three orders of magnitude higher than for the completely protonated or deprotonated species (cf. Fig. 27.6).

This result can be explained by assuming a depression of intermolecular charge interactions due to charge screening. This leads to the growth of rod-like micelles. The increased viscosity is therefore a sign for the generation of a network existing of rod-like micelles.

The interfacial tension between the pure surfactant solution and n-decane passes through a minimum when plotted as function of the degree of protonation, which is a hint

27.4 Results and Discussion

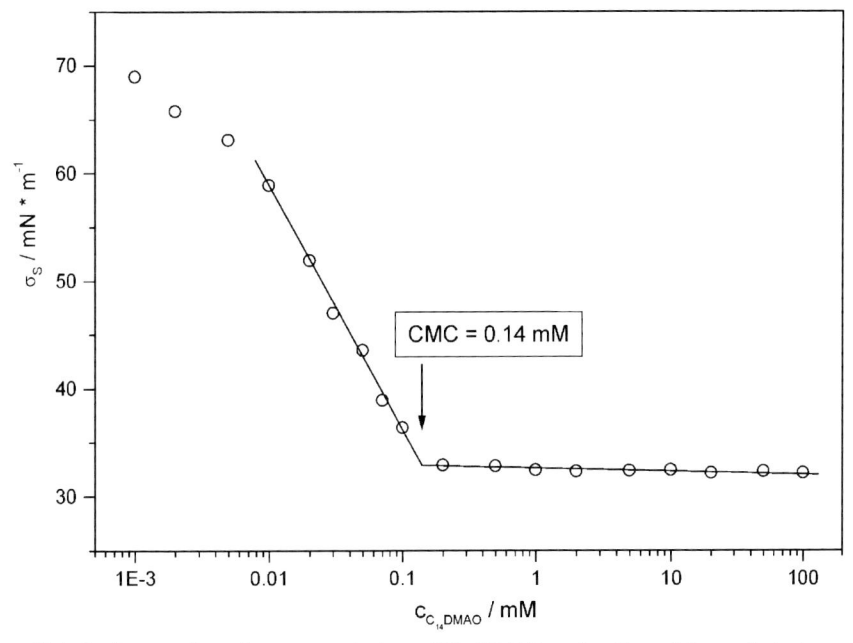

Figure 27.4: Surface tension of aqueous solutions of C_{14}DMAO as function of the surfactant concentration.

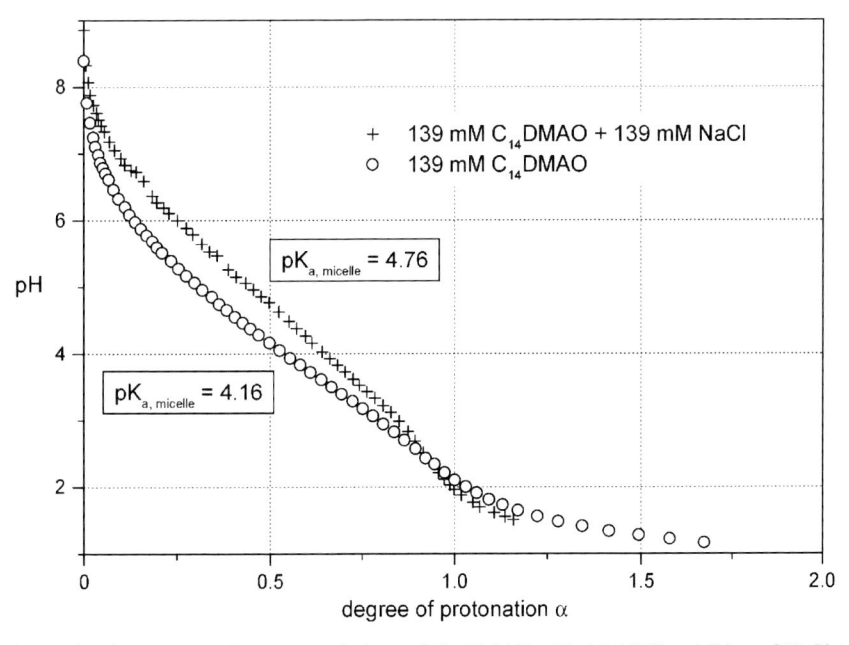

Figure 27.5: Titration curves of aqueous solutions of C_{14}DMAO with 1 M HCl; addition of NaCl to increase the pH.

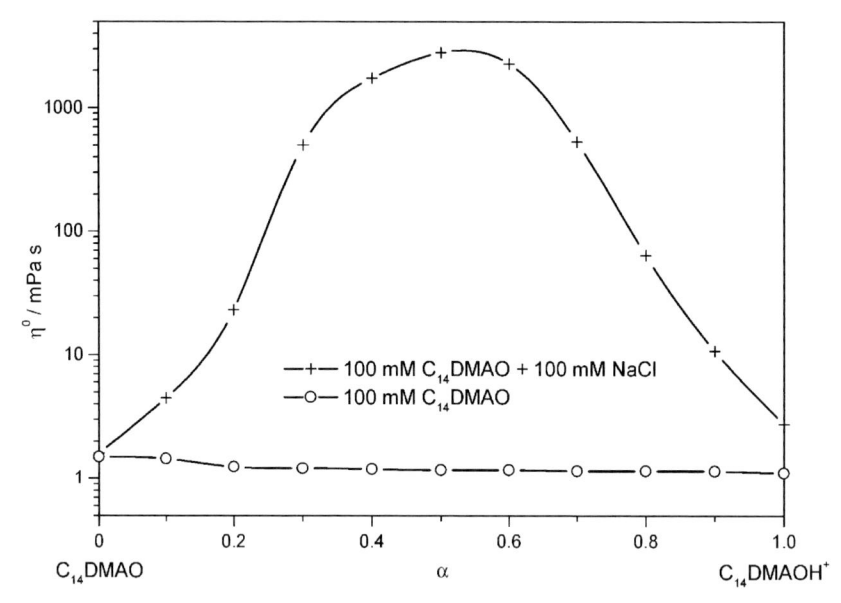

Figure 27.6: Zero shear viscosity of aqueous solutions of C_{14}DMAO (100 mM) without and with 100 mM NaCl as function of the degree of protonation.

for a decrease of the space requirement of the surfactant head group and the formation of synergistical effects between protonated and deprotonated species (cf. Fig. 27.7).

27.4.3 Interactions between PAA and C_{14}DMAO in Aqueous Solution

The titration curve of an aqueous solution of a mixture of PAA and C_{14}DMAO runs at pH < 8 above the titration curve of the aqueous solution of PAA, i.e., the addition of C_{14}DMAO increases the pH of a PAA solution (cf. Fig. 27.8). It can be concluded that C_{14}DMAO is protonated by PAA. At pH < 6.9 a white precipitate is formed which was identified as a polyelectrolyte-surfactant complex.

The course of the zero shear viscosity as function of the surfactant concentration in aqueous solutions of PAA + C_{14}DMAO mixtures indicates that at pH ≥ 8 no interactions between PAA and C_{14}DMAO occur. The zero shear viscosity increases at 30 mM C_{14}DMAO, which is only a result of the formation of rod-like micelles (cf. Fig. 27.9).

In the same way the Kerr constant B which is determined by electric birefringence measurements increases at 30 mM C_{14}DMAO (cf. Fig. 27.10).

At pH ≥ 8 the results of surface tension measurements are the same, both for the mixtures and the pure surfactant solutions (cf. Fig. 27.11). This indicates that there aren't any interactions neither between the monomer surfactant nor the micelles and the polyelectrolyte.

At pH < 7.5 highly viscoelastic solutions are observed. That is a hint for strong interactions between PAA and C_{14}DMAO. Figure 27.12 shows the results of rheological measure-

27.4 Results and Discussion

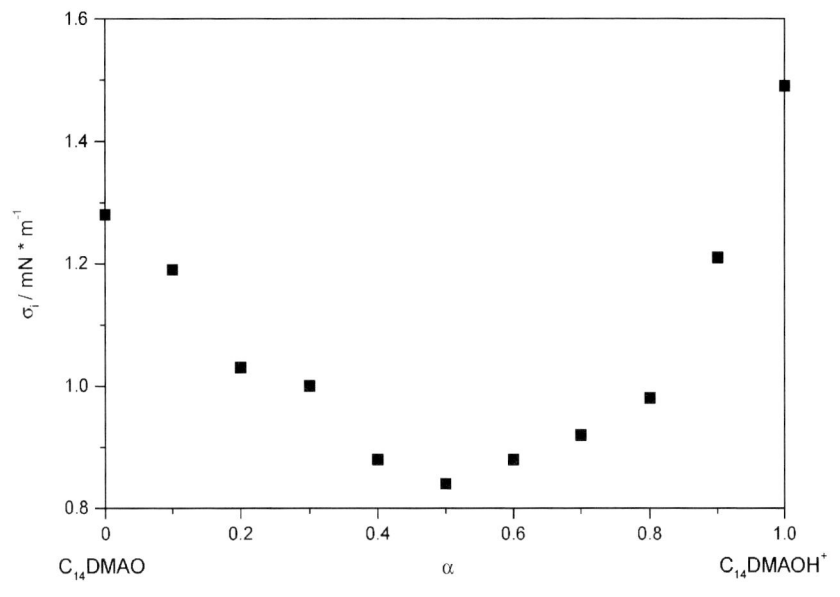

Figure 27.7: Interfacial tension between an aqueous solution of C_{14}DMAO (100 mM) and *n*-decane as function of the degree of protonation.

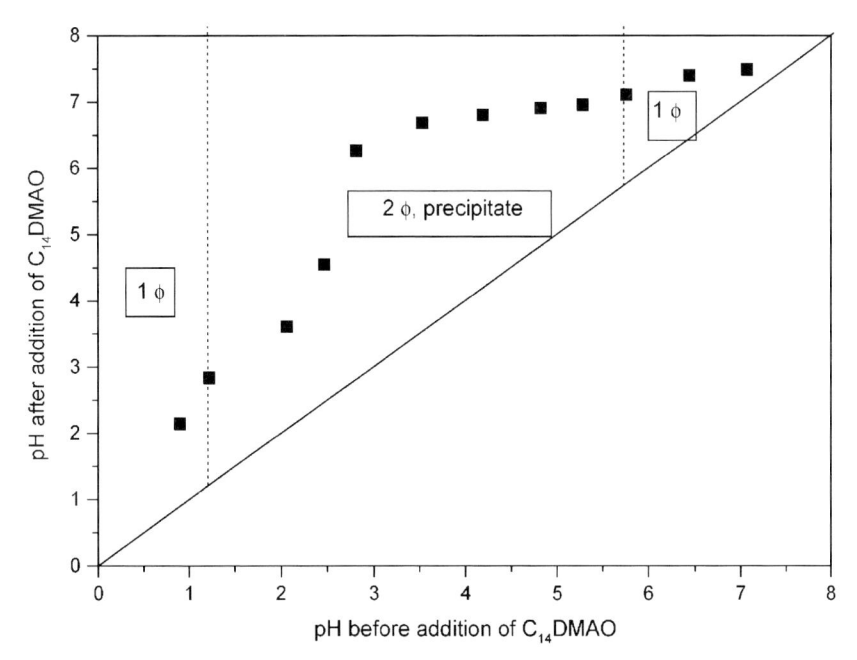

Figure 27.8: pH-shift caused by the addition of 100 mM C_{14}DMAO to a 1 mass% aqueous solution of PAA 230.

639

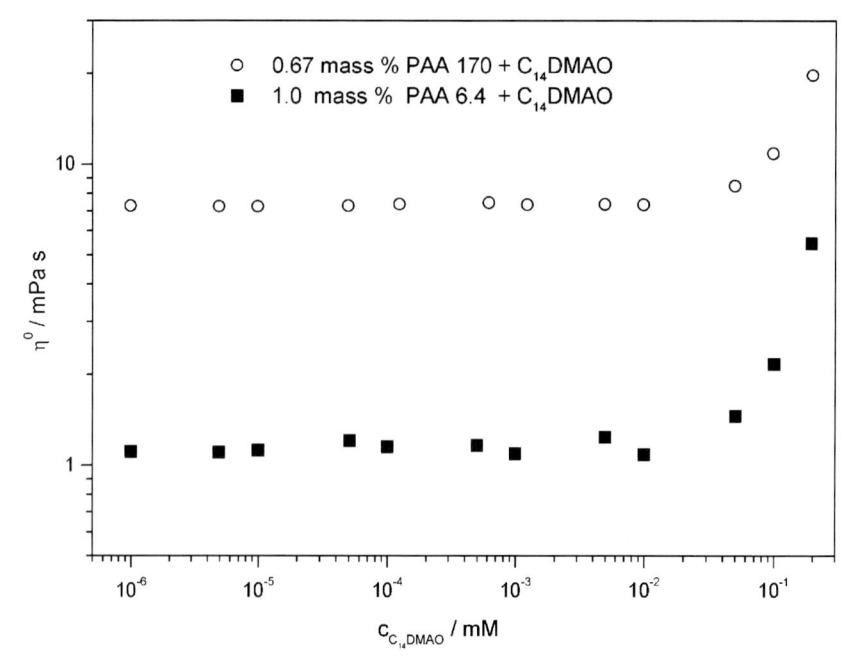

Figure 27.9: Zero shear viscosity of aqueous solutions of PAAs and C$_{14}$DMAO as function of the surfactant concentration at pH = 8.0.

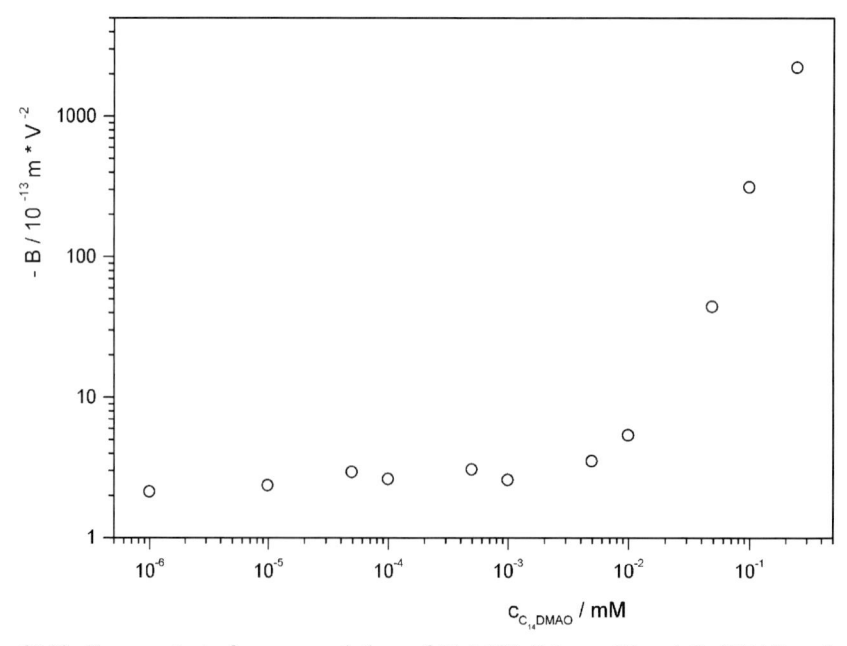

Figure 27.10: Kerr constant of aqueous solutions of PAA 170 (0.1 mass%) and C$_{14}$DMAO as function of the surfactant concentration at pH = 9.0.

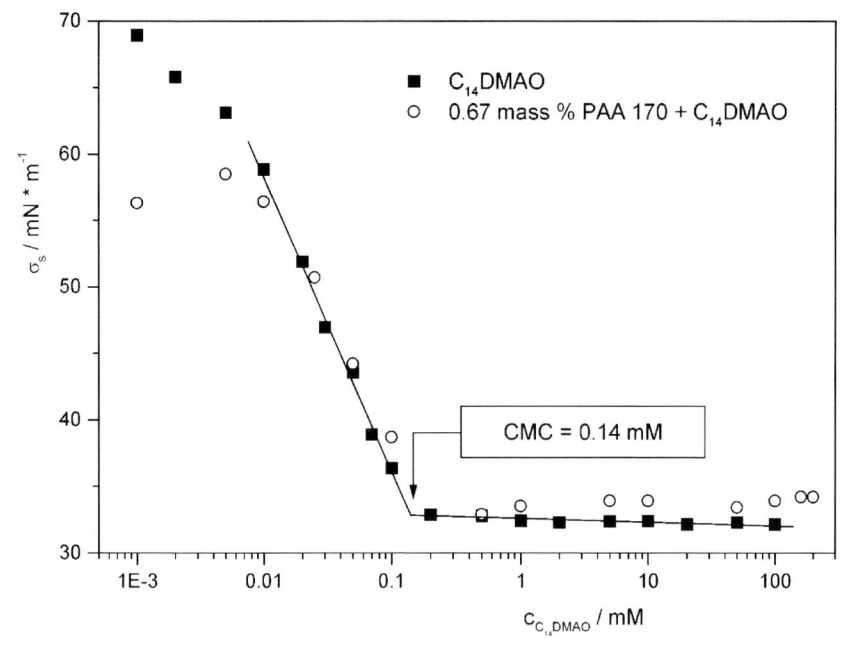

Figure 27.11: Influence of PAA 170 on the surface tension of aqueous solutions of C_{14}DMAO at pH = 8.0.

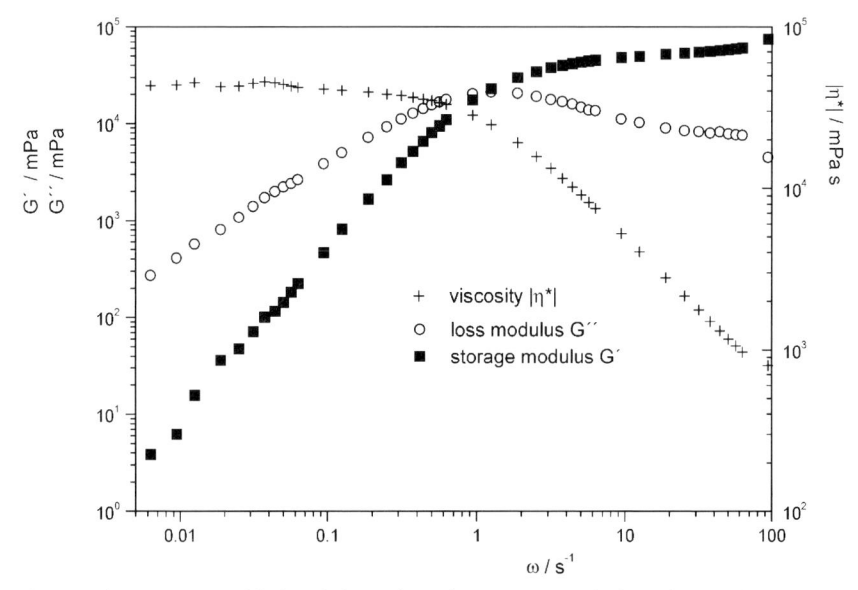

Figure 27.12: Rheogram at oscillating deformation of an aqueous solution of PAA 230 (1 mass%) and 100 mM C_{14}DMAO at pH = 7.05.

ments with oscillating deformation of an aqueous solution containing 1 mass% PAA 230 and 100 mM C$_{14}$DMAO at pH = 7.05.

When the pH is decreased from 7.5 to 6.9 the zero shear viscosity increases four orders of magnitude to 100 Pa s, whereas the viscosity of the surfactant-free polymer solution remains constant (cf. Fig. 27.13).

Interactions develop between the negatively charged PAA and the partly protonated C$_{14}$DMAO at these pH values. Therefore a network of polyelectrolyte and rod-like micelles is formed which leads to the strong viscosity enhancement.

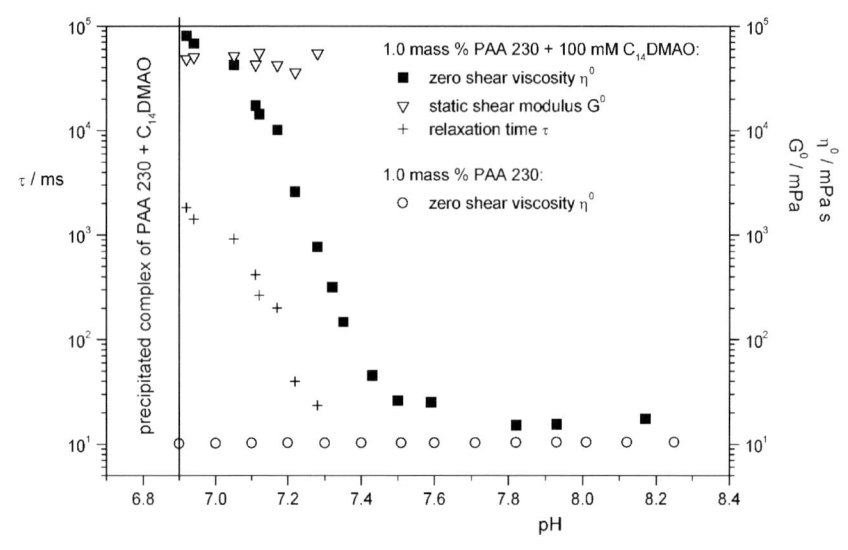

Figure 27.13: Rheological data of an aqueous solution of PAA 230 (1 mass%) and 100 mM C$_{14}$DMAO as function of pH; data for the surfactant-free solution are displayed for comparison.

27.4.4 Modified PAAs in Aqueous Solution

Examinations on Sokalan solutions (and on completely neutralized PAA) showed that the pH value first increases, runs through a maximum and finally decreases again with increasing polyelectrolyte concentration (cf. Fig. 27.14).

The pH increase is based on hydrolysis of the Na-salts of the weak carboxylic acid. The decrease of pH at higher concentrations can be understood by an increased coiling of the macroions at a characteristic concentration, caused by the influence of intermolecular repulsive interactions. This leads to stronger binding of the Na-ions to the carboxylate groups and to decreasing hydrolysis.

Rheological experiments show that a low concentrated polyelectrolyte solution behaves as a Newtonian liquid whose zero shear viscosity is in the range of water. With increasing polyelectrolyte concentration there is a sudden increase in viscosity at two characteristic concentrations c_1 and c_2. The slope at $c > c_2$ is very high and the same for all examined polyelectrolytes (cf. Fig. 27.15).

27.4 Results and Discussion

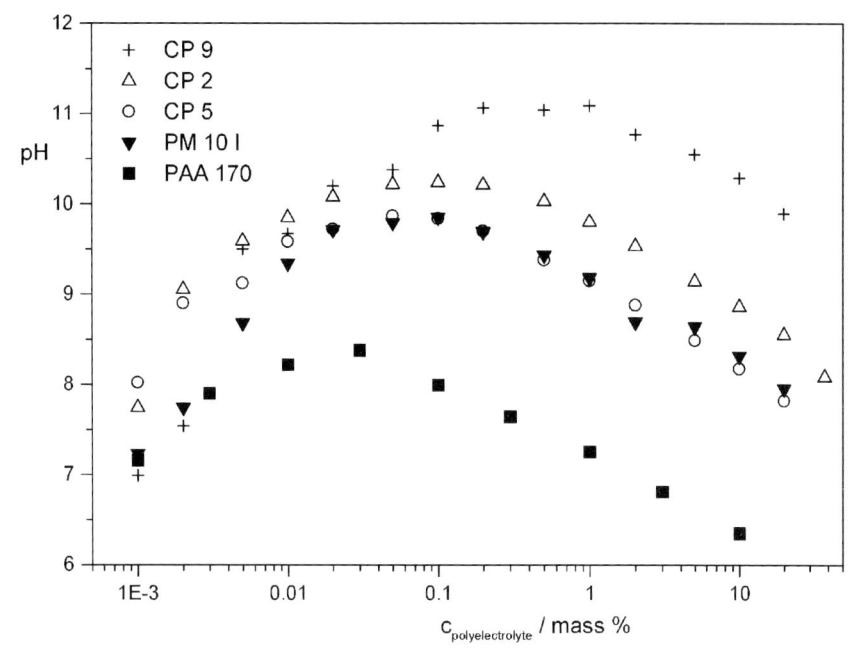

Figure 27.14: pH of aqueous solutions of different polyelectrolytes as function of the concentration.

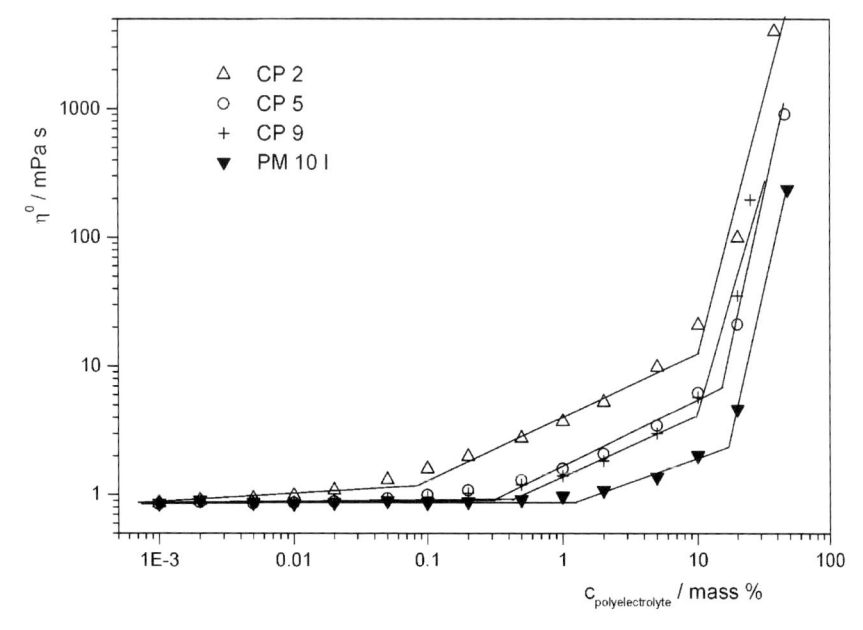

Figure 27.15: Zero shear viscosity of aqueous solutions of different Sokalanes as function of the polyelectrolyte concentration.

As mentioned in Section 27.4.1 c_1 can be identified as the overlap concentration of the ion clouds of the stretched macroions, while c_2 is the overlap concentration of the coiled polyelectrolytes. These characteristic concentrations are given in Table 27.2 for the examined polyelectrolytes.

Table 27.2: Overlap concentrations c_1 and c_2 of Sokalanes in aqueous solution.

Polyelectrolyte	M	c_1/mass%	c_2/mass%
Sokalan PM 10 I	4,300	1.3	17
Sokalan CP 5	55,000	0.3	15
Sokalan CP 2	92,000	0.09	10.1
Sokalan CP 9	100,000	0.4	9.6

Both overlap concentrations decrease with increasing molecular mass. Comparing the Sokalanes CP 2 and CP 9 – which have about the same molecular mass – the lower charge density of CP 9 leads to an increase of c_1. The higher the charge density of the macroion, the stronger is the expansion of the ion cloud and the lower is c_1. Between c_1 and c_2 the slope of the zero shear viscosity has a value of 0.5 in a double logarithmic plot. This is characteristic for semi-dilute polyelectrolyte solutions (Fuoss, 1951). Above c_2 the slope of the curves is 4.0. This is close to the value of 5.5 for neutral polymers (Cates, 1987, 1988; de Gennes, 1971). It shows that in the concentrated region hard-core interactions dominate the behavior of the systems.

Electric birefringence measurements (EBF) did not show an effect for Sokalan PM 10 I. The used field strengths were not high enough to orientate these small molecules. For CP 9 there is detected EBF without the often seen anomaly for polyelectrolytes (Krämer and Hoffmann, 1991; Yamamoto et al., 1982) and the Kerr law is valid up to high concentrations. The overlap concentration c_1 can be seen as a kink in the plot of the Kerr constant against the concentration (Lauer, 1999); this number corresponds well with the number determined by rheology. The two other polyelectrolytes CP 2 and CP 5 show a clear anomaly of EBF. The Kerr constant changes sign from positive to negative and again to positive with increasing concentrations (cf. Fig. 27.16). At high concentrations it decreases to zero. This anomaly can be explained by changes in conformation of the macroions (Krämer, 1990) but also by the existence of a permanent and induced dipolmoment (Oppermann, 1988) with different directions regarding to the axes of the macroion molecule. The loss of EBF signal at higher concentrations is probably due to coiling of the macroions.

Surface tension measurements did not show any interfacial activity for CP 5 because it has no hydrophobic parts. With the other polyelectrolytes a more or less marked decrease in surface tension with a kink at a certain concentration can be seen (cf. Fig. 27.17). This is similar to surfactants.

The decrease of surface tension indicates an adsorption of the macroions to the phase border. It is not clear if the kink of the curves is due to aggregation of the polyelectrolyte molecules above a CMC as it is with surfactants or due to another fact. Light scattering, conductivity and rheological measurements (Lauer, 1999) did not show an aggregation of the macroions. Therefore the kink is probably due to saturation of the surface with adsorbed molecules; this would also explain the decrease of the surface tension above the kink.

27.4 Results and Discussion

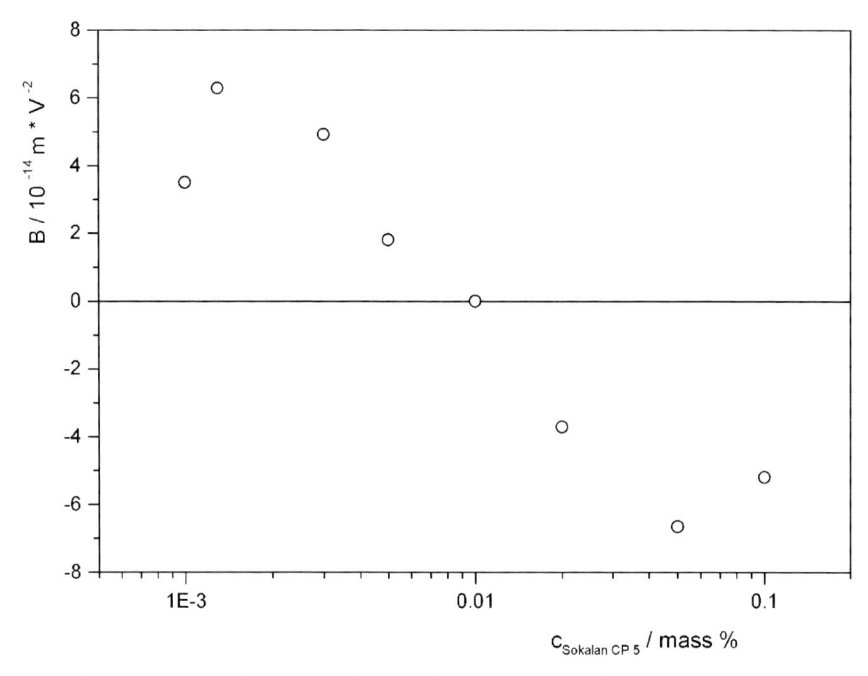

Figure 27.16: Kerr constant of aqueous solutions of Sokalan CP 5 as function of the polyelectrolyte concentration.

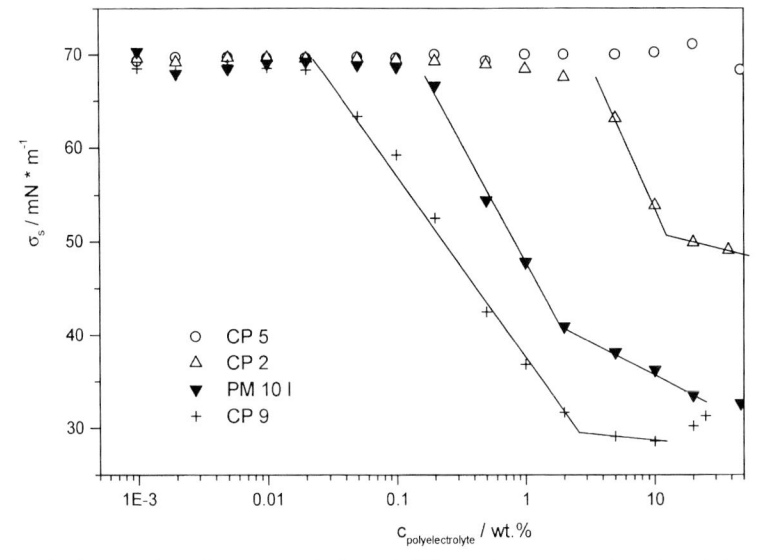

Figure 27.17: Surface tension of aqueous solutions of different Sokalanes as function of the polyelectrolyte concentration.

27.4.5 Interactions Between Modified PAAs and C$_{14}$DMAO in Aqueous Solution

The addition of C$_{14}$DMAO to a 1 mass% aqueous solution of Sokalanes which exist as Na-salts doesn't result in a pH-shift (Lauer, 1999). Also the viscosity remains constant up to the concentration of the sphere-rod transition which occurs at 30 mM C$_{14}$DMAO as for the pure surfactant solution. Above this concentration the expected increase of viscosity can be detected (cf. Fig. 27.18). For CP 9 there is even a slight decrease of viscosity just before the viscosity increases. This indicates repulsive interactions between macroions and micelles.

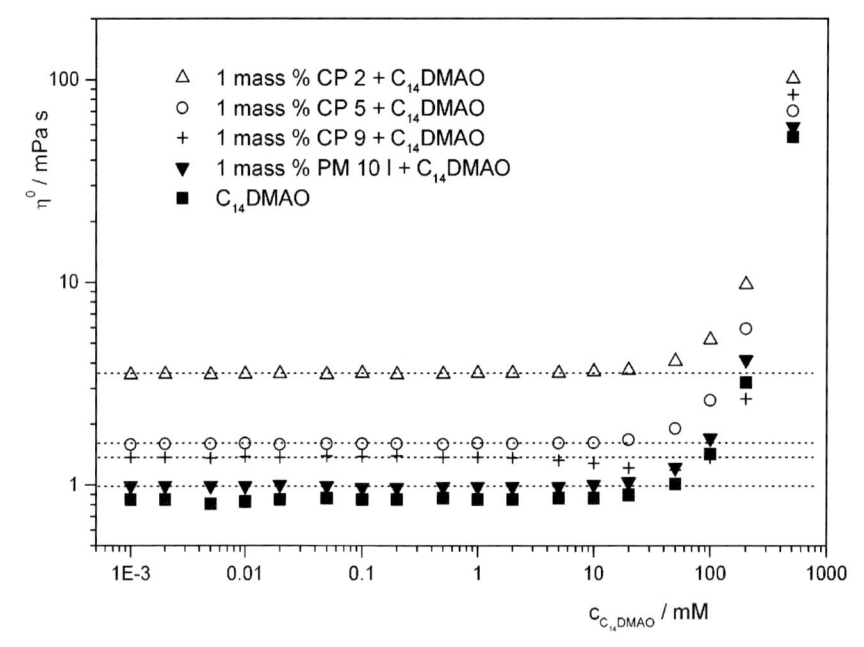

Figure 27.18: Zero shear viscosity of mixtures of Sokalanes and C$_{14}$DMAO as function of the surfactant concentration; the dot lines represent the viscosity of the surfactant-free solutions; data of the pure surfactant solution are displayed for comparison.

Experimental results for the surface tension (as function of the surfactant concentration) do not show much change by addition of CP 5 and CP 2 compared with the curve of the pure surfactant. Though the more surface active CP 9 and PM 10 I decrease the surface tension, especially below the CMC of the surfactant (cf. Fig. 27.19). But also with these polyelectrolytes the CMC of the surfactant and the number for the constant surface tension above the CMC is not influenced compared to the pure surfactant solution. These results indicate that there isn't any binding to the macroions, neither of monomeric surfactant molecules nor of micelles. The same result was found with completely neutralized polyacrylic acids (cf. Section 27.4.3).

27.4 Results and Discussion

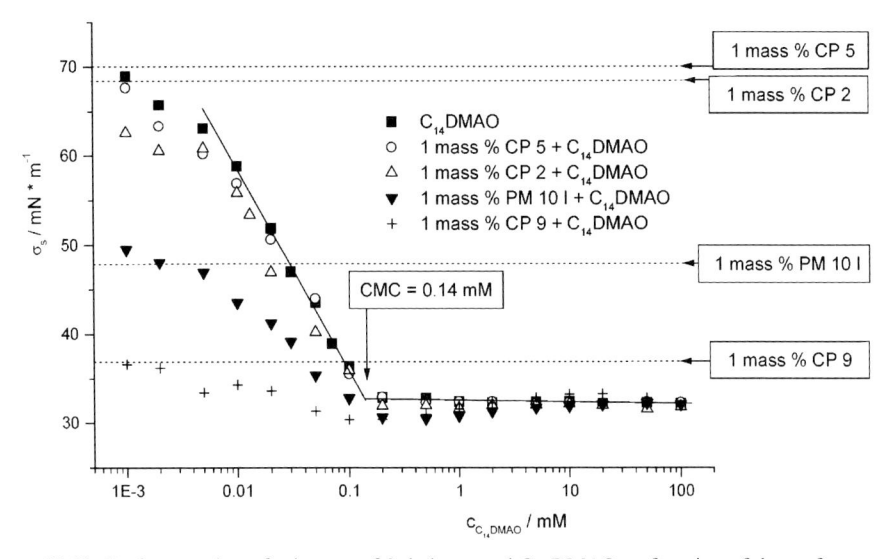

Figure 27.19: Surface tension of mixtures of Sokalanes and $C_{14}DMAO$ as function of the surfactant concentration; the dot lines represent the surface tension of the surfactant-free Sokalan solutions; data of the pure surfactant solution are displayed for comparison.

The last aim of this project was to examine if there are interactions between polyelectrolytes and $C_{14}DMAO$ if the charge density of the macroions is reduced. This was reached by addition of 30 mM HCl to the 1 mass% polyelectrolyte solution.

Figure 27.20 shows the phase behavior of the polyelectrolyte solutions with 30 mM HCl as function of surfactant concentration. It can be seen that with all systems the pH stays constant at low surfactant concentrations. At a certain surfactant concentration – which is lower for CP 5 than for PM 10 I and CP 9 – the system becomes biphasic by forming a precipitate. This can be explained by protonation of the aminoxide headgroup and the electrostatic binding of the cationic surfactant to the anionic polyelectrolyte. In the two-phase area the CMC of the surfactant can be found. Because of the precipitation of the surfactant-polyelectrolyte complex it is of course more than one order of magnitude above the CMC of the pure zwitterionic $C_{14}DMAO$.

Above the CMC there is a strong increase of the pH and the system gets completely miscible again above a second characteristic surfactant concentration. In case of CP 5 there is an additional concentration range above the two-phase area where streaks can be observed. This indicates inhomogeneous parts in this region in spite of clear solutions.

Rheological measurements prove that in the concentration range below the two-phase area the flow behavior of the systems is not influenced by addition of surfactant. The solutions are Newtonian. Its viscosity is in the range of water. In the one-phase region above the two-phase region the solutions are strongly viscoelastic and their zero shear viscosities are up to five orders of magnitude higher than those of the surfactant-free solutions (cf. Fig. 27.21). The highest zero shear viscosity is observed immediately after reaching the one-phase area; with increasing surfactant concentration the viscosity decreases again but remains about two orders of magnitude above water viscosity.

27 *Interactions of Polyelectrolytes and Zwitterionic Surfactants in Aqueous Solution*

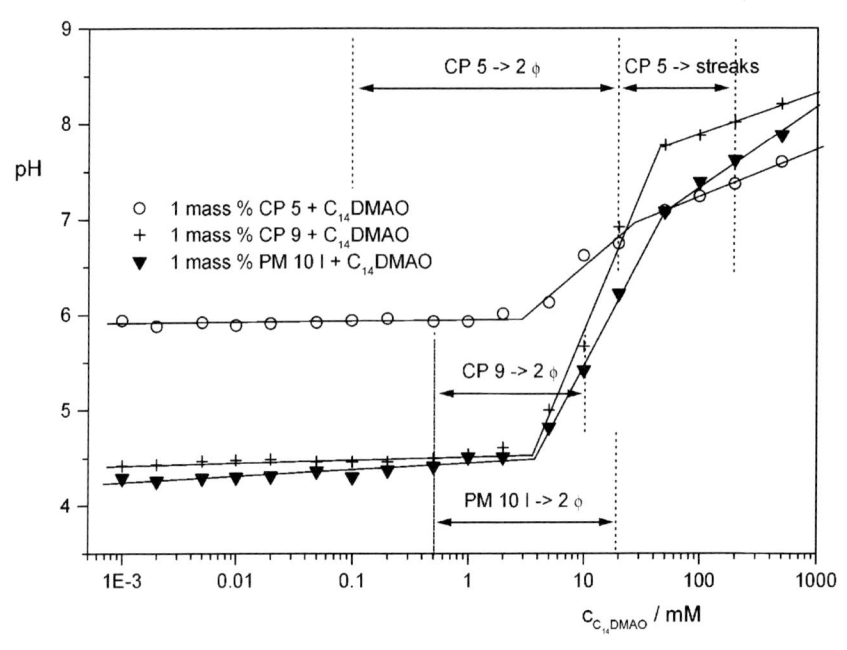

Figure 27.20: pH of mixtures of different Sokalanes with $C_{14}DMAO$ in 30 mM HCl as function of the surfactant concentration.

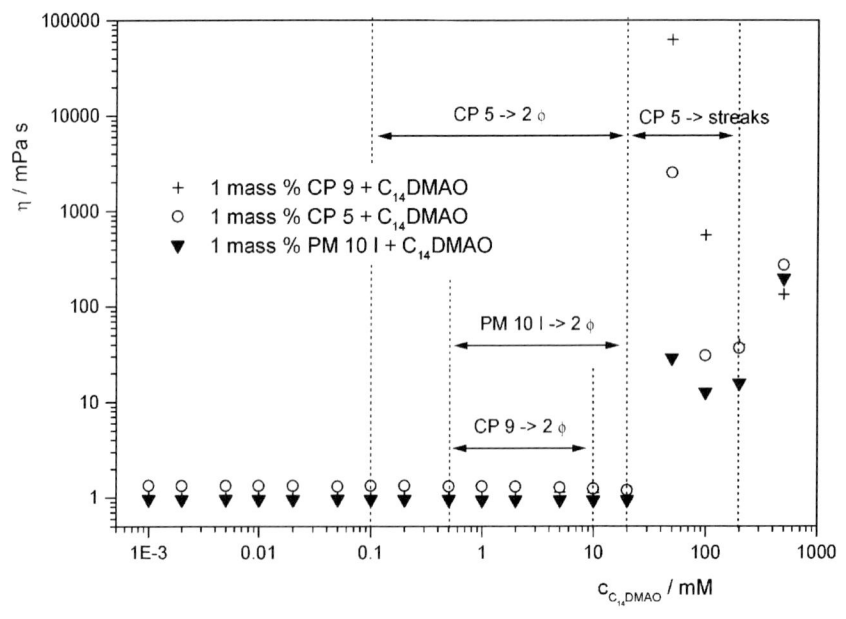

Figure 27.21: Zero shear viscosity of aqueous solutions of different Sokalanes with $C_{14}DMAO$ in 30 mM HCl as function of the surfactant concentration.

27.4 Results and Discussion

In most cases the viscoelastic solutions can be described by a Maxwell model with a structure relaxation time (Lauer, 1999). Only in case of CP 5 there is a more complicated rheogram in the streak range which is shown in Figure 27.22. It indicates the existence of a yield stress. In the whole frequency range there is a decrease of complex viscosity with increasing frequency. Therefore no zero shear viscosity can be found. On the other hand storage and loss modulus are of the same magnitude and increase with increasing frequentcy. This is different from other systems with a yield stress (Hoffmann and Rauscher, 1993).

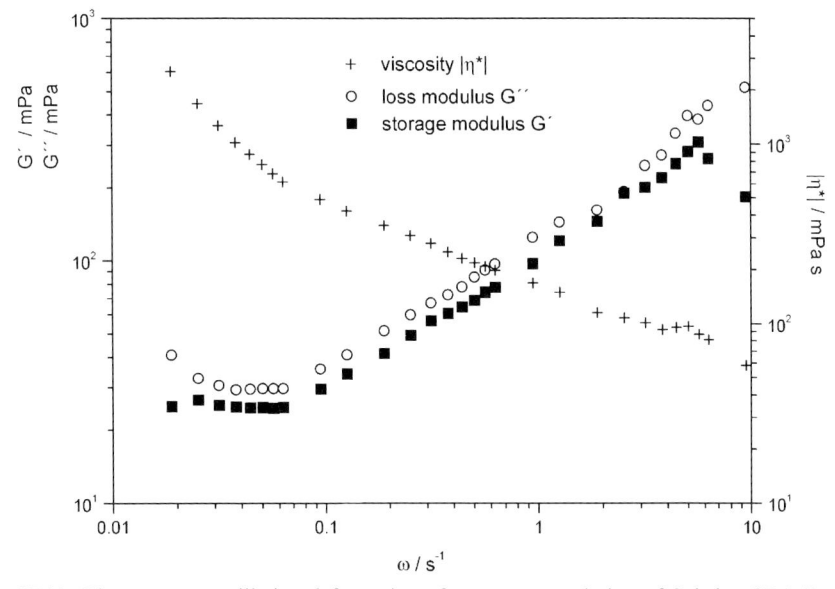

Figure 27.22: Rheogram at oscillating deformation of an aqueous solution of Sokalan CP 5 (1 mass%) and 50 mM C_{14}DMAO in 30 mM HCl.

These results show that above the two-phase region micelles of the zwitterionic surfactant exist which have positive surface charges due to the protonation by HCl. They can build up a temporary network with the anionic polyelectrolytes. Therefore similar structures are formed as with unmodified PAAs and C_{14}DMAO-micelles. They are schematically shown in Figure 27.23 and described as "decorated micelles" (Cabane, 1977). With increasing surfactant concentration the pH increases. Therefore the amount of positive charges on the micelles decreases. This leads to a decrease of cross links in the network and therefore to a decrease of viscosity and elasticity of the system.

Surface tension measurements are not able to give further information about the systems because the build-up of the networks can only occur above the CMC and therefore has no influence on the surface tension (Lauer, 1999).

In addition the macroions with hydrophobic parts (namely CP 9 and PM 10 I) are protonated at the carboxyl groups by addition of HCl. This leads to a decrease of charge density and therefore to a strong enhancement of surface activity of the polyelectrolyte molecules

PAA + C₁₄DMAO

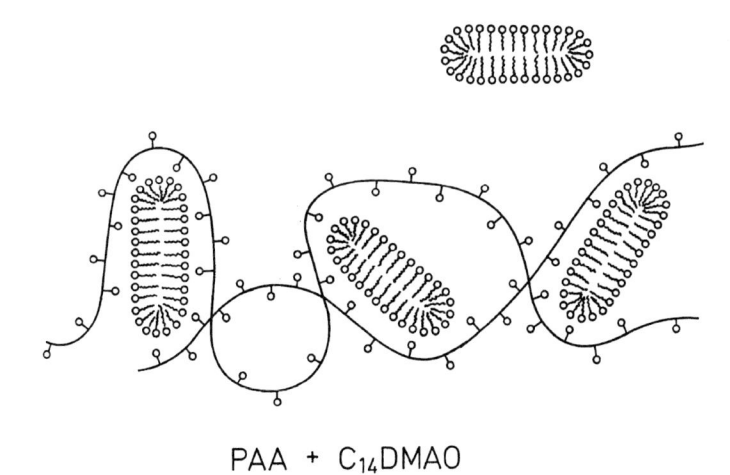

Figure 27.23: Network structure of polyelectrolyte and surfactant micelles ("decorated micelles").

themselves. The surface tension of 1 mass% aqueous solutions of CP 9 or PM 10 I with 30 mM HCl is therefore lower or only slightly higher than the surface tension of the aqueous solution of C_{14}DMAO above the CMC. Therefore the surface tension remains unchanged or is only changed little when the surfactant is added. CP 5 does not have any hydrophobic molecule parts and is (neither without nor with added HCl) surface active. By addition of HCl a continuous decrease of the surface tension is observed until the two-phase region. The curves are slightly and parallel shifted to lower surface tension values compared to the aqueous solutions of the pure surfactant.

27.5 Conclusion

Interactions of polyacrylic acid and modified polyacrylic acids with the zwitterionic surfactant C_{14}DMAO in aqueous solutions strongly depend on the pH of the solution. At high pH values no interactions can be observed. At $6.9 < pH < 7.5$ strong interactions can be observed. The partly protonated surfactant forms a transient network with the negatively charged polyelectrolyte. This results in viscoelastic and highly viscous systems. At $pH < 6.9$ the systems get biphasic with a solid, white precipitate which was identified as a polyelectrolyte-surfactant complex.

Acknowledgement

We thank BASF/Germany for providing the modified polyacrylic acids (Sokalanes).

References

Cabane, B. (1977): Structure of some polymer detergent aggregates in water. J. Phys. Chem. 81, 1639–1645.
Cates, M.E. (1987): Reptation of living polymers – Dynamics of entangled polymers in the presence of reversible chain-scission. Macromolecules 20, 2289–2296.
Cates, M.E. (1988): Dynamics of living polymers and flexible surfactant micelles – Scaling laws for dilution. J. Phys. France 49, 1593–1600.
de Gennes, P.G. (1971): Reptation of a polymer chain in presence of fixed obstacles. J. Chem. Phys. 55, 572–579.
Fuoss, R.M. (1951): Polyelectrolytes. Discuss. Faraday Soc. 11, 125–134.
Hoffmann, H.; Rauscher, A.K. (1993): Aggregating systems with a yield stress value. Colloid Polym. Sci. 271, 390–395.
Krämer, U. (1990): Elektrodoppelbrechungsmessungen an Tensid- und Polyelektrolytlösungen. PhD Thesis, University of Bayreuth.
Krämer, U.; Hoffmann, H. (1991): Electric birefringence measurements in aqueous polyelectrolyte solutions. Macromolecules 24, 256–263.
Lauer, H. (1999): Polyelektrolyte in wässriger Lösung und ihre Wechselwirkungen mit Tensid. PhD Thesis, University of Bayreuth.
Maeda, H.; Muroi, S.; Kakehashi, R. (1997): Effects of ionic strength on the critical micelle concentration and the surface excess of dodecyldimethylamine oxide. J. Phys. Chem. B 101, 7378–7382.
Oppermann, W. (1988): Electro-optical properties and molecular flexibility of polyelectrolytes in dilute solution. Makromol. Chem. 189, 927–937.
Yamamoto, T.; Mori, Y.; Ookubo, N.; Hayakawa, R.; Wada, Y. (1982): Relaxational behavior of birefringence of aqueous carboxymethylcellulose under an alternating electric field at frequencies ranging from 0.1 Hz to 100 kHz. Colloid Polym. Sci. 260, 20–26.

Nomenclature

B	Kerr constant detected by electric birefringence measurements
c	concentration
CMC	critical micellar concentration
C_{14}DMAO	tetradecyldimethylaminoxide
C_{14}DMAOH$^+$	tetradecyldimethylaminoxide in the protonated state
EBF	electric birefringence

G'	storage modulus
G''	loss modulus
G⁰	static shear modulus
M	molecular mass
PAA	polyacrylic acid
pK_a	chemical equilibrium constant

Greek letters

α	degree of neutralization/protonation
σ_s	surface tension
σ_i	interfacial tension
η^0	zero shear viscosity
$|\eta^*|$	complex viscosity
τ	relaxation time
ω	frequency
1φ	single phase
2φ	two phases